D1612864

WITHDRAWN

LIVERPOOL JMU LIBRARY
3 1111 01491 2511

Textbook of Drug Design and Discovery

Fifth Edition

Textbook of Drug Design and Discovery

Fifth Edition

Edited by

Kristian Strømgaard
Povl Krogsgaard-Larsen
Ulf Madsen

CRC Press is an imprint of the
Taylor & Francis Group, an **informa** business

CRC Press
Taylor & Francis Group
6000 Broken Sound Parkway NW, Suite 300
Boca Raton, FL 33487-2742

Printed on acid-free paper
Version Date: 20160510

International Standard Book Number-13: 978-1-4987-0278-2 (Hardback)

Visit the Taylor & Francis Web site at
http://www.taylorandfrancis.com

and the CRC Press Web site at
http://www.crcpress.com

Contents

Preface

Textbook of Drug Design and Discovery is meant to be used primarily in teaching at undergraduate and postgraduate level courses, where an insight into the complex process of the early drug design and discovery process is central. The chapters in the book are written by experts within their topic of expertise, and it is assumed that readers have a basic knowledge of organic and physical chemistry, biochemistry, and pharmacology.

The textbook covers a broad range of aspects related to the early drug design and discovery process, presented in an up-to-date review form, with an underlying and fundamental focus on the educational aspects. The first part of the book covers general aspects, methods, and principles within drug design and discovery and the second part exemplify important and recent medicinal chemistry developments for a number of specific targets and diseases.

To perform both academic and industrial medicinal chemistry research at the highest level, it is required to attract the attention of bright students, interested in the creative and fascinating nature of drug design. In order to reach this goal, it is of utmost importance to maintain focus on the integration of the scientific disciplines of chemistry and biology. Interesting developments in this regard is the increased interactions between academic and more industrial medicinal chemistry efforts, as seen in a number of public–private partnerships, often involving "big pharma" companies and universities around the world. Regardless of the setting, students should be taught that the conversions of hits into lead structures and further into drug candidates require the integration of a number of related scientific disciplines, such as advanced synthetic chemistry, computational chemistry, biochemistry, structural biology, and molecular pharmacology.

Clearly, the early processes of drug design and development are constantly undergoing changes, which call for a regular update of a textbook like this and are reflected in the current edition. This update includes chapters describing the particular challenges and aspects of developing peptide-based drugs, as well as a broader description of protein-based drugs. In addition, a chapter is devoted to the medicinal chemistry considerations in the pharmaceutical industry (Chapter 5), when developing hit compounds into lead compounds and later clinical candidates. These considerations are important for students to learn and understand and highlight putative distinctions between more academic medicinal chemistry endeavors and those ongoing in the pharmaceutical industry.

Kristian Strømgaard
Povl Krogsgaard-Larsen
Ulf Madsen

Editors

Kristian Strømgaard is a professor of chemical biology in the Department of Drug Design and Pharmacology, Faculty of Health and Medical Sciences, at the University of Copenhagen and a director of the Center for Biopharmaceuticals, Copenhagen, Denmark. He received his master's degree in chemical research from the Department of Chemistry at the University College London under the supervision of Professor C. Robin Ganellin. In 1999, he received his PhD in medicinal chemistry from the Royal Danish School of Pharmacy (now School of Pharmaceutical Sciences). Subsequently, Dr. Strømgaard carried out his postdoctoral studies with Professor Koji Nakanishi in the Department of Chemistry at Columbia University. During this period, his main focus was on medicinal chemistry studies of neuroactive natural products, with a particular emphasis on polyamine toxins and ginkgolides. In 2002, he returned to the Faculty of Pharmaceutical Sciences as an assistant professor and became an associate professor in 2004 and a professor in 2006. He is currently heading the Center for Biopharmaceuticals, which is an interdisciplinary center focusing on protein medicinal chemistry. The Strømgaard lab explores membrane-bound proteins and their downstream signaling proteins using a combination of chemical and biological approaches. In addition, Dr. Strømgaard is cofounder of a biotech company, Avilex Pharma, which explores peptide-based inhibitors of protein–protein interactions as a novel treatment for stroke.

Povl Krogsgaard-Larsen is a professor of medicinal chemistry in the Department of Drug Design and Pharmacology, Faculty of Health and Medical Sciences, at the University of Copenhagen, Copenhagen, Denmark. In 1970, he received his PhD in natural product chemistry from the former Royal Danish School of Pharmacy. As an associate professor, he established a research program focusing on the conversion of naturally occurring toxins into specific pharmacological tools and therapeutic agents. Key lead structures in this research program were the *Amanita muscaria* constituents, muscimol and ibotenic acid, and the *Areca* nut alkaloid, arecoline, all of which interact nonselectively with GABA, glutamate, and muscarinic receptors, respectively. The redesign of muscimol resulted in a variety of specific GABA agonists, notably THIP and isoguvacine, and specific GABA uptake inhibitors, including nipecotic acid and guvacine. Ibotenic acid was converted into a broad range of subtype-selective glutamate receptor agonists, including AMPA, from which the AMPA receptor subgroup was named. Arecoline was redesigned to provide a variety of subtype-selective muscarinic agonists and antagonists. Whereas nipecotic acid was subsequently developed into the antiepileptic agent tiagabine, THIP is currently used in advanced clinical trials.

In 1980, Dr. Krogsgaard-Larsen received his DSc. He has published nearly 460 scientific papers and edited a number of books, and during the period 1998–2013, he was editor of the *Journal of Medicinal Chemistry*. He has been awarded honorary doctorates at the universities of Strasbourg (1992), Uppsala (2000), and Milan (2008), apart from receiving numerous other scientific awards and prizes. He is a member of a number of academies, including the Royal Danish Academy of Sciences and Letters. In 2002, he founded the Drug Research Academy as an academic/industrial research training center. He is currently chairman of the Brain Prize Foundation under the Lundbeck Foundation, member of the board of the Lundbeck Foundation, and deputy chairman of the Benzon Foundation. During the period 2003–2011, he was chairman of the Carlsberg Foundation.

Ulf Madsen is currently director at the School of Pharmaceutical Sciences, Faculty of Health and Medical Sciences, University of Copenhagen, Copenhagen, Denmark. In 1988, he received his PhD in medicinal chemistry from the Royal Danish School of Pharmacy, where he later became an associate professor in the Department of Medicinal Chemistry. Dr. Madsen has been a visiting scientist at the University of Sydney, Australia; at Johann Wolfgang Goethe University, Frankfurt, Germany; and at Syntex Research, California. He has extensive research experience with design and synthesis of glutamate receptor ligands. This includes structure activity studies and development of selective ligands for ionotropic and metabotropic glutamate receptors, work that has led to a number of important pharmacological tools. Compounds with antagonist activities have shown neuroprotective properties in animal models and are consequently leads for potential therapeutic candidates. Projects involving biostructure-based drug designs have resulted in the development of important pharmacological agents with high subtype selectivity. The work generally involves the synthesis of heterocyclic compounds, the use of bioisosteric principles, and molecular pharmacology on native and recombinant receptors. The work has led to 120 scientific papers. Before becoming school director in 2012, Dr. Madsen was the associate dean at the former Faculty of Pharmaceutical Sciences and before that head of the Department of Medicinal Chemistry for nine years.

Contributors

Louise Albertsen
Steno Diabetes Center
Gentofte, Denmark

Thomas Balle
Faculty of Pharmacy
The University of Sydney
Sydney, New South Wales, Australia

Benny Bang-Andersen
H. Lundbeck A/S
Valby, Denmark

Bo Hjorth Bentzen
Department of Biomedical Sciences
University of Copenhagen
Copenhagen, Denmark

Fredrik Björkling
Department of Drug Design and Pharmacology
University of Copenhagen
Copenhagen, Denmark

Klaus P. Bøgesø
H. Lundbeck A/S
Valby, Denmark

P. Ann Boriack-Sjodin
Epizyme, Inc.
Cambridge, Massachusetts

Hans Bräuner-Osborne
Department of Drug Design and Pharmacology
University of Copenhagen
Copenhagen, Denmark

Lennart Bunch
Department of Drug Design and Pharmacology
University of Copenhagen
Copenhagen, Denmark

Erica S. Burnell
School of Physiology, Pharmacology and Neuroscience
University of Bristol
Bristol, United Kingdom

Ib C. Bygbjerg
Department of Public Health
University of Copenhagen
Copenhagen, Denmark

Guy T. Carter
Carter-Bernan Consulting, LLC
New City, New York

Søren B. Christensen
Department of Drug Design and Pharmacology
University of Copenhagen
Copenhagen, Denmark

Rasmus P. Clausen
Department of Drug Design and Pharmacology
University of Copenhagen
Copenhagen, Denmark

Robert A. Copeland
Epizyme, Inc.
Cambridge, Massachusetts

Iwan de Esch
Department of Chemistry and Pharmaceutical Sciences
Vrije University Amsterdam
Amsterdam, the Netherlands

Anastassios Economou
Rega Institute
KU Leuven
Leuven, Belgium

Bente Frølund
Department of Drug Design and Pharmacology
University of Copenhagen
Copenhagen, Denmark

Steen Gammeltoft (emeritus)
Department of Clinical Biochemistry
Rigshospitalet
Copenhagen, Denmark

Ulrik Gether
Department of Neuroscience and Pharmacology
University of Copenhagen
Copenhagen, Denmark

Bo Falck Hansen
Novo Nordisk A/S
Bagsvaerd, Denmark

Harald S. Hansen
Department of Drug Design and Pharmacology
University of Copenhagen
Copenhagen, Denmark

Piet Herdewijn
Rega Institute
KU Leuven
Leuven, Belgium

Matthias M. Herth
Department of Drug Design and Pharmacology
University of Copenhagen
Copenhagen, Denmark

David E. Jane
School of Physiology, Pharmacology and Neuroscience
University of Bristol
Bristol, United Kingdom

Anders A. Jensen
Department of Drug Design and Pharmacology
University of Copenhagen
Copenhagen, Denmark

Flemming Steen Jørgensen
Department of Drug Design and Pharmacology
University of Copenhagen
Copenhagen, Denmark

Morten Jørgensen
H. Lundbeck A/S
Valby, Denmark

Jette Sandholm Kastrup
Department of Drug Design and Pharmacology
University of Copenhagen
Copenhagen, Denmark

Jan Kehler
H. Lundbeck A/S
Valby, Denmark

Jesper L. Kristensen
Department of Drug Design and Pharmacology
University of Copenhagen
Copenhagen, Denmark

Povl Krogsgaard-Larsen
Department of Drug Design and Pharmacology
University of Copenhagen
Copenhagen, Denmark

Krista Laine
School of Pharmacy
University of Eastern Finland
Kuopio, Finland

Jesper Lau
Novo Nordisk A/S
Måløv, Denmark

Rob Leurs
Division of Medicinal Chemistry
Vrije University Amsterdam
Amsterdam, the Netherlands

Tommy Liljefors (deceased)
Department of Drug Design and Pharmacology
University of Copenhagen
Copenhagen, Denmark

Claus J. Løland
Department of Neuroscience and Pharmacology
University of Copenhagen
Copenhagen, Denmark

Ulf Madsen
Department of Drug Design and Pharmacology
University of Copenhagen
Copenhagen, Denmark

José Moreira
Department of Veterinary Disease Biology
University of Copenhagen
Copenhagen, Denmark

Søren-Peter Olesen
Department of Biomedical Sciences
University of Copenhagen
Copenhagen, Denmark

Søren Østergaard
Novo Nordisk A/S
Måløv, Denmark

Smitha Rao C.V.
Rega Institute
KU Leuven
Leuven, Belgium

Lars Kyhn Rasmussen
LEO Pharma A/S
Ballerup, Denmark

Jarkko Rautio
School of Pharmacy
University of Eastern Finland
Kuopio, Finland

Jan Stenvang
Department of Veterinary Disease Biology
University of Copenhagen
Copenhagen, Denmark

Kristian Strømgaard
Department of Drug Design and Pharmacology
University of Copenhagen
Copenhagen, Denmark

Lena Tagmose
H. Lundbeck A/S
Valby, Denmark

Robert J. Thatcher
School of Physiology, Pharmacology and Neuroscience
University of Bristol
Bristol, United Kingdom

Henk Timmerman
Division of Medicinal Chemistry
Vrije University Amsterdam
Amsterdam, the Netherlands

Daniel B. Timmermann
Novo Nordisk A/S
Måløv, Denmark

1 Introduction to Drug Design and Discovery

Ulf Madsen and Povl Krogsgaard-Larsen

CONTENTS

1.1 MEDICINAL CHEMISTRY: AN INTERDISCIPLINARY SCIENCE

Therapeutic agents are chemical entities that prevent disease, assist in restoring health to the diseased, or alleviate symptoms associated with disease conditions. Medicinal chemistry is the scientific discipline that makes such drugs available either through discovery or design processes. Throughout history, drugs were primarily discovered by empirical methods, investigating substances or preparations of materials, such as plant parts or plant extracts, found in the local environment. Over the previous centuries, chemists developed methods for the isolation and purification of the active principles in medicinal plants. The purification and structure determination of natural products like morphine, hyoscyamine, quinine, and digitalis glycosides represent milestones in the field of drug discovery and the beginning of medicinal chemistry as a fascinating independent field of research (Figure 1.1).

In the twentieth century, a very large number of biologically active natural products were structurally modified in order to optimize their pharmacology and drug properties in general, and novel drugs were prepared by an increasing use of advanced synthetic methods. Moreover, the rapidly growing understanding of the nature of disease mechanisms, how cells function, and how drugs interact with cellular processes has led to the rational design, synthesis, and pharmacological evaluation of new drug candidates. Most recently, new dimensions and opportunities have emerged from a deeper understanding of cell biology, genetics, and biostructures.

Modern medicinal chemistry draws upon many scientific disciplines, with organic chemistry, physical chemistry, and pharmacology being of fundamental importance. But other disciplines such as biochemistry, molecular biology, toxicology, genetics, cell biology, biophysics, physiology, pathology, and computer modeling approaches play important roles. The key research objective of medicinal chemistry is to investigate relationships between chemical structure and biological effects. When the chemical structure of a particular drug candidate has been optimized to interact

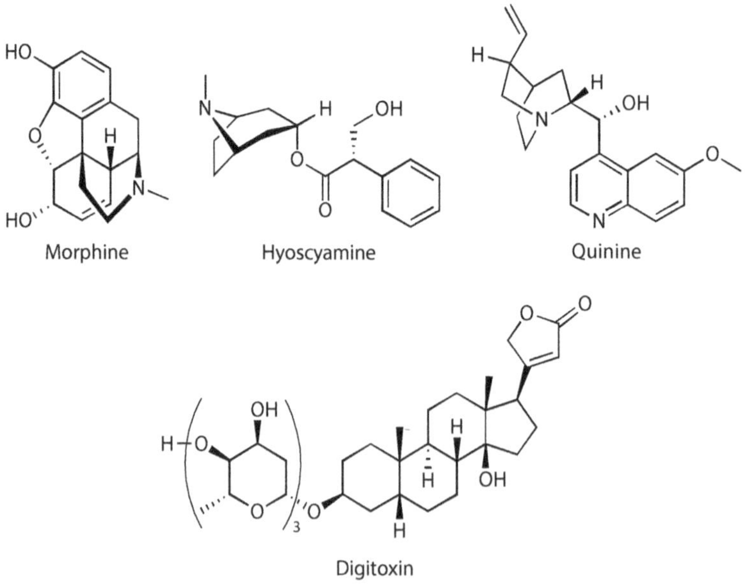

Morphine

Hyoscyamine

Quinine

Digitoxin

FIGURE 1.1 Chemical structures of four naturally occurring classical therapeutic agents.

with the biological target, the compound further has to fulfill a multifaceted set of criteria before it can be safely administered to patients. Absorption, distribution, metabolism, excretion (ADME), and toxicology studies in animals and humans are time-consuming research tasks which often call for redesign of the chemical structure of the potential therapeutic agent investigated. It is an iterative process which in reality ends up in an overall compromise with respect to multiple desired properties.

1.2 DRUG DISCOVERY: A HISTORICAL PERSPECTIVE

In early times, there was no possibility of understanding the biological origin of a disease. Of necessity, progress in combating disease was disjointed and empirical. The use of opium, ephedra, marijuana, alcohol, salicylic acid, digitalis, coca, quinine, and a host of other drugs still in use long predate the rise of modern medicine. These natural products are surely not biosynthesized by plants for our therapeutic convenience, but they normally have survival value to the plants in dealing with their own ecological challenges.

The presence of biologically active substances in nature, notably in certain plants, was in medieval times interpreted more teleologically. In the early sixteenth century, the Swiss-Austrian medical doctor and natural scientist Paracelsus formulated the "Doctrine of Signatures":

> Just as women can be recognized and appraised on the basis of their shape; drugs can easily be identified by appearance. God has created all diseases, and he also has created an agent or a drug for every disease. They can be found everywhere in nature, because nature is the universal pharmacy. God is the highest ranking pharmacist.

The formulation of this doctrine was in perfect agreement with the dominating philosophies at that time, and it had a major impact on the use of natural medicines. Even today, remanences of this doctrine can be observed in countries where herbal medical preparations are still widely used. Although the "Doctrine of Signatures" evidently is out of the conception of modern medicinal natural product research, the ideas of Paracelsus were the first approach to rational drug discovery.

More than 100 years ago, the mystery of why only certain molecules produced a specific therapeutic response was rationalized by the ideas of Emil Fischer and further elaborated by John Langley and Paul Ehrlich that only certain cells contained receptor molecules that served as hosts for the drugs. The resulting combination of drug and receptor created a new super molecule that had properties producing a response of therapeutic value. One extension of this conception was that the drug fits the target specifically and productively like "a key into its corresponding lock." When the fit was successful, a positive pharmacological action (agonistic) followed, analogous to opening the door. On the contrary, when the fit prevented the intrinsic key to be inserted an antagonist action resulted—i.e., the imaginative door could not be opened. Thus, if one had found adventitiously a ligand for a receptor, one could refine its fit by opportunistic or systematic modifications of the ligand's chemical structure until the desired function was obtained.

This productive idea hardly changed for the next half century and assisted in the development of many useful drugs. However, a less fortunate corollary was that it led to some limitations of creativity in drug design. The drug and its receptor (whose molecular nature was unknown when the theory was formulated) were each believed to be rigid molecules precrafted to fit one another precisely. Today, we know that receptors are highly flexible transmembrane glycoproteins accessible from the cell surface that often comprise more than one drug compatible region. Further complexities have been uncovered continually. For example, a number of receptors have been shown to consist of clusters of proteins either preassembled or assembled as a consequence of ligand binding. The component macromolecules may be either homo- or heterocomplexes. The challenge of developing specific ligands for systems of this complexity may readily be imagined (Chapters 4 and 12).

The opposite extreme to "the lock and key model" is "the zipper model." In this view, a docking interaction takes place (much as the end of a zipper joins the talon piece) and, if satisfactory complementarity is present, the two molecules progressively wrap around each other and adapt to the steric and electrostatic needs of each other. A consequence of accepting this mutual adaptation is that knowledge of the receptor ground state may not be particularly helpful as it adjusts its conformation to ligand binding. Thus, in many cases, one now tries to determine the 3D structure of the receptor–ligand complex. In those cases where X-ray analysis remains elusive, modeling of the interactions involved is appropriate. This is the subject of Chapters 2 through 4.

Earlier, it was also noted that enzymes could be modulated for therapeutic benefit. Enzymatic proteins share many characteristics with receptors, although enzymes catalyze biochemical reactions. Receptor ligands interact with the receptor glycoproteins or with the interfaces between the macromolecular subunits of di- or polycomponent receptor complexes and modify the conformation and dynamics of these complexes. Thus, neither receptor agonists nor antagonists directly interfere with chemical reactions and generally are dissociated from the receptor recognition sites structurally unchanged.

The reaction mechanisms underlying the function of the vast majority of enzymes have been elucidated in detail, and based on such mechanistic information, it has been possible to design a variety of mechanism-based enzyme inhibitors, notably k_{cat} inactivators and transition-state analogs, many of which are in therapeutic use (Chapter 11). Until very recently, it was usually only possible to inhibit enzyme action rather than facilitate it. Actually, diseases frequently result from excessive enzymatic action, making selective inhibition of these enzymes therapeutically useful.

Much later, further classes of receptors were disclosed, explored, and exploited as therapeutically relevant pharmacological targets. This heterogeneous group of receptors comprises nuclear receptors operated by steroid hormones and other lipophilic biochemical mediators, a broad range of membrane-ion channels (Chapter 13), DNA or RNA (Chapter 22), and a number of other biostructures of known or unknown functions. These aspects will be discussed in different chapters of this book.

1.3 DRUG DEVELOPMENT PROCESS: AN OUTLINE

The stages through which a drug discovery/development project proceeds from inception to marketing and beyond are illustrated in Figure 1.2 and described briefly in the following text. The discovery and development process can be described by a number of individual steps, but is also a continuous and iterative process not necessarily performed in a strict stepwise process. From this outline, the complexity of the task of finding new therapeutic agents is evident:

- Target discovery comprises identification and validation of disease-modifying targets. Two major strategies are used for target identification and validation: (1) the molecular approach, with focus on the cells or cell components implicated in the disease and the use of clinical samples and cell models, and (2) the systems approach based on target discovery through the study of diseases in whole organisms.
- Before or after identification of target disease, establishment of a multidisciplinary research team, selection of a promising approach, and decision on a sufficient budget. Initiation of chemistry normally involves synthesis based on available chemicals, in-house chemical libraries, or collection of natural product sources. Start of pharmacology includes suitable screening methods and choice of receptor or enzymatic assays.
- Confirmation of potential utility of initial class(es) of compounds in animals, focusing on potency, selectivity, and apparent toxicity.
- Analog syntheses of the most active compounds, planned after careful examination of literature and patents. More elaborated pharmacology in order to elucidate the mode of action, efficacy, acute and chronic toxicity, and genotoxicity. Studies of ADME characteristics. Planning of large-scale synthesis and initiation of formulation studies. Application for patent protection.

These first project phases which typically last 4–5 years, are followed by highly time- and resource-demanding clinical, regulatory, and marketing phases which normally last 6–10 years:

- Very-large-scale synthesis in parallel or before clinical studies
- Phase I clinical studies which include safety, dosage, and blood level studies on healthy volunteers
- Phase II clinical studies focusing on efficacy and side effects on delimited groups of patients
- Phase III clinical studies which involve studies of range of efficacy and long-term and rare side effects on large patient groups
- Regulatory review
- Marketing and phase IV clinical studies focusing on long-term safety
- Distribution, advertisement, and education of marketing and information personnel

After these project stages from initiation to successful therapeutic application after approval, the patent protection expires, normally after 17–25 years, and generic competition becomes a reality.

This outline of a drug discovery and development process illustrates that, it takes many years to introduce a new therapeutic agent, and it must be kept in mind that most projects are terminated before marketing, even at advanced stages of clinical studies. The later a project fails the more expensive, and many efforts are done in order to consider as many potential failure problems as early as possible in the process. Especially forward translation of preclinical data to possible clinical outcome and back translation of clinical data to "humanized" preclinical data of more predictive value are important issues in the desire to avoid late failures.

Some of the aspects of drug discovery phase are described in more detail in the following sections.

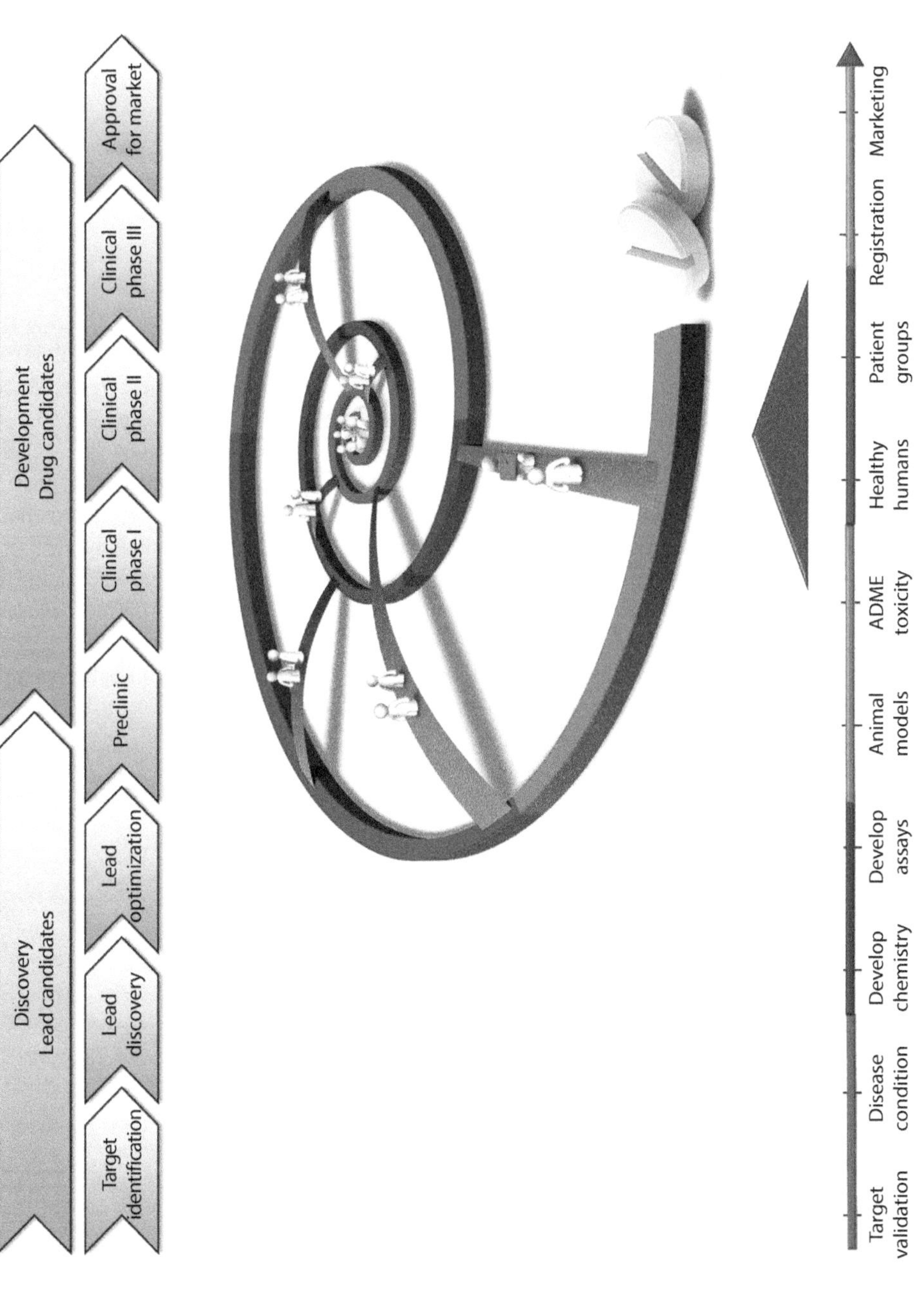

FIGURE 1.2 Outline of the drug discovery and development process with indication of individual steps (blue arrows). The lower multicolored timeline shows the process as a continuous flow which also involves iterative processes at all stages. The spiral representation illustrates such forward and backward translation of knowledge by bridges between the different stages of development.

1.4 DISCOVERY OF DRUG CANDIDATES

Prehistoric drug discovery started with higher plant and animal substances, and this continues today to be a fruitful source of biologically active molecules frequently belonging to unanticipated structural types. Adding to the long list of classical plant products that are still used in modern medicine, one can list many substances of more recent origin, including antibiotics such as penicillins, cephalosporins, tetracyclines, aminoglycosides, various glycopeptides, and many others (Chapter 23). Anticancer agents of natural origin comprise taxol, camptothecin, vinca alkaloids, doxorubicin, and bleomycin (Chapter 21). Among immunosuppressant agents, cyclosporine and tacrolimus deserve special mention.

Other sources of lead structures and drug candidates include endogenous compounds and other compounds with known activity at the target(s) in question, as well as screening programs. The role of natural products in target identification and as lead structures is further described in the following sections, whereas examples of use of other sources are described in different chapters throughout the textbook.

1.4.1 Natural Products: Role in Target Identification

Many naturally occurring compounds have potent and/or selective activity on different biological targets and are of potential therapeutic value. Most often these activities are toxic effects, since these compounds are either animal venoms (e.g., snake poison, spider, or wasp toxins) which can paralyze or kill prey or plant toxins preventing animals to eat the plants. However, toxicity is generally a matter of dose, and in some instances a toxin can be used as a drug in the appropriate dose.

Various biologically active natural products have played a key role in the identification and characterization of receptors, and such receptors are often named after these compounds (Chapter 12). Morphine is a classical example of a natural product used for receptor characterization. Radiolabeled morphine was shown to bind with high affinity to receptors in the nervous system, and these receptors are known as opiate receptors. More than three decades ago, the physiological relevance of these receptors was documented by the findings that endogenous peptides, notably enkephalins and endorphins, served as receptor ligands (agonists). Analogs of morphine have been useful tools for the demonstration of heterogeneity of opiate receptors (Chapter 19) (Figure 1.3).

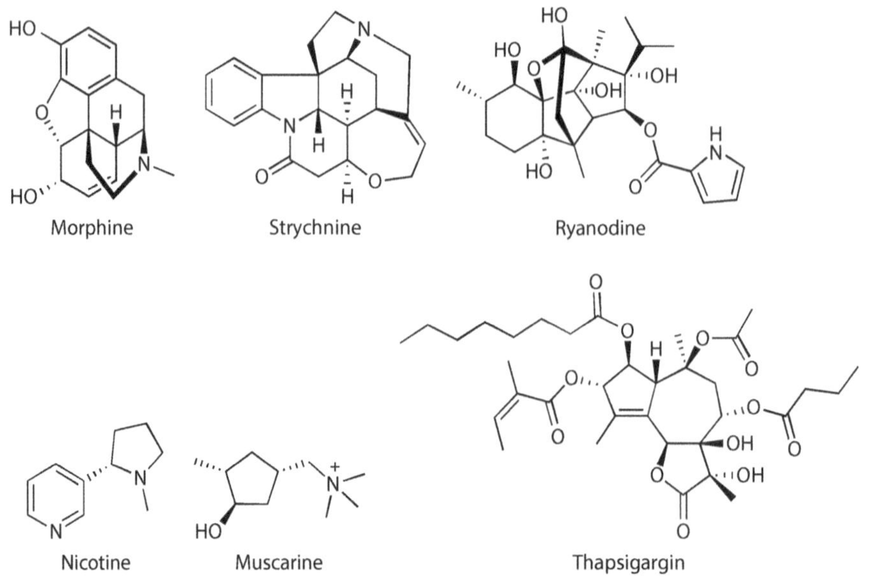

Morphine Strychnine Ryanodine

Nicotine Muscarine Thapsigargin

FIGURE 1.3 Chemical structures of morphine, strychnine, ryanodine, nicotine, muscarine, and thapsigargin.

The very toxic and convulsive alkaloid, strychnine, has been extensively studied pharmacologically. Using electrophysiological techniques and tritiated strychnine for binding studies, strychnine was shown to be an antagonist for the neuroreceptor mediating the inhibitory effect of glycine, through the glycine$_A$ receptor located primarily in the spinal cord.

Acetylcholine is a key transmitter in the central and the peripheral nervous system. Acetylcholine operates through multiple receptors, and the original demonstration of receptor heterogeneity was achieved using the naturally occurring compounds, nicotine and muscarine. Whereas the ionotropic class of acetylcholine receptors binds nicotine with high affinity and selectivity, muscarine specifically and potently activates the metabotropic class of these receptors. Using molecular biological techniques, a number of subtypes of both nicotinic and muscarinic acetylcholine receptors have been identified and characterized (Chapters 12 and 16).

The ryanodine receptor is named after the insecticidal naturally occurring compound, ryanodine. Extensive studies have disclosed that ryanodine interacts with high affinity and in a calcium-dependent manner with its receptor which functions as a calcium release channel. There are three genetically distinct isoforms of the ryanodine receptor which play a role in the skeletal muscle disorder, central core disease.

The sesquiterpene lactone, thapsigargin which is structurally unrelated to ryanodine, also interacts with an intracellular calcium mechanism. Thapsigargin has become the key pharmacological tool for the characterization of the sarco(endo)plasmic reticulum Ca^{2+} ATPase (SERCA). Thapsigargin effectively inhibits this ATPase, causing a rise in the cytosolic calcium level which eventually leads to cell death. Although the SERCA pump is essential for all cell types, attempts to target thapsigargin toward prostate cancer cells have been made based on a prodrug approach (see Chapter 10).

1.4.2 Natural Products as Lead Structures

Although a number of biologically active natural products have been indispensable as tools for identification and characterization of pharmacological and potential therapeutic targets, these compounds normally do not satisfy the demands on drugs for therapeutic use (Chapter 7).

Thus, although morphine is used therapeutically, it is not an ideal drug and has, to some extent, been replaced by a number of analogs showing lower side effects and higher degrees of selectivity for subtypes of opiate receptors (Chapter 19). Prominent examples are the μ-selective opiate agonist fentanyl and the experimental tool U50,488 which selectively activates the κ-subtype of opiate receptors (Figure 1.4).

The main psychoactive constituent of *Cannabis sativa*, the highly lipophilic tetrahydrocannabinol (THC), has been a useful tool for the identification of the two cannabinoid receptors, CB1 and CB2 receptors, operated by endocannabinoids. Since different preparations of *C. sativa* have psychoactive effects, health authorities in most countries have been reluctant to accept THC and analogs as therapeutic agents for the treatment of pain and other disease-related conditions. This may change with time, as medicinal chemists have synthesized a number of cannabinoid receptor ligands, including the receptor agonist CP55,940 which is markedly less lipophilic than THC (Chapter 19).

The nicotine acetylcholine receptors (nAChRs) have become important targets for therapeutic approaches to treat pain, cognition disorders, depression, schizophrenia, and nicotine dependence. For several reasons, nicotine has limited utility as a therapeutic agent, and a wide variety of nAChR agonists have been synthesized and characterized (Chapter 16). (–)-Cytisine is a naturally occurring toxin acting as a partial nAChR agonist. Using (–)-cytisine as a lead structure, varenicline was developed as a partial nAChR agonist showing a balanced agonist/antagonist profile for smoking cessation. Muscimol is another example of a naturally occurring toxin which has been extensively used as a lead for the design of specific GABA receptor agonists and GABA uptake inhibitors (Chapter 15). Muscimol which is a 3-isoxazolol bioisostere (see Section 1.4.3.1) of GABA, is a

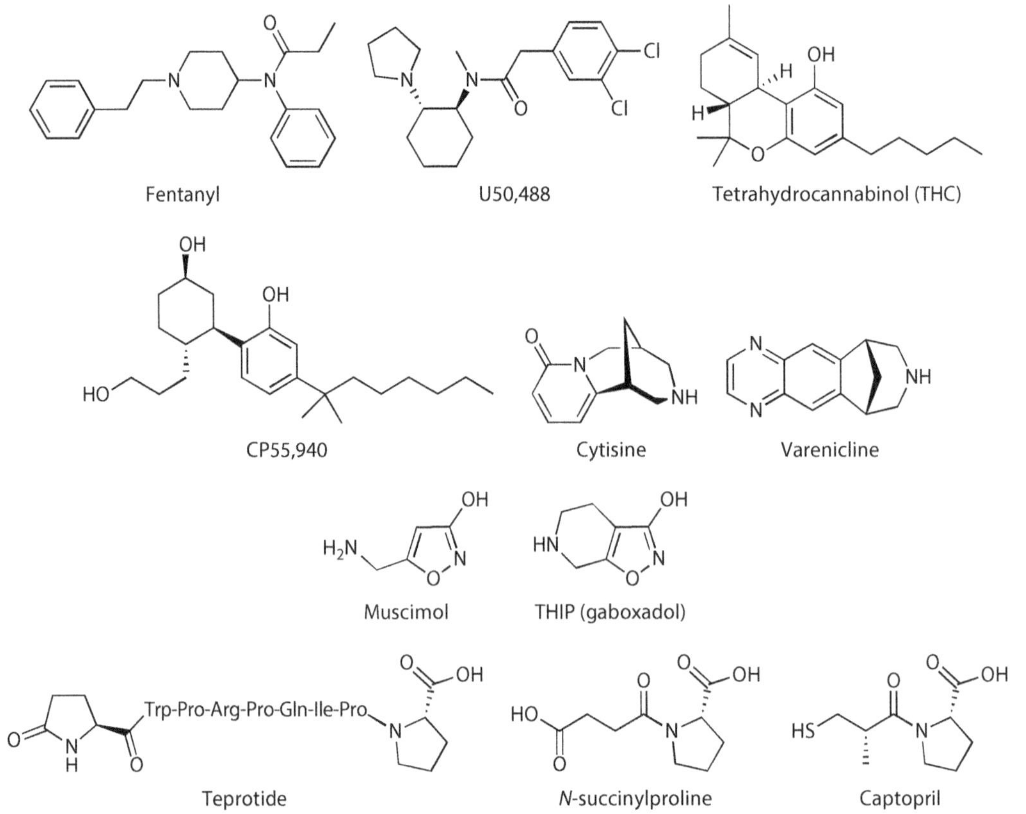

Fentanyl U50,488 Tetrahydrocannabinol (THC)

CP55,940 Cytisine Varenicline

Muscimol THIP (gaboxadol)

Teprotide *N*-succinylproline Captopril

FIGURE 1.4 Chemical structures of fentanyl, U50,488, tetrahydrocannabinol (THC), CP55,940, cytisine, varenicline, muscimol, THIP (gaboxadol), teprotide, *N*-succinylproline, and captopril.

constituent of the mushroom *Amanita muscaria*. Muscimol is toxic, it is metabolically unstable, and it interacts with the different GABA synaptic mechanisms and with a broad range of $GABA_A$ receptor subtypes. The cyclic analog of muscimol, THIP (gaboxadol), is highly selective for the therapeutically interesting extrasynaptic $GABA_A$ receptors. Gaboxadol is a clinically active nonopioid analgesic and a nonbenzodiazepine hypnotic which at present is in clinical trials (see also Chapter 15).

The angiotensin-converting enzyme (ACE) is a zinc carboxypeptidase centrally involved in the regulation of blood pressure and is an important target for therapeutic intervention. Peptide toxins from the Brazilian pit viper, *Bothrops jararaca*, and the synthetic peptide analog, teprotide, are inhibitors of ACE (Figure 1.4), but are not suitable for therapeutic use. Systematic molecular dissection of teprotide led to the nonpeptide ACE inhibitor, *N*-succinylproline which was converted into the structurally related and much more potent analog, captopril, that is now marketed as an effective antihypertensive drug.

1.4.3 Basic Principles in Lead Development and Optimization

Potency, efficacy, and selectivity are essential but certainly not the only parameters to fulfill for a pharmacologically active compound to become a therapeutic drug. A large number of additional requirements have to be met, and the most important ones have been summarized in the acronym, ADME or ADME-Tox (ADME and toxicity). Obviously, the drug must reach the site of action in a timely manner and in sufficient concentration to produce the desired therapeutic effect.

After oral administration, one of many routes of administration, the drug must survive the acidic environment of the stomach. In the small intestine, the bulk of absorption takes place. Here, the pH is neutral to slightly acidic. In the gastrointestinal system metabolism can take place. The presence of digestive enzymes creates particular problems for polypeptide drugs which may call for other routes of administration, as the gut wall is rich in oxidative enzymes.

Unless the drug acts as a substrate for active energy-requiring uptake mechanisms which normally facilitate uptake of, for example, amino acids and glucose, it must be significantly unionized to penetrate cell membranes in order to enter the blood stream. Following absorption, the blood rapidly presents the drug to the liver, where Class I metabolic transformations (oxidation, hydrolysis, reduction, etc.) and in some cases phase II transformations (glucuronidation, sulfation, etc.) take place. The polar reaction products from these reactions are typically excreted in the urine or feces.

The rate of absorption of drugs, their degree of metabolic transformation, their distribution in the body, and their rate of excretion are collectively named pharmacokinetics. This is in effect the influence of the body on a drug as a function of time. The interaction of the drug with its targets, and the consequences of this interaction as a function of time are pharmacodynamics.

Both of these characteristics are alone governed by the drug's chemical structure. Thus, the medicinal chemist is expected to remedy any shortcomings by structural modification. In addition to ADME-Tox, a number of other characteristics must also be satisfactory, such as

- Freedom from mutagenesis
- Freedom from teratogenicity
- Chemical stability—shelf stability
- Synthetic or biological accessibility
- Acceptable cost
- Ability to patent
- Clinical efficacy
- Solubility
- Satisfactory taste (per oral administration)
- Ability to formulate satisfactorily for administration
- Freedom from idiosyncratic problems

A number of strategies are used by the medicinal chemists in order to optimize lead compounds in order to fulfill all these requirements related to optimization of desired activities and minimization of undesired effects:

- Variation of substituents—change of size, shape, and polarity
- Extension/contraction of structure—change chain size or ring size
- Ring closure/ring variation/ring fusion
- Simplification of structure
- Rigidification of structure

Examples of such modification are presented especially in Chapters 15 through 19, and generally these efforts aim toward optimizing the active conformation and physicochemical properties of the drugs with the essential and necessary pharmacophoric groups present. A very versatile principle for variation of molecules, functional groups, and substituents with focus on optimizing biological activity is the use of bioisosteres (see Section 1.4.3.1). Furthermore, stereochemical control of drug interactions with the chiral environment is essential as described in Section 1.4.3.2 and subsequent sections.

These challenges emphasize the key importance of scientists trained in interdisciplinary medicinal chemistry in drug discovery projects.

1.4.3.1 Bioisosteres

Bioisosteric replacement, also named molecular mimicry, is one of the most widely used principles for optimization of drug molecules. Bioisosteres are molecules in which atoms or functional groups are modified in order to obtain new molecules with a biological activity related to the parent molecule. The purpose is to obtain drug molecules with improved biological properties. Bioisosteric replacement can change a number of physicochemical properties of the resulting molecules compared to the parent molecule: size, shape, electronic distribution, solubility, pK_a, and chemical reactivity. These changes may lead to changes in the pharmacodynamics as well as pharmacokinetic properties, e.g., changes in potency, selectivity, bioavailability, and metabolism.

Bioisosteres have been classified as either classical or nonclassical. In classical bioisosterism, similarities in certain physicochemical properties have enabled investigators to successfully exploit several monovalent isosteres. These can be divided into the following groups: (1) fluorine versus hydrogen replacements; (2) amino-hydroxyl interchanges; (3) thiol-hydroxyl interchanges; and (4) fluorine, hydroxyl, amino, and methyl group interchanges (Grimm's hydride displacement law, referring to the different number of hydrogen atoms in the isosteric groups to compensate for valence differences). The nonclassical bioisosteres include all those replacements that are not defined by the classical definition of bioisosteres. These isosteres are capable of maintaining similar biological activity by mimicking the spatial arrangement, electronic properties, or some other physicochemical properties of the molecule or functional group that are of critical importance. A number of classical and nonclassical bioisosteres are shown in Table 1.1 representing only a selection of more commonly used bioisosteres. The concept of nonclassical bioisosterism, in particular,

TABLE 1.1
Examples of Bioisosteric Replacements

Classical	
Monovalent atoms and groups	$-F$ $-OH$ $-NH_2$ $-CH_3$ $-Cl$ $-SH$ $-Br$ $-i-Pr$ $-I$ $-t-Bu$
Bivalent atoms and groups	$-CH_2-$ $-O-$ $-S-$ $-Se-$ $-NH-$ $-COCH_2R$ $-CONHR$ $-COOR$ $-COSR$
Trivalent atoms and groups	$-N=$ $-P=$ $-CH=$ $-As=$
Ring equivalents	
Nonclassical	
Hydroxyl group	$-OH$ $-CH_2OH$ $-NHCOR$ $-NHSO_2R$ $-NHCONH_2$
Carbonyl group	CO $C{=}C(CN)_2$ $-SO_2NRR'$ $CONRR'$ $=CHCN$
Carboxylic acid	$-COOH$ $-SO_2NHR$ $-SO_2OH$ $-PO_3H_2$ $-CONHOH$
Halogen	$-X$ $-CF_3$ $-CN$ $-N(CN)_2$ $-C(CN)_3$
Spacer group	$-(CH_2)_3-$

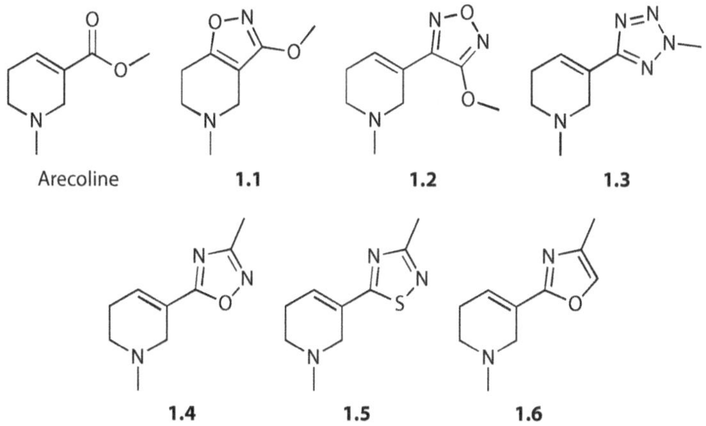

FIGURE 1.5 Chemical structures of arecoline and analogs.

is often considered to be qualitative and intuitive, and there are numerous and a constantly growing number of such bioisosteres used effectively in drug design projects (see, e.g., Chapters 15 and 16).

The conversion of the muscarinic acetylcholine receptor agonist arecoline, containing a hydrolysable ester group, into different hydrolysis-resistant heterocyclic bioisosteres is illustrated in Figure 1.5. The annulated (**1.1**) and nonannulated (**1.2** and **1.3**) bicyclic bioisosteres are potent muscarinic agonists. Similarly, compounds **1.4** and **1.5** interact potently with muscarinic receptors as agonists, whereas **1.6**, in which the 1,2,4-oxadiazole ester bioisosteric group of **1.4** is replaced by an oxazole group, shows reduced muscarinic agonist effects. Thus, the electronic effects associated with these heterocyclic rings appear to be essential for muscarinic activity.

It must be emphasized that a bioisosteric replacement strategy which has been successful for a particular group of pharmacologically active compounds, may not necessarily be effectively used in other groups of compounds active at other pharmacological targets.

1.4.3.2 Stereochemistry

Receptors, enzymes, and other pharmacological targets which by nature are protein constructs, are highly chiral. Other targets like DNA and other macromolecules in the human body are all build up of chiral building blocks. Thus, it is not surprising that chirality in the drug structures generally plays an important role in pharmacological responses. In racemic drug candidates, the desired pharmacological effect typically resides in one enantiomer, whereas the other stereoisomer(s) are pharmacologically inactive or possess different pharmacological effects. Thus, chiral drugs should preferentially be resolved into stereochemically pure isomers prior to pharmacological examination. Since many, especially of older date, synthetically prepared chiral biologically active compounds have been described pharmacologically as racemates, much of the older pharmacological literature should be read and interpreted with great care. At best the nonactive stereoisomer in a racemate is inactive and can be looked upon as chemical waste. However, with the introduction of an "inactive enantiomer" by therapeutic application of a racemic drug, one always runs the risk of introducing undesired and unknown activities and side effects. The most significant example of undesired effects of a racemic drug is the tragic case of thalidomide (Figure 1.6a). Racemic thalidomide was developed in the 1950s and was used as a mild sleeping agent and to treat morning sickness in pregnant women. Unfortunately, the drug was teratogenic and gave serious fetal abnormalities. Later it was discovered that the (*S*)-enantiomer possessed the teratogenic effect and the (*R*)-enantiomer possessed the desired activity. However, the studies also revealed that the enantiomers of thalidomide racemize under physiological conditions; thus the use of pure (*R*)-form was not a solution. This caused major changes for the legislative requirements for the introduction of chiral

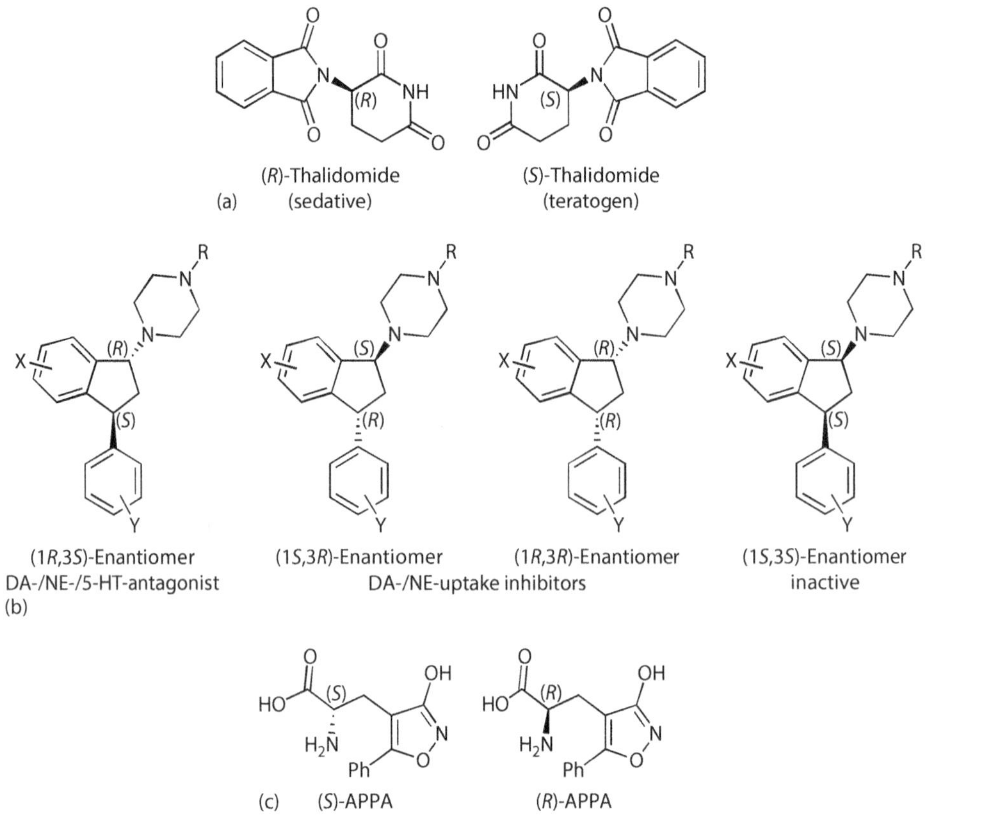

(*R*)-Thalidomide (sedative) (*S*)-Thalidomide (teratogen)
(a)

(1*R*,3*S*)-Enantiomer DA-/NE-/5-HT-antagonist
(1*S*,3*R*)-Enantiomer (1*R*,3*R*)-Enantiomer DA-/NE-uptake inhibitors
(1*S*,3*S*)-Enantiomer inactive
(b)

(c) (*S*)-APPA (*R*)-APPA

FIGURE 1.6 Chemical structures of (a) (*R*)- and (*S*)-thalidomide, (b) the four stereoisomers of 1-piperazino-3-phenylindans, and (c) the two enantiomers of the phenyl analog of AMPA (APPA).

drugs subsequently and mandatory test for teratogenic activity. Thalidomide was off the market for many years, but has been introduced again for treatment of leprosy and other diagnoses, but under very strict guidelines.

Figure 1.6 exemplifies the importance of stereochemistry in studies of the relationship between structure and pharmacological activity (SAR studies). Figure 1.6b shows four stereoisomers which are two pairs of enantiomers of two diastereomeric compounds. These 1-piperazino-3-phenylindans were synthesized, resolved, structurally analyzed, and pharmacologically characterized as part of a comprehensive drug research program in the field of central biogenic amine neurotransmission. Whereas one of these stereoisomers turned out to be inactive, two of them were inhibitors of dopamine (DA) and norepinephrine (NE) uptake, and one isomer showed antagonist effects at DA, NE, and serotonin (5-HT) receptors. It is evident that a pharmacological characterization of a synthetic mixture of these compounds would be meaningless.

The 3-isoxazolol amino acid, APPA (Figure 1.6c), is an analog of the standard agonist, AMPA, for the AMPA subtype of excitatory glutamate receptors (Chapter 15). APPA was tested pharmacologically as the racemate which showed the characteristics of a partial agonist at AMPA receptors. Subsequent pharmacological characterization of the pure enantiomers quite surprisingly disclosed that (*S*)-APPA is a full AMPA receptor agonist, whereas (*R*)-APPA turned out to be an AMPA antagonist. This observation prompted intensive pharmacological studies, and as a result it was demonstrated that administration of a fixed ratio of an agonist and a competitive antagonist always provides a partial agonist response at an efficacy level dependent on the administered ratio of compounds and their relative potencies as agonist and antagonist, respectively. This phenomenon was named "functional partial agonism." An interesting aspect of this pharmacological concept is that administration of an antagonist drug inherently establishes functional partial agonism together with the endogenous agonist at the target receptor.

1.4.3.3 Membrane Penetration: Including the Lipinski Rule of Five

In drug discovery projects, an issue of major importance is the design of drug molecules capable of penetrating different biological membranes effectively and that allow effective concentrations to build up at the therapeutic target. The structure and physiochemical properties of the drug molecule obviously are of decisive importance, and it is possible to establish the following empirical rules:

- Some small and rather water-soluble substances pass in and out of cells through water-lined transmembrane pores.
- Other polar agents are conducted into or out of cells by membrane-associated and energy-consuming proteins. Polar nutrients that the cell requires, such as glucose and many amino acids, fit into this category. More recently, drug resistance by cells has been shown to be mediated in many cases by analogous protein importers and exporters.
- The blood–brain barrier (BBB) normally is not easily permeable by neutral amino acids. However, such compounds with sufficiently small difference between the pK_a values will have a relatively low I/U ratio (the ratio between the ionized/zwitterionic form and the unionized form of the amino acid in solution). As an example, THIP (Figure 1.4) has pK_a values of 4.4 and 8.5 and a calculated I/U ratio of about 1000. Thus, 0.1% of THIP in solution is unionized, and this fraction apparently permits THIP to penetrate the BBB quite easily. Other neutral amino acids typically have I/U ratios around 500,000 and thus have much lower fractions of unionized molecules in solution, and such compounds normally do not penetrate into the brain after systemic administration.
- Molecules that are partially water soluble and partially lipid soluble can pass through cell membranes by passive diffusion and are driven in the direction of the lowest concentration.
- In cells lining the intestinal tract, it is possible for molecules with these characteristics to pass into the blood through the cell membrane alone.
- Finally, it is also possible for molecules with suitable water solubility, small size, and compact shape to pass into the blood between cells. This last route is generally not available for passage into the CNS, because the cells forming the BBB are organized much closer together and thus prevent such entry into the brain.

Whereas there are no guarantees and many exceptions, the majority of effective oral drugs obey the Lipinski rule of five:

- The substance should have a molecular weight of 500 or less.
- It should have fewer than five hydrogen-bond donors.
- It should have fewer than 10 hydrogen-bond acceptors.
- The substance should have a calculated log P (clog P) between approximately −1 and +5.

The Lipinski rule of five is thus an empirical rule, where the number five occurs several times. The rule is a helpful guide rather than a law of nature.

1.4.3.4 Structure-Based Drug Design

During the early 1980s, the possibility to rationally design drugs on the basis of structures of therapeutically relevant biomolecules was an unrealized dream for many medicinal chemists. The first projects were underway in the mid-1980s, and today, even though there are still many obstacles and unsolved problems, structure-based drug design now is an integral part of many drug discovery programs. Major breakthroughs are represented by the publication of high-resolution 3D X-ray crystallographic structures of neurotransmitter receptors and transporters, e.g., determination of full-length $GABA_A$ receptor and full-length dopamine transporter structures, both obtained after many years of extensive research. In Chapter 4, a number of examples of this impressive drug design approach are described.

As structural genomics, bioinformatics, and computational power continue to almost explode with new advances, further successes in structure-based drug design are likely to follow. Each year, new targets are being identified, and structures of those targets are being determined at an amazing rate, and our capability to capture a qualitative picture of the interaction between macromolecules and ligands is accelerating.

1.5 INDIVIDUALIZED MEDICINE AND CONCLUDING REMARKS

The mapping of the human genome leads us to the identification of new targets for therapeutic interventions, and even allows us to dream of the possibility of correcting genetic defects, enhancing our prospects for a longer and more healthy life, and for devising drugs for specific individuals. Presuming that individual variations in therapeutic response may have a genetic origin, and thus dividing populations into subgroups with similar genetic characteristics, might allow us to prescribe drugs and even dosages within these groups. This form of individual gene typing is already possible but still very resource demanding as per day's techniques. It is likely that perplexing species differences in response to, for example, chemotherapy, that complicates drug development, may also be understood, when the individual genome mapping becomes more elaborate and cheap.

The new biological capabilities raise many new prospects and problems for drug companies and, in general, for the society, not only scientifically but also morally. Scientific knowledge by itself is morally neutral, but how it is used, is not.

In conclusion, there has never been a more exciting time to take up the study of medicinal chemistry. The technological developments and the amount of information will grow with increasing speed, and scientists may eventually risk to be drowned in this multitude of possibilities. However, the intelligent, intuitive, and skilled medicinal chemist will be able to maneuver in this ocean of multiplicity and to continue the series of brilliant achievements by the pioneers in drug discovery during the past century.

FURTHER READING

Anderson, A.C. 2003. The process of structure-based drug design. *Chem. Biol.* 10:787–797.

Harvey, A.L., Edrada-Ebel, R.A., and Quinn, R.J. 2015. The re-emergence of natural products for drug discovery in the genomics era. *Nat. Rev. Drug Discov.* 14:111–129.

Hughes, J.P., Rees, S., Kalindjian, S.B., and Philpott, K.L. 2011. Principles of early drug discovery. *Br. J. Pharmacol.* 162:1239–1249.

Lindsay, M.A. 2003. Target discovery. *Nat. Rev. Drug Discov.* 2:831–838.

Lipinski, C.A., Lombardo, F., Dominy, B.W., and Feeney, P.J. 2001. Experimental and computational approaches to estimate solubility and permeability in drug discovery and development settings. *Adv. Drug Deliv. Rev.* 46:3–26.

Meanwell, N.A. 2011. Synopsis of some recent tactical application of bioisosteres in drug design. *J. Med. Chem.* 54:2529–2591.

Morgan, S., Grootendorst, P., Lexchin, J., Cunningham, C., and Greyson, D. 2011. The cost of drug development: A systematic review. *Health Policy* 100:4–17.

Patani, G.A. and LaVoie, E.J. 1996. Bioisosterism: A rational approach in drug design. *Chem. Rev.* 96:3147–3176.

2 Molecular Recognition

*Thomas Balle and Tommy Liljefors**

CONTENTS

2.1 INTRODUCTION

Molecular recognition is the foundation for the function of virtually any biological system. It relies on the existence of favorable interactions between two (or more) molecules which could be a neurotransmitter molecule and its protein receptor or an enzyme and its substrate. Recognition of one molecule by another is driven by energetics. If two molecules attract each other, the total free energy of the two molecules and their surroundings will be lower compared to the situation where the two molecules are far apart. Therefore, based on "the lowest potential energy principle," the two molecules will tend to stick together in a noncovalent complex. Understanding the basic principles of noncovalent interactions is essential to understand how biological systems work at the molecular level, and it is a prerequisite for understanding how drugs interact with their target macromolecules and how they obtain selectivity or even specificity. It is also the foundation of "structure-based drug design," a discipline where medicinal chemists seek to optimize the strength of molecular interactions exploiting the knowledge of the three-dimensional structure of the host molecule combined with an understanding of energetic contributions to binding from different parts of the molecules (see Chapter 4).

Molecular recognition is often highly selective and the first attempt to understand this selectivity was made in 1894 by Emil Fisher. Fisher formulated the "lock-and-key principle" (Figure 2.1a) to explain why certain enzymes would only degrade certain substrates. The principle in this hypothesis is that the substrate has to fit like a key in a lock to trigger an enzymatic reaction. The lock-and-key principle was also applied in the broader context of medicinal chemistry, and it makes perfect sense that a drug has to bind to its receptor in order to exert an action—"if it doesn't bind it doesn't work."

* Deceased.

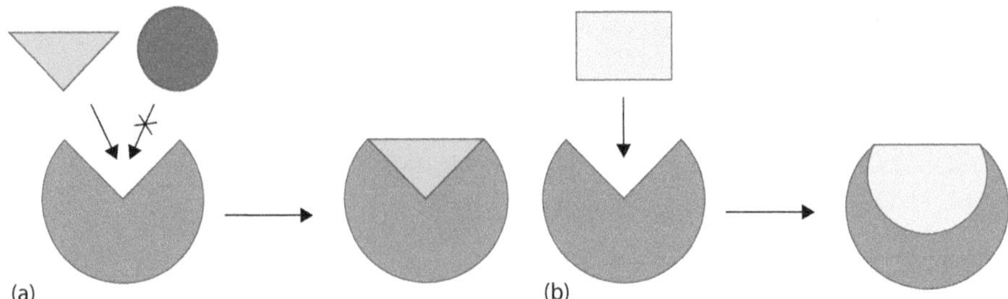

FIGURE 2.1 (a) Lock and key and (b) induced fit hypothesis.

However, one of the limitations of the lock-and-key principle as a general model was its inability to explain why certain inhibitors could block enzyme activity without affecting binding of the substrate and why some enzymes were extremely specific toward their substrate while others were not. To accommodate these scenarios, Daniel E. Koshland in 1958 introduced the "induced fit theory," which is illustrated in Figure 2.1b. In the induced fit model which is often referred to as the "hand in glove" model, the substrate and the enzyme are mutually allowed to adapt to each other upon binding just like the glove changes its conformation when a hand slips into it. The model is in perfect agreement with the notion that small molecules are flexible and can change conformation by rotation around single bonds. Likewise, proteins such as enzymes and receptors are flexible and can adapt to small molecule ligands by changing amino acid side chain conformations or by movement of the protein backbone. Today, the three-dimensional structures of a huge number of proteins have been determined by X-ray crystallography in the presence of different drug molecules. These structures support the induced fit hypothesis which today is generally accepted.

The understanding of molecular recognition and specificity in drug binding has been greatly aided by X-ray crystallographic structures of macromolecules in complex with various ligands. Combined with an understanding of the attractive and repulsive physical forces acting between the two interacting molecules, one can understand molecular recognition and learn how to optimize molecules toward better affinity and specificity. The aim of this chapter is to understand, in terms of basic physical chemistry and examples, the nature and magnitude of molecular interactions. This is one of the most fundamental aspects of medicinal chemistry and will provide a framework for many of the other chapters in this book where affinity, selectivity, and structure–affinity relationships are being discussed.

Throughout this book, protein X-ray structures are referred to by a four-letter code which is the same identifier that is used in the protein data bank (PDB), a repository for experimentally determined structures of protein and protein–ligand complexes. All the structures can be accessed from the web page www.rcsb.org or via apps for mobile devices, e.g., "Pymol" from App Store or "RCSB PDB Mobile" from App Store and Google Play. The apps can be downloaded free of charge, and once installed, you can tap in the four-letter pdb code and the app will automatically download and display a three-dimensional model of the protein complex so you can explore the fascinating world of protein structures on your own.

2.2 DETERMINATION OF AFFINITY

The affinity of a ligand for its receptor is defined as the equilibrium (i.e., the difference in free energy, ΔG) between the bound ligand–protein complex on one side and the "free" unbound ligand and unbound protein on the other as illustrated in Figure 2.2. This means that it is a basic thermodynamic equilibrium that underlies the principles of molecular recognition.

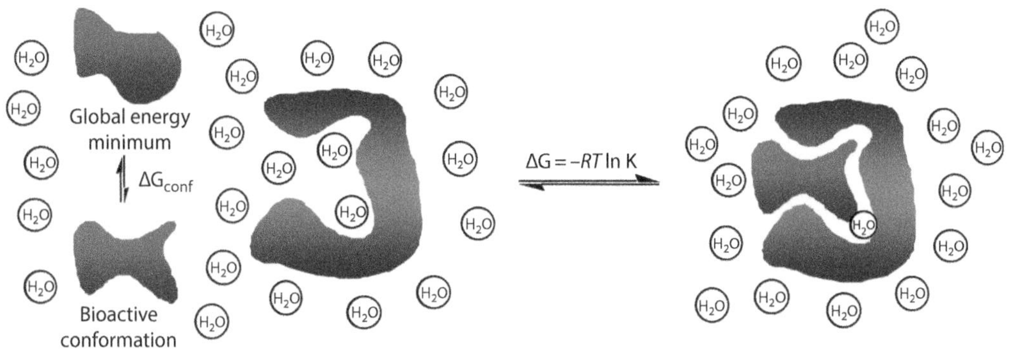

FIGURE 2.2 The equilibrium determining the affinity of a ligand.

Before the ligand can bind to the receptor it may have to change its conformation and so may the receptor. Water also plays an important role in the binding process and a change in the network of water molecules surrounding ligand and receptor may occur as illustrated in Figure 2.2.

The free energy difference is related to the equilibrium constant K by

$$\Delta G = -RT \ln K \tag{2.1}$$

where

R is the gas constant (8.315 J/K/mol)
T is the temperature in kelvin
K is (complex)/(ligand) × (receptor)

A higher affinity is the result of the equilibrium in Figure 2.2 being pushed toward the receptor–ligand complex on the right. This implies a larger positive value of K and, thus, a larger negative value of ΔG. In medicinal chemistry, the affinity of a ligand is most often given as an inhibition constant K_i or by an IC_{50}-value which is the concentration of an inhibitor that is required to displace 50% of the specific binding of a radioactively labeled ligand in a radioligand binding assay. Since $K = 1/K_i$, the free energy difference in terms of K_i can be written as

$$\Delta G = RT \ln K_i \tag{2.2}$$

The ΔG term can be dissected into an enthalpic component (ΔH) and an entropic component (ΔS) according to

$$\Delta G = \Delta H - T\Delta S \tag{2.3}$$

A higher affinity (a more negative ΔG) corresponds to a smaller value of the inhibition constant K_i (most often given in nM or μM). A K_i of 1 nM corresponds to a ΔG of −53.4 kJ/mol at 310 K and a K_i of 1 μM to −35.6 kJ/mol. Furthermore, using Equation 2.2 it can be calculated that an increase in ΔG by 5.9 kJ/mol corresponds to a 10-fold loss in affinity (10-fold increase in K_i). An example of the size of this energy in terms of molecular structural change is shown in Figure 2.3. A conformational change of the ethyl group in ethylbenzene from a perpendicular conformation

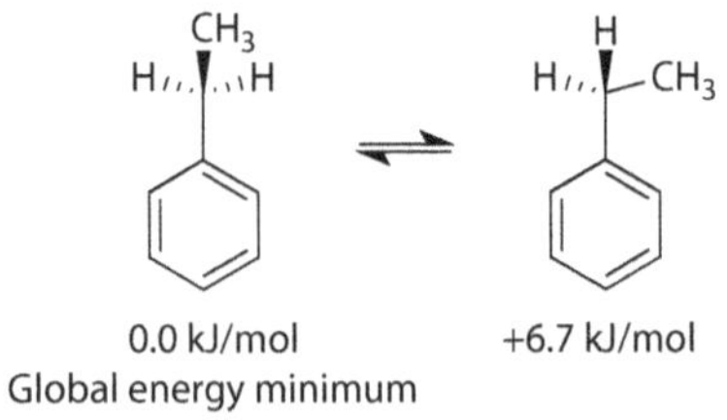

FIGURE 2.3 Conformational energies of ethylbenzene.

with respect to the phenyl ring (the lowest energy conformation) to a conformation with a coplanar carbon skeleton increases the energy by 6.7 kJ/mol. Thus, even modest changes in the conformation (the shape) of a ligand can result in a significant decrease in the affinity, a fact that should carefully be taken into account in ligand design (for further discussions on ligand conformations, see Section 2.3.2). As mentioned earlier, IC_{50} expresses the concentration of an inhibitor which displaces 50% of the specific binding of a radioactively labeled ligand in a radioligand displacement experiment. The IC_{50} value can be converted to an inhibition constant K_i by the Cheng–Prusoff equation:

$$K_i = \frac{IC_{50}}{\left(1 + \frac{[L]}{K_D}\right)} \tag{2.4}$$

where

[L] is the concentration of the radioligand used in the assay

K_D is the affinity of the radioligand for the receptor

It should be noted that IC_{50} values are dependent on the concentration and the affinity of the radioligand used to determine it. Therefore, care should be taken when comparing IC_{50} values across different experiments unless the same radioligand and assay conditions were used. If not, one needs to convert the IC_{50} value to K_i using Equation 2.4 to get comparable values. K_i is namely independent of radioligand and radioligand concentration and can be compared across assays.

2.3 PARTITIONING OF ΔG

To get a better understanding of the driving forces underlying molecular recognition, it is helpful to express the free energy difference ΔG as a sum of the individual components from which it may be considered to be made up. The main drivers of recognition are

- Electrostatic interactions including ion–ion, ion–dipole, and dipole–dipole interactions
- Lipophilicity/hydrophobicity
- Shape complementarity

Other factors such as entropy and conformational changes also influence the recognition process, and summing up all these contributions leads to Equation 2.5. Several different partitioning schemes have been suggested in the literature and differ in their level of detail. The one used here is essentially as suggested by Williams (see Further Reading):

$$\Delta G = \Delta G_{transl+rot} + \Delta G_{conf} + \Delta G_{polar} + \Delta G_{hydrophob} + \Delta G_{vdW} \quad (2.5)$$

where

$\Delta G_{transl+rot}$ accounts for the restrictions of translational movements (movements in x-, y-, and z-directions) and restrictions of rotations (about the x-, y-, and z-axes) of the "whole" molecule from the unbound to the bound state

ΔG_{conf} is the difference in the conformational free energies between the unbound and bound states due to conformational restrictions in the ligand–protein complex

ΔG_{polar} is the free energy change due to interactions of polar functional groups in the binding cavity of the protein

$\Delta G_{hydrophob}$ accounts for the binding free energy due to the hydrophobic effect

ΔG_{vdW} gives the difference in free energy due to van der Waals (vdW) interactions in the bound and unbound states

In the following sections, the different terms in Equation 2.5 and their magnitudes will be discussed in more detail and illustrated in terms of ligand–protein recognition.

2.3.1 $\Delta G_{TRANSL+ROT}$: The Freezing of the Overall Molecular Motion

In the "free" unbound state, both receptor and ligand are able to rotate and translate freely around in the aqueous solution. When bound in the cavity, the freedom of the ligand to tumble is lost because the small ligand will now have to follow the motions of the big protein. The loss of freedom is unfavorable in a thermodynamic sense and is a cost associated with ligand binding that must be overcome by the favorable binding forces to enable formation of a complex. The freezing or restriction of the motions of the ligand is a decrease in entropy resulting in a more negative ΔS and consequently a more negative TΔS. According to Equation 2.3, this results in a more positive ΔG. The magnitude of this free energy cost has been much debated in the literature. Explicit calculations show that it varies only slightly with molecular weight, but an important problem for the estimation of $\Delta G_{transl+rot}$ is that it depends on the "tightness" of the ligand–protein complex. A tighter complex leads to a greater loss of freedom of movement and thus to a more negative TΔS. Most estimates of $\Delta G_{transl+rot}$ range from 12 kJ/mol for a "loose" complex to 45 kJ/mol for a tightly bound complex. Whatever the exact magnitude of $\Delta G_{transl+rot}$ is in a particular case, it is a very significant energy to overcome by the favorable binding forces. Consider a ligand with an affinity (K_i) of 1 nM corresponding to ΔG of −53.4 kJ/mol at 310 K (Section 2.2). In order to end up with this free energy difference between the bound and unbound states, the favorable binding forces must produce not only 53.4 kJ/mol of ligand–protein binding energy but in addition 12–45 kJ/mol of free energy is required to compensate for the loss of entropy associated with binding. It should be noted that this free energy cost of ligand–protein association is always present and cannot be reduced by ligand design. However, the exact value of $\Delta G_{transl+rot}$ is only important for predictions of "absolute" ΔG values. To a first approximation it cancels out when comparing the affinities of different ligands to the same receptor.

2.3.2 ΔG_{CONF}: Conformational Changes of Ligand and Receptor

The restriction of motions that are accounted for in $\Delta G_{transl+rot}$ described in Section 2.3.1 refers to the "overall" motion of the molecule relative to its surroundings. However, there is an additional internal motion which is more or less frozen upon ligand binding. Most ligand molecules are flexible which means that in the aqueous phase outside the binding cavity, the ligand constantly undergoes conformational changes by rotation around single bonds. For example, the dihedral angle in a hydrocarbon chain alters between gauche and anticonformations resulting in a mixture of ligand conformations of different ligand shapes. A ligand generally binds to a protein in a single well-defined conformation that positions functional groups used for binding in appropriate locations in space for interactions with their binding partners in the protein. This implies that the motions corresponding to the conformational freedom in aqueous solution are to a large extent frozen in the binding site. As discussed earlier for $\Delta G_{transl+rot}$, this leads to a decrease in the entropy (conformational entropy) giving a more negative ΔS and $T\Delta S$ and thus a free energy cost for binding. The magnitude of ΔG_{conf} due to $T\Delta S_{conf}$ has been estimated to be 1–6 kJ/mol per restricted internal rotation and depends on the "tightness" of the ligand–protein complex as in the case of $\Delta G_{transl+rot}$ (Section 2.3.1). To minimize the cost associated with freezing of internal motion, one may consider constraining the ligand by introduction of double bonds or by incorporating ring structures.

A second and very important energy contribution to ΔG_{conf} comes from the cost associated with the ligand adopting its bioactive conformation. The ligand conformation in aqueous solution is not necessarily the same as the protein bound conformation and this difference in energy is reflected directly in the binding affinity. Comparisons of ligand conformations observed in X-ray structures of ligand–protein complexes and ligand conformations in aqueous phase (as calculated by state-of-the-art computational methods) show that a ligand in general does not bind to the protein in its preferred conformation (lowest energy conformation) in aqueous solution. An example of this is shown in Figure 2.4. Palmitic acid prefers the well-known all-anti (zigzag) conformation of the hydrocarbon chain in aqueous solution, but binds to the adipocyte lipid-binding protein with an affinity (K_i) of 77 nM in a significantly folded conformation. The energy required for palmitic acid to adopt the binding conformation has been calculated to be 10.5 kJ/mol. This conformational energy penalty is detrimental to binding and has the effect of increasing the K_i value in comparison to a case in which the ligand binds in its preferred conformation in aqueous solution. As shown in Section 2.2, a conformational energy penalty of 5.9 kJ/mol corresponds to a decrease in affinity (increase of K_i) by a factor of 10. The 10.5 kJ/mol penalty for palmitic acid thus corresponds to an almost 100-fold loss of affinity. It is consequently of high importance to avoid introducing significant conformational energy penalties when designing new ligands. Calculations of the conformational energy penalties for ligands in a series of X-ray structures of ligand–protein complexes indicate that these energy penalties in general are below 13 kJ/mol. This may be used as a rule of thumb in ligand design. In this context, it is important to note that in calculations of conformational energy penalties, the conformational properties of the unbound ligand "in aqueous phase" must be used as the reference state

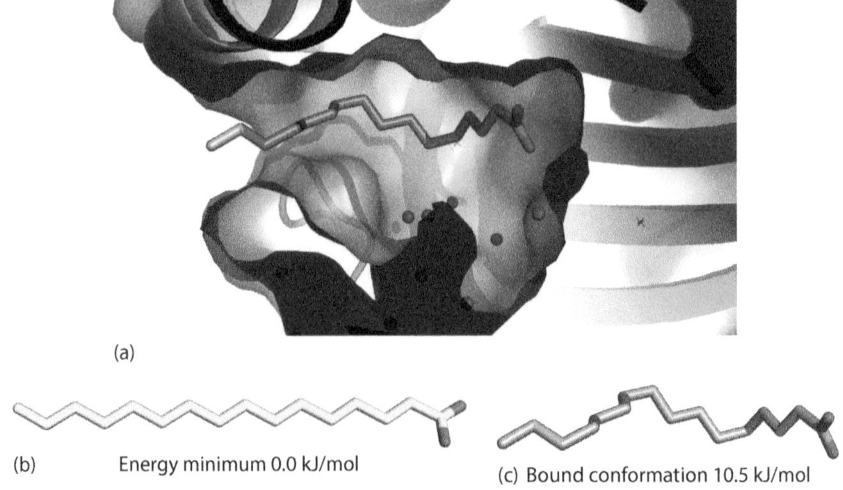

FIGURE 2.4 (a) Palmitic acid bound to the adipocyte lipid-binding protein (pdb-code 1LIE), (b) the preferred conformation of palmitic acid in aqueous solution and (c) the conformation bound to the protein.

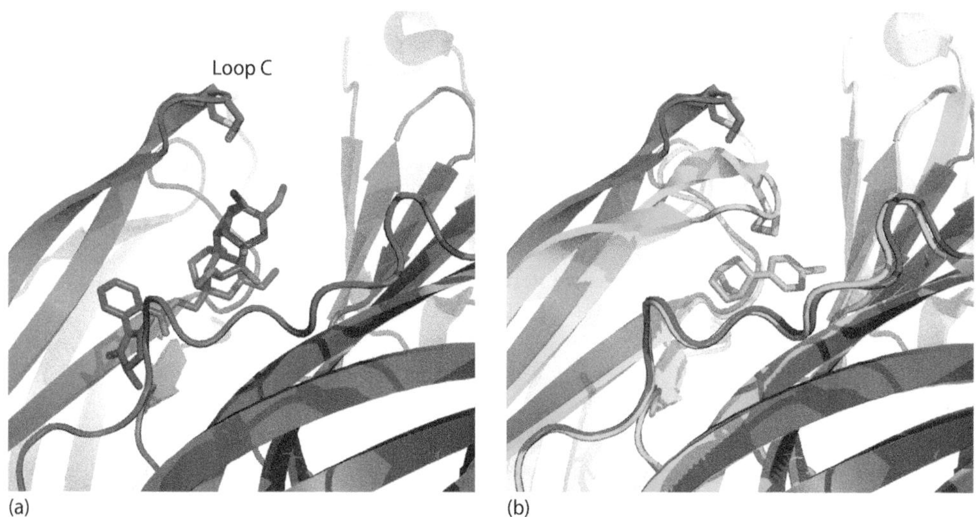

FIGURE 2.5 Backbone movements in acetylcholine binding protein (AChBP). (a) Methyllycaconitine (nicotinic acetylcholine receptor [nAChR] antagonist, purple carbons) bound to AChBP (pdb-code 2BYR). (b) Epibatidine (nAChR agonist, cyan carbons) bound to AChBP (pdb-code 2BYQ). The protein is superimposed on the methyllycaconitine bound structure from Figure 2.5a to illustrate the large-scale backbone movements of loop C.

(see Further Reading). In terms of ΔG_{conf}, rigid molecules have an advantage relative to more flexible ligands. The binding of a rigid ligand does not result in a loss of conformational entropy and if the ligand has only one possible conformation (or has one strongly preferred conformation corresponding to the binding conformation), the conformational energy penalty is zero. Although highly rigid molecules are ideal as ligands, it is a great challenge to design such ligands. For instance, functional groups taking part in interactions in the binding cavity have to be designed to occupy precisely correct positions in space, as no (or very small) adjustments of their positions are possible in a rigid ligand.

So far only conformational changes in the ligand have been discussed, but conformational changes most often also occur in the protein. When comparing structurally similar ligands, the conformational changes of the protein can, to a first approximation, be expected to be similar and will thus cancel out when comparing relative affinities. For structurally diverse ligands, this may not be the case. Not only the amino acid side chains adjust their conformations to optimize their interactions with the ligand, but also the protein backbone conformation may change. In some cases, this may result in major movements of, for example, loops or even entire protein domains. Examples of such movements are shown in Figure 2.5. The acetylcholine binding protein (AChBP) which is a soluble protein, is used as surrogate for structural studies of nicotinic acetylcholine receptors, because these cannot easily be crystallized. The AChBP has a flexible loop (loop C) that adjusts its position in response to the size of the ligand (Figure 2.5). This has implications for the pharmacological profile of the ligand. Protein flexibility is a major challenge in structure-based drug design and is currently the focus of much research.

2.3.3 ΔG_{POLAR}: Electrostatic Interactions and Hydrogen Bonding

ΔG_{polar} is the change in free energy due to interactions between polar functional groups in the ligand and polar amino acid side chains and/or C=O and NH backbone groups in the binding cavity of the protein. Indirect ligand–protein interactions mediated via bridging water molecules in the binding cavity also play an important role. Electrostatic interactions include salt bridges between positively and negatively charged residues (ion–ion), ion–dipole, and dipole–dipole interactions which are well described in books on physical chemistry to which the reader is referred for details. The attraction between opposite charges or antiparallel dipoles plays an important role in ligand–protein recognition. The strength of any electrostatic interaction (E_{polar}) is given by Coulomb's law:

$$E_{polar} = \frac{q_i q_j}{\varepsilon r_{ij}} \tag{2.6}$$

where

- q_i and q_j are integer values for ion–ion interactions or partial atomic charges (summed over the participating atoms) for other polar interactions
- ε is the dielectric constant
- r_{ij} is the distance between the charges

In Equation 2.6, it is important to note that the electrostatic energy E_{polar} depends on the dielectric constant (ε) which is a measure of the shielding of the electrostatic interactions by the environment. The dielectric constant in vacuum is 1 (no shielding) and in water it is 78.4 (25°C). E_{polar} is difficult to quantify in proteins as ε is not uniform throughout the protein but depends on the microenvironment in the protein. A value of about 4 is often used for a lipophilic environment in the interior of a protein.

The relative strength of the different types of electrostatic interactions is ion–ion > ion–dipole > dipole–dipole. Ion–ion interactions do not depend on the relative orientation of the interacting partners, whereas ion–dipole and dipole–dipole interactions are strongly dependent on geometry. For instance, the interaction between antiparallel dipoles is attractive, whereas that between parallel dipoles is repulsive.

2.3.3.1 Hydrogen Bonds

A hydrogen bond X–H–A may be described as an electrostatic attraction between a hydrogen atom bound to an electronegative atom X (in ligand–protein interactions most often nitrogen or oxygen) and an additional electronegative hydrogen bond acceptor atom A. The typical hydrogen bond distance is 2.5–3.0 Å and is measured as the distance between the heavy atoms X and A. This convention is used because the position of hydrogens in most cases is not well determined at the resolution typically achieved in protein X-ray crystallography. Hydrogen bonds are highly orientation dependent with an optimal X–H–A angle of 180°. Examples of different types of hydrogen bonds commonly observed in ligand–protein complexes are shown in Figure 2.6.

Figure 2.7 displays the binding of (*S*)-glutamate to the ligand-binding domain of the ionotropic glutamate receptor GluA2 (see Chapters 12 and 15) featuring a "salt bridge" and a number of other charge-assisted hydrogen bonds between the ligand and the receptor, and also between the ligand and water molecules in the active site.

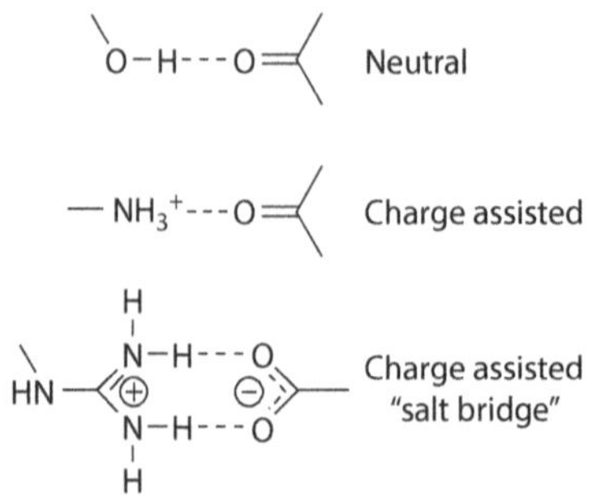

FIGURE 2.6 Examples of different types of hydrogen bonds observed in ligand–protein complexes.

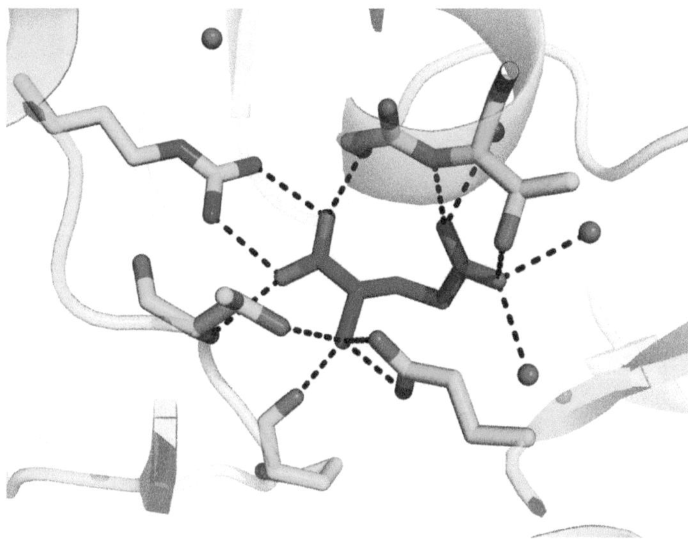

FIGURE 2.7 The binding of (*S*)-glutamate (purple carbons) to the ligand-binding domain of the ionotropic glutamate receptor GluA2 (pdb-code 1FTJ).

In order to understand the contribution of hydrogen bonding or other polar interactions to ligand binding, it is crucial to keep in mind that the ligand–protein interaction is an equilibrium process (Figure 2.2) and that hydrogen bonding is an exchange process. Before formation of the ligand–protein complex (left-hand side in Figure 2.2), the polar functional groups of the ligand, the polar amino acid residues, and C=O and NH backbone groups in the protein are engaged in hydrogen bonding with surrounding solvent water molecules. In the ligand–protein complex (right-hand side in Figure 2.2), these hydrogen bonds to the solvent are replaced by hydrogen bonds between the ligand and the protein. The net effect of this hydrogen bond exchange process is the difference in free energy between hydrogen bonding to water and to the protein. As a consequence of this exchange process, a substituent, which is hydrogen, bonded to water molecules in the aqueous phase, but is buried in the binding cavity but not hydrogen bonded to the protein (an unpaired hydrogen bond), is strongly unfavorable for binding. It has been shown that the energy cost for an unpaired hydrogen bond is ca. 4 kJ/mol for a neutral substituent and ca. 16 kJ/mol for a charged substituent. This is equivalent to a loss of affinity by a factor of 5 and 500, respectively. The successful formation of a hydrogen bond in the binding cavity has been estimated to contribute to the binding affinity by 2–6.5 kJ/mol (corresponding to an affinity increase by a factor of 2–13) for a neutral bond and by 10–20 kJ/mol (equivalent to a 50–500-fold increase in affinity) for a charge-assisted hydrogen bond or a salt bridge.

2.3.3.2 Polar Interactions Involving Aromatic Ring Systems

Other types of polar interactions often observed in ligand–protein complexes are π–π and cation–π interactions. The exact nature of these interactions is quite complex, but qualitatively they can be easily understood in terms of electrostatics. Figure 2.8 displays the calculated molecular electrostatic potential of benzene. It is obtained by calculating the energies of interaction between a benzene ring and a cation placed in different positions around the aromatic ring. The electrostatic potential in Figure 2.8 is color-coded on the vdW surface of the molecule. A red color indicates a strong attraction between the cation and the aromatic ring and a blue color indicates a strong repulsion. Thus, a cation, for example, an ammonium ion, may favorably interact with the face of the benzene ring as shown in Figure 2.8. In π–π interactions, the edge of one benzene ring interacts with the face of the other.

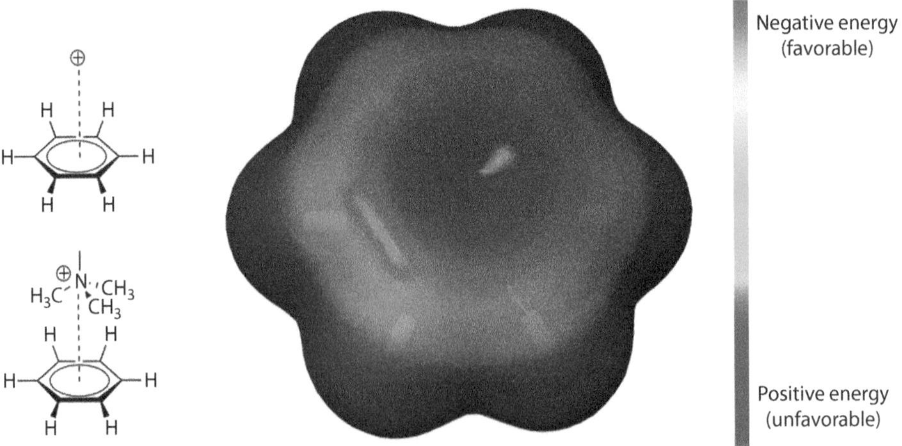

FIGURE 2.8 The molecular electrostatic potential of benzene, representing the amino acid phenylalanine.

Other aromatic rings such as phenol and indole display similar electrostatic potentials as benzene. Thus, the aromatic rings of phenylalanine, tyrosine, and tryptophan side chains may favorably interact with positively charged functional groups of the ligand. It has been estimated that a cation–π interaction may contribute by 8–17 kJ/mol to the overall binding of the ligand which is equivalent to a 23–760-fold increase in affinity. Figure 2.9 shows the binding of acetylcholine bound to the AChBP. The binding displays cation–π interaction between the ammonium group acetylcholine and the indole ring system of Trp143. Cation–π interactions are also observed to the aromatic rings of Tyr185, Tyr192, and Trp53.

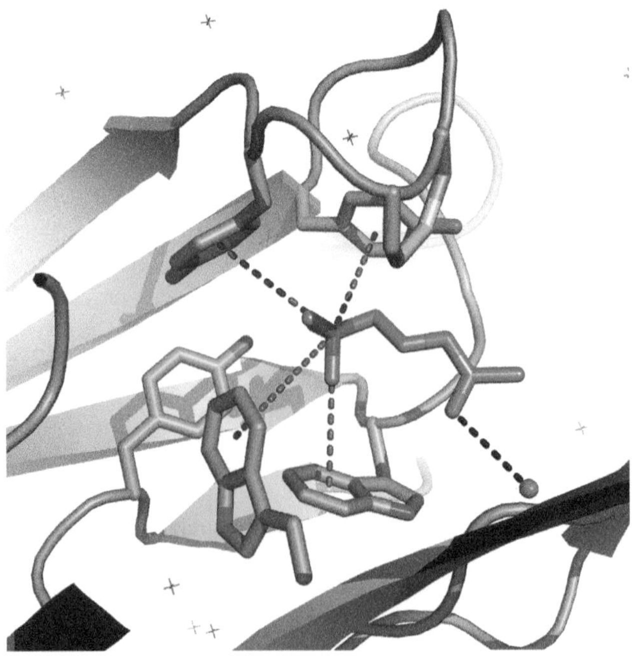

FIGURE 2.9 Acetylcholine (blue carbons) in the binding pocket of the acetylcholine binding protein (pdb code 3WIP). Cation–π interactions are indicated by gray dashed lines. The black dashed line represents a hydrogen bond from the carbonyl oxygen atom of acetylcholine to a tightly bound water molecule.

2.3.4 $\Delta G_{HYDROPHOB}$: The Hydrophobic Effect

The hydrophobic effect is a concept used to describe the tendency of nonpolar compounds to transfer from water to an organic phase, for example, a lipophilic region of a protein. When a lipophilic compound interacts with water, it changes the dynamic network of hydrogen bonds between water molecules in pure liquid. It creates a new interface in which water molecules around the lipophilic compound assume a more ordered arrangement than bulk water. This results in a decrease in entropy which according to Equation 2.3 leads to a decreased ΔG. Formation of a ligand–protein complex displaces the ordered water from the ligand and the protein by bulk water as shown in Figure 2.10. The increased "freedom" of movement of the released water molecules gives an increase in entropy (ΔS) and, according to Equation 2.3, a more negative ΔG—an increase in affinity. The magnitude of the hydrophobic effect is related to the area of hydrophobic surface that is buried in the binding cavity. Estimates, based on measurements of solvent transfer and ligand binding, range between 0.1 and 0.24 kJ/Å^2 mol. The burial of a methyl group of ca. 25 Å^2 is thus expected to result in an affinity increase by a factor of 3–10. In cases of a more perfect fit between a methyl group and the protein, the affinity increase may be even larger.

Analysis of ligand–protein interactions and attempts at ligand design often focus on hydrogen bonding and other electrostatic interactions. However, in many cases, even strong hydrogen bond interactions may favorably be replaced by hydrophobic interactions. An example of this is shown in Figure 2.11. The influenza neuraminidase inhibitor **2.1** binds to the protein with an affinity (IC_{50})

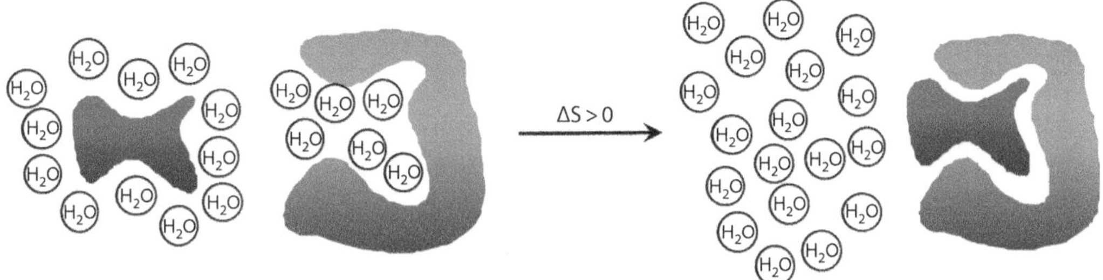

FIGURE 2.10 The hydrophobic effect.

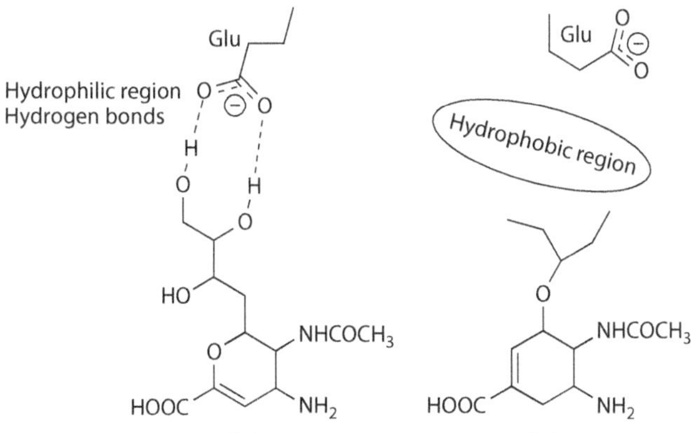

Glu
Hydrophilic region
Hydrogen bonds
HO
NHCOCH$_3$
HOOC
NH$_2$
2.1
$IC_{50} = 150$ nM

Glu
Hydrophobic region
NHCOCH$_3$
HOOC
NH$_2$
2.2
$IC_{50} = 1$ nM

FIGURE 2.11 Influenza neuraminidase inhibitors.

of 150 nM. According to the ligand–receptor X-ray structure, the binding displays a bidentate, charge-assisted hydrogen bond between the terminal glycerol hydroxyl groups and a glutamate side chain. Removal of the glycerol side chain as in analog **2.2** and replacement with a hydrophobic alkyl group increases the affinity to 1 nM (the minor modifications of the six-membered ring are not expected to influence the affinity significantly). An X-ray structure of the protein complex (**2.2**) shows that the glutamate side chain is folded back, opening up a large hydrophobic pocket as indicated in Figure 2.11.

2.3.5 ΔG_{vdW}: Attractive and Repulsive vdW Interactions

Nonpolar interactions between atoms may be calculated according to Equation 2.7. When the vdW energy is calculated and plotted as a function of distance (radius), it results in the energy curve shown in Figure 2.12. As evident from the figure, vdW interactions may be attractive as well as repulsive depending on the distance between the two interacting atoms/molecules. At short atom–atom distances, the vdW interaction is repulsive due to overlap of the electron clouds. The repulsion rises steeply with decreasing atom–atom distance in this region of the energy curve. When a part of the ligand clashes with atoms in the binding site, this steric repulsive vdW interaction is responsible for the often dramatic reduction in affinity which is observed. At a longer distance, there is a region of attraction between the atoms. This attraction is due to the so-called dispersion forces. These are basically of electrostatic nature and due to interactions between temporary dipoles induced in two adjacent atoms. For a single atom–atom contact, the strength of the interaction is small, ca. 0.2 kJ/mol. However, as the total number of such interactions may be large, the dispersion interaction may be significant in cases of a close fit between ligand and protein. Note also that E_{vdW} is inversely correlated to distance (r) to the power of 6

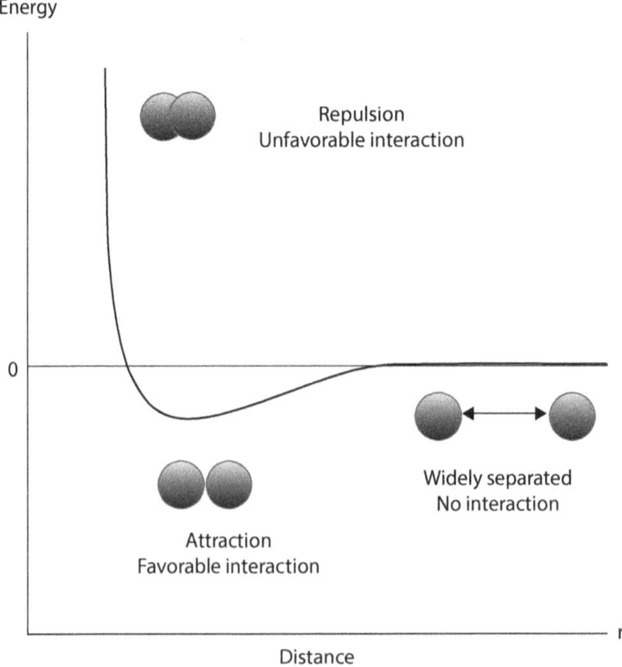

FIGURE 2.12 The van der Waals energy curve.

(Equation 2.7). This means that unlike the electrostatic energy (Equation 2.6), vdW interaction energy decays rapidly with increasing distance:

$$E_{vdW} = \frac{A}{r^{12}} - \frac{B}{r^6} \quad (2.7)$$

where
r is the distance between two interacting atoms
A and B are atom-specific repulsive and attractive term coefficients, respectively

As discussed in Section 2.3.5, a methyl group may be expected to increase the affinity of a compound by a factor of 3–10 due to the hydrophobic effect. This shows that the methyl group can be accommodated in the binding cavity. If it cannot be accommodated, vdW repulsion may instead give a significant decrease in affinity. Thus, the effect on the affinity of introducing a methyl group at different positions in a ligand may be a useful strategy to map out the dimensions of a receptor cavity in the absence of experimental information on three-dimensional protein structures. Providing that the introduced methyl group does not change the conformational properties of the parent molecule, the changes in affinity may be interpreted exclusively in terms of hydrophobic interactions and vdW interactions. An example of effects, which may be observed when introducing methyl groups in a ligand, is shown in Figure 2.13 for a series of flavones that binds to the benzodiazepine site of the $GABA_A$ receptor (see Chapter 15). The effects on the affinity of methyl groups in different positions of the parent flavone compound (**2.3**) may be used to gain indirect information of the shape and properties of the binding cavity.

When a methyl group is introduced in the 6-position (**2.4**), the affinity is increased by a factor of 23. This indicates that the methyl group can be very well accommodated in a lipophilic cavity in the binding site and that the affinity increase is due to hydrophobic interactions including dispersion interactions. An introduction of a methyl group to the 4′-position in **2.4** to give **2.5** results in a 24-fold decrease in the affinity. This is most likely due to repulsive vdW interactions between the

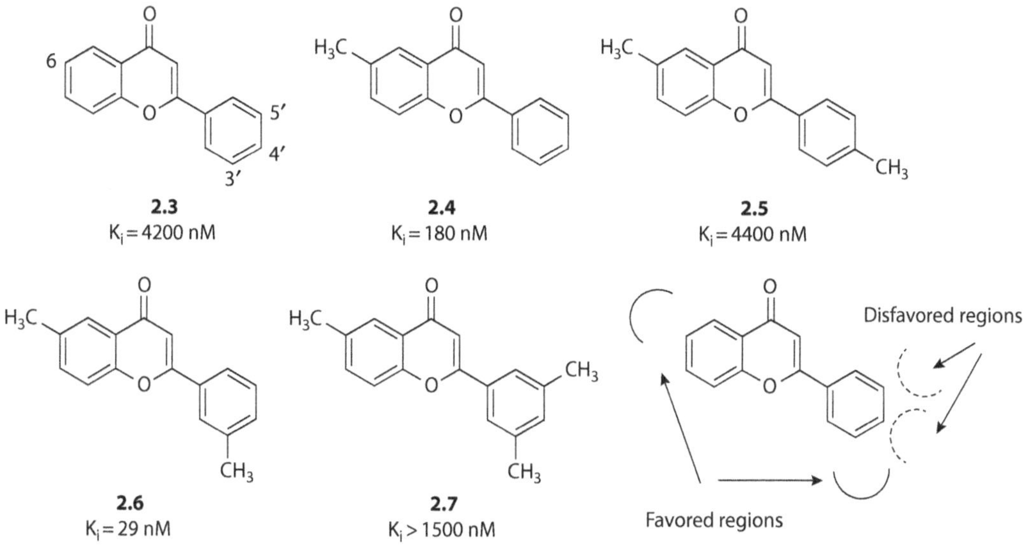

FIGURE 2.13 Affinities of methyl-substituted flavones binding to the benzodiazepine site of $GABA_A$ receptors.

4′-methyl group and the receptor giving an indication of the dimensions of the binding site in this region. An introduction of a 3′-methyl group in **2.4** to give **2.6** increases the affinity by a factor of 6. This is a significantly lower affinity increase than shown by the 6-methyl group in **2.4**. Thus, the two receptor regions in the vicinities of the 6′- and 3′-positions clearly have different properties.

The region in the vicinity of the 3′-position can accommodate a methyl group but the fit between the methyl group and the receptor is not as good as in the case of the 6′-methyl compound **2.4**. Finally, introduction of a 5′-methyl group in **2.6** giving **2.7** strongly decreases the affinity by a factor of more than 52. This is undoubtedly due to strong repulsive vdW interactions with the receptor. This identifies another steric repulsive receptor region adjacent to that identified by compound **2.5**. This example shows that conclusions drawn on the basis of a few compounds may provide valuable information on the properties of the protein binding site. Such information may be used to draw up a sketch of the binding site based on information gained from the ligands alone. In the absence of three-dimensional protein structures, this is a powerful way to gain insight into the binding pocket to aid in the design of new compounds.

2.4 CONCLUDING REMARKS

Molecular recognition is a prerequisite for drug action in virtually any biological system. Understanding the underlying physical chemistry is important in order to understand medicinal chemistry. Recognition is driven by three main factors: electrostatic interactions, lipophilicity/hydrophobicity, and shape complementarity. Together with the energies associated with restricting translation and rotation in the receptor bound complex, the energy associated with desolvation of receptor and ligand and the energy associated with conformational changes of ligands and receptors to assume their respective bioactive conformations, the interaction energies determine the binding free energies we measure in the form of binding affinities.

FURTHER READING

Ajay, A. and Murcko, M.A. 1995. Computational methods to predict binding free energies in ligand–receptor complexes. *J. Med. Chem.* 38:4953–4967.

Andrews, P.R., Craig, D.J., and Martin, J.L. 1984. Functional group contributions to drug–receptor interactions. *J. Med. Chem.* 27:1648–1657.

Bissantz, C., Kuhn, B., and Stahl, M. 2010. A medicinal chemist's guide to molecular interactions. *J. Med. Chem.* 53:5061–5084.

Boström, J., Norrby, P.-O., and Liljefors, T. 1998. Conformational energy penalties of protein–bound ligands. *J. Comput. Aided Mol. Des.* 12:383–396.

Davies, A.M. and Teague, S.J. 1999. Hydrogen bonding, hydrophobic interactions, and failure of the rigid receptor. *Angew. Chem. Int. Ed.* 38:736–749.

Jencks, W.P. 1981. On the attribution and additivity of binding energies. *Proc. Natl. Acad. Sci. USA* 78:4046–4050.

Murray, C.W. and Verdonk, M.L. 2002. The consequences of translational and rotational entropy lost by small molecules on binding to proteins. *J. Comput. Aided Mol. Des.* 16:741–753.

Olsen, J.A., Balle, T., Gajhede, M., Ahring, P.K., and Kastrup, J.S. 2014. Molecular recognition of the neurotransmitter acetylcholine by an acetylcholine binding protein reveals determinants of binding to nicotinic acetylcholine receptors. *PLOS ONE* 9(3):e91232.

Williams, D.H., Cox, J.P.L., Doig, A.J. et al. 1991. Toward the semiquantitative estimation of binding constants. Guides for peptide–peptide binding in aqueous solution. *J. Am. Chem. Soc.* 113:7020–7030.

3 Ligand-Based Drug Design

Erica S. Burnell, David E. Jane, and Robert J. Thatcher

CONTENTS

3.1 INTRODUCTION

It is no exaggeration to state that the advances in understanding of molecular structure and organic chemistry in the twentieth century brought about a renaissance in medicine. For the first time, chemists were able to isolate biologically active compounds from their sources and, crucially, identify their molecular structure. Thus, it became possible to correlate the structure of compounds with their therapeutic effect, the concept that has underpinned the field of medicinal chemistry for the last 150 years.

With the knowledge of structures of active compounds, the synthetic chemist has the opportunity to design and synthesize related compounds in order to assess how their biological activity changes with structure. This correlation of structure and activity is termed the structure–activity relationship (SAR) and the iterative process of modifying chemical structures in order to increase their biological activity is termed ligand-based drug design (LBDD).

An advantage of LBDD is that no knowledge of the drug's specific mechanism of action at the molecular level is required, as long as there is some measure of the biological activity. For example, Nicholson and Adams developed ibuprofen for treatment of rheumatoid arthritis a decade before the action of NSAIDs was linked to inhibition of prostaglandin biosynthesis and well before the cyclooxygenase enzymes were structurally characterized by X-ray diffraction methods. Indeed, before the emergence of structural models, generated by protein crystallography and molecular modeling (see Chapter 4), LBDD was the only method of rational drug design.

In the last 50 years, profound advances in the understanding of biology at the molecular level have led to a change in focus in drug discovery from symptom-based evaluation of drug activity to the specific targeting of validated biological molecules. In the majority of cases, drugs are developed to target proteins, and, for clarity, this chapter will discuss drug development with respect to protein receptors. The same principles, however, are relevant for the development of drugs that bind to other biological targets such as enzymes, nucleic acids, lipids, and structural proteins.

3.2 PROCESS OF LIGAND-BASED DRUG DESIGN

3.2.1 Determining "Hit" and "Lead" Compounds

In the initial stages of LBDD, two things are required.

3.2.1.1 A Suitable Biological Assay That Provides a Measure of Drug Activity

In order to optimize a drug structure, it is necessary to have a relevant method of measuring compound activity in an appropriate biological assay. Classically, a primary assay may have been an animal model of the target disease or live tissue preparations. For example, in the development of ibuprofen, Adams screened compounds for anti-inflammatory activity using an ultraviolet light-induced erythema model in guinea pigs. Nowadays, in vivo methods have largely been superseded by in vitro models utilizing isolated enzymes or cells expressing the target receptor. The latter techniques have the advantages of being amenable to automation and are considerably less expensive, although there remains the risk that specific activity on a particular protein target may not translate to a therapeutic effect. In addition, this approach cannot identify drugs with a novel mechanism of action, hence some drug companies are returning to phenotypic screening.

3.2.1.2 "Lead" Compound That Shows Activity in the Assay

With a suitable assay in place, LBDD requires a starting point from which to investigate the structure–activity landscape. Suitable lead molecules may come from a variety of sources such as natural products or chemical libraries held by drug companies. In practice, finding an initial "hit" compound can also be a matter of luck, with drug companies commonly performing initial screens of hundreds of thousands of compounds in appropriate pharmacological assays to find a handful of active molecules.

Hit compounds that exhibit activity in the assay are not only ranked according to their potency, but also by their potential to be developed into a suitable drug-like molecule. For example, it may well be easier to develop a structurally simple hit with lower activity than attempt SAR development on a synthetically challenging molecule. The medicinal chemist will consider synthetic tractability so that analogs can be obtained easily, and potential bioavailability and inherent toxicity when selecting a starting point for drug development. Once a hit compound is selected for investigation, it is termed a "lead" compound. The lead compound is a structure that exhibits the best drug profile of the compounds tested to date and is the starting point for all SAR development until it is succeeded by a better compound.

3.2.2 Building a Model of the Binding Site: The Pharmacophore

During the process of LBDD, a model of the ligand binding site in the target protein is determined indirectly by establishing which structural features are necessary to maintain biological activity. The minimum structural framework required to elicit the desired biological effect is termed the pharmacophore, and can be extrapolated by comparison of a number of active structures.* The pharmacophore model is usually an abstract arrangement of structural features inferred by considering what features might complement the functional groups of the lead compound. For example,

* The IUPAC definition is more elaborate, defining a pharmacophore as "an ensemble of steric and electronic features that is necessary to ensure the optimal supramolecular interactions with a specific biological target and to trigger (or block) its biological response."

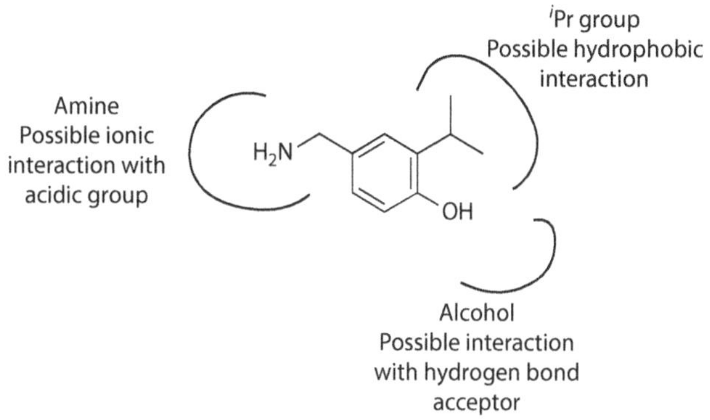

FIGURE 3.1 Inferring a pharmacophore model from an active compound.

given the hypothetical active compound in Figure 3.1, it can be postulated that the amino group may be protonated and form an ionic bond with a charged residue in the binding site, the isopropyl group may be interacting with a hydrophobic pocket, etc.

The pharmacophore model itself is continually refined as more SAR data are obtained and, in the process, more active compounds are developed to validate and fit the model. The question is as follows: "how should a structure be modified to increase activity?"

3.2.3 Increasing the Affinity of Drugs for Their Protein Target

Given a lead compound, the design of analogs is limited only by the imagination of the chemist and, perhaps more practically, by their ease of synthesis. That said, there are some common guiding principles that underpin drug modification. A major goal of drug design is to increase the binding affinity of the lead compound. To rationalize the modifications made to a lead structure, it is worth considering the thermodynamics of drug binding (see also Chapter 2).

At a very simple approximation, binding affinity is directly related to the free energy of binding of the drug with the target receptor assuming that the drug is binding in a specific manner that instigates the therapeutic effect (Often a molecule may bind very strongly to a receptor without triggering the desired response). The free energy of binding is related to the equilibrium binding constant (K_{eq}) as follows:

$$\Delta G_{bind} = -RT \ln K_{eq}$$

The Gibb's free energy, ΔG_{bind}, of a binding event is given by

$$\Delta G_{bind} = \Delta H_{bind} - T\Delta S_{bind}$$

The enthalpy term, ΔH_{bind}, includes the sum of all favorable and unfavorable molecular interactions that are formed or broken by the binding event. The relative strengths of common intermolecular interactions are shown in Table 3.1. Clearly the synthetic chemist needs to maximize these favorable interactions to obtain a potent drug, however, it should be emphasized that favorable interactions such as drug solvation are broken during binding and that these will count against the net enthalpy gain. As such, van der Waals interactions and the hydrophobic effect make a disproportionately high contribution to the overall binding energy for the majority of ligands. Other potential penalties include perturbation of the receptor structure from the ground state upon ligand binding.

The entropy term, ΔS_{bind}, describes the change in disorder over the course of the binding event. This term encompasses entropy changes due to solvent reorganization both inside and outside the binding site and penalties associated with configurational change in the ligand and receptor.

TABLE 3.1
Strength of Intermolecular Interactions

Interaction	Example	Strength (kJ/mol)
Dispersion forces	R-^{i}Pr···^{n}Bu-R	0.05–40
Ion or dipole-induced dipole	H_2O···O_2	2–15
Dipole–dipole	C=O···C=O	5–25
Hydrogen bonding	R_2NH···O=C	10–40
Ion–dipole	Na^+···O=C	40–600
Ion–ion	RNH_3^+···^-OOCR	400–4000
Covalent bond (for reference)	C—O	150–1200

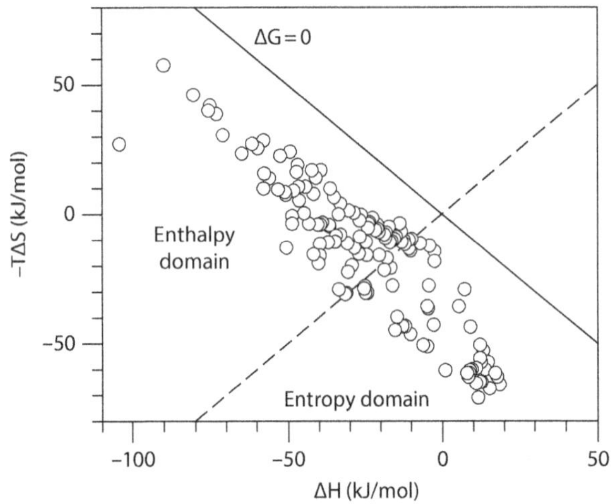

FIGURE 3.2 Relative contributions of ΔH_{bind} and $-T\Delta S_{bind}$ for 156 small molecules binding to various proteins. The enthalpy and entropy domains are labeled where ΔH_{bind} or $-T\Delta S_{bind}$ makes the greater contribution to ΔG_{bind}, respectively. (Graph constructed using data obtained from the SCORPIO project http://scorpio.biophysics.ismb.lon.ac.uk.)

A summary of experimental data for the relative contributions of ΔH_{bind} and ΔS_{bind} for various ligands and receptors is shown in Figure 3.2. It can be seen that both enthalpy and entropy can make a significant contribution to the free energy of binding. Thus, any LBDD study should attempt to address both components of ΔG_{bind}.

While binding affinity is a major consideration when developing a hit compound, there are other characteristics that need to be considered in drug development. Parameters such as solubility, pharmacokinetics, metabolic susceptibility, and toxicity must be taken into account. Drug design is a process of optimizing several parameters at once and ultimately, compromises will have to be made such that the overall drug profile is optimal for clinical use.

In practice, medicinal chemists will usually attempt to develop the most active compound possible, and then seek to improve the drug profile while maintaining the desired level of biological activity. That said, depending on the assay used, solubility and metabolism can complicate SAR data. For example, a compound may be the perfect fit for a binding site of a receptor, but if it is metabolized before it can reach the binding site, it will return a negative result for activity. The lesson is that the medicinal chemist must consider the broader picture when interpreting results of SAR analysis. Low activity may not necessarily mean a bad fit to the protein target! (See also Section 1.4.3.)

3.2.4 Functional Groups and Bioisosteres

As previously shown, maximizing the favorable intermolecular interactions between drug and receptor is the primary method of minimizing ΔH_{bind}. A hit compound will already bind to the receptor by a number of these interactions depending on which functional groups are present. The initial challenge in LBDD is to deconvolute how specific moieties of the lead compound interact with the receptor. To do this, the medicinal chemist generalizes structural groups into their possible modes of interaction: ionic bonding groups, hydrogen bond acceptors/donors, hydrophobic groups, etc., as introduced in Table 3.1. These groups are then systematically varied to assess if they are indeed interacting as expected. For example, if swapping an alkyl group for a polar group of similar size causes a significant decrease in activity, it suggests that the moiety occupies a hydrophobic pocket in the receptor. The process of swapping like-for-like functional groups is termed isosteric replacement and is an important concept in SAR development.

An isostere is an atom or functional group that presents similar chemical or physical properties as another group. Classically the term was used to describe groups with the same number of valence electrons such as CH_2, NH, O, S; however, modern usage also encompasses larger functionalities that are similar in terms of steric bulk and/or polarity. For example, oxazoles can be considered isosteres of amide groups on account of their similar size and propensity to act as H-bond acceptors. In contrast to amides, oxazoles will not act as a hydrogen bond donors thus isosteric replacement of the two groups would be one method to determine the importance of an H-bond donor at that position of the molecule. Aside from helping to validate a pharmacophore model, isosteres can present subtly different steric and electronic qualities when binding to a receptor, and thus isosteric replacement is an important method of optimizing drug affinity.

Bioisosteres are specific examples of isosteres and are groups that can replace other moieties while maintaining the biological activity of the compound. Aside from their effect on the interaction between drug and receptor, bioisosteric replacement can improve other drug characteristics. Examples are as follows:

- Deuterium is an isostere of hydrogen but is metabolized slower as a result of the kinetic isotope effect. Consequently, replacement of a hydrogen atom by deuterium in the antibacterial fludalanine significantly improves the in vivo half-life.
- Fluorine is also a bioisostere of hydrogen, albeit less benign. Depending on its position in a molecule, the incorporation of fluorine can significantly increase or decrease the lipophilicity of a drug. The greater strength of the C—F bond also leads to greater metabolic stability.
- The replacement of cycloalkyl rings with heterocycles can greatly improve water solubility and reduce metabolic lability. Similarly, ethylene glycol chains are isosteric with long-chain hydrocarbons but are significantly more soluble in aqueous conditions. For further description of bioisosteres, see Section 1.4.3.1 and Table 1.1 with examples of common bioisosteric replacements.

3.2.5 Steric Occlusion

An important contribution to drug affinity is a molecule's ability to displace organized water from the active site of the target receptor. This results in a net entropy gain leading to a lower ΔG_{bind}. Further, increasing surface area of contact between the drug molecule and receptor provides a greater potential for favorable van der Waals interactions. In LBDD, it is common for medicinal chemists to investigate the available space for a drug by incrementally increasing the size of the structure until a drop in potency is observed. Increasing the size of a lead can be accomplished by extending aliphatic chains, homologation, ring expansion, or addition of polycyclic aromatic rings (Figure 3.3).

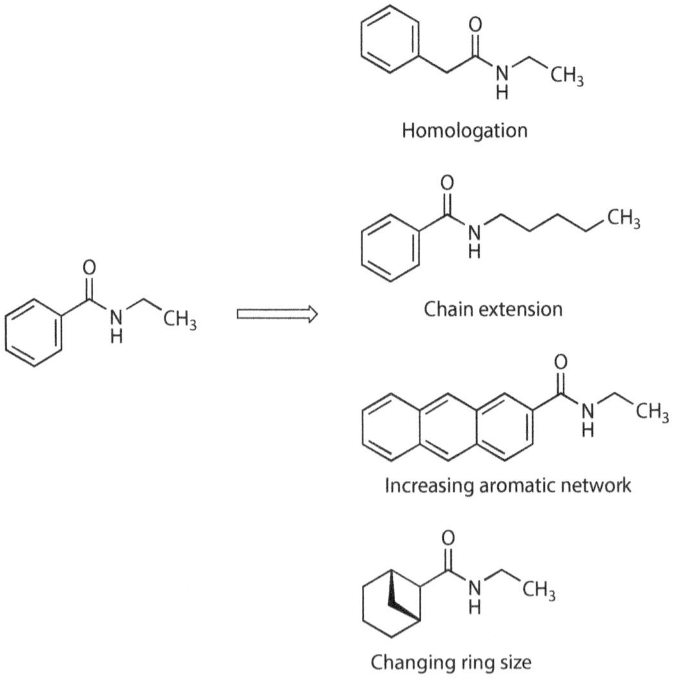

FIGURE 3.3 Structural modifications to increase steric occlusion.

3.2.6 Conformational Constraint and Chirality

It is no coincidence that many potent drugs are relatively rigid molecules. For a conformationally flexible molecule, there may only be one biologically active conformation among many inactive conformers. Therefore, for the flexible drug to bind to the receptor, there is an entropic penalty associated with the conformational change required. This penalty can be significant, for instance, the calculated conformational entropy penalty for amprenavir (Figure 3.4) binding to HIV protease is ~105 kJ/mol.

As such, it can be highly advantageous to constrain a structure into an active conformation by addition of rings and/or unsaturation into conformationally flexible sites in the drug molecule. For example, a long, relatively flexible hydrocarbon chain can be made more rigid by including unsaturated groups such as alkenes and alkynes, to confer a specific conformation (Figure 3.5a). Similarly, the incorporation of ring structures can also lock a molecule into a conformationally active state (Figure 3.5b). Perhaps more subtly, steric hindrance can also influence the shape of molecules. For example, the addition of methyl groups *ortho* to a biphenyl moiety introduces restricted rotation about the exocyclic bond and causes the second ring to deviate out of plane to the other (Figure 3.5c). Stereoelectronic effects can also stabilize particular molecular conformations but in general are far weaker than steric repulsion (Figure 3.5d).

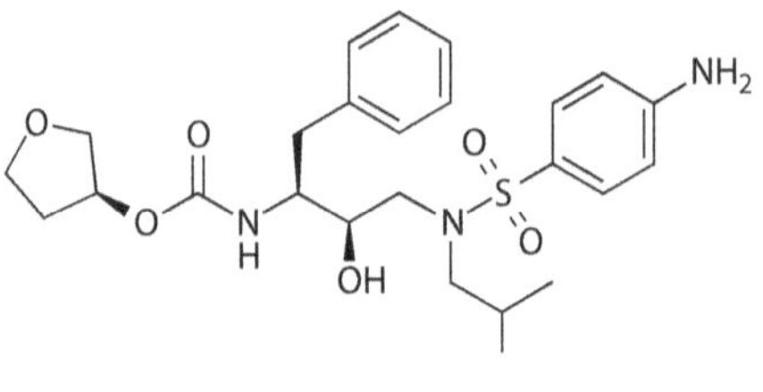

FIGURE 3.4 Amprenavir—an HIV protease inhibitor.

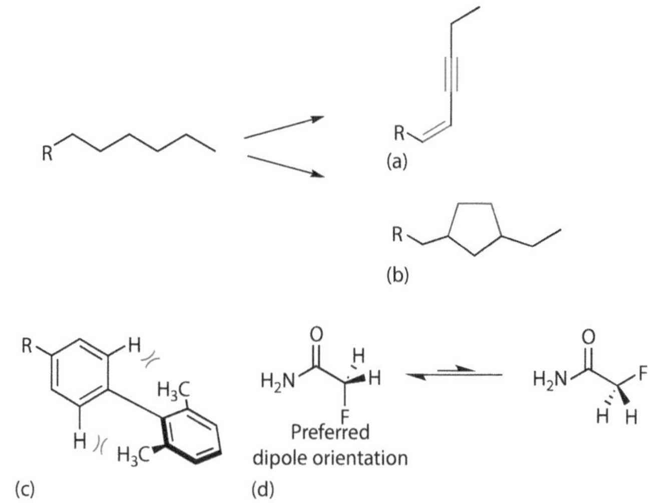

FIGURE 3.5 Examples of chemical modification to enhance conformational constraint. (a) Use of rigid unsaturated groups; (b) cyclisation; (c) restricted rotation by addition of sterically encumbering groups; (d) stereo-electronic effects.

Chirality can also be considered a type of spatial constraint. All receptors in nature are inherently chiral and present an asymmetric binding site. Thus, introducing a suitable stereogenic center can grant a drug greater selectivity and potentially provide a more complementary fit than achiral analogs. It follows that enantiomers and diastereoisomers of drug structures typically show different biological activities. Indeed, of the top 50 selling small molecule pharmaceuticals in the United States, 33 are, or contain, single enantiomers.

3.3 INTRODUCTION TO *N*-METHYL-D-ASPARTATE RECEPTOR CASE STUDIES

The concepts introduced earlier are, at best, a very brief introduction to a large and complex subject and the practical method of LBDD is best demonstrated with reference to real-world examples. Outlined in the following are three case studies that demonstrate the development of agonists and antagonists of the *N*-methyl-D-aspartate receptor (NMDAR) from initial SAR of simple amino acids.

The NMDAR belongs to a family of ligand-gated ion channels activated by the excitatory neurotransmitter L-glutamate and the co-agonist glycine (see also Chapter 15). This family also includes (*S*)-2-amino-3-hydroxy-5-methylisoxazole-4-propionic acid receptors (AMPARs) and kainate receptors (KARs). The NMDAR family consists of seven subtypes: GluN1, GluN2A-D, and GluN3A-B. NMDARs are tetramers and in most areas of the central nervous system (CNS) are comprised of two GluN1 and two GluN2 subunits. The GluN2 subunits contain binding sites for L-glutamate, whereas the GluN1 and GluN3-B subunits contain binding sites for the co-agonist glycine. In these case studies, the development of NMDA and other more potent NMDAR agonists will be discussed, along with competitive NMDAR antagonists.

3.3.1 Case Study 1: Development of NMDAR Agonists

A SAR study centered on the structures of the amino acids aspartate and glutamate was conducted in the early 1960s by Jeff Watkins. In the initial studies, and with no knowledge of the structure of glutamate receptors and their subunits, compounds were tested for their propensity to cause depolarization of spinal neurons and their activity was compared to that of the individual enantiomers of glutamate and aspartate.

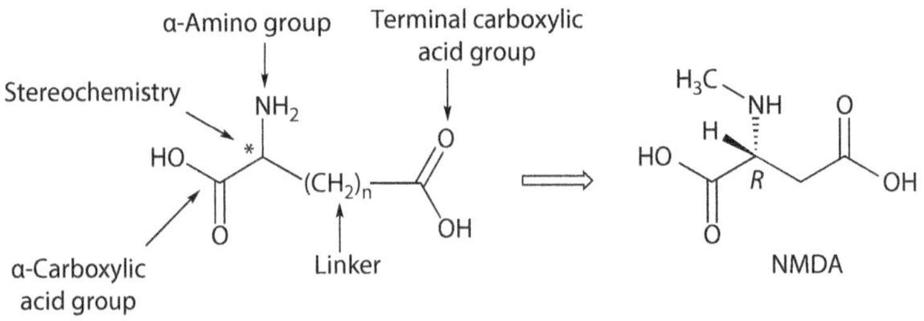

α-Amino group
Terminal carboxylic acid group
Stereochemistry
α-Carboxylic acid group
Linker
NMDA

FIGURE 3.6 Structure–activity relationship study noting the effect of a number of structural changes to aspartate (n = 1) and glutamate (n = 2) that led to the identification of *N*-methyl-D-aspartate.

In this study, the effect of changing the structure of glutamate and aspartate on agonist activity was investigated as follows: (1) role of stereochemistry, (2) nature of alpha and terminal acidic groups, (3) chain length of linker connecting amino acid moiety and terminal acidic group, and (4) substitution of the α-amino group with an alkyl group (Figure 3.6). This study led to the identification of NMDA as a potent agonist for glutamate receptors. It was later found to be highly selective for a subtype of glutamate receptor that was eponymously named the NMDAR.

A number of conclusions regarding structural requirements for optimal agonist activity for the NMDAR were made following this SAR study:

1. For simple open-chain glutamate analogs, *S* stereochemistry is preferred, whereas for some aspartate analogs such as NMDA, *R* stereochemistry is preferred.
2. An alpha and terminal acidic group is required. Replacement of the α-CO_2H group with a phosphonate group lowers NMDAR affinity. With regard to the terminal acidic group, the rank order of affinity for the NMDAR is CO_2H and $SO_2H > SO_3H > PO_3H_2$ for *R* or *S* aspartate analogs and for (*S*)-glutamate analogs is $CO_2H > SO_2H$ and $SO_3H \gg PO_3H_2$. An analog of glycine with a tetrazole substituent on the α-carbon, (*RS*)-TetGly (Figure 3.7), has a high affinity for the NMDAR and is among the most potent agonists tested to date. Due to the delocalization of the negative charge in the deprotonated tetrazole ring, it probably represents an isosteric replacement of CH_2COO^-,and therefore, (*RS*)-TetGly is an aspartate analog.
3. The optimal number of CH_2 groups in the linker is either 1 or 2.
4. With the exception of NMDA, *N*-alkyl substitution of the amino group is detrimental for aspartate and glutamate, and NMDAR affinity becomes lower with increasing size of the alkyl chain.

Conformational restriction of glutamate has led to some potent agonists such as homoquinolinic acid and (2*S*,3*R*,4*S*)-CCG, the latter being one of the most potent NMDAR agonists described to date (Figure 3.7).

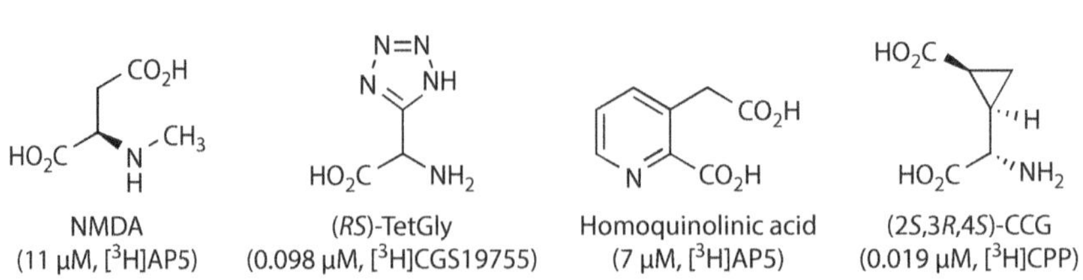

NMDA (11 μM, [^{3}H]AP5)
(*RS*)-TetGly (0.098 μM, [^{3}H]CGS19755)
Homoquinolinic acid (7 μM, [^{3}H]AP5)
(2*S*,3*R*,4*S*)-CCG (0.019 μM, [^{3}H]CPP)

FIGURE 3.7 Structures of *N*-methyl-D-aspartate receptor agonists identified in structure–activity relationship studies. K_i values obtained from competition binding assays using rat brain membranes are given in parenthesis together with the radioligand used.

3.3.2 Case Study 2: SAR Studies on α-Aminoadipic Acid: Development of Competitive NMDAR Antagonists

A breakthrough in the development of competitive NMDAR antagonists came about when a chain extended form of glutamate, (*RS*)-α-aminoadipic acid (α-AA) (Figure 3.8), was shown to selectively antagonize NMDA-induced responses in an electrophysiological assay. In a similar fashion to the agonist SAR study, the effects of the stereochemistry, terminal acidic group, and chain length were investigated.

Testing of the individual enantiomers of α-AA revealed that the antagonist activity resided in the *R* form whereas the *S* form was a NMDAR agonist with moderate affinity. SAR studies on α-AA revealed that the linker chain length and the nature of the terminal acidic group played an important role in determining the affinity of the NMDAR antagonist.

With regard to the terminal acidic group the rank order of affinity of α-AA analogs was PO_3H_2 > CO_2H ≫ SO_3H. The phosphonate analog (*R*)-AP5 (Figure 3.8) was the first NMDAR antagonist to be widely used as a tool to block NMDAR activity in functional studies. Schild analysis of the antagonism by (*R*)-AP5 of the NMDA-induced response in an electrophysiological assay on the neonatal rat spinal cord showed that it had a competitive mode of action. One explanation for the potent NMDAR antagonist activity observed upon phosphono substitution was the possibility of two interactions with the receptor via the two ionizable OH groups of the phosphonate group. The pK_a values for the ionization of the hydroxyl groups on the phosphonate group are <1 and 7.8. When the experimental pH was increased from 7.3 to 8.2, the potency of the antagonist activity of (*R*)-AP7 was increased approximately fourfold, suggesting that the doubly charged phosphonate group was optimal for antagonist activity. In support of this theory, when the phosphonate group of AP5 was replaced by a phosphinate group (with only one ionizable OH group), the NMDAR antagonist activity was reduced considerably.

Changing the inter-acidic group chain length of (*R*)-AP5, by adding or removing CH_2 groups, revealed an interesting relationship between chain length and NMDAR antagonist affinity. A chain of three CH_2 groups, as observed in (*R*)-AP5, or five CH_2 groups, as observed in (*R*)-AP7 (Figure 3.8)

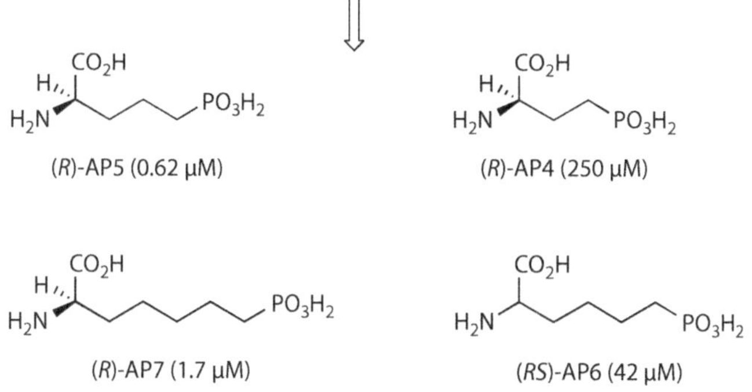

FIGURE 3.8 Structure–activity relationship studies on (*R*)-AA led to the development of (*R*)-AP5 and (*R*)-AP7, which had the optimal inter-acidic group chain length. (*R*)-AP4 and (*RS*)-AP6 had lower *N*-methyl-D-aspartate receptor affinity. IC_{50} values from competition binding assays on rat brain membranes using [^{3}H] AP5 are given in parenthesis.

was optimal for NMDAR antagonist activity, whereas inter-acidic group chain lengths of one, two, four, or six CH_2 groups produced compounds that were at best only weak to moderate antagonists.

The NMDAR antagonists described up to this point are open-chain analogs with a very high degree of conformational flexibility. Conformational restriction of these open-chain compounds produced NMDAR antagonists of much higher affinity. A number of methods of producing conformational restriction were attempted (see Figure 3.9 for structures of compounds):

1. Incorporation of a double bond into the inter-acidic group chain of (*R*)-AP5 (e.g., CGP 40116).
2. Formation of carbocyclic rings from the inter-acidic group chain, e.g., ACPEB (cyclobutyl ring) and NPC 17742 (cyclohexyl ring).

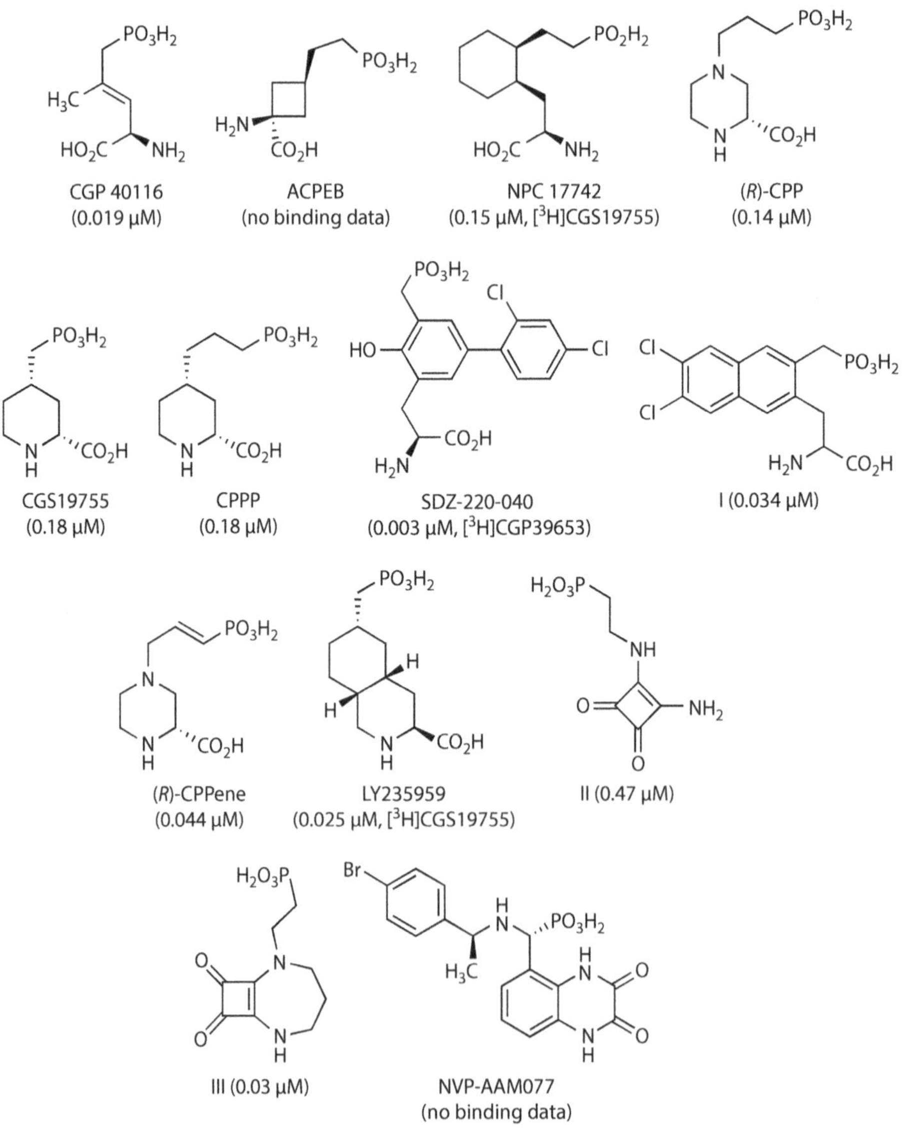

CGP 40116 (0.019 μM)
ACPEB (no binding data)
NPC 17742 (0.15 μM, [3H]CGS19755)
(*R*)-CPP (0.14 μM)
CGS19755 (0.18 μM)
CPPP (0.18 μM)
SDZ-220-040 (0.003 μM, [3H]CGP39653)
I (0.034 μM)
(*R*)-CPPene (0.044 μM)
LY235959 (0.025 μM, [3H]CGS19755)
II (0.47 μM)
III (0.03 μM)
NVP-AAM077 (no binding data)

FIGURE 3.9 Structures of conformationally restricted AP5 and AP7 derivatives. IC_{50} values for inhibition of [3H]CPP binding to rat brain membranes are given in parenthesis (unless an alternative radioligand is indicated).

3. Incorporation of the α-amino group and inter-acidic group chain of AP5 or AP7 into a heterocyclic ring, e.g., piperazine (e.g., CPP) and piperidine rings (e.g., CGS19755 and CPPP).
4. Introduction of an aryl or heteroaryl ring into the inter-acidic group chain to produce phenylalanines (e.g., SDZ 220-040) or quinoxaline derivatives (e.g., compound I, Figure 3.9).
5. Combination of strategies 1 and 3 to produce heterocyclic compounds with unsaturated side chains bearing the terminal phosphonate group (e.g., CPPene) (in the case of the double bond in CPPene, *trans* (*E*) geometry was optimal).
6. Formation of bicyclic ring systems by incorporating the phosphonoalkyl side chain of heterocyclic analogs into another ring (e.g., conversion of CPPP to the bicyclic analog LY235959).
7. Bioisosteric replacement of the amino acid moiety with the 3,4-diamino-3-cyclobutene-1,2-dione group (e.g., compounds II and III, Figure 3.9)—another strategy of bioisosteric replacement was to replace the amino acid moiety with a quinoxaline-2,3-dione unit (e.g., NVP-AAM077).

The *cis* isomer of the heterocyclic analog of AP5, CGS19755, had much higher affinity for the NMDAR than the *trans* isomer. However, this *cis* selectivity was not observed with CPPP which is an analog of CGS19755 with the same chain length as AP7.

The carbocyclic derivative ACPEB and the quinoxaline derivative (Compound I, Figure 3.9) are anomalous as they are conformationally restricted analogs of AP6 rather than AP5 or AP7. However, in modeling studies it was possible to obtain a good overlay of the key interacting groups of these compounds with the corresponding groups of the template structure CGS19755 which is a conformationally restricted analog of AP5.

Compound II (Figure 3.9) is an analog of AP5 in which the amino acid has been replaced with a squaric acid moiety and showed approximately the same affinity for the NMDAR as the parent compound. Compound III is a conformationally restricted analog of II and displayed a greater than 10-fold higher affinity for the NMDAR. Thus, bioisosteric replacement of the amino acid moiety can be used as a novel strategy to produce high-affinity NMDAR antagonists.

Many of the conformationally restricted NMDAR antagonists described earlier showed the same preference for *R* stereochemistry to that observed with the open-chain analogs (*R*)-AP5 and (*R*)-AP7. However, there are exceptions to this rule; the *S* enantiomer of biphenylalanine derivative SDZ 220-040 (Figure 3.9) which is a conformationally restricted analog of AP7, was much more potent than the *R* enantiomer. A similar preference for *S* stereochemistry at the amino acid stereogenic center was observed with the decahydroisoquinoline derivative LY235959 (Figure 3.9).

In a number of cases NMDAR antagonists obtained by conformational restriction had high affinity with K_i values in the low nanomolar range (e.g., D-CPPene) and were used to uncover roles for NMDARs in a number of neurodegenerative and neurological disorders such as stroke, Alzheimer's disease, Parkinson's disease, schizophrenia, epilepsy and chronic pain. Indeed, D-CPPene made it as far as Phase III clinical trials as a neuroprotectant for the treatment of ischemic stroke. One issue with the competitive NMDAR antagonists used in clinical trials was poor oral bioavailability, i.e., some did not get into the systemic circulation at high enough concentrations and showed poor brain penetration. Novartis developed a series of biphenyl-substituted AP7 analogs as potent competitive NMDAR antagonists, such as SDZ 220-040 which showed improved oral bioavailability. The biphenyl substituent would improve lipid solubility and likely enhance biomembrane penetration but the main reason for the improved oral bioavailability was that these compounds were substrates of the large neutral amino acid transporter in the intestine.

3.3.3 Case Study 3: Development of GluN2 Subunit Selective Competitive NMDAR Antagonists

The different GluN2 subunits likely play distinct physiological roles, but little is known of their functions in the CNS, primarily due to the lack of subunit-selective antagonists. In addition to their usefulness as tool compounds to understand brain function, selective antagonists could also be of therapeutic value as outlined earlier. Increasing subunit selectivity is also likely to minimize multitarget interaction and consequent side effects. During the development of most of the competitive NMDAR antagonists described earlier, compounds were characterized either in radioligand binding assays on isolated rat brain membranes or in electrophysiological assays using the neonatal rat spinal cord or cortical wedge preparation on native NMDARs. It was only later when the subunits that make up the NMDAR were cloned that some of these compounds were evaluated on individual NMDAR subtypes. When tested in a two-electrode voltage clamp electrophysiological assay on NMDARs comprised of GluN1 and an individual type of GluN2 subunit, (*R*)-AP5, (*R*)-AP7 (Figure 3.8), and their analogs displayed a rank order of antagonist potency of GluN2A > GluN2B > GluN2C > GluN2D (Table 3.2). One compound NVP-AAM077 (Figure 3.9), a conformationally restricted AP5 analog, showed approximately 10-fold selectivity for GluN2A *vs* GluN2B but much lower selectivity for GluN2A *vs* GluN2D. However, AP7 and its conformationally restricted analogs showed greater GluN2A *vs* GluN2D subunit selectivity. For example, (*R*)-CPP (Figure 3.9) displayed a 50-fold higher affinity for GluN2A compared to GluN2D (Table 3.2). Thus, despite the high amino acid sequence homology in the glutamate ligand binding domains of GluN2A-D it was deemed possible to design GluN2 subunit selective competitive antagonists.

TABLE 3.2
Summary of the Activity of Piperazine-2,3-Dicarboxylic Acid Derivatives

	GluN2A	GluN2B	GluN2C	GluN2D
(*R*)-AP5	0.28 ± 0.02	0.46 ± 0.14	1.64 ± 0.14	3.71 ± 0.67
(*R*)-AP7	0.49 ± 0.18	4.14 ± 0.36	6.42 ± 1.08	17.10 ± 0.65
NVP-AAM077	0.0054 ± 0.0004	0.067 ± 0.003	0.012 ± 0.001	0.037 ± 0.004
(*R*)-CPP	0.041 ± 0.003	0.27 ± 0.02	0.63 ± 0.05	1.99 ± 0.20
PBPD	**15.79 ± 0.43**	**5.01 ± 0.25**	**8.98 ± 0.18**	**4.29 ± 0.11**
UBP130	>100	>100	>100	>100
UBP112	13.2 ± 0.89	18.0 ± 2.4	27.0 ± 3.1	59.7 ± 8.5
PPDA	**0.55 ± 0.15**	**0.31 ± 0.02**	**0.096 ± 0.006**	**0.13 ± 0.04**
PPDA-(−)	**0.21 ± 0.02**	**0.22 ± 0.04**	**0.07 ± 0.01**	**0.09 ± 0.01**
PPDA-(+)	17.6 ± 2.1	13.5 ± 1.0	3.4 ± 0.3	4.6 ± 0.2
UBP142	0.32 ± 0.01	0.25 ± 0.02	0.10 ± 0.01	0.15 ± 0.01
UBP135	1.54 ± 0.11	0.76 ± 0.11	0.29 ± 0.02	0.57 ± 0.03
UBP129	0.84 ± 0.08	0.3 ± 0.06	0.85 ± 0.19	1.94 ± 0.14
UBP131	9.6 ± 1.1	18.2 ± 1.3	3.89 ± 0.09	9.26 ± 0.43
UBP141	**14.19 ± 1.11**	**19.29 ± 1.38**	**4.22 ± 0.52**	**2.78 ± 0.16**
UBP160	7.2 ± 0.7	15.2 ± 2.6	3.2 ± 0.2	1.7 ± 0.1
UBP145	**11.5 ± 0.8**	**8.0 ± 0.4**	**2.8 ± 0.1**	**1.2 ± 0.1**
UBP161	7.4 ± 0.9	3.8 ± 0.9	1.9 ± 0.2	1.0 ± 0.1
UBP140	11.18 ± 1.29	10.25 ± 1.87	2.67 ± 0.13	3.18 ± 0.19
UBP138	10.48 ± 1.23	9.08 ± 1.02	4.88 ± 0.57	6.03 ± 2.01

Note: K_i values (μM) of antagonists for inhibiting the responses of recombinant rat NMDAR subtypes expressed in *Xenopus* oocytes (mean ± SEM, n = 3–10). Lead compounds are shown in bold type.

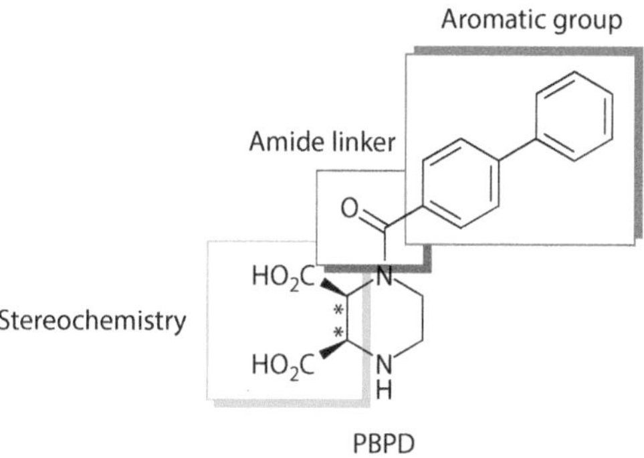

FIGURE 3.10 Structure–activity relationship study on PBPD derivatives to investigate structural features necessary for GluN2 subunit selectivity.

Although some of the compounds described earlier discriminated between the subunits more effectively than others, they still followed the same selectivity pattern that had been observed for the majority of "traditional" antagonists. Examining the existing pool of competitive antagonists, a series of NMDAR antagonists were found to go against this traditional selectivity pattern. These were derivatives of piperazine-2,3-dicarboxylate (PzDA) which are conformationally restricted derivatives of NMDA. Some of these PzDA derivatives showed selectivity for GluN2C/D *vs* GluN2A/B (Table 3.2). Most NMDAR competitive antagonists had 5 or 7 bonds between the acidic groups; unusually, the PzDA analogs had a short inter-acidic group chain length more commonly seen in agonists. It was hypothesized that the amino acid end of these compounds was binding to the same site as agonists such as NMDA and that the bulky aromatic ring was causing the antagonist activity. In particular, the biphenyl derivative PBPD (Figure 3.10) stood out as an interesting lead as it had highest affinity for GluN2D and lowest affinity for GluN2A (Table 3.2) which is the opposite of that observed for (*R*)-CPP. Therefore, a SAR study was carried out on a series of PBPD derivatives with variations to the biphenyl group to investigate the size and shape of the predicted hydrophobic pocket and investigate whether these changes would improve the GluN2D subunit selectivity. In addition to the aromatic group, the amide linking group and the stereochemistry were also examined for their effect on binding affinity and subunit selectivity (Figure 3.10). The affinity and selectivity of these compounds were evaluated at recombinant NMDARs using two-electrode voltage clamp electrophysiology. Each of the GluN2 subunits were individually co-expressed with GluN1a in *Xenopus* oocytes.

It was hypothesized that the aromatic ring was crucial for the antagonist activity and GluN2 subunit selectivity pattern observed in PBPD and that it occupied a hydrophobic pocket within the binding site. Without knowing the structure of the binding site, this could be attributed to a number factors relating to the size and shape of the pocket or the various interactions that aromatic groups can be involved in, such as hydrophobic interactions, cation–π interactions or hydrogen bonding interactions between the dense π system of the ring with strong hydrogen bond donors such as quaternary ammonium ions.

A number of PBPD analogs were synthesized, where the biphenyl group was systematically altered (Figure 3.11). Removing the second ring to leave a phenyl ring (UBP130) resulted in a loss of affinity across all the subunits. Replacing the first ring with an (*E*)-ethylene linker (UBP112) reduced affinity at GluN2B-D, but slightly increased affinity for GluN2A, showing a similar selectivity pattern to that of (*R*)-AP7 (Table 3.2). These alterations suggested that both rings were important in conferring GluN2D *vs* GluN2A selectivity.

As it appeared that the general size of the biphenyl group was favorable, its geometry was examined next. In a biphenyl ring system, the aromatic rings are twisted out of plane. To force a coplanar arrangement, the biphenyl group was substituted for a phenanthrene ring (PPDA) (Figure 3.11).

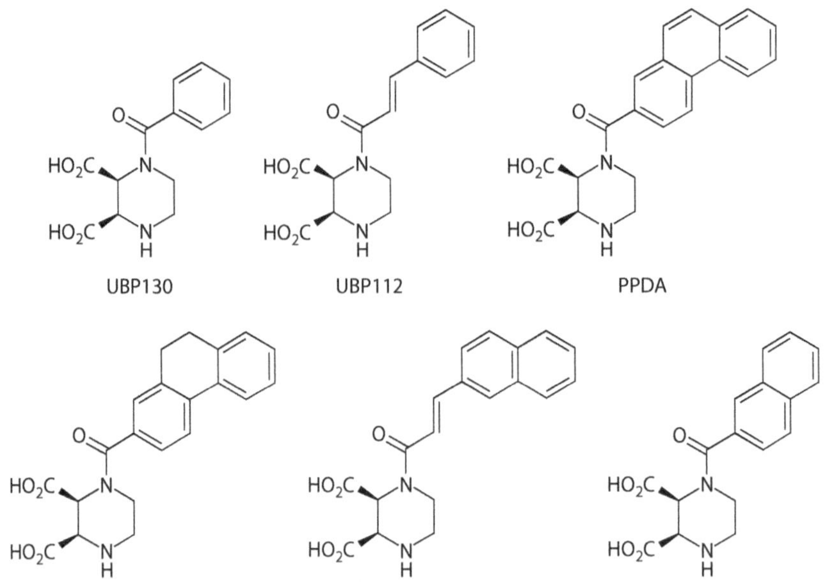

FIGURE 3.11 Compounds synthesized to examine optimal number of aromatic rings for high GluN2 subunit affinity and selectivity.

This modification resulted in a dramatic increase in affinity and, in addition, increased selectivity for GluN2C and GluN2D *vs* GluN2A and GluN2B (Table 3.2). When phenanthrene was replaced with dihydrophenanthrene (UBP135), the affinity was reduced across all subunits, confirming that a co-planar arrangement was favorable. This improvement in affinity could be attributed to a better fit within the receptor, a favorable interaction or a decrease in entropy loss brought about by conformational constraint.

Replacing the first phenyl ring of PPDA with an (*E*)-ethylene linker (UBP129) (Figure 3.11), as was done with PBPD, had a similar effect on the selectivity pattern, i.e., conferred GluN2A *vs* GluN2D selectivity. This confirmed the importance of this ring for GluN2C/D *vs* GluN2A/B selectivity. Removal of the last benzene ring of phenanthrene (UBP131) reduced affinity across all GluN2 subunits.

Having established that a tricyclic aromatic ring was important, the position of attachment to the piperazine ring and the effect of adding substituents to the aromatic ring were examined. A positional isomer of PPDA (UBP141) with the carbonyl linker attached to the 3-position of the phenanthrene ring was less potent than PPDA, but more GluN2C/2D subunit selective. Of a series of 9-halo-substituted derivatives of UBP141 (Figure 3.12), the 9-iodo-derivative (UBP161) was the most potent across all subtypes but lacked GluN2C/2D subunit selectivity. The 9-chloro-derivative (UBP160) was slightly more potent than UBP141, but slightly less selective. The 9-bromo-derivative (UBP145) was the most promising halo-substituted compound as it had a greater affinity for GluN2D than UBP141, and retained selectivity for this subunit over GluN2A and GluN2B (Table 3.2). A 7-bromo-derivative of PPDA (UBP142) was also tested; although it showed high affinity, it was completely non-selective.

The effect of changing the carbonyl group linking the piperazine ring to the aromatic ring was also briefly examined. Replacing the carbonyl functionality of PPDA with a methylene group (UBP140) led to a 30-fold drop in affinity (Figure 3.12, Table 3.2).

The configuration of the carboxylic acid groups attached to the piperazine ring was found to have a significant influence on subunit selectivity as well as antagonist potency; the *cis* isomer of

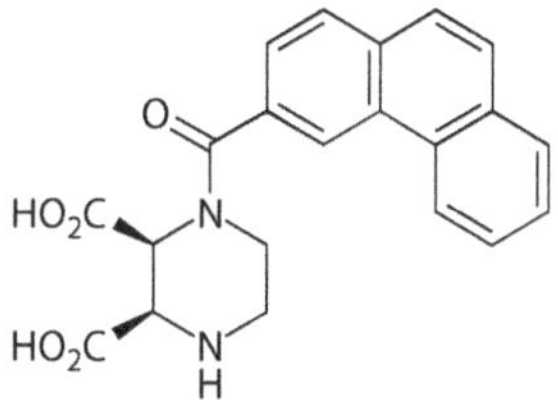

UBP141

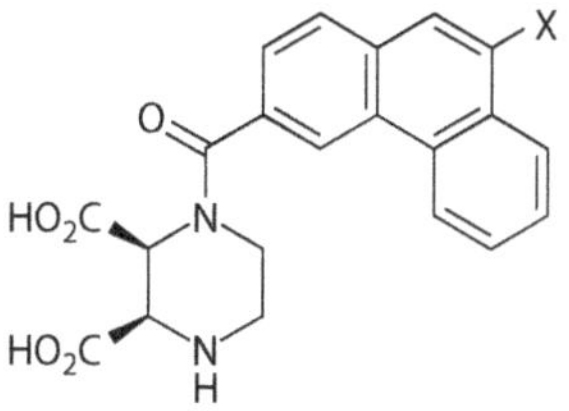

X = Br (UBP145)
X = Cl (UBP160)
X = I (UBP161)

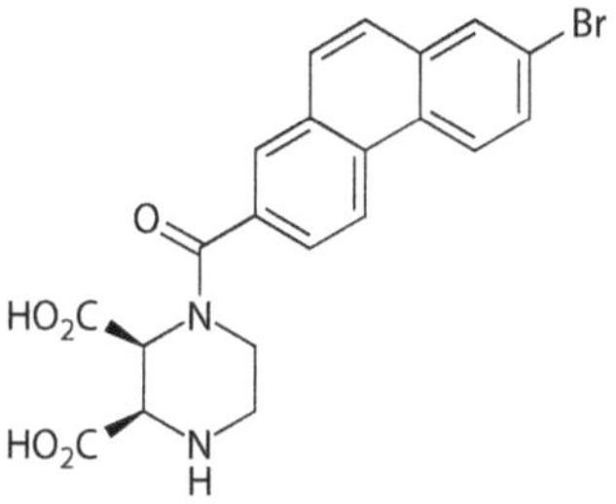

UBP142

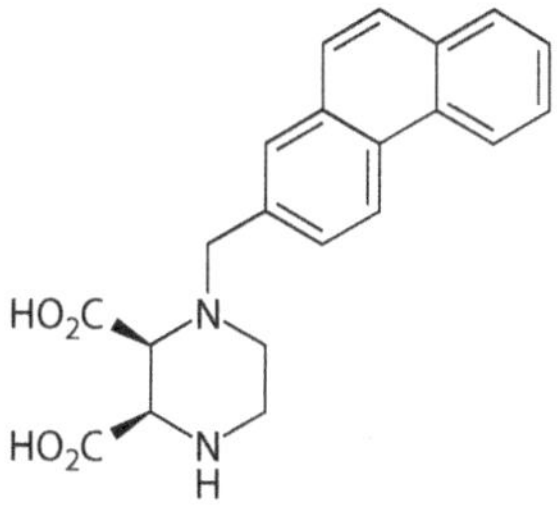

UBP140

FIGURE 3.12 Compounds synthesized to examine the effect of ring position and substituents.

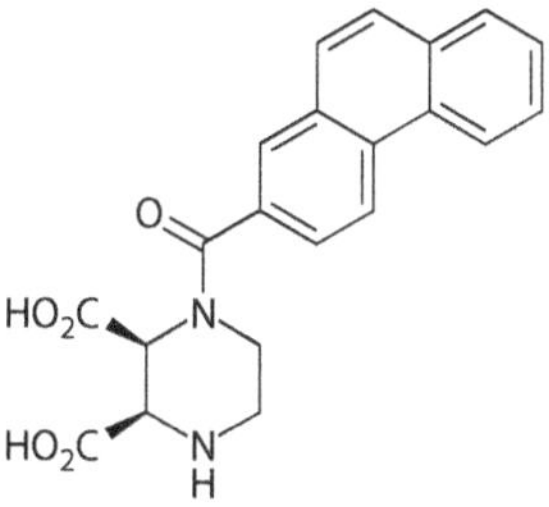

PPDA

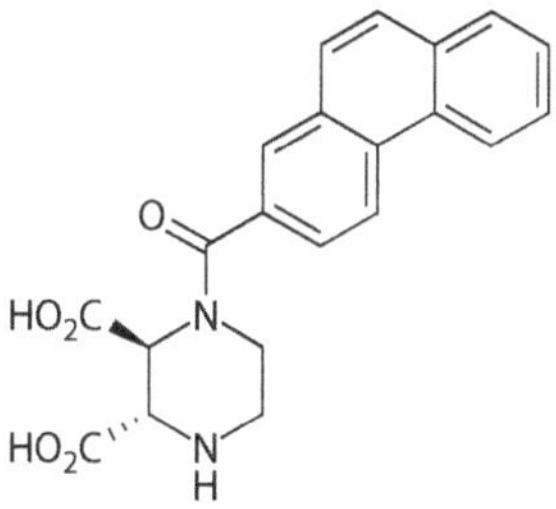

UBP138

FIGURE 3.13 Structures synthesized to study the effect of configuration on *N*-methyl-D-aspartate receptor antagonist activity.

PPDA displayed a high affinity for all the subunits, with a GluN2C/GluN2D selectivity, while the *trans* isomer (UBP138) was 60 times less potent at GluN2C and GluN2D, and showed very little GluN2 subunit selectivity (Figure 3.13, Table 3.2). All of the PzDA analogs were tested initially as racemic mixtures. As it is often the case that only one enantiomer will be active, it was important to test the individual enantiomers of one of the more potent antagonists. When the enantiomers of PPDA were separated and tested, it was found that the (–)-isomer had 50 times higher affinity for GluN2D than the (+)-isomer, although there was no difference in their GluN2 subunit selectivity (Table 3.2).

It was concluded that all three rings of the phenanthrene contributed to the high affinity and GluN2C/D selectivity, but that there was an opportunity to manipulate the selectivity and affinity by changing the position at which the carbonyl was attached to the piperazine ring and by adding new substituents to the phenanthrene ring. Both the stereochemistry and the amide linker were also crucial for high affinity and GluN2C/2D subunit selectivity. PPDA, UBP141, and UBP145 are now leads for the development of more potent and GluN2 subunit selective competitive antagonists.

3.4 CONCLUDING REMARKS

In this chapter, we have described some of the many techniques that can be used in LBDD and shown examples where these techniques were put into practice to enhance the biological activity of lead compounds. The case studies demonstrate that by investigating SARs, considerable detail of the binding site can be inferred even in the absence of molecular modeling and crystal structures of target proteins. For example, since the initial publication of these case studies, the crystal structure of the ligand binding domain of GluN2A in complex with PPDA has been determined, largely confirming the predictions and conclusions made by these studies.

The last 50 years has seen the emergence of new strategies for the design of drugs such as quantitative SAR, computer-aided pharmacophore modeling, and structure-based methods (Chapter 4). While these techniques provide new ways of designing molecules, they complement LBDD rather than replace it, influencing the design of ligands and helping to rationalize the data obtained.

FURTHER READING

Brown, N., Mannhold, R., Kubinyi, H., and Folkers, G. 2012. *Bioisosteres in Medicinal Chemistry*. Weinheim, Germany: Wiley-VCH.

Feng, B., Morley, R.M., Jane, D.E., and Monaghan, D.T. 2005. The effect of competitive antagonist chain-length on NMDA receptor subunit selectivity. *Neuropharmacology* 48:354–359.

Feng, B., Tse, H.W., Skifter, D.A., Morley, R.M., Jane, D.E., and Monaghan, D.T. 2004. Structure-activity analysis of a novel NR2C/NR2D-preferring NMDA receptor antagonist: 1-(phenanthrene-2-carbonyl) piperazine-2,3-dicarboxylic acid. *Br. J. Pharmacol.* 141:508–516.

Irvine, M.W., Costa, B.M., Dlaboga, D. et al. 2012. Piperazine-2,3-dicarboxylic acid derivatives as dual antagonists of NMDA and GluK1-containing kainate receptors. *J. Med. Chem.* 55:327–341.

Jane, D.E., Olverman, H.J., and Watkins, J.C. 1994. Agonists and competitive antagonists: Structure-activity and molecular modelling studies. In *The NMDA Receptor*, 2nd ed., eds. G.L. Collingridge and J.C. Watkins, pp. 31–104. Oxford, U.K.: Oxford University Press.

Morley, R.M., Tse, H.W., Feng, B., Miller, J.C., Monaghan, D.T., and Jane, D.E. 2005. Synthesis and pharmacology of N^1-substituted piperazine-2,3-dicarboxylic acid derivatives acting as NMDA receptor antagonists. *J. Med. Chem.* 48:2627–2637.

Watkins, J.C. and Jane, D.E. 2006. The glutamate story. *Br. J. Pharmacol.* 147:S100–S108.

4 Biostructure-Based Drug Design

Flemming Steen Jørgensen and Jette Sandholm Kastrup

CONTENTS

4.1 INTRODUCTION

The idea behind biostructure-based drug design is to utilize the information on shape and properties of the binding site of a target molecule (e.g., enzyme or receptor) to design compounds which possess complementary properties. Thus, biostructure-based drug design requires methods for determination of the 3D structure of the target molecules as well as knowledge of which molecular interactions are important for obtaining the desired binding characteristics.

Examples of ligands (drug molecules) binding to proteins are shown in Figure 4.1. The two ligands have been selected to illustrate different types of molecular interactions between the ligand and the target protein.

The 3D structure of a target protein can be determined experimentally by methods such as X-ray crystallography and NMR spectroscopy or predicted by computational methods such as homology modeling (comparative model building). Out of ca. 100,000 experimentally determined protein structures, 94,000 have been determined by X-ray crystallography (Protein Data Bank, March 2015). An X-ray crystallographic structure determination requires protein crystals, and irradiation with a high-energy X-ray source generates a diffraction pattern by the scattering of X-rays from organized molecules in a continuous arrangement in the crystal. Based on the diffraction pattern, an electron density map of the protein can be derived, and subsequently a molecular model reflecting the electron density, the 3D structure, can be determined. Presently, the data collection, data processing, model building, and refinement are highly automated and computerized processes. The present limiting factor for determining the 3D structure of a protein is to get sufficient amounts of pure and stable protein and proper diffracting crystals.

It is important to consider the quality of an X-ray structure before using it for biostructure-based drug design. The resolution is a measure of how detailed the electron density map is and thereby how accurately the positions of the individual atoms can be determined (Figure 4.2a). Structures based on electron densities at ca. 1.2 Å resolution are referred to as atomic-resolution structures and, e.g., hydrogen-bonding networks can unambiguously be identified. Generally, a resolution of ca. 2 Å provides an accurate structure, and accordingly such a structure is also suitable for biostructure-based drug design. Structures based on 3 Å resolution electron densities should be used with caution. Therefore, the smaller the resolution number the more accurate the structure will usually be.

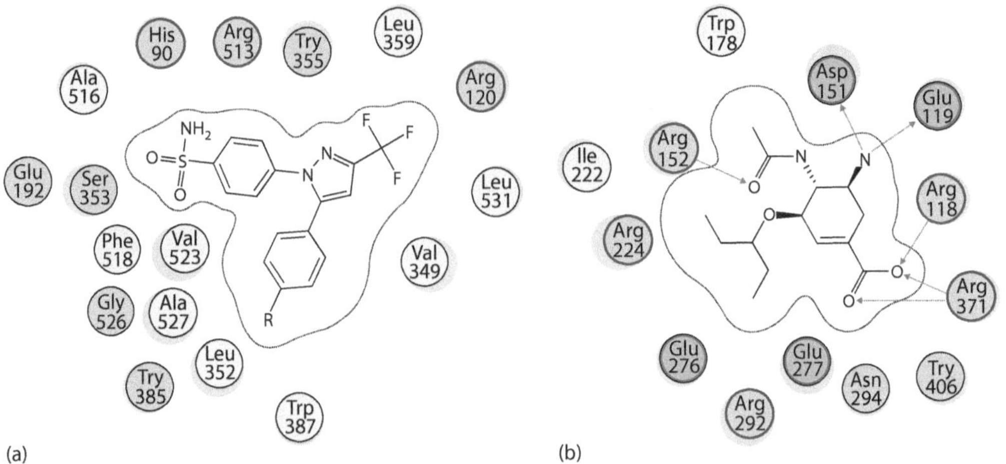

FIGURE 4.1 Example on a nonpolar (a) and a polar (b) binding site. In (a), the ligand, which is an analog (R=Br) of the COX-2 inhibitor celecoxib (R=CH_3), binds to cyclooxygenase-2 in a pocket primarily formed by nonpolar amino acid residues (pdb-code 1CX2). In (b), the ligand, which corresponds to the active part of the anti-influenza drug oseltamivir (Tamiflu), binds to an influenza virus neuraminidase in a pocket formed by polar residues (pdb-code 2QWK). Green arrows indicate hydrogen bonds. Green and red circles represent nonpolar and polar residues, respectively. Blue and red interior rings are highlighting basic and acidic residues, respectively. The dotted lines illustrate the shape of the binding sites. The sizes of the turquoise halos reflect the solvent accessible surface area of the residues.

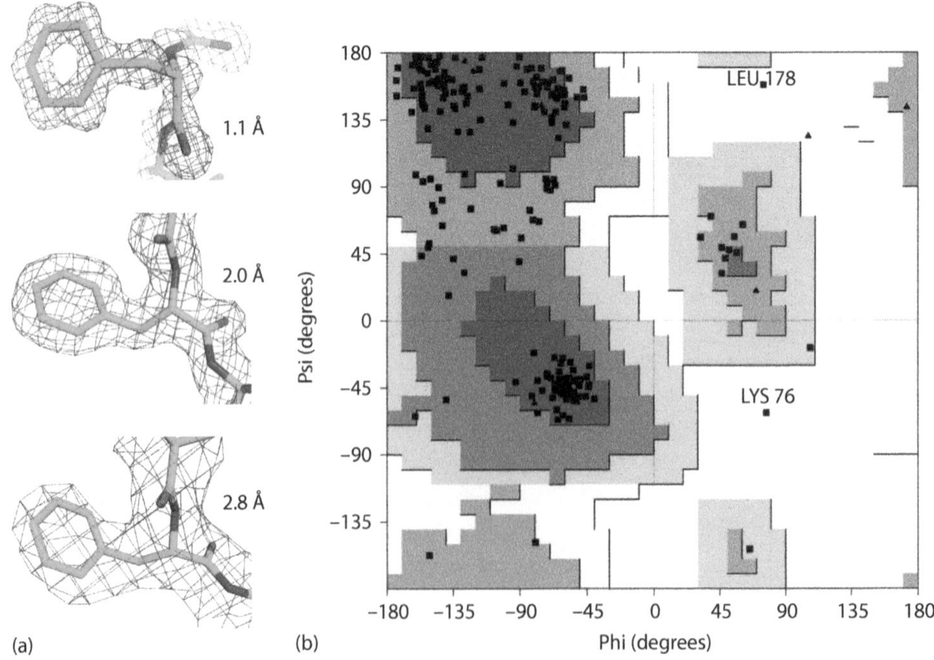

FIGURE 4.2 (a) Electron density of a phenylalanine side chain at 1.1, 2.0, and 2.8 Å resolution, respectively. (b) Ramachandran plot of a protein. Two residues (Lys76 and Leu178) adopt unfavorable backbone conformations. The majority of the residues are located in the dark-gray regions corresponding to favorable backbone conformations.

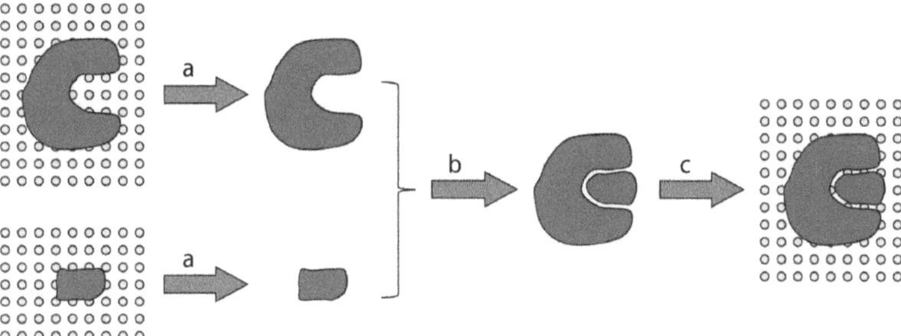

FIGURE 4.3 Simplified binding process. (a) Desolvation of protein and ligand (large unfavorable energy change). (b) The binding process (small favorable energy change). (c) Solvation of the protein–ligand complex (large favorable energy change).

The stereochemical quality of a protein structure should be carefully evaluated prior to using it for biostructure-based drug design. Planarity of peptide bonds, bond lengths, bond angles, and torsion angles should not deviate from average values. The protein backbone geometry is usually validated by a Ramachandran plot which is a plot of the two variable torsion angles (phi and psi) (Figure 4.2b). Special attention should be on the part of the protein involved in direct interactions with ligands, e.g., flexible loops and binding site residues.

Binding of a drug molecule to a protein is a complex process that often is considered as composed of a number of discrete steps in order to simplify the understanding of the process: (1) The protein may contain a cavity where the drug may bind, but prior to binding the water molecules occupying the cavity has to be removed. Similarly, the drug molecule may be surrounded by water molecules that have to be removed before binding. These two desolvation processes are associated with an unfavorable energy change (Figure 4.3a) that needs to be taken into account in biostructure-based design. (2) The next step is the actual binding between the drug molecule and the protein. The drug molecule may change conformation in order to fit into the binding site of the protein, and this conformational change requires energy. The protein may also change conformation (see example in paragraph 4.4), but this is unfortunately often ignored. The binding process requires that the drug molecule and protein are complementary not only with respect to shape, but also with respect to molecular properties. Positive parts of the drug molecule will bind in negative parts of the binding cavity and vice versa, and hydrogen-bond donors will bind to hydrogen-bond acceptors and vice versa. The binding process usually leads to a small favorable energy change (Figure 4.3b). (3) Finally, the complex between the drug molecule and the protein has to be surrounded by water molecules (solvated), and this process is associated with a favorable energy change (Figure 4.3c). Thus, the actual binding energy is a relatively small difference (typically 5–20 kcal/mol) between processes representing large destabilizing versus stabilizing forces (for further details see Chapter 2).

4.2 ANTI-INFLUENZA DRUGS

One of the first examples on biostructure-based drug design was the discovery of the anti-influenza drug zanamivir (Relenza). Together with the subsequent discovery of the orally active anti-influenza agent oseltamivir (Tamiflu) they now constitute the classical example on a successful drug design story.

Two glycoproteins, hemagglutinin and neuraminidase, present on the surface of the virus are important for the replication cycle of the virus. The enzyme neuraminidase, or sialidase, is a glycohydrolase responsible for cleavage of the terminal sialic acid residues from carbohydrate moieties on the surface of the host cells and the influenza virus envelope. This process facilitates the release of newly formed virus from the surface of an infected cell. Inhibition of neuraminidase will leave uncleaved sialic acid residues that may bind to viral hemagglutinin, causing viral aggregation and thereby a reduction of the amount of virus that may infect other cells.

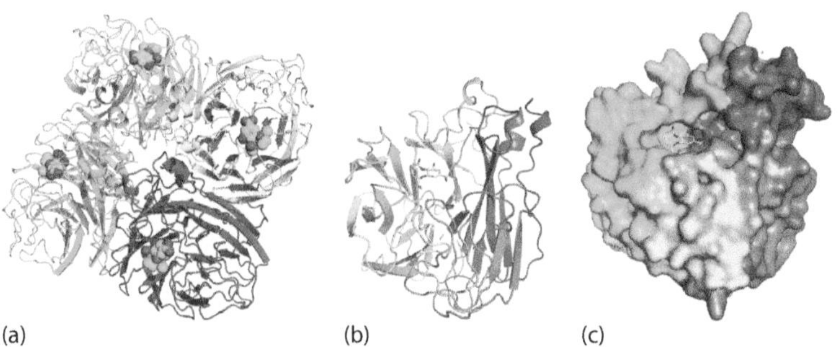

FIGURE 4.4 Three-dimensional structure of influenza virus neuraminidase (pdb-code 2QWK). (a) Homotetramer shown as ribbon representation. Four inhibitor molecules bound to the enzyme are shown as space-filling models. (b) Ribbon representation of one monomer colored continuously from blue (*N*-terminus) to red (*C*-terminus). An inhibitor is shown as a stick model. (c) Surface representation of monomeric neuraminidase with an inhibitor occupying the active site.

Thus, selective inhibitors of the enzyme neuraminidase are potential drugs against influenza. The determination of a 3D structure of neuraminidase by X-ray crystallography in 1991 yielded a breakthrough in the discovery of neuraminidase inhibitors being sialic acid analogous by biostructure-based methods.

It was shown that neuraminidase forms a homotetramer and that each monomer contains six four-stranded antiparallel beta-sheets arranged as blades on a propeller. The active site was identified from complexes with inhibitors of the enzyme and is located in a deep cavity on the surface of the enzyme (Figure 4.4). The active site is primarily formed by charged and polar residues, reflecting that the substrate also is a polar compound.

The mechanism for hydrolysis of glycoconjugates (Figure 4.5a) by neuraminidase yielding sialic acid (Figure 4.5c) is proposed to proceed via a flat oxonium cation transition state (Figure 4.5b). Sialic acid is also a weak inhibitor of neuraminidase (IC_{50} ~ 1 mM) and binds to the enzyme in a half-boat conformation. The weak, nonselective inhibitor 2-deoxy-2,3-didehydro-D-*N*-acetylneuraminic acid (Neu5Ac2en, DANA, Figure 4.5d) was developed as a sialic acid analog resembling the transition state of the enzymatic process (IC_{50} = 1–10 μM). Based on the structure of the neuraminidase–sialic acid complex, the complex with DANA and computational studies using the program GRID which determines favorable binding sites for probes resembling various functional groups, it could be shown that replacement of the hydroxyl group at the 4-position on DANA by an amino or a guanidinyl group would enhance binding. It was predicted that a substantial increase in binding could be obtained by introducing an amino group because a salt bridge was formed to a negatively charged side chain in the enzyme. Further replacement of the amino group with the more basic guanidinyl group was as anticipated yielding an even tighter binding inhibitor due to the formation of salt bridges to two negatively charged side chains in the enzyme. Both the predicted increase in affinity and the binding mode were subsequently confirmed experimentally. The 4-guanidinyl analog (Figure 4.5e) was named zanamivir and was the first neuraminidase inhibitor approved for treatment of influenza in humans. It is marketed under the trade name Relenza.

The low oral bioavailability and rapid excretion of zanamivir clearly showed that further improvements were needed in order to obtain a successful anti-influenza drug. Based on the 3D structures of several neuraminidase-inhibitor complexes and computer-based studies, the characteristics of the active site and thereby the properties of the optimal inhibitor were deduced. Replacing the pyranose ring with a benzene ring reduced the affinity, showing that the half-boat conformation of the six-membered ring was important for obtaining a proper orientation of the substituents. Inhibitors with the pyranose ring replaced by a carbocyclic ring system showed promising affinities, and by introducing more lipophilic substituents compounds with significantly better oral bioavailability

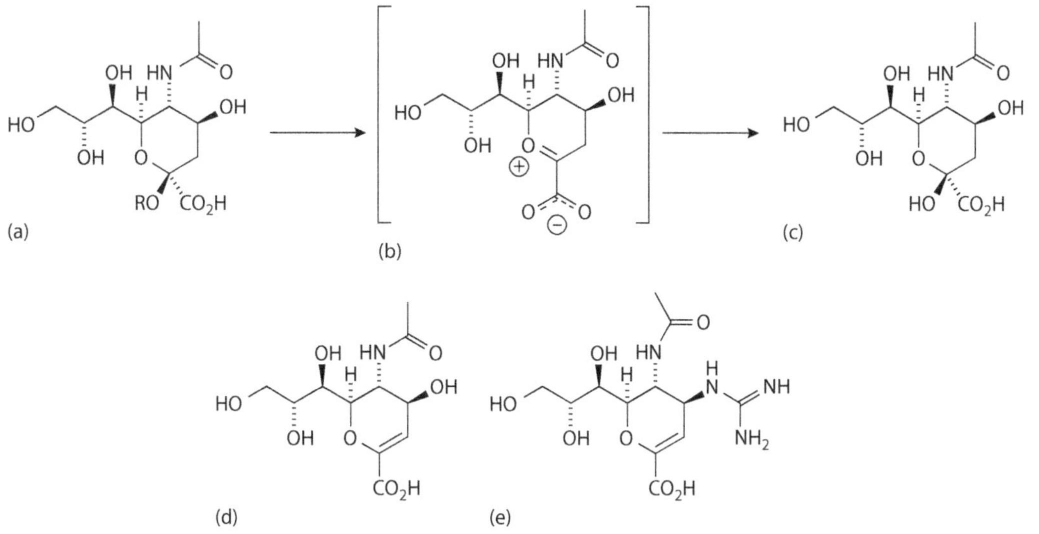

FIGURE 4.5 The biostructure-based design process leading to zanamivir (Relenza): (a) chemical structure of the glycoconjugates (R = sugar), (b) the proposed transition state, (c) sialic acid, (d) DANA, and (e) zanamivir.

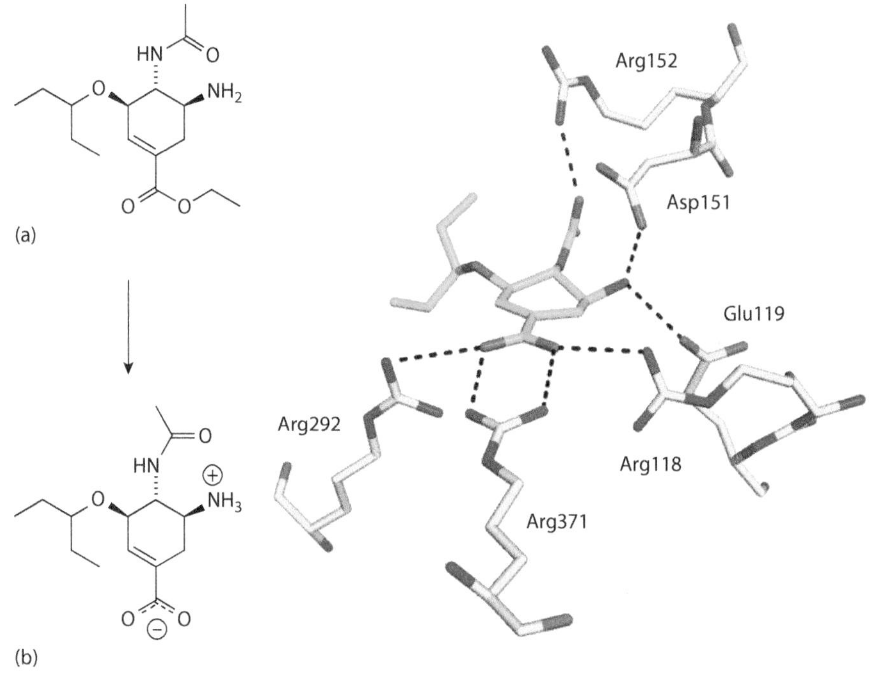

FIGURE 4.6 Chemical structures of (a) oseltamivir (oseltamivir phosphate salt = Tamiflu) and (b) GS4071, the active component formed by enzymatic ester cleavage. Right: hydrogen-bonding network between (b) and neuraminidase (pdb-code 2QWK).

were obtained. In the final active drug candidate GS4071 (IC_{50} ~ 1 nM), the carbocyclic ring adopts the half-boat conformation, and the polar substituents are all involved in hydrogen bonding to polar residues in the neuraminidase active site (Figure 4.6). The ethyl ester was named oseltamivir and its phosphate salt (a prodrug; see Chapter 10) is the first orally administered neuraminidase inhibitor and is marketed under the trade name Tamiflu.

4.3 HIV PROTEASE INHIBITORS

When it was discovered in 1984 that the human immunodeficiency virus (HIV) caused acquired immunodeficiency syndrome (AIDS), it was the start of an intensive drug hunting process. The DNA in the virus encodes for a number of enzymes, e.g., a reverse transcriptase, an integrase, and a protease. Each of these represented a potential drug target, and drugs have subsequently been developed for each of the enzymes. Presently, cocktails of inhibitors against at least two of these enzymes are used therapeutically. In the following, we will concentrate on the HIV-1 protease as biostructure-based design has been applied extensively to this target.

The first 3D structure of a HIV-1 protease in complex with an inhibitor, MVT-101, was reported shortly after it had been shown that inhibition of the HIV-1 protease prevented the virus in producing new virions. MVT-101 binds to the enzyme in an extended conformation and forms a network of hydrogen bonds between the ligand and enzyme (Figure 4.7). Hydrophobic substituents on

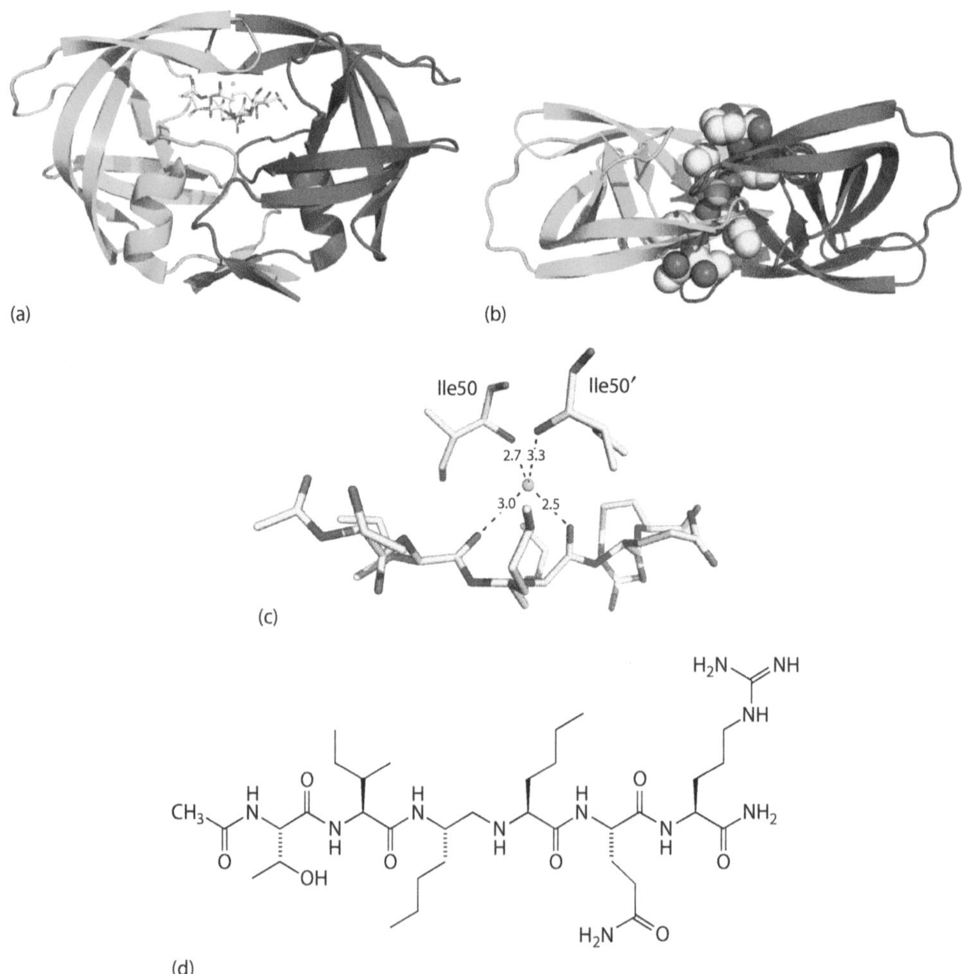

FIGURE 4.7 Three-dimensional structure of the HIV-1 protease with bound MVT-101 (pdb-code 4HVP). (a) Side view showing that the active form of the HIV-1 protease is a homodimer (colored red and green, respectively) and that the inhibitor MVT-101 binds between the two monomers. (b) Top view showing the extended form of the inhibitor and that the inhibitor via a structural water molecule (cyan) binds to the flaps. (c) The structural water molecule makes four hydrogen bonds to the inhibitor and to Ile50 in the flaps. (d) Chemical structure of MVT-101.

the inhibitor occupy hydrophobic pockets in the enzyme, and a water molecule mediates contact between the inhibitor and two residues in two flexible beta-sheets, normally referred to as the flaps. Today, more than 500 structures of complexes between HIV-1 protease and ligands have been determined which makes HIV-1 protease one of the structurally most extensively studied proteins.

The first HIV-1 protease inhibitors such as indinavir (Crixivan), nelfinavir (Viracept), and saquinavir (Invirase, Fortovase) (Figure 4.8) were derived from the polypeptide sequences cleaved by the protease in the HIV. Accordingly, they were very peptide-like and had poor bioavailability.

Unfortunately, the HIV rapidly developed resistance against these first-generation inhibitors. The shapes of the hydrophobic pockets are sensitive to mutations in the enzyme and accordingly, the virus could easily prevent an inhibitor from binding by mutation of residues forming the hydrophobic pockets.

In order to circumvent this problem, inhibitors binding not primarily due to hydrophobic effects but also by hydrogen bonding to the enzyme backbone atoms (amide NH group and amide carbonyl oxygen atom) were designed. These hydrogen-bonding contacts are not sensitive to mutations, since they do not involve the side chains but only the backbone atoms. These second-generation inhibitors such as amprenavir (Agenerase) and KNI-764 (Figure 4.8) showed significantly different binding characteristics.

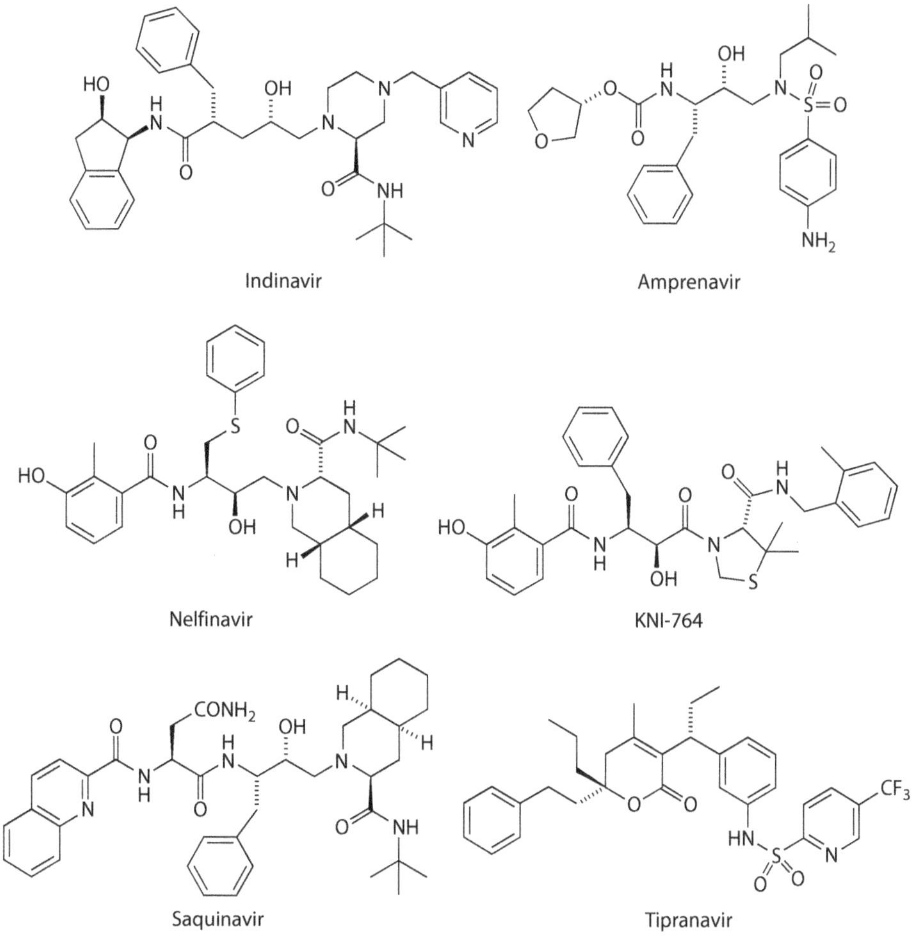

FIGURE 4.8 Examples on HIV-1 protease inhibitors. Indinavir, nelfinavir, and saquinavir are examples of first-generation, amprenavir and KNI-764 of second-generation, and tipranavir of third-generation HIV-1 protease inhibitors.

Thermodynamic determination of the enthalpy and entropy components to the free energy of binding ($\Delta G = \Delta H - T\Delta S$) can be performed by isothermal titration calorimetry (ITC), and this method has led to a much more detailed understanding of the energetics associated with the process of binding a ligand to a protein. By using ITC to guide the design of new inhibitors, it has been possible to optimize the binding characteristics of these.

For the first-generation HIV-1 protease inhibitors, the majority of the free energy of binding is due to an entropy gain associated with filling hydrophobic pockets in the HIV-1 protease with hydrophobic substituents on the inhibitor. The second-generation inhibitors were characterized by both enthalpy and entropy now contributing to the free energy of binding, making the enzyme less likely to develop resistance. Third-generation HIV-1 protease inhibitors have been developed based on careful optimization of the structural and energetic contributions to binding. Tipranavir (Aptivus) (Figure 4.8) is an example of a HIV-1 protease inhibitor with unique binding characteristics. It is a highly potent inhibitor (K_i = 19 pM) that primarily binds to the wild type HIV-1 protease due to entropy effects. The unusually high binding entropy is most likely caused by release of buried water molecules from the active site of the HIV-1 protease. When binding to the multidrug-resistant HIV mutants, tipranavir only loses little in potency, because the reduction in binding entropy is compensated by a gain in binding enthalpy.

In 1994, researchers at the DuPont Merck Pharmaceutical Company reported an important observation. They had realized that in most of the complexes between HIV-1 protease and the peptide-like inhibitors, a structural water molecule bridged the ligand and enzyme. They concluded that by designing an inhibitor that would displace this water molecule, a favorable entropy contribution to the binding energy would be obtained. In addition, selectivity should be gained since this structural water molecule was unique for the viral proteases.

The DuPont Merck scientists defined a pharmacophore (see Section 3.2.2) from known dihydroxyethylene inhibitors (Figure 4.9a), used this pharmacophore for searching a database, and obtained a hit (Figure 4.9b) which subsequently gave the idea to the six-membered cyclic ketone as lead structure (Figure 4.9c). Ring expansion to the seven-membered cyclic ketone (Figure 4.9d) enabled incorporation of two hydrogen-bonding donor/acceptor groups and synthetic reasons led

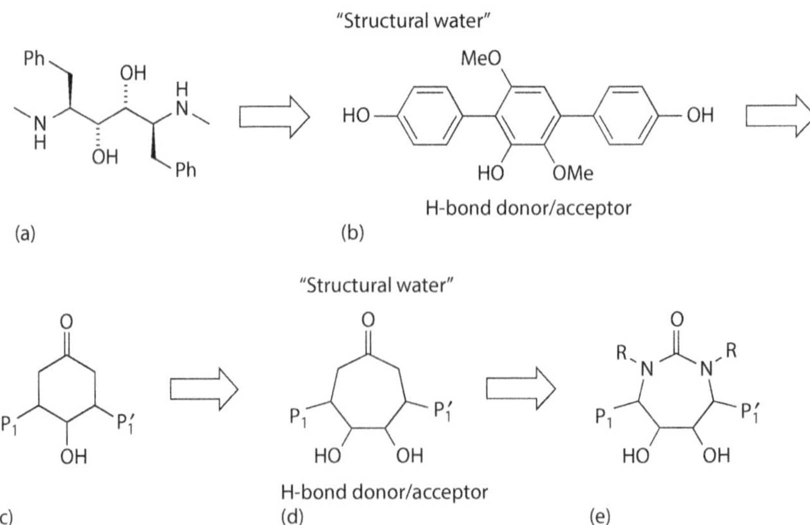

FIGURE 4.9 The DuPont Merck design process leading to the seven-membered urea HIV-1 protease inhibitors. (a) Symmetric dihydroxyethylene inhibitor used to define a pharmacophore. (b) Initial hit from 3D database search. (c) Initial synthetic scaffold. (d) Scaffold modified to accommodate two hydrogen-bond donors/acceptors. (e) Final scaffold optimized for synthetic feasibility and improved hydrogen bonding.

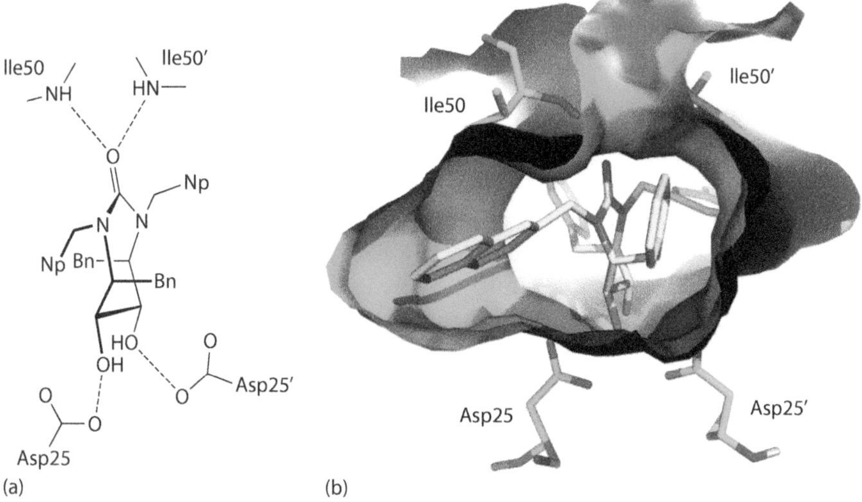

FIGURE 4.10 Schematic representations of the binding of the seven-membered urea HIV-1 protease inhibitor XK-263 to HIV-1 protease (pdb-code 1HVR). In (a), Bn and Np refer to benzyl and 2-naphthyl, respectively. In (b), the two Asp25 and Ile50 residues making hydrogen bonds to the inhibitor are shown as stick models. Hetero-atoms are colored red and blue for oxygen and nitrogen, respectively. The surface is colored accordingly.

to the series of seven-membered cyclic urea HIV-1 protease inhibitors (Figure 4.9e). Using the structural information available from the many peptide-like inhibitors, the nature and stereochemistry of the substituents on the cyclic urea could be designed so they were preorganized for binding to the enzyme (Figure 4.10). By preorganization of a ligand for binding, the conformational energy penalty often associated with ligand binding is reduced and the ligand may be more potent.

Although the DuPont Merck cyclic urea inhibitors were based on a brilliant structural idea and potent inhibitors were designed, the inhibitors generally had low bioavailability and the HIV quickly developed resistance against the compounds. Thus, none of these inhibitors are among the drugs used today. Other companies have subsequently adopted the same idea in their design processes. One example is the previously mentioned HIV-1 protease inhibitor tipranavir (Aptivus) where the oxygen atom in the carbonyl group replaces the structural water molecule.

4.4 MEMBRANE PROTEINS

The largest group of biological targets for therapeutic intervention comprises the membrane proteins. These proteins perform various functions in the cell, serving as enzymes, pumps, channels, transporters, and receptors. To underline the importance of considering structure-based drug design on membrane proteins, ca. 40% of all drugs target G-protein-coupled receptors and additionally ca. 20% other membrane proteins.

The first membrane structure was reported in 1985 of the photosynthetic reaction center from *Rhodopseudomonas viridis*. Today, the number of known membrane protein structures is still limited but steadily increasing. Currently, 3D structures are only known for ca. 530 unique membrane protein structures, including proteins of the same type from different organisms (Membrane Proteins of Known 3D Structure website, March 2015). The number of membrane protein structures and the change in the number of structures as a function of time are similar to the state of soluble protein structures approximately 25 years ago. The reason for this is primarily that expression, purification, and crystallization of membrane proteins still are non-trivial and require substantial time and resources. This means that structure determination of and biostructure-based drug design on full-length membrane proteins represent one of the most challenging areas of modern drug research.

Another strategy taken on membrane proteins is to produce soluble constructs of parts of the proteins. The resolution (quality) of structures usually are much higher of soluble proteins compared to membrane proteins and this allows detailed information on, e.g., ligand–receptor interactions and water networks in the binding site.

One example of important membrane proteins is the ionotropic glutamate receptors (iGluRs). These receptors mediate most fast excitatory synaptic transmission within the central nervous system. The glutamate receptors are involved in various aspects of normal brain function but are also implicated in a variety of brain disorders and diseases. For example, the drug memantine targeting the iGluRs is used in the treatment of Alzheimer's disease and the drug perampanel (Fycompa) was approved by the European Commission in 2012 for treatment of epilepsy. Hence, iGluRs are potential targets for biostructure-based drug design (for further details on glutamate receptor structures, see Chapters 12 and 15).

In 1998, the first structure of a soluble ligand-binding domain construct of an iGluR was reported, and 10 years later the first structure of a full-length iGluR, confirming that the receptor is composed of four subunits, each containing an *N*-terminal domain, a ligand-binding domain where glutamate binds, a transmembrane region and a C-terminal region, the latter not seen in the structure (Figure 4.11). In 2014, new full-length iGluR structures appeared in the literature, both of different types of iGluRs and in different conformational states. Presently, more than 300 structures (mostly domain constructs) of iGluRs with bound glutamate, agonists, antagonists, or allosteric modulators have been reported.

The binding of ligands to the ligand-binding domain of iGluRs can be described as a "Venus flytrap" mechanism. In the resting state, the ligand-binding domain is present in an open form

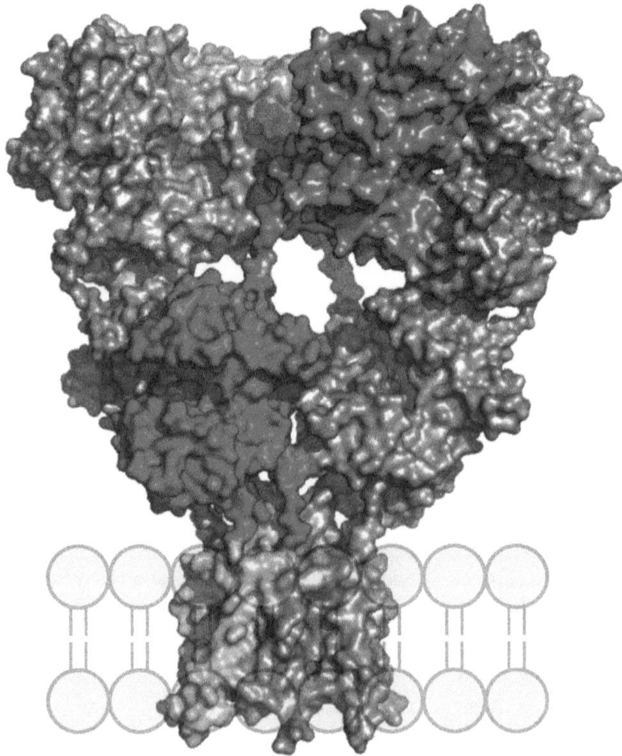

FIGURE 4.11 Structure of a full-length tetrameric iGluR (pdb-code 3KG2). The domains in one subunit are shown in different colors: The N-terminal domain in blue, the ligand-binding domain where glutamate binds in red, and the ion channel part in orange. The location of the cell membrane is indicated.

and it is this form that is stabilized by competitive antagonists (Figure 4.12). When glutamate or an agonist binds to the ligand-binding domain, a change in conformation occurs, resulting in a closed form of the ligand-binding domain. In full-length receptors, this domain closure is thought to lead to the opening of the ion channel (receptor activation). It has been observed that different agonists and antagonists can induce a range of domain movements, from domain opening of ca. 5°

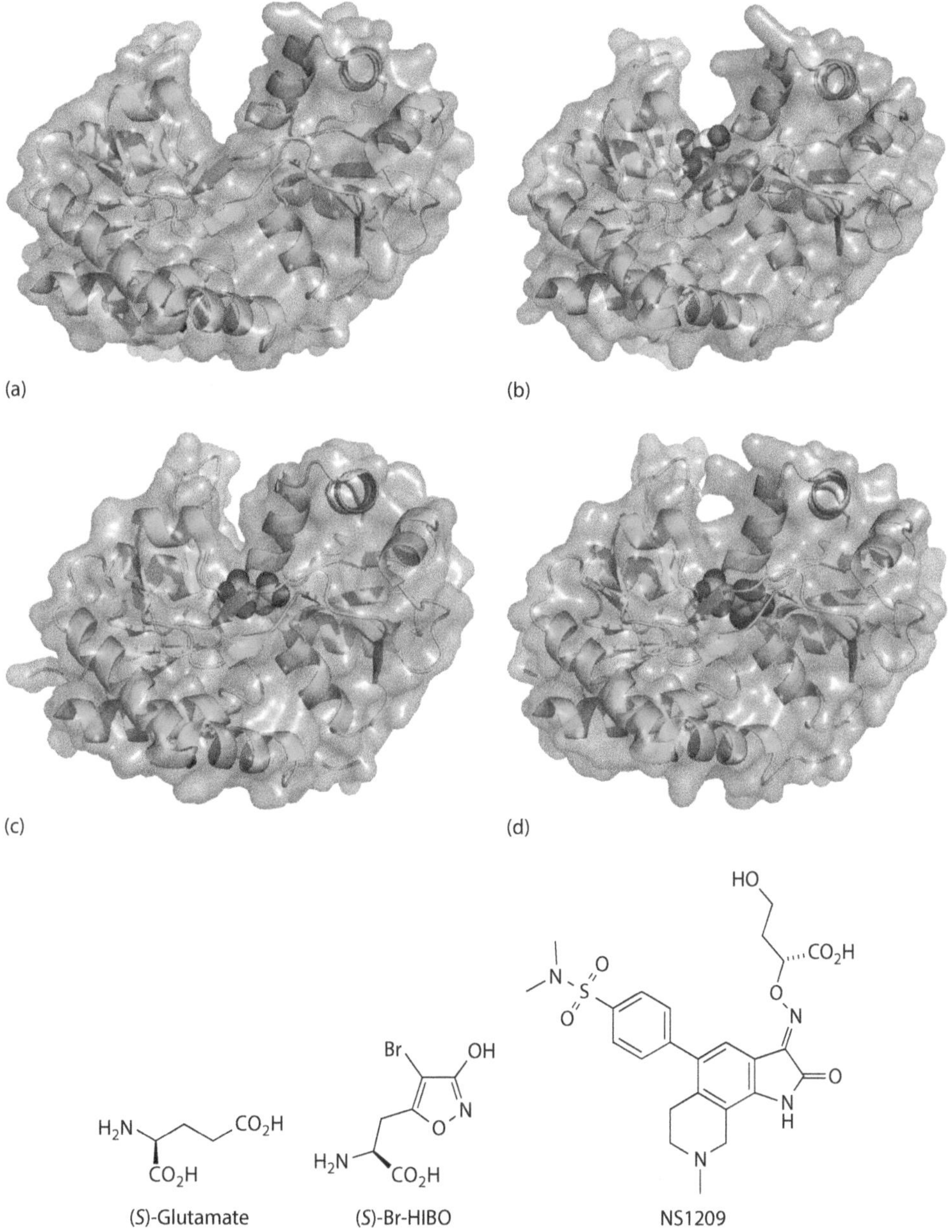

FIGURE 4.12 Structures of the soluble ligand-binding domain of the ionotropic glutamate receptor GluA2. (a) The open, unbound form of GluA2 (pdb-code 1FTO). (b) The NeuroSearch compound NS1209 stabilizes GluA2 in the open form (pdb-code 2CMO). (c) The endogenous ligand (*S*)-glutamate introduces domain closure of GluA2 by a "Venus flytrap" mechanism (pdb-code 1FTJ). (d) Various synthetic agonists (here (*S*)-Br-HIBO) also introduce domain closure in GluA2 (pdb-code 1M5C). Below: chemical structures of (*S*)-glutamate, the agonist (*S*)-Br-HIBO, and the antagonist NS1209.

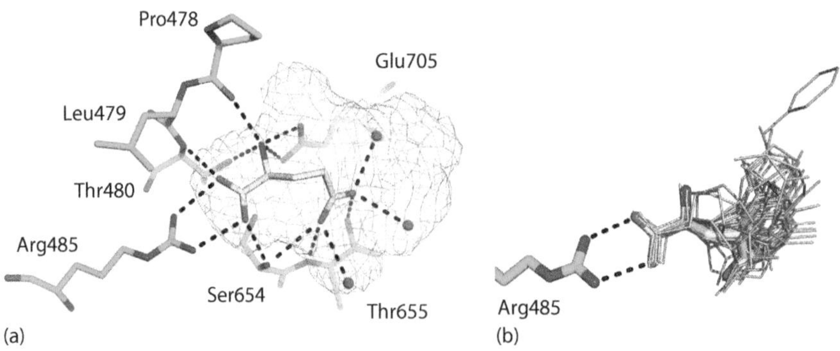

FIGURE 4.13 Zoom on the binding site of GluA2. (a) Binding mode of (*S*)-glutamate (pdb-code 1FTJ, yellow carbon atoms). Hydrogen bonds from glutamate to GluA2 residues (green carbon atoms) and water molecules (red spheres) are shown as dotted lines (within 3.2 Å). Oxygen atoms are red and nitrogen atoms blue. The gray mesh illustrates that glutamate does not entirely fill out the binding site. (b) Overlay of GluA2 structures with different agonists bound (glutamate in yellow stick representation; 24 other agonists in gray line representation). Only the agonists and the essential residue Arg485 in GluA2 is shown as well as the salt bridge from (*S*)-glutamate to Arg485.

by antagonists to closure of ca. 25° by agonists, compared to the unbound form of the iGluR. The presence of several conformations of the ligand-binding domain of iGluRs clearly underlines the importance of knowing more than one structure of the receptor as fundamental for biostructure-based drug design.

Glutamate forms different types of polar contacts to binding-site residues and water molecules (Figure 4.13a). The α-carboxylate group of glutamate makes a salt bridge with the essential Arg485, and the α-ammonium group of glutamate a salt bridge to Glu705. In addition, charge-assisted hydrogen bonds from glutamate to Pro478, Thr480, Arg485, Ser654, Thr655 and three water molecules are present. As seen in Figure 4.13a, glutamate only partly fills up the binding site. In structure-based design, it might be advantageous to make compounds that, compared to glutamate, either protrude into unoccupied regions of the receptor or displace water molecules by substituents introduced into the compound. Comparison of the binding mode of various agonists shows that the α-amino acid moiety of the agonists is essentially located at the same position in the binding site, whereas much larger variability is seen for the compound's distal functionalities (Figure 4.13b).

4.5 FAST-ACTING INSULINS

Biostructure-based drug design is not limited to design of low molecular weight compounds based on knowledge of the structure of their biological targets. In the following, we present an example on biostructure-based design of macromolecular drug molecules, i.e., insulin analogs. This design was only made possible by a detailed insight into the structure of insulin and the intermolecular interactions between insulin molecules in the crystalline phase.

Insulin is a hormone produced in the pancreas and it is responsible for regulation of glucose uptake and storage. Insulin is most often associated with diabetes mellitus which is a disease causing hyperglycemia. Healthy people have a basal level of insulin in the bloodstream, but in response to the intake of food or to cover glucose clearance from the blood, peaks of larger insulin concentrations appear throughout the 24 hours of a day. Patients with diabetes may have difficulties in maintaining the proper insulin concentrations, basal as well as peak concentrations, and accordingly regulation of their insulin level is essential. Type I diabetes patients need insulin to supplement endogenously produced insulin, whereas Type II diabetes patients often are getting insulin in order to improve glycemic control.

Insulin has been available for treatment of diabetes since 1923, and the major form for administration is still by subcutaneous injection. Insulin therapy typically involves multiple doses of different forms of insulin to maintain near-physiological insulin (and thereby glucose) levels. Long-acting insulin maintain the basal insulin level over 24 hours with a single administration, and fast- and short-acting insulin analogs which are instantaneously absorbed, are used to meet the insulin requirements associated with food intake. Thus, the development of insulin formulations with tailored properties, e.g., a prolonged effect or a faster onset, has always had a high priority in insulin research.

With the biosynthesis of recombinant human insulin in the 1980s, it became possible to optimize the insulin therapy by designing insulin analogs with optimal pharmacokinetic properties by changing the amino acid sequence of the insulin molecule. A rational design of insulin analogs was only possible because a large number of insulin structures were determined experimentally by NMR spectroscopy and X-ray crystallography. Insulin was one of the first proteins whose 3D structure was determined by X-ray crystallography, and today ca. 300 insulin structures are available in the Protein Data Bank.

Insulin exists in the crystalline phase as hexamers, dimers, and monomers. Actually, several different forms of hexamers, T6, T3R3, and R6, exist depending on the presence of zinc ions and phenol (Figure 4.14). The presence of several hexameric forms of insulin reflects that even in the crystalline form insulin exerts some kind of conformational flexibility. Both zinc ions and phenol are stabilizing the hexameric form of insulin and are accordingly added to insulin formulations in order to improve their stability. After subcutaneous injection, the insulin hexamer dissociates to dimeric insulin and by further dissociation to monomeric insulin which represents the bioactive form. Thus, from the early beginning it was believed that shifting the equilibrium toward the monomeric form would lead to faster-acting insulins, whereas stabilization of the hexameric form would lead to longer-acting insulins.

The insulin molecule consists of two peptide chains, an A-chain of 21 residues and a B-chain of 31 residues. The A- and B-chains are connected by two disulfide bridges, linking A7-B7 and A20-B19. With the introduction of recombinant DNA techniques, the residues involved in interaction with the insulin receptor or involved in the hexamer versus dimer stabilization were identified.

From the X-ray structure of hexamer insulin, it was evident that the side chain of the HisB10 residue was involved in zinc binding and thereby in stabilizing the hexamer (Figure 4.14). Mutation of the B10 residue from His to Asp yielded an insulin analog being absorbed twice as rapidly as normal insulin. Unfortunately, this analog turned out to be mitogenic and thus not suitable for clinical use.

Based on various structural studies of insulin, it could be concluded that the flexibility of the C-terminus is crucial for the binding of insulin to its receptor (Figure 4.15). It also became evident

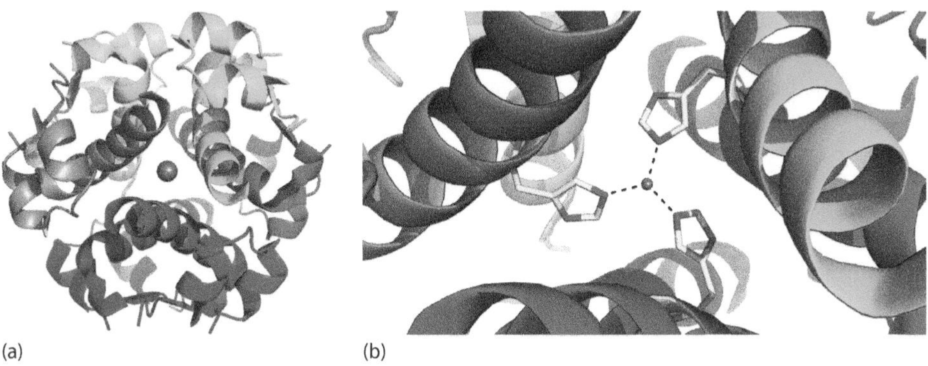

FIGURE 4.14 (a) Insulin R6 hexamer (pdb-code 1EV6) showing the threefold symmetry of the three dimers (red-green, cyan-orange, and magenta-blue). (b) Three HisB10 residues coordinate to a zinc ion.

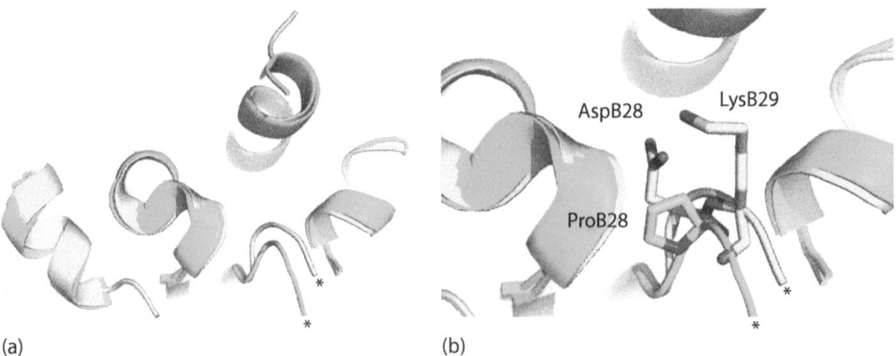

FIGURE 4.15 (a) Human insulin (pdb-code 1ZNJ, green) and insulin aspart (pdb-code 1ZEG, yellow), showing the change in conformation of the C-terminus (marked by an asterisk) by the mutation ProB28Asp. (b) Human insulin ProB28 and for insulin aspart AspB28 and the neighboring LysB29 are shown as stick models. Hetero-atoms are colored red and blue for oxygen and nitrogen, respectively.

that the B24-B26 residues are stabilizing the dimer by making an intermolecular anti-parallel beta-sheet and at the same time burying the nonpolar side chains. Removal of the B24-B30 residues yields monomeric insulin and eliminates the association of monomers into dimers and hexamers. Furthermore, molecular modeling studies showed that ProB28 was important for dimer formation, but apparently not involved in receptor binding. Mutation of B28 from Pro to Lys or Asp has a profound effect on dimerization. The ProB28Asp mutant is presently marketed as rapid-acting insulin named aspart (Novorapid).

The knowledge of the interactions, which stabilize the hexamer and dimer, has made it possible to engineer the properties of insulin to yield therapeutic insulin analogs with tailored properties. Homology modeling indicated that inversion of the residues B28 and B29 should give faster acting insulins (Figure 4.16). The double mutant, ProB28Lys + LysB29Pro, indeed had a faster onset than normal insulin and it was the first genetically engineered insulin analog to become available for clinical use. Generally, modifications of B29 have a less pronounced effect on dimerization than modification of ProB28. The double mutant is especially interesting because it has the same isoelectric point (pI) as human insulin as their amino acid compositions are identical, and thereby the same solubility. The double mutant called insulin lispro is marketed under the brand name Humalog.

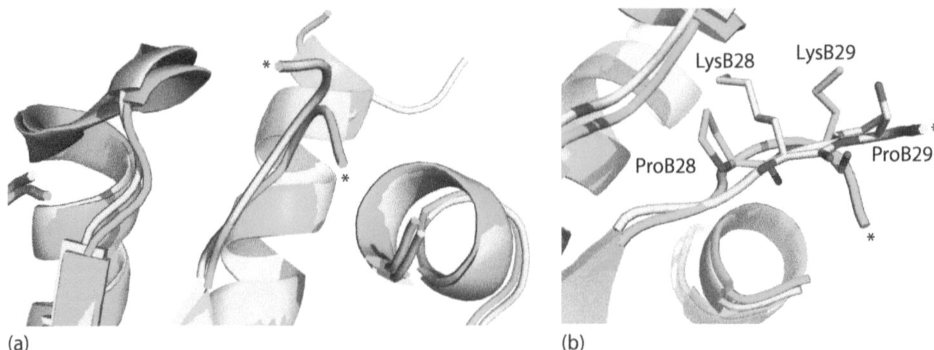

FIGURE 4.16 (a) Human insulin (pdb-code 1ZNJ, green) and insulin lispro (pdb-code 1LPH, yellow), showing the change in conformation of the C-terminus (marked by an asterisk). (b) The ProB28-LysB29 in human insulin and the LysB28-ProB29 in insulin lispro are shown as stick models. Hetero-atoms are colored red and blue for oxygen and nitrogen, respectively. The side-chain nitrogen atom in LysB29 could not be observed in the X-ray structure.

Insulin analogs with changes of ProB28, as the aforementioned insulin analogs aspart and lispro, primarily exist in the monomeric form which is known to be more vulnerable for unfolding and subsequent fibril formation. To avoid this problem, mutations were limited to the neighboring LysB29 and to residues in the N-terminus of the B-chain. It was already known that changes of the B1-B8 residues primarily affected the stabilization of the dimer form of insulin. The basic LysB29 was exchanged for the acidic, polar, and hydrophilic Glu, and the neutral, polar, and hydrophobic AsnB3 was exchanged for the basic, polar, and hydrophilic Lys. This double mutant has a slightly lower pI of 5.1 compared with human insulin with a pI of 5.5, and accordingly its solubility is also enhanced. The double mutant, AsnB3Lys + LysB29Glu, is especially interesting because higher-order forms dominate relative to the monomeric form, i.e., it is more stable, while still maintaining rapid dissociation to monomeric insulin. This double mutant is normally referred to as insulin glulisine and is marketed as Apidra.

Thus, the design of rapidly absorbed, fast-acting insulin analogs must be characterized as a clear success. The design of the aforementioned insulin analogs were made possible because the large number of insulin structures provided the researchers with detailed information about the molecular interactions responsible for receptor binding and the hexamer–dimer–monomer equilibrium. Analogously, it has also been possible to develop long-acting insulin analogs by minor structural modifications based on the structural knowledge. Insulin degludec (Tresiba) is an ultralong-acting insulin analog where the C-terminal residue in the B-chain (ThrB30) has been removed and LysB29 has been conjugated via a spacer to a fatty acid moiety (see also Chapter 9).

4.6 NEW METHODS

The experimental and computational methods used in biostructure-based drug design are constantly evolving. Here, we will only mention a few of the methods which have and are expected to have a great impact on the design process.

The availability of an increasing number of experimentally determined structures of relevant targets is crucial. Crystallization, data collection at synchrotron radiation centers, and structure determination processes are today partially or fully automated. Centers for high-throughput determination of 3D structures have been established at several places in the world. These developments and initiatives are reflected in an increase in the number of experimentally determined structures deposited at the Protein Data Bank. Many pharmaceutical companies have also in-house groups doing structure determinations, but the number of structures determined are not available and is difficult to estimate.

Computationally, the developments are also considerable. Improved methods for determination of models of structures based on structures of related proteins, i.e., homology modeling or comparative modeling, are being developed. Faster computers allow longer and thereby more realistic simulations of proteins where flexibility and solvation can be considered. Ligand docking programs are being improved by considering not only the flexibility of the ligand but also the flexibility of the protein. The prediction of binding energies has always been a problem in docking methods, but improved quantum mechanics based or quantum mechanics derived methods combining speed and accuracy are being developed. Better description and handling of the solvation versus desolvation processes are also crucial for correct prediction of binding affinities.

4.7 CONCLUSION

Biostructure-based drug design is being used to efficiently develop new therapeutic candidates and incorporates multiple scientific disciplines, including medicinal chemistry, pharmacology, structural biology, and computer modeling.

The examples described earlier illustrate that biostructure-based drug design in cases where structural information of the targets are available is a powerful method for the design of new or improved drugs.

The DuPont Merck example on biostructure-based design of the cyclic urea HIV-1 protease inhibitors nicely illustrates that in many cases it is possible based on the 3D structures of a target to design ligands to control or regulate a biological system. Unfortunately, the example also illustrates that several other features have to be considered in order for a ligand to become a drug.

FURTHER READING

Acharya, K.R. and Lloyd, M.D. 2005. The advantages and limitations of protein crystal structures. *Trends Pharmacol. Sci.* 26:10–14.

Brange, J., Ribel, U., Hansen, J.F. et al. 1988. Monomeric insulins obtained by protein engineering and their medical implications. *Nature* 333:679–682.

Davis, A.M., Teague, S.J., and Kleywegt, G.J. 2003. Applications and limitations of X-ray crystallographic data in structure-based ligand and drug design. *Angew. Chem. Int. Ed.* 42:2718–2736.

Hardy, L.W. and Malikayil, A. 2003. The impact of structure-guided drug design on clinical agents. *Curr. Drug Discov.* 3:15–20.

Klebe, G. 2013. *Drug Design: Methodology, Concepts and Mode-of-Action*. Berlin, Germany: Springer-Verlag.

Lam, P.Y.S., Jadhav, P.K., Eyermann, C.J. et al. 1994. Rational design of potent, bioavailable, nonpeptide cyclic ureas as HIV protease inhibitors. *Science* 263:380–384.

McCusker, E.C., Bane, S.E., O'Malley, M.A., and Robinson, A.S. 2007. Heterologous GPCR expression: A bottleneck to obtaining crystal structures. *Biotechnol. Prog.* 23:540–547.

Pøhlsgaard, J., Frydenvang, K., Madsen, U., and Kastrup, J.S. 2011. Lessons from more than 80 structures of the GluA2 ligand-binding domain in complex with agonists, antagonists and allosteric modulators. *Neuropharmacology* 60:135–150.

Russell, R.J., Haire, L.F., Stevens, D.J. et al. 2006. The structure of H5N1 avian influenza neuraminidase suggests new opportunities for drug design. *Nature* 443:45–49.

Sobolevsky, A.I., Rosconi, M.P., and Gouaux, E. 2009. X-ray structure, symmetry and mechanism of an AMPA-subtype glutamate receptor. *Nature* 462:745–756.

Thornton, J.M., Todd, A.E., Milburn, D., Borkakoti, N., and Orengo, C.A. 2000. From structure to function: Approaches and limitations. *Nat. Struct. Biol.* 7:991–994.

von Itzstein, M., Wu, W.-Y., Kok, G.B. et al. 1993. Rational design of potent sialidase-based inhibitors of influenza virus replication. *Nature* 363:418–423.

5 Drug-Like Properties and Decision Making in Medicinal Chemistry

Jan Kehler, Lars Kyhn Rasmussen, and Morten Jørgensen

CONTENTS

5.1 INTRODUCTION

The discovery and invention of new chemical entities that will prove to be clinically useful drugs is challenging. First and foremost, therapeutic drugs must observe affinity to the intended target(s). Then, drugs must be able to be administered by a reasonable route, and importantly, they must be able to distribute to the tissue where the target receptor resides with a relevant exposure and half-life. Physicochemical properties determine oral absorption, drug metabolism, toxicity, and in the context of molecular architecture determine the pharmacological properties. In this chapter, we will discuss principles for drug design aimed at increasing the chances of success. First, we will introduce the concept of drugability followed by methods that can assist decision making during drug discovery.

The concept of drug-likeness is useful to medicinal chemists as a guide for molecule design in hit and lead optimization. Orally administered small molecules that have appropriate properties in terms of absorption-distribution-metabolism-excretion (ADME) and acceptable toxicity properties to advance through to the completion of human Phase I clinical trials have been coined "drug-like." Another, yet still related application of the term "drug-like" is broadly to encompass small molecules that possess physicochemical properties that fall within the range specified by marketed drugs. Numerous analyses of the physicochemical characteristics of marketed drugs have shown that drug-like properties fall within a relatively narrow range. Indeed, lipophilicity and hydrogen bond donor (H-bond; HBD) count have not changed significantly in drugs over the last three decades, indicating that they are important properties of oral drugs; in contrast, the number of O and N atoms, H-bond acceptors (HBA), rotatable bonds, molecular weight (MW), and number of rings increased in drugs between 1983 and 2002. Physicochemical properties determine important drug parameters like solubility, permeability, metabolic stability, transporter effects, and protein binding. These fundamental properties dictate oral bioavailability and exposure at the physiological site of action. The drug-likeness terminology, therefore, can be understood as a way to relate physicochemical properties to ADME and toxicological (ADME-T) properties and pharmacological profile. Drug-like properties are intrinsic molecular properties and simultaneous optimization of physicochemical properties along with pharmacological properties is critical to success in drug discovery.

During the multidimensional optimization process, the medicinal chemist needs methods to monitor the progress, as well as get feedback from the iterative learning process, of establishing the structure–activity relationships (SAR). The concepts of ligand efficiency (LE) and lipophilic ligand efficiency (LLE) are introduced in this chapter as simple measures to monitor and ultimately balance the pharmacological profile and the physicochemical properties. These concepts intend to integrate the need for drug-like properties with the hunt for affinity to the target. They can be used by project teams to make informed decisions and ultimately drive projects toward higher quality compounds that are less likely to fail in development due to compound-related issues. Importantly, compounds with properties that fail the criteria of drug-likeness have higher attrition. By nature, the concept of drug-likeness has to do with probabilities, so is not an absolute science. The goal is to increase the *probability* of prospectively designing molecules that are more likely to survive preclinical safety studies and that possess optimal pharmacokinetic and pharmacodynamic properties to test hypotheses in the clinic.

5.2 DRUG-LIKE PROPERTIES OF ORAL DRUGS

The discussion in this chapter relates to small-molecule drugs. The meaning of "drug-likeness" depends on the mode of administration. The highest patient compliance (i.e., the degree to which a patient correctly follows medical advice) is observed for the oral route of drug administration; this is also the most cost-effective and most convenient. Consequently, most drugs are administered as tablets or capsules. Oral drug space is a generic term that describes the region of chemical space in which oral drugs have a high probability of being found. It can in principle be defined using any number of physicochemical or theoretical descriptors. A descriptor is a structural or physicochemical property of a molecule or part of a molecule. Examples include lipophilicity (assessed by the log P and log D values), molecular weight, number of hydrogen bond donors and acceptors, and topological polar surface area (TPSA).

A drug-like molecule must have some solubility in both water and fat, since an orally administered drug needs to pass through the intestinal lining after ingestion, be carried in aqueous blood and penetrate the lipid-based cell membrane to reach the inside of a cell. A model compound for the lipophilic cellular membrane is 1-octanol (a lipophilic hydrocarbon). Lipophilicity is measured as the partition of a small molecule between octanol and water (log P) or aqueous buffer (log D, often at pH = 7.4) as depicted in Figure 5.1. For small-molecule drugs log P is typically 3–5 meaning that the molecule has a preference for octanol over water by a factor of 1,000–100,000. A hydrogen bond is the electrostatic attractive force between a polar hydrogen bound to an electronegative atom such as nitrogen or oxygen of one molecule and an electronegative atom of a different molecule. The name hydrogen bond is in fact a misnomer; as it is not a true bond but in fact a strong dipole–dipole attraction. The topological polar surface area (TPSA) of a molecule is defined as the surface sum over all polar atoms, primarily oxygen and nitrogen, and is therefore roughly a count of the number of polar atoms, each of which contributes with approximately 14 Å^2. The TPSA is not always "easy" to deduce directly from the chemical structure as not all polar heteroatoms contribute equally and due to the fact that the 3D conformation of a molecule is not

Octanol

Water/aq. buffer

$$P = \text{Partition coefficient} = \frac{[\text{Molecule}]_{\text{octanol}}}{[\text{Molecule}]_{\text{water}}}$$

$$D = \text{Partition coefficient} = \frac{[\text{Molecule}]_{\text{octanol}}}{[\text{Molecule}]_{\text{aq. buffer}}}$$

If $\log P = 5$ then $[\text{molecule}]_{\text{octanol}} = 100{,}000 \cdot [\text{molecule}]_{\text{water}}$

FIGURE 5.1 Definition of the partition coefficient P or D for the equilibrium distribution of a molecule between octanol/water and octanol/aqueous buffer, respectively.

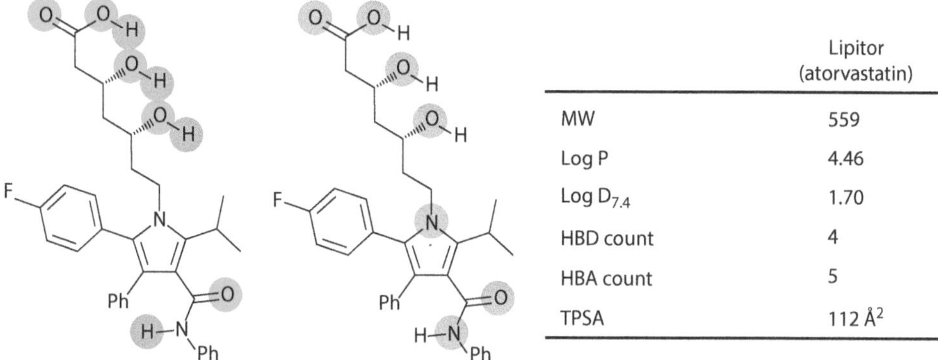

	Lipitor (atorvastatin)
MW	559
Log P	4.46
Log $D_{7.4}$	1.70
HBD count	4
HBA count	5
TPSA	112 Å^2

FIGURE 5.2 Physicochemical properties for atorvastatin. The atoms contributing to the number of HBD (orange), HBA (green), and TPSA (gray) are indicated.

known a priori but requires a computational analysis. Drawing tools such as ChemDraw are the most convenient tools to obtain the values for each of these descriptors directly from the chemical structure, but several public websites also offer these calculations for free. The fundamental properties are illustrated for the cholesterol-lowering drug Lipitor (atorvastatin) in Figure 5.2. The significantly lower log $D_{7.4}$ value as compared to log P is a consequence of deprotonation of the carboxylic acid at physiological pH (7.4) and the accompanying higher hydrophilicity of the carboxylate anion.

Most commonly the assessment of drug-likeness is performed using guidelines, the original and most well-known of which is Lipinski's rule of five (Ro5). In the following we first discuss two of these rules of drug-likeness, and then turn our attention to a measure of the so-called molecular beauty based on scoring functions to avoid hard cutoffs.

5.2.1 The Lipinski Rule of Five (Ro5)

In 1997, Lipinski et al. reported a seminal paper in which they analyzed the physicochemical properties of drugs in a database of about 2500 orally active small molecules that had entered at least phase II clinical trials. From this analysis, the "Rule of five" (Ro5) was developed. The Ro5 owes its name to the fact that it is based on four essential physical properties, each of which is constrained by an upper value related to the number five (MW ≤ 500, log P ≤ 5, HBD ≤ 5, HBA ≤ 10). Nine out of ten orally active small-molecule drug candidates that achieved phase II clinical status are found within these boundaries. This also applied broadly to oral small-molecule *drugs* as we shall learn later in this chapter. These physicochemical parameters are associated with acceptable aqueous solubility and intestinal permeability that are the essential first two steps toward oral bioavailability. Although it is termed the "Rule of five," it is in fact an "anti-rule" that states that it is unlikely that compounds with two or more "violations" can be orally absorbed. The Ro5 does not

TABLE 5.1
Three Oral Drugs and Their Lipinski Parameters

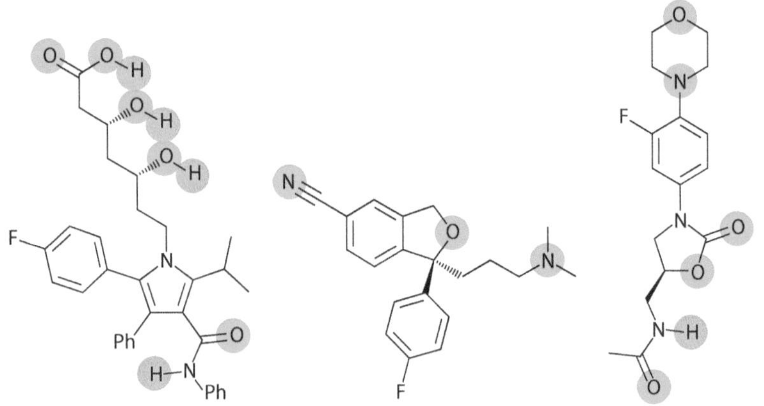

Lipinski Criteria	Lipitor (Atorvastatin)	Lexapro (Escitalopram)	Zyvox (Linezolid)
MW < 500	559	324	337
Log P < 5	4.46	3.13	0.17
HBD count < 5	4	0	1
HBA count > 10	5	3	5
Number of Ro5 violations	1	0	0

Note: The H-bond donors and acceptors are indicated by the orange and green colors, respectively.

apply to compounds that are subject to active up-take across the gut by transport proteins. Indeed, Pfizer's blockbuster drug Lipitor (atorvastatin) is an example of such a compound. The antidepressant Lexapro (escitalopram) and the antibacterial drug Zyvox (linezolid) are examples of CNS and peripherally acting Ro5-compliant oral drugs, respectively. These compounds and their Lipinski profiles are illustrated in Table 5.1.

5.2.2 The Golden Triangle

Since the publication of the Ro5, a large number of additional analyses have been performed by scientists from many different organizations on a variety of small and large datasets and generally have confirmed the central role of physicochemical properties. One of these analyses led to the so-called Golden Triangle model. This guideline was developed from in vitro ADME data (permeability and metabolic stability) and computational data with the objective of aiding medicinal chemists to identify permeable and metabolically stable compounds. In a plot of MW versus log $D_{7.4}$, the Golden Triangle appears with its apex at MW = 450 and log $D_{7.4}$ = 1.5 and a baseline from log $D_{7.4}$ = −2.0 to log $D_{7.4}$ = 5.0 at MW = 200. Compounds that reside inside the Golden Triangle are more likely to be both metabolically stable and to possess good membrane permeability than those outside. In general, compounds placed in the lower left part of the plot tend to exhibit low permeability and good stability toward hepatic (oxidative) clearance, and they are often excreted unchanged via the urine. Conversely, compounds placed in the upper right part of the plot beyond the Golden Triangle often have poor properties and in order to be useful as oral drugs, special features have to be "build in" to the chemical structure, particularly, problems with metabolic stability are often noticed, so a special attention needs to be paid to secure the compound to be very resistant to liver microsomal clearance (a measure of hepatic clearance). This is illustrated in Figure 5.3.

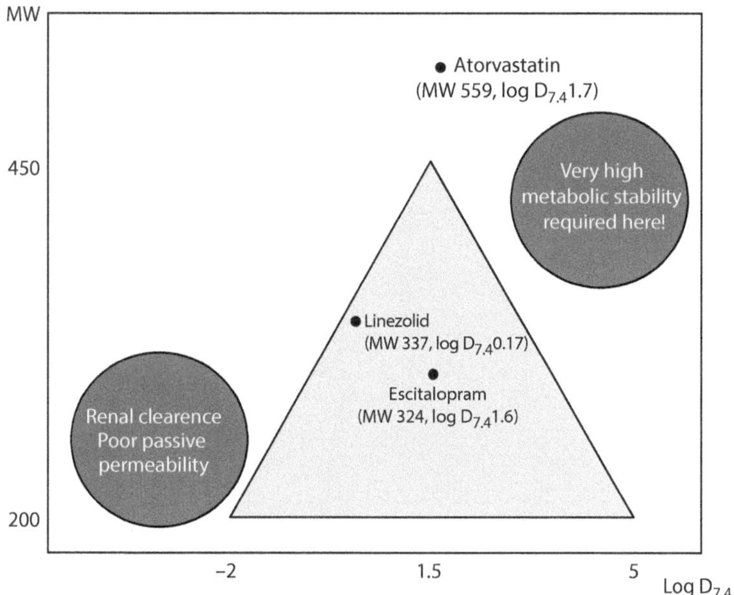

FIGURE 5.3 The Golden Triangle. Compounds with combined good liver microsomal stability and good permeability are typically found within the triangle. The three drugs atorvastatin, linezolid, and escitalopram are indicated in the plot.

5.2.3 Multiparameter Optimization (CNS MPO-Score)

Neither the Ro5 nor the Golden Triangle was intended to build "fixed walls" or "forbidden areas" or "uncompromising rules." However, in reality they are typically applied as strict cutoff filters to, e.g., *in silico* designed compounds or corporate screening collections. When using these kinds of rules of thumb as absolute selection criteria, compounds are discriminated solely qualitatively: either they pass or they fail. All compounds that comply with the rules are considered equal, as are all that breach them. The fact that there are some successful drugs on the market that are not compliant with these rules and the inability of a single descriptor to account for the quality of a small molecule as an oral drug has prompted "moving beyond rules" and paved the way for more sophisticated descriptors of drug-likeness based on scoring schemes. In the following, we will discuss one of these "softer guidelines": the CNS multi-parameter optimization algorithm (MPO).

Instead of operating with hard cutoff values, the MPO scoring function allows compounds to "go outside" one or more of the optimal property ranges by assigning a gradually decreasing sub-score for that parameter depending on the deviation from the optimal range as illustrated in Figure 5.4. As the name implies, CNS MPO-score was developed for oral CNS drugs, but it is equally applicable to oral peripheral small-molecule therapeutics. The CNS MPO-score for the three drugs discussed so far in this chapter are listed in Table 5.2. From these examples, it is evident that there are significant differences between the three drugs with atorvastatin displaying the lowest score and the other two drugs have near-optimal profiles despite falling short of the ideal for one or more properties. In general, a CNS MPO-score of above 4 is considered good. As we shall see in the next section there is no difference between centrally acting small-molecule drugs and their peripheral counterparts in terms of the MPO-score.

5.2.4 Analysis of Small-Molecule Oral Drugs

Following the publication of the numerous rules and guidelines for best practices in medicinal chemistry, we performed an analysis of a large set of small-molecule oral drugs to evaluate the validity of some of these mnemonics for a large set of successful drugs. In this study we

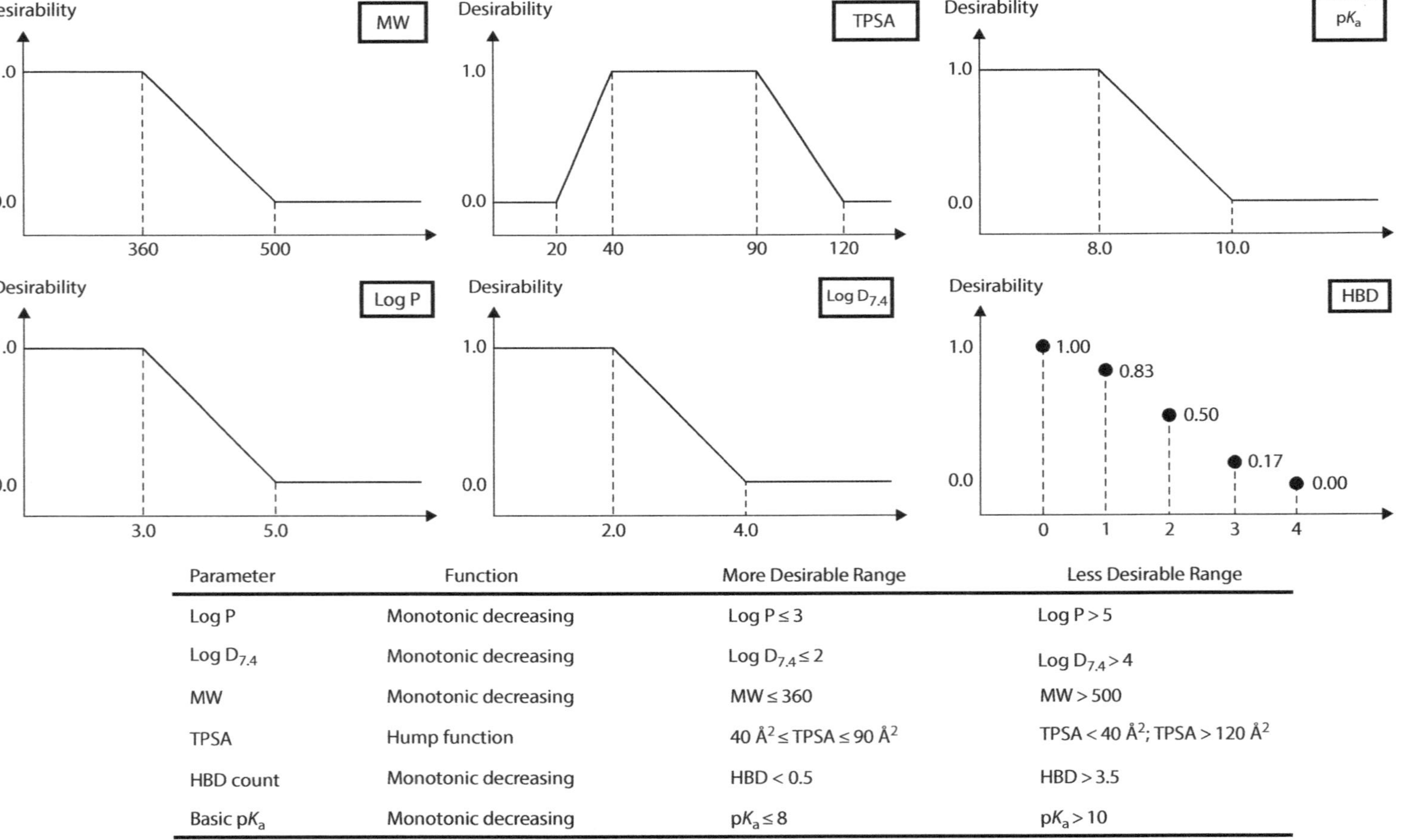

Parameter	Function	More Desirable Range	Less Desirable Range
Log P	Monotonic decreasing	Log P ≤ 3	Log P > 5
Log $D_{7.4}$	Monotonic decreasing	Log $D_{7.4}$ ≤ 2	Log $D_{7.4}$ > 4
MW	Monotonic decreasing	MW ≤ 360	MW > 500
TPSA	Hump function	40 $Å^2$ ≤ TPSA ≤ 90 $Å^2$	TPSA < 40 $Å^2$; TPSA > 120 $Å^2$
HBD count	Monotonic decreasing	HBD < 0.5	HBD > 3.5
Basic pK_a	Monotonic decreasing	pK_a ≤ 8	pK_a > 10

FIGURE 5.4 The six parameters that underlie the CNS MPO-score. With desirability scores for each parameter of ranging from zero to one, the maximal (best) possible MPO-score is 6.

TABLE 5.2
Calculation of the CNS MPO-Score for Atorvastatin, Escitalopram, and Linezolid

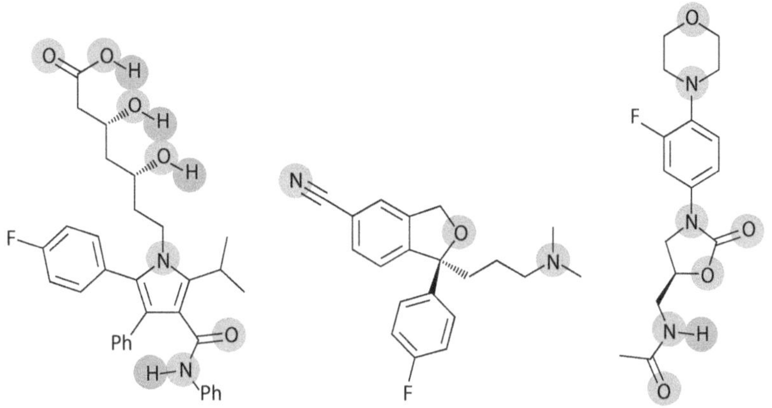

Property	Lipitor (Atorvastatin) Value	Score	Lexapro (Escitalopram) Value	Score	Zyvox (Linezolid) Value	Score
MW	559	0.00	324	1.00	337	1.00
TPSA	112	0.27	36	0.80	71	1.00
pK_a	—	1.00	8.91	0.55	2.45	1.00
Log P	4.46	0.27	3.13	0.94	0.17	1.00
Log $D_{7.4}$	1.70	1.00	1.60	1.00	0.17	1.00
HBD count	4	0.00	0	1.00	1	0.83
CNS MPO score	Σ	2.54	Σ	5.28	Σ	5.83

Note: The H-bond donors are indicated in orange and the polar atoms contributing to TPSA in gray.

compared 591 peripheral drugs with 273 CNS drugs in terms of Ro5 compliance, location inside/outside the Golden Triangle, and the MPO-score. The data are summarized in Figure 5.5. It is evident that the oral small-molecule drugs, in particular the CNS drugs, are indeed highly compliant with the Lipinski Ro5. Although it was developed for CNS drug discovery, the MPO algorithm is equally applicable to peripheral drugs with 70% of both drug classes having a score above 4. Almost 80% of the CNS drugs are found within the Golden Triangle, whereas nearly half of the peripheral drugs fall outside. This analysis illustrates that CNS drugs are found in a more constrained physicochemical space than their peripheral counterparts. This difference is most likely a consequence of the requirement for passage across the blood–brain barrier (BBB), including the possibility for a CNS drug to escape the many protective protein efflux pumps which are integrated in the BBB.

5.3 DECISION MAKING IN MEDICINAL CHEMISTRY

A number of metrics have appeared in the medicinal chemistry literature over the last two decades that relate the target affinity to a fundamental descriptor of molecular size (for example, MW or TPSA) or lipophilicity (typically log P). Herein, we discuss two of them: ligand efficiency (LE) and ligand lipophilic efficiency (LLE). These key metrics allow informed optimization of the potency averaged for molecular size and normalized for lipophilicity, respectively. Increasing the size of the molecular footprint and especially the lipophilicity is known to be associated with increased attrition in clinical development of drug candidates. The ADME properties of compounds typically

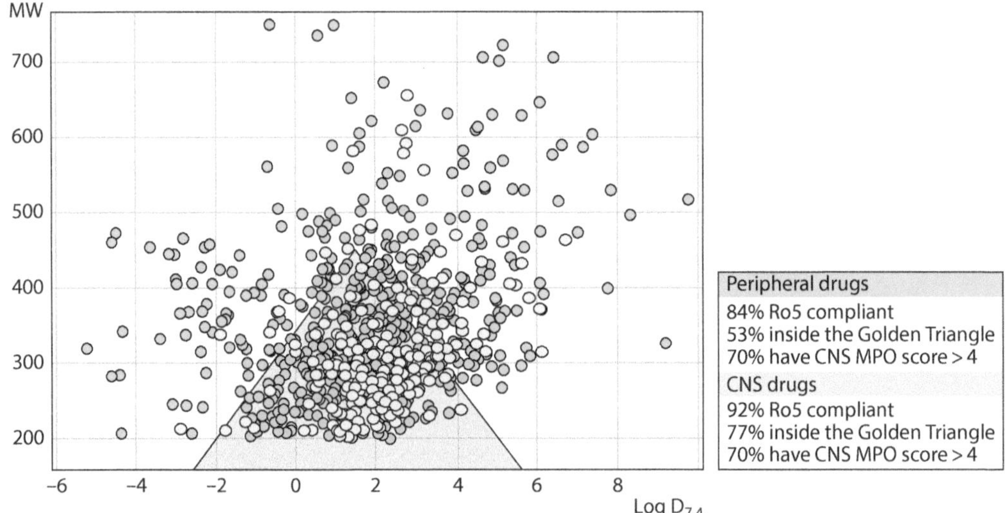

FIGURE 5.5 Plot of MW vs. log $D_{7.4}$ for 591 peripheral drugs (cyan) and 273 CNS drugs (yellow). The Golden Triangle is indicated. The insert summarizes the degree of Ro5 compliance, location with respect to the Golden Triangle, and CNS MPO-score for the two drug sets.

deteriorate with either increasing MW and/or log P. With this in mind it is important to select the best possible starting point and to adopt a suitable strategy for lead optimization. It is of utmost importance to keep track of MW and log P in relation to the target affinity to ensure that the improvement in potency does not come at the cost of deteriorating compound quality (quality here measured as the ADME-T properties). Furthermore, lipophilicity is a major driver for off-target pharmacology, often referred to as promiscuity. Increasing affinity without increasing log P increases the probability for obtaining a selective drug. Ligand efficiency and ligand lipophilicity efficiency are simple means of monitoring if a change in potency of a molecule is based on specific strong interactions with the target macromolecule or if it is a consequence of lipophilicity-driven affinity.

5.3.1 Ligand Efficiency (LE)

Ligand efficiency is a measure of Gibbs free energy of binding in relation to the molecular size (represented by the number of non-hydrogen atoms, called the heavy-atom-count, HAC). For interactions between a ligand and its target the following equilibrium exists and K_i is defined as in Equation 5.1.

$$\text{Free ligand} + \text{Free target} \underset{k_{off}}{\overset{k_{on}}{\rightleftarrows}} \text{Ligand-target complex}$$

$$K_i = \frac{k_{on}}{k_{off}} \quad \text{and} \quad \Delta G = -RT \ln K_i \tag{5.1}$$

The ligand efficiency (LE) is defined as the binding energy (ΔG) per HAC as shown in Equation 5.2:

$$LE = \frac{\Delta G}{HAC} = \frac{-RT \ln K_i}{HAC} \tag{5.2}$$

The majority of oral drugs have a HAC between 10 and 30. Equation 5.2 can be simplified by changing the natural logarithm function to the base 10 logarithm, using T = 300 K and the gas constant yielding Equation 5.3:

$$\Delta G = -RTLn\ K_i = 300K \cdot 1.987 \cdot 10^{-3} \frac{kcal}{mol \cdot K} \cdot LnK_i \leftrightarrow$$

$$\Delta G = -0.5961 \frac{kcal}{mol} \cdot 2.303 \cdot \log K_i = -1.4 \log K_i$$

$$LE = -1.4 \frac{\log K_i}{HAC} \tag{5.3}$$

or if K_i is not available it can be replaced with IC_{50}:

$$LE = -1.4 \frac{\log IC_{50}}{HAC} \tag{5.4}$$

When comparing and evaluating early hits from different chemical classes as starting points for optimization, for example, from a high-throughput screen (HTS), LE is a useful parameter that relates potency to the size/MW. All other parameters being equal, one should start working from the hit offering the highest LE. Off course it is necessary to make a holistic evaluation of hits taking all other parameters into account as well. As a rule of thumb a good starting point for a lead would be a LE above 0.3.

In the process of optimizing hits and leads, LE is a metric which can be used to monitor how structural modifications influence the efficiency of the binding to the biological target. By monitoring LE and trying to increase the value or keeping it close to the starting point, one can evaluate the influence of adding a new atom to the molecule. If LE drops and no improvement of properties is observed, the applied change does not improve the compound quality. It is a sign indicating that the particular added chemical substituent or moiety does not have any attractive intermolecular interaction with the target. Conversely, if a change in the lead structure solves an identified critical issue such as poor metabolic stability or insufficient membrane permeability it will be completely acceptable with lower ligand efficiency. Optimally, a structural modification will simultaneously increase LE and address a non-potency-related problem.

A plot of pIC_{50} versus HAC is typically used when evaluating a compound series as shown in Figure 5.6. This analysis illustrates what affinity is needed to keep LE constant when the number of heavy atoms is increased. For example, a small fragment hit with 17 heavy atoms and an IC_{50} of 10 μM has a ligand efficiency of 0.4. When adding 10 heavy atoms HAC increases to 27 and by following the dotted line it is evident that a potency of just above 10 nM is required to maintain the same ligand efficiency.

Optimization of a structure in terms of ligand efficiency means moving toward the green area in the upper left corner (Figure 5.6). However, staying on one of the dotted lines is acceptable because the LE is maintained. In contrast, a change toward the lower right corner must be accompanied by improvement of other properties of the molecule to be acceptable and to offset the reduction in LE.

Despite the utility of the underlying thinking behind the LE, some important weaknesses should be mentioned. One problem with using the LE metric is that HAC treats all heavy atoms equally, meaning that the introduction of CH_3 is identical to the introduction of, for example, NH_2, OH, F, Cl, or Br and these changes will lead to the same ligand efficiency if the compounds are equipotent. However, it has been demonstrated that C, N, O, S or halogens, on average, do not contribute

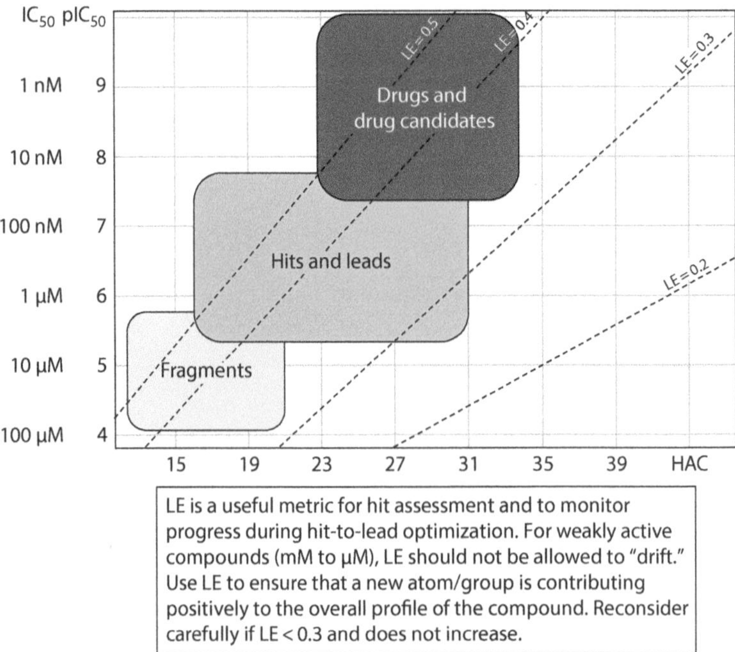

FIGURE 5.6 Plot of pIC_{50} vs. HAC. Lines for compounds with equal LE values are indicated. The typical location ranges for fragments, hits and leads, and drug candidates are illustrated. The insert summarizes the use of LE during the compound optimization process.

to potency identically. Another problem is that LE is not independent of HAC. A 10-fold change in potency per heavy atom does not result in constant LE, for instance, a 15 HAC compound with a pIC_{50} of 3 does not have the same LE as a 16 HAC compound with a pIC_{50} of 4 ($\Delta pIC_{50} = 1$, $\Delta HAC = 1$, $\Delta LE = 0.07$). Because of this it is advisable not to focus solely on the metric LE but keep other factors in mind such as log P, solubility, and metabolic stability. Another significant drawback of LE is that it does not take log P into account. A way of addressing this issue will be discussed in the next section.

5.3.2 Lipophilic Ligand Efficiency (LLE)

High lipophilicity increases the likelihood that a compound will bind to multiple targets and result in off-target pharmacology that may limit the therapeutic window due to adverse events or dose-limiting toxicity. Lipophilicity is also known to be related to low solubility and increased susceptibility to oxidative metabolism, typically in the liver. Medicinal chemists are pursuing optimization of potency at a specific target without compromising physicochemical properties or the safety profile. Thus, an approach to identify high-quality compounds suitable for further development is to focus on increasing potency without increasing the lipophilicity excessively.

LLE is a way of balancing potency normalized with respect to lipophilicity. This concept is useful because an increase in log P can lead to decreased selectivity, and a suboptimal ADME profile. LLE is commonly referred to as a measure of the compound's target affinity versus its non-specific binding to fatty tissues; a compound with a LLE = 0 is unselective while a compound with LLE = 5 would 100,000 times rather bind to the target than bind unspecifically. LLE is a measurement of binding in relation to unit of grease:

$$\mathrm{LLE} = -\log \mathrm{K_i} - \log \mathrm{P} = \mathrm{pK_i} - \log \mathrm{P} \tag{5.5}$$

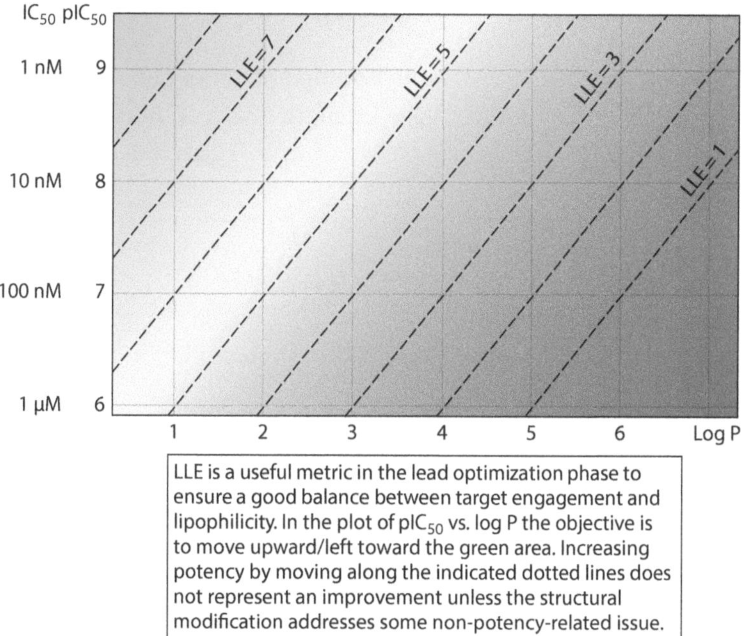

FIGURE 5.7 LLE plot/LiPE plot. Lines for compounds with equal LLE values are indicated. The insert describes the practical application of this plot.

or when K_i is not available:

$$\text{LLE} = -\log \text{IC}_{50} - \log \text{P} = \text{pIC}_{50} - \log \text{P} \quad (5.6)$$

As a rule of thumb, compounds that have the potential to become development candidates should have LLE in the range of 5–7 or greater and compounds attractive as leads in the range of 3–5 (or greater). The calculation of LLE from formula (5.6) is straightforward. A plot of pK_i or pIC_{50} versus log P is a powerful way to monitor the progression of a compound series. This is illustrated in Figure 5.7 that shows how changing target affinity and/or log P influence LLE. A very important insight and a key message from this analysis is that two different molecules that reside on the same dotted line have the same quality in terms of LLE despite the common misperception that the more potent compound is the better compound. In general, it is desirable to increase LLE as much as possible, with the caveat that negative log P can make essentially inactive compounds look promising. However, as mentioned in the discussion of LE, a decrease in LLE can be accepted when the change is solving a given issue in the series. Using LLE for small and weakly binding ligands such as fragments is not recommended.

Increasing affinity without increasing lipophilicity is normally desirable for a medicinal chemist in order to make high-quality compounds. In Figure 5.7, the dotted lines represent compounds that have equal quality in terms of LLE, meaning that the potency of these compounds is the same when they have been corrected for the lipophilicity (Equation 5.6). A structural change that moves compounds toward the upper left corner of the graph would represent a compound of higher potency that is not originating from increased lipophilicity which would hence have a better chance of making it through development.

5.3.3 Expected Potency Improvements

Sometimes it can be difficult to get a sense of what changes in target affinity one would or should expect from the addition of a given functionality to a chemical structure. Table 5.3 summarizes this information for a number of common ligand–target interactions and, also for the addition of a

TABLE 5.3
Expected Potency Improvements upon the Addition of Some Common Functional Group in Relation to Specific Interactions and the Strength of a C—C Bond in Comparison

Group/Interaction	ΔG kcal/mol	Expected X-Fold Improvement in Potency
H-bond	−1.4	10
Salt bridge (ion-pair)	−3.4	300
Methyl from lipophilicity	−0.7	3
Buried methyl	−1.4	10
Cl from lipophilicity	−1.0	5
Phenyl from lipophilicity	−2.5	60
Buried phenyl	−3.4	300
Ethane rotation barrier	3.0	−140
C—C bond energy	80	$1 \cdot 10^{57}$

methyl or a phenyl group as well as a chlorine atom. Not surprisingly the strongest interaction is a salt bridge (for example, between a protonated ligand nitrogen and a deprotonated aspartic acid amino acid residue). In contrast, it is notable that a "buried methyl" (i.e., a methyl group positioned deeply in a binding pocket or in a sub-pocket) is "worth as much" as a hydrogen bond between a small molecule and the biological target. It is equally important to note that the addition of a phenyl group should result in a massive 300-fold boost in potency. This also implies that if, for example, the removal of a phenyl group only leads to a potency loss around 60-fold then that structural change was a very good idea. The understanding of this type of SAR analysis can be further expanded if you make an exercise to calculate the changes in LLE and relate it to the change in log P (Δlog P) versus the change in pIC_{50} (ΔpIC_{50}). Restriction of the otherwise free rotation of an ethyl group is energetically unfavorable and causes a reduction in potency. Thus, the loss in conformational freedom of an ethane rotor due to a locked bioactive conformation is reflected in huge potency loss. Finally, it is pertinent to note that these non-bonded interactions are insignificant when compared to the strength of a carbon–carbon single bond.

5.3.4 Calculation of Ligand Efficiency and Ligand Lipophilicity Efficiency

As exemplified for the antidepressant escitalopram in Table 5.4, it is straightforward to calculate the LE and the LLE from the target affinity and the number of non-hydrogen atoms and calculated

TABLE 5.4
Calculation of LE and LLE for the Antidepressant Escitalopram

Lexapro
(escitalopram)

Parameter	Value	Calculation of LE and LLE
Target affinity	2.1 nM	$LE = \frac{-1.4 \times \log(2.1 \times 10^{-9})}{36} = 0.36$
HAC	36	
log P	3.13	$LLE = -\log(2.1 \times 10^{-9}) - 3.13 = 5.55$

lipophilicity, respectively. For larger datasets, it is recommended to perform these calculations using appropriate software such as Microsoft Excel or Spotfire.

5.4 CASE STORY: APPLICATION IN DRUG DISCOVERY

The concepts discussed so far in this chapter are illustrated below for a phosphodiesterase 2 (PDE2) inhibitor project. The objective of this drug discovery work was to identify orally active PDE2 inhibitors with potential applications as therapeutics for cognitive impairment associated with schizophrenia.

5.4.1 Classical Strategy Based on High-Throughput Screening

The project started with an HTS screening of the Lundbeck corporate compound collection. A number of structurally related tricyclic compounds were identified as hits. One of these hits was the bis-chloro-triazolophthalazine (**5.1**). The medicinal chemistry team worked primarily on a sub-series derived from **5.1** and synthesized and evaluated several hundred compounds before identifying the ether-linked phenylpiperazine Lu AF64280 (**5.2**) as the lead compound (see Figure 5.8). This compound was benchmarked versus the corresponding amino-linked phenyl morpholine **5.3**, a competitor compound known from the patent literature.

The change in potency from the hit compound **5.1** to the lead compound Lu AF64280 (**5.2**) corresponded to a less than 10-fold improvement and came at the price of increased lipophilicity and molecular footprint. Despite the overall similar structures, Lu AF64280 (**5.2**) did represent a substantial improvement over the Altana Pharmaceuticals compound (**5.3**) in that it exhibited good brain exposure and showed strong effects in animal models of cognition and memory/learning.

The physicochemical profile and ADME properties of Lu AF64280 (**5.2**) and related compounds precluded them from being developed as oral drugs. Indeed, these compounds have a relatively poor profile when assessed by the "drug-likeness" descriptors discussed previously. Both the Altana Pharmaceuticals compound (**5.3**) and Lu AF64280 (**5.2**) are compliant with Lipinski's Ro5 with no and one violation, respectively (see Table 5.5). However, their physicochemical properties are relatively poor as assessed by the CNS MPO-score and the fact that both compounds reside well outside the Golden Triangle. In spite of high target affinities, neither compound is characterized by good

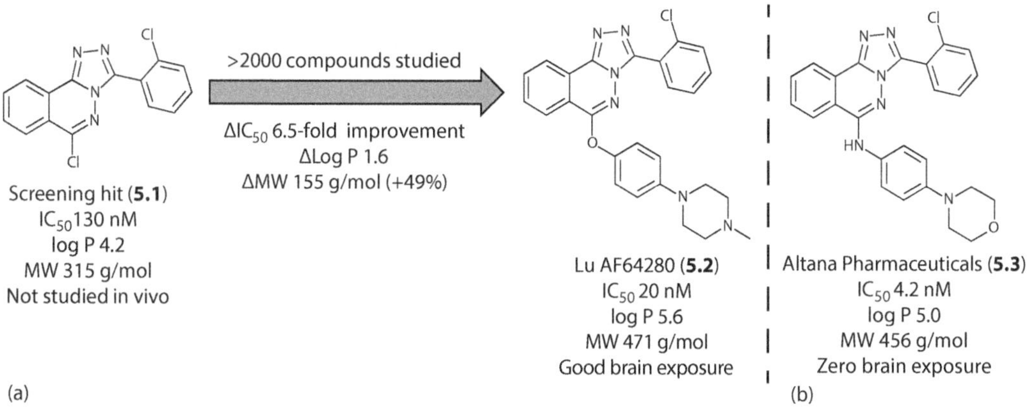

FIGURE 5.8 Discovery of the selective PDE2 inhibitor Lu AF64280 (a). A reference compound (**5.3**) from Altana Pharmaceuticals (b).

TABLE 5.5
Comparison of Lu AF64280 (5.2) and the Altana Pharmaceuticals Compound 5.3

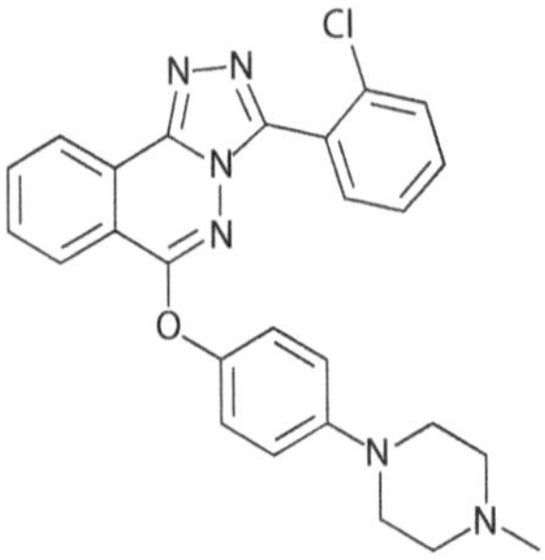

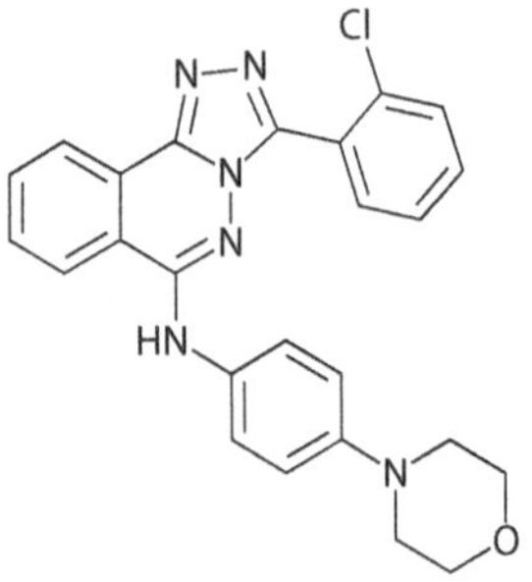

Descriptor/Metric	Lu AF64280 (5.2)	Altana Pharmaceuticals Compound 5.3
Lipinski Ro5	1 violation (clog P > 5)	0 violations
MPO	2.96 of 6.00	3.45 of 6.00
Golden Triangle	Outside	Outside
LE	0.32	0.36
LLE	2.96	3.67
Actual potency, IC_{50}	20 nM	4.2 nM
IC_{50} required to reach LE > 0.45	<0.02 nM	<0.05 nM
IC_{50} required to reach LLE > 5	<0.06 nM	<0.2 nM

Note: The potencies required for the two compounds to reach the "optimal ranges" of LE and LLE are indicated at the bottom of the table.

LE or LLE scores. In other words, they are too large and too lipophilic; if they were to exhibit good numbers (LE > 0.45 and LLE > 5) they would have had to be around 100-fold more potent which would put them in the generally unrealistic double-digit picomolar potency range. This example illustrates that medicinal chemists should exhibit great care when optimizing lead compounds in such a challenging physicochemical property space.

5.4.2 Ligand Efficiency-Driven Approach

A fundamentally different strategy was applied in a second-generation of the PDE2 inhibitor project where ligand efficiency metrics (LE and LLE) were applied. Aided by crystallographic information, this approach was guided by structure-based drug design (SBDD) principles. The starting point was the X-ray crystal structure of the moderately active triazolo-quinoline **5.4** bound to the enzyme target (see Figure 5.9).

This structural information prompted an exploration of a small pocket near the 4-position of the scaffold leading to compound **5.6** which was much more potent than the corresponding des-methyl analog **5.5** (see Table 5.6). Methyl groups such as this one are sometimes referred to as being "magic" because they contribute significantly to the target affinity. Compounds **5.4–5.6** were relatively lipophilic; this issue was resolved by adding one more polar heteroatom into the ring system thereby changing to the triazolo-quinoxaline scaffold in compounds **5.7– 5.9**. This lead to a very significant improvement in LLE due to both an increase in potency and a substantial reduction in lipophilicity of more than three orders of magnitude (see Table 5.6). This analysis pinpointed compounds **5.8** and **5.9** as the overall best within this structural class.

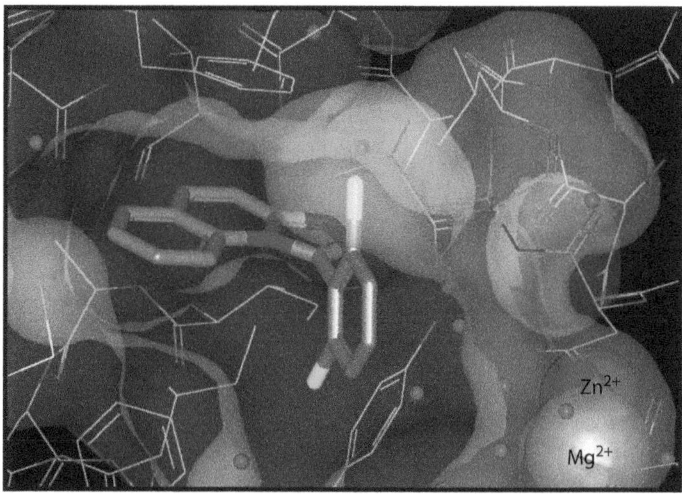

FIGURE 5.9 The crystal structure of screening hit **5.4** bound to the PDE2 enzyme.

Conclusions from these LE and LLE analyses are further supported by the fact that all of these compounds are Lipinski Ro5 compliant and that the most optimized of them (**5.7**– **5.9**) have good MPO-scores and are situated inside the Golden Triangle as illustrated in Figure 5.10. It is further evident that the optimization of the first-generation compounds despite a similarly good starting point in compound **5.1** toward Lu AF64280 (**5.2**) was not undertaken with appropriate consideration of physicochemical properties; also the Altana compound displays a suboptimal profile as judged from this analysis.

When exploring SAR trends and monitoring the lead optimization phase of a project, it is often advisable to use as little numerical SAR tables as possible, especially for ligand efficiency metrics. Instead it is highly recommended to work with graphical tools in addition to the plot of MW versus log $D_{7.4}$ as illustrated above. As representative examples, it is instructive to compare the original lead compound **5.2** with the second-generation lead series. For this purpose, the plots of pIC_{50} versus heavy atom count (HAC) or log P are illustrative (see Figures 5.11 and 5.12). It is evident that despite being a better starting point, the optimization of compound **5.1** to Lu AF64280 (**5.2**) was much less efficient than the SBDD-guided route from compound **5.5** to compounds **5.8** and **5.9**. Introduction of the phenoxy-*N*-methyl-piperazine group in going from **5.1** to Lu AF64280 (**5.2**) yielded a modest increase in potency at the expense of a significantly lower ligand efficiency. The removal of one chlorine atom in **5.4** led to the less potent, but similarly ligand efficient **5.5**. The subsequent optimization via **5.6**–**5.7** to **5.8** and **5.9** more than compensated for this initial loss of potency in **5.4**.

The SBDD approach was also superior in terms of LLE as illustrated in Figure 5.12. Again compound **5.1** is a better place to start than compound **5.4** or **5.5**. Nevertheless, the potency-driven optimization led to Lu AF64280 (**5.2**) which is significantly more lipophilic and has a significantly lower LLE than its predecessor **5.1**. The initial SBDD work around compounds **5.4** and **5.5** saw virtually no improvement in the LLE score. The introduction of the "magic methyl" resulted in an increase in LLE for compound **5.6**, but it was the change of the scaffold to the triazolo-quinoxaline core in **5.7** that really made the difference with an LLE increase of more than 2.5. The final addition of the methoxy group ensured a second hydrogen bond (HBA) to the enzyme and explains the further increase in LLE for compounds **5.8** and **5.9**.

TABLE 5.6
Optimization of Screening Hit 5.4 Using LE and LLE

	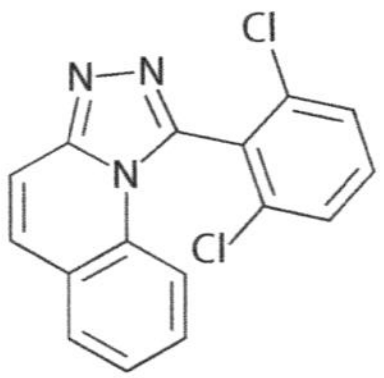 **5.4**	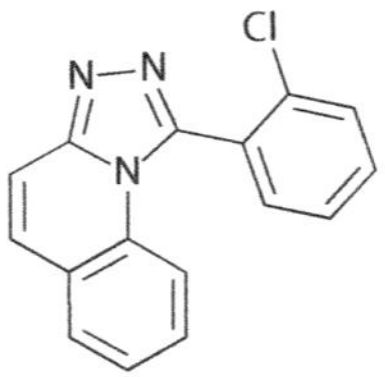 **5.5**	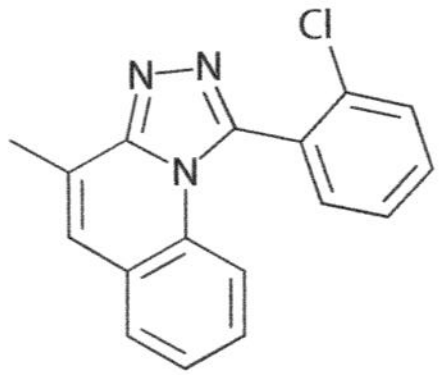 **5.6**	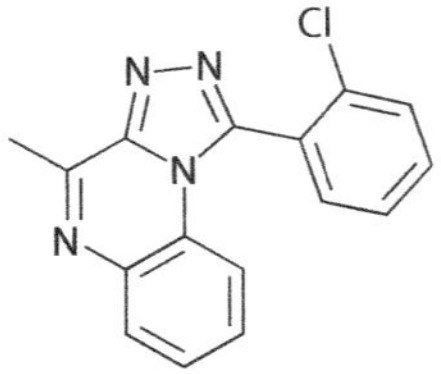 **5.7**	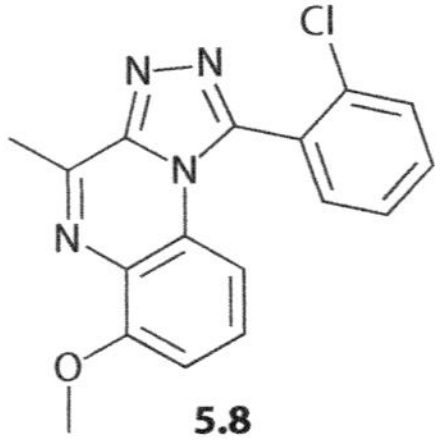 **5.8**	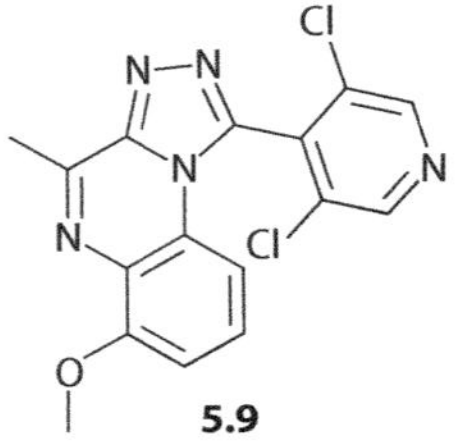 **5.9**
IC_{50}	660 nM	2400 nM	140 nM	55 nM	13 nM	14 nM
HAC	21	20	21	21	23	25
Log P	4.9	4.2	4.8	2.6	2.2	2.4
LE	0.40	0.39	0.47	0.48	0.48	0.44
LLE	1.5	1.4	2.1	4.7	5.7	5.5
MPO	3.68	3.91	3.66	4.46	4.30	4.41

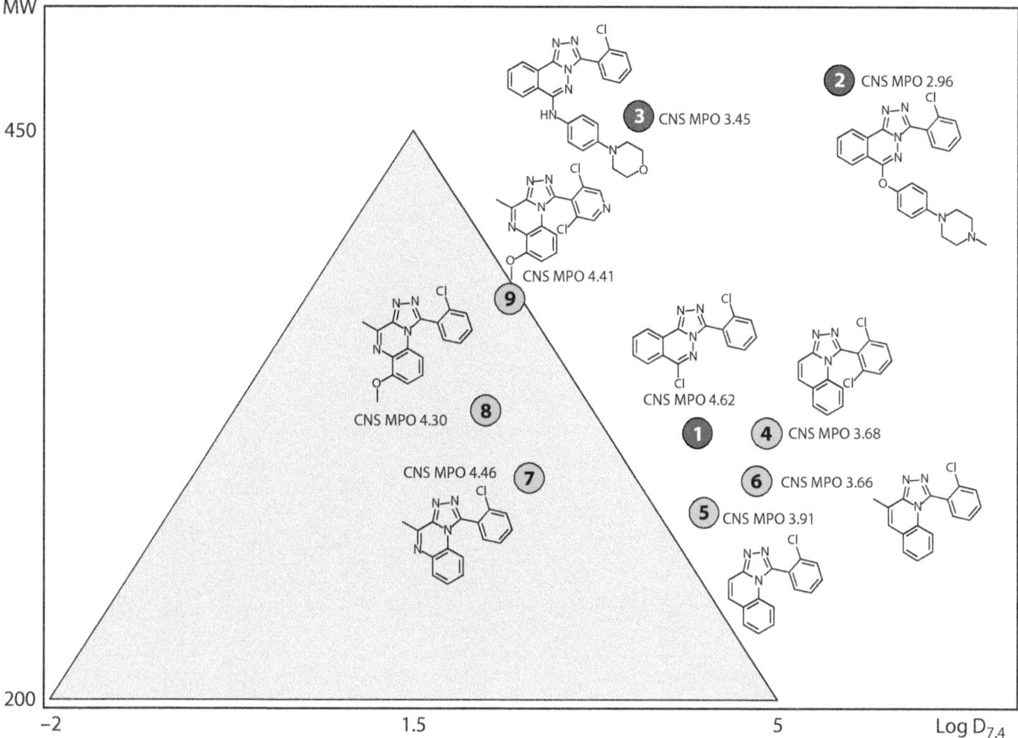

FIGURE 5.10 Plot of MW vs. Log $D_{7.4}$ for compounds **5.1– 5.9**. The Golden Triangle is indicated.

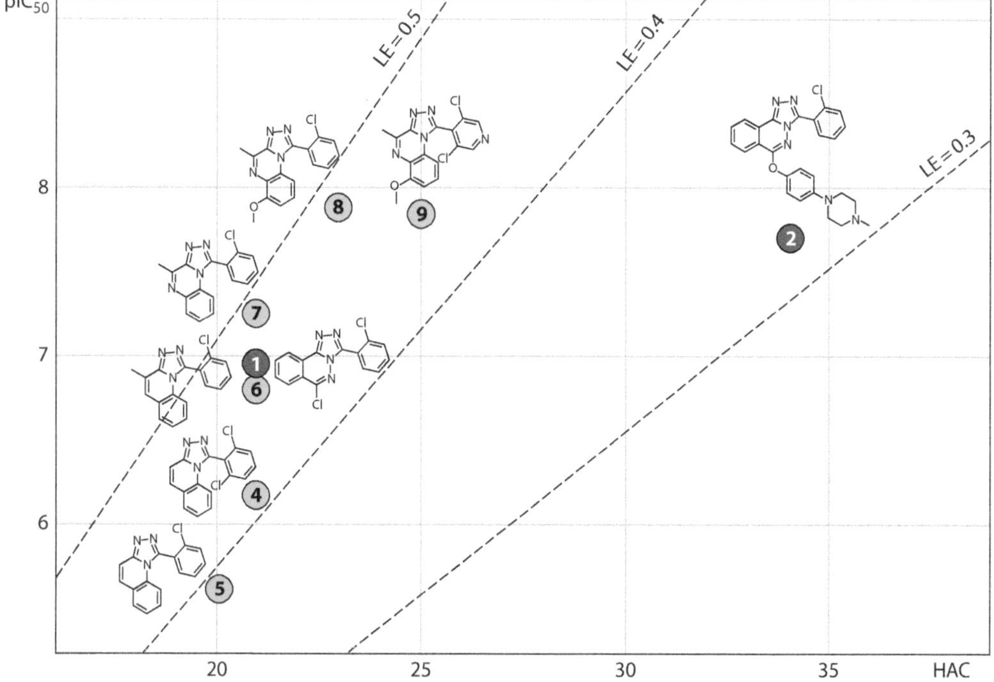

FIGURE 5.11 Plot of pIC_{50} vs. heavy atom count for the PDE2 inhibitor project. The compounds **5.1– 5.2** and **5.4–5.9** are highlighted. The lines for LE = 0.3, 0.4, and 0.5 are indicated.

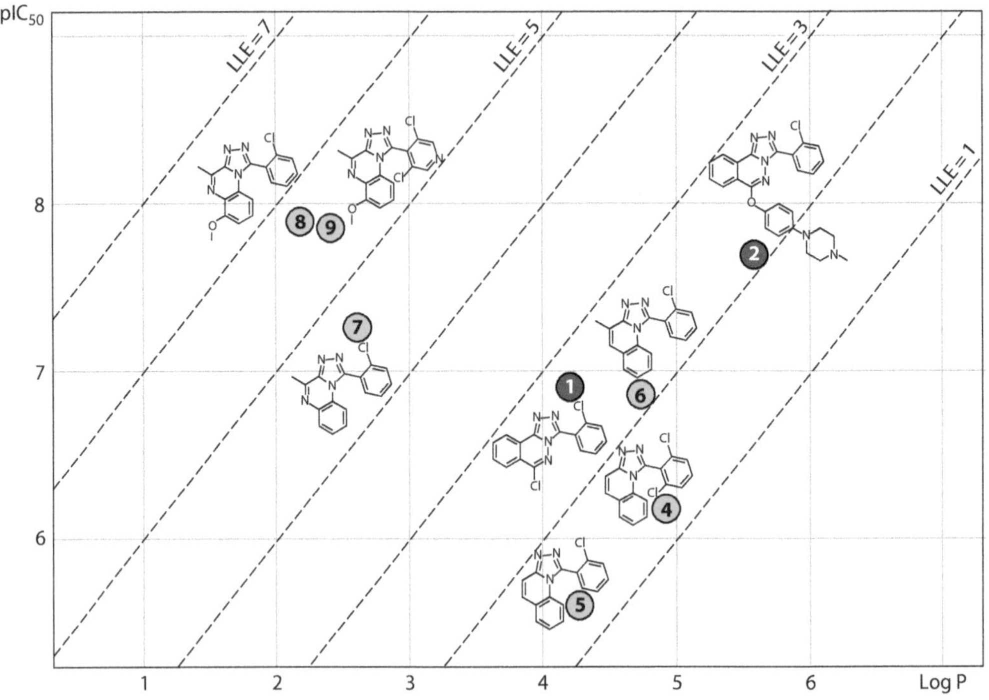

FIGURE 5.12 Plot of pIC_{50} vs. lipophilicity for the PDE2 inhibitor project. The compounds **5.1–5.2** and **5.4–5.9** are highlighted. The lines for LLE = 1, 2, 3, 4, 5, 6, and 7 are indicated.

5.5 CONCLUDING REMARKS

Despite the utility of the concept of drug-likeness in decision making in medicinal chemistry, a final word of warning and a strong recommendation for a "holistically thinking mind" in drug discovery is relevant. Much molecular information is neglected in the descriptors used by medicinal chemists to assess drug-likeness and specifically, the dynamic nature of molecules is a missing dimension when encoding chemical information. Furthermore, one of the limitations of drug-like evaluation is the inability to incorporate dosage. Nevertheless, physicochemical properties certainly have a major impact on the efficacy and safety of drug candidates. And the concept of drug-likeness has won a legitimate place as a help to optimize pharmacokinetic and pharmaceutical properties. Tools that estimate drug-likeness are valuable in the early stages of lead discovery, and are useful to prioritize hits from primary screens and in decision making during the compound design phase and the lead optimization process. Lipophilicity and MW have consistently emerged in study after study relating measurable physical properties with molecular behavior.

By improving potency relative to HAC and lipophilicity, it has been shown that the quality of the compound, and therefore the probability of success, will be increased. Successful optimizations are highly correlative with improved LLE and it has been reported that in successful drug discovery programs, LLE improves relative to the lead. Importantly, the physicochemical parameters are the parameters directly controlled by medicinal chemists in multidimensional optimizations striving toward the goal of developing oral drugs.

FURTHER READING

Hopkins, A.L., Keserü, G.M., Leeson, P.D., Rees, D.C., and Reynolds, C.H. 2014. The role of ligand efficiency metrics in drug discovery. *Nat. Rev. Drug Discov.* 13:105–121.

Johnson, T.W., Dress, K.R., and Edwards, M. 2009. Using the Golden Triangle to optimize clearance and oral absorption. *Bioorg. Med. Chem. Lett.* 19:5560–5564.

Lipinski, C.A., Lombardo, F., Dominy, B.W., and Feeney, P.J. 1997. Experimental and computational approaches to estimate solubility and permeability in drug discovery and development settings. *Adv. Drug Deliv. Rev.* 23:3–25.

Wager, T.T., Chandrasekaran, R.Y., Hou, X., Troutman, M.D., Verhoest, P.R., Villalobos, A., and Will, Y. 2010. Defining desirable central nervous system drug space through the alignment of molecular properties, in vitro ADME, and safety attributes. *ACS Chem. Neurosci.* 1(6):420–434.

Wager, T.T., Hou, X., Verhoest, P.R., and Villalobos, A. 2010. Moving beyond rules: The development of a central nervous system multiparameter optimization (CNS MPO) approach to enable alignment of drug-like properties. *ACS Chem. Neurosci.* 1(6):435–449.

6 Chemical Biology

Louise Albertsen and Kristian Strømgaard

CONTENTS

6.1 INTRODUCTION

Modern drug design and discovery is an increasingly complex process, where a wide range of actors with many different backgrounds work together toward the unifying goal of developing novel drugs. Their success is highly dependent on insights and advances from the combined use of particularly chemical and biological research, and hence it is pivotal to have researchers from chemistry and biology to work in a concerted effort, in order to succeed. Similarly, research at academic institutions is becoming more and more interdisciplinary which is driven by the increasing complexity that modern research is facing. Multi- or inter-disciplinary research disciplines are having a growing impact and boundaries between traditional natural sciences are vanishing. Chemical biology is just one example of such a cross-disciplinary science; others include nanosciences and systems biology which comprises research from physics, chemistry, biology, and computer sciences.

Chemical biology is a contemporary scientific discipline which integrates several chemical and biomedical sciences such as organic chemistry, peptide chemistry, pharmacology, biochemistry, and cell biology. Chemical biology is generally defined as the application of chemical approaches to modulate biological systems to study biological problems or create new biological functions. In other words, chemical biology is the scientific discipline for chemists who are applying the principles, language, and tools of chemistry to biological systems and for biologists who are interested in understanding biological processes at the molecular level. Although research in chemical biology is characterized by covering a very broad range of both chemical and biological approaches, it can generally be classified into two main classes of research:

1. Applying small molecules to perturb biomacromolecules, such as proteins, carbohydrates, lipids, DNA, or RNA
2. Manipulating biomacromolecules by chemical methods to interrogate biological phenomena

The application of small molecules in biological studies traces back to the isolation of individual natural products which is still today one of the cornerstones of conventional medicinal chemistry. In the context of chemical biology the application of small molecules is often called chemical genetics, where small molecules are used in a systematic manner to probe biology. Although there is no strict definition of a small molecule, it typically is nonoligomeric and has a molecular weight between 300 and 700 Da. In contrast, biomacromolecules are oligomeric molecules composed of building blocks, and the three major classes are proteins, nucleic acids, and polysaccharides that are built up of amino acids, nucleotides, and carbohydrates, respectively. Chemical biology studies of proteins and nucleic acids have the great advantage of robust automated methodologies for the synthesis of peptides and nucleic acids, respectively. More recently, a range of chemically inspired technologies have been developed to modify and study proteins and nucleic acids, opening new and exciting possibilities for probing structure and function of these biomacromolecules.

6.2 SMALL MOLECULES

The use of small molecules has played a key role in biological research as well as in drug discovery (see Section 6.1) for many years. Historically, the most important source of small molecules has been nature, which has delivered a wealth of compounds to be applied in both basic research and drug discovery (see Chapters 1 and 7). In some cases, the natural product could be used directly, but often medicinal chemistry efforts were required to optimize the properties of the natural product and generate an improved small molecule. In the 1980s and 1990s combinatorial chemistry emerged as a response to the development of high-throughput (HTP) technologies in biology, and was widely applied in a more systematic search for bioactive small molecules which often were synthesized using solid-phase chemistry. However, it was realized that combinatorial chemistry in itself did not succeed in providing better starting points for drug development than other available methods.

There is an urge to find chemistry-based methods and principles that in a most efficient manner can provide pharmacological tools and drug candidates. Besides medicinal chemistry and natural products chemistry which still play pivotal roles in any small-molecule development, novel approaches such as diversity-oriented synthesis (DOS) and fragment-based approaches are being investigated and will be discussed in the following. These developments follow along progress in related areas to the drug discovery process, such as screening technologies, constitution of compound libraries and systems biology approaches, but a discussion of these areas is beyond the scope of this chapter.

6.2.1 Diversity-Oriented Synthesis

Synthetic chemists have generally strived to prepare predesigned molecules or challenging compounds from nature. Similarly, medicinal chemists are initially designing target molecules which subsequently are prepared in the laboratory. Such approaches can be classified as target-oriented synthesis (TAS, Figure 6.1). This mindset was modified with the advent of combinatorial chemistry, where compounds were designed and prepared in either parallel or combinatorial manner, instead of preparing a target molecule(s) in a linear fashion. Thus, combinatorial chemistry could provide a significantly increased number of compounds and it was believed that this increased number eventually would lead to an increased number of drugs on the market. However, it appears as if combinatorial chemistry has not led to the expected revolution in the number of new drug leads, and it is believed that one of the primary issues is the lack of "diversity," i.e. that in the pursuit of generating large number of compounds the design and diversity of the compounds might have been compromised.

As a consequence, a different concept for preparing biologically relevant compounds has been proposed, where compounds are generated with the focus on maximizing "diversity" of the compounds generated (Figure 6.1). These compounds are typically not directed toward a single biological target, but can be used to identify new ligands for a number of targets. The concept of diversity-oriented synthesis (DOS) furthermore strives to apply a relatively limited number of chemical reactions, but

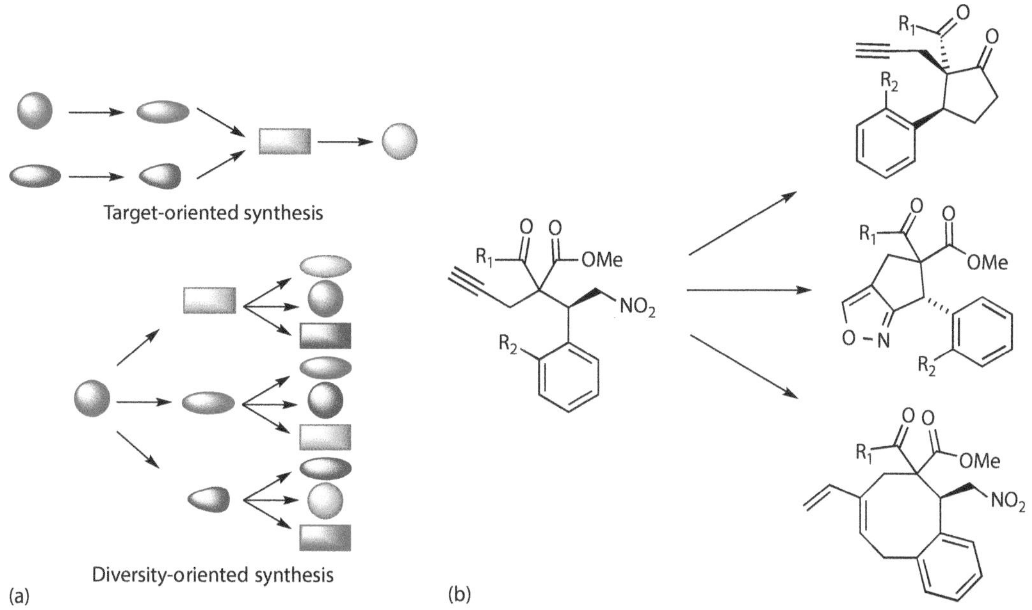

FIGURE 6.1 (a) Illustration of the principles of target-oriented synthesis (TAS) and diversity-oriented synthesis (DOS). In TAS, the synthesis of a single target molecule is convergent, and the synthetic strategy is devised by a retrosynthetic analysis. The synthesis of multiple target molecules in DOS is divergent and the synthetic strategy planned by forward synthetic analysis. (b) Example of DOS, where different pairings of the functional groups (ester, alkyne/alkene, and nitro group) lead to three different scaffolds.

each of these reactions should generate maximal diversity. When applying DOS, a number of principally different strategies are being used; in particular, strategies distinguishing between building block diversity and skeleton diversity. In the former a variation of building blocks are attached to the same scaffold, whereas performing DOS with skeleton diversity leads to generation of compounds with fundamental changes in the skeleton.

A number of successful examples of the application of DOS for the synthesis of diverse combinatorial libraries, phenotypic screening, and subsequent elucidation of the biological target have been carried out and will be discussed in Section 6.2.3.

6.2.2 Fragment-Based Approaches

The application of fragments of chemical compounds has been used for some time to simplify computational analysis of ligand binding and in analysis of pharmacophore elements. However, more recently, a similar approach has been used in drug screening; the concept being that by screening low molecular weight compounds or fragments, low-affinity binders are found, and ideally combining or evolving fragments should then lead to chemical compounds with much improved affinity. One of the advantages compared to high-throughput screening (HTS) is that hits from fragment-based screening generally have lower molecular weight and more efficient binding affinity (ligand efficiency) and provide better starting points for lead generation.

A requirement for fragment-based screening is that a sufficiently sensitive technology for screening is available, i.e., a technology that allows identification of low-affinity (in the mM range) binders. In addition, structural information is required to generate potent lead compounds from fragments. Therefore, screening of fragments is typically carried out using bioassays such as surface plasmon resonance (SPR) in combination with biophysical techniques such as nuclear magnetic resonance (NMR) technologies or X-ray crystallography, the latter providing structural guidance for further development

Once one or more fragments have been identified, iterative cycles of medicinal chemistry, ideally guided by three-dimensional structural data, are performed to optimize the weakly binding fragment hits into potent and selective lead compounds. Principally, there are two different approaches for optimization: (1) fragment linking and (2) fragment evolution (Figure 6.2). If two fragments have been identified which bind in separate binding sites that are sufficiently close to each other, those fragments can be chemically linked and provide a lead. This is called fragment linking (Figure 6.2). Alternatively, knowledge of the identified fragments is used to develop hypotheses on how to expand the fragment by additional interactions in the active site of the protein. This "evolution" then leads to compounds with improved binding which can be further optimized (Figure 6.2). The fragment "evolution" principle probably is most often applicable and is an elegant demonstration of

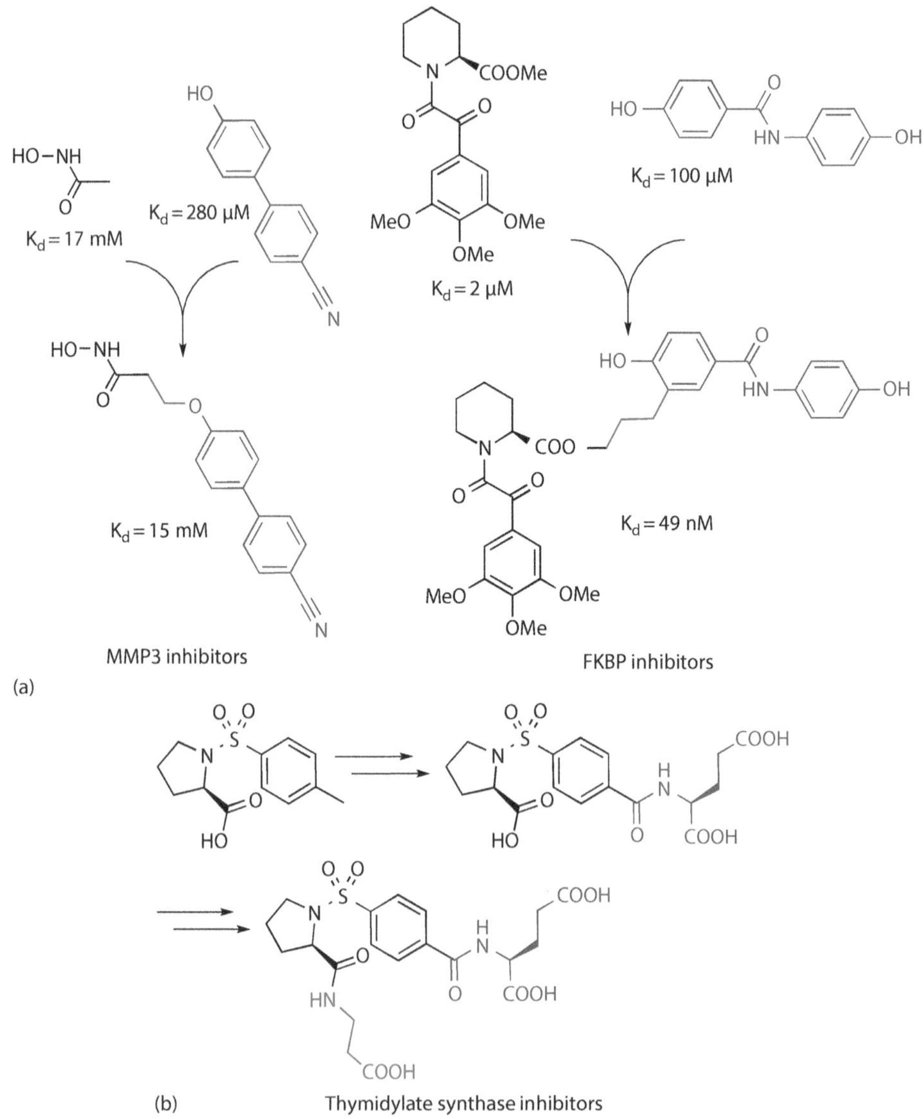

$K_d = 17$ mM

$K_d = 280$ μM

$K_d = 2$ μM

$K_d = 100$ μM

$K_d = 15$ mM

$K_d = 49$ nM

MMP3 inhibitors

FKBP inhibitors

(a)

(b) Thymidylate synthase inhibitors

FIGURE 6.2 Illustration of the two principles applied in fragment-based screening. (a) Fragment linking: Two fragments were identified as very weak inhibitors of matrix metalloproteinase 3 (MMP3), but by linking the two fragments a highly potent lead compound was identified. Similarly, a potent inhibitor of the FK506-binding protein (FKBP) was achieved by linking two weaker inhibitors. (b) Fragment evolution: A very potent inhibitor of thymidylate synthase was identified.

structure-based drug design (see Chapter 4) and bear many resemblances to conventional medicinal chemistry. Fragment-based drug design is a promising strategy that is widely applied both in industry and at academic institutions. It is still primarily focused on studies of soluble proteins as easy access to detailed structural information is required.

6.2.3 Chemical Genetics

Extracts from natural sources have been used in the treatment of diseases since ancient times, but in the early nineteenth century, isolated natural products became available and have been extensively used as biological probes and drugs. The first isolated natural products that were used as drugs were alkaloids such as morphine, quinine, and codeine (see Chapters 1 and 7), and later derivatives of natural products such as acetylsalicylic acid have been applied.

Chemical genetics is basically a research method that uses small molecules to perturb the function of proteins and to do this directly and in real time, rather than indirectly by manipulating their genes. It is used to identify proteins involved in pertinent biological processes, to understand how proteins perform their biological functions, and to identify small molecules that may be of therapeutic value. The term "chemical genetics" indicates that the approach uses chemistry to generate the small molecules, and that it is based on principles which are similar to classical genetic screens. In genetics two kinds of genetic approaches, forward and reverse, are applied, depending on the starting point of the investigation, and the same is true for chemical genetics, except that small molecules are used to perturb protein function. In forward chemical genetics, a ligand that induces a phenotype of interest is selected and the protein target of this ligand is identified.

In reverse chemical genetics, small molecules are screened for effects at the protein of interest, and subsequently a ligand is used to determine the phenotypic consequences of perturbing the function of this protein. Chemical genetics can therefore be regarded as a fruitful and complementary alternative to classical genetics or to the use of RNA-based approaches, such as RNA interference (RNAi) technology (see Section 6.4.2).

A now classical example illustrating the power of small molecules in elucidating protein targets is FK506 (or Fujimycin, Figure 6.3) which is a macrolide natural product structurally related to

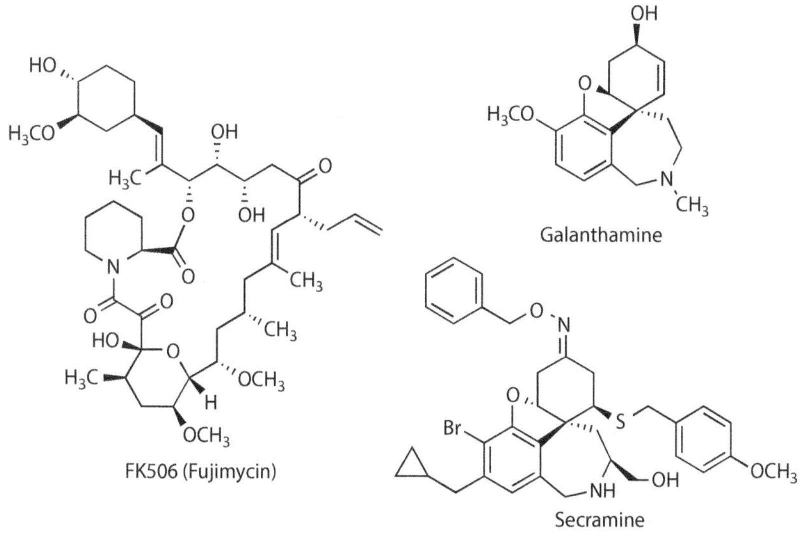

FK506 (Fujimycin)

Galanthamine

Secramine

FIGURE 6.3 Examples of compounds used in chemical genetic studies. FK506 was used to identify its protein target, FK506-binding protein (FKBP). Galanthamine was employed as a template for the DOS of structurally diverse analogs, which lead to the identification of secramine, as an inhibitor of vesicular traffic out of the Golgi apparatus.

rapamycin (see Chapter 7) and currently used as an immunosuppressant after organ transplantation. The target of FK506 was not known, but using FK506 as molecular bait to fish its binding protein from biological samples, a protein was identified, called FK506 binding protein (FKBP). It was found that binding of FK506 to FKBP inhibits immune function by shutting down a specific molecular signaling pathway. Another natural product, galanthamine (Figure 6.3, see Chapter 16), which is an inhibitor of acetylcholinesterase (AChE) isolated from certain species of daffodil and used for the treatment of mild to moderate Alzheimer's disease, was used as a template for DOS. Specifically, a 2527 compound library of galanthamine-like structures was prepared and underwent phenotypic screening, identifying secramine (Figure 6.3) as an inhibitor of vesicular traffic out of the Golgi apparatus by an unknown mechanism. After an extensive effort, it was discovered that secramine inhibits activation of the Rho GTPase Cdc42, a protein involved in membrane traffic.

6.3 MODIFYING PROTEINS

Proteins constitute up to 50% of the dry weight of cells, and are thereby the most abundant biomacromolecules in cells. In eukaryotes, proteins are produced in the ribosome, where a messenger RNA (mRNA) carries the code for the primary sequence of the protein and is read by aminoacylated transfer RNA (aa-tRNA). The code, the genetic code, contains 64 triplet codons, of which 61 codes for 20 different amino acids which we call cognate, canonical, or proteinogenic amino acids that are the building blocks for protein biosynthesis. The three remaining triplet codons, UAG (amber), UAA (ochre), and UGA (opal) are stop codons, also known as nonsense codons. Thus, eukaryotic proteins are generally made up of the 20 amino acids, although in recent years two to three extra amino acids have been added to this repertoire (Figure 6.4). The 21st amino acid, selenocysteine, which is found in nuclear proteins in eukaryotes and other organisms, is a cysteine analog in which sulfur has been replaced by selenium. The 22nd amino acid is pyrrolysine which is predominantly found in Achaea and where the ε-amino group of lysine is derivatized with β-methylpyrroline. A putative 23rd amino acid is *N*-formylmethionine, an *N*-terminal formylated methionine analog which is present in mitochondrial proteins that upon tissue damage is released and activates immunity. Importantly, nature has substantially increased the diversity of proteins by a large number of posttranslational modifications (PTMs, Figure 6.4) that are added by enzymatic processing after translation.

In general, the use of chemical, as a complement to conventional genetic methods, to alter protein structure and function offers exciting possibilities. Genetic methods are generally limited to the use of the 20 proteinogenic amino acids comprising a finite number of functional groups. However, by combining the principles and tools of chemistry with the synthetic strategies and processes of living organisms, it is possible to generate proteins with novel functions. Such proteins can be applied in structural and functional studies of proteins, in ways previously considered unattainable.

A key feature for generating novel proteins has been site-specific incorporation of PTMs which uniquely allow addressing the structural and functional importance of such modifications in great detail. In principle, any PTM can be introduced including phosphorylation, acetylation, hydroxylation, glycosylation and lipidation, as well as more subtle chemical alterations of parent amino acids including deamidation and citrullination (Figure 6.4). Another class of modifications includes those that incorporate biophysical probes or reactive handles for further derivatization. Examples include site-specific labeling with ^{13}C- or ^{15}N-labeled amino acids for biological NMR studies, incorporation of fluorescent amino acids or amino acids containing photolabile groups such as benzophenone (Figure 6.5) which can be used to identify and map intra- and intermolecular interactions of proteins. Amino acids with reactive groups such as azides or alkyne groups, for selective derivatization via the Huisgen 1,3-dipolar cycloaddition, also known as "click chemistry," are also of increasing interest (Figure 6.5). Another example is introduction of ketone functionalities (Figure 6.5) that can be selectively modified, for example, with polyethylene glycol (PEG) linkers containing a hydrazide or oxyamine functionality. Finally, very subtle changes of proteins, such as incorporation of

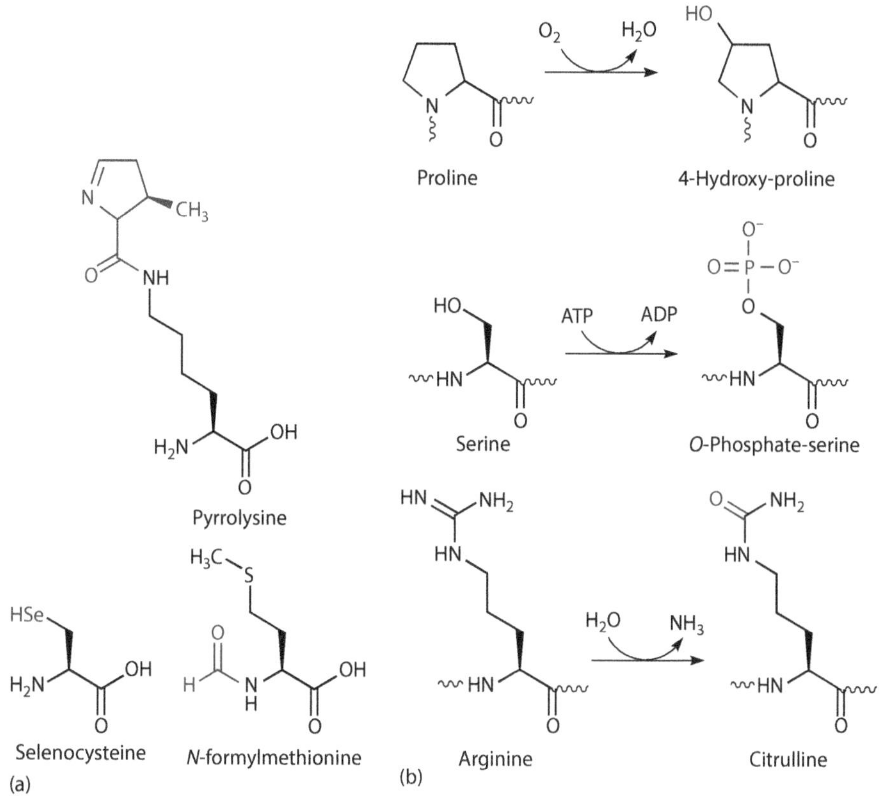

FIGURE 6.4 The three nonproteinogenic amino acids and examples on posttranslational modifications. (a) The 21st (selenocysteine), 22nd (pyrrolysine), and 23rd (*N*-formylmethionine) amino acids are obtained by conversion of serine, lysine, and methionine, respectively. (b) Proline, serine, and arginine can undergo post-translational hydroxylation, phosphorylation, and deimination (or citrullination), respectively.

D-amino acids, close analogs of encoded amino acids, and modification of the amide backbone can also be introduced (Figure 6.5). This allows very fine-tuned studies of, for example, ligand–receptor interactions and protein function in general and has been described as "protein medicinal chemistry."

A number of technologies have been developed to achieve this objective, and it is now possible to generate proteins containing, in principle, any functionality. Genetic methods for residue-specific incorporation of close analogs of natural amino acids, involving depletion of one amino acid and addition of the structural-related analog, have existed for many years. A typical example is the incorporation of selenomethionine (Se-Met), in place of methionine, which is used in structural studies of proteins, as the heavy atom, selenium, can help solving the phase problem in X-ray crystallography. In the following, we will focus on approaches for site-specific introduction of unnatural amino acids (UAAs) into proteins using the cells' own protein synthesis machinery (unnatural mutagenesis) or through full- or semi-synthesis by ligation-based strategies, as well as approaches for chemical or enzymatic conjugation.

6.3.1 Unnatural Mutagenesis

In 1989, a biosynthetic in vitro method that allowed site-specific incorporation of UAAs into proteins was introduced based on earlier work on nonsense suppression. The term "nonsense suppression" refers to the use of stop (nonsense) codons and suppressor transfer RNA (tRNA) which

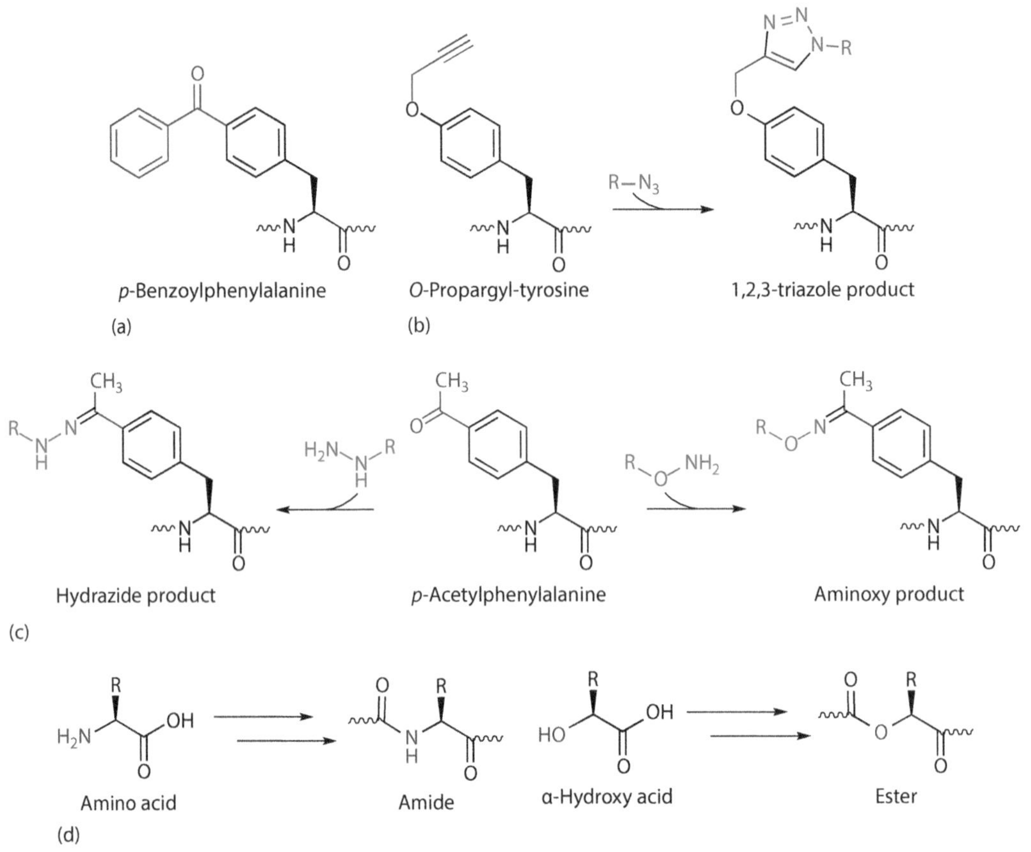

FIGURE 6.5 Unnatural amino acids and backbone modifications that can be incorporated into proteins using chemistry-based methods. (a) The biophysical probe, *p*-benzoylphenylalanine, contains a photolabile benzophenone group. (b) *O*-Propargyl-tyrosine contains an alkyne group that can react with azides (red) in "click chemistry" to form 1,2,3-triazole products (R can, for example, be polyethylene glycol). (c) *p*-Acetylphenylalanine contains a ketone group that can react with hydrazines (red, left) or oxyamine (red, right) in a hydrazone and oxime reaction, respectively. (d) The backbone amide can be changed into an ester by introducing an α-hydroxy acid in place of an amino acid. This can be used to evaluate the electronic effects of backbone amides in H-bonding interactions.

recognize stop codons. The method is based on the fact that only one of three stop codons in the genetic code is necessary for termination of protein synthesis, and the two unused stop codons can then be exploited for introduction of UAAs.

The primary challenge in this technology is the generation of the modified suppressor tRNA with the UAA (Figure 6.6). Once generated the aminoacylated-tRNA (aa-tRNA) is recognized by the mRNA carrying the specific stop codon, which allows the ribosome to read through (suppress) the stop codon, resulting in incorporation of the UAA into the protein at the specific position. The principle is applied in two different versions: One method applies tRNAs that are chemically aminoacylated in vitro with the UAA of interest, and the aminoacylated-tRNA (UAA-tRNA) is subsequently applied in an expression system to generate the protein of interest (Figure 6.6). The other method employs development of pairs of orthogonal (i.e., not used natively by the host) tRNA and aminoacyl-tRNA synthetases (aaRS), where the latter is evolved so that it selectively recognizes and aminoacylates an UAA inside living cells (Figure 6.6).

In the chemical aminoacylation of tRNA, a truncated tRNA lacking a dinucleotide at the 3′ terminus is enzymatically ligated to a chemically synthesized aminoacylated dinucleotide carrying the

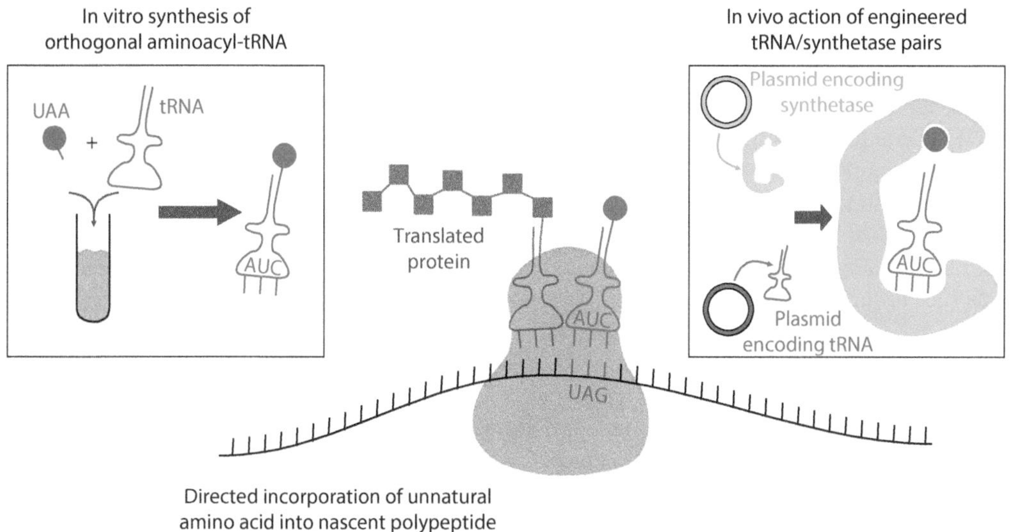

FIGURE 6.6 Incorporation of unnatural amino acids (UAAs) into proteins in living cells using nonsense suppression. An orthogonal tRNA loaded with the UAA directs the incorporation of the UAA into a nascent polypeptide (gray squares) at the site of a stop codon (usually UAG) that has been introduced into the gene of interest. Loading of the tRNA with the UAA can be done synthetically (top left panel) by ligating a UAA with a synthetic tRNA in vitro, followed by injection of the resulting aminoacylated-tRNA into an expression system along with mRNA for the gene of interest. Alternatively, cells can be engineered to generate the required tRNA in vivo by introducing plasmids coding for an orthogonal tRNA/aminoacyl-tRNA synthetase pair (top right panel). The synthetase that has been evolved to recognize the foreign tRNA and the desired UAA (usually added to growth media) catalyze the ligation to generate the aminoacylated-tRNA. (From Pless, S.A. et al.: Atom-by-atom engineering of voltage-gated ion channels: Magnified insights into function and pharmacology. *The Journal of Physiology*. 2015. 593: 2627–2634. Copyright Wiley-VCH Verlag GmbH & Co. KGaA. Reproduced with permission.)

UAA of interest. If a cell-free expression system is used, the UAA-tRNA is simply added to the media, and when intact cell expression systems are used, the UAA-tRNA is injected into the cell. In practice, the most widely used expression system for this methodology is *Xenopus* oocytes which is generally used for electrophysiological studies of ion channels, receptors, and transporters. The oocytes are microinjected with the modified mRNA encoding for the target protein and the synthetic UAA-tRNA which results in synthesis and surface expression of the target protein containing the UAA. This has enabled a large number of structurally diverse UAAs to be incorporated into proteins. In most cases the unnatural amino acids have been α-amino acids, but also non-α-amino acids and most notably α-hydroxy acids have been incorporated, the latter introducing an amide-to-ester mutation in the protein backbone (Figure 6.5).

In studies of ligand-gated ion channels, such as γ-aminobutyric acid (GABA), nicotinic acetylcholine (nACh) and serotonin ($5\text{-}HT_3$) receptors (see Chapters 15, 16, and 18), the technology has proven particularly valuable. These studies were pioneered by Dougherty and Lester, who have explored the molecular details of the cation–π interaction between the quaternary ammonium group of acetylcholine and aromatic residues in the nACh receptor. Similar studies have also provided unique details of ligand binding to different nACh receptor subtypes, and paved the way for unraveling the basis for stronger binding of nicotine to brain versus muscle nACh receptors which is a crucial aspect of nicotine addiction. More recently, the methodology has also been applied in studies of voltage-gated ion channels to obtain detailed mechanistic information about the regulation of channel activation and inactivation.

Although the technology has proven exceptionally useful in studies of ion channels, it comes with a number of limitations. First, the generation of UAA-tRNAs requires highly skilled persons in both chemistry and molecular biology. Secondly, the amount of protein generated is very low, thus exceptionally sensitive detection systems, such as electrophysiology or fluorescence, are required in these studies. In order to overcome some of these limitations, a modified method was developed. In this approach, a custom-made pair of orthogonal tRNA and aaRS is genetically introduced into a cell. The aaRS is engineered such that it only recognizes the UAA and efficiently acylates the corresponding tRNA. Subsequently, the UAA, which has to be nontoxic and cell permeable, is added to the cell culture medium, taken up by the host organism, and incorporated into the protein by the specific tRNA/aaRS pair. This approach was first used for incorporation of UAAs into bacteria, but has since been optimized for use in yeast, mammalian cells, primary neuronal cultures, and even in animals such as the nematode *Caenorhabditis elegans* and the fruit fly *Drosophila melanogaster.* These studies showed that increased incorporation efficiencies were obtained, leading to substantially higher yields of modified proteins. The primary challenge of this technology is that the specific aaRS has to be generated for each UAA which is done by extensive mutational studies and rounds of positive and negative selections. However, the approach is developing rapidly and recent advancements include identification of a broader number of orthogonal tRNA/aaRS pair as well as genetic engineering of the ribosome and the termination release factor to allow more efficient UAA incorporation.

The methodology has been applied to a variety of proteins including ion channels, G protein-coupled receptors (GPCRs) and histones to examine protein–protein interactions, conformational changes, signaling processes, and the role of posttranslational modifications.

A specific and widespread application is the incorporation of photo-cross-linking UAAs such as benzoyl-phenylalanine or azido-phenylalanine which upon UV radiation cross-link with nearby reactive species. This was used to map the interaction of peptide and small-molecule ligands to GPCRs (Figure 6.7) and also to map the intracellular interaction to a G-protein. Another application is the introduction of photo-caged UAAs that enables the researcher to control particular protein functions. This approach has, for example, been used to generate a light-activatable potassium channel functioning as an optical switch for suppressing neuronal firing (Figure 6.7). Very recently, incorporation of a fluorescent UAA has enabled in situ imaging of protein dynamics of the Shaker Kv channel. The methodology is also used to incorporate reactive UAAs that can subsequently be specifically chemically modified which is also being exploited commercially by a biopharmaceutical company to develop improved protein-based therapeutics.

A general constraint of these technologies is that the genetic code only contains three stop codons which limits the number of different UAAs that can be incorporated into a single protein. To potentially overcome this limitation, extended codons (quadruplets) have been employed. In this approach, an mRNA containing the quadruplet codon is being read by a modified UAA-tRNA containing the corresponding extended anti-codon. The quadruplet codon approach has successfully been used to incorporate UAAs into proteins, and also to incorporate two different UAAs into two different sites of a single protein, showing that the quadruplet codons are not only orthogonal to their host organism but also to each other. In order to use the quadruplet codon strategy to incorporate multiple UAAs by the orthogonal tRNA/aa-RS methodology, orthogonal tRNA/aa-RS pairs for each UAAs have to be developed. This has recently been achieved by evolving orthogonal tRNA/aa-RS from two different widely used tRNA/aa-RS pairs which has enabled simultaneous incorporation of two different UAAs in response to a stop codon and a quadruplet codon. More recent advances in this field include the de novo design of orthogonal tRNA/aa-RS pairs and engineering of an orthogonal ribosome that reads quadruplet codons more efficiently.

6.3.2 Peptide/Protein Ligation

A conceptually different strategy for modification of proteins is to employ methods based on solid-phase peptide synthesis (SPPS) for generation of proteins. This would allow incorporation of principally any amino acid. SPPS has in a few cases been applied for the synthesis of proteins, although yields are generally rather low. The first example was the synthesis of ribonuclease A (124-residues)

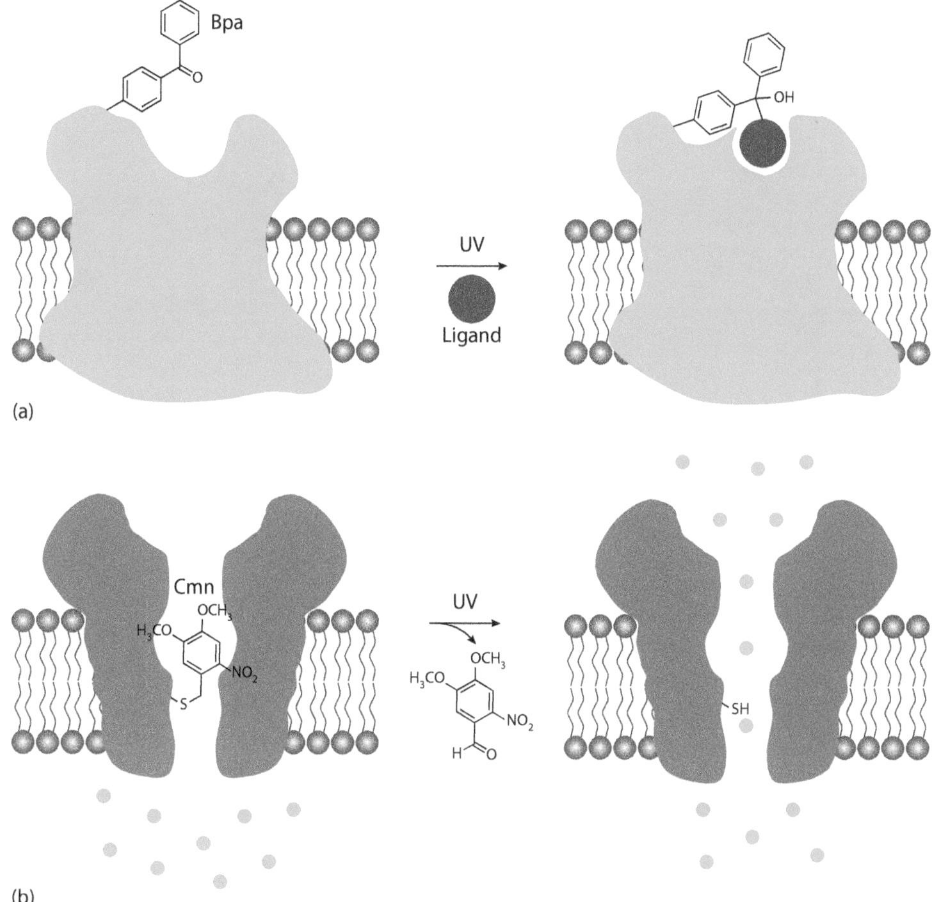

FIGURE 6.7 Biophysical probes can be used to map ligand-binding sites and control function of membrane proteins after incorporation via unnatural mutagenesis. (a) Light is used to cross-link *p*-benzoylphenylalanine (Pba) in G protein-coupled receptors upon binding of the ligand (gray sphere) which result in a conformational change into the ligand-bound state. (b) When the photo-caged amino acid, 4,5-dimethoxy-2-nitrobenzyl-cysteine (Cmn), is incorporated into the pore of an ion channel, it blocks the channel. Exposure to UV light results in irreversible removal of the dimethoxynitrobenzyl group from cysteine, rendering the channel conductive.

by Bruce Merrifield in 1966 and since then a few other proteins have been prepared by this approach; most notably, HIV protease (99-residues) which enabled structural characterization of the protein.

However, SPPS is generally limited to the preparation of up to 40–60 amino acid peptides, whereas most proteins are considerably larger. Therefore, there has been a considerable interest in developing methods that are not confined to these restrictions and in 1994, a strategy for preparation of proteins from peptide fragments was introduced, called native chemical ligation (NCL, Figure 6.8). In NCL, two or more unprotected peptide fragments can be ligated together, generating a (native) cysteine residue in the ligation site. The ligation requires a peptide with a *C*-terminal thioester and a peptide with an *N*-terminal cysteine residue: the thiolate of the *N*-terminal cysteine attacks the *C*-terminal thioester to affect transthioesterification, followed by formation of an amide bond after *S*→*N* acyl transfer (Figure 6.8). The reaction takes place in aqueous buffer and generally proceeds in good to excellent yield.

Thus, NCL is a very useful approach for the total chemical synthesis of proteins and has been used for the preparation of numerous proteins, including glycoproteins and proteins with fluorescent labels. An example is the synthesis of an analog of erythropoietin (EPO) which was derivatized

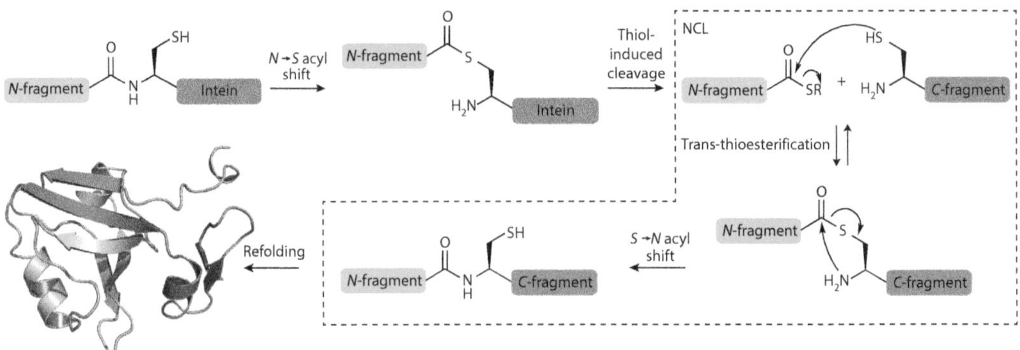

FIGURE 6.8 Principles of expressed protein ligation (EPL) and native chemical ligation (NCL). In NCL, an *N*-terminal peptide thioester fragment undergoes trans-thioesterification with a *C*-terminal peptide fragment containing an *N*-terminal cysteine. This is followed by a spontaneous *S*-to-*N* acyl shift, resulting in the formation of a native amide bond and a cysteine at the ligation site. In EPL, an *N*-terminal protein thioester is generated from a recombinant intein fusion construct by reaction with an exogenous thiol. The protein thioester is then reacted in an NCL reaction with a synthetic peptide, enabling semi-synthesis of larger proteins. After both NCL and EPL, the protein has to be refolded into its native structure.

with monodisperse polymer moieties in order to improve duration of action in vivo. The 166-residue protein was prepared by ligation of four peptide fragments, two of which was modified with the polymer and the EPO analog displayed improved properties in vivo compared to EPO.

In 1998, an extension of the NCL principles was introduced, called expressed protein ligation (EPL). The technology applies the same reaction as in NCL, but in contrast to NCL, one of the components is a recombinantly expressed protein, rather than a peptide (Figure 6.8). Protein thioesters is prepared by expressing the protein as a so-called intein fusion construct which can be converted into the thioester and subsequently reacted with a peptide with an *N*-terminal cysteine generating a full-length protein (Figure 6.8). Thus, the EPL methodology combines the advantages of molecular biology with chemical peptide synthesis and enables the addition of unnatural functionalities to a recombinant protein framework.

EPL has, similar to NCL, been applied in studies of a plethora of proteins. A prominent example is studies of histones which are important for storage of DNA and have flexible *N*-terminal tails that are heavily modified by PTMs. EPL has been applied to prepare full length, ubiquitylated histones which were used to demonstrate a direct cross-talk between PTMs on different histones. EPL has also been applied to integral membrane proteins, specifically the potassium channel KcsA. EPL was used to prepare truncated KcsA subunits (122-residues) which were then refolded and reconstituted into lipid membranes. This allowed incorporation of, for example, D-alanine and an amide-to-ester mutation (Figure 6.5) in the selectivity filter of the channel which has revealed important information of the mechanism behind ion conductance. In subsequent studies, a modular EPL approach, involving three fragments and two consecutive ligation steps, was used to generate full-length KcsA subunits (160-residues) that were refolded into the native channel. EPL has also been used to address the importance of backbone hydrogen bonds in intracellular signaling processes. Specifically, protein domains, the so-called PDZ domains, were generated by EPL, and amide-to-ester mutations (Figure 6.5) were introduced in the central recognition site to demonstrate the general importance of specific hydrogen bonds. Clearly, such backbone interactions cannot be addressed by conventional mutagenesis.

The EPL technology also holds commercial prospective and is exploited to generate recombinantly derived protein *C*-terminal thioesters that can be converted into the corresponding hydrazide or oxyamine derivatives, enabling subsequent chemical modification and thus, allowing a wide range of *C*-terminal modifications of proteins.

6.3.3 Peptide/Protein Conjugation

In addition to the technologies just described, a large number of chemistry-based methods allow derivatization of the parent peptide or protein structure, something that has been successfully exploited for generation of peptide and protein therapeutics.

The most common way of modifying an endogenous protein is by reacting at cysteine residues which can often be achieved with either none or minimal changes to the parent protein. The advantage is that the thiol of cysteine allows for selective modification, relative to the other proteinogenic amino acids, and the frequency by which cysteine occurs in proteins is relatively low, thus often allowing selective modification of specific cysteine residues. Even if a protein contains more than one cysteine residue, these might have different accessibility which can allow selective modification of certain residues.

When peptides and proteins are being developed as drugs, the pharmacokinetic (PK) and pharmacodynamic (PD) properties of proteins often need to be improved, and this is often attained by chemical derivatization. A particularly popular strategy is introduction of PEG moieties, known as PEGylation, which can help reducing immunogenicity, increasing the circulatory time by reducing renal clearance, and also provides water solubility to hydrophobic drugs and proteins. PEGylation is generally performed by reaction of a reactive derivative of PEG with the target protein, typically with side-chains of amino acids such as lysine or cysteine, or by reaction at the *C*- or *N*-terminal of the protein or peptide. Today, a number of therapeutic proteins are marketed as PEGylated derivatives including PEGylated α-interferons, which are used in the treatment of hepatitis C and injected only once a week, compared to three times a week for conventional α-interferon.

An alternative way of improving protein and peptide therapeutics is by adding lipids to the protein, which can increase half-life. An example of this is the long-acting insulin analog, insulin detemir (Levemir®), where the side chain N^{ε}-amino group of a terminal lysine in the B-chain of insulin has been modified with tetradecanoic acid (myristic acid, C_{14} fatty acid chain). This modification increases self-association and binding to albumin, leading to stable insulin supply for up to 24 hours. Similarly, liraglutide (Victoza®) is an analog of glucagon-like peptide-1 (GLP-1), where a lysine side-chain has been modified; a palmitic acid (hexadecanoic acid, C_{16} fatty acid chain) was added through a glutamate linker. The modification leads to a substantial increase in half-life due to increased binding to serum albumin and does not compromise the biological activity.

Selective modification of proteins by chemical reactions similar to those used in organic synthesis requires that reactions are compatible with the aqueous (buffer) conditions, and a number of robust and water-compatible reactions have evolved. To achieve exquisite selectivity it is often required to introduce selective reactive "handles," as previously described. The examples of "click chemistry," i.e., a 1,3-dipolar cycloaddition between an azide and an alkyne providing a 1,2,3-triazole have already been mentioned. Another prominent example is the Staudinger reaction which is a phosphine-mediated reduction of an azide to an amine, also known as an aza-Wittig reaction that has been used particularly in protein glycosylation studies.

Finally, enzymes can be used to selectively modify proteins. Enzymes have the inherent advantage that they efficiently add, or remove, groups to proteins, and they are often highly specific for certain sequences (consensus motifs) of amino acids, so modifications are typically site-specific. Enzymes are generally highly substrate-specific, i.e., kinases add only phosphate groups to serine, threonine, and tyrosine. Thus, modification of the enzyme is required if other groups have to be introduced. However, some enzymes such as glycosyltransferases which transfer carbohydrates to serine or asparagine, have broader substrate specificity, and here it can be desirable to modify the enzyme to achieve increased reactivity for specific carbohydrates. A particularly powerful method to develop enzymes with desired properties is directed evolution which basically consist of two steps: (1) generation of a library of mutants of the enzymes and (2) rounds of screening/selection for the desired properties. Enzymes are particularly useful for furnishing proteins with specific PTMs which are often essential for regulation and dynamics of biological activity. For example, most proteins are glycosylated, and controlling glycosylation patterns of proteins is a key challenge. Glycosyltransferases catalyze the transfer of a monosaccharide to a protein and through directed evolution it was possible to modify the transferases, so that monosaccharides of interest could be selectively added to a protein framework. Another example is using transglutaminase (TGase) to obtain selective PEGylation. TGase catalyzes transfer reactions between the γ-carboxamide group of glutamine residues and primary amines, resulting in the formation of γ-amides of glutamic acid and ammonia. Thus, by using an amino-derivative of PEG (PEG-NH_2) as substrate for the enzymatic reaction it is possible to covalently bind the PEG polymer to a therapeutic protein.

6.4 EXPLORING NUCLEIC ACIDS

The central dogma of molecular biology, as proposed by Francis Crick, describes the process of converting DNA to proteins via RNA. Crick also hypothesized that RNA has more functions than just being a passive carrier of information. This was confirmed by the discovery of the first RNA enzymes (ribozymes), whereas the tremendous functional versatility of RNA we know today came much later. It is now evident that of all our genes, only 1.5% are coding for proteins. Hence the remaining 98.5% of genes must have other functions, and a large number of noncoding RNAs (ncRNAs) have been discovered. These are emerging as key regulators of gene expression at the transcriptional and post-transcriptional level and have been divided into two groups: small ncRNAs (<200 nucleotides) and large ncRNA (>200 nucleotides). Although, the majority of ncRNA transcripts belong to the latter group, very little is known about their biological function. In contrast, studies of small ncRNAs have revealed a wealth of novel properties of RNA. Small ncRNAs are typically involved in processing or regulation of other RNAs and are classified into a continuously increasing number of subgroups. The best characterized classes are micro RNA (miRNA), short interfering RNA (siRNA), and piwi-interacting RNA (piRNA) which all regulate gene expression through base-pairing to target nucleic acids. Probably the best known example is gene silencing through RNA interference (RNAi), where protein translation is inhibited after binding of the ncRNA to a specific mRNA (see Section 6.4.2).

The realization that the functions of RNA are much more diverse than first anticipated has inspired the application of in vitro directed evolution methods to further expand the properties of nucleic acids. In these studies, large libraries of RNA or DNA are used to screen for novel functionalities which has resulted in notable breakthroughs such as the discovery of oligonucleotides with high-affinity ligand binding abilities (aptamers) and the generation of the first DNA enzyme (DNAzyme) that can catalyze the cleavage of RNA phosphodiesters. The properties of ribozymes and DNAzymes are constantly being expanded, for example, to catalyze chemical transformations such as Diels–Alder cycloadditions, Michael addition, aldol condensation, as well as hydrolysis, ligation, and deglycosylation of nucleic acid substrates. At the same time extensive efforts have been invested into developing nucleic acid analogs, in which the tripartite chemical structure is modified. Recently, this led to the discovery of the unnatural polymer with catalytic activity (see Section 6.3.1). Thus, the study of nucleic acids is a core area in chemical biology and in the following, chemical biology studies of nucleic acids and RNAi as a novel therapeutic strategy will be discussed in more detail.

6.4.1 Modification of Nucleic Acids

The motivation for modification of nucleic acids is primarily related to increasing the stability and also exploring novel chemical properties of DNA and RNA. This is achieved by modifying the three principal components of nucleic acids: nucleobases, sugar moieties, and the phosphodiester backbone.

In particular, modifications of the backbone, leading to the so-called xeno nucleic acids (XNAs), have provided important tools and compounds with therapeutic perspectives. A fundamental characteristic of XNAs is their ability to form regular Watson–Crick base pairs, as seen in DNA, and prominent examples include peptide nucleic acid (PNA), 2′-substituted nucleic acids and locked nucleic acid (LNA) (Figure 6.9). In PNA, the deoxyribose-phosphate backbone of DNA is replaced with *N*-(2-aminoethyl)-glycine units which confer several advantages over regular DNA: The backbone is linked by peptide bonds, and can be easily synthesized, and the stability of PNA is significantly increased compared to DNA. In LNA, the ribose moiety has been locked into the bioactive 3′-endo conformation via an extra carbon–oxygen bond connecting the 2′-oxygen and 4′-carbon. This locked conformation enhances base pairing affinities and backbone pre-organization, as well as increases the stability. Finally, replacement of the 2′-hydroxy on ribose with substituents such as 2′-fluoro

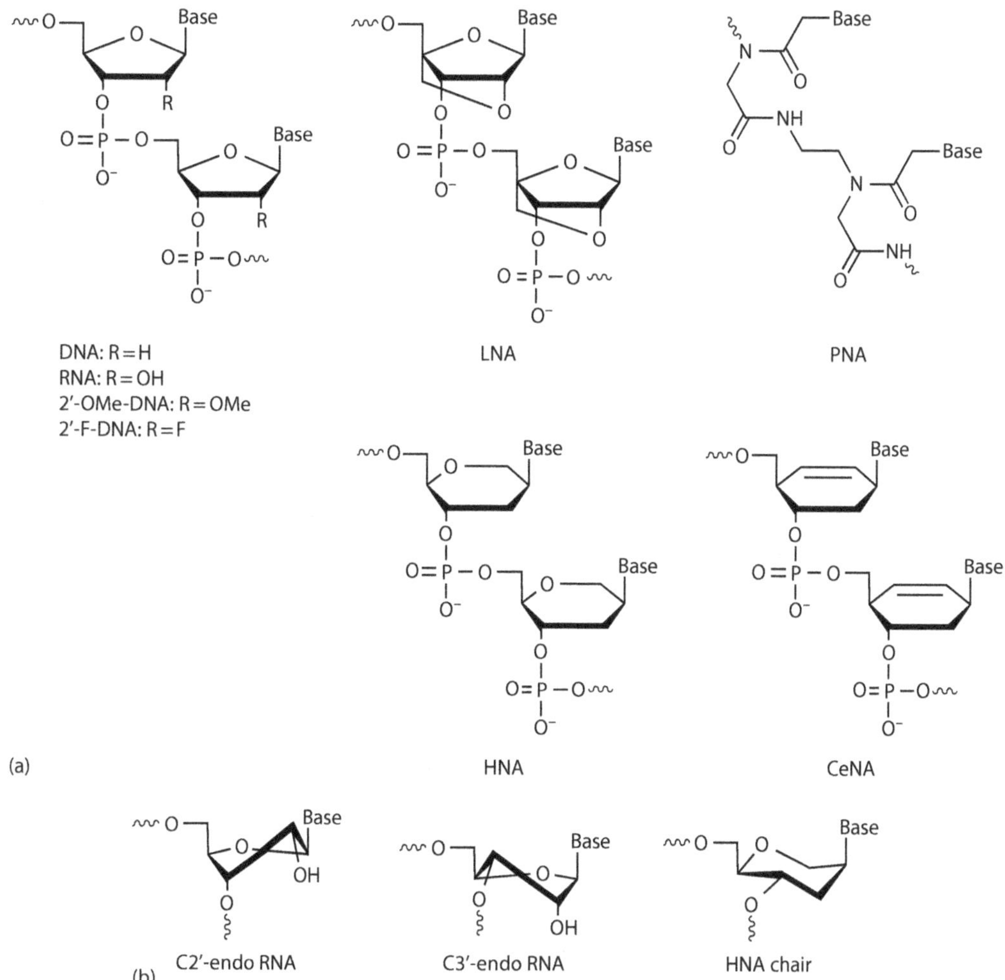

FIGURE 6.9 Examples on backbone-modified nucleic acid analogs. (a) Substitution of DNA/RNA on the 2′ position with a fluoro or methoxy group results in 2′-F-DNA and 2′-MeO-DNA, respectively. In locked nucleic acid (LNA), the ribose moiety of RNA is modified with an extra bridge connecting the 2′ oxygen and 4′ carbon. In peptide nucleic acid (PNA), the deoxyribose-phosphate backbone of DNA is replaced with *N*-(2-aminoethyl)-glycine units. In hexitol nucleic acids (HNA) and cyclohexene nucleic acids (CeNA), the sugar moiety is altered to anhydrohexitol and cyclohexenyl, respectively. (b) The electronegative 2′-hydroxy substituent of RNA influences the conformation of the sugar moiety so that RNA is predominantly found in the bioactive C3′-endo conformation. The anhydrohexitol chair conformation found in HNA resembles the C3′ endo conformation of ribose.

(2′-F) or 2′-methoxy (2′-OMe) has also been carried out. These electronegative substituents induce a preference for the 3′-endo conformation of ribose (Figure 6.9) which is the structural conformation in A-type duplexes. PNA, LNA, and 2′-substituted nucleic acid analogs have all been pursued as potential drug candidates, as well as diagnostic tools.

A range of other analogs in which the ribose is replaced by six-membered ring structures includes hexitol nucleic acids (HNA) and cyclohexene nucleic acids (CeNA) (Figure 6.9). The hexitol moiety of HNA adopts an anhydrohexitol chair conformation which is a faithful mimic of the ribose C3′-endo conformation in RNA (Figure 6.9). Thus, HNA can form duplexes that are highly similar to the A-type helical structure and the bases of HNAs are oriented in an axial position leading to stronger base pairing. In contrast, CeNAs are more flexible around the C2′–C3′ bond as the cyclohexenyl moiety easily adopts different conformations. Importantly, both HNAs and CeNAs are resistant to nuclease degradation.

Generation of XNA polymers was until recently, generally achieved by chemical synthesis since most unnatural nucleotide analogs are poor polymerase substrates. Pioneering work on re-design of RNA polymerases identified enzyme variants that were capable of synthesizing XNAs nearly fully substituted with nucleotides harboring 2′-OMe modified sugars. More recently, a combination of rational design and directed evolution of DNA polymerases enabled the synthesis of eight different XNAs including HNA, CeNA, LNA, and 2′-F substituted DNA. Notably, the XNAs could be inversely transcribed into DNA and evolved into high-affinity aptamers, demonstrating that the XNAs comprise the capacity for both heredity and evolution. Subsequently, six different XNA enzymes (XNAzymes) were discovered including HNAzymes and CeNAzymes capable of cleaving and ligating RNA.

Chemical modification of nucleobases generally does not modulate stability to nucleases, but allows for functional versatility, and thus for complementary storage of genetic information. With only four different bases or nucleotides, compared to the 20 different amino acids in proteins, the functional diversity of nucleic acids is confined. In an attempt to address this, a substantial number of unnatural base pairs (UBPs) have been developed. This has been successfully achieved, for example, by reshuffling of the hydrogen-bond donor and acceptor patterns (Figure 6.10). Even though some of these lack the complementary hydrogen bonds that underlie Watson–Crick base pairing, it has been demonstrated that they can be replicated in vitro with efficiencies approaching the natural bases. Similarly, other UBPs have been efficiently incorporated and replicated in vitro, but incorporating unnatural bases into DNA in vivo is a substantial bigger challenge. A number of criteria need to be fulfilled: first, the unnatural nucleoside triphosphate building blocks have to be introduced into the cells; second, the endogenous polymerases must incorporate the unnatural building block within a complex intracellular environment; third, the UBP needs to be formed during replication enabling duplex formation and finally, the UBP must be resistant toward the endogenous DNA repair mechanisms. Remarkably, the incorporation of an unnatural base pair in *Escherichia coli* was achieved by importing the unnatural building blocks via an exogenously expressed nucleotide triphosphate transporter into cells transformed with an artificial plasmid coding for the UBP. In this way, a semi-synthetic organism with an expanded genetic alphabet was generated. This could enable the site-specific replacement of a wide range of chemical functionalities in standard DNA, thereby enhancing the functional diversity of, for example, DNA aptamers and DNAzymes.

6.4.2 RNA Interference

The first evidence for small regulatory RNAs was demonstrated with the discovery of Lin-4 which is a miRNA that controls the temporal development of the nematode *Caenorhabditis elegans*. It was later demonstrated that Lin-4-mediated gene silencing was triggered by double-stranded RNA (dsRNA) precursors that inhibited protein translation by base-pairing to complementary mRNA in a process termed RNA interference (RNAi) (Figure 6.11). Subsequently, small synthetic dsRNAs were used to demonstrate that RNAi pathways are also present in mammalian cells.

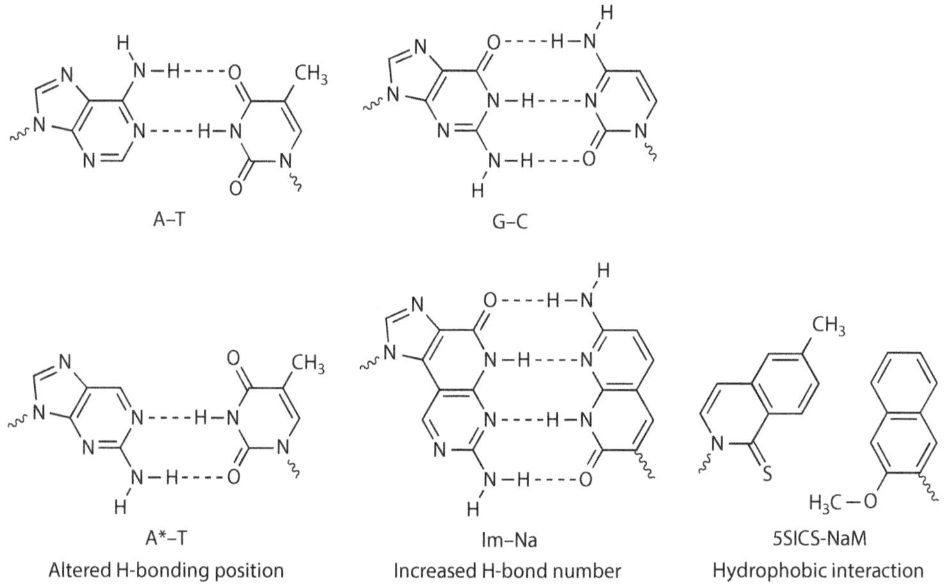

FIGURE 6.10 Three examples of unnatural nucleobase pairs (UBPs) developed to form extra base pairs beyond the two found in nature: A–T (adenine–thymine) and G–C (guanine–cytosine). The A*–T base pair involves an artificial adenine analog with altered positioning of the H-bonding donor, resulting in a novel H-bond pattern between A* and thymine. The Im–Na (Imidazopyridopyrimidine–Naphthyridine) base pair packs through four H-bonds, resulting in a more stable duplex structure than observed for the native A–T and G–C pairs. The hydrophobic 5SICS-NaM base pair does not contain any H-bonds and stacking is caused by steric complementarity.

These discoveries have had a tremendous impact on basic and applied science and opened entirely new areas of research. Today, RNAi tools have widespread applications in gene knockdown studies in cell cultures and in vivo (model organisms such as nematodes, fruit flies, and rodents) to study gene function, dissect complex biological pathways, and uncover novel therapeutic targets. In addition, RNAi have also been pursued as putative drug candidates for a range of diseases.

The therapeutic potential of RNAi was first demonstrated when small synthetic dsRNAs was used to target a specific mRNA sequence from hepatitis C virus in mice. Later, it was demonstrated that small dsRNAs, targeting a specific gene, protected from liver fibrosis in a mouse model of autoimmune hepatitis. Quickly, RNAi drug candidates, such as Bevasiranib for aged-related macular degeneration, entered clinical development (Phase I clinical trials). Although several other clinical studies using RNAi followed, to date none have made it to the market, generally due to lack of efficacy or stimulation of immune responses. The development of RNAi therapeutics is further challenged by inefficient delivery and safety issues caused by nonspecific binding or saturation of the endogenous RNAi pathway.

Disease-related genes are most commonly targeted by short siRNA duplexes (usually 18–25 nucleotides) that are either chemically synthetized and introduced into the cells or generated in vivo through the RNAi pathway from short hairpin RNAs (shRNAs) introduced to the cell via a small gene cassette (Figure 6.11). The majority of the RNAi-based drug candidates that have reached clinical trials were synthetic siRNA duplexes, but these naked siRNAs are inherently associated with poor enzymatic stability and inefficient cell permeation. Instead, chemically modified siRNAs have therefore been explored and short double-stranded nucleotides based on a LNA or 2′-OMe-DNA backbone (see Section 6.4.1) have entered clinical trials. Synthetic siRNAs have also been encapsulated into nanocarriers such as cyclodextrin and lipid nanoparticles (LNPs), both of which simultaneously protect against enzymatic degradation and improve cell penetration. The latter

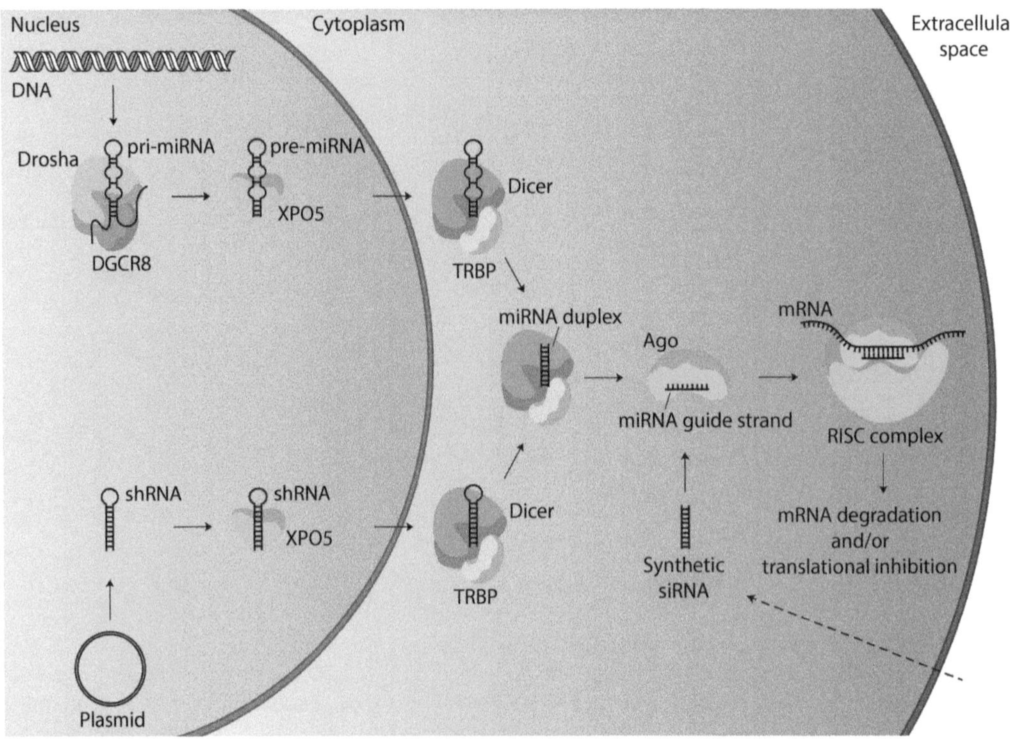

FIGURE 6.11 RNA interference (RNAi) is a cellular pathway of gene silencing. In eukaryotic cells, miRNA is generated from genome-encoded long single-stranded RNA that is folded into a dsRNA structure (pri-miRNA). The pri-miRNA is processed in the nucleus by the RNase III ribonuclease Drosha, and its cofactor DGCR8, into a shorter imperfectly base-paired hairpin structure called pre-miRNA (60–80 nucleotides). The pre-miRNA is then exported to the cytoplasm by exportin 5 (XPO5). In the cytoplasm, the pre-miRNA is further processed by Dicer, another RNase III ribonuclease, acting in association with Tar RNA-binding protein (TRBP) into the mature miRNA duplex (19–25 nucleotides). The guide strand is loaded into an Argonaute (Ago) protein that together with other proteins and co-factors form the RNA-induced silencing complex (RISC) which is guided to its mRNA target by the miRNA strand. This results in gene silencing through either translational repression or site-specific cleavage and degradation of the mRNA. Artificially designed siRNA duplexes which are used to target disease-related genes, can be delivered to the cells via small gene cassettes like plasmids that are transcribed into a short hairpin RNA (shRNA) capable of entering the RNAi pathway. Alternatively, the siRNA duplex is synthesized in vitro and delivered directly into the cytoplasm by a suitable method such as transfection, nanocarrier encapsulation, or viral gene transfer.

strategy has resulted in the development of RNAi therapeutics that has recently entered late-stage clinical trials for the treatment of transthyretin (TTR)-mediated amyloidosis, specifically TTR-mediated cardiomyopathy and TTR-mediated polyneuropathy.

Although RNAi-based therapeutics has shown great promise in recent years, a number of important challenges remain in order for widespread application of this therapeutic principle. These challenges include improvements in nuclease resistance, reduced immune activation, and development of efficient delivery methodologies.

6.5 CONCLUDING REMARKS

In this chapter, we have discussed how small molecules can be generated and can be used to probe and discover biology, both in an academic setting, but also in the initial drug design and development process. We also discussed the application of chemical biology technologies in studies of proteins and nucleic acids and how this has opened up new avenues in protein and nucleic acid engineering, paving the way for groundbreaking discoveries and new therapeutic principles. Chemical biology is a scientific discipline that has emerged primarily from chemical sciences to apply chemical tools and principles in studies of biological phenomena. However, chemical biology has evolved in recent years, to include scientists from many other disciplines, particularly those that emanate from biological sciences, making chemical biology a truly interdisciplinary playing ground. Like other interdisciplinary sciences, such as nanoscience and synthetic biology, chemical biology will most likely have an increasing impact on several aspects related to the early phases of drug design and discovery.

FURTHER READING

Chin, J.W. 2014. Expanding and reprogramming the genetic code of cells and animals. *Annu. Rev. Biochem.* 83:379–408.

Coyne, A.G., Scott, D.E., and Abell, C. 2010. Drugging challenging targets using fragment-based approaches. *Curr. Opin. Chem. Biol.* 14:1–9.

Davis, L. and Chin, J.W. 2012. Designer proteins: Applications of genetic code expansion in cell biology. *Nat. Rev. Mol. Cell Biol.* 13:168–182.

Fire, A., Nirenberg, M., and Appasani, K. 2011. *RNA Interference Technology: From Basic Science to Drug Development.* Cambridge, U.K.: Cambridge University Press.

O'Connor, C.J.O., Beckmann, H.S.G., and Spring, D.R. 2012. Diversity-oriented synthesis: Producing chemical tools for dissecting biology. *Chem. Soc. Rev.* 41:4444–4456.

Pinheiro, V.B. and Holliger, P. 2014. Towards XNA nanotechnology: New materials from synthetic genetic polymers. *Trends Biotechnol.* 32:321–328.

Pless, S.A. et al. 2015. Atom-by-atom engineering of voltage-gated ion channels: Magnified insights into function and pharmacology. *J. Physiol.* 593:2627–2634.

7 Natural Products in Drug Discovery

Guy T. Carter

CONTENTS

7.1 INTRODUCTION

Natural products are loosely defined as secondary metabolites produced by living organisms. These compounds are called "secondary" because they are generally not directly involved in the primary processes of growth and development of the host. In the majority of cases, the intrinsic biological functions of these secondary metabolites remain unknown. It has been postulated that such secondary metabolites must have a vital role, otherwise evolutionary pressure would have purged their encoding genes. One case where the role of the secondary metabolites seems reasonably clear occurs in the microbial world. Antibiotics are natural products produced by microorganisms that adversely affect other microbes, either through outright killing or inhibition of growth. The producer microbe maintains a resistance mechanism that renders it insensitive to its own antibiotic. Although the production of antibiotics for self-defense or to eliminate competition for resources seems reasonable, there are still many unanswered questions about this process.

Natural products are conveniently classified into families of compounds according to their biosynthetic origin. For the purpose of this chapter, we will consider the most prevalent biosynthetic types as examples. Representative examples of the major biosynthetic classes are illustrated in Figure 7.1. Perhaps the most biogenetically diverse family of natural products is the alkaloids, originally described as basic compounds found predominantly in plants. Alkaloids have varied biosynthetic

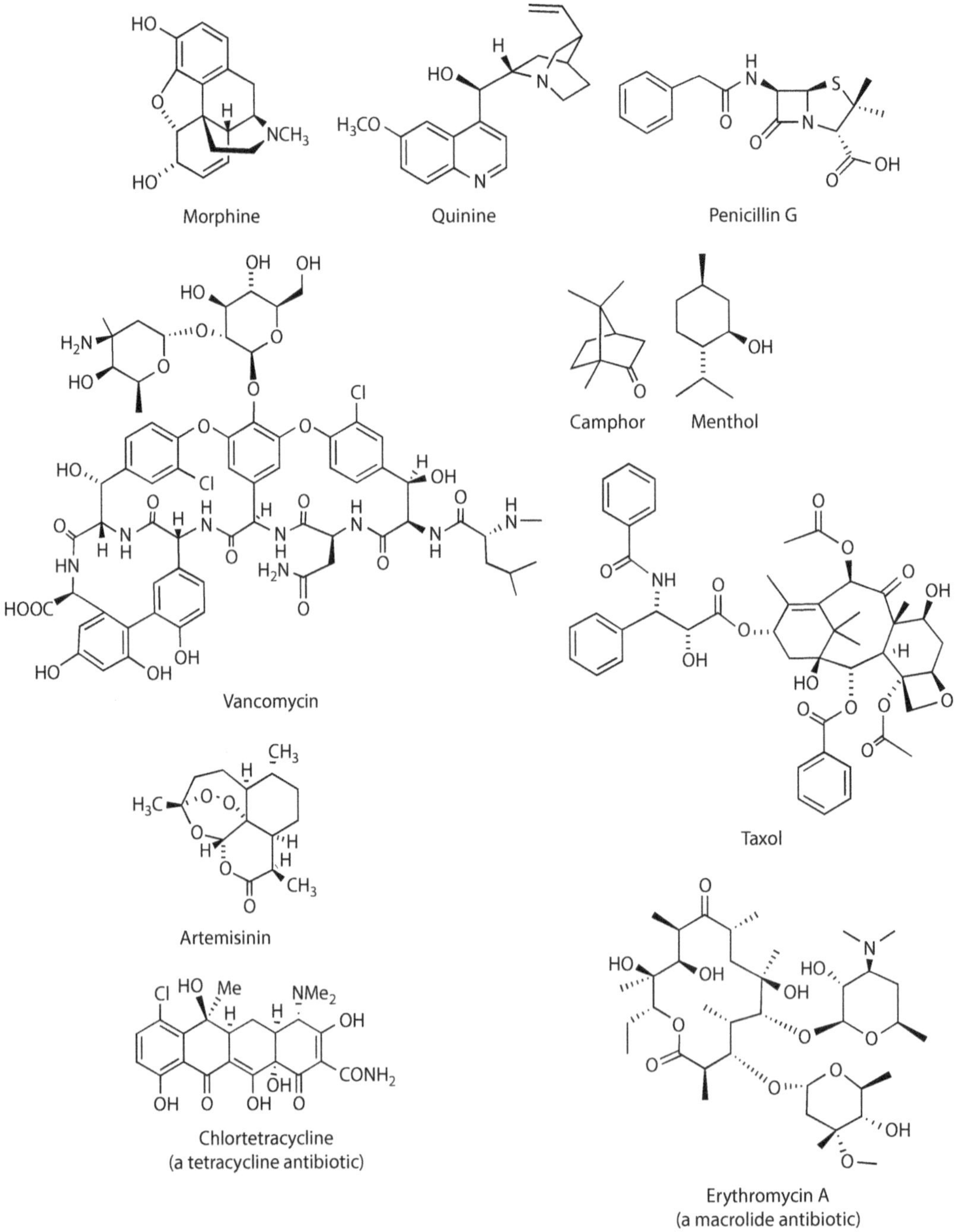

Morphine

Quinine

Penicillin G

Vancomycin

Camphor

Menthol

Taxol

Artemisinin

Chlortetracycline
(a tetracycline antibiotic)

Erythromycin A
(a macrolide antibiotic)

FIGURE 7.1 Examples from the major classes of natural products.

origins, frequently based on an amino acid core; the unifying structural aspect remains the existence of a basic nitrogen function. Examples of well-known alkaloids that have been used in medicine are morphine (from the opium poppy, *Papaver somniferum*), and quinine (from *Chincona* species).

Closely related to alkaloids are the peptides, which are composed largely of amino acids, often with unusual structural modifications. These peptides are characteristically much smaller than proteins, generally less than 2000 Da in molecular weight, and are often not assembled on the ribosome.

As shown in Figure 7.1, the powerful antibiotics penicillin and vancomycin are representatives of nonribosomally synthesized peptide natural products that have great utility in modern medicine.

Terpenoids comprise a highly diverse class of secondary metabolites, whose members are constituted by combinations of five-carbon units. These segments are commonly referred to as "isoprenoid units" that are biogenetically derived from one of two alternative pathways, either through mevalonic acid or deoxyxylulose phosphate. Traditionally, these compounds have been characterized as monoterpenes (C10), e.g., menthol and camphor; sesquiterpenes (C15), e.g., artemisinin; diterpenes (C20), e.g., paclitaxel (Taxol); and so on to sesterterpenes (C25), and triterpenes (C30). The obvious structural diversity is derived from cation-induced rearrangements of the nascent isoprenyl chain to form a large variety of cyclic frameworks.

Polyketides are the result of an exceptionally versatile biosynthetic pathway that assembles polyfunctional compounds by sequential condensation of small carboxylic acid units, followed by a variety of other steps, such as reductive processing, "tailoring" reactions, and cyclization. The range of structural diversity of polyketides is so vast that it is difficult to summarize with only a few examples, so for illustration two of the best known antibiotic polyketides, tetracycline and erythromycin, are shown in Figure 7.1.

These biosynthetic origins remain basically the same across the phylogenetic spectrum; however, the distribution of various pathways is highly dependent on the type of organism. Polyketides and nonribosomally synthesized peptides, particularly those with antimicrobial activity have primarily been isolated from bacteria and fungi. Higher plants are historically the most prolific sources of terpenes and alkaloids. Since higher plants are readily accessible they were the first sources to be explored for medicinal properties.

7.2 HISTORICAL PERSPECTIVE

Plants have been used as medicines for centuries, according to folklore, often without any perceptible efficacy. Plant materials were processed for medicinal use by chopping and grinding into powders or through aqueous extraction to make teas, smoking, or chewing. Often mixtures of plant products were combined ostensibly to create the most beneficial medicines. While the majority of these products were not effective, it was accepted practice, and other than faith healing there was no real alternative.

Despite the difficulties encountered by administering mixtures of plant products in these crude preparations, several have yielded medicinally useful products upon purification. Opium, a dried concentrate of the milky latex derived from the poppy, *P. somniferum*, has been used for thousands of years as an analgesic. Recreational smoking of opium for its euphoric effect became popular in Europe in the early nineteenth century. Morphine (see Figure 7.1), which is the major active alkaloid in opium, is a powerful pain medicine, but owing to its addictive properties is mainly prescribed for the management of pain in terminal illnesses.

Malaria is an infectious disease caused by protozoan parasites of the genus *Plasmodium*. Symptoms of malaria have been treated with a number of plant preparations. Cinchona bark was originally discovered to have antimalarial properties in South America in the early 1600s and was soon imported to Europe where it was widely prescribed. Chemical investigations in the early 1800s by Pelletier and others led to the isolation of purified alkaloids possessing the antimalarial properties of the plant materials. Quinine (Figure 7.1), the major active principle obtained in this work, soon became the preferred treatment for symptoms of malaria and was manufactured through large-scale isolation from the bark. The advent of substituting purified chemicals for crude plant preparations marked a turning point for pharmaceutical discovery. Quinine continues to have practical use in the treatment of certain resistant forms of *Plasmodium falciparum*.

In Chinese traditional medicine, *Artemisia annua* or qinghao has been employed as an antimalarial agent for many centuries. In this case, the sesquiterpene lactone artemisinin (Figure 7.1), containing a rare peroxide bridge, was isolated from the plant material and shown to possess effective

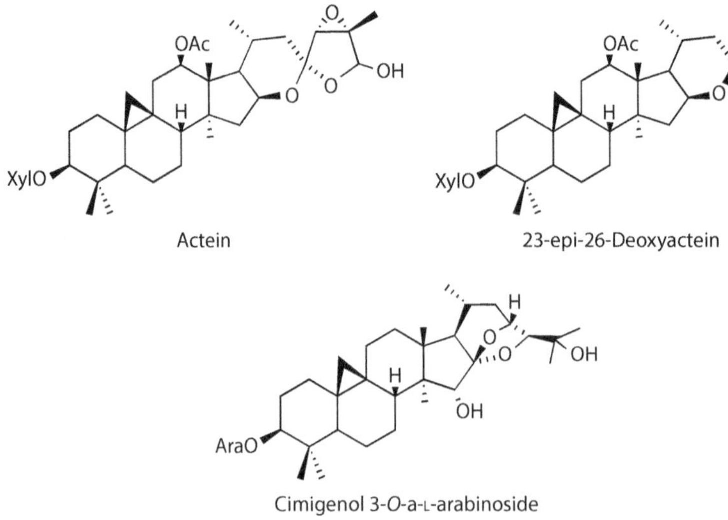

Actein

23-epi-26-Deoxyactein

Cimigenol 3-*O*-a-L-arabinoside

FIGURE 7.2 Some common triterpenoid constituents of black cohosh.

antimalarial activity. Owing to the development of resistance to synthetic antimalarial drugs, artemisinin has become an effective alternative therapy. The yield of artemisinin in *A. annua* is relatively low making its commercial production an expensive process. Considering that the majority of people suffering from malaria live in underdeveloped regions of the world, a practical and cost-effective method for the production of artemisinin is highly desirable. This situation is a prime example of what has become known as the "supply issue" for the practical production of natural products. The issue is focused on the difficulty of producing commercial quantities of complex natural products.

Herbal remedies continue to play a significant role in human medicine. Chemical investigations have identified many of the active principles in many commonly used products. These products are often sold as dietary supplements rather than ethical pharmaceutical products. Because these products are complex mixtures of many natural products, there is a need to establish criteria for their standardization. This situation is complicated by the natural variation in secondary metabolites produced by closely related species of medicinal plants. Owing to the possibility that variations in the composition of the products will result in unpredictable potency, the herbal products industry has been developing quality control standards. Black cohosh, for example, which is taken for the relief of menopausal symptoms, has a number of signature triterpenoid constituents, including actein, 23-*epi*-26-deoxyactein, and cimigenol-3-*O*-arabinoside (Figure 7.2). These compounds can be identified by coupled high-performance liquid chromatography—mass spectrometry (LC/MS), in comparison with authentic reference compounds. Our ability to quantify the amount of the key constituents in these herbal products will facilitate a better understanding of their efficacy and permit greater confidence in their use in medicine.

7.3 ANTIBIOTICS

The development of antibiotics for the treatment of bacterial infections, which was critical during the war years of the early 1940s, changed the course of drug discovery efforts in the pharmaceutical industry. Following the pioneering experiments of Selman Waksman at Rutgers University on soil actinomycetes, pharmaceutical companies began the systematic evaluation of antibiotics produced by bacteria isolated from the soil. During the succeeding quarter century, often referred to as the "Golden Age" of antibiotic discovery, all of the major classes of life-saving antibiotics were found.

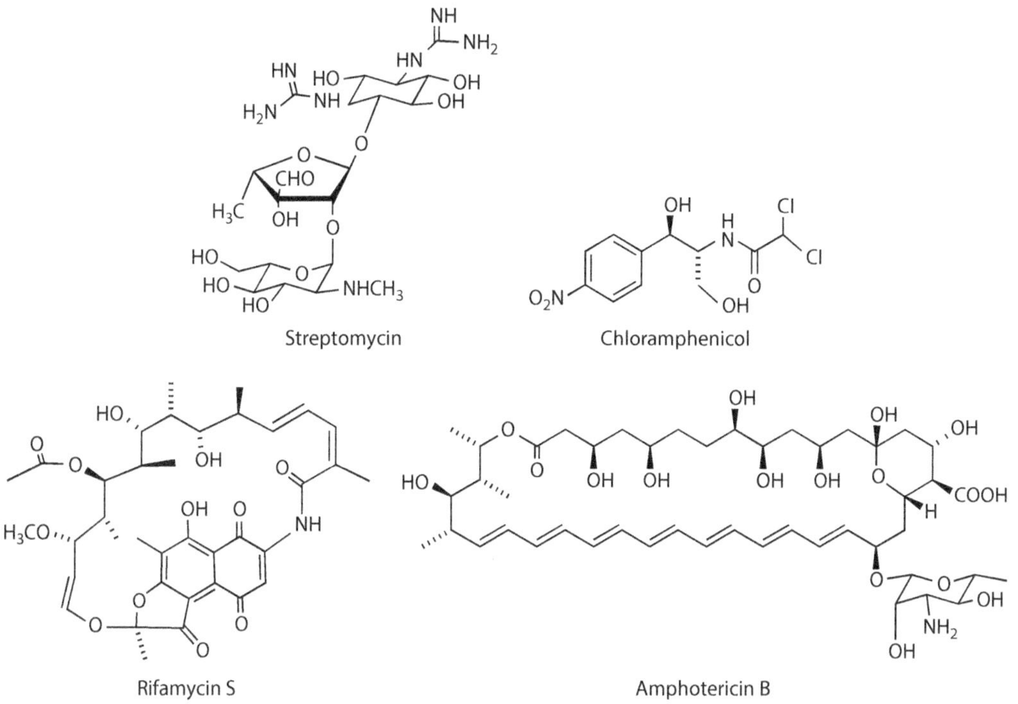

Streptomycin

Chloramphenicol

Rifamycin S

Amphotericin B

FIGURE 7.3 Additional examples of antibiotics found during the "Golden Age of Antibiotics Discovery."

The previously illustrated penicillin, vancomycin, tetracycline, and erythromycin represent the progenitors of the most important classes of antibacterial agents still in use today. Other examples of these wonderfully complex compounds are illustrated in Figure 7.3. Streptomycin, isolated by Waksman from *Streptomyces griseus*, was the first of the class of aminoglycoside antibiotics to be introduced into therapy. Chloramphenicol, rifamycin, and amphotericin were also discovered during this time and each has a valuable niche in modern chemotherapy.

In addition to antibiotics used for the treatment of microbial diseases, microbial products have been explored for a number of other therapeutic uses. Owing to the relative ease with which new organisms could be isolated from the environment and grown in culture, these provided versatile sources of new chemistry. Beginning in the 1960s these sources were employed for screening against other diseases, such as parasitic and fungal infections, as well as for the ability to differentially kill cancer cells. Notable among the antiparasitic compounds discovered in this way are the milbemycins. These polyketide-derived macrolides, produced by *Streptomyces* species, are exceptionally effective against several types of parasites that infect livestock. Compounds in the milbemycin class, e.g., ivermectin and moxidectin (Figure 7.4) have also found utility against the devastating human disease of river blindness caused by filarial worms, which is endemic to sub-Saharan Africa, and other tropical areas of the world.

Actinomycete-derived antibiotics with efficacy as anticancer agents have also been a major focus of screening programs (see Figure 7.5). Waksman once again discovered the first of these, actinomycin D, from *Streptomyces parvullus*. Today, actinomycin has quite a limited use, but it served as a prototype for the discovery of other antitumor antibiotics. Doxorubicin, which interacts similarly with DNA, was isolated in the 1960s and remains an important component of typical chemotherapy regimes. Another early discovery from the Golden Age that remains in use today for chemotherapy is bleomycin. Bleomycin is a complex glycopeptide antibiotic produced by *S. verticillus* that induces DNA damage through oxidative reactions.

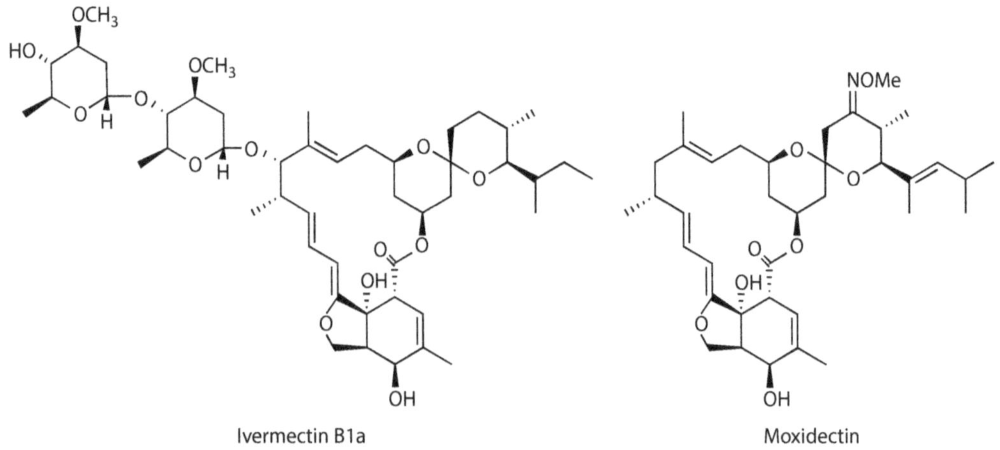

Ivermectin B1a

Moxidectin

FIGURE 7.4 Antiparasitic milbemycin analogs.

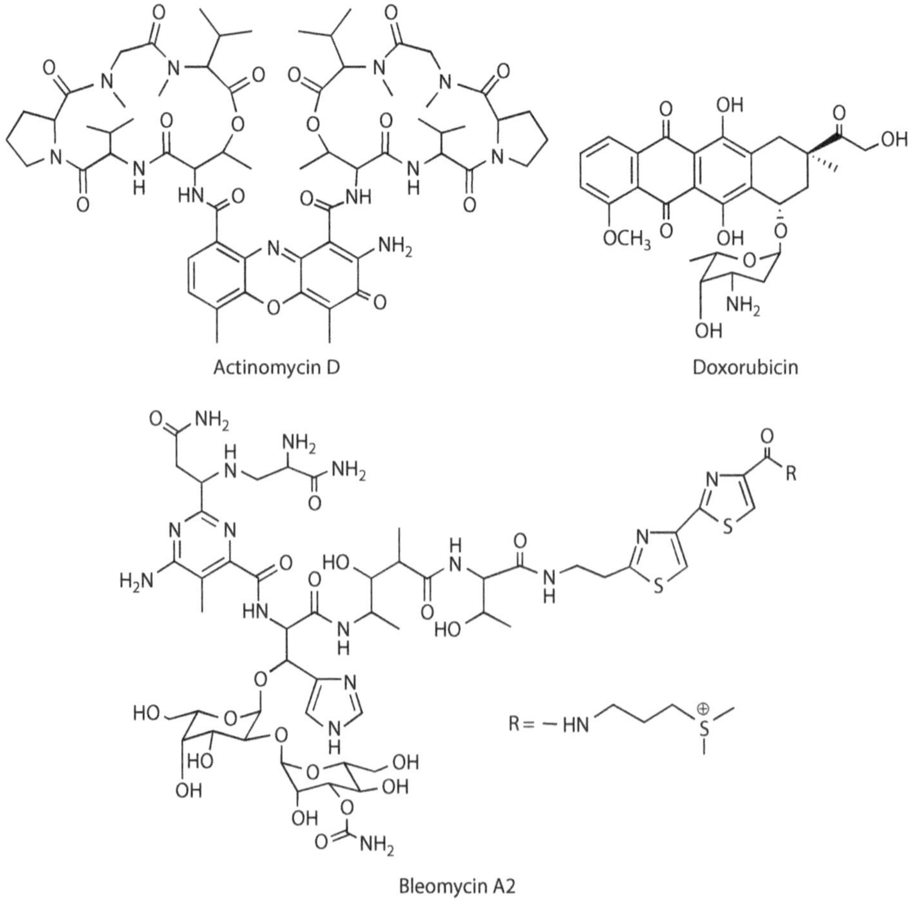

Actinomycin D

Doxorubicin

Bleomycin A2

FIGURE 7.5 Cytotoxic antibiotics.

7.4 SCREENING

7.4.1 General Concepts

In the most general sense, screening refers to the process of investigating sources of compounds that exhibit a particular type of property or biological activity. In this chapter we explore ways of using natural products for drug discovery, and therefore the "investigations" are typically linked to a biological assay. A positive response in the assay (a "hit") is determined by the intrinsic potency of a given compound and its concentration in the screening sample. The sources of natural products used in the screening process can be quite diverse, ranging from bacterial products to higher plants and animals; however, the processes involved are similar. Once a sample, which is typically an extract of an organism, or a part of an organism (e.g., fermentation broth, fruiting body of a mushroom, leaves or roots of plants, etc.) has shown a positive response in a given assay, the process of "bioassay-guided fractionation" begins. This process is shown as a loop diagram in Figure 7.6. Resolution of the active principle(s) in these materials is a highly experimental process. The ease of resolution is dependent upon such parameters as the concentration of the active compound in an extract, the overall constitution of the extract, in terms of interferences (e.g., tannins, fatty acids and other lipids, and complex carbohydrates), as well as the chemical properties of the compound of interest. Trial and error is the operational mode of these processes and is highly dependent on the preferences and experience of the individual investigator. As indicated in Figure 7.6, the original crude material is initially split into fractions by a rough process such as differential solubility in solvents with different polarities, or by liquid–liquid partitioning between immiscible phases, usually aqueous versus organic. Subsequent steps are generally of higher resolution often with different forms of chromatography, perhaps using a normal phase high-capacity technique like silica gel chromatography in organic mobile phases first, followed by a reversed phase system with a hydrophobic stationary phase eluted with an aqueous-organic mixture. It is usually the case that a suite of structurally related compounds is isolated in this process, each having some activity in the bioassay of interest. Subtle differences in the potency or selectivity shown by these congeners form the basis for the "natural structure–activity relationship" of the series that may be useful in designing improved compounds by synthetic or biosynthetic methods during subsequent optimization of the lead. Once a compound is shown to have the activity of interest and passes a criterion of purity, it can proceed for resolution of its chemical structure and further biological evaluation.

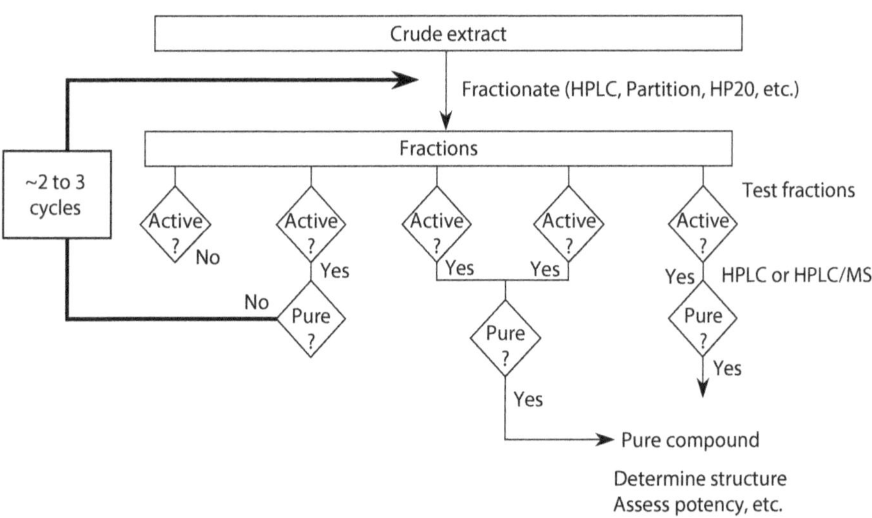

FIGURE 7.6 Bioassay-guided fractionation of natural products.

7.4.2 Whole Organism Screening

Often referred to as "phenotypic screening," as well, the screening of natural products in whole organisms was historically the original mode, and one that still has a great deal of potential. One of the most robust examples of this approach is screening for antimicrobial activity by inhibiting the growth of selected strains of bacteria on seeded agar plates. A positive response in such an assay is observed as a zone of inhibition of growth around a test sample deposited on the Petri plate. In most cases, the diameter of the zone of inhibition is directly related to the concentration of the antibacterial agent in the test sample, so that some degree of quantification of the response is also possible in this simple test. Such assays provide no indication of the mode of action of the active principle, but are generally quite effective tools for following fractionation processes that lead to pure compounds. A crucial extension of this in vitro whole organism assay is another phenotypic test in which the isolated material is administered to rodents that have been challenged with a lethal infection of the target bacterium. The positive end point of the in vivo rodent test is survival beyond that of the control animals. Positive results in the in vivo model of infection provides the critical information that the compound has sufficient drug-like properties to penetrate the normal xenobiotic defenses of the host animal and reach the target population of infecting bacteria.

There are many such whole organism models that have been used for drug screening. Among the simplest of these are those related to infectious diseases, including the aforementioned antibacterial system with various classes of pathogenic agents, as well as those designed for antifungal, antiviral, and antiparasitic agents. In the quest to find new effective agents against cancer, animal models of disease remain a mainstay of the process. Similar models are the norm for advancing the development of drugs in terms of understanding the efficacy, tolerability, metabolism, and long-term effects. These systems are rarely used for high-throughput screening of crude natural products because they require substantial resources to maintain. Therefore, most live animal models are used to verify the efficacy of compounds that have been isolated with the aid of a simpler in vitro assay.

As will be discussed in the following section, screening against isolated target biomolecules, such as enzymes or receptors is now favored for high-throughput screening operations. In the case of screening mixtures of natural products, however, the whole organism approach offers tremendous advantages. The discovery of a novel secondary metabolite that confers a positive response in a whole organism screen provides the opportunity to discover a new target, and potentially a new mechanism of action. In current parlance, these studies are often referred to as "chemical biology" or "chemical genetics" (see also Chapter 4). Specifically, in forward chemical genetics, a small molecule, in our case a natural product, is employed to probe for their cellular targets. Typical experiments include the creation of affinity binding reagents or affinity matrices that include the small molecule of interest and these systems are used to fish out target macromolecules from cellular components. Molecular targets for rapamycin and geldanamycin were found by such methods. Once the target macromolecules are verified, additional mechanistic studies are developed to understand the relationship between the binding partners and the disease process.

7.4.3 Target-Based Screening

Molecular biology has provided the tools to engineer and produce macromolecular targets of drug action. If it is believed that the inhibition of a particular cell-signaling process will mediate the development of disease, then the isolated enzyme or receptor that is responsible for the signaling can be used as a target for screening. Alternatively, a selective whole cell screen can be employed which is designed to respond by providing some measurable signal as a result of the interaction with a particular target. Owing to developments in automation for such assay systems, hundreds of thousands of compounds can be conveniently tested for activity in a short period of time. Natural product mixtures may also be tested in these highly automated systems. With mixtures, particularly crude extracts, there is the potential for significant interference with the assay. Such interferences include nonspecific inhibition of targets by ubiquitous classes of natural products like fatty acids, or the presence of a highly potent cytotoxic agent that kills the host cell designed for the specific assay. These issues are not insurmountable, but require diligence in evaluating screening results. Some solutions for these issues are presented in Section 4.5.

7.4.4 Dereplication

Dereplication is a term that when used by natural products chemists refers to the rapid identification of a compound (or class of compounds). How is this different from the usual process encountered in the isolation of natural products? It is different in that the process refers to the identification of expected (or nuisance) compounds. These nuisance compounds will vary depending upon the particular assay system that is being followed, and therefore it takes some experience with a given bioassay to identify the classes of interfering compounds. Before the advent of high-throughput LC/MS systems, specific tests were developed to identify the nuisances. In the early days of antibiotic discovery paper chromatography was used, as were thin layer chromatography with specific detection, and liquid chromatography with diode-array detection, and so on. This process has been greatly expedited by the use of LC/MS such as the system diagrammed in Figure 7.7. The system is based upon the separation of a mixture by reversed phase HPLC, the continuous recording of UV/visible absorption spectra and mass spectra throughout the chromatogram, and the correlation of these data with biological activity. Once the active wells in the bioassay plate are related to a retention time, the optical and mass spectral data are correlated and used for querying suitable databases that provide matches of known compounds. Once an active compound is identified in this way, one can simply rely on the chromatographic and spectral data for dereplication of future samples.

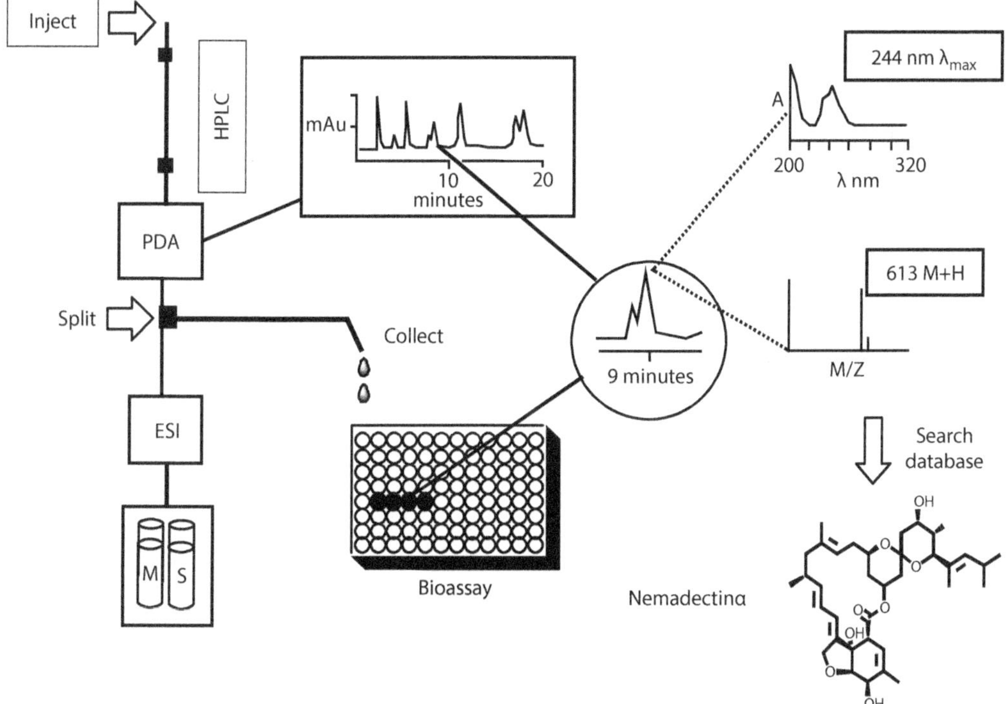

FIGURE 7.7 Schematic of HPLC/UV/MS system for dereplication of natural products.

7.4.5 Screening Improvements

In order to enhance the quality of screening data generated from highly valuable natural products, the nature of the extracts can be improved. There are a number of approaches to this problem, but the common goal has been to simplify the materials being tested. When the offending materials are well understood, such as the presence of tannins in plant extracts, pretreatment of the crude extracts may suffice to remove the tannins from the screening samples. A broader approach that is applicable to samples derived from a variety of sources is diagrammed in Figure 7.8. This HPLC-based method seeks both, to remove offending nonspecific materials, as well as simplify the actual screening samples. As diagrammed in Figure 7.8, the natural products are concentrated, often in the form of a solvent extract, and then subjected to separation by reversed phase HPLC. The compounds are separated employing gradient elution and only the material contained in the shaded area is retained

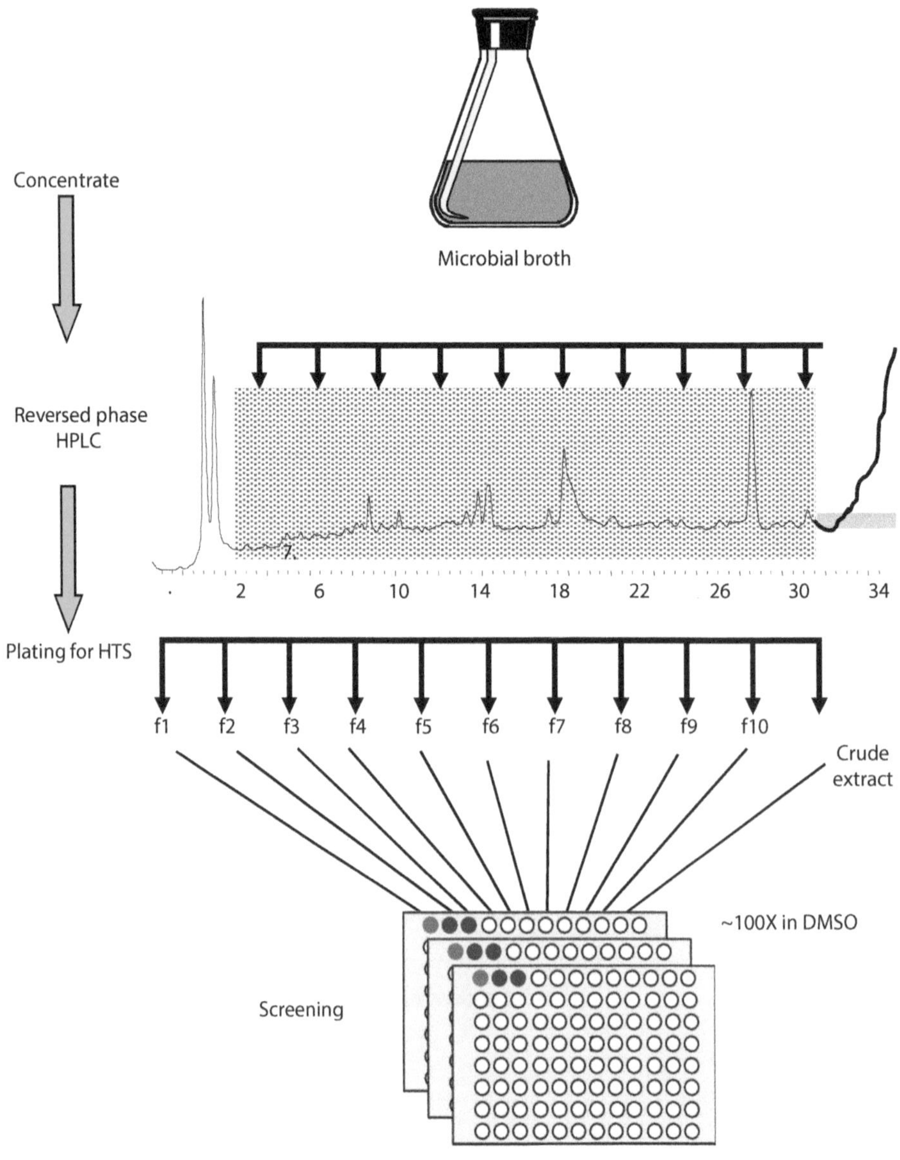

FIGURE 7.8 Prefractionation of natural products to generate improved samples for screening.

for testing. This area contains components with polarities consistent with drug-like properties, having eliminated the early-eluting highly polar compounds such as saccharides and amino acids and the highly retained lipophilic materials that elute at the end of the run. The area is divided into 10 fractions that are concentrated and plated for use in high-throughput screening.

Fractionated screening samples derived from chromatographic separations such as indicated in Figure 7.8 offer several advantages over crude extracts. The major one is the ability to independently evaluate diverse components produced by a particular source organism. In cases where one component in an extract is toxic, its adverse effect on the test organism in a phenotypic assay may obscure the positive response of a second component. Actinomycetes are particularly notable in their ability to produce numerous families of compounds. For example, it has been well documented that *Streptomyces* species that produce the milbemycin class of macrolides, such as the previously mentioned nemadectin, generally also produce oligomycins. The oligomycin macrolides are known respiration inhibitors that are highly toxic to eukaryotic cells. Therefore, if one were screening in a rodent model for antiparasitic activity, a crude extract containing both the milbemycin and oligomycin would likely only show the toxicity. Whereas if the components were resolved chromatographically prior to screening, the cryptic antiparasitic effect of the milbemycin would be observed.

Apart from unmasking activities, there are benefits that accrue from enhancing the concentration of minor components present in a complex mixture. This is particularly true if in the preparation of fractionated screening samples the effort is made to normalize the concentration of the samples. The benefits of obtaining the maximum positive responses from these samples, often representing precious material collected under unique conditions, argue for expending the extra effort required.

In the course of resolving "hits" from natural product screening through bioassay-guided fractionation, as shown in Figure 7.6, it was emphasized that this is an empirical and highly experimental process, each new extract requiring an individual strategy for the isolation of its biologically active principles. If one has prefractionated the extract prior to initial screening, and the activity falls into a neat cluster of fractions, then one has valuable information on how to begin the purification process. This information will facilitate the resolution of the hit and lead to greater efficiency of the entire isolation and purification process.

7.5 OPTIMIZATION OF NATURAL PRODUCT LEADS

Nature has preserved the ability of a given organism to make these fascinating secondary metabolites, although their inherent biological roles remain obscure. As scientists seek to co-opt these metabolites as medicinal agents, attempts are typically made to enhance their pharmaceutical effectiveness. Such enhancements may be to improve the spectrum of activity against a range of targets, as in the case of antibiotics where broad-spectrum activity in inhibiting the growth of both gram-positive and gram-negative bacteria is important. In other cases, it may be crucial to enhance the specificity to a narrower range of targets, such as the ability to selectively inhibit a particular kinase reaction in a signaling cascade. Furthermore, it may be necessary to improve the drug-like properties of a natural product lead. Here, improvements in solubility, chemical stability in biological matrices or metabolic stability may be crucial. These and a host of other reasons drive the process to make structural modifications of the core natural product, which may be effected by chemical or biosynthetic means.

7.5.1 SEMISYNTHESIS

Semisynthesis refers to the process of performing synthetic chemical transformations starting with a natural product, for the purpose of enhancing the pharmaceutical performance of the natural product. This approach has been most effectively used with complex microbial products, owing to the ready availability of the starting material through fermentation of highly productive variants of the parent organism. The challenge in these experiments is twofold: one, to achieving adequate selectivity in the chemical process and two, the subsequent purification of reaction mixtures. A good example

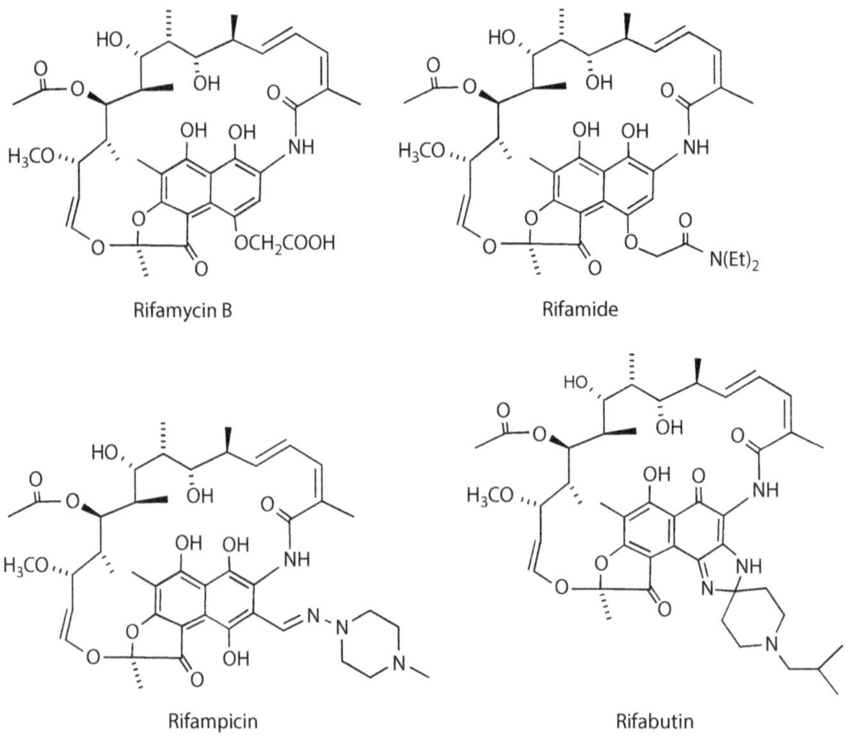

Rifamycin B

Rifamide

Rifampicin

Rifabutin

FIGURE 7.9 Rifamycin B and semisynthetic analogs.

of the successful application of semisynthesis is in the case of the rifamycin antibiotics, shown in Figure 7.9. Rifamycin B is the originally isolated natural product derived from *Nocardia mediterranei*. Rifamycins have potent activity against gram-positive bacteria and are of greatest importance to inhibit the growth of the tuberculosis causing organism *Mycobacterium tuberculosis*. Rifamycin B was only modestly effective when administered to infected animals and this led to investigation of derivatives in search of improved potency. Greater potency was achieved through substitutions on the aromatic portion as exemplified by rifamide, rifampicin, and rifabutin. The latter two compounds continue to be important drugs for the treatment of re-emerging epidemics of tuberculosis.

7.5.2 Improvements in Natural Products through Total Synthesis

Although the total synthesis of natural products has been the forte of many prominent academic laboratories, only a few totally synthetic analogs of natural products have been introduced into commerce. The continued development of efficient and selective synthetic methods could provide alternative supply routes for simpler natural products of the future. Regardless of the issue of practical scalability, total synthesis enables the production and testing of analogs that often illuminate key features of the structure that are critical for biological activity. Paul Wender's research on the bryostatins, potent cytotoxic principles isolated from marine invertebrates, illustrates some of the key insights that can be revealed through total synthesis.

7.5.3 Biosynthetic Modifications

Genetic engineering of biosynthetic pathways to create specific modifications in the chemical structure of secondary metabolites is now a practical reality in bacterial systems. In the simplest cases, a single enzymatic function is eliminated by inactivating the respective gene, resulting in an

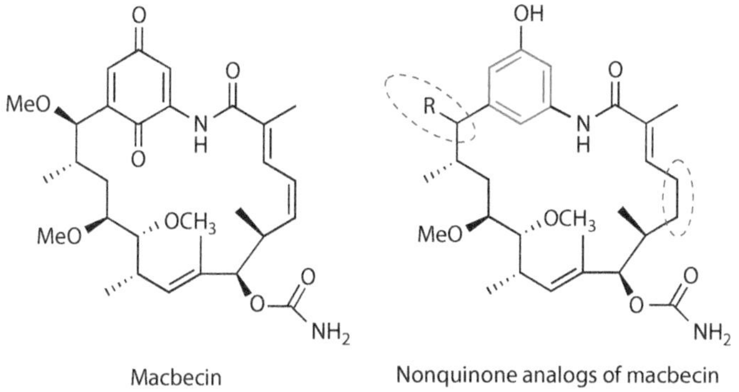

Macbecin Nonquinone analogs of macbecin

FIGURE 7.10 Macbecin and nonquinone analogs derived from the knock out of a key oxidative function.

altered product. One such example is shown in Figure 7.10 from the work of scientists at Biotica Technologies, Cambridge, UK, where the oxidation state of the aromatic ring in the ansamycin antibiotic macbecin is reduced from quinone to phenol. This was accomplished by inactivation of the gene *macM*, which was found through genetic analysis to code for the specific enzyme responsible for addition of the *para* oxygen to the phenol ring that is further oxidized to the quinone. The macbecins are promising HSP-90 inhibitors with potential in cancer chemotherapy whose off-target effects have been linked to the reactive quinone moiety. The new products lack this reactive unit and are expected to have reduced side effects. This precise alteration in structure was made possible by the identification of the functions associated with the key genes found in the macbecin biosynthetic gene cluster.

7.5.3.1 Mutasynthesis

Another technique that relies upon the knockout of an enzymatic function is known as mutasynthesis. In mutasynthesis, a key step in a biosynthetic sequence is knocked out such that no product is made without the addition of a suitable precursor. In the past, these processes were done by random mutagenesis followed by screening of the resultant mutants for the desired phenotype. Today, it is a straightforward process to obtain the fully annotated genetic map of a biosynthetic pathway and to specifically design experiments to knock out the targeted function. One such example is shown in Figure 7.11 for the microbial product rapamycin. This work was pioneered by Peter Leadlay at the University of Cambridge, England who mapped the biosynthetic gene cluster for rapamycin. As illustrated, knock out of the gene rapL results in the organism's inability to make pipecolic acid which is the usual amino acid incorporated into the rapamycin macrocycle. Supplementing the fermentation medium of the knockout strain with alternative cyclic amino acids, such as substituted proline analogs, results in efficient incorporation of these units yielding selectively modified rapamycin analogs.

7.5.3.2 Polyketide Synthase (PKS) Engineering

In the case of polyketide-derived compounds, the biosynthetic modules that are responsible for the iterative addition of two-carbon units to the nascent chain can be exchanged within sequences to alter both the substitution and oxidation state of the resultant unit. The most studied case of this biosynthetic class is erythromycin. The polyketide assembly of this macrolide antibiotic is illustrated in Figure 7.12, where three multifunctional enzymes DEBS 1, 2 and 3, encoded by *eryAI, II* and *III* genes, assemble a starter unit propionate residue with six propionate extender units to produce the substituted polyketide chain. The chain is indicated in the figure growing as the successive condensations add the propionate units. The resulting keto groups are reduced to alcohols by ketoreductase functions (KR, modules 1, 2, 5, and 6), not reduced at all as in module 3 (note the lack of a reductive loop), or fully reduced to the bare methylene by ketone reduction, enolization, and

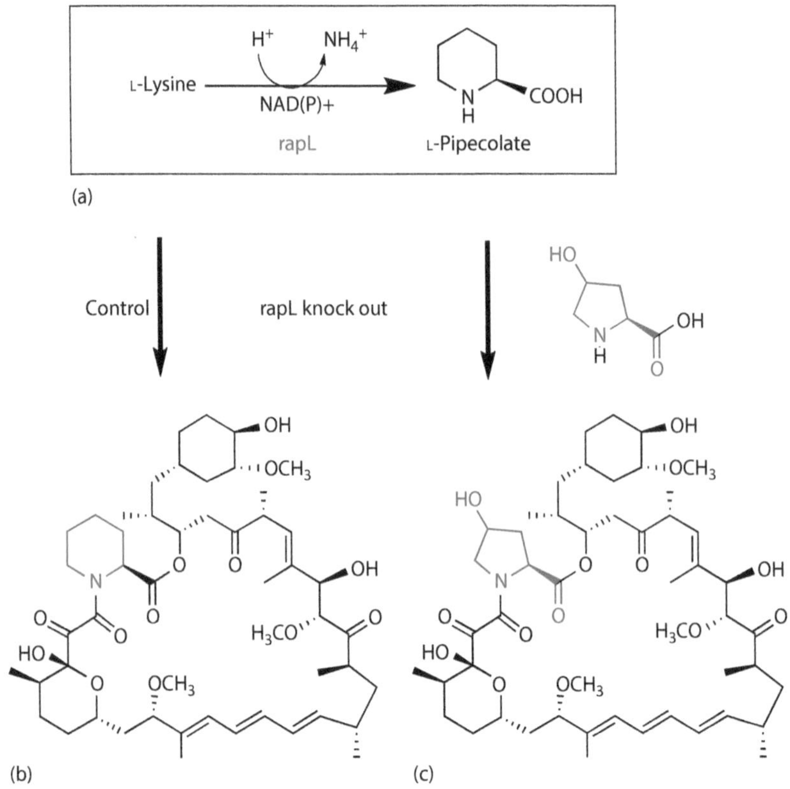

FIGURE 7.11 Mutasynthesis of rapamycin analogs. (a) Conversion of lysine to pipecolic acid mediated by rapL. (b) Incorporation of pipecolate in control strain leads to rapamycin. (c) Feeding 4-hydroxyproline to the rapL knockout stain yields a novel analog.

hydrogenation (KR, ER, DH, module 4). The chain is terminated and cyclized through the action of the final active site in DEBS 3, the thioesterase (TE). Additional tailoring enzymes further modify 6-deoxyerythronolide B, by oxidation, glycosylation, and methylation processes to yield the fully functionalized erythromycin A.

In order to modify the alkyl substitution of the macrocycle, one would alter the specificity of the acyltransferase moiety (AT) that is responsible for recruiting these units to join the growing chain. As shown in Figure 7.13, by swapping the genes that encode for the propionyl group to be added in module 4 for another AT domain that is specific for an acetyl group, it is possible to encode for an erythromycin analog that lacks methyl substitution at position 6. In theory, similar modifications in alkyl substitution and oxidation state can be engineered for the other positions on the macrocycle.

7.5.4 Structure–Activity Relationships

Such alterations in structure, whether they are done by chemical synthesis or biosynthetically, result in variations in the biological properties of the compound in comparison with the parent. Such structure–activity relationships or SAR reveal the areas of a given structure type that are optimal for driving the potency or selectivity of the series. A very simple example of SAR observed in naturally occurring congeners is illustrated in Figure 7.14 for the antibiotic mannopeptimycin. The mannopeptimycins are ordinarily produced as a mixture of components, some of which contain an isovaleryl ester group on the terminal unit of the di-mannose side chain. The bioassay data in the

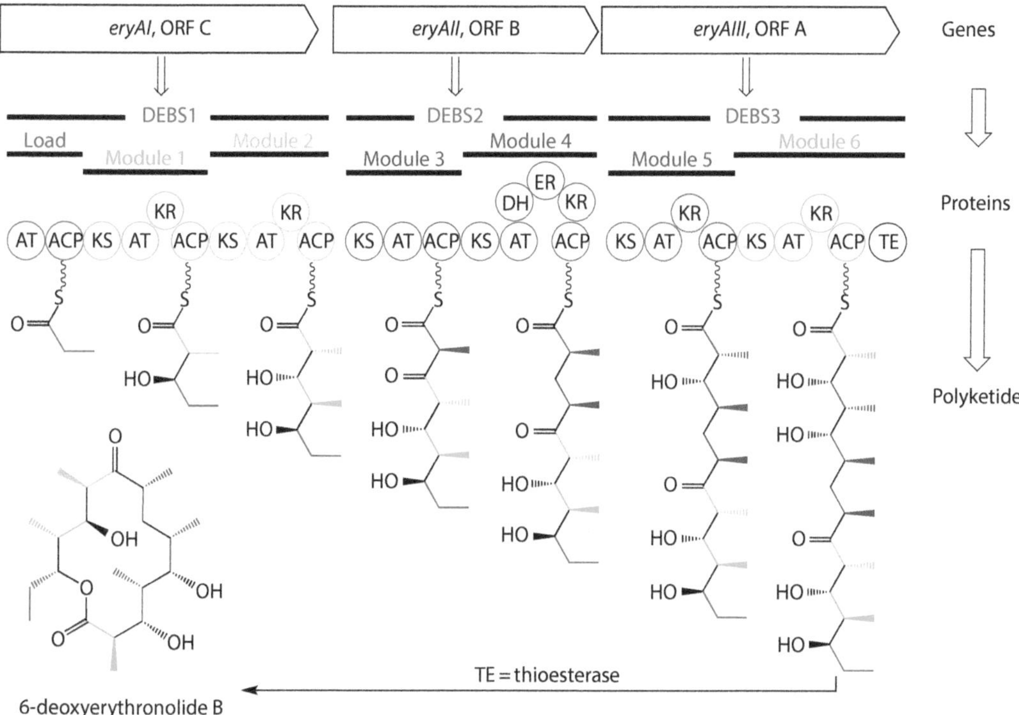

FIGURE 7.12 Polyketide pathway for the biosynthesis of erythromycin.

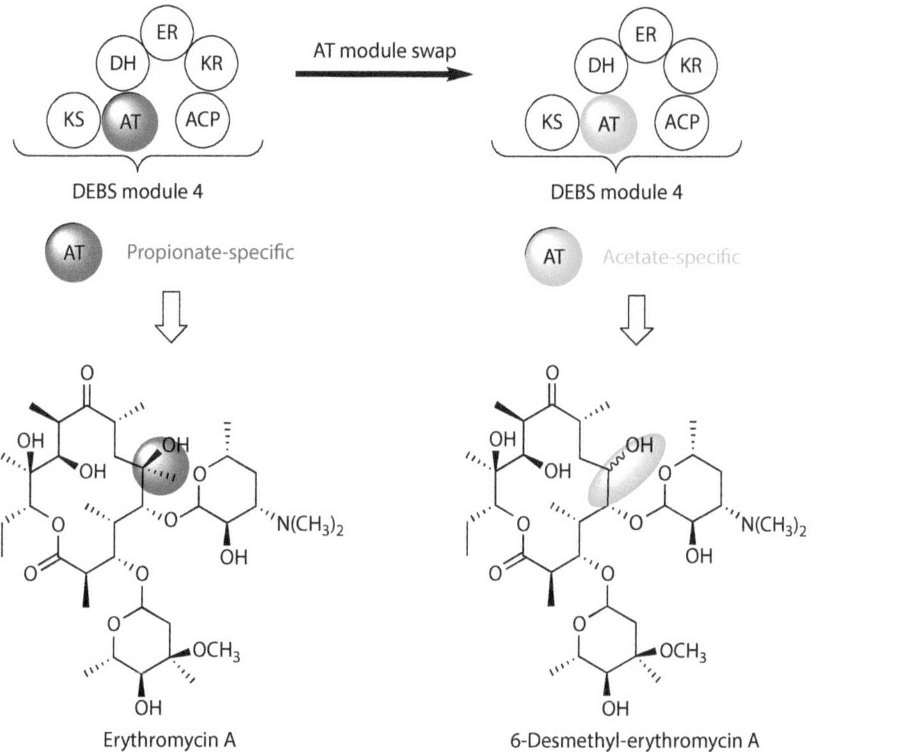

FIGURE 7.13 Substitution of an acetate-specific AT for the propionate-specific AT in DEBS module 4 leads to 6-desmethyl analogs.

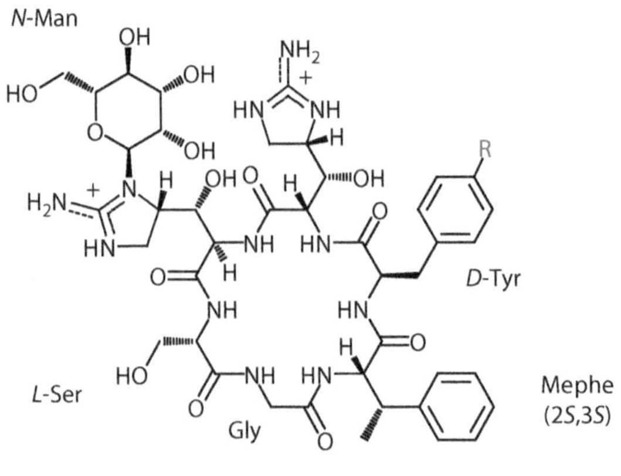

Congener	R Group	MIC (μg/mL) *S. aureus*	ED_{50} (iv, mg/kg) Mouse, *S. aureus*
α	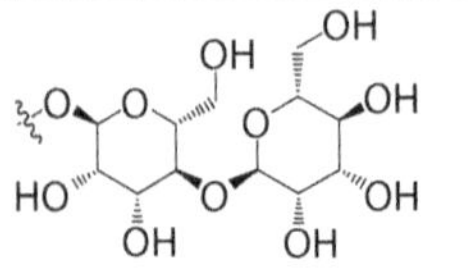	>64	20
γ	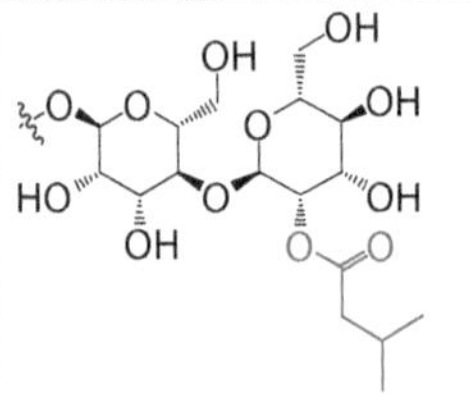	8	3.5
δ	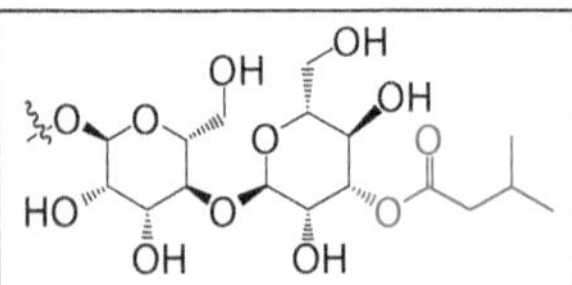	4–8	2.6
ε	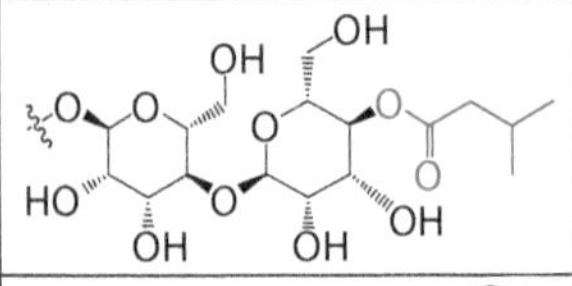	4	0.6
708	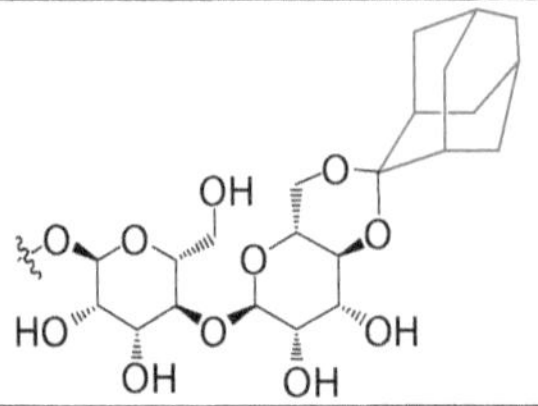	0.5–1	0.04

FIGURE 7.14 Structure–activity relationships in the mannopeptimycin antibiotics.

table are measures of antibiotic effectiveness. Minimal inhibitory concentration values (MIC) are measures of antibiotic potency in vitro that are determined by serial dilution of a solution of antibiotic substance to the point where no inhibition of growth is obtained. This value is reported in terms of a concentration, in this case in micrograms per milliliter of the final test solution. The ED_{50} value refers to the potency of the compound against an infection induced in a mouse model. The lowest concentration of the dose that was still effective in protecting the mouse is shown. The units are in milligrams of antibiotic per kilogram weight of a mouse. What we are able to discern from these data is that the ester function is responsible for conferring a great deal of the potency to the compounds, as alpha is considerably less potent than any of the esterified components. In addition, the position of the ester group on the terminal mannose unit also has a significant effect on potency with the epsilon component having the ester at the 4-position being the most potent. These data provide key insight for the design of semisynthetic and biosynthetically derived analogs in a lead optimization program. In the case of mannopeptimycin, this initial natural SAR led to the semisynthesis of numerous lipophilic derivatives on the terminal di-mannose moiety. One of the most potent is shown as compound 708 in Figure 7.14. This compound is one of a series of cyclic acetals that showed remarkably enhanced potency as well as presented excellent chemical and metabolic stability.

7.6 CONCLUDING REMARKS

One of the most exciting developments in the study of secondary metabolites, not touched on in this chapter, is the sequencing and annotation of bacterial genomes. Genome mining has revealed a plethora of potential chemistry waiting to be revealed. Challenges also await as scientists endeavor to elucidate the mechanisms that regulate the expression of these cryptic pathways. Issues of supply of these precious materials have frequently plagued the efforts of drug discovery from higher organisms, such as marine sponges or plants. But there is hope. Jay Keasling's work on engineering of a high-producing terpene cyclase pathway in bacteria promises to enable large-scale economical production of the antimalarial drug artemisinin and thereby unlock the supply chain for these lifesaving chemicals.

Natural products remain a fascinating and incredibly rich source of leads for drug discovery. Owing to developments in chemical and biosynthetic technologies, the moment is right for the exploration for new chemistry and further exploitation of known secondary metabolites. Advances in molecular biology will enable experiments aimed at a more fundamental understanding of the intrinsic biological roles for secondary metabolites and this knowledge can be expected to illuminate future applications in drug research.

FURTHER READING

Chang, M.C.Y. and Keasling, J.D. 2006. Production of isoprenoid pharmaceuticals by engineered microbes. *Nat. Chem. Biol.* 2:674–681.

Dewick, P.M. 2009. *Medicinal Natural Products: A Biosynthetic Approach*, 3rd edn. West Sussex, U.K.: John Wiley & Sons Ltd.

Harvey, A.L., Edrada-Ebel, R., and Quinn, R.J. 2015. The re-emergence of natural products for drug discovery in the genomics era. *Nat. Rev. Drug Discov.* 14:111–129.

Hesse, M. 2002. *Alkaloids: Nature's Curse or Blessing.* Weinheim, Germany: Wiley-VCH.

Hopwood, D.A. 2007. *Streptomyces in Nature and Medicine: The Antibiotic Makers.* New York: Oxford University Press.

Koehn, F.E. and Carter, G.T. 2005. The evolving role of natural products in drug discovery. *Nat. Rev. Drug Discov.* 4:206–220.

Osbourne, A., Goss, R.J., and Carter, G.T. (eds.). 2014. *Natural Products: Discourse, Diversity and Design.* Oxford, U.K.: John Wiley & Sons, Inc.

Weissman, K.J. and Leadlay, P.F. 2005. Combinatorial biosynthesis of reduced polyketides. *Nat. Rev. Microbiol.* 3:925–936.

Wender, P.A., Baryza, J.L., Hilinski, M.K., Horan, J.C., Kan, C., and Verma, V.A. 2007. Beyond natural products: synthetic analogues of bryostatin 1. In *Drug Discovery Research: New Frontiers in the Post-Genomic Era*, ed. Z. Huang, pp. 127–162. Hoboken, NJ: John Wiley & Sons, Inc.

8 In Vivo Imaging in Drug Discovery

Jesper L. Kristensen and Matthias M. Herth

CONTENTS

8.1 INTRODUCTION

At an early stage of the drug discovery process, it is important to advance the evaluation of potential drug candidates from in vitro to in vivo investigations. The initial development of compounds is guided by their ability to interact with the chosen biological target using parameters such as affinity and efficacy, in combination with physiochemical parameters such as solubility and lipophilicity.

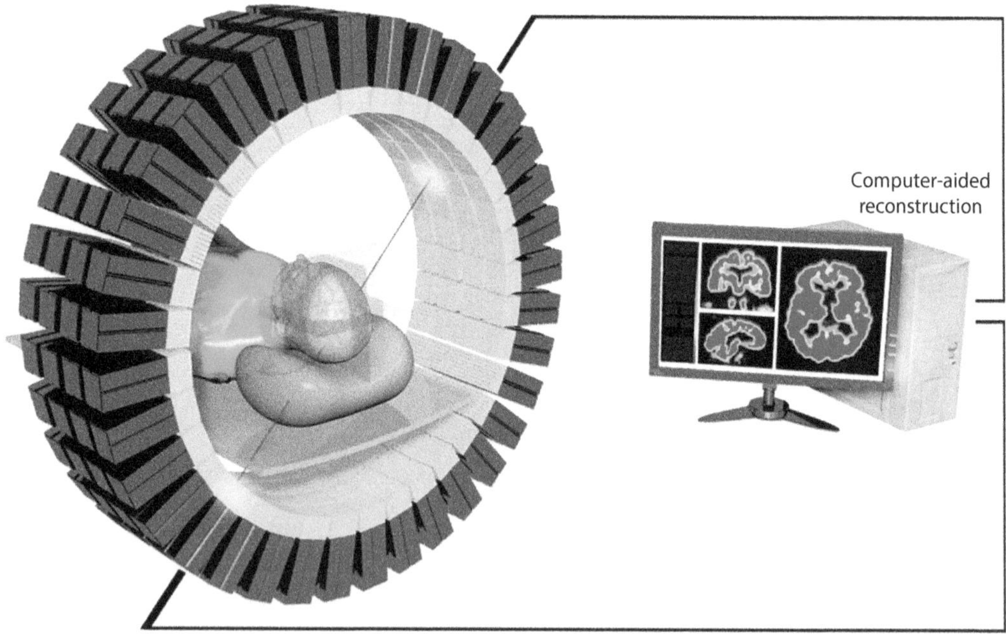

FIGURE 8.1 Positron emission tomography (PET): illustration of an acquisition process. (Reprinted with permission from Piel, M., Vernaleken, I., and Rösch, F., Positron emission tomography in CNS drug discovery and drug monitoring, *J. Med. Chem.*, 57, 9232–9258. Copyright 2014 American Chemical Society.)

Although a compound may show a very favorable profile in vitro, the ability of the given molecule to reach and engage the desired target in vivo is very difficult to predict, given the complexity of an entire human being. Thus, it is crucial to be able to investigate and quantify the distribution of a compound within a complex biological system when evaluating and selecting compounds that is to be advanced further in a drug discovery program. Figure 8.1 illustrates a setup that can be used to image in vivo biochemical parameters or pharmacological effects of a specific drug using a technique called positron emission tomography (PET).

8.2 IN VIVO IMAGING

A number of different anatomical imaging techniques—most notably magnetic resonance imaging (MRI) and computed tomography (CT)—are available for generating three-dimensional reference pictures (often referred to as a map or atlas) of an animal or human being. These techniques provide anatomical information such as size and localization of organs. In contrast, functional imaging technologies can provide more detailed information about particular drug targets like receptors, ion channels, transporters, and enzymes. For example, biochemical parameters such as receptor availability, enzymatic reaction rates, and metabolic pathways can be visualized and quantified in vivo. All these parameters are very important when selecting compounds to be advanced in the drug discovery process.

Positron emission tomography (PET) can be used to determine these parameters in a noninvasive and reliable way. It is highly sensitive, reproducible, and can be used to detect organ accumulation accurately, regardless of tissue depth, and finds widespread use in the drug discovery and development process. Single photon emission computed tomography (SPECT) offers some of the same possibilities but is less versatile. For that reason, this chapter will focus on PET and its application in drug discovery and development.

8.3 IN VIVO IMAGING WITH PET: CONCEPTS AND METHODS

In PET, a ligand is labeled with an unstable, radioactive isotope which emits a detectable amount of radiation. When this radioactive ligand is injected into a living being, the distribution of that ligand can be followed via the detection of the radiation in a noninvasive way. If the radioligand binds selectively to a biological target, the interaction of the radioligand with its target, for example, a receptor, can be visualized. The primary information that can be gained from such experiments is the biodistribution of the radioligand–target complex in a timely fashion (dynamic PET data). However, secondary biochemical parameters such as in vivo binding affinities or reaction rates can be quantified via simple kinetic modeling models of the dynamic PET data.

In general, there are two ways to utilize molecular imaging in the drug discovery and development process: Either a PET ligand (often referred to as a tracer) can be used to investigate a specific target directly, such as a receptor or enzyme (direct approach) or a tracer can be used to determine secondary effects like the proliferation rate of a tumor before and after treatment (functional response studies).

8.3.1 Direct Approach

The direct approach often employs competition studies where the binding of a PET ligand is challenged with a novel drug (blocking studies). This can be done with another ligand (normal block) or the unlabeled version of the PET ligand (self-block). The outcome of such an intervention study provides information about the in vivo selectivity of the drug, but can also be used to study a drug's receptor occupancy. This is helpful to determine the maximum and optimal clinical dose. The receptor occupancy can be determined from the relative changes in the binding potential with increasing doses of the prospective drug molecule. Thus, the direct approach can be used to study the ability of a specific drug to engage a particular receptor or transporter in vivo. Pharmacokinetic and pharmacodynamic parameters of the potential new drug molecules can also be evaluated.

8.3.2 Functional Response Studies

This approach relies on a test–retest PET study of the same subject before and after treatment. The PET tracer is not necessarily targeting the same enzyme or receptor as the compound used in the investigation. Usually, established PET ligands like [^{18}F]2-fluoro-2-deoxy-*d*-glucose (FDG) are used to determine a functional response.

8.4 POSITRON EMISSION TOMOGRAPHY (PET)

8.4.1 Basic Principles

The basic imaging principle in PET makes use of the unique decay characteristics of positron emitting radionuclides: a neutron-deficient isotope converts a proton into a neutron with subsequent emission of a positron (β^+ particle). Once emitted, this positron travels up to a few millimeters until it encounters an electron—typically from an adjacent water molecule. Upon contact, a positron and an electron merge into a positronium which annihilates almost instantaneously into γ-photons moving in opposite directions. The coincident detection of several of these photon pairs in dedicated scanners form the basis of PET imaging, since computational reconstruction along straight lines between detector pairs allows the determination of the photon's source of origin in a three-dimensional space (Figure 8.2).

An example of how a PET image may look like is shown in Figure 8.3, which displays a PET image of phosphodiesterase 10A, a drug target for several neurological diseases in the CNS.

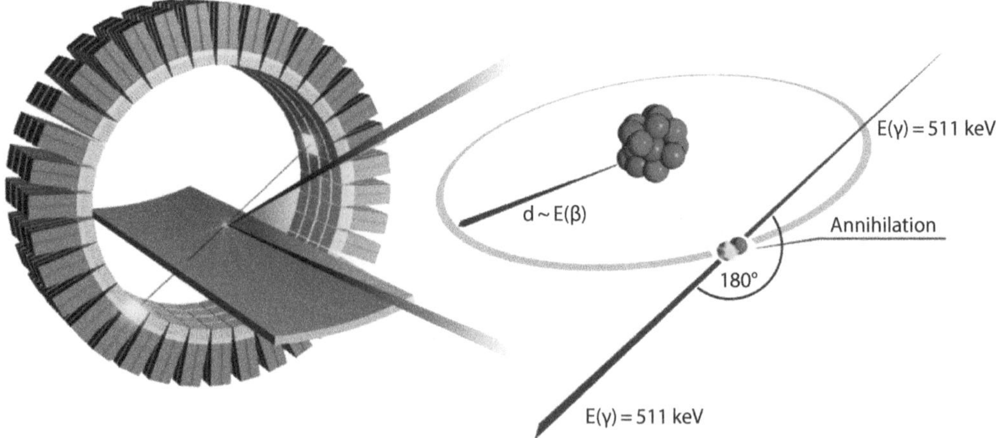

FIGURE 8.2 Basic physical principle of positron emission tomography (PET). Coincident detection of numerous photon pairs (after annihilation) in a PET scanner provides the raw data for the PET images. (Reprinted with permission from Piel, M., Vernaleken, I., and Rösch, F., Positron emission tomography in CNS drug discovery and drug monitoring, *J. Med. Chem.*, 57, 9232–9258. Copyright 2014 American Chemical Society.)

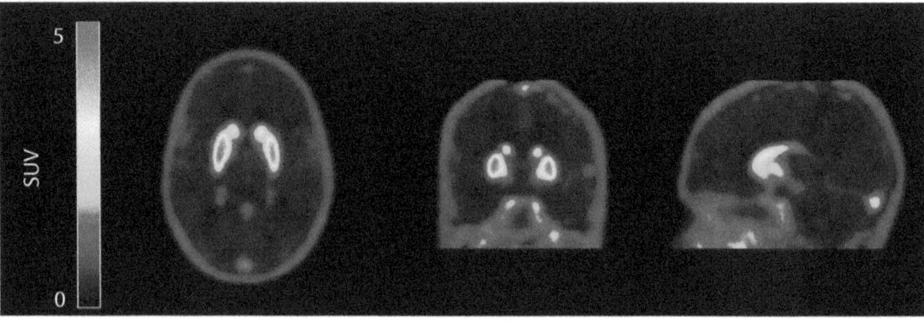

FIGURE 8.3 PET image of [^{11}C]Lu AE92686 in the human brain visualizing phosphodiesterase 10A. This research was originally published in JNM. (From Kehler, J. et al., *J. Nucl. Med.*, 55(9), 1513, 2014. Figure 6. Copyright by the Society of Nuclear Medicine and Molecular Imaging, Inc.)

The PET ligand used in these experiments [^{11}C]Lu AE92686 was developed in the pharmaceutical industry to be able to quantify the occupancy of potential drug candidates.

8.4.2 General Requirements for a PET Ligand

The ideal PET ligand has to fulfill other criteria than those that apply to drug molecules. The efficacy of a drug (i.e., its ability to evoke a pharmacological/biological response) is crucial for its success, whereas the relative affinity/selectivity (i.e., a compound's ability to bind to a given target in preference to others) is much more important for a PET ligand used for imaging of a specific target. Drugs that interact with several targets may be very effective in the clinic, but for imaging purposes, the ability of a PET ligand to bind to a single target is important to simplify the interpretation of the final images. An effective drug does not necessarily make a good PET ligand and vice versa. Therefore, separate approaches have to be taken when developing compounds into either drugs or PET ligands.

8.4.3 Nonspecific and Unspecific Binding

In general, a ligand's nonspecific binding is a crucial factor in PET ligand development since it has a great impact on the contrast in the final PET images. Nonspecific binding should not be confused with unspecific binding. Nonspecific binding refers to the compound's propensity to bind to membranes, proteins, lipids, or other cell components without a specific and selective target in a nondisplaceable way, whereas unspecific binding refers to interactions with other well defined targets (e.g., receptors or enzymes). If a given region of interest that contains the targeted receptor or transporter is saturated with uniformly distributed radioligand, it is not possible to detect minute changes in the binding to the target. In fact, nonspecific binding is one of the main reasons why ligands with a promising in vitro profile fail to be successful PET ligands in vivo.

Nonspecific binding is usually evaluated in a blocking experiment. Here, the binding of the radioligand is challenged with a pharmacological dose of a compound that is known to reach and interact with the target. If it is not possible to displace the binding of the radioligand, further development is usually discontinued. Nonspecific binding is often correlated with the lipophilicity of the ligand and the octanol/water distribution coefficient at physiological pH (log $D_{7.4}$) can be used as a rough indicator. For PET ligands used in neuroimaging, a log $D_{7.4}$ in the range of 2–3 is considered to be optimal, but there are several successful PET ligands that do not meet that criteria and many radioligands within this range still show high nonspecific binding.

8.4.4 In Vivo Selectivity and B_{MAX}

Even if a PET ligand is not completely selective toward a given target, it may still be effective as a PET ligand as the relative receptor density (B_{max}) of the target in different regions is an important parameter to consider. The B_{max} value is a measure of how many receptors are present in a given region. Low selectivity of a compound toward a certain receptor may be compensated by a high B_{max} value of the desired target in a particular region. Thus, the observed PET images are a function of the relative affinity of the ligand toward a target and the B_{max} value of that target in a specific region. Usually, PET ligands should be at least 10-fold selective when the relative selectivity and receptor density are taken into consideration.

8.4.5 Tracer Principle

In PET studies using the direct approach, a ligand interacts with the target in question, but it is important to keep in mind that the radioligand is given in very small doses—usually <5 μg for human applications. Thus, the ligand will not be able to influence any physiological process—i.e., to evoke a pharmacological response via the activation of a receptor or inhibition of an enzyme, as the injected dose is typically ~1000-fold lower than the pharmacological dose. This means that it is possible to investigate the function of drug targets, like a transporter in a cell membrane, without disturbing the function of that transporter, since PET is sufficiently sensitive to detect trace amounts of a labeled compound. This concept is often referred to as "the tracer principle." In order to make this possible, it is important that the radioligand is produced with a high specific activity (A_s), which is determined as the ratio between labeled and unlabeled ligand.

8.4.6 Dosimetry

A test subject involved in a PET experiment is exposed to potentially harmful ionizing radiation. Dosimetry relates to the quantification and risk assessment of that exposure. The radiation burden for a PET scan depends on the amount of injected radioactivity, the radionuclide itself and the tissue in which the radioactivity is released. One should strive to keep the exposure to an absolute minimum. The dosage is measured in Sievert [Sv]. [^{18}F]FDG is the most frequently used PET tracer in nuclear medicine and has an effective radiation dose of 6–8 mSv. For comparison, the effective dose from background radiation when living in Europe is in the range of 2–7 mSv/year. Radiation dosage for a chest X-ray is ca. 0.02 mSv and for a CT scan of the chest 6.5–8 mSv.

8.5 SELECTION OF RADIONUCLIDES: SHORT- OR LONG-LIVED ISOTOPES?

A number of parameters need to be considered before a suitable nuclide can be selected for the task at hand, for example:

1. The half-life ($T_{1/2}$) of the nuclide in relation to the kinetics of ligands interaction with the target (How long does it take for the ligand to locate and bind to the target; minutes or hours?)
2. The β^+-range: The distance from the originally decayed radionuclide to the annihilation point, which depends on the energy of the emitted positron, which in turn depends on the nature of the parent radionuclide. The β^+-range influences the spatial resolution of the final PET images; the lower the range the higher the resolution.
3. The β^+-branching ratio, which will influence the radiation burden of the patient: a lower branching ratio will lead to a higher radiation burden. It can be seen as the percentage of the emitted radiation that adds to the final image.
4. The target localization (Will the PET ligand have to pass over cell membranes or perhaps even the blood–brain barrier? This can influence the choice between nonmetal and metal-based ligands).

Table 8.1 summarizes some important physical properties of commonly applied positron-emitting nuclides.

PET nuclides can be divided into short-lived (minutes to 2 hours) and long-lived (several hours to days), as well as into nonmetal and metal nuclides. In the following, the advantages and disadvantages of these nuclides will be discussed.

TABLE 8.1
Selected Radionuclides for PET Imaging

Nuclide	Half-Life ($T_{1/2}$)	Branching Ratio (%)		Mean β^+-Range in H_2O (mm)	Specific Activity (GBq/µmol)
^{11}C	20.4 minutes	β^+	99.8	1.1	20–100
^{18}F	110 minutes	β^+	96.9	0.6	200–1000
^{68}Ga	68 minutes	β^+	89.1	2.9	Carrier-free[a]
^{64}Cu	12.7 hours	β^+	17.8	0.6	Carrier-free[a]
^{89}Zr	78.4 hours	β^+	22.7	1.2	Carrier-free[a]

Note: $T_{1/2}$ is the amount of time required for a radioactive quantity to decay to half of its original value; the branching ratio is the fraction of particles which decay by an individual decay mode with respect to the total number of particles which decay; the β^+-range specifies the distance until the β^+-particle annihilates with an electron and is thus one factor that controls the spatial resolution of a PET tracer.

[a] Carrier-free indicates that the nuclide is essentially free from stable isotopes of the element in question (carrier). In reality, it is almost impossible to achieve carrier-free samples, but for radiometals the level is assumed to be very low.

8.5.1 ^{11}C AND ^{18}F

The short-lived positron emitting radionuclides that have had the greatest impact in PET are ^{11}C and ^{18}F. Carbon is a ubiquitous building block of life, and the most abundant isotope of carbon (^{12}C) can be substituted for ^{11}C without influencing the bioactivity of the molecule. Thus, the number of ligands that could be ^{11}C-labeled is in principle unlimited. Fluorine containing compounds are not abundant in nature, but popular in drug discovery due to its ability to engage in hydrogen bonding and influence the metabolic rate of compounds. Nevertheless, known ligands may often have to be modified and re-evaluated before an ^{18}F-moiety can be introduced and the ligand be applied to PET studies.

The short half-life of ^{11}C is both an advantage and a challenge. Chemical modifications have to be conducted as fast as possible. A rule of thumb suggests that a radioactive synthesis should not exceed 2–3 half-lives which in the case of ^{11}C provides 1 hour for all the necessary manipulations. Thus, the final application has to be in close proximity to the site of synthesis and the labeled product cannot be transported to other facilities. On the other hand, the half-life is long enough to investigate many biological processes like drug–receptor interactions. In addition, test–retest experiments are feasible in 1 day using the same animal.

^{18}F has a significant longer half-life (110 minutes) than ^{11}C (20 minutes), providing additional time to perform more complex synthetic manipulations and biological experiments with the final PET ligand. Additionally, it is possible to synthesize the radioligand at one facility and subsequently transport it to another site. ^{18}F possesses a relatively low positron energy resulting in a mean β^+-range of 0.6 mm in water, which leads to PET images with high spatial resolution in combination with a relatively low radiation burden for the patient. All of these factors taken together means that ^{18}F is very useful and very widely applied in PET.

8.5.2 ^{68}GA, ^{64}CU, AND ^{89}ZR

In contrast to ^{11}C and ^{18}F, which typically are covalently linked to the PET ligand, metal nuclides rely on chelating groups that have to be attached to the target molecule. These entities bind a single metal ion by two or more separate coordinate bonds (see Figure 8.4 for an example). The coupling of the chelator to the original ligand obviously influences the pharmacodynamic and pharmacokinetic behavior and ultimately, changes the affinity, selectivity, biodistribution and metabolism of the compound, but in some cases, these changes are tolerated. ^{68}Ga-DOTA-Tyr3-octreotide, a somatostatin analog targeting neuroendocrine tumors, is a good example of a metal-based tracer, see Figure 8.4.

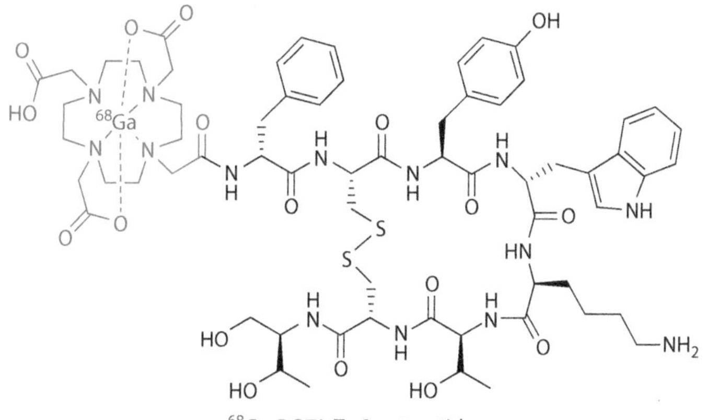

^{68}Ga-DOTA-Tyr3-octreotide

FIGURE 8.4 Example of ^{68}Ga-PET ligand. The DOTA-chelator (in red) is appended to the active peptide (in black) without disturbing its ability to interact with the desired target. The PET ligand is synthesized simply by adding a solution of $^{68}Ga^{3+}$ to the peptide-DOTA precursor.

Chelator approaches are often applied when labeling large molecules like peptides, proteins, antibodies, and nanoparticles, where the ligand's active site can be separated from the chelator via a spacer group. Many small molecules do not accommodate such manipulations and metal-based PET ligands usually cannot be applied within CNS research as they do not cross the blood–brain barrier. Compared to ^{18}F and ^{11}C, ^{68}Ga-based ligands have an inferior β^+-branching ratio, spatial resolution, and dosimetry. Nevertheless, ^{68}Ga-based ligands are routinely applied in PET scans. The popularity of ^{68}Ga stems from the fact that it is readily accessible via a simple generator system that easily can be set-up in a standard laboratory, whereas the use of ^{11}C and ^{18}F requires access to a cyclotron—a particle accelerator requiring very extensive dedicated infrastructure and know-how.

The application of longer-lived metal nuclides like ^{64}Cu and ^{89}Zr is also chelator-based in analogy to ^{68}Ga-based ligands. The advantages and disadvantages discussed before are also valid for these nuclides. The main advantage of ^{64}Cu and ^{89}Zr is the longer half-life, but their β^+-branching ratios are substantially lower meaning that a major fraction of the decay of these nuclides does not contribute to the PET image. This results in a higher radiation burden for the patients.

In conclusion, shorter-lived isotopes are preferable with respect to dosimetry. But many important targeting processes like the accumulation of monoclonal antibodies are slow (usually up to several days) and cannot be imaged with short-lived nuclides. Therefore, the choice of nuclide is strongly dependent on the biological process that is investigated, target localization, the nature of the ligand, and the pharmacokinetics of that ligand's interaction with the target.

8.6 LABELING CHEMISTRY

Besides the obvious decaying characteristics, there is a fundamental difference between the synthesis of radioactive materials and classical organic synthesis. In the synthesis of PET ligands a very large excess (×10^6) of the precursor is reacted with a minute amount of radioactive material (like [^{11}C]MeI or [^{18}]F^-, see below). Consequently, radioactive reactions usually follow pseudo-first reaction order kinetics whereas standard organic or inorganic reactions often follow binary reaction kinetics. Therefore, it is often necessary to substantially alter and modify standard reaction conditions when applied to radioactive procedures. In addition, special precautions have to be considered while working with radioactive material, and there is a strong focus on automation in the development of radiochemical procedures to limit exposure and increase reproducibility.

8.6.1 ^{11}C-Labeling

One of the biggest challenges when labeling molecules with ^{11}C is the short half-life of 20.4 minutes. ^{11}C is primary accessible from cyclotrons as [^{11}C]CO_2 or [^{11}C]CH_4. All of the chemical transformations known for CO_2 can in principle be applied to [^{11}C]CO_2, as long as the critical time perspective is addressed. In Figure 8.5, a representative ^{11}C-labeling strategy starting from [^{11}C]CO_2 is shown. The sequence starts with the addition of an alkyl Grignard reagent giving [^{11}C]propionic acid. Conversion to the corresponding acid chloride and addition of the appropriate amine, followed by a reduction gives [^{11}C]PHNO which is used to image dopamine $D_{2/3}$ receptors, see Chapter 18.

[^{11}C]CH_3I is another very popular reagent that can be accessed from [^{11}C]CH_4. Although this is an effective alkylating agent it is often converted into the even more reactive [^{11}C]CH_3OTf. Both [^{11}C]CH_3I and [^{11}C]CH_3OTf can be used to perform standard alkylation of a range of different nucleophiles—typically amines and phenols. Figure 8.5 shows representative PET ligands that have been labeled via alkylation of a suitable precursor with either [^{11}C]CH_3I or [^{11}C]CH_3OTf.

It is also possible to perform more complex transformations—like cross-coupling reactions. [^{11}C]CH_3I can be trapped by a Pd-catalyst and reacted with a suitable organometallic reagent—usually based on Sn or B. Vortioxetine and Stavudine (both marketed drugs) have been ^{11}C-labeled using this approach, see Figure 8.5. The examples in Figure 8.5 are far from exhaustive and a plethora of other PET ligands, labeling procedures, and ^{11}C-synthons have been developed.

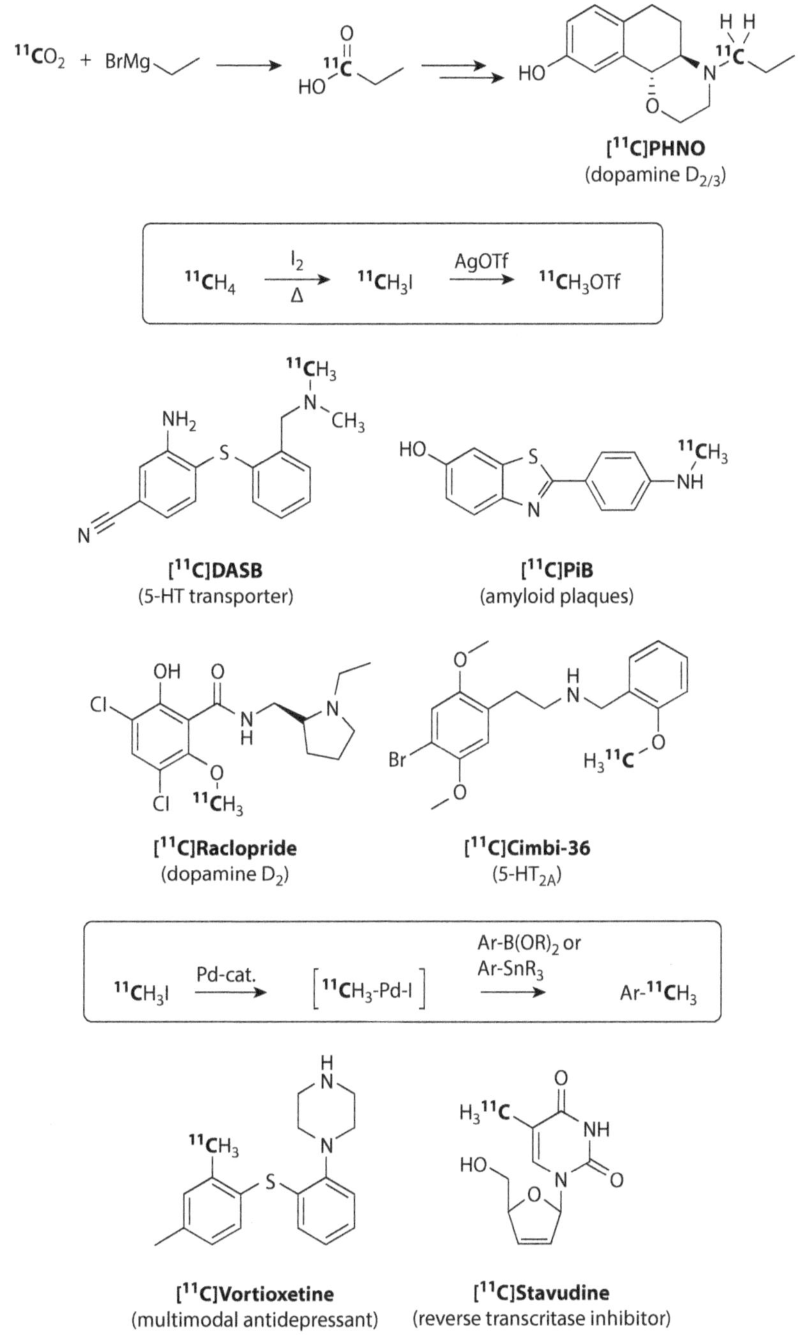

$^{11}CO_2$ + BrMg

[^{11}C]PHNO
(dopamine $D_{2/3}$)

$^{11}CH_4$ $\xrightarrow[\Delta]{I_2}$ $^{11}CH_3I$ $\xrightarrow{AgOTf}$ $^{11}CH_3OTf$

[^{11}C]DASB
(5-HT transporter)

[^{11}C]PiB
(amyloid plaques)

[^{11}C]Raclopride
(dopamine D_2)

[^{11}C]Cimbi-36
(5-HT_{2A})

$^{11}CH_3I$ $\xrightarrow{\text{Pd-cat.}}$ [$^{11}CH_3$-Pd-I] $\xrightarrow{\text{Ar-B(OR)}_2 \text{ or Ar-SnR}_3}$ Ar-$^{11}CH_3$

[^{11}C]Vortioxetine
(multimodal antidepressant)

[^{11}C]Stavudine
(reverse transcritase inhibitor)

FIGURE 8.5 Representative examples of ^{11}C-labeled PET ligands and their biological targets.

8.6.2 ^{18}F-Labeling

At present, nucleophilic aromatic and aliphatic fluorination using [^{18}F]F$^-$ is the most reliable way to produce ^{18}F-labeled PET ligands with high specific activity. Either direct nucleophilic substitution on the target molecule or on secondary precursors which are then coupled to the molecule of interest, is therefore still the method of choice. In Figure 8.6, some examples of tracers prepared either by

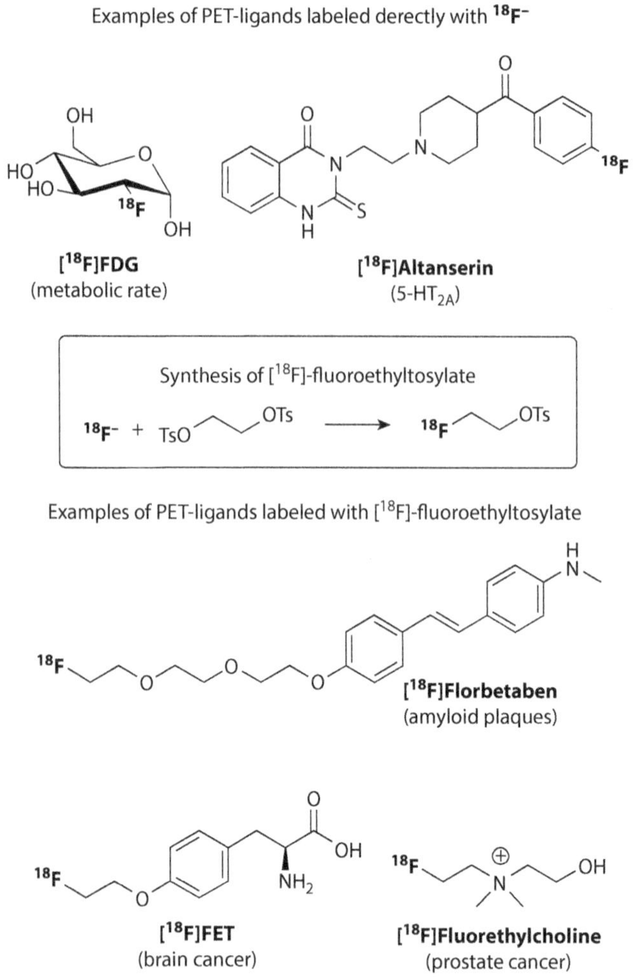

FIGURE 8.6 Representative structures of ^{18}F-labeled PET ligands.

direct or indirect fluorination are presented. [^{18}F]2-fluoro-2-deoxy-*d*-glucose ([^{18}F]FDG) is one of the most widely used PET ligands used primarily in cancer diagnostics. [^{18}F]altanserin, also labeled via a nucleophilic labeling of a suitable precursor, can be used to image the 5-HT_{2A} receptor in the CNS.

[^{18}F]Fluoroethyl tosylate is readily prepared and can be used in analogy with [^{11}C]CH_3I to alkylate a wide range of precursors. [^{18}F]Florbetaben is used to image amyloid plaques in analogy to [^{11}C]PiB in Figure 8.5; [^{18}F]FET and [^{18}F]Fluorethylcholine are used to image specific types of cancers. Again, a plethora of other ^{18}F-based PET ligands, synthons, and procedures are available.

8.7 METABOLISM OF PET LIGANDS

The site of labeling in a given molecule needs to be considered carefully. The most obvious and easiest way of labeling a molecule may not be the optimal solution as the metabolism of the PET ligand needs to be considered. As the ligand is metabolized, one (or more) derivatives of the original PET ligand can compromise the outcome of the experiments. Changing the site of labeling may alleviate that problem, see Figure 8.7.

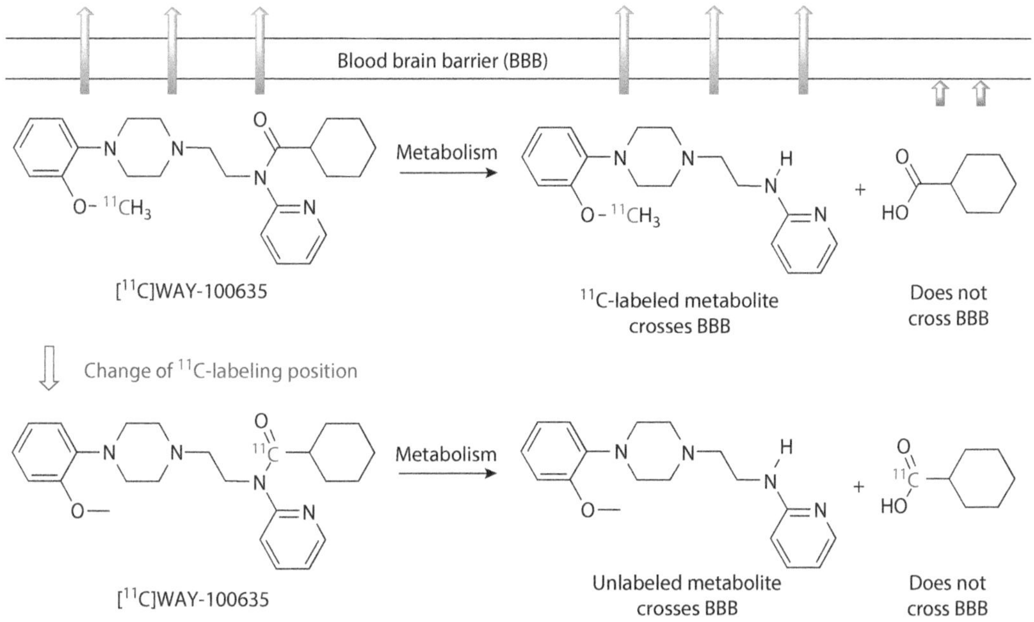

FIGURE 8.7 Metabolism of a PET ligand may lead to a metabolite that compromises the PET images as shown for [^{11}C]WAY-100635—used to image the 5-HT$_{1A}$ receptor. Judicious choice of labeling position can circumvent the problem.

[^{11}C]WAY-100635, a PET ligand used to image 5-HT$_{1A}$ receptors in the CNS, may most conveniently be labeled at the methoxy group as shown at the top of Figure 8.7. However, the primary pathway for its degradation in vivo is via hydrolysis of the amide bond in plasma to give the shown amine and cyclohexyl carboxylic acid. The formed ^{11}C-labeled metabolite is able to cross the BBB and enter the CNS, thereby adding to the combined PET image. Simply moving the site of labeling to the amide carbonyl leads to the formations of [^{11}C]cyclohexyl carboxylic acid as the radiolabeled metabolite that does not cross the BBB, avoiding the contamination of the final images. Even though the unlabeled metabolite is able to cross the BBB, the amount of compound present is so low that it does not impact the binding of the PET ligand to the target.

8.8 IMAGING IN DRUG DISCOVERY AND DEVELOPMENT

Molecular imaging methods such as PET are increasingly involved in the development of novel drugs since they are able to identify a biological target associated with a specific disease, to determine the drug mechanism of action, to examine the drug's biological characteristics such as target engagement, nonspecific binding or metabolism, to determine the optimal drug dosage, and thereby to improve the efficiency of selecting the appropriate drug candidate for clinical trials.

8.8.1 Application of Molecular Imaging Probes

There are many different phases of the drug discovery and development process where molecular imaging can be utilized (Figure 8.8). Table 8.2 lists some examples of PET ligands that have been used to study different parameters.

In the following, some examples of how PET can be used in the different phases of the drug discovery and development are discussed.

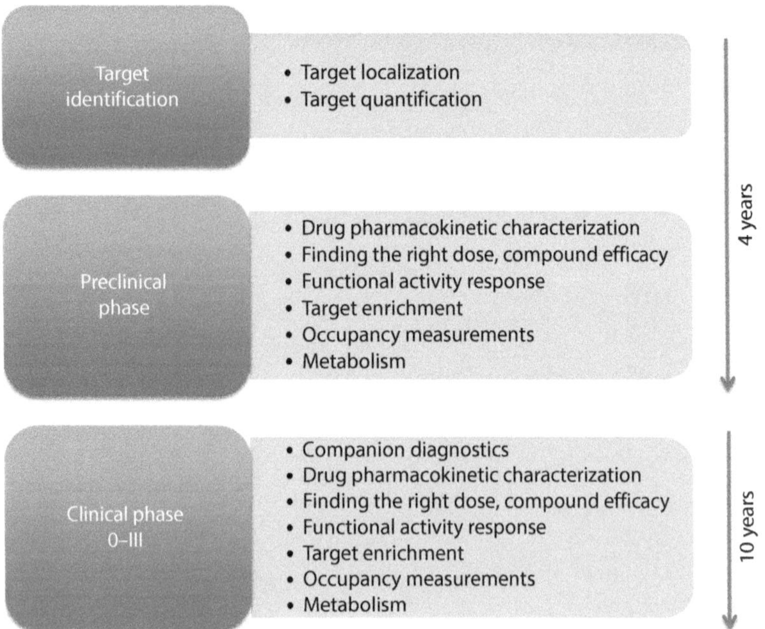

FIGURE 8.8 Overview of the different applications of imaging in the drug discovery and development process.

TABLE 8.2
Selected PET Ligands Applied within the Drug Discovery and Development Process

Drug	Applied Tracer	Measurement	Phase	Application
Cisplatin	FLT	Pharmacokinetics	Preclinical/clinical	Lung cancer
Fluorouracil	FLT	Method of action	Clinical	Colorectal cancer
Tamoxifen	[^{18}F]Tamoxifen	Pharmacokinetics	Clinical	Breast cancer
Gefitinib	FDG	Tumor metabolism	Preclinical/clinical	EGFR inhibitor
Lu AE92686	[^{11}C]Lu AE92686	Pharmacokinetics	Preclinical	PDE10A
Chemothera-peutica	FDG	Tumor metabolism	Clinical	Breast cancer
Raclopride	[^{11}C]raclopride	Occupancy	Preclinical/clinical	D_2 receptors
Herceptin	[^{89}Zr]Herceptin	Pharmacokinetics	Preclinical/clinical	Breast cancer
Selegiline	FDG	CNS metabolism	Clinical	Cocaine addiction
GR205171	[^{11}C]GR205171	Pharmacokinetics	Preclinical	NK1 receptor
Bevacizumab	FET	AA transport	Preclinical/clinical	Brain tumors
Bapineuzumab	PiB	Aβ plaques	Preclinical/clinical	Alzheimer's
MK-4232	[^{11}C]MK-4232	Target	Preclinical	CGRP-R

8.8.2 Target Identification

Recent findings indicated that calcitonin gene-related peptide receptors (CGRP-Rs) are involved in the pathogenesis of migraine and that the application of CGRP-R selective ligands result in pain relief. However, it was uncertain if central or peripheral CGRP-Rs mediate migraine attacks. A PET study with [^{11}C]MK-4232 and the clinically effective drug telcagepant as a blocking agent revealed only low to moderate occupancies at central CGRP-Rs implying that peripheral CGRP-Rs are more likely to be involved in the relief of migraine pain. Such finding can be used to direct the future development of more effective drugs.

8.8.3 Drug Pharmacokinetic Characterization

Labeled versions of drug molecules can be very useful when evaluating their pharmacokinetic profile. But as mentioned earlier, an effective drug does not necessarily make a good PET ligand. Nonspecific binding, irreversible/slow binding kinetics, or a receptor-rich profile may compromise the success of such investigations. Nevertheless, if possible, such investigations may provide very valuable information and PET has been used to profile [^{18}F]tamoxifen—an estrogen receptor antagonist and [^{11}C]vortioxetine—an atypical antidepressant.

Parameters such as BBB passage, metabolism, biodistribution, and reversibility of target binding of a drug can be investigated and quantified using the isotopically labeled drug. An illustrative example is the preclinical evaluation of two different neurokinin 1 (NK1) receptor antagonists, GR203040 and GR205171. Both displayed a promising in vitro profile to treat migraine, emesis, and pain. ^{11}C-labeling and subsequent nonhuman primate PET evaluation studies of both compounds revealed that [^{11}C]GR205171 showed superior in vivo binding characteristics (for example, higher brain uptake and target binding, lower nonspecific and unspecific binding) than [^{11}C]GR203040. Based on those results, GR205171 was selected for further investigations and eventually evaluated in clinical trials.

8.8.4 Finding the Right In Vivo Dose

Incorrect dosage of potential drug candidates in clinical trials is one of the major reasons why compounds fail in the development process. Thus, it is very important to be able to determine target occupancy levels of the drug molecules at different doses and correlate these data with the in vivo potency. The ideal dose of a drug is one that is high enough to have the desired effect, but not so high that possible side-effects begin to appear. Identification of the appropriate dose range in vivo is perhaps the most important application of PET in the drug discovery process.

One such example is from the dopaminergic receptor system, see Figure 8.9. PET studies with [^{11}C]raclopride and [^{18}F]desmethoxyfallypride have defined a narrow and optimal therapeutic window of 65%–78% D_2 receptor blockade using D_2 antagonists. Most antipsychotics show optimal clinical efficacy within this therapeutic window with minimal side effects. Increasing striatal D_2

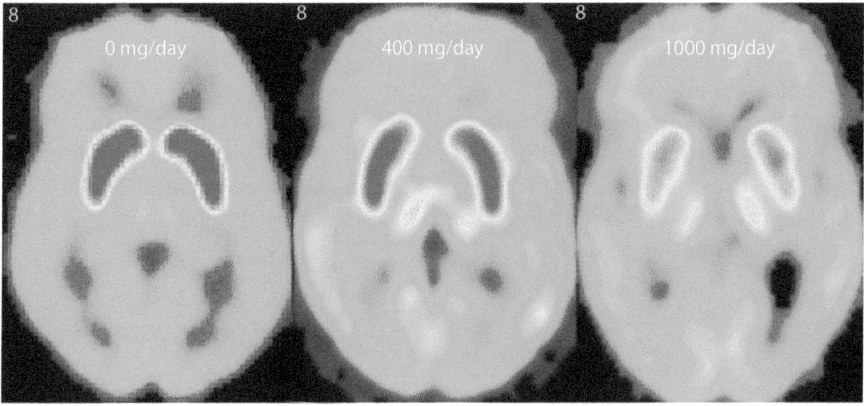

FIGURE 8.9 Increasing receptor occupancy at increasing doses of amisulpride (an atypical antipsychotic) manifests itself by decreasing [^{18}F]desmethoxyfallypride binding to $D_{2/3}$ receptors in striatum. (Reprinted with permission from Piel, M., Vernaleken, I., and Rösch, F., Positron emission tomography in CNS drug discovery and drug monitoring, *J. Med. Chem.*, 57, 9232–9258. Copyright 2014 American Chemical Society.)

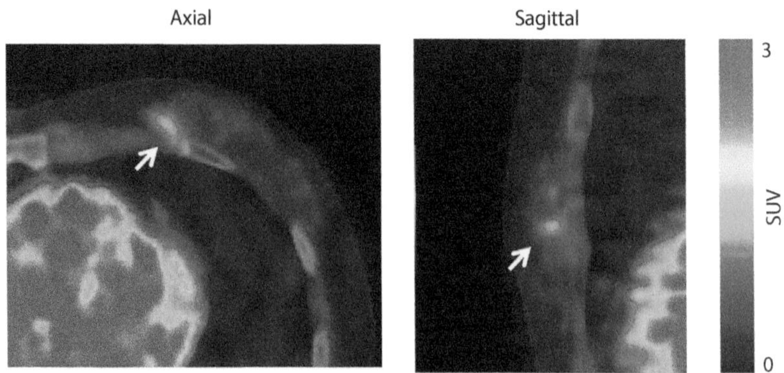

FIGURE 8.10 ^{64}Cu-DOTA-trastuzumab PET images of HER2-positive primary breast tumor. Arrows show primary breast tumor in patient. (Red regions indicate high uptake ^{64}Cu-DOTA-trastuzumab in heart and blood vessels.) This research was originally published in JNM. (From Tamura, K. et al., *J. Nucl. Med.*, 54(11), 1869, 2013. Figure 8.3. Copyright by the Society of Nuclear Medicine and Molecular Imaging, Inc.)

blockade leads to extrapyramidal motor symptoms, whereas lower striatal D_2 blockade diminishes the therapeutic effect drastically.

8.8.5 Companion Diagnostics

Companion diagnostics help tailor treatment schemes to individual patients and monitor the success of the selected treatment. In the drug development, the ability to identify and select patients that will benefit from the treatment enables smaller and more cost-effective clinical trials.

A prominent example for in vivo monoclonal antibodies (mAbs) companion diagnostic imaging is human epidermal growth factor receptor 2 (HER2)-positive breast cancer imaging with ^{64}Cu-labeled Herceptin (trastuzumab). The labeled mAbs can be used for patient selection, characterization of the target binding, and determination of the mAb's fate in vivo. In general, HER2 is a very attractive drug target since an outstanding therapy response rate up to 86% has been observed—unfortunately, only 20%–25% of all breast cancers are HER2-positive (Figure 8.10).

8.9 DETERMINATION OF FUNCTIONAL ACTIVITY RESPONSE

The ability to image a drug's effects on a biological process in vivo provides a direct readout of the efficacy of the potential drug. Molecular imaging of functional responses can validate whether a prospective drug is able to modulate the desired target in vivo or not. The metabolism of glucose, amino acids, and lipids can be imaged with PET. These measurements appear to be superior when assessing tumor response to targeted drugs which predominantly result in no or minor tumor size changes early in a treatment cycle.

8.9.1 [^{18}F]FDG

[^{18}F]FDG (FDG) is the most widely used tracer for PET imaging in oncology. FDG is transported into cells by the glucose transporter, and then phosphorylated by a hexokinase to form FDG-6-phosphate. This phosphorylated product of FDG (in contrast to the phosphorylated product of glucose) is not a substrate for further glycolysis and accumulates in cells that have increased activity of hexokinase and increased glucose transporter levels. Tumor cells typically display such a characteristics (the Warburg effect) and FDG is routinely used in cancer-related investigations.

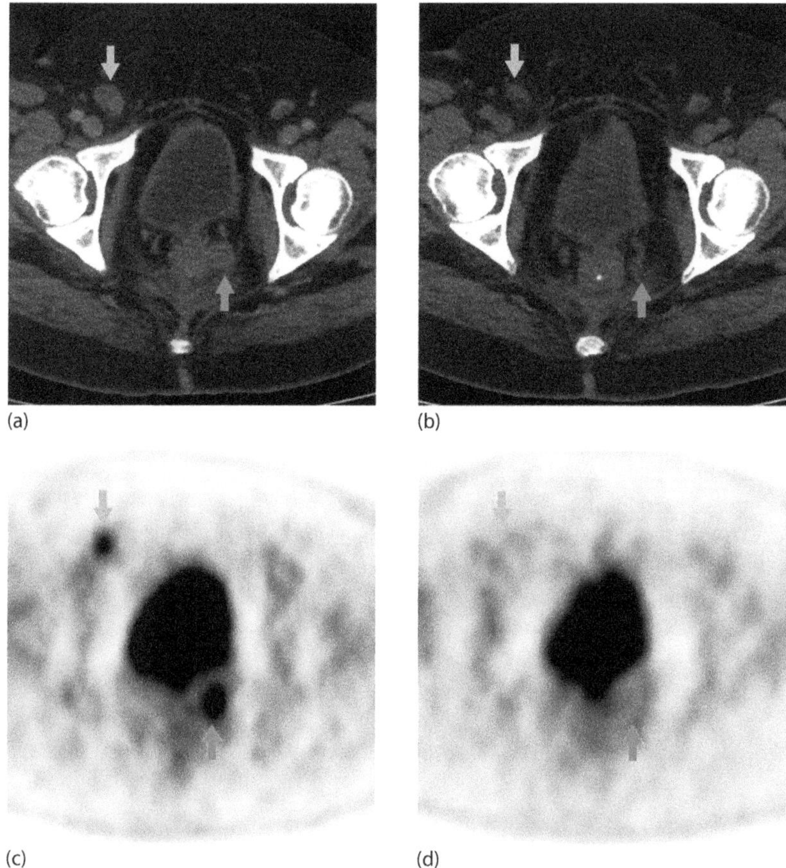

FIGURE 8.11 Cross-sectional axial CT (a and b) and [^{18}F]FDG PET (c and d) images from cancer patient responding to imatinib mesylate treatment. a and c are prior to and b and d are after 1 month of therapy. (Reprinted with permission from Hodi, F. et al., *J. Clin. Oncol.*, 31(26), 3182–3190. Copyright 2013 American Society of Clinical Oncology.)

FDG can also be used to detect a functional response after a treatment with anticancer agents providing information on drug efficacy much earlier than, e.g., survival data of patients. In patients with gastrointestinal stromal tumors, a response as early as 24 hours can be observed after the first application of imatinib mesylate (Glivec), a tyrosine-kinase inhibitor used in the treatment of multiple cancers, see Figure 8.11.

FDG can also be applied to CNS studies and it has been shown that cocaine consumption reduces cerebral glucose metabolism in the brain. Based on that result it was speculated that FDG could be used as a surrogate marker for the treatment efficacy of cocaine addiction. Indeed, FDG-PET investigations revealed that the monoamine oxidase B inhibitor selegiline which is used as a pharmacological adjunct in the treatment of cocaine addiction, altered glucose metabolism in most brain regions.

8.9.2 [^{18}F]FLT

3′-Deoxy-3′-[^{18}F]fluorothymidine (FLT) is a substrate of the intracellular mammalian thymidine kinase which is part of the DNA synthesis machinery. FLT is trapped within the cells (after phosphorylation) at levels proportional to the thymidine kinase activity. FLT–PET is superior to FDG-PET for the imaging of changes in the proliferation rate after treatment with the anti-proliferative drug cisplatin.

8.9.3 [^{18}F]FET

Amino acid transport is another interesting imaging target used to measure functional activity responses. Usually, the amino acid transport is increased in tumor cells. [^{18}F]fluoro-ethyl-tyrosine (FET, see Figure 8.6) has been used to study brain tumors since FDG-PET diagnosis is unreliable in relatively benign brain tumors due to high normal FDG brain uptake. The response of tumors treated with bevacizumab, an angiogenesis inhibitor, could be assessed using FET and these PET experiments were predictive for treatment failure.

8.9.4 [^{11}C]PiB and [^{18}F]Florbetaben

Imaging with [^{11}C]PiB and [^{18}F]Florbetaben (see Figures 8.5 and 8.6) in Alzheimer's disease makes it possible to quantify the regional Aβ plaque load in the living brain. Aβ plaques are abundant in brains of Alzheimer's patients and the depositions begin a decade (or earlier) before clinical symptoms of Alzheimer's become apparent.

Bapineuzumab is a humanized monoclonal antibody, which has been reported to enhance the clearance of amyloid plaques in the brains of overexpressing transgenic mice and [^{11}C]PiB, was subsequently used to evaluate the effectiveness of Bapineuzumab in humans. Treatment of Alzheimer's patients for 78 weeks did indeed significantly reduce PiB-PET binding in cortical brain regions when compared to placebo, see Figure 8.12.

Despite these promising results, Pfizer and Johnson & Johnson later reported that bapineuzumab failed to give significant cognitive improvements in patients, suggesting that amyloid reduction cannot reverse the cognitive effects of Alzheimer's disease.

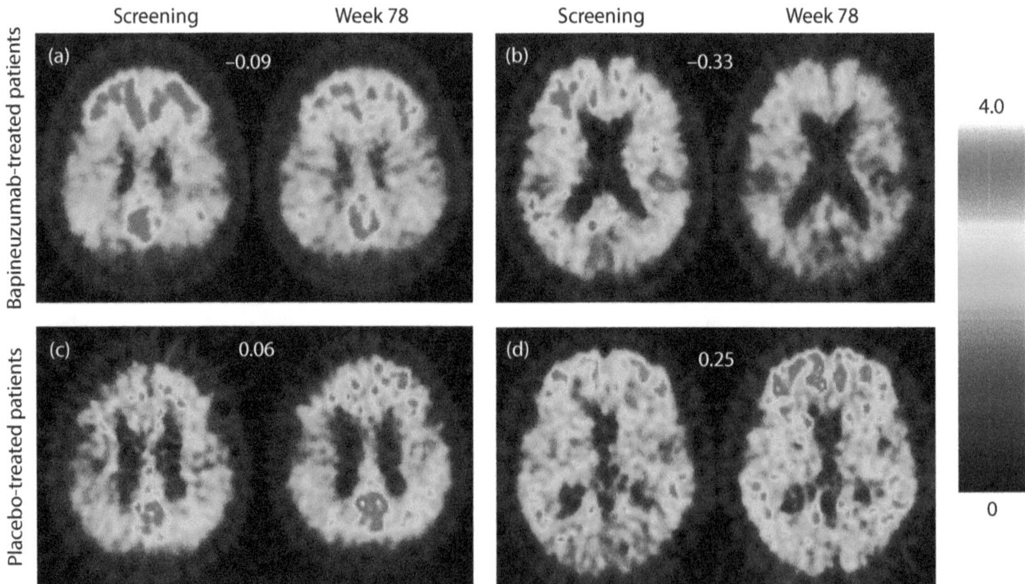

FIGURE 8.12 [^{11}C]PiB PET images from patients treated with bapineuzumab (a and b) and placebo (c and d). Mean [^{11}C]PiB PET changes are shown at the top of each picture for individual patients, indicating a significant change in the amount of amyloid plaques for both bapineuzumab (decrease) and placebo (increase). (Reprinted from *Lancet Neurol.*, Rinne, J. et al., ^{11}C-PiB PET assessment of change in fibrillar amyloid-β load in patients with Alzheimer's disease treated with bapineuzumab: A phase 2, double-blind, placebo-controlled, ascending-dose study, 9, 368. Copyright 2010, with permission from Elsevier.)

8.10 CONCLUDING REMARKS

In vivo molecular imaging methods such as PET are increasingly involved in the drug discovery and development process. The ability to quantify key parameters in vivo makes them a very powerful tool when transitioning from in vitro to in vivo investigations.

FURTHER READING

Cunha, L., Szigeti, K., Mathé, D., and Metello, L.F. 2014. The role of molecular imaging in modern drug development. *Drug Discov. Today* 19:936–948.

Hodi, F., Corless, C.L., Jonathan, A. et al. 2013. Imatinib for melanomas harboring mutationally activated or amplified KIT arising on mucosal, acral, and chronically sun-damaged skin. *J. Clin. Oncol.* 31(26):3182–3190.

Kehler, J., Kilburn, J.P., Estrada, D. et al. 2014. Discovery and development of 11C-Lu AE92686 as a radioligand for PET imaging of phosphodiesterase10A in the brain. *J. Nucl. Med.* 55(9):1513–1518.

Miller, P.W., Long, N.J., Vilar, R., and Gee, A.D. 2008. Synthesis of 11C, 18F, 15O, and 13N radiolabels for positron emission tomography. *Angew. Chem. Int. Ed.* 47:8998–9033.

Piel, M., Vernaleken, I., and Rösch, F. 2014. Positron emission tomography in CNS drug discovery and drug. *J. Med. Chem.* 57:9232–9258.

Pike, W. 2009. PET radiotracers: Crossing the blood–brain barrier and surviving metabolism. *Trends Pharmacol. Sci.* 30:431–440.

Rinne, J., Brooks, D.J., Rossor, M.N. et al. 2010. ^{11}C-PiB PET assessment of change in fibrillar amyloid-β load in patients with Alzheimer's disease treated with bapineuzumab: A phase 2, double-blind, placebo-controlled, ascending-dose study. *Lancet Neurol.* 9:368.

Tamura, K., Kurihara, H., Yonemori, K. et al. 2013. ^{64}Cu-DOTA-trastuzumab PET imaging in patients with HER2-positive breast cancer. *J. Nucl. Med.* 54(11):1869–1875.

Willmann, J.K., Bruggen, N., Dinkelborg, L.M., and Gambhir, S.S. 2008. Molecular imaging in drug development. *Nat. Rev. Drug Discov.* 7:591–607.

9 Peptide and Protein Drug Design

Jesper Lau and Søren Østergaard

CONTENTS

9.1 INTRODUCTION

9.1.1 A Historical Perspective

Peptides and proteins have been used for decades as pharmaceuticals to treat life-threatening diseases, and 95 years ago insulin was among the first plasma-derived proteins used to effectively treat a fatal disease. In 1920, Frederick Banting and Charles Best isolated and purified insulin while working in the laboratory of Professor John Macleod at University of Toronto, and saved the life of 14-year-old Leonard Thomson, a type I diabetic, by injecting him with insulin. Other proteins like factor VIII (hemophilia) and XI, INF-β1 (multiple sclerosis), anti-TNFα (inflammation), and growth hormone have since demonstrated great therapeutic value. With the entrance of recombinant technology in the late 1970s it became possible to design analogs of natural proteins in expression systems like *E. coli*, yeast, or various mammalian cells. It was now possible to make mutations in a protein, or, native-like post-translational modifications such as glycosylation, giving rise to protein analogs with altered and/or improved properties. Rapid acting insulin analogs like insulin aspart or insulin lispro were the first protein analogs to enter the market. They were engineered so one or two mutations in the dimer interface destabilized the dimer formation leading to preferentially monomeric insulin which in turn led to a more rapid acting analog than native insulin.

In the early 1990s, high-throughput expression technologies such as the phage technology that could display billions of peptides and later even larger proteins such as antibodies were developed and this accelerated the discovery of protein analogs with improved or completely new pharmaceutical properties.

Protein–protein interactions (PPIs) are fundamental to any living organism and modulating these interactions can have profound importance. With the introduction of various humanized antibody screening technologies it quickly became feasible to identify antibodies that could interfere as PPI inhibitors and many blockbuster protein drugs today are based on this approach. However, although we recognize that antibodies and other domain scaffolds are some of the most important biopharmaceuticals on the market, they are beyond the scope of this Chapter. Rather, we will focus on ways to improve efficacy of naturally occurring peptides and proteins. The success of today's biopharmaceuticals is reflected by the fact that in 2013 among the top 20 top selling drugs 11 were proteins or peptide-based drugs.

9.1.2 Aspects of Peptide and Protein Drug Design

The design of a peptide or protein drug candidate needs to take a variety of different aspects into consideration in order to lead to a successful drug candidate. Thus, during the design process, a number of important factors that modulate the protein properties must be considered such as

1. Potency
2. Selectivity
3. Distribution
4. Elimination
5. Route of administration
6. Formulation
7. Toxicity and immunogenicity
8. Production process

In general, highly potent proteins and peptides that circulate in a low, picomolar range are much more specific than small molecules to the same target. This is of particular importance if homologous receptors are present, as in the case of the insulin and the insulin-like growth factor 1 (IGF-1) receptor, or, in the case of the neuropeptide Y (NPY) receptor family where four receptors Y1, Y2, Y4, and Y5 necessitate an even greater awareness concerning selectivity issues, since these are distributed across many organs and in the central nervous system (CNS) as well.

The half-life or rate of elimination is also an important parameter that affects efficacy. Elimination may be caused by a rapid receptor clearance, enzymatic instability, or renal clearance which rapidly removes peptides and proteins that are less than 60 kDa. Increasing half-life by modifying the size of the peptide or protein, or implementing other changes to make them less prone to proteolysis, is a major focus area in the engineering of therapeutic peptides and proteins.

The distribution and route of delivery of biopharmaceuticals versus small molecules is also very different. Intracellular targeting is generally reserved for small molecules since peptides and proteins do not readily pass the cell membrane, and most proteins likely do not cross the blood–brain barrier although some exceptions exist. Importantly, most small molecules show oral bioavailability which peptides and proteins do not. Thus, one disadvantage of protein and peptides compared with small molecules is that they need to be injected. Administration by injection necessitates that the drug is stable in formulation which may be challenging as many endogenous peptides and proteins lack sufficient chemical or physical stability to be manufactured in an aqueous formulation. Alternatively, they may be provided as lyophilized products that must be dissolved prior to use which is less convenient when, for example, daily injections are needed.

Likewise, in contrast to small molecules, both proteins and peptides may form insoluble aggregates or fibrils upon quiescent storage which in the long run may provoke an immune response and the formation of antibodies. Antibody formation is also a concern when a protein sequence is changed from that of the native peptide or protein which may lead to a loss of drug's effect or reduced efficacy.

9.2 PEPTIDE AND PROTEIN PROTRACTION

As previously discussed, major challenges in the development of biopharmaceuticals as therapeutics are rapid clearance and thus a very short blood residence time. The reason for this is partly due to fast renal elimination of peptides and small proteins, and also proteolytic plasma clearance as well as receptor clearance.

9.2.1 Polymer Extension

The most important determinant of renal clearance is the size or the hydrodynamic volume of the peptide or protein. While there are several ways to increase the volume of a peptide or protein, one common approach is to add a random polymer that simply increases the size of the resulting polymer–protein conjugate. In addition to reducing clearance, such modification appears to provide a shielding effect that lowers the antigenicity of the conjugate and protect against proteolysis. It is important to keep in mind, however, that such shielding may also affect access to the target protein and thus impact the potency.

9.2.1.1 PEGylation

The most successful polymer has so far been polyethylene glycol (PEG) (Figure 9.1a). PEGylation of peptides and proteins originated in the 1970s and was one of the very first protein engineering strategies to be used. Pegademase bovine, a modified enzyme used for enzyme replacement therapy, was the first to reach the market after approval in 1990, and several other PEGylated proteins have since been approved, including pegaspargase, 1994, peginterferon alfa-2b, 2000, peginterferon alfa-2a, 2002, pegfilgrastim, 2002, pegvisomant, 2002, pegaptanib, 2004, methoxy polyethylene glycol-epoetin beta, 2007, and certolizumab pegol, 2008.

Most of these drugs were randomly PEGylated through amidation to surface exposed lysines. The main reason for PEGylation was to increase the hydrodynamic volume and thereby reduce renal clearance, but in many cases it also increased stability toward proteolysis which also contributed to a longer circulation time. The molecular weight cut-off for glomerular filtration is approximately 60 kDa, but interestingly renal clearance of PEGylated peptides had a cut-off at about 20–30 kDa. This was explained by the tendency of PEG to bind water and the tendency to form a random coiled noncharged polymer with a larger hydrodynamic radio compared to a charged compact and structured protein of similar molecular weight. By attachment of PEG several other benefits include reduced immunogenicity, antigenicity, and in several cases also improved solubility which was of importance for drug product formulation. The first applications used PEGs of less than 12 kDa and with rather high degree of polydispersity due to the polymerization process in manufacturing the PEGs. The process technology has since improved and today various PEGs are available that offer higher molecular weights and a narrower polydispersity index. The conjugation technology has also

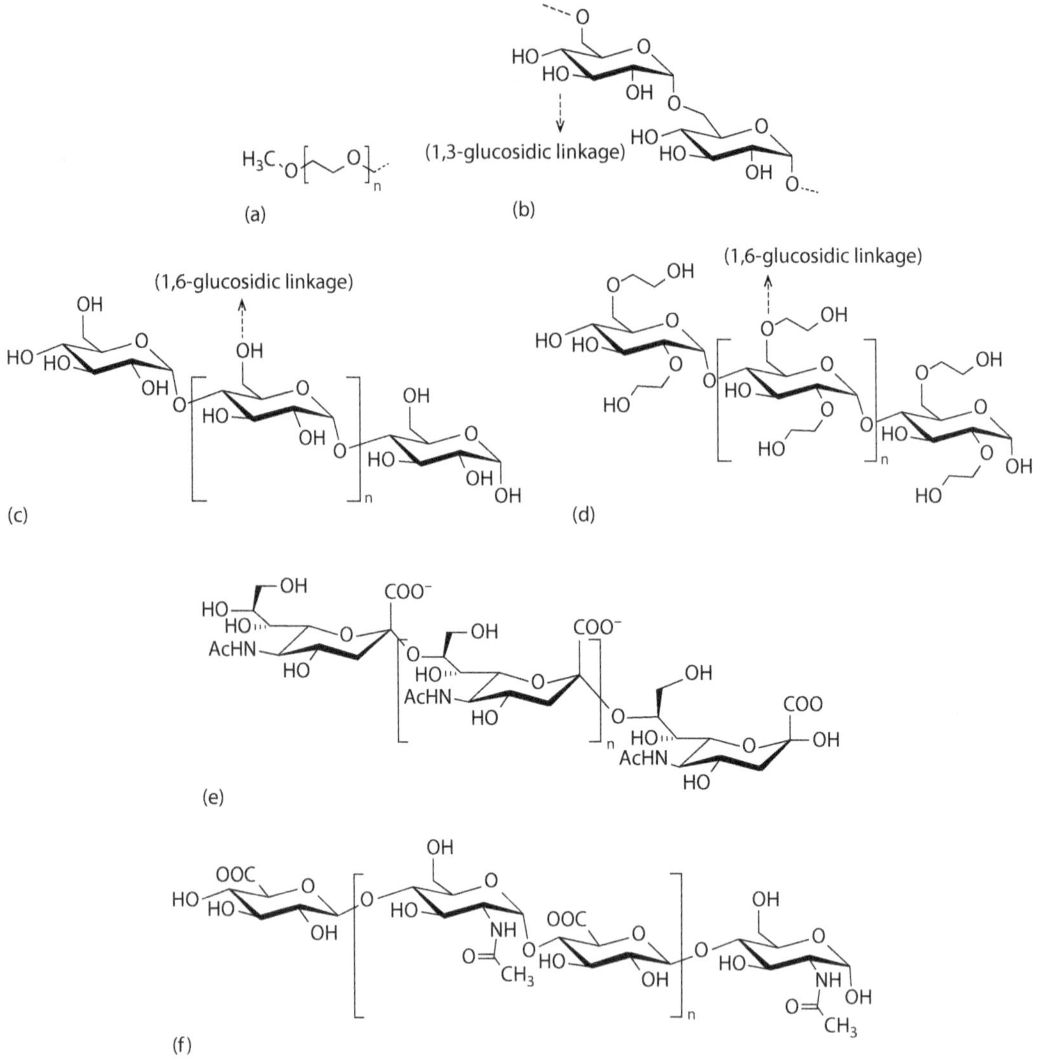

FIGURE 9.1 Structure of polymers that have been used in the modification of proteins: polyethylene glycol—PEG (a); dextran (b); dextrin (c); hydroxyethyl starch (d); polysialic acids (PSA) (e); and heparosan (f).

been refined to site selective PEGylation using amber codon technology (as discussed in Chapter 6) that allow for a well-defined product with a high degree of homogeneity.

The most versatile (but less selective) PEG reagents were based on activated esters that reacted with the primary amino groups of lysine residues. More selective methodologies have since been developed. Such linkage chemistries include the coupling of a free cysteine residue of the protein via a maleimide or haloacetyl group of the PEG reagent, and also reductive alkylation of *N*-terminals of the protein with PEG aldehydes in the presence of a reductive agent. Enzymatic reactions have also been explored for site-selective attachment of PEG such as the use of transglutaminase to PEGylate growth hormone. Another technology was recently introduced whereby the glycans of glycoproteins are applied as selective handles for glycoPEGylation using various enzymes to trim the glycan structure and attach tailored PEG reagents.

One of the major downsides of PEGylation is that the active protein or peptide is shielded significantly diminishing bioactivity. Releasable PEGylation has therefore been investigated in which the protein is reversibly attached to PEG and released after dosing. This strategy allows for a prodrug approach where the fully active protein or peptide is slowly released from a systemic inactive deposition that can then be tailored to specific release profiles. This technology is still in the early stages of development and so far no pharmaceuticals using this technology have been approved.

9.2.1.2 Polypeptide Modification

Other attractive polymer conjugation and fusion technologies are available that are either based on random polypeptides or polymers of carbohydrates. The polypeptide approach was based on the design of a random coil structure composed of a minor set of small polar amino acids. The size of the polypeptide impacts the half-life by reduction of renal clearance. Although the polymers are stable in plasma and nonimmunogenic, they are degraded in endosomes liberating only endogenous amino acids, that is in contrast to PEGylation which is not degradable. In addition, the polypeptides are all of a defined size in contrast to the polydisperse nature of PEG making the analysis of the drug protein much easier. Two types of polypeptides, the XTEN and the PASylation technologies, have attracted attention in the recent years. The XTEN technology developed by Amunix is a polypeptide comprising of Ala, Ser, Thr, Glu, and Gly in more or less random sequence. This ensures a highly hydrophilic and flexible polymer that offers both low immunogenicity and is also stable against plasma degradation. The PASylation technology developed by XL-Protein is based on the same principle but uses only three amino acids: Pro, Ala, and Ser (PAS). Both technologies offer the opportunity to express the drug target as a fusion partner, or, chemically link the drug target to the polymer via suitable and site-specific linkages. In the latter case, not just one but several drug molecules may be attached to the polymer raising the avidity of the active molecule. An example of this technology is a once-monthly growth hormone developed by Versatis and Amunix which is currently entering phase II clinical trials.

9.2.1.3 Carbohydrate-Based Polymer Modification

Initially used as a plasma expander, *dextran* (Figure 9.1b) is one of the most studied classes of carbohydrate-based polymers conjugated to proteins. It is produced by bacteria and consists of repeating units of glucose monomers linked by α-1,6-glucosidic linkages. The polymer also displays varying degrees of branching via 1,3-glucosidic linking. Coupling of dextran is accomplished by oxidation with periodate yielding aldehyde groups that react with lysine amino groups or the *N*-terminal of the protein. This procedure results in a relatively inhomogeneous conjugate since the aldehyde groups are randomly distributed along the polymer and the attachment to the protein is nonspecific.

Dextrin (Figure 9.1c) is a similar polysaccharide to dextran and is composed of D-glucose units linked by a 1,4-glucoside linkage that forms a linear polymer with branches via 1,6-glucosidic linkages. This biodegradable polymer is produced by hydrolysis of starch and degraded in vivo by α-amylase which has hampered its use. *Hydroxyethyl starch* (*HES*) (Figure 9.1d) represents a class of polysaccharides that have been engineered to be more stable against α-amylase by the chemical modification of the hydroxy groups to hydroxyethyl. By this chemical modification, the rate of degradation can be modulated by controlling the extent of hydroxyethylation. This polymer is approved as a blood expander and site-specific conjugation has been achieved by the reductive amination of one distal aldehyde in the polysaccharide to the *N*-terminal of the protein target. *Polysialic acid* (*PSA*) (Figure 9.1e) is a linear polymer chain which is linked together by *N*-acetylneuraminic acid (sialic acid) acid. Sialic acid is found on the external membrane of a number of cell types in the body, and polymers of sialic acids are widely expressed on the external membrane on a number of bacterial types. When used for therapeutic protein and peptide drug delivery, PSA provides an increased size which seems to provide a shielding effect similar to the one obtained by PEGylation. Comparable to the HESylation method, PSA can be conjugated using the distal three vicinal hydroxyl groups, which after oxidation to aldehyde by periodate can be coupled to the protein, or to a bifunctional handle, to allow other types of coupling chemistries. PSAs have been chemically conjugated to a few clinically relevant therapeutic proteins like interferon and shown to improve their circulating half-life without adversely affecting efficacy. A similar negatively charged polymer is heparosan (Figure 9.1f) which is composed of repeating units of *N*-acetylglucosamine and glucuronic acid.

9.2.2 Albumin as Protractor

Human serum albumin (HSA) with a molecular weight of 66.5 kDa is the most abundant plasma protein (Figure 9.2). Like most other plasma proteins it is synthesized in the liver where it is produced at a rate of approximately 0.7 mg/hour for every gram of liver. This translates to roughly 10–15 g of albumin per day that is exported into circulation reaching a concentration in the blood of 30–50 g/L. HSA has several important physiological functions that include stabilizing plasma pH, heat and other denaturing conditions and is also a radical scavenger that counts as an important antioxidant. Another key function is its ability to bind various ligands thereby providing a depot for a wide variety of compounds that goes well beyond their solubility in plasma including fatty acids and steroids. Several small molecule drugs are also known to bind to HSA, which makes it an important factor in the pharmacokinetic behavior of many drugs, affecting their efficacy and rate of delivery. Multiple binding sites for drug ligands have been described since the first structure of HSA was published in 1992. The multiple binding sites and the high plasma concentration of albumin give an extraordinarily high capacity for transport of fatty acids and other ligands. Albumin has been used as a drug carrier for small molecules for several decades, and has also been used in peptide and protein engineering. Several technology platforms have been established and shown success in clinical trials. The main technologies will be described in the following paragraphs.

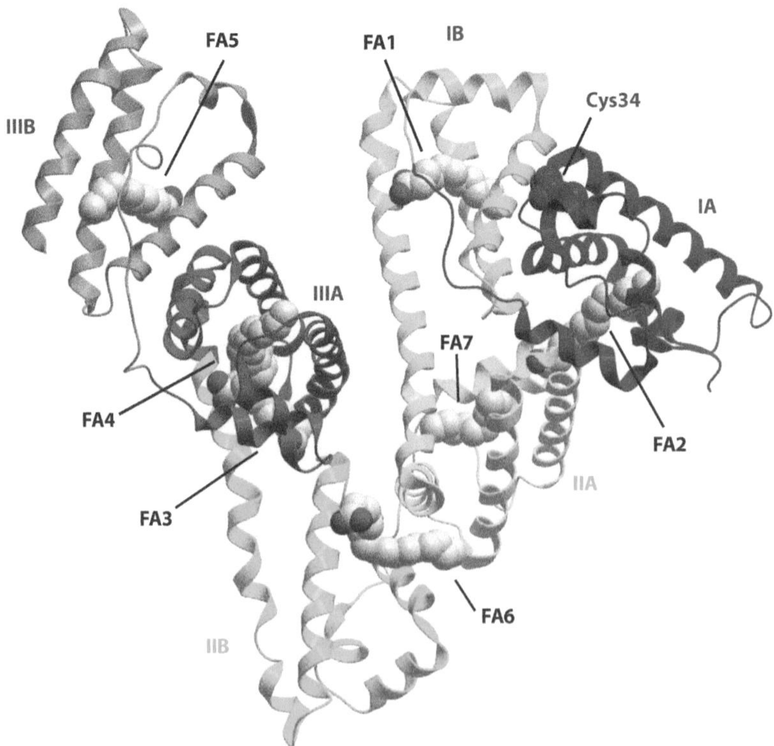

FIGURE 9.2 HSA structure with palmitic acid in light yellow (ball and stick models) occupying all seven fatty acid (FA) binding pockets. In domain IA the free cysteine 34 is shown in purple. The six subdomains of HSA are colored as follows: subdomain IA: blue; subdomain IB: light blue; subdomain IIA: light green; subdomain IIB: dark green; subdomain IIIA: red; subdomain IIIB: orange (pdb entry 1E7H).

Albumin and antibodies have the largest plasma stability with half-lives up to 3 weeks. They both bind simultaneously to the neonatal receptor (FcRn) with high affinity at low pH in endosomes. This protects against degradation during endocytosis and recycles both proteins back to plasma, where they are rapidly released due to low affinity to the FcRn at neutral pH.

9.2.2.1 Reversible Binding by Fatty Acid Acylation

The high capacity of albumin to bind fatty acids was used in the molecular design of insulin detemir that was approved in Europe in 2004. The idea was to develop soluble and protracted insulin by linking a fatty acid to an insulin analog that could then bind to albumin in a reversible manner after injection. The concept was to link myristic acid to the lysine in B29 of desB30 insulin mimicking that of a free fatty acid (Table 9.1). Whereas the half-life of human insulin is only 4–6 minutes, the extended pharmacokinetic profile of insulin detemir makes it possible to manage blood glucose levels using a once-daily subcutaneous (s.c.) injection due to a combination of a *sub cutis* deposition and low albumin affinity. Recently, a second insulin analog, insulin degludec, was approved in Europe in 2013 and US in 2015 that is based on the same technology, but here myristic acid was replaced by hexadecandionyl attached to DesB30 human insulin via a L-γ-glutamic acid spacer (Table 9.1). Insulin degludec binds to albumin, but the main effect of protraction is due to self-assembly to multi-hexamers after subcutaneous injection, and this molecular construct extends the pharmacokinetic profile beyond that of insulin detemir (Figure 9.3).

Liraglutide (Table 9.1) is a glucagon-like peptide 1 (GLP-1) analog and was the very first human GLP-1 analog to be approved for treatment of diabetes (Table 9.1) and later for obesity. Native GLP-1 has a half-life of approximately 2 minutes due to rapid cleavage of GLP-1(7–37) to GLP-1(9–37) by dipeptidyl peptidase-4 (DPPIV). Liraglutide was designed based on reversible binding to albumin through the attachment of palmitic acid via a L-γ-glutamic linker to lysine 26 of Arg34

TABLE 9.1
Insulin and GLP-1 Analogs Modified with Fatty Acid Acylation That Are on the Market or in Phase III

Drug	Modification	Position	Protractor
Detemir	desB30 insulin	B29Lys	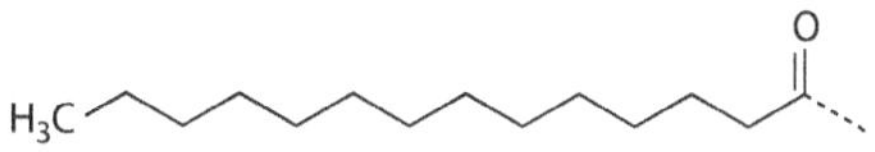
Degludec	desB30 insulin	B29Lys	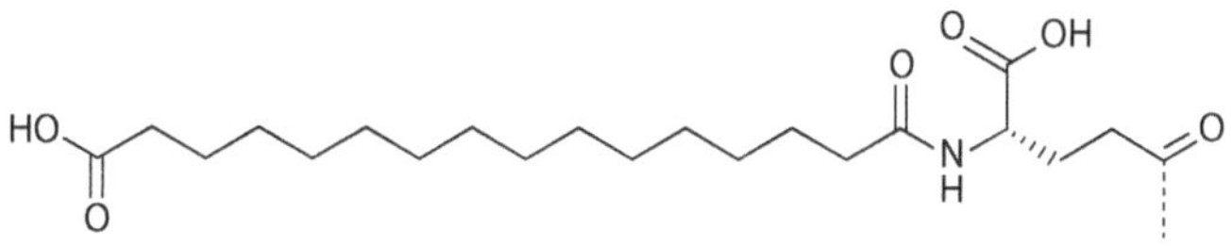
Liraglutide	[34Arg] GLP-1	26Lys	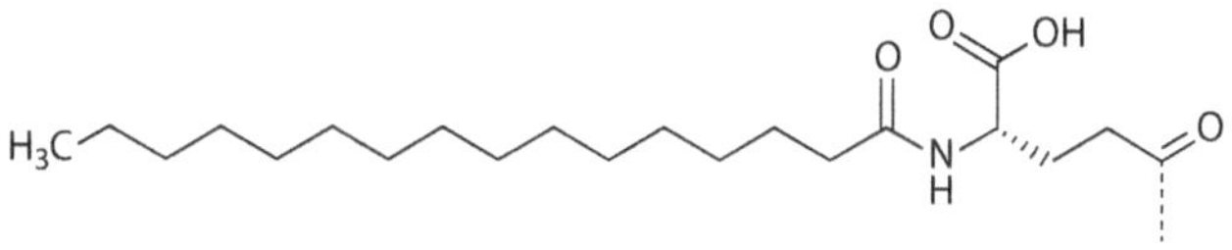
Semaglutide	[8Aib,34Arg] GLP-1	26Lys	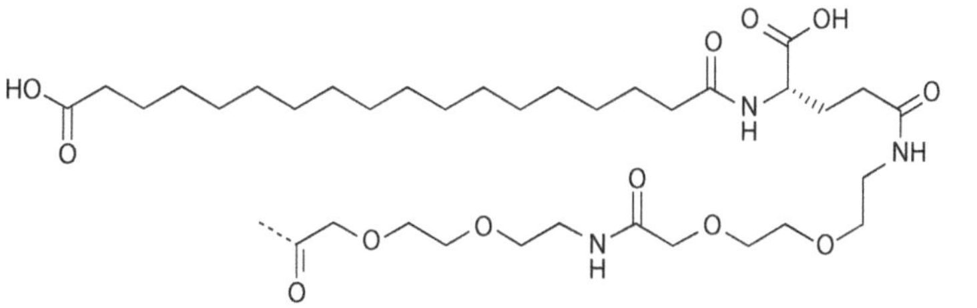

FIGURE 9.3 Depletion of phenol from injection site can be mimicked by buffer exchange of pharmaceutical formulation (a). The subcutaneous depot is composed of long strands of multi-hexamers (b) as observed by transmission electron microscopy (c).

GLP-1 (7–37). The modification of Lys34 to Arg34 made it possible to produce Arg34 GLP-1(7–37) in yeast followed by acylation of Lys26 in a similar way as was used for the production of detemir and degludec. Liraglutide has a half-life of about 11 hours after s.c. dosing in humans combined with a delayed absorption from *sub cutis* which gives a pharmacokinetic profile suitable for once-daily dosing. The reason for extended circulation is due to reversible albumin binding, which protects liraglutide from DPPIV degradation and glomerular filtration, whereas the delayed absorption is explained by the ability of liraglutide to form heptamers by self-assemble controlled by the fatty acid side chain at position 26. The next generation of GLP-1 analogs using the reversible albumin affinity is represented by semaglutide which is currently in phase III clinical trials. This analog Aib8, Arg36 GLP-1 (7–37) is derivatized with a side chain composed of hexadecandinoyl attached to Lys26 via a L-γ-glutamic acid linker and a small hydrophilic spacer (Table 9.1). The Aib8 was introduced to improve the DPPV stability beyond that of liraglutide, and the new side chain increased the albumin affinity resulting in a half-life in man sufficient for once-weekly dosing.

9.2.2.2 Reversible Binding by Other Moieties

Several alternative ligands that bind to albumin have been discovered including peptides, as well as antibody fragments, and other protein-based ligands. Phage display technology has been used to identify high-affinity peptide ligands for a variety of protein targets. Genentech used this technology to discover small cyclic peptides by screening against rat, rabbit and human albumin and found a series of albumin ligands with the core sequence DICLPRWGCLW that bound to albumin with a 1:1 stoichiometry at a unique binding site. These peptides were used to fuse to a fab fragment resulting in 37-fold half-life extension in rabbits and 26-fold extension in mice. It was also shown that it is possible to identify albumin ligands based on small domain antibody fragments (dAbs) that are composed of the variable parts of either heavy (HV) or light (VL) chains of human antibodies, and have a molecular weight of approximately 11–13 kDa. These albumin binding dAbs were found by phage display and have albumin affinities of 34–660 nM depending on the species.

Several applications were investigated using this platform including interleukin-1 receptor antagonist (IL-1ra) and interferon-α2b which were fused to the dAb and both constructs were active in vivo with extended half-lives. Finally, the Swedish company, Affibody, was able to generate high-affinity small proteins that bind to albumin. The proteins have a molecular weight of approximately 6 kDa and are based on a stable α-helical bacterial receptor domain Z, derived from staphylococcal protein. The lead peptide, SDFYKRLINKAKTVEGVEALKLHILAALP (ABD035), has human serum albumin affinity in the femto molar range, and has been used to extend the circulating half-life of several proteins and peptide that are in preclinical or clinical phases.

9.2.2.3 Fusion Proteins and Conjugates

The first approved albumin fused protein was the GLP-1 HSA fusion protein, albiglutide, discovered by Genentech and developed by GlaxoSmithKline. As a free *N*-terminal is important for GLP-1 activity the peptide was fused via the *C*-terminal to HSA. The construct is composed of a tandem repeat of Ala8 to Gly8 GLP-1 (7–36) that is fused to the *N*-terminal of HSA. Gly8 GLP-1 was used in order to protect for degradation of the *N*-terminal by DPPIV. The plasma half-life was extended from approximately 2 minutes to about 6–8 days which made this fusion construct applicable for once-weekly dosing. One drawback is the reduced potency that is most likely due to a combination of the Gly8 modification and the covalent attachment to the larger HSA protein. The tandem repeat of GLP-1 was used to obtain greater distance between the albumin and the distal GLP-1 peptide, but the GLP-1 receptor potency of albiglutide is 0.61 nM compared to 0.02 nM for native GLP-1. The protein was approved for treatment of diabetes in 2014.

Aiming for an alternative to PEGylated interferon α, Genentech developed a fusion protein composed of albumin and interferon-α2b. This fusion protein had a half-life of about 6 days which gave potential for dosing every two to four weeks.

It has also been attempted to generate long action insulin by direct fusion of single chain insulin to HSA. The fusion protein, called albulin, has greatly extended the half-life in mice and a sustained blood glucose lowering effect, but has apparently not entered clinical trials.

Chemical conjugation to albumin has been investigated by several groups. One unique feature of albumin is that it has a free cysteine in position 34 that can be coupled to electrophiles like maleimide groups (Figure 9.2). CJC-1134-PC is an exendin-4 (a GLP-1 agonist) developed by ConjuChem that is chemically coupled to Cys34 in HSA via a maleimide containing linker. CJC-1134-PC has a half-life of approximately 8 days in humans which is significantly longer than exendin-4 itself. This chemical conjugate was tested to be effective in the regulation of blood glucose in several rodent models and is currently in phase II clinical trials. The same technology platform has been used to explore other pharmaceutical peptides such as insulin.

9.2.3 Fc-Fusions and Conjugates

It is the Fc domain of the IgG antibody that binds to FcRn and ensures that antibodies are recycled and prolonged. Obviously, this protein domain has been used as a fusion partner to extend the plasma half-life of various approved biopharmaceuticals. The first Fc-fusion of this type was enternacept (Amgen/Pfizer) for treatment of rheumatoid arthritis, it was approved by the FDA in 1998 and several others have since been launched including alefacept (Astellas Pharma, 2003), abatacept (BMS, 2005), romiplostim (Amgen/Pfizer, 2008), rilonacept (Regeneron, 2008), aflibercept (Regeneron, 2011), belatacept (BMS, 2011), and more recently GLP-1 Fc-fusion dulaglutide (Eli Lilly, 2014). Most of these proteins act as antagonists to block receptor binding and only a few have agonistic activities (e.g., alefacept, romiplostim, and dulaglutide). It is of critical importance not to introduce immunogenic features in such fusion proteins. Often, fusion proteins are produced as fully recombinant proteins in mammalian host cells in order to control the glycosylation pattern to be fully humanized.

9.3 PEPTIDE AND PROTEIN ENGINEERING

As discussed earlier, the renal clearance can be reduced significantly by increasing the hydrodynamic size of the drug target. Another important parameter influencing elimination beyond receptor clearance is the inactivation of the peptide or protein due to proteolysis. It is therefore very important that peptides and proteins are resistant enough against plasma degradation to provide a sufficiently long half-life.

9.3.1 Increasing the Enzymatic Stability

Proteolytic enzymes in plasma may cleave and inactivate the peptide or protein, and, in general, peptides are much more prone to enzymatic degradation than larger proteins. The rigid structure of the protein helps to stabilize against proteolytic cleavage since proteases require an extended form of the backbone in order to bind and cleave the peptide substrate. In contrast, peptides may possess secondary structure such as an α-helix that is also partly unfolded making them more susceptible to cleavage. This is very easily determined with the help of liquid chromatography mass spectrometry (LC-MS) of plasma samples and helps the elucidation of the cleavage site as it occurs in vivo. Once the degradation pattern of the metabolites is identified, the design of more proteolytically stable peptide analogs can take place.

Since proteases are highly promiscuous, subtle changes of the amino acid side chain in the proximity of the cleavage site, such as changing a residue to a homologous residue, e.g., leucine to a valine, may have a very limited stabilizing effect. On the other hand, radical changes such as the exchange of lysine with a glutamic acid or proline residue, may indeed introduce enzymatic stabilization, but may also compromise biological activity. It is the balance of keeping the potency of the target peptide intact and at the same time stabilizing the peptide against cleavage, which in some cases involves several enzymes, that is important in drug design. In the stabilization of peptides, the most conventional approach is to substitute with another proteinogenic amino acid which ensures the production process can be made recombinant. Alternatively, nonproteinogenic amino acids can be employed, and as an example in semaglutide, where Aib8 in GLP-1(7–37) helps to stabilize against DPPIV degradation.

9.3.1.1 Global Modifications

Since the protein fold assists in stabilizing against proteolytic cleavage, other approaches that increase the stability of three-dimensional structure even further may be employed. This could be the introduction of additional disulfide bridges or salt bridges. The α-helix is the most abundant secondary structure in protein structures comprising of ca. 30%, and while the α-helixes are stabilized by the hydrophobic core of the protein structure itself, the helix is much more flexible in peptides. An α-helix has 3.4 residues per turn and therefore on average 3–4 residues will face the same side of the helix. An often used approach in increasing the enzymatic stability of peptides is the stabilization of the α-helix which in peptides can be accomplished by the formation of a linkage between the side chain that face the same side, e.g., residues i to a residue i+3 or i+4. One popular approach to achieve this has been the introduction of salt bridges between, such as, a lysine or arginine and a glutamic or aspartic acid. A linkage could be the formation of a disulfide bridge between Cys residues i and i+3 (Figure 9.4a) and another approach is to covalently link the side chains of lysine and glutamic acid (Figure 9.4b). While the salt bridge still allows for flexibility of the peptide, the covalent bond is much more rigid, and has a more substantial effect on the α-helix stabilization. Often structural information about the conformation of the peptide while bound to the receptor is not known and a salt bridge may allow for the flexibility needed to accommodate the right active structure. This may not be the case when the peptide has less structural freedom due to a linkage. Not surprisingly the position and size of this linkage may therefore be very critical. On the other hand the covalent linkage of side chains allows for a larger space of opportunities, since the size of

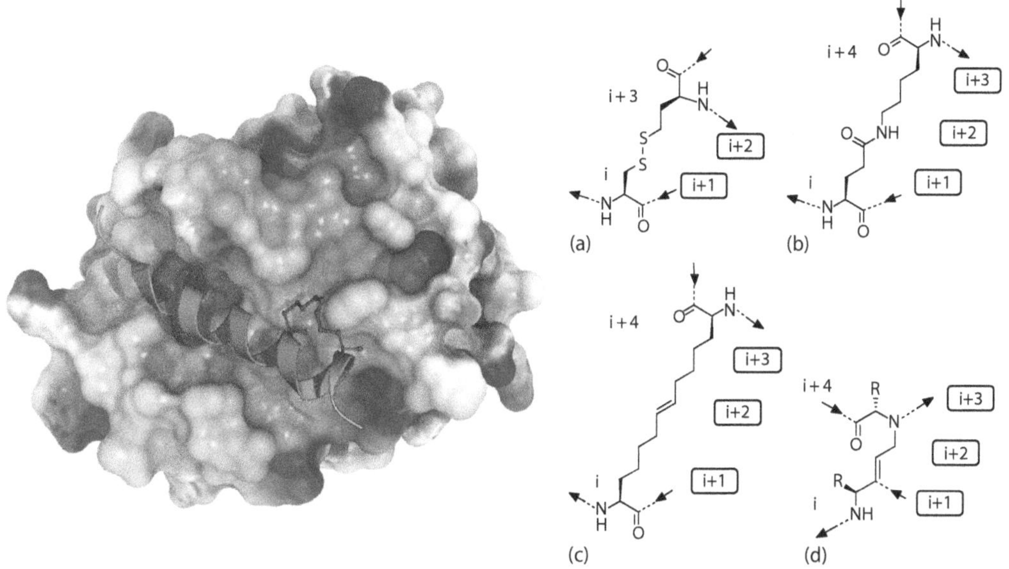

FIGURE 9.4 A structural representation of the interaction of a stabilized (hydrocarbon type of stabling) alpha-helix peptide (MCL-BH3) interacting with the myeloid cell leukemia 1 (MCL-1) receptor (Pbd ID: 3MK8). The linkage itself interacts with the receptor. On the right part of the figure is represented four types of linkages disulfide bridge (a), lactam bridge between Glu and Lys (b), a hydrocarbon type of stabling (c) and a hydrogen surrogate type of linkage (d).

this linkage may be varied, increasing the possibilities to adjust parameters such as receptor potency and/or selectivity. The linkage itself may also be a part of the interaction with the receptor and not just a stabilization of the α-helix as visualized in Figure 9.4.

Various types of chemistries exist for the covalent formation of linkages. Besides the popular lactam formation between a lysine and glutamic acid, linkages that are based on the carbon–carbon bond type (Figure 9.4c) have gained some interest in the recent years. These types of linkages are completely stable against degradation in vivo. Besides making them more stable toward enzymatic degradation, they may also play a direct part in the receptor interaction. In 2005, Aileron Therapeutics was founded which utilized a variety of different forms of linkages to identify potent peptide-based protein–protein inhibitors. In addition, it was claimed that these stabled peptides possess improved properties such as better oral bioavailability and even cell penetration properties which could lead to highly efficacious peptide drugs. In addition, another type of linkage termed hydrogen bond surrogate mimics that of the hydrogen bonding pattern normally observed between i and i+4 peptide bonds in α-helixes (Figure 9.4d).

9.3.1.2 Peptide Bond Modification

The stabilization of the labile peptide bond itself is a more direct and efficient approach. The amide bond can be stabilized in a number of ways. Popular approaches include *N*-methylation, *N*-substituted glycine (peptoid), Cα-methyl, and reduced peptide bond (Figure 9.5). The labile peptide bond may very well be a part of the receptor interaction and thus the risk of losing too much potency is likely. The introduction of *N*-methyl amino acids or a D-amino acid are also not compatible with preserving a stable secondary structure such as an α-helix, so even if these amino acid substitutions are outside of the direct receptor interaction they may have negative effect on the potency. On the other hand both D- and *N*-methyl amino acids are generally more accepted in, e.g., β-turns or smaller cyclic peptides. In smaller, notably, cyclic peptides, the effect of altering the peptide bond may be large and the structural space of conformations may lead to biologically inactive peptides, but in

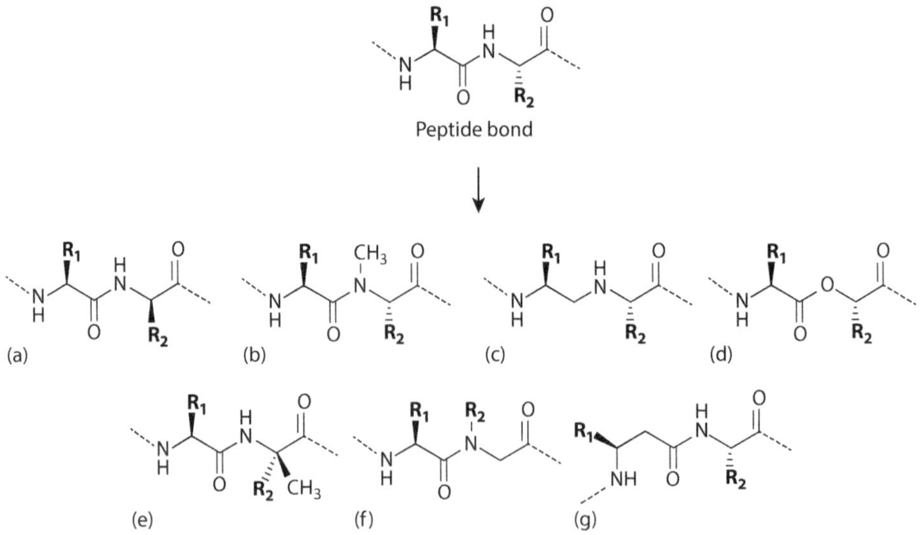

FIGURE 9.5 Examples of various structures that can be introduced in place for the labile peptide bond; D-amino acid (a), *N*-methyl-amino acid (b), reduced peptide bond (c), ester bond (d), Cα-methyl (e), *N*-substituted glycine (peptoid monomer) (f), and β-amino acid (g).

some cases very selective and potent compounds can be identified. Either way it often requires an empirical approach in which a larger set of peptides are synthesized and screened for receptor activity. Other consideration in the design phase is ease of synthesis of these analogs and again later on the cost-effectiveness in a larger scale production. As an example, the introduction of an ester bond requires the synthesis of building blocks that are not readily available. This naturally impedes the synthesis of a larger collection of peptides in a screening process. In contrast, the introduction of *N*-methyl amino acids (or D-amino acids) where a larger number of building blocks are commercially available or otherwise readily synthesized from relative inexpensive starting materials, makes it feasible to create large collections (arrays) of peptides by standard solid phase peptide synthesis (SPPS) that can be tested in various high-throughput receptor assays.

9.3.2 Methods for Structure–Activity Relationship

Getting usable information concerning the structure–function–activity relationship in a quantitative manner (Q-SAR) is an important part of the drug discovery process. Molecular modeling plays a pivotal role in this design process of biopharmaceuticals and is discussed in Chapter 4. While design of experiment (DoE) approaches are arguably gaining more and more attention in many disciplines, the way to intelligently reduce the number of analogs and still gain useful information about the SAR is a challenge, since residues in close proximity may influence each other's activity. In cases where there is no structural knowledge of the ligand–receptor interaction, a more empirical approach, e.g., the synthesis and screening of many analogs in a parallel or combinatorial fashion can prove highly advantageous.

9.3.2.1 Replacement and Analysis

For decades, simple truncation and replacement analysis in the peptide or protein have yielded very useful information about the receptor interaction. By replacing the residues with alanine (Ala-scan) which only displays a methyl group as the side chain, the significance of the individual side chains can be assessed since any peptide bond interaction is preserved (Figure 9.6). The key residues are often referred as hot spots in the ligand and are considered to be the most essential for

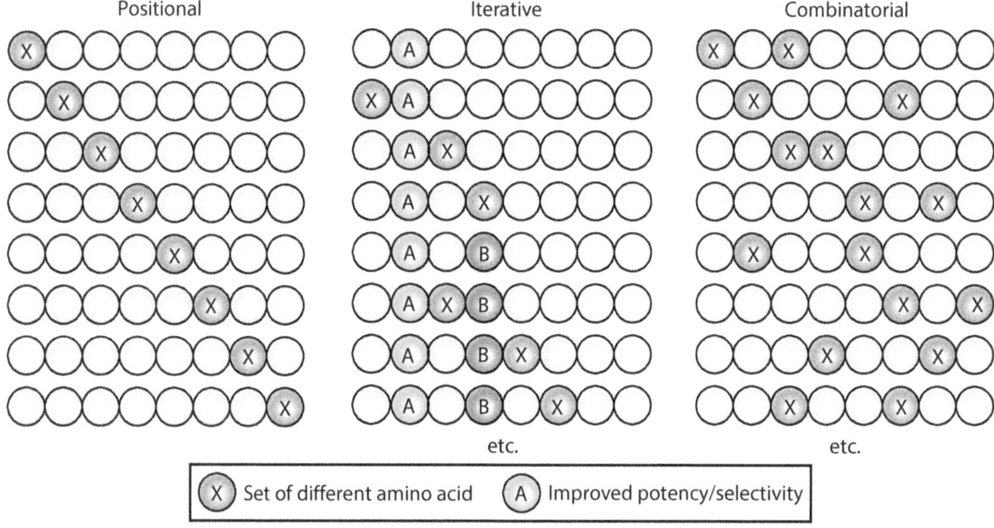

FIGURE 9.6 Schematic representation of different approaches for replacement analysis. In positional scanning (e.g., Ala-scan) each position is replaced in the context of the native sequence. In the iterative approach positions are replaced in the context of a new optimized sequence and in the combinatorial approach replacement analysis is done in a random approach allowing structural changes from more than one residue.

the binding activity. It is however challenging to unambiguously determine if a particular residue is responsible for the direct interaction with the receptor, or if it is more important for maintaining the correct fold and positioning of other residues. In situations where the peptide or protein is an agonist which is most often the case when investigating endogenous peptides and proteins, it is important to determine both the potency (agonist efficacy) and the binding affinity. Although these tend to follow the same trend, some amino acid replacements may affect the binding very little, but may have a dramatic impact on potency, whereas changes that have a hugely negative effect on the affinity also negatively affect potency as well. The in vitro data acquired from an Ala-scan can point to the regions of the peptide or protein where the interaction with the receptor occurs, however, it is important to realize that the Ala-scan only reveals the effect when the side chain is absent and not *per se* the type of surroundings that the side chains are engaged in. However, once the positions of critical residues are known in the peptide or protein, then more dedicated efforts can be initiated in order to enhance potency or selectivity by the introduction of other residues by an iterative approach (Figure 9.6). Peptide chemists have hundreds of amino acids that are commercially available to work with. But it is also important to mention that backbone modification, including the introduction of rigidity to the backbone, can be an important part of this molecular toolbox.

9.3.2.2 Multi-Parallel Synthesis and Peptide Arrays

The SPPS approach in plate format is an attractive tool for the parallel synthesis of hundreds or even thousands of different peptide analogs. It is much less labor intensive to synthesize many analogs in parallel as compared to recombinant expressions of proteins, and by integrating the SPPS format with in vitro pharmacology screening assays vast in vitro biological data sets can be obtained that facilitate the drug design process. Since residues in close proximity can affect each other, combinatorial replacement analysis may reveal particular combinations that are not readily identified by other means (Figure 9.6). For certain ligand–receptor interactions, even more complex synthesis and screening methods can be used such as paper (SPOT array) and chip-based arrays. In situations where the target receptor exists as a free soluble and stable receptor, peptide arrays offer the opportunity to screen thousands of peptide analogs with respect to binding affinity. The former spot array method using paper sheets is a low cost methodology that has gained success notably within the area of proteomics and epitope mapping due to simple synthesis equipment and standard SPPS conditions. It allows for the rapid synthesis of several hundred to a few thousand peptides and a simple screening procedure that can identify the ones with highest binding affinity. The latter chip-based method utilizes photolithographic techniques and noncommercially available building blocks to construct the arrays. Due to this large number of peptides, both technologies offer the opportunity to use simple positional scanning or even combinatorial design strategies (Figure 9.6).

9.3.3 De Novo Design

The term "de novo design" essentially describes a design process leading to a biologically active, novel, peptide or protein without an active structural template to begin from. While most biopharmaceuticals (excluding antibodies) to date are modified versions of nature's repertoire of peptides and proteins, the erythropoietin mimetic peginesatide (hematide) from Affymax, and a peptide-Fc fusion thrombopoietin mimetic, romiplostim by Amgen, were the first de novo designed protein mimetics to be approved by FDA. Hematide binds and activates the erythropoietin receptor leading to red blood cell production, and romiplostin stimulates the formation of blood platelets with a prolonged effect due to FcRn recycling. Despite the subsequent withdrawal from market of hematide, both cases are hallmarks of what can be accomplished by combining de novo *design* using large combinatorial libraries to identify an active peptide, and using peptide or protein drug design to make a suitable drug based on the peptide.

9.3.3.1 Biological Approaches

In the very early 1990s, the random peptide phage display approach appeared and literally revolutionized de novo design. It is a technology where hundreds of millions to billions of random (or semi-random) peptides are displayed on the surface of bacteriophages. A simple screening scheme then allows for the identification of peptides binding to the immobilized protein target. Various formats have since appeared that do not employ phages but are done directly on ribozymes as an example. In addition to peptides, larger proteins may be expressed. Indeed, antibody-based phage display has revolutionized the way pharmaceutical companies identify protein–protein interaction inhibitors using various protein scaffolds. Serving as vast source of potential drug molecules, these huge libraries are screened against the target of interest. Notably, within the area of cancer and inflammation, a number of blockbuster antibody drugs have reached the market demonstrating the success of this approach. Less attention has been given to using peptide display technology as the direct drug. One reason for this is that protein scaffolds often offer the identification of binders with a higher affinity than linear peptide-based binders. However, cyclic peptides as well as bicyclic variants may also offer ligands that can bind to protein surfaces with good affinity. Another reason is that many receptors of drug interest are G-protein-coupled receptors (GPCR or 7TM-type), and those specific receptors are not readily screened in these phage libraries. Finally, it is believed that although high-affinity peptides may be found against a target receptor, the half-life is much shorter than the corresponding half-life of, e.g., a humanized antibody.

9.3.3.2 Synthesis Approach

One approach first published in 1991 that is similar to the phage method was termed one-bead-one-peptide library or one-bead-one-compound (OBOC), since the random peptides were displayed on small beads. The method relies on the synthesis of millions of random (or semi-random) compounds synthesized by a method termed split-and-mix synthesis. This ensures that each and every bead in the library only displays one unique sequence. The whole set of beads comprising the library can then be screened against a biotin or fluorescent-labeled target receptor and only the active beads are retrieved from the library and analyzed. As opposed to the phage technology the synthetic bead library approach only allows the synthesis of shorter peptides, and not larger proteins. On the other hand, in addition to the proteinogenic amino acids that are used, a vast source of nonproteinogenic amino acids can be incorporated as well. In addition, other types of compounds, including other types of peptidomimetics and small molecules can be displayed on the beads. Since the screening is based on the direct binding of the soluble purified receptor to the immobilized ligand on the surface of the bead, and not the detection of a functional readout, the risk of identifying ligands that are nonspecific or of no relevance is quite high. In some cases this has led to disappointing results, but in cases where specific binding motifs were identified the method was very efficacious in identifying de novo peptides that were used as a starting point for further development. Disappointingly, GPCR and other integral membrane receptors are not readily suited for this type of solid combinatorial library approach, and the use of the split-and-mix methodology as a general drug discovery platform has had limited application in the pharmaceutical industry in the past decade. However, for specific protein targets, it still represents a vital source for de novo design of ligands and also as a tool for identifying enzyme specificity using on-bead analysis.

9.3.4 Chemical and Biophysical Stability

The chemical stability of peptides and proteins concerns a change within the molecule itself while biophysical stability describes the ability to self-associate into a diverse set of oligomers or aggregates that is different from the native structure. While chemical stability is easier to predict and quantify by analysis, foreseeing aggregation problems is more difficult due to the fact that this has many causes.

Of the most common events concerning chemical instability is the deamidation of asparagine residues that can cyclize to a five-membered succinimide ring that re-opens to aspartic acid, D-aspartic acid, L-iso-aspartic acid, or D-isoaspartic acid (Figure 9.7). The neighboring residue to the *C*-terminal side (n+1) significantly affects deamidation rate. Increasing pH also accelerates this process. A secondary structure, such as an α-helix, lowers the rate of formation since the flexibility of the peptide chain is reduced. Therefore, peptides are generally more prone to deamidation

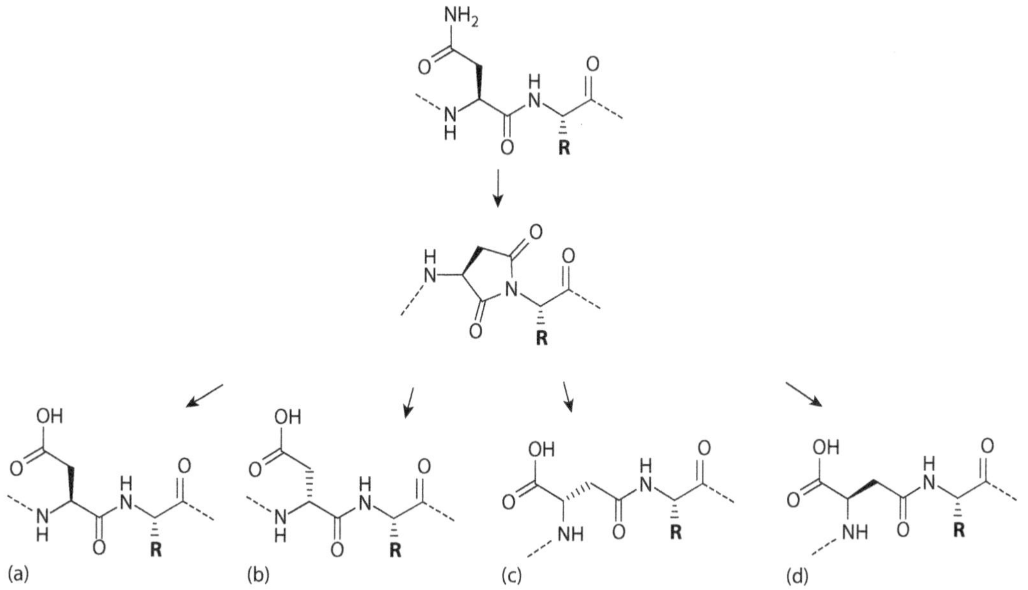

FIGURE 9.7 Deamidation and isomerization of asparagine and aspartic acid, respectively, occurs via a succinimide formation and results in the formation of L-aspartic acid (a), D-aspartic acid (b), L-isoaspartic acid (c), and D-isoaspartic acid (d).

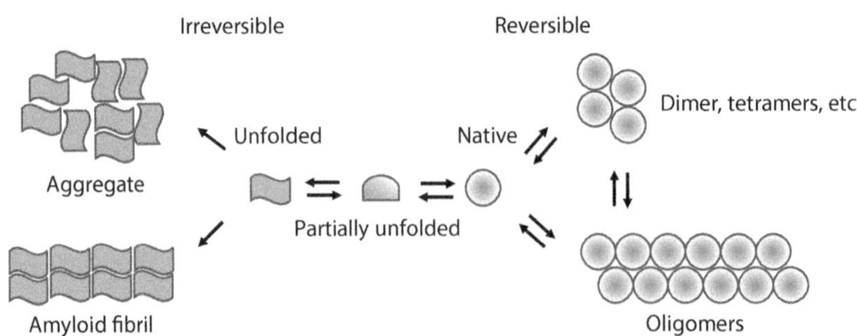

FIGURE 9.8 The conformational changes of a protein in formulation that may lead to reversible or irreversible self-assembly.

than proteins where deamidation more frequently occurs in loops or flexible regions. In addition, aspartic acid can undergo cyclization via succinimide formation but the rate is favored at low pH. Glutamine can deamidate via glutarimide; however, the rate is roughly 100 times lower compared to asparagine, except when glutamine is the *N*-terminal residue, where the rate is very rapid leading to pyroglutamine. Other residues may also pose problems and free cysteines in a peptide or protein in particular impair stability due to the formation of disulfide bridges. In peptides and proteins containing disulphide bridges, these may re-open and form pairs with other cysteines at high pH > 10, a phenomenon termed disulfide scrambling or shuffling that may lead to misfolding and loss of activity. Other sensitive residues are methionine that oxidizes to methionine sulfoxide upon storage, serine (dehydration), and tryptophan (oxidation). The significance, however, may not be as critical as deamidation and disulfide shuffling. It is worth pointing out that the mentioned side reactions also occur when the peptide or protein is lyophilized, albeit at a lower rate.

Peptides and proteins may oligomerize or aggregate and this process is extremely complex. While the former may refer to the association of native structures and be reversible, the latter refers to the association of unfolded forms and most often an irreversible event (Figure 9.8). Aggregation may be insoluble fibrous proteins (fibrils), which are often linear ordered aggregates, or, amorphous and disordered structures. Both may be found in a range of different sizes. It is widely accepted that irreversible aggregation also increases the risk of immunogenicity, so it is of great importance to include biophysical analysis early in the design phase. Factors that influence the oligomerization or aggregation potential includes change in primary structure (mutations), attachment of side chain moieties, the expression system (e.g., altered type of glycosylation), excipients, liquid–air interface, pH, and temperature.

9.3.5 Immunogenicity

The formation of anti-drug antibodies is a challenge unique to biopharmaceuticals, as small molecules very rarely cause an antigenic response. The formation of peptide or protein aggregates may lead to anti-drug antibodies even if the peptide or protein is identical to the endogenous version. B-cell activation and subsequent antibody formation may be T-cell independent and are thought to occur by the repetitive nature of aggregates, and in that perspective the quality of protein drug formulation is of high concern. Another way to be immunogenic is when the biopharmaceutical has a critical number or type of amino acid substitutions that alter existing or create new T-cell epitopes and interacts with the immune cells. This will lead to an IgG antibody response due to B-cell activation mediated by T-cell stimulation. In some cases, the antibody response may even lead to loss of function of the endogenous protein. In at least a few cases, patients treated with human erythropoietin developed anti-erythropoietin antibodies resulting in pure red cell aplasia. Within the field of hemophilia, neutralizing antibodies are common and roughly 15%–30% of congenital hemophilia

A patients develop neutralizing antibodies against FVIII. In the case of interferon-β 1a (INF-β) used in the treatment of multiple sclerosis, approximately ca. 25% of the patients develop antibodies that neutralize the function of INF-β. On the other hand, treatment with the GLP-1 agonist exenatide derived from the Gila monster causes an antibody response in 30%–40% of the patients without any significant impact on the efficacy of the drug, although in some of the high antibody titer patients less efficacy was found indicating that the antibody response may have had an impact.

9.3.5.1 Predicting Immunogenicity

No reliable experimental methods for predicting immunogenicity currently exist, but some *in silico* predictions may provide useful information with respect to the consequences of mutations in the primary structure between analogs of relevance. As such, these methods could be of relevance in the *ranking* of otherwise equal protein analogs but are of no value for peptides with unnatural building blocks, conjugations to various molecular moieties, or other post-translational modifications since the models are based on the natural T-cell epitope repertoire. Ex vivo methods may be an alternative way to predict the level of T-cell stimulation. One method is to determine the T-cell response after incubation of representative pools of T cells from humans with the drug analogs and to then rank potential risk of similar drug candidates.

9.4 CONCLUDING REMARKS

Biopharmaceuticals will gain more market shares in various disease areas in the years to come. Predominantly, in the fields of inflammation, hemophilia, cancer, diabetes, inflammation, as well as infectious diseases, peptides and proteins have shown their value and huge potential. New methodologies for extending half-life will likely result in new analogs among peptides and proteins that possess even greater efficacy not only in existing fields but also in new types of diseases. Progress has also been made in the area of oral formulation of larger peptides and may offer new future opportunities within the oral delivery of peptide pharmaceuticals.

FURTHER READING

Azzarito, V., Long, K., Murphy, N.S., and Wilson, A.J. 2013. Inhibition of α-helix-mediated protein–protein interactions using designed molecules. *Nat. Chem.* 5:161–173.

Elsadek, B. and Kratz, F. 2012. Impact of albumin on drug delivery–new applications on the horizon. *J. Control. Release* 157(1):4–28.

Frøkjær, S. and Otzen, D.E. 2005. Protein drug stability: A formulation challenge. *Nat. Rev.* 4:298–306.

Giorgi, M.E., Agusti, R., and de Lederkremer, R.M. 2014. Carbohydrate PEGylation, an approach to improve pharmacological potency. *Beilstein J. Org. Chem.* 10:1433–1444.

Kang, J.S., Deluca, P.P., and Lee, K.C. 2009. Emerging PEGylated drugs. *Expert Opin. Emerg. Drugs* 14(2):363–380.

Ratanji, K.D., Derrick, J.D., Derman, R.J., and Kimber, I. 2014. Immunogenicity of therapeutic proteins: Influence of aggregation *J. Immunotoxicol.* 11(2):99–109.

Robinson, N.E. and Robinson, A.B. 2004. *Molecular Clocks: Deamidation of Asparaginyl and Glutaminyl Residues in Peptides and Proteins.* Althouse Press, Cave Junction, OR (ISBN 1-59087-250-0).

Sleep, D., Cameron, J., and Evans, L.R. 2013. Albumin as a versatile platform for drug half-life extension. *Biochim. Biophys. Acta* 1830(12):5526–5534.

Walsh, G. 2014. Biopharmaceutical benchmarks 2014. *Nat. Biotechnol.* 32:992–1000.

10 Prodrugs in Drug Design and Development

Jarkko Rautio and Krista Laine

CONTENTS

10.1 INTRODUCTION

Drug discovery is an exceedingly complex and demanding enterprise aiming to select a drug candidate that displays promising efficacy, safety, and marketing potential for clinical trials. The initial phase of the drug discovery process is devoted to increase pharmacological potency with high affinity and selective ligands at the target. This initial phase often results in lead compounds with less than optimal biopharmaceutical and pharmacokinetic properties, or properties more frequently referred to as ADME (absorption, distribution, metabolism, and excretion) properties. Designing prodrugs, when used appropriately, has the potential to enable a suitable drug candidate with the optimal ADME properties and pharmacological potency to be selected for development faster and with less overall costs. Fortunately, prodrugs are becoming an integral part of many drug discovery efforts.

10.2 PRODRUG CONCEPT

Prodrugs are derivatives of active parent drugs that require in vivo conversion, either an enzymatic and/or a chemical, to release the pharmacologically active drug. Because prodrugs are considered to be inactive or at least significantly less active than the released drugs, salt forms and drugs, whose metabolites contribute to the overall pharmacological response, are not considered as prodrugs. Typically, prodrugs are chemical derivatives consisting of a promoiety which is released in one or two chemical or enzymatic (or both) steps to release the active parent drug. Bioprecursor prodrugs

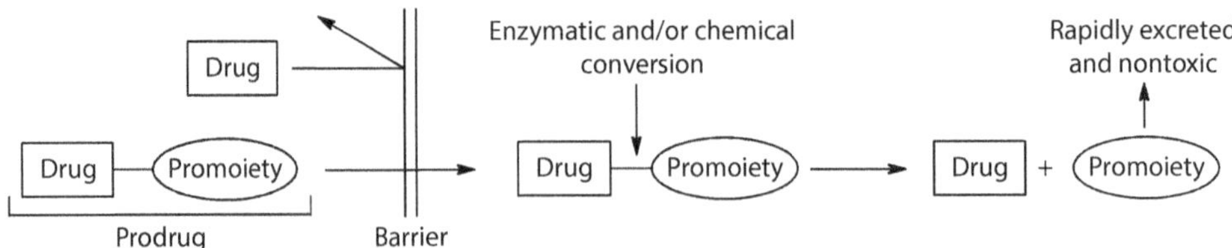

FIGURE 10.1 Simplified illustration of the prodrug concept.

do not contain an obvious promoiety, but these prodrugs form the active compound through one, or with today's increasing levels of technical sophistication, through a cascade of metabolic reactions such as oxidation, reduction, hydrolysis, or phosphorylation in vivo. In some rare cases, a prodrug may consist of two active drugs which are merged together in a single molecule. These derivatives, in which each drug acts as a promoiety for the other, are called codrugs. Finally, also soft drugs rely upon bioconversion to dictate their course of action. However, soft drugs are not a class of prodrugs. Opposite of prodrugs, soft drugs are active molecules that become deactivated by the bioconversion process after exerting their therapeutic effect. The prodrug concept is illustrated in Figure 10.1.

10.3 HISTORY AND CURRENT PREVALENCE OF PRODRUGS

The term "prodrug" was first introduced by Adrien Albert in 1958. However, the prodrug concept had been implemented long before Albert's publication. Early examples of prodrugs (Figure 10.2) are methenamine (or hexamine) and acetylsalicylic acid (or aspirin) which were introduced in 1899. Methenamine was intentionally designed to release antibacterial formaldehyde along with ammonium ions in acidic urea for the treatment of urinary tract infection. In the similar way, acetylsalicylic acid was designed to be a less-irritating and better-tasting replacement for the anti-inflammatory drug salicylic acid. Unlike methenamine and acetylsalicylic acid, prontosil that was introduced in 1935 and isoniazid that was introduced in 1952 were not intentionally designed as prodrugs, but their prodrug nature was revealed in hindsight. The discovery of sulfanilamide being the active antibacterial metabolite released from prontosil by reductive enzymes gave rise to the era of sulfonamide antibiotics. The prodrug nature of the antituberculosis drug isoniazid was discovered in hindsight more than 40 years after its launch. Bioconversion of isoniazid is catalyzed by the mycobacterial catalase-peroxidase called KatG. The reactive species generated by bioconversion form adducts with NAD^+ and $NADP^+$ which are potent inhibitors of biosynthesis of mycolic acid required for the mycobacterial cell wall.

Since the 1960s, there has been an explosive increase in the use of prodrugs in drug discovery and development. Today, the interest in prodrugs is evident based on published journal articles,

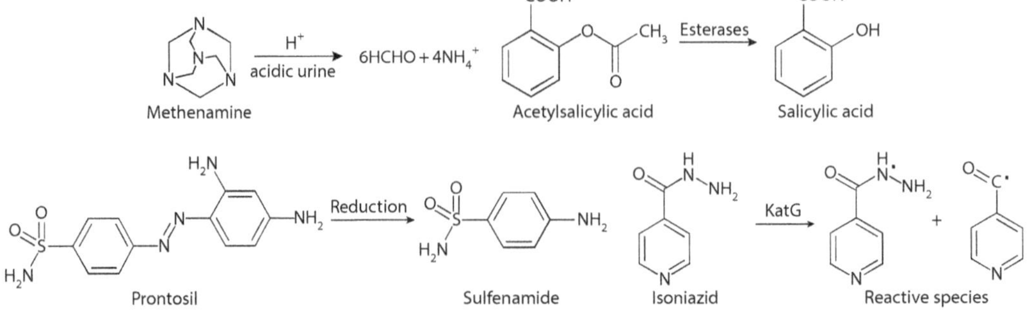

FIGURE 10.2 Early prodrug examples in clinical use.

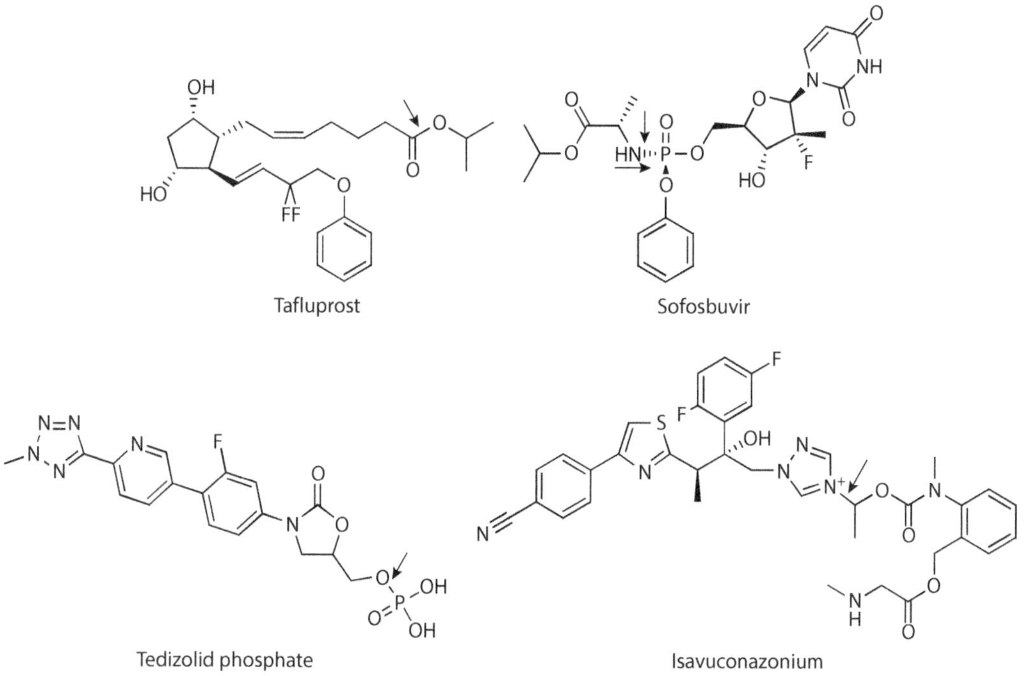

Tafluprost

Sofosbuvir

Tedizolid phosphate

Isavuconazonium

FIGURE 10.3 Recent prodrugs in clinical use. Bonds between an active drug and a promoiety are indicated by arrows.

reviews, and patents, as well as on the number of clinical studies and marketed prodrugs. From 2010 to 2014, 127 small molecular weight drugs were approved by the FDA. Of these, at least 12 drugs, or 9.4%, were prodrugs. Some of the recently approved prodrugs are tafluprost (in 2012), an ocular isopropyl ester for the treatment of glaucoma and ocular hypertension, an oral phosphoramidate prodrug sofosbuvir (2013) for the treatment of hepatitis C, as well as orally and/or intravenously given tedizolid phosphate (2014) and isavuconazonium (2015) for the treatment of skin and fungal infections, respectively (Figure 10.3). It is estimated that approximately 10% of all worldwide marketed drugs can be classified as prodrugs.

10.4 DESIGNING OF PRODRUGS

The design of a prodrug structure should always be considered at the early stages of preclinical development, keeping in mind that prodrugs may alter the tissue distribution, metabolism, efficacy, and even toxicity of the parent drug. For example, the intended hydrolysis of the inactive lactone form to the pharmacologically active hydroxyl acid of a HMGCoA reductase inhibitor simvastatin is relatively fast, but the more lipophilic lactone also predisposes the drug to unwanted metabolism by cytochrome P450 enzymes. Many times, the prodrug design can, however, represent a comparable smaller challenge and be more feasible than the alternative of searching for a new therapeutic agent that also inherently possesses the desired properties. The prodrug strategy can also enable the selection of a suitable drug candidate faster and with lesser costs. There are several factors that should be carefully considered when designing prodrug structures:

- What is the limitation of the parent drug that needs prodrug intervention?
- Which functional groups on the parent drug are suitable for chemical modification?
- Can the prodrug be readily synthesized?

- Chemical modification must be reversible and allow the parent drug to be released by an in vivo chemical and/or enzymatic reaction at the rate that is greater than the elimination rate of the prodrug or parent drug.
- The promoiety should ideally be safe and rapidly excreted from the body. The selection of promoiety should be considered with respect to the disease, the dose, and the duration of therapy.
- The absorption, distribution, metabolism, and excretion (ADME) properties of both the parent drug and prodrug need to be comprehensively understood.
- Can bioavailability in humans be predicted, with a high degree of certainty, using preclinical animal models?
- Is the final prodrug formulation sufficiently chemically and physically stable with a reasonable shelf life?

Some of the most common functional groups on parent drugs that are amenable to chemical modification to form prodrugs include hydroxyl (—OH), carboxyl (—COOH), as well as basic and acidic NH-groups. Prodrugs typically produced via the modification of these groups include esters, carbonates, carbamates, amides, phosphates, as well as various *N*-acyl derivatives and *N*-Mannich bases. Also phosphate (—OPO(OH)$_2$) and phosphonate (—C—PO(OH)$_2$) groups present in various nucleoside analogs have been very popular targets in prodrug design resulting in many different types of experimental and clinically approved phosphate and phosphonate nucleoside prodrugs. Figure 10.4 illustrates prodrug structures for the most common functional groups of parent drugs. Not surprising that the most common prodrugs are those requiring a hydrolytic bioconversion in vivo by ubiquitous hydrolases. Less frequently prodrugs are designed to be bioconverted by oxidative or reductive processes mediated by enzymes such as cytochrome P450, monoamine oxidase, nitroreductase, or azoreductase. The use of these enzymes is often sought out for liver or cancer targeting prodrug approaches, and some of the examples are represented in more detail in Section 10.5.

Besides utilization of enzymes to carry out necessary bioconversion of prodrugs to their respective active drugs, a wide variety of different prodrug bonds are designed to undergo a spontaneous nonenzymatic reaction. For example, lower pH in tumor environment has prompted the exploitation of acid-sensitive prodrug bonds such as acetals, ketals, hydrazones, and imines. *Cyclo*SAL phosphate and phosphonate prodrugs have recently been extensively studied to deliver nucleoside analog monophosphates. These prodrugs are stable in acidic media but undergo a chemical hydrolysis to generate nucleoside analog monophosphates and salicyl alcohol under basic conditions. Hydroxyalkyl and especially hydroxymethyl prodrugs of carboxylic acids and NH-acidic groups are known to undergo rapid spontaneous conversion at pH 7.4 resulting in the formation of an aldehyde (e.g., formaldehyde from hydroxymethyl derivative) and a respective parent drug. While the hydroxymethyl esters of carboxylic acids are highly unstable, the pK_a value of NH-acid influences on the rate of conversion of *N*-hydroxymethyl prodrugs with the more acidic parent drugs giving the fastest conversion. For *N*-hydroxymethyl prodrugs, a conversion half-life is less than 1 hour for NH-acids having the pK_a less than 13. While hydroxymethyl prodrug approach offers very little design flexibility, its practical utility can be extended to double prodrug approach where a terminal hydroxyl group is further phosphorylated or acetylated. These phosphoryloxy or acyloxyalkyl prodrugs undergo a two-step cleavage mechanism where the first rate-determining step is the enzymatic hydrolysis of the terminal acyl group with a subsequent spontaneous chemical reaction releasing the active drug. As a representative example (Figure 10.5), after cleavage of the phosphate group, dehydroxymethylation has been shown to be very rapid (less than 2 seconds at pH 7.4 and 37°C) for a relatively strong NH-acid, phenytoin. In comparison, a conversion half-life of *N*-hydroxymethyl intermediate of a weaker NH-acid, aripiprazole, should be in hours.

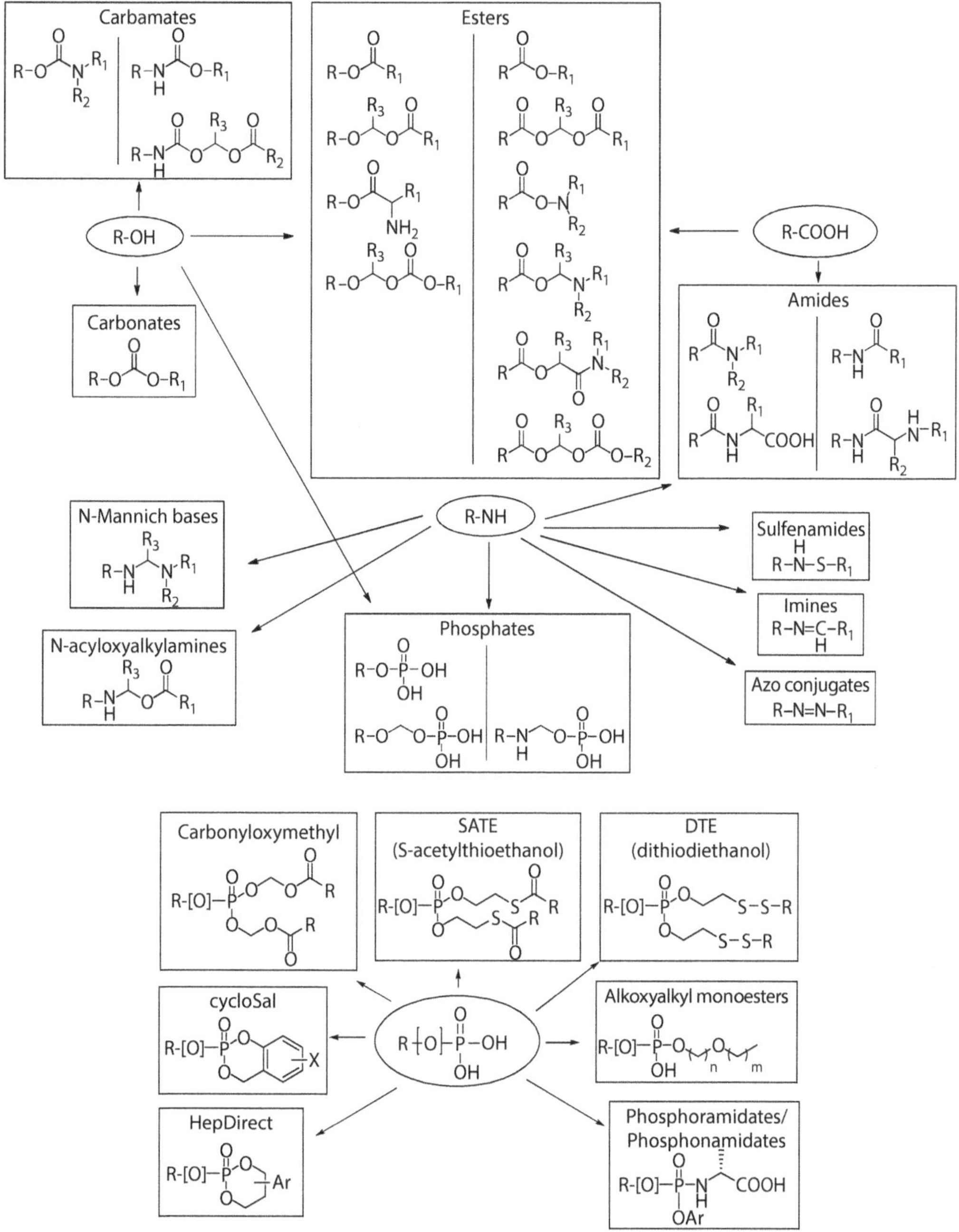

FIGURE 10.4 Functional groups amenable to prodrug design and representative prodrugs.

10.5 APPLICATIONS OF PRODRUGS

Prodrugs have offered a versatile approach to enhance the clinical usefulness of existing drugs by improving their physicochemical, formulation, or pharmacokinetic properties. Additionally, prodrugs have been designed to increase the selectivity of drugs to their intended target. Design of prodrugs with improved properties may also provide an opportunity for commercial life cycle

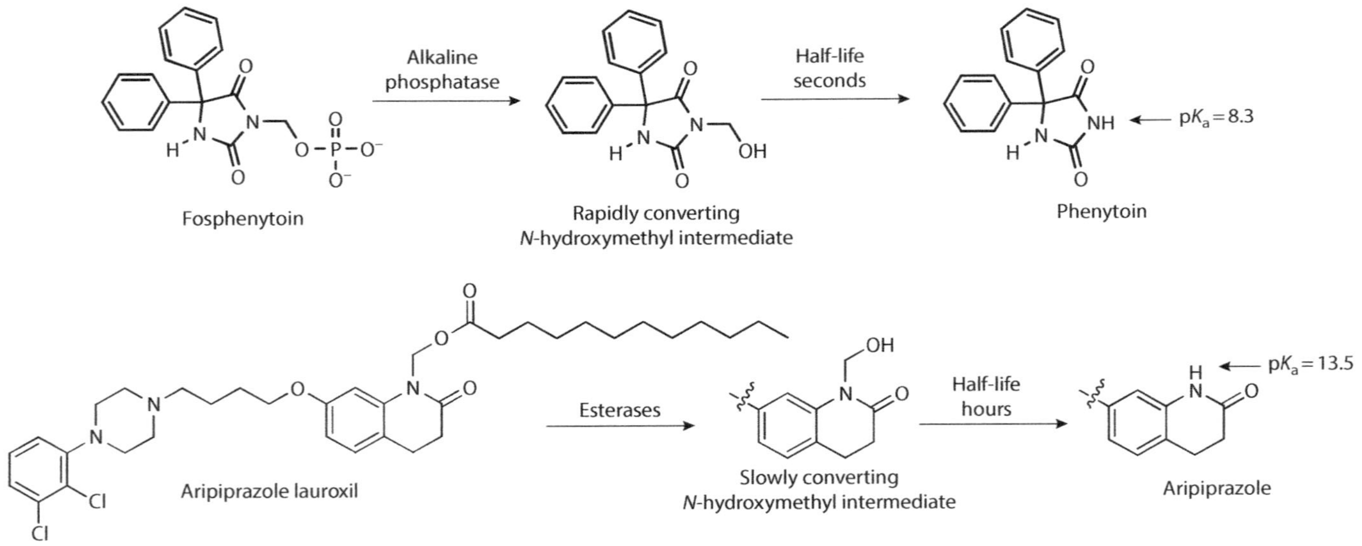

Fosphenytoin

Alkaline phosphatase

Rapidly converting *N*-hydroxymethyl intermediate

Half-life seconds

Phenytoin

$pK_a = 8.3$

Aripiprazole lauroxil

Esterases

Slowly converting *N*-hydroxymethyl intermediate

Half-life hours

Aripiprazole

$pK_a = 13.5$

FIGURE 10.5 Influence of a pK_a value of a NH-acid on the conversion rate of its *N*-hydroxymethyl intermediate.

TABLE 10.1
Major Limitations to Drug's Usefulness That May Be Overcome by Prodrugs

Formulation and administration
Insufficient aqueous solubility
Insufficient shelf life for liquid and solid dosage forms
Irritation or pain after local administration
Unpleasant taste or odor
Absorption
Insufficient dissolution rate due to low aqueous solubility
Poor membrane permeability and low oral or topical bioavailability due to poor lipophilicity
Insufficient stability during first-pass metabolism or in acidic gastric juices
Insufficient availability due to efflux mechanisms
Distribution
Lack of site specificity (e.g., poor brain or tumor targeting)
Excessively strong protein binding in plasma or disposition in lipophilic compartments of body
Metabolism and excretion
Lack or need of site-specific bioactivation
Short duration of action
Toxicity
Lack of site specificity
Lack or need of site-specific bioactivation
Irritation or pain after local administration
Need to temporarily mask a reactive, inherently active, functional group
Life cycle management
Development of a prodrug with improved properties that may represent a life cycle management opportunity for an existing drug

management of an existing drug. The major barriers which limit drug's usefulness and may be overcome by prodrug modification, are listed in Table 10.1, and some of the issues are also briefly discussed in the following sections.

10.5.1 Improved Formulation and Administration

Inadequate water solubility of a drug is a prerequisite for the formulation of aqueous parenteral or injectable dosage forms. Prodrug approaches have become a worthy opinion, when conventional formulation techniques, such as salt formation, particle size reduction, solubilizing excipients, and complexation agents have proved unsuccessful and failed. The most common prodrug approach for increased water solubility has been the introduction of ionizable or polar neutral group, such as phosphate, amino acid, or sugar moiety, to the poorly soluble parent drug. Phosphate esters are widely used prodrug strategy for improving the aqueous solubility of orally and parenterally administered drugs which contain hydroxyl or amine functionalities. Their synthesis is quite straightforward and their chemical stability is usually good or at least adequate. Bioconversion of phosphate prodrugs back to their active parent drugs usually occurs rapidly in the intestinal brush border epithelium before or during the absorption, or during first-pass metabolism in the liver, and is catalyzed by phosphatases. However, the premature enzymatic cleavage of a very insoluble parent drug may result in its precipitation in intestinal lumen yielding to the reduced bioavailability.

One successful example of phosphate prodrugs is fosamprenavir, an orally administered phosphate ester of antiretroviral protease inhibitor amprenavir. Amprenavir is poorly soluble in water (approximately 0.04 mg/mL at 25°C), and its dosage regimen is as high as 1200 mg twice a day.

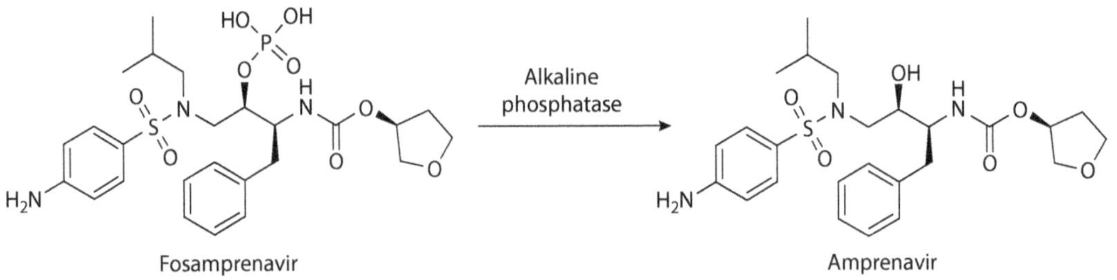

Fosamprenavir Alkaline phosphatase Amprenavir

FIGURE 10.6 Structure and bioconversion of fosamprenavir to its parent drug amprenavir.

Amprenavir was marketed as 50 and 150 mg capsules and its oral formulation contained a large amount of solubilizing excipients, particularly vitamin E TPGS (d-alpha tocopherol polyethylene glycol succinate) which caused potential toxicity issues. To reduce the excipient intake and the burden of 8 capsules twice a day, approximately 10-fold more water soluble fosamprenavir which required only 2700 mg tablets twice a day, was developed (Figure 10.6). In fosamprenavir, a single phosphate group is directly attached to the free hydroxyl group of amprenavir. During its absorption, orally administered fosamprenavir is rapidly and quantitatively hydrolyzed to yield amprenavir and inorganic phosphate by gut epithelial alkaline phosphatases. Amprenavir and its phosphate prodrug fosamprenavir possess comparable therapeutical efficacy and safety profiles, but fosamprenavir provides more simplified and patient-compliant dosage regimen for HIV patients. This clinical advantage has served also as a life cycle management tool for amprenavir.

A good example of a prodrug which can overcome parenteral formulation problems of sparingly water soluble parent drug is fosphenytoin, an injectable phosphate ester of poorly water soluble anticonvulsant phenytoin, in which the phosphate group is attached to an acidic amine of the parent drug via oxymethylene spacer (Figure 10.7). Due to its limited aqueous solubility (20–30 μg/mL), the injectable formulation of phenytoin sodium salt consisted of 40% propylene glycol and 10% ethanol at a pH of 12. This formulation caused several administration problems, including severe irritation and pain at the injection site, precipitation of drug, or even death of the patient, if injected too quickly. In contrast, the significantly increased aqueous solubility of fosphenytoin (142 mg/mL) allows its formulation in purely aqueous solution with pH of about 8.5. Thus, fosphenytoin enables more safe, convenient, and rapid intravenous administration with markedly lower potential for adverse effects at the injection site. When in systemic circulation, fosphenytoin is rapidly and almost completely converted back to the phenytoin by alkaline phosphatases of blood and tissues through a chemically unstable *N*-hydroxymethyl intermediate with half-lives ranging from 7 to 15 minutes in humans. The by-products of the bioconversion of fosphenytoin include formaldehyde and inorganic phosphate.

Other examples of prodrugs with improved aqueous solubility include sulindac, miproxifene phosphate, extramustine phosphate, prednisolone phosphate, irinotecan, and fludarabine phosphate.

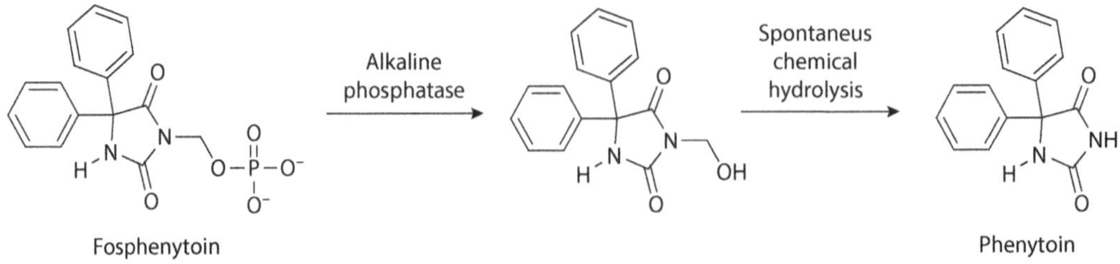

Fosphenytoin Alkaline phosphatase Spontaneus chemical hydrolysis Phenytoin

FIGURE 10.7 Bioconversion of fosphenytoin to its parent drug phenytoin.

10.5.2 Improved Passive Permeability

During its absorption, the molecule has to pass through several cell membranes that represent an "impermeable" barrier which only relatively lipophilic uncharged molecules can cross by passive diffusion. However, the permeability across the biological membranes remains one of the major obstacles for polar and charged drugs. Their poor permeability tends to result in low and often variable oral absorption, low oral bioavailability, and low exposure for specific target organs. The improvement of passive drug permeation has been the most successful area of prodrug research so far, and most frequently the lipophilicity of the parent drug has been enhanced by masking its polar and ionized functionalities. Typically, short hydrocarbon moieties have been attached to the polar or ionized groups of the hydrophilic parent drug to increase its lipophilicity. Hydrophilic hydroxyl, carboxyl, phosphate, or amine groups have been converted to the more lipophilic alkyl or aryl esters, which are rapidly hydrolyzed back to their parent drugs in the body by the ubiquitous esterase activity. The majority of the lipophilic prodrugs have been developed for improved absorption from the gastrointestinal tract. Additionally, this prodrug strategy has been employed to improve topical administration of parent drugs through the skin or come in the eye.

Oseltamivir is a dosed active prodrug of oseltamivir carboxylate which is a selective inhibitor of neuraminidase enzyme of the influenza viruses A and B (Figure 10.8). As a more lipophilic ethyl ester, oseltamivir is rapidly and readily absorbed from the gastrointestinal track. In fact, almost 80% of an oral dose of oseltamivir reaches systemic circulation as oseltamivir carboxylate, whereas the oral bioavailability of hydrophilic oseltamivir carboxylate is only 5% in humans. Once absorbed, oseltamivir undergoes rapid bioconversion to its parent drug and ethanol by the catalytic action of human carboxylesterase 1, and the maximal plasma levels are reached within 3–4 hours after oral dose.

Sofosbuvir is a novel pyrimidine nucleotide analog that inhibits nonstructural protein 5B (NS5B) polymerase of hepatitis C virus and is used for the treatment of chronic hepatitis C infection in adults as a once-daily oral dose regimen. Initially, 2′-deoxy-2′-fluoro-2′-C-methyluridinemonophosphate was discovered as a precursor for the highly potent NS5B polymerase inhibitor 2′-deoxy-2′-fluoro-2′-C-methyluridinetriphosphate, but as a very hydrophilic compound it was unable to sufficiently cross biological membranes. Thus, a more lipophilic phosphoramidate prodrug of 2′-deoxy-2′-fluoro-2′-C-methyluridinemonophosphate, sofosbuvir which enable oral dosing, was developed. Sofosbuvir is metabolized at its desired site of action, in liver, to yield active tri-phosphorylated nucleotide. In fact, the release of an active triphosphate nucleotide is proposed to occur via four enzymatic steps and two spontaneous reaction (Figure 10.9). The key enzymes involved in the enzymatic steps include human catepsin A, carboxylesterase 1, and histidine triad nucleotide-binding protein 1.

Prostaglandin analogs latanoprost, bimatoprost, travoprost, and isopropyl unoprostone represent a class of lipophilic ocular prodrugs for the treatment of high intraocular pressure in glaucoma (Figure 10.10). Latanoprost, travoprost, and isopropyl unoprostone are isopropyl ester and bimatoprost is ethanolamine amide prodrugs, respectively, which are hydrolyzed in intraocular tissues to generate their biological active prostaglandins. Their active carboxylic acids are poorly permeable and cause eye irritation, whereas an improved ocular absorption and safety is achieved with their more lipophilic prodrugs.

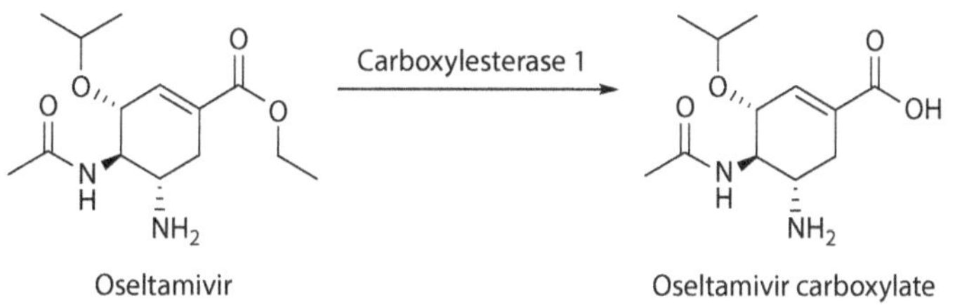

Carboxylesterase 1

Oseltamivir

Oseltamivir carboxylate

FIGURE 10.8 Bioconversion of oseltamivir to more hydrophilic oseltamivir carboxylate.

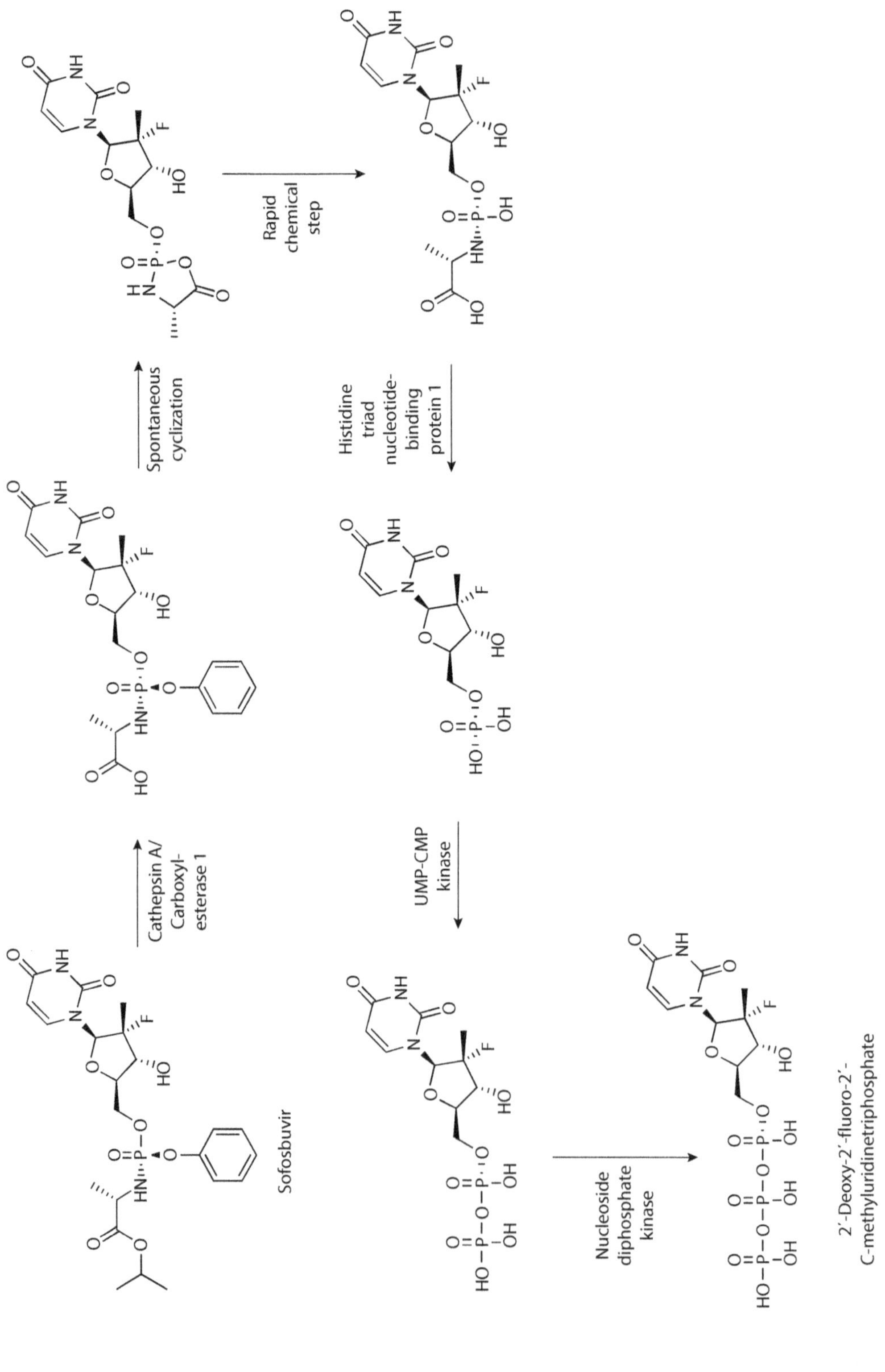

Sofosbuvir

Cathepsin A/ Carboxyl-esterase 1

Spontaneous cyclization

Rapid chemical step

Histidine triad nucleotide-binding protein 1

UMP-CMP kinase

Nucleoside diphosphate kinase

2′-Deoxy-2′-fluoro-2′-C-methyluridinetriphosphate

FIGURE 10.9 Metabolic pathway of sofosbuvir to its active triphosphate nucleoside derivative.

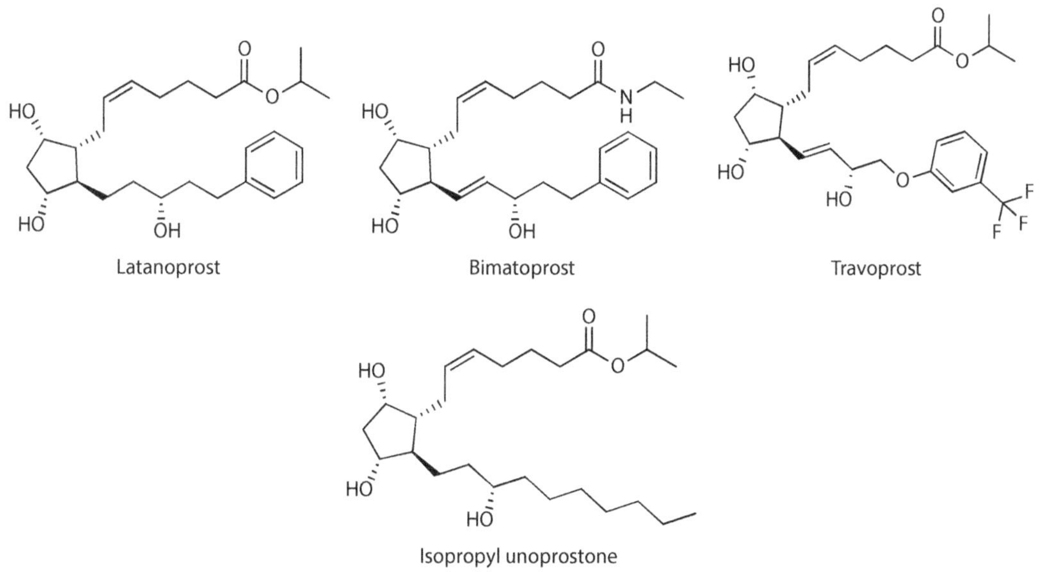

Latanoprost Bimatoprost Travoprost

Isopropyl unoprostone

FIGURE 10.10 Chemical structures of latanoprost, bimatoprost, travoprost, and isopropyl unoprostone.

Other examples of prodrugs with improved lipophilicity or permeability include enalapril, pivampicillin, fenofibrate, olmesartan medoxomil, adefovir dipivoxil, tenofovir disoproxil, tenofovir alafenamide, famciclovir, dabigatran, and tazarotene.

10.5.3 Improved Transporter-Mediated Permeability

Transporters are membrane proteins that act as gatekeepers for cells, controlling the intake and efflux of crucial polar endogenous compounds. The specificity of these transporters is, however, not limited to their endogenous substrates, and other molecules that bear a close structural resemblance can be transported across cell membranes by these transporters as well. Transporter-mediated permeability is particularly important for polar and charged drugs which have negligible passive absorption. While surprisingly many drugs already exploit the transporters of gastrointestinal tract during their absorption, a number of prodrugs have designed as substrates of these transporters. Successful design of transporter-targeted prodrugs requires good knowledge about the structure–activity features of the transporter in question.

Valacyclovir and valganciclovir, the L-valine amino acid esters of acyclovir and ganciclovir, respectively, are probably the first examples of commercially available prodrugs that utilize intestinal peptide transporter 1 (PepT1) to overcome the limited and variable oral bioavailability of their highly polar parent drugs (Figure 10.11). PepT1 is an active influx mechanism for dietary di- and tripeptides located in the small intestine, and its utilization by L-valyl prodrug strategy increased the intestinal permeation of acyclovir and valganciclovir by 3–10-fold. After their absorption, both prodrugs are rapidly converted to their parent drugs by intracellular hydrolytic enzymes. Actually, both valacyclovir and valganciclovir act as double prodrugs, since like other nucleosides, they require triphosphorylation prior to formation of the active antiviral agents.

Gabapentin enacarbil is an acyloxyalkylcarbamate prodrug of analgesic and anticonvulsant drug gabapentin which has problematic pharmacokinetic properties, including short half-life, saturable absorption, high inter-patient variability, and lack of linear dose–response relationship. Gabapentin enacarbil was designed to be absorbed throughout the entire length of the gastrointestinal tract, and its absorption is mediated by high-capacity nutrient transporters,

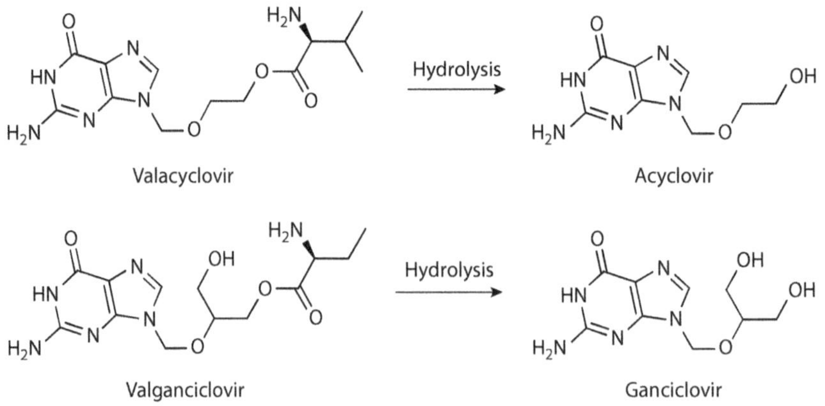

FIGURE 10.11 Hydrolysis of valacyclovir and valganciclovir to their parent drugs.

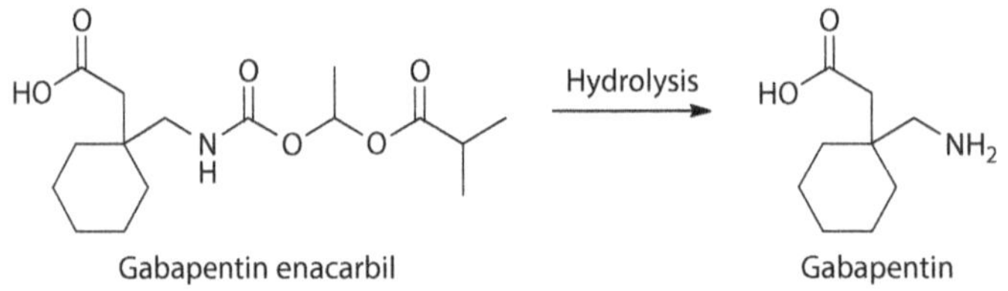

FIGURE 10.12 Structure and hydrolysis of gabapentin enacarbil to the active gabapentin.

including monocarboxylate transporter 1 (MCT-1) and sodium-dependent multivitamin transporter (SMVT). Prodrug modification produced an extended release of gabapentin with twofold improved, more predictable, and dose-proportional oral bioavailability in humans. During and after its absorption, gabapentin enacarbil is efficiently hydrolyzed by nonspecific esterases to yield gabapentin (Figure 10.12). Currently, gabapentin enacarbil is commercially available for the treatment of restless legs syndrome and post-herpetic neuralgia of adults.

Other examples of prodrugs with improved transporter-mediated permeability include midodrine and enalapril.

10.5.4 Improved Targeting

Targeting drug action into specific organs, tissues, or cells is an attractive strategy for more effective therapeutics with less adverse effects. In a prodrug approach, the targeted drug action can be achieved either by site-directed drug delivery or site-specific drug bioactivation. In site-directed drug delivery, intact prodrug is selectively transported to its site of action, and thus constitutes a very challenging task after oral or systemic administration. In site-specific bioactivation, a prodrug can be widely distributed throughout the body, but undergoes bioactivation and exerts pharmacological action only at the desired site. The site-specific bioactivation of prodrugs can be achieved either by: (1) exploiting endogenous transporters, enzymes or physiological conditions of target tissue, such as pH or hypoxia, (2) delivering genes that encode prodrug-activating enzyme into target tissue (e.g., virus-directed enzyme prodrug therapy [VDEPT] and gene-directed enzyme prodrug therapy [GDEPT]), or (3) delivering prodrug-activating enzyme into target tissue via monoclonal antibodies (e.g., antibody-directed enzyme prodrug therapy [ADEPT]). In this chapter, only examples of prodrugs, which exploit physiological differences between target and other tissues, are discussed, but it should be noted that some ADEPT and GDEPT systems are currently under clinical investigation.

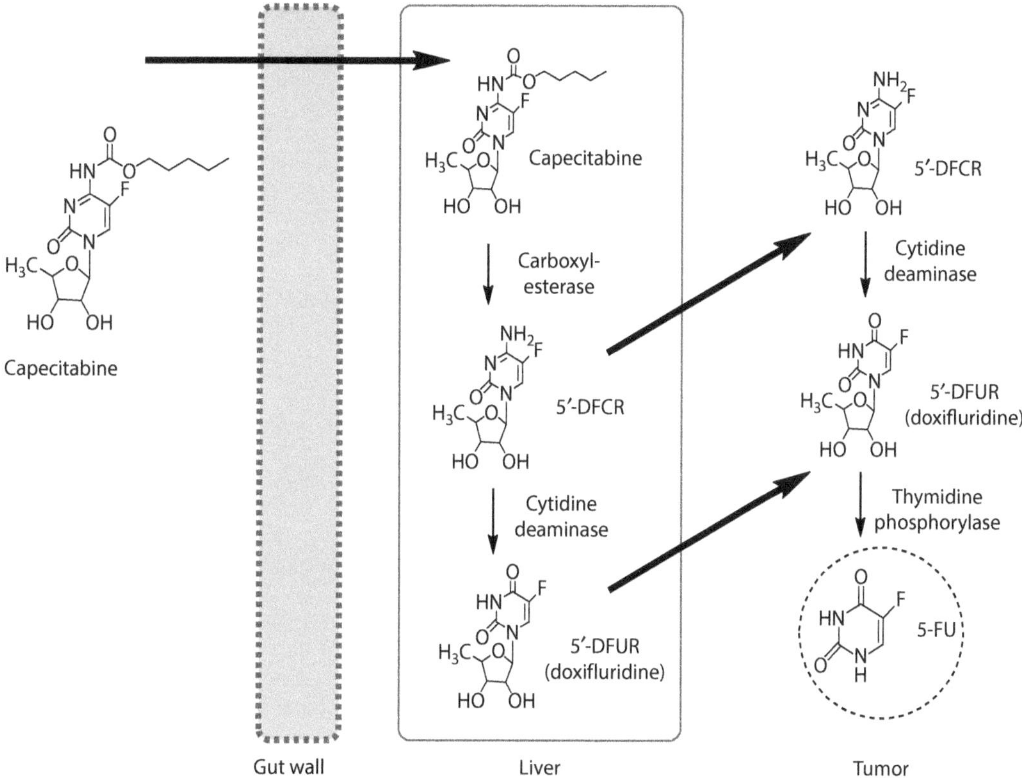

FIGURE 10.13 Tumor-selective activation pathway of capecitabine to cytotoxic 5-FU.

A need for more tumor selective delivery and reduced healthy tissue concentrations of the cytotoxic drug, 5-fluorouracil (5-FU), led to the rational design of capecitabine. Capecitabine is an orally administered chemotherapeutic prodrug, of which bioconversion to 5-FU requires three reaction steps that all are enzymatic (Figure 10.13). After its administration, capecitabine is first hydrolyzed to 5′-deoxy-5′-fluorocytidine (5′-DFCR) by carboxylesterase activity in the liver. Next, 5′-DFCR is converted into 5′-deoxy-5′-fluorouridine (5′-DFUR) by cytidine deaminase either in liver or in tumor tissue. Finally, the release of an active 5-FU occurs selectively at the tumor site by the catalytic activity of thymidine phosphorylase. The oral bioavailability of 5-FU after capecitabine administration is nearly 100% and the maximal concentrations of 5-FU are reached within 1.5–2 hours. In patients with advanced breast cancer, the 5-FU levels in tumor tissue were 2.5-fold higher compared to healthy tissue and 14-fold higher compared to plasma after capecitabine administration. Capecitabine enables more convenient and safer oral treatment for cancer patients with reduced adverse effects in healthy tissue.

The thienopyridine antiplatelet agents, clopidogrel and prasugrel, are excellent examples of liver-targeted prodrugs, which release their parent drugs at or close to their site of action (Figure 10.14). As a metabolizing organ, the liver possesses a wide variety of metabolizing enzymes that are capable of prodrug activation. Clopidogrel and prasugrel are both bioactivated through the hepatic cytochrome P450 enzymes to produce their active but unstable thiol-compounds which bind to the $P2Y_{12}$ class of platelet adenosine triphosphate receptors to inhibit platelet aggregation. The fact that about 85% of clopidogrel dose is hydrolyzed by esterases to yield an inactive carboxylic acid has led to the design of over 10-fold more potent prasugrel. Prasugrel undergoes bioconversion through a cascade of events which involve carboxylesterase 1-catalyzed hydrolysis and cytochrome

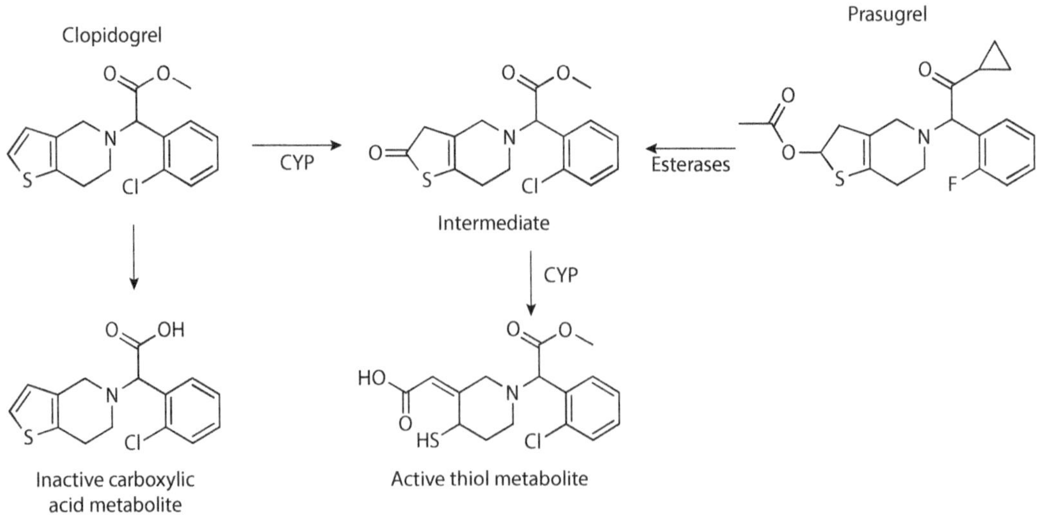

FIGURE 10.14 Bioconversion of clopidogrel and prasugrel to their active thiols.

P450-mediated oxidation. In 2010, clopidogrel has held the position of the second most prescribed drug in the world with over $9 billion in sales.

The increased knowledge about cytochrome P450-mediated bioconversion pathways has led to the design of the liver-specific cyclic phosphate or phosphonate prodrugs called HepDirect®-prodrugs. These prodrugs are cyclic 1,3-propanyl esters containing a ring substituent that render them sensitive to cytochrome P450-mediated oxidation. This oxidation results in ring opening and formation of transient negatively charged intermediate which cannot diffuse across cell membranes and is thus retained in hepatic cells. A consequent β-elimination reaction releases the phosphate or phosphonate drug, and aryl vinyl ketone as a by-product (Figure 10.15). Pradefovir is a HepDirect®-prodrug of adefovir which is currently under clinical development for the treatment of hepatitis B. In fact, pradefovir is developed to improve the therapeutical potential of adefovir dipivoxil which kidney toxicity limits its use. In contrast, pradefovir produces 12-fold liver/kidney and 84-fold liver/intestine targeting ratios compared to adefovir dipivoxil, and after its administration the systemic adefovir levels remain low. Pradefovir undergoes a cytochrome P450-mediated oxidation predominantly in the hepatocytes of liver (Figure 10.15).

Although the proton pump inhibitors, such as omeprazole, lansoprazole, pantoprazole, rabeprazole, esomeprazole and tenatoprazole, were not originally developed as prodrugs, they provide a good example of site-selective prodrugs. Due to the basic pyridine group (pK_a of omeprazole is 3.97), these drugs are protonated and accumulated in the acidic secretory parietal cells (Figure 10.16). In the acidic conditions of parietal cells (pH of 1–2), prazoles undergo spontaneous chemical reaction to their active sulfenamide metabolites followed by their irreversible binding to a cysteine group of H^+/K^+ ATPase. This irreversible binding inhibits the ability of parietal cells to secrete gastric acid. The fact that proton pump inhibitors are only effective on H^+/K^+ ATPases, which contain highly acidic compartments that nongastric H^+/K^+ ATPases lack, corresponds their excellent safety profiles. Therefore, proton pump inhibitors are converted to their active species only under highly acidic conditions at their site of action.

Examples of other site-selective prodrugs include ticlopidine, simvastatin, lovastatin, and hypoxia-activated prodrugs, such as AQ4N, PR104, and evofosfamide which are currently under clinical investigation.

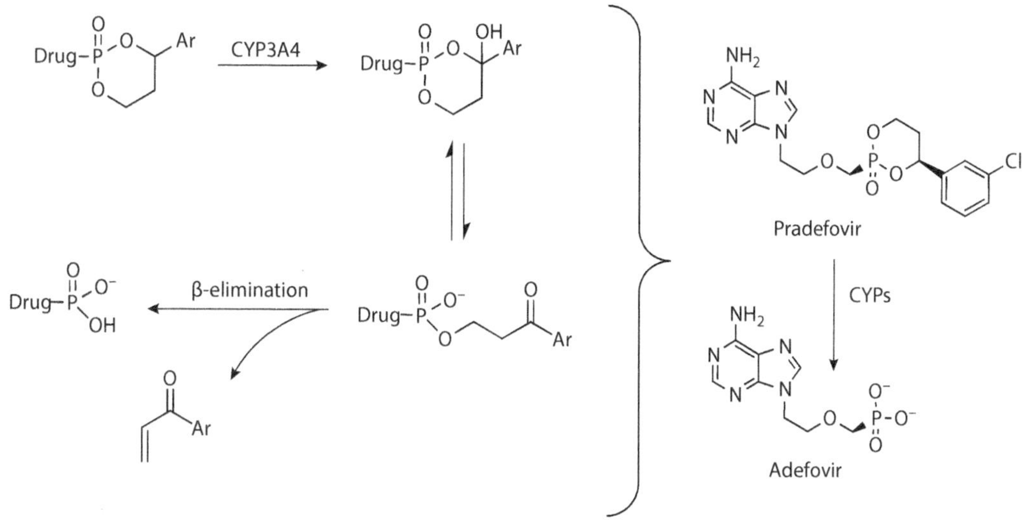

FIGURE 10.15 Bioconversion of HepDirect®-prodrugs to corresponding phosphates or phosphonates.

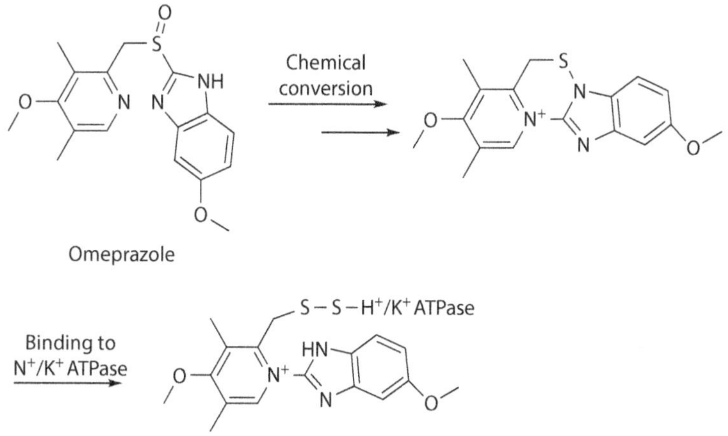

FIGURE 10.16 Bioconversion of omeprazole to its active sulfonamide and following irreversible binding to the cysteine group of H^+/K^+ ATPase.

10.5.5 Improved Metabolic Stability

The therapeutic activity of a drug can be prolonged either by slowing its enzymatic hydrolysis rate or by diminishing its rate of metabolism. Even though various pharmaceutical formulations or the synthesis of new drug analogs have most frequently been exploited to avoid rapid metabolism of a drug, a few examples of prodrugs also exist. In these prodrugs, the metabolically labile but pharmacologically essential functional group(s) of parent drug has generally been masked to avoid rapid elimination. In the case of bambuterol, which is a long-lasting bis-dimethylcarbamate prodrug of the bronchodilator and β_2-receptor agonist terbutaline, the metabolically susceptible phenolic hydroxyl groups have been protected to avoid rapid and extensive first-pass metabolism in the gut and the liver. After administration, bambuterol is slowly converted to terbutaline via its monocarbamate metabolite mainly outside the lungs by nonspecific butyrylcholinesterase, but other hydrolytic and oxidable enzymes may also be involved in the release of an active drug (Figure 10.17). As a result

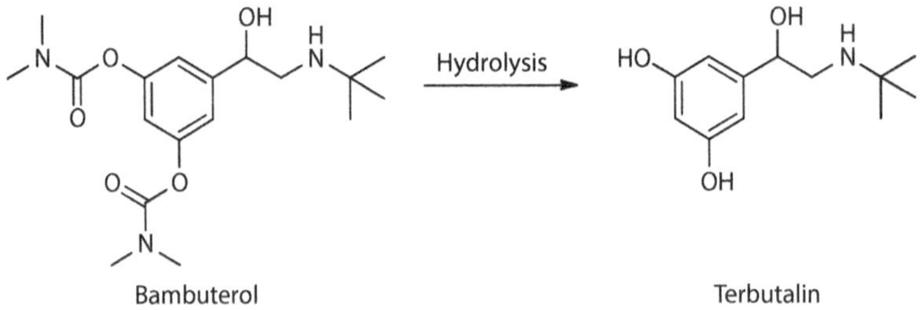

FIGURE 10.17 Structure and butyrylcholinesterase catalyzed hydrolysis of bambuterol to active terbutaline.

of the sustained release, a once-daily bambuterol treatment provides the relief of asthma symptoms with a lower incidence of side effects than terbutaline taken three times a day.

10.5.6 Prolonged Duration of Drug Action

Prolonged duration of drug action, sustained plasma levels, and reduced dosing frequency are typically achieved by controlled-release formulations, such as suspensions, osmotic pumps, and polymeric matrixes. Using prodrug approaches, the controlled release of an active drug can be achieved by modifying its aqueous solubility, partition and dissolution properties in a way that affects the release rate of the active drug, the rate of absorption from injection site, or alter its tissue distribution. Prodrug approaches have been very successful in the development of several parenteral sustained release depot injections which maintain therapeutic plasma levels for weeks to months. In this approach, the free hydroxyl group of parent drug is esterified with a long-chain fatty acid to form respective ester prodrugs, such as decanoates, palmitates, cypionates, or valerates. These highly lipophilic prodrugs are formulated in a vegetable oil and slowly released in the systemic circulation from the site of subcutaneous or intramuscular injection resulting in sustained plasma levels of 2–8 weeks. This approach has resulted in many commercially available oily depot injection products from several drugs including estrogens (e.g., estradiol), neuroleptics (e.g., fluphenazine, flupentixol haloperidol, pipotiazine, and zuclopenthixol), contraceptive (e.g., hydroxyprogesterone and norethisterone), and steroids (e.g., nandrolone and testosterone).

Examples of other prodrugs with prolonged duration of action include lisdexamfetamine.

10.6 CHALLENGES AND CONSIDERATIONS IN PRODRUG DISCOVERY AND DEVELOPMENT

Prodrug strategies have been successful in a number of drug discovery cases based on clinically approved prodrugs and as illustrated earlier in the text. However, embarking on a prodrug strategy can bring some additional complications to the drug discovery and development processes. For example, synthesis difficulties, more complex analytical profiling, bioconversion, further metabolism and pharmacokinetic studies requiring the analysis of both the prodrug and parent drug as well as concerns about the toxicity of not only the prodrug and drug, but also the released promoieties or by-products bring an additional challenge to prodrug discovery and development strategy. In addition, navigating the regulatory environment with prodrugs is far from straightforward, particularly when prodrugs of already marketed drugs are developed.

10.6.1 Bioconversion of Prodrugs

Efficient and site-controlled conversion of a prodrug to the parent drug is critical for the prodrug approach to be successful. Typically, conversion involves metabolism by enzymes that are distributed throughout the body. The most common prodrug approaches rely on metabolic bioconversion to the active drug by functionally prominent and nonspecific hydrolases such as peptidases, phosphatases, and esterases including acetylcholinesterase, paraoxonase, butyrylcholinesterase, and in particular, carboxylesterase. Carboxylesterases (CESs) are predominantly involved in hydrolase activity in the liver and the small intestine of human and various preclinical species. CESs efficiently catalyzes the hydrolysis of a variety of esters and amides such as cocaine, heroin, irinotecan, and temocapril. CESs are ubiquitously distributed and their potential to become saturated or for their substrates to become involved in drug–drug interactions is considered to be negligible, although not impossible. Therefore, CESs are attractive targets for various ester and amide prodrugs of hydroxyl, phenolic, carboxyl and amino group containing compounds. Several prodrugs also rely on cytochrome P450 (CYP) enzymes in their bioactivation process. The rest of the text in this chapter focuses on hydrolases, particularly, carboxylesterase, and to lesser extent on CYPs.

The majority of CESs fall within two isozyme families, CES1 and CES2 which are characterized by their substrate specificities, tissue distribution, and gene regulation. In humans, CES1 is highly expressed in the liver, and distributed throughout many tissues with the notable exception of the intestine. Human CES2, in contrast, is highly expressed in the small intestine, the kidney, and the colon, but at much lower levels than CES1 in the liver. Although there is 40%–50% sequence homology between CES1 and CES2, differences in hydrolysis of their substrates have been observed. CES1 prefers compounds with large acyl and small alcohol groups, whereas CES2 prefers hydrolyzing compounds with large alcohol and small acyl groups. For example, oseltamivir is predominantly hydrolyzed by CES1, while irinotecan, with its bulky alcohol group, is hydrolyzed almost exclusively by CES2 (Figure 10.18).

In addition to substrate specificities, significant differences in CES expression between mammalian species should be considered when selecting animal models for preclinical studies of prodrugs and translating preclinical data in humans. Similarly, with dog and monkey, human plasma does not contain CES which is highly expressed in rabbit, mouse, and rat plasma. The esterase expression and hydrolyzing pattern of dog plasma has been found to be closest to that of human plasma which are both ubiquitously expressing other hydrolyzing enzymes paraoxonase and butyrylcholinesterase. Hydrolase activity in the small intestine of humans and rats is similar with exclusive expression of CES2 and similar expression along the whole length of the intestine. Interestingly, no hydrolase activity has been found in dog small intestine. The human liver contains both CES isozymes like various preclinical species with similar or less hydrolase activity depending on species compared. Using human recombinant enzymes or human tissue extracts can reduce the interspecies variability and identify tissue-specific hydrolysis.

While ubiquitous distribution of hydrolase enzymes can be utilized in effective bioconversion of ester prodrugs, premature hydrolysis during absorption process of hydrolytically susceptible prodrugs can significantly diminish their bioavailability. Reduced absorption has been observed with several ester prodrugs of penicillins, cephalosporins, and angiotensin-converting enzyme inhibitors. In general, these ester prodrugs have bioavailabilities of around 40%–60% which can be

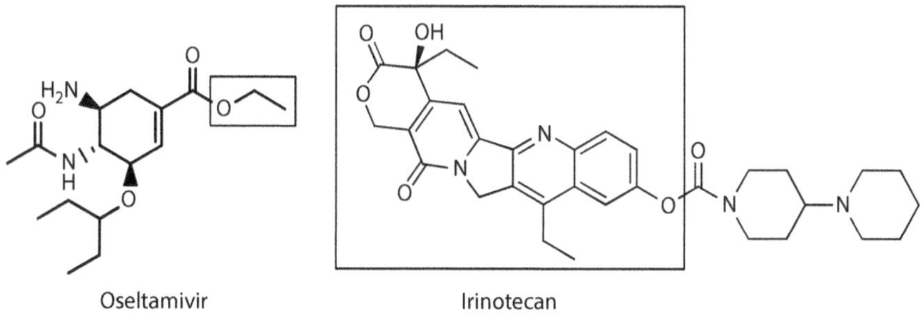

Oseltamivir

Irinotecan

FIGURE 10.18 Human CES1 mainly hydrolyses a substrate with a small alcohol group (oseltamivir) while human CES2 mainly hydrolyses a substrate with a large alcohol group (irinotecan).

explained by premature hydrolysis in the esterase-rich enterocytes during the absorption process. This indicates that complete oral bioavailability of ester prodrugs is unlikely.

The enzymes involved in prodrug bioconversion can also be polymorphic, and therefore, subject to interindividual variability in activity. Although, in general, CESs do not appear to be subject to high genetic polymorphisms in a manner that might compromise their exploitation in prodrug design, scattered evidence observed with prodrugs such as angiotensin-converting enzyme inhibitors, capecitabine, and irinotecan has nevertheless demonstrated that polymorphisms of CES genes can have some impact on the clinical variability of CES-activated prodrugs. Genetic variation in CYPs, on the other hand, is known to contribute substantially to variability in exposure as well as clinical efficacy and safety of prodrugs activating by these enzymes. CYPs that are expressed in several tissues being prominent in the liver and to lesser extent in the intestine are capable of oxidation and reduction reactions of several various, especially bioprecursor prodrugs, into their active species. Clopidogrel, codeine, tamoxifen, tegafur, and cyclophosphamide are examples of prodrugs which have demonstrated interindividual variation in prodrug exposure because of polymorphisms of CYPs. Therefore, an appropriate selection of dose of genetically defined patient subsets should enhance the efficacy and safety of prodrugs undergoing bioconversion by CYPs.

10.6.2 Safety Evaluation of Prodrugs

Discovery and development of prodrugs present many challenges in the safety assessment. Potential toxicity of both prodrug and promoiety as compared to the parent drug needs to be carefully evaluated. In many cases, a comparative toxicological analysis can assist in determining which toxicity is contributed by the prodrug, its intermediates, or the active drug. Two specific sources of possible toxicity in prodrugs are promoiety itself and by-products released during the bioconversion process. For example, the toxicity concern associated with the ethylene sulfide by-product released from S-acyl-2-thioethyl (SATE) prodrugs of phosphates/phosphonates has limited the advancement of these prodrugs into development. Another and more frequently discussed by-product is formaldehyde that is released during bioconversion of various double prodrugs. However, considering the normal daily formaldehyde levels in humans, it is unlikely that formaldehyde from a prodrug could adversely affect normal physiological functions. A third example of promoieties associated with possible toxicity concern is pivalic acid used as a promoiety in prodrugs such as adefovir dipivoxil, and pivampicillin. Pivalic acid is shown to interrupt carnitine homeostasis which can lead to depletion of carnitine in humans. In many cases, exposure to the pivalic acid has no or only minor toxicological impact, and in extended treatment with high doses simultaneous carnitine supplementation can be administered to avoid its deficiency. Therefore, in cases where prodrugs raise any toxicity concern, daily dose and duration of treatment should be carefully taken into consideration in the overall risk evaluation process.

10.6.3 Regulatory Aspects of Prodrugs

The development of a prodrug from an existing drug may represent an opportunity for life cycle management and add several years to the life of a patent. If the prodrug structure has not been disclosed previously, it can be considered as a new chemical entity (NCE) and is likely to have the added benefit of being considered as intellectual property (IP). However, for an inventive step to be acknowledged, it has to be demonstrated that the prodrug's design was not obvious to the skilled person. Difficulties in overcoming an obviousness objection may arise if the structural difference between the prodrug and parent active drug is trivial such as forming an ester of an existing drug. As of today, other widely used prodrug strategies such as amides and phosphates have overcome obviousness criteria and respective prodrugs of existing drugs have been granted NCE status. In scenarios where both the active drug and its prodrug are discovered in the same time and included in the same patent filing ensuring adequate protection of both is more straightforward. Finally, in cases where a prodrug of an existing drug has been developed, existing documentation and clinical experience have been, at least in part, accepted by regulatory agencies, therefore, resulting in simplified development path for the prodrug.

10.7 CONCLUDING REMARKS

Using prodrugs is a versatile and powerful strategy to temporarily improve the problematic characteristics of drug molecules. Prodrug strategies have traditionally been embarked to address ADME properties and risks of marketed drugs or as a tool in late-stage problem solving for drug candidates in development phases. However, prodrug design is now increasingly being integrated into early drug discovery. Admittedly, embarking a prodrug strategy can certainly present its own challenges, but depending on the chemical nature of the parent drug and the therapeutic target, many times the prodrug design can represent a comparable smaller challenge than the alternative of searching for a new therapeutically active molecule that also inherently possesses the desired ADME properties.

FURTHER READING

Pradere, U., Garnier-Amblard, E.C., Coats, S.J., Amblard, F., and Schinazi, R.F. 2014. Synthesis of nucleoside phosphate and phosphonate prodrugs. *Chem. Rev.* 114:9154–9218.

Rautio, J. (ed.). 2011. *Prodrugs and Targeted Delivery: Towards Better ADME Properties.* Weinheim, Germany: Wiley-VCH.

Rautio, J., Kumpulainen, H., Heimbach, T., Oliyai, R., Oh, D., Järvinen, T., and Savolainen, J. 2008. Prodrugs: Design and clinical applications. *Nat. Rev. Drug Discov.* 7:255–270.

Stella, V.J., Borchardt, R.T., Hageman, M.J., Oliyai, R., Maag, H., and Tilley, JW. (eds.). 2007. *Prodrugs: Challenges and Rewards Part 1 and 2*, 1st ed. New York: Springer.

11 Enzyme Inhibitors
Biostructure- and Mechanism-Based Design

Robert A. Copeland and P. Ann Boriack-Sjodin

CONTENTS

11.1 INTRODUCTION

Inhibition of disease-associated enzyme targets by small molecular weight drugs is a well-established modality for pharmacologic intervention in human disease. Indeed, a 2007 survey of the FDA Orange Book showed that more than 300 marketed drugs work through enzyme inhibition. Among orally dosed drugs in clinical use, nearly half function by inhibition of specific enzyme targets. Likewise, much of the current preclinical drug discovery efforts in biotechnology and pharmaceutical companies—as well as those in government and academic laboratories—are focused on the identification and optimization of small molecules that function by inhibition of specific enzyme targets. The reasons for the popularity of enzymes as targets for drug discovery have been reviewed a number of times (see, e.g., Copeland 2013). In brief, enzymes make good drug targets for two significant reasons. First, the catalytic activity of specific enzymes is often critical to the pathophysiology of the disease, such that inhibition of catalysis is disease-modifying. Second, the binding pockets for natural ligands of enzymes play a crucial role in catalytic activity, and these pockets are often uniquely well-suited for interactions with small molecule drugs. Thus, the very nature of the chemistry of enzyme catalysis makes these proteins highly vulnerable to inactivation by small molecule inhibitors that have the physicochemical characteristics of oral drugs.

Enzyme catalysis involves the conversion of a natural ligand (the substrate) into a different chemical species (the product), most often through a process of chemical bond breaking and formation steps.

The chemical transformation of substrate to product almost always involves the formation of a sequential series of intermediate chemical species along the reaction pathway. Paramount in this reaction pathway is the formation of a short-lived, high energy species referred to as the transition state. To facilitate this sequential process of intermediate species formation, the ligand binding pocket(s) of enzymes must undergo specific conformational changes that induce strains at correct locations and align molecular orbitals to augment the chemical reactivity of the appropriate functionalities on the substrate molecule(s), at defined moments during the reaction cycle. The basis of mechanistic enzymology includes understanding the chemical nature of the various intermediate species formed, and their interactions with those elements of the enzyme binding pocket that facilitate chemical transformations. When these studies are coupled with structural biology methods, such as X-ray crystallography and multi-dimensional nuclear magnetic resonance (NMR) spectroscopy, a rich understanding of the structure–activity relationships (SAR) that attend enzyme catalysis can be obtained. What is germane to the present discussion is that this structural and mechanistic understanding can be exploited to discover and design small molecule inhibitors—mimicking key structural features of reaction intermediates—that form high-affinity interactions with specific conformational states of the ligand binding pocket of the target enzyme. In this chapter, we describe the application of mechanistic and structural enzymology to drug discovery efforts with an emphasis on the evolution of structural changes that attend catalysis and the exploitation of these various conformational forms for high-affinity inhibitor development.

11.2 MODES OF INHIBITOR INTERACTION WITH ENZYMES

The simplest enzyme-catalyzed reaction that one can envisage is that of a single substrate (S) being converted by the enzyme (E) to a single product (P). This reaction can be summarized by the following equation:

$$E + S \underset{k_2}{\overset{k_1}{\rightleftarrows}} ES \xrightarrow{k_{cat}} E + P \tag{11.1}$$

As summarized by Equation 11.1, enzyme and substrate combine to form a reversible initial encounter complex (ES) that is governed by a forward rate constant for association (k_1) and a reverse rate constant for dissociation (k_2). The equilibrium dissociation constant for the ES complex is given the symbol K_S and is mathematically equivalent to the ratio of the rate constants k_2/k_1. Subsequent to initial complex formation, a series of chemical steps ensue that are collectively quantified by the cumulative rate constant k_{cat}. Thus, k_{cat} is not a microscopic rate constant, but rather summarizes all of the intermediate states that must be formed during the chemical transformation of substrate to product (see Section 11.3 for more details on the individual intermediate steps that may contribute to k_{cat}).

Three modes of inhibitor interaction with an enzyme target can be defined, based on their effects on the catalytic steps summarized in Equation 11.1. Competitive inhibitors bind to the free enzyme in a manner that blocks the binding of substrate so that they increase the apparent value of K_S, but have no effect on the apparent value of k_{cat}. Noncompetitive inhibitors can bind to both the free enzyme and to the ES complex (or intermediate species that follow formation of the ES complex). Such inhibitors can have some effect on the value of K_S but show the greatest effect on k_{cat}, as they inhibit by blocking catalytic steps subsequent to substrate binding. Finally, uncompetitive inhibitors have no affinity for the free enzyme and only bind subsequent to formation of the ES complex. These inhibitors decrease the apparent value of K_S (i.e., increasing the apparent affinity of the enzyme for substrate) and also decrease the apparent value of k_{cat} (i.e., diminishing the ability of the enzyme to catalyze chemical steps subsequent to substrate binding). Among drugs in current clinical use, one finds multiple examples of each of these three modalities of enzyme inhibition.

11.3 PROTEIN DYNAMICS IN ENZYME CATALYSIS AND INHIBITOR INTERACTIONS

The catalytic pathway summarized in Equation 11.1 is a gross over-simplification of even the simplest of enzymatic reactions. At minimum, this reaction requires the formation of two additional forms of enzyme–ligand binary complex, these being the enzyme–transition state complex and the enzyme–product complex. In practice, one often finds that additional intermediate states are accessed during the catalytic cycle of an enzyme. Thus, one can say that the catalytic cycle of an enzyme is a sequential series of protein–ligand complexes, each representing a unique chemical form of the ligand with attendant changes in the protein conformation of the ligand binding pocket. Each conformational state of the ligand binding pocket that is accessed during this catalytic cycle is a potential target for small molecule drug interactions. Hence, one can think of the ligand binding pocket of an enzyme not as a single target for drug intervention, but rather a collection of targets that evolve and interconvert over the time course of catalytic turnover.

To illustrate these concepts, let us consider the reaction cycle of an aspartyl protease. The aspartyl proteases constitute a family of protein/peptide hydrolyzing enzymes that use a pair of aspartic acid residues within the enzyme active site to facilitate peptide bond cleavage. Figure 11.1 provides a schematic representation of the canonical reaction cycle of an aspartyl protease, illustrating the changes in active site structure that attend catalysis. Before substrate binding, the enzyme is in a resting form (E) in which the two active site aspartic acid residues are bridged by a water molecule. One of the aspartates is present in the protonated acid form while the other is present as the conjugate base form, and the two residues share the acid proton through a strong hydrogen bond.

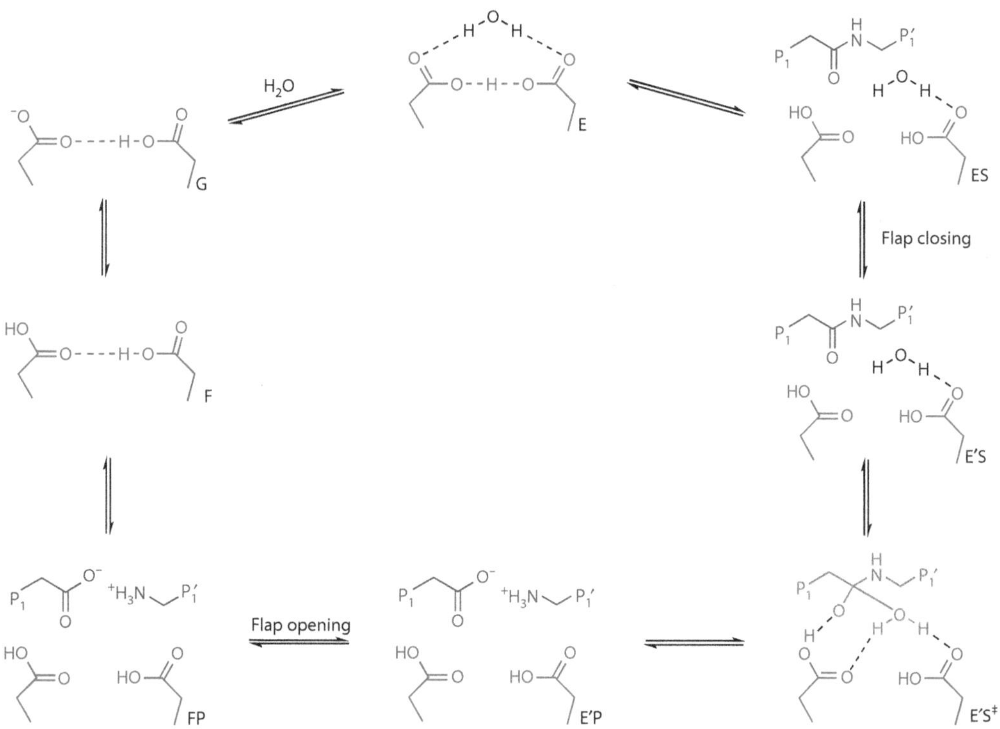

FIGURE 11.1 The reaction pathway for an aspartyl protease. Protein catalytic residues are red, substrate peptide is blue, product peptides are purple, water molecules are black, and the transition state is green. (Adapted and modified from Copeland, R.A., *Evaluation of Enzyme Inhibitors in Drug Discovery: A Guide for Medicinal Chemists and Pharmacologists*, 2nd ed., John Wiley & Sons, Hoboken, NJ, 2013.)

Initial substrate binding causes disruption of the hydrogen bonding interactions and displacement of the water molecule in species ES. After initial substrate binding, a flexible loop of the protein, referred to as the "flap" closes down over the active site to occlude the active site groups from bulk solvent. The substrate-bound enzyme in this altered conformation is referred to as form E′S in Figure 11.1. Subsequently, the water of the enzyme active site attacks the carbonyl carbon of the scissile peptide bond to form a dioxy, tetrahedral carbon center on the substrate. This constitutes the enzyme-bound transition state of the reaction and is symbolized as $E'S^{\ddagger}$ in the figure. Bond rupture then occurs, leading to a species with both active site aspartates protonated and with both the anionic and cationic peptide products bound (form E′P). After that, the flap retracts from the active site to generate a new conformational state of the active site, referred to as FP. With the flap out of the way, the product peptides can now dissociate from the enzyme, forming enzyme state F. Deprotonation of one of the active site aspartates occurs next to form state G. Finally, addition of a water molecule returns the enzyme back to the original conformational state E, thus completing the catalytic cycle.

It is clear from Figure 11.1 that protein dynamics is an important and integral component of the catalytic cycles of enzyme reactions. As stated earlier, the importance of this concept to drug discovery is that each intermediate state accessed along the reaction pathway provides unique opportunities for inhibitor interactions. For example, in the case of the aspartyl proteases, there are three ligand-free, conformationally distinct forms of the enzyme (states E, F, and G); small molecule inhibitors are known that preferentially bind to each of these individual states.

The protein structures that are represented along the reaction pathway of an enzyme reflect a collection of conformational microstates that interconvert among themselves through rotational and vibrational excursions. Thus, any particular conformation of the enzyme can be represented as a manifold of conformational substates (microstates), each stabilized to different degrees by specific interactions with ligands. Isomerization of the enzyme from one structure to another therefore involves the potential energy stabilization of certain microstates at the expense of others.

Similarly, high-affinity inhibitor interactions often involve isomerization of the enzyme from an initial structure resembling the unliganded enzyme to a new conformation in which a particular microstate(s) is highly stabilized by interactions with the inhibitor. Kinetically, this requires two distinct steps in overall binding of high-affinity inhibitors. As illustrated in Figure 11.2a, the first step involves formation of a reversible encounter complex between the enzyme and inhibitor (EI) which often displays only modest affinity. Forward binding of the inhibitor to the enzyme is dictated by the association rate constant k_3 and dissociation of the initial EI complex is dictated by the dissociation rate constant k_4. Once the EI complex is formed, enzyme isomerization can occur to form the much higher affinity complex E*I. The forward conversion of EI to E*I is dictated by the isomerization rate constant k_5 and the reverse conversion of E*I back to EI is dictated by rate constant k_6. All three enzyme forms (E, EI, and E*I) can also be represented as potential energy diagrams, as illustrated in Figure 11.2b. It should be noted that the affinity of the enzyme–inhibitor complex is related to the potential energy stabilization of the system which is reflected in the depth of the potential well in the energy diagram. The deeper the potential energy well for the inhibited form, the more energy that is required to escape this well and thus access the other conformational microstates required for continued catalysis. The potential energy stabilization of the inhibitor-bound form is mediated by productive interactions between the inhibitory ligand and the binding pocket of the protein, in the form of hydrogen bonds, electrostatic interactions, hydrophobic interactions, van der Waal forces, and the like. These concepts can also be conceptualized in terms of an induced fit between the binding pocket and the inhibitor, as illustrated in cartoon form in Figure 11.2c.

The state E*I thus represents a state of high-affinity interactions between the enzyme and the inhibitor. As long as the inhibitor is bound to the enzyme, either in the form of EI or E*I, the biologic activity of the enzyme is blocked. Dissociation of the inhibitor from the enzyme can occur for any reversible inhibition process; once the enzyme is free of the inhibitor, catalytic activity is restored. In the case of tight-binding inhibitors that induce enzyme isomerization, the overall rate

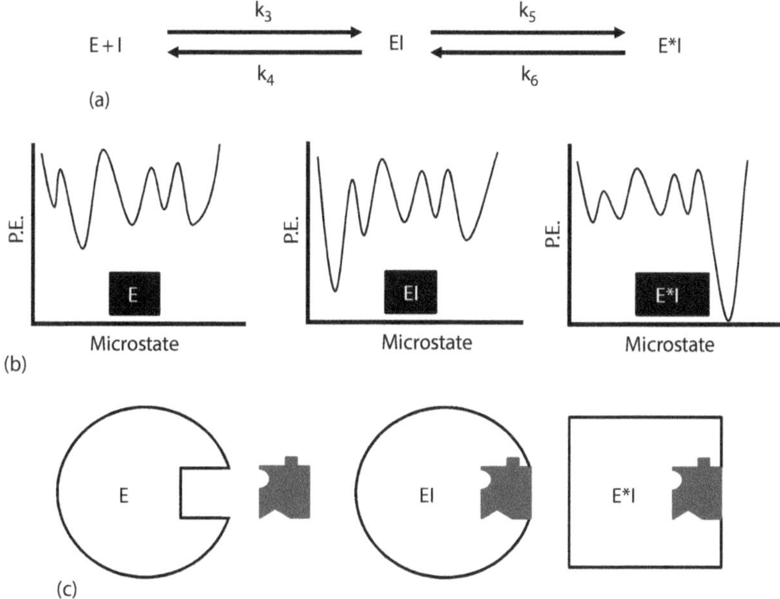

FIGURE 11.2 A two-step inhibitor binding mechanism involving initial binding of the inhibitor to the enzyme in one conformation and a subsequent isomerization of the enzyme to a new conformation. (a) Reaction sequence illustrating the forward and reverse kinetic steps of binding and enzyme isomerization. (b) Potential energy diagrams representing the three conformational states of the enzyme: E, EI, and E*I. (c) Cartoon representation of the inhibitor binding and enzyme isomerization steps in this mechanism.

constant for inhibitor dissociation, k_{off}, must take into account reversal of the isomerization step, re-isomerization via k_5 and dissociation of the inhibitor from EI via k_4. Mathematically, the value of k_{off} is given by:

$$k_{off} = \frac{k_4 k_6}{\left(k_3 + k_5 + k_6\right)} \tag{11.2}$$

For this two-step binding mechanism, it is almost always the case that the reverse isomerization step, mediated by rate constant k_6, is by far the slowest step in overall inhibitor dissociation. Thus, the lower the value of k_6, the longer the duration of potent inhibition by the drug. There are a large and growing number of examples of highly efficacious drugs that demonstrate tight-binding interactions with their target enzyme through a two-step enzyme isomerization mechanism as described here. In some cases, the slowness of the reverse isomerization step leads to prolonged duration of inhibition that may translate into an extended duration of pharmacodynamic activity in vivo; this concept is considered further in Section 11.6.

11.4 MECHANISM-BASED INHIBITOR DESIGN

Enzymes are designed by nature to catalyze a specific chemical reaction. As described earlier, every enzyme accesses a sequential series of intermediate states along the reaction pathway, thus providing unique opportunities for inhibitor interaction. Consequently, enzyme inhibitors can be effectively designed based on an understanding of the mechanistic and structural details of the catalyzed reaction pathway. The majority of known enzyme inhibitors are structurally related to natural ligands of the enzymatic reaction; one survey suggested that more than 60% of marketed drugs that target enzymes are either analogs of substrates or enzyme cofactors, or they undergo catalyzed structural conversion within the active site of an enzyme. Substrate, cofactor, and product mimicry, however, is not the only method for the design of high-affinity, selective enzyme inhibitors. Advances in transition state theory during the past four decades have helped to establish an alternative approach for mechanism-based design: intermediate state-based design (sometimes also referred to as transition state-based design). In this latter approach, inhibitors that mimic the steric and electronic features of high-energy reaction intermediate states are designed to capitalize on the specific interactions of active site residues with the reaction intermediate. In the next two sections, cases for substrate structure-based design and intermediate state-based design will be discussed to exemplify inhibitor design strategies that have led to successfully marketed products or clinical candidates.

11.4.1 Substrate Structure-Based Design

11.4.1.1 Nucleoside and Nucleotide Inhibitors of HIV Reverse Transcriptase

HIV reverse transcriptase (RT) is one of two main targets for anti-acquired immunodeficiency syndrome (AIDS) therapy (the second target being the HIV protease; *vide infra*). The RT enzyme catalyzes the synthesis of double-stranded proviral DNA from single-stranded genomic HIV RNA. Drugs targeting HIV RT can be divided into two categories: (1) nucleoside and nucleotide RT inhibitors (NRTIs) which are competitive with respect to the natural deoxynucleotide triphosphates (dNTPs) and serve as alternative substrates for catalysis (resulting in chain termination) and (2) non-nucleoside RT inhibitors (NNRTIs) which are allosteric, noncompetitive inhibitors that bind at a site distal to the RT active site. NRTIs were the first class of chemotherapeutic agents to be utilized in the clinic to treat AIDS patients and offer excellent examples of inhibitor design based on substrate mimicry. The first NRTI, Zidovudine (AZT) was approved by the FDA in 1987 (Figure 11.3).

AZT is a thymidine analog with an azido group in place of the hydroxyl group at the 3′ position of the ribose. Since the advent of AZT-based therapy, a number of NRTIs have joined the anti-AIDS treatment armamentarium. Most of these are nucleoside analogs with the exception of tenofovir

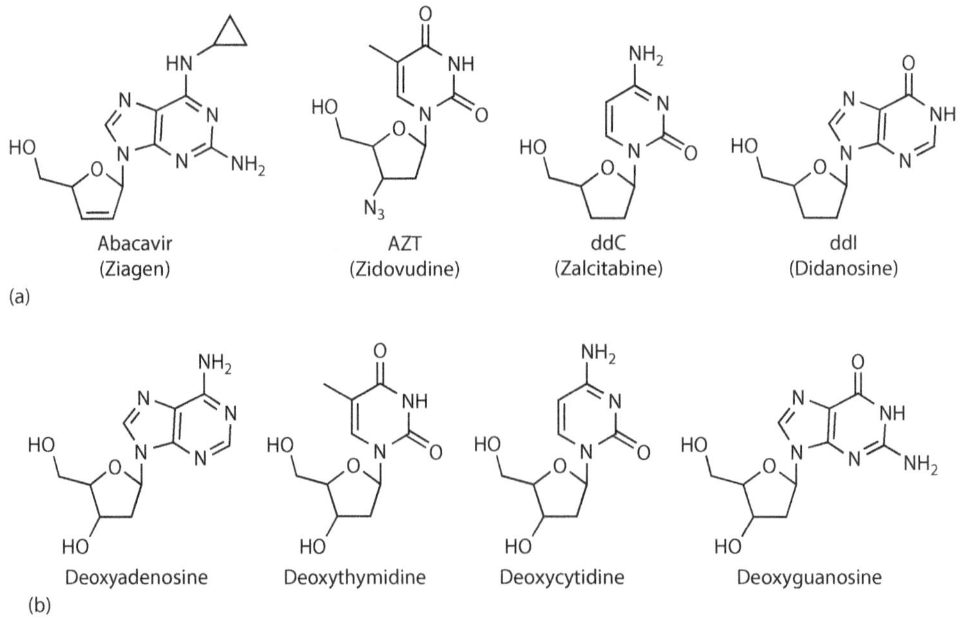

Abacavir (Ziagen) AZT (Zidovudine) ddC (Zalcitabine) ddI (Didanosine)

(a)

Deoxyadenosine Deoxythymidine Deoxycytidine Deoxyguanosine

(b)

FIGURE 11.3 Representative FDA-approved nucleoside/nucleotide reverse transcriptase inhibitors (a) that closely mimic the natural deoxynucleotides (b).

disoproxil fumarate (TDF) which is a nucleotide analog of adenosine phosphate. The NRTIs are administered as unphosphorylated prodrugs. Upon entering the host cell, these prodrugs are recognized by cellular kinases and further converted to the tri-phosphorylated form. The tri-phosphorylated NRTIs then bind to the active site of RT and are catalytically incorporated into the growing DNA chain. The incorporated NRTIs block the further extension of the chain since the NRTIs lack the 3′ hydroxyl group on their ribose or pseudo-ribose moiety and thus cannot form the 3′–5′ phosphodiester bond needed for DNA extension. NRTIs are one of the major classes of inhibitors used in all combination therapies for the treatment of HIV-infected patients. However, the clinical successes of these agents are limited by viral resistances to NRTIs, arising through mutations in the coding region of RT. These mutations confer viral resistances through improved discrimination of a nucleotide analog relative to the natural substrate, or by increased phosphorolytic cleavage of an analog-blocked primer. To overcome these acquired resistances, the design of the next generation of NRTIs has been mainly focused on two fronts: (1) nucleoside analogs possessing a 3′ hydroxyl group that can induce delayed polymerization arrest and (2) nucleotide analogs that are designed to be incorporated into the viral genome during replication. These nucleotide analogs can introduce mutations into the HIV genome through mispairing and blockade of the replication process.

11.4.1.2 Human Steroid 5α-Reductase Inhibitors

The human enzyme steroid 5α-reductase is responsible for the conversion of testosterone (T) to the more potent androgen, dihydrotestosterone (DHT). It has been shown that abnormally high 5α-reductase activity in humans leads to excessively high DHT levels in peripheral tissues. Inhibition of 5α-reductase thus offers a potential treatment for DHT-associated diseases, such as benign prostate hyperplasia, prostate cancer, acne, and androgenic alopecia. In humans, there are two types of steroid 5α-reductase: type I and type II. The type I 5α-reductase is mainly expressed in the sebaceous glands of skin and the liver, while the type II enzyme is most abundant in the prostate, seminal vesicles, liver, and epididymis. The first 5α-reductase inhibitor approved for clinical application in the United States was finasteride; it is currently employed in the treatment of benign prostatic hyperplasia (BPH) in men. This compound is approximately 100-fold more potent toward the type II than the type I isozyme of 5α-reductase. In humans, finasteride decreases prostatic DHT levels by 70%–90%, resulting in reduced prostate size. The detailed biochemical characterization of finasteride inhibition suggested that finasteride is a mechanism-based inhibitor. It is proposed that by closely mimicking the substrate (testosterone), finasteride is accepted as an alternate substrate and forms an NADP-dihydrofinasteride adduct at the enzyme active site (Figure 11.4). This covalent NADP-dihydrofinasteride adduct represents a bisubstrate analog with extremely high affinity ($K_i \leq 1 \times 10^{-13}$ M) to the type II 5α-reductase. Interestingly, finasteride is also a mechanism-based inhibitor of the human type I 5α-reductase. However, the NADP-dihydrofinasteride adduct formation rate at the type I 5α-reductase active site is reduced by more than 100-fold compared to that for the type II isozyme. This difference in NADP-dihydrofinasteride adduct formation rate accounts for the isozyme selectivity of finasteride both in vitro and in vivo. Knowledge of the mechanism of inhibition of 5α-reductase by 4-azasteroids (represented by finasteride) and of the SAR for dual 5α-reductase inhibition led to the discovery of a potent, dual inhibitor of 5α-reductase, known as dutasteride. Dutasteride is equipotent versus type I and type II 5α-reductase and demonstrates exceptional in vivo potency. This compound has also been approved for clinical use in the treatment of BPH.

11.4.2 Intermediate State-Based Design

11.4.2.1 Inhibitors of Hydroxymethylglutaryl-CoA Reductase (HMG-CoA Reductase)

The biosynthetic pathway for cholesterol involves more than 25 different enzymes. The enzyme 3-hydroxy-3-methylglutaryl coenzyme A (HMG-CoA) reductase catalyzes the conversion from HMG-CoA to mevalonate, the rate-limiting step of the entire pathway. Inhibition of HMG-CoA reductase provides a very attractive opportunity to inhibit cholesterol biosynthesis because no

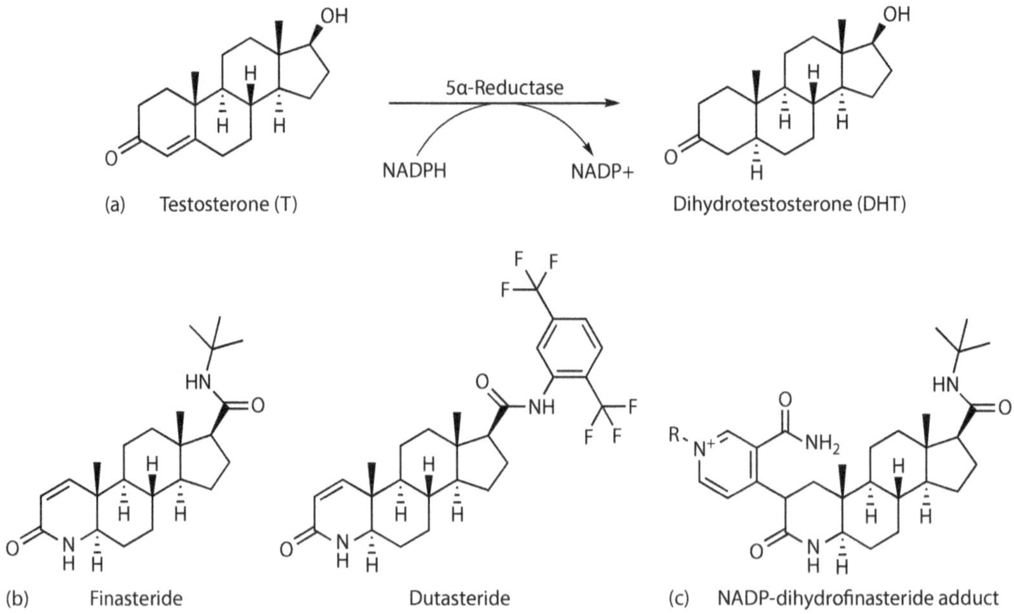

FIGURE 11.4 (a) 5α-Reductase catalyzed conversion of testosterone (T) to dihydrotestosterone (DHT); (b) chemical structures for finasteride and dutasteride; and (c) the proposed structure of the NADP-dihydrofinasteride adduct. R = phosphoadenosine diphosphoribose.

buildup of potentially toxic precursors occurs upon inhibition. In 1976, Japanese microbiologist Akira Endo isolated a series of compounds from *Penicillium citrinum*, including ML236B (compactin), with powerful inhibitory effect on HMG-CoA reductase. Since then, several HMG-CoA reductase inhibitors have become marketed drugs for lowering cholesterol levels (Figure 11.5).

These HMG-CoA reductase inhibitors, commonly referred to as statins, have accounted for the majority of prescriptions for cholesterol-lowering drugs worldwide. All the statins in clinical use are analogs of the substrate HMG-CoA with an HMG-like moiety which may be present in an inactive lactone form in the prodrugs (Figure 11.5). Statins are classified into two groups according to their molecular structures. Type I statins, including lovastatin and simvastatin, are lactone prodrugs originally isolated from fungi. They are enzymatically hydrolyzed in vivo to produce the active drug. Type II statins are all synthetic products with larger groups attached to the HMG-like moiety. All the statins are competitive with respect to HMG-CoA and noncompetitive with respect to NADPH, a cosubstrate of the reaction. Crystal structures of HMG-CoA reductase complexed with six different statins showed that the statins occupy the HMG-binding region, but do not extend into the NADPH site. The orientation and bonding interactions of the HMG-like moiety of the statins resemble those of the substrate complex. However, from a combination of crystal structures, binding thermodynamics, and SAR studies it is clear that the 5′-hydroxyl group of the acidic side chain acts as a mimetic of the tetrahedral intermediate of the reduction reaction. The multiple hydrogen bonds between the C5–OH of the statins and the HMG-CoA reductase active site contribute significantly to the tight binding of the statin inhibitors. Strictly speaking, the HMG-CoA reductase inhibitors are not products of rational design; rather they were identified through natural product screening and analoguing of the natural product hits. Nevertheless, it is quite clear that all statins share a common strategy for inhibiting their target: tetrahedral intermediate state mimicry.

11.4.2.2 Inhibitors of Purine Nucleoside Phosphorylase

Purine nucleoside phosphorylase (PNP) catalyzes the phosphorolysis of 6-oxypurine nucleosides and deoxynucleosides. In humans, the PNP pathway is the only route for deoxyguanosine degradation,

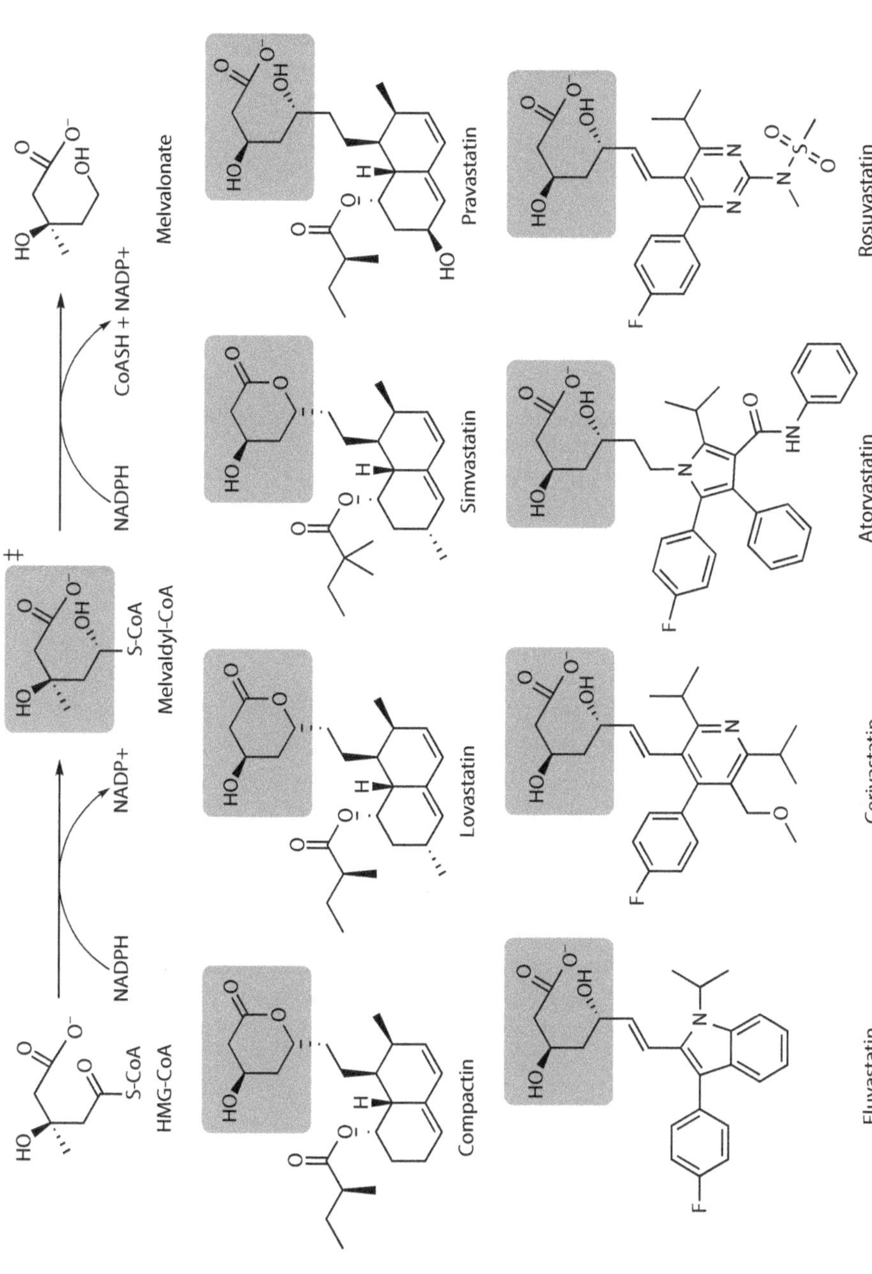

FIGURE 11.5 Structures of HMG-CoA reductase reaction substrate, tetrahedral intermediate, product, and selected statin inhibitors. Compactin, lovastatin, and simvastatin are type I statins. All other statins are type II statins. The mevaldyl tetrahedral intermediate that is mimicked in all statins is identified within brackets.

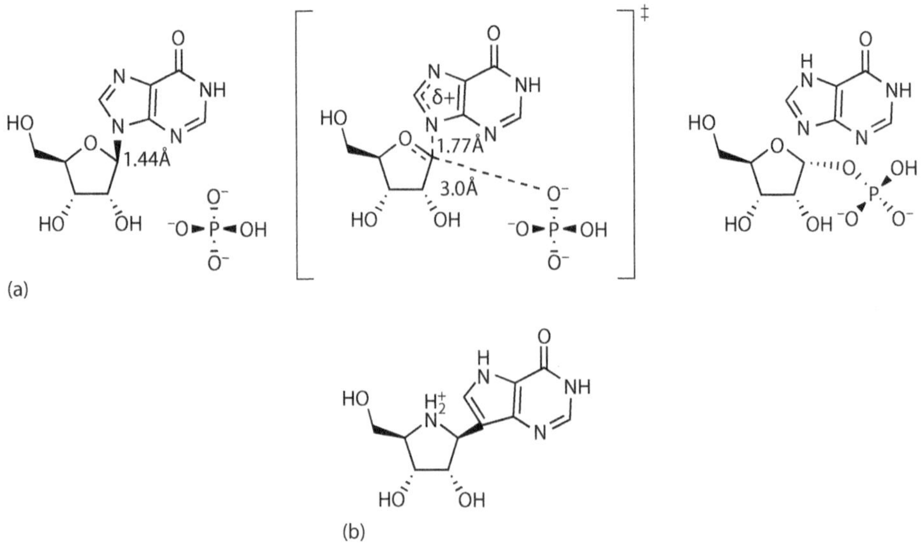

FIGURE 11.6 The structures of PNP reaction substrate inosine, transition state and reaction product (a) and transition state-based inhibitor Immucillin H/BCX-1777 (b).

and genetic deficiency in this enzyme leads to profound T-cell-mediated immunosuppression. Inhibition of PNP has applications in treating aberrant T lymphocyte activity which is implicated in T-cell leukemia and autoimmune diseases. The challenge to inhibitor design for PNP arises from the abundance of the enzyme in human tissues. It has been shown that near complete inhibition of PNP (>95%) is required for significant reduction in T-cell function. Structure-based inhibitor design produced some inhibitors with K_d values in the nanomolar range. However, clinical evaluations showed that these inhibitors did not produce sufficient inhibition of PNP to be effective anti-T-cell therapies. Much more potent PNP inhibitors were later designed with the aid of transition state analysis. In theory, a perfect transition state inhibitor of PNP should bind with a K_d value of approximately 10^{-17} M (10 attomolar). The structure of the transition state for human PNP was determined by Schramm and coworkers in 1995 by measuring kinetic isotope effects. Their studies revealed a transition state with significant ribooxycarbenium character (Figure 11.6). On the basis of the features of this transition state, compounds with picomolar affinity to PNP were synthesized. Among them, Immucillin H/BCX-1777 was a 56 pM inhibitor of human PNP with good potency against cultured human T-cell lines in the presence of deoxyguanosine. Early-stage clinical trials of Immucillin H/BCX-1777 showed moderate efficacy in peripheral and cutaneous T-cell lymphomas and the molecule is currently under investigation for recurrent refractory peripheral T-cell lymphoma.

11.5 BIOSTRUCTURE-BASED DESIGN

In the preceding section of this chapter, we established the fundamental importance to drug discovery of a deep, mechanistic understanding of the reaction mechanism of an enzyme target. While this can be accomplished by the application of mechanistic enzymology, it can be facilitated greatly by the knowledge of the three-dimensional structure of the protein, ideally obtained via experimental techniques such as X-ray crystallography and NMR spectroscopy or by computational methods such as homology modeling. Visualization of the detailed architecture of an enzyme's active site, in complex with a small-molecule inhibitor, can be an important driver in the optimization of a medicinal chemistry effort. The structural insights obtained allow for improvements in target potency, selectivity, and inhibitor physicochemical properties, all of which are paramount in establishing inhibitor SAR.

The term "rational drug design" is often used to describe the application of structure-guided drug discovery approaches. Over the past three decades, several drugs have been made available to patients as a result of advances in protein crystallography and other structural methods. In 2007, at least 10 compounds whose discovery was reliant upon a structure-based approach had been approved by regulatory agencies. As improvements have been made to the technologies and methodologies needed for structure-based efforts, the impact of these approaches on drug development has increased. A review of the literature revealed that 9 of the 27 drugs approved by the FDA in 2013 had information on the structure of the protein target or the protein–inhibitor complex available, and at least 4 of these compounds utilized structure-based approaches during early development (Wasserman 2014). In this section, we exemplify how a detailed understanding of the topography of a ligand–enzyme complex can provide a basis for the design of better inhibitors and can complement enzymologic studies to rationalize their biochemical mode of action.

11.5.1 Structure-Based Design of Protein Kinase Inhibitors

Owing to their central roles in mediating cellular signaling pathways, protein kinases are increasingly important targets for treating a number of diseases. In particular, many of the over 500 kinases encoded by the human genome function to regulate tumor cell proliferation, migration, and survival, rendering them attractive targets for chemotherapeutic intervention in the treatment of cancer. Despite their diversity, all protein kinases catalyze the identical chemical reaction, the transfer of the γ-phosphate of ATP to the hydroxyl group of serine, threonine, or tyrosine residues on specific proteins. Their catalytic domains reflect this singular function in that they share a common feature called the protein kinase fold; this includes a highly conserved ATP-binding pocket, formed between the interface of the *N*-terminal and *C*-terminal domains. The ATP-binding site has been the major focus of inhibitor design; owing to its high degree of conservation, however, selectivity has been a major challenge for inhibitors that target this binding site of protein kinases. The use of biostructure-based approaches has therefore been of great importance in the optimization of kinase-targeted anticancer therapies.

X-ray crystallographic studies have indicated that the catalytic activity in most kinases is controlled by an "activation loop" which adopts different conformations depending upon the phosphorylation state of serine, threonine, or tyrosine residues within the loop. In kinases that are fully active, the loop is thought to be stabilized in an open conformation as a result of phosphorylation, allowing a β-strand within the loop to serve as a platform for substrate binding. While the "active" conformation of the loop is very similar in all known structures of activated kinases, there is great variability in the loop conformation in the inactive state of kinases. In this inactive-like conformation, the loop places steric constraints which preclude substrate binding.

One of the first protein kinase inhibitors developed as a targeted cancer therapy is imatinib (Gleevec®; Novartis Pharmaceuticals, Basel, Switzerland, see also Chapters 22 and 23). Imatinib has been used with remarkable success to treat patients with chronic myelogenous leukemia (CML), a malignancy resulting from the deregulated activity of the kinase Abl due to a chromosomal translocation which gives rise to the breakpoint cluster region-abelson tyrosine kinase oncogene (Bcr-Abl). Imatinib inhibits the tyrosine kinase activity of Bcr-Abl and it is considered as a frontline treatment for CML by virtue of its high degree of efficacy and kinase selectivity. Together with biochemical analyses, crystallographic studies of the interaction of imatinib with the Abl kinase domain have revealed that imatinib binds to the Bcr-Abl ATP-binding site preferentially when the centrally located activation loop is not phosphorylated, thus stabilizing the protein in an inactive conformation (Figure 11.7). In addition, imatinib's interactions with the *N*-terminal lobe of the kinase appear to involve an induced-fit mechanism, further adding to the unique structural requirements for optimal inhibition. One of the most interesting aspects of this interaction is that the specificity of inhibition is achieved despite the fact that residues that contact imatinib in Abl kinase are either identical or very highly conserved in other Src-family

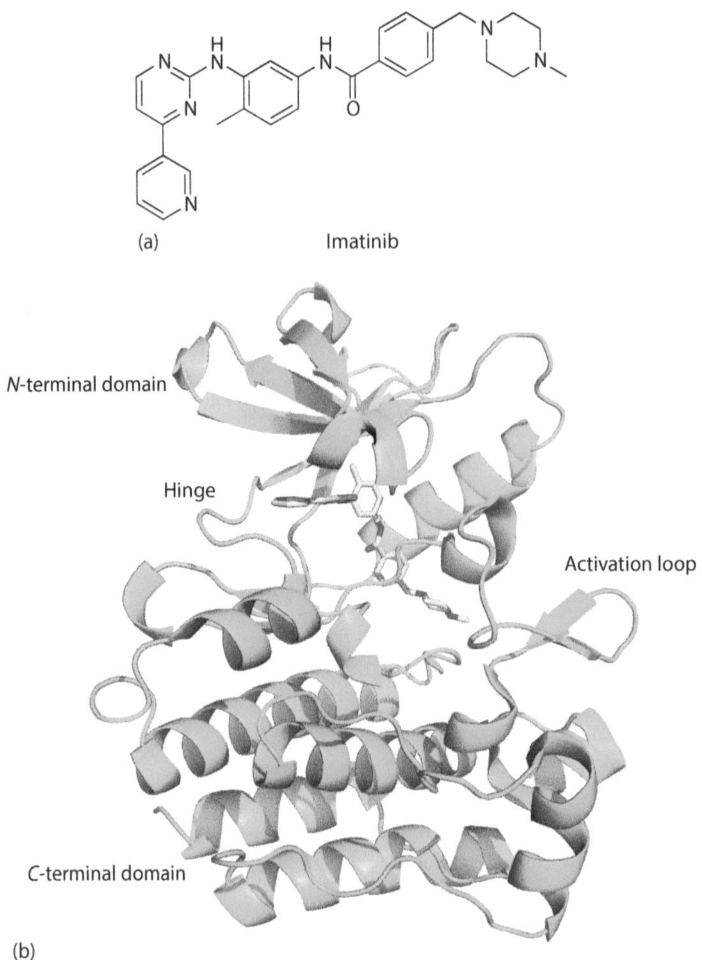

FIGURE 11.7 The structure of Imatinib (Gleevec; a) and the compound in complex with the catalytic domain of cAbl (PDB 1IEP; b).

tyrosine kinases. Thus, despite targeting the relatively well-conserved nucleotide-binding pocket of Abl, studies have shown that imatinib achieves its high specificity by recognizing the distinctive inactive conformation of the Abl activation loop. Biostructure-based methods have had a further impact on more recent efforts to design second-generation therapies targeting imatinib-resistant mutations in Bcr-Abl kinase that have been identified in CML patients. It is likely that these new inhibitors will have substantial clinical utility in the treatment of imatinib-resistant CML; continued exploration of the structural details of the interactions between these compounds and the mutant kinase are still necessary, as resistance remains an inevitable consequence of such drug treatment regimens.

The three catalytically active receptor tyrosine kinases (RTKs) of the ErbB family represent another attractive target group for the treatment of a variety of cancers: epidermal growth factor receptor (EGFR, also known as ErbB1), ErbB2 (also known as HER2/*neu*), and ErbB4. These RTKs are large, multi-domain proteins that contain an extracellular ligand binding domain, a trans-membrane domain, and a cytoplasmic domain responsible for the tyrosine kinase activity. Ligand binding to the extracellular domain induces the formation of receptor homo- and hetero-dimers which leads to the activation of the tyrosine kinase activity and subsequent phosphorylation of the cytoplasmic tail. A number of ErbB-targeted molecules have already reached the market, with a number of others in various stages of clinical investigation. Two of these molecules, erlotinib (Tarceva®, Genetech, Inc. and OSI Pharmaceuticals, Inc. San Francisco, USA) and lapatinib (Tykerb®, GlaxoSmithKline plc. Brentford, U.K.), share a common 4-anilinoquinazoline core, yet their ErbB inhibition profiles and mechanisms of action are clearly differentiated on the basis of biochemical and crystallographic studies. For example, while erlotinib is a potent and selective inhibitor of EGFR only (K_i^{app} = 0.4 nM), lapatinib exhibits potent activity against both EGFR and ErbB2, with estimated K_i^{app} values of 3 and 13 nM, respectively. In addition to its dual kinase activity profile, lapatinib can be distinguished further from erlotinib in that it has a prolonged off-rate from its kinase targets compared to the very fast off-rate from EGFR of erlotinib. This translates to a half-life of dissociation of 300 minutes for the lapatinib–EGFR complex. Importantly, in cellular washout experiments this slow off-rate correlates with a prolonged inhibition of receptor tyrosine phosphorylation in tumor cells (see Section 11.6 for additional information).

An evaluation of the binding mode of lapatinib, based on the crystal structure of the compound in complex with EGFR, suggests a rationale for its long target residence time compared to other 4-anilinoquinazoline inhibitors. Not surprisingly, the quinazoline ring was observed to be hydrogen-bonded to the flexible hinge region between the *N*- and *C*-terminal lobes of the kinase, but there are variations in the key hydrogen-bonding interactions compared to those revealed in the erlotinib–EGFR structure. These differences indicate that lapatinib binds to a relatively closed form of this binding site, whereas erlotinib binds to a more open form. In addition, the ATP-binding pocket of the lapatinib–EGFR complex has a larger back pocket than the apo-EGFR or erlotinib-EGFR structures owing to a shift in one end of the C-helix. This enlarged back pocket accommodates the 3-fluorobenzyloxy group of lapatinib (Figure 11.8). The structural change is significant because it results in the loss of a highly conserved Glu738–Lys721 salt bridge which is an important regulatory mechanism of kinases, functioning to ligate the phosphate groups of ATP. The net result of these structural differences is that the activation loop in the lapatinib–EGFR structure adopts a conformation that is reminiscent of that found in inactive kinases. In contrast, the erlotinib–EGFR structure in Figure 11.8 displays the activation loop in an active conformation. These effects provide a potential molecular rationale for the prolonged residence time of lapatinib on its target which in turn may result in the observed duration of drug activity in cells. In total, these elegant structural and biochemical studies have important implications for the discovery of novel, targeted signal transduction inhibitors and suggest that subtle differences in kinase inhibitor structure can have a profound impact on the binding mode, kinetics, and cellular activity.

11.5.2 Structure-Based Design of HIV Protease Inhibitors

Perhaps the greatest impact of structure-based design on the identification of novel medicines has been in the treatment of AIDS, the etiologic agents of which are human immunodeficiency virus type 1 and type 2 (HIV-1 and HIV-2). These retroviruses encode relatively simple genomes consisting of three open reading frames (ORFs): *gag*, *pol*, and *env*. The *gag* gene encodes the structural capsid, nucleocapsid, and matrix proteins, while the *env* gene is processed by multiple alternative splicing events to yield regulatory proteins. The *pol* ORF encodes the essential viral enzymes necessary for viral replication: reverse transcriptase (RT), integrase, and protease (PR). HIV-1 PR is an aspartyl protease which is required for proteolytic processing of the Gag and Gag-Pol polyprotein precursors to yield the viral enzymes and structural proteins, and it is absolutely indispensable for proper virion assembly and maturation. For this reason it has been an important target for the discovery of anti-HIV therapeutics, and indeed there are at least eight drugs in current clinical use whose antiviral mode of action is by potent inhibition of the HIV protease (Figure 11.9).

One of the major driving forces behind the rapid progress in the identification of HIV protease inhibitors to combat AIDS has been the intense investigation of the structure of the enzyme, particularly in complex with a number of different inhibitors. HIV-1 PR is a homo-dimer comprising

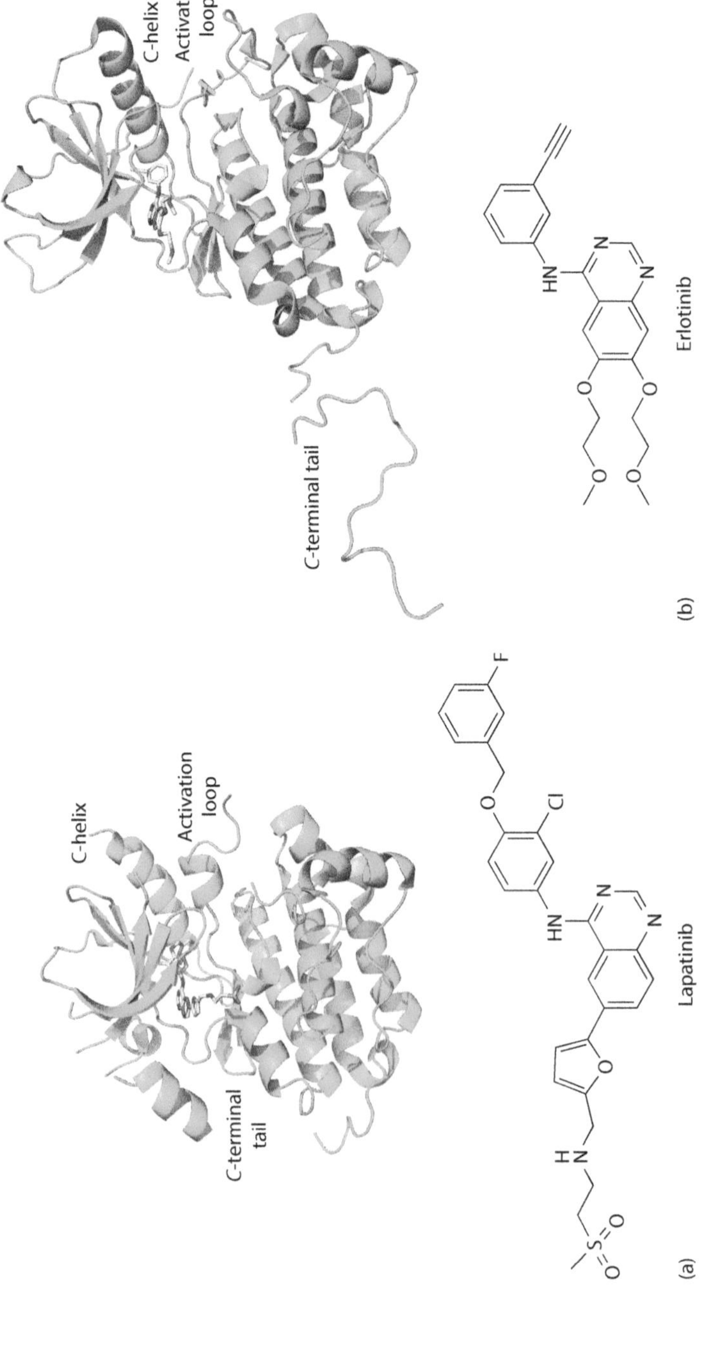

FIGURE 11.8 Protein-ligand binding modes of lapatanib (PDB 1XKK) (a) and erlotinib (PDB 1M17) (b) in complex with EGFR.

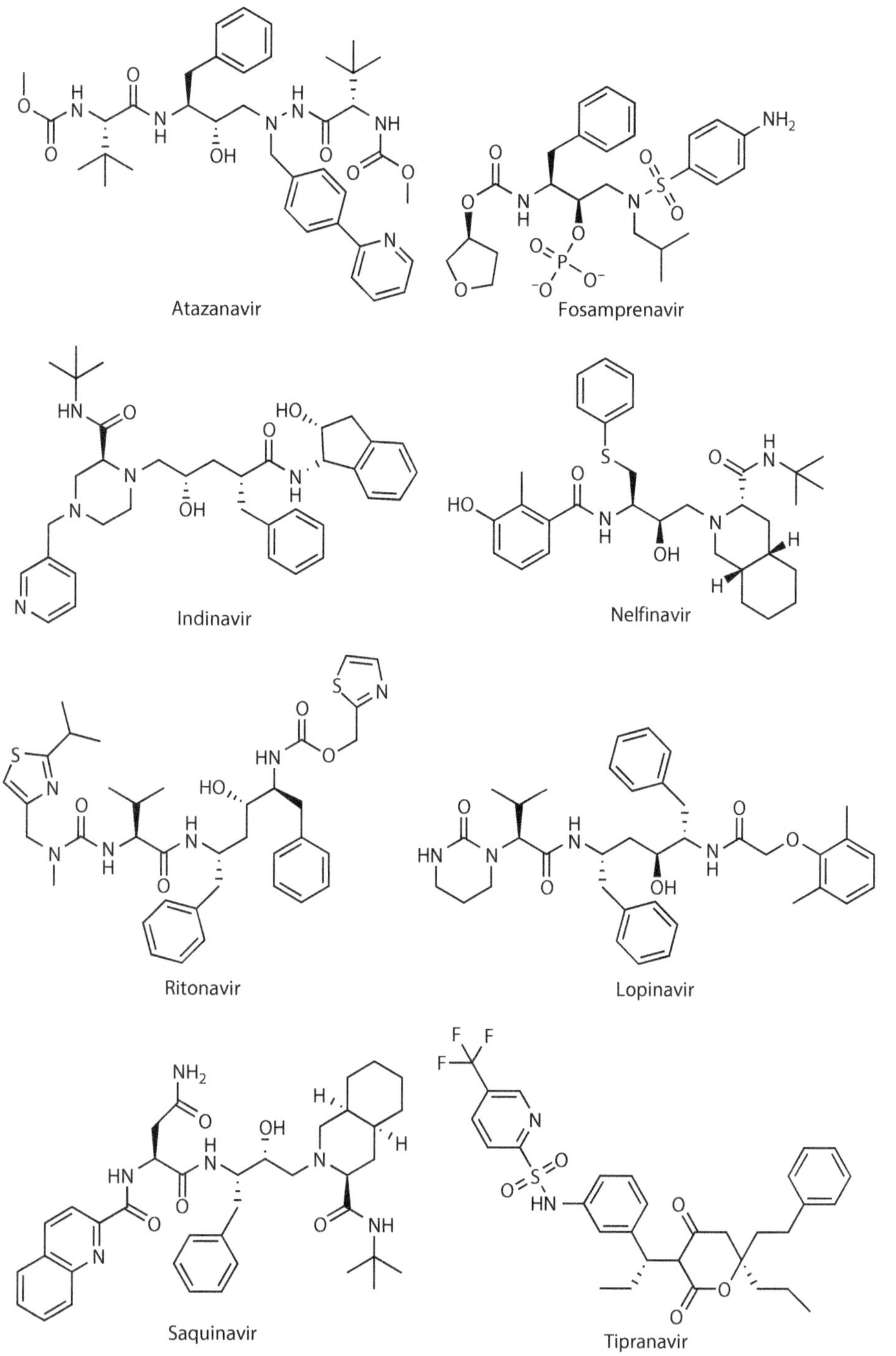

Atazanavir

Fosamprenavir

Indinavir

Nelfinavir

Ritonavir

Lopinavir

Saquinavir

Tipranavir

FIGURE 11.9 Clinically approved HIV-1 protease inhibitors for the treatment of AIDS.

two polypeptide chains of 99 amino acids, each contributing a single catalytic aspartate residue within the active site that lies at the dimer interface. This active site is covered by two symmetric flaps whose dynamic motions allow entry and exit of polypeptide substrates. For each of the different substrates, three to four amino acids on each side of the scissile bond are thought to be involved in binding to the substrate cavity. Since there is little similarity in the primary sequence of the cleavage sites of each of the protease substrates, binding specificity is thought to be driven by the conservation in the secondary structure surrounding the cleavage sites. All of the inhibitors currently used to treat HIV infection targeting HIV-1 PR are competitive in nature and bind to the protease active site.

Saquinavir was the first HIV protease inhibitor available for the treatment of AIDS, and its design was based on a strategy using a transition state mimetic. A distinguishing feature of HIV-1 PR is its ability to cleave Tyr-Pro and Phe-Pro sequences found in the viral substrates, as mammalian endopeptidases are unable to cleave peptide bonds followed by a proline. A rational inhibitor design approach based on this property offered hopes of identifying inhibitors selective for the viral enzyme. Since reduced amides and hydroxyethylamine isosteres most readily accommodate the amino acid moiety of Tyr-Pro and Phe-Pro in the HIV substrates, they were chosen for further interrogation. Systematic substitutions were explored on a minimum peptidic pharmacophore, and one compound containing an (S,S,S)-decahydro-isoquinoline-3-carbonyl (DIQ) replacement for proline exhibited a K_i value of 0.12 nM at pH 5.5 for HIV-1 PR and <0.1 nM for HIV-2 PR. The interactions of this compound, later named saquinavir, with HIV-1 PR were studied using X-ray crystallography (Figure 11.10). The compound was shown to bind to the enzyme in an extended conformation with the carbonyl of the DIQ group binding to a water molecule that connects the inhibitor with the flap regions. These studies shed much light on the binding mode of the first HIV PR inhibitors and set the stage for further exploration of novel compounds with improved properties.

The availability of new HIV protease inhibitors represented a great triumph in the fight against AIDS, but it was only a matter of time before the selective pressure of antiretroviral therapy led to the emergence of HIV strains harboring drug-resistant mutations against protease inhibitors. One of the primary mutations first noted in protease inhibitor-resistant strains was in Val82 of HIV-1 PR. Crystallographic and modeling studies suggested that the binding of protease inhibitors like ritonavir might be compromised due to the loss of hydrophobic interactions between the isopropyl

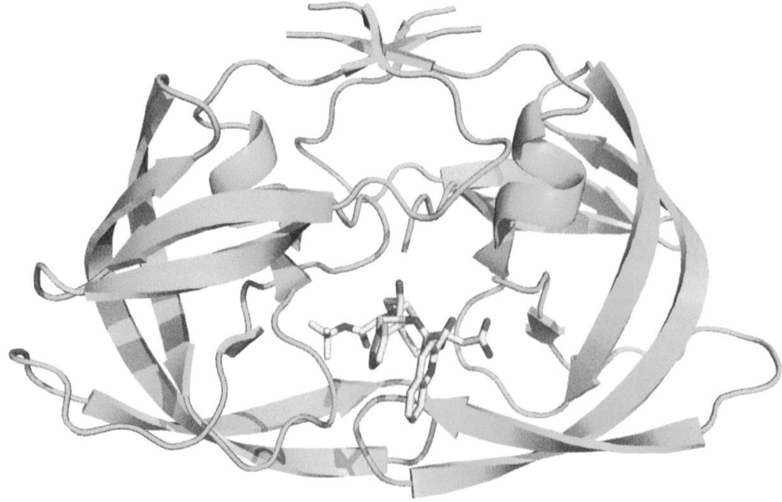

FIGURE 11.10 The crystal structure of HIV-1 protease in complex with the inhibitor saquinavir (PDB 3OXC). For clarity, only one of two orientations of the inhibitor is shown.

side chain of Val 82 of the enzyme and the isopropyl substituent projecting from the 2 position of the P3 thiazolyl group of ritonavir. This functionality was substituted to identify an inhibitor whose activity was less dependent on interaction with Val 82 and the optimization, supported by modeling studies, led to the identification of ABT-378, later named lopinavir. This novel inhibitor had extraordinary potency against wild-type and mutant HIV-1 PR (K_i = 1.3–3.6 pM) in vitro, and maintained activity against ritonavir-resistant mutants of HIV-1. Lopinavir, as a combination drug with ritonavir, is known as Kaletra, and is an important salvage drug for patients who have failed primary therapy with other protease inhibitors.

11.6 CONCLUDING REMARKS

The structural details of drug-binding pocket interactions, gleaned from crystallographic and other biophysical methods, can provide a rich source of information for inhibitor optimization. Historically, SAR through structure-based methods had been limited by the time and protein demands of X-ray crystallography. These limitations, however, are rapidly diminishing due to significant advances in the technologies of protein expression and protein crystallography. In particular, robotic methods for crystallization trials have significantly reduced the amount of protein required for experiments and the time required to obtain crystals of a target protein in complex with multiple inhibitory compounds. Likewise, the more routine use of high-energy beam sources has facilitated structure determinations from crystals that would otherwise be insufficient for diffraction studies and has significantly decreased the time required to perform the needed data collection. Advances in software development and computational capabilities have further decreased the time required to solve each structure, resulting in a faster turnaround time for each structure-based design iteration. These advances provide the basis for greater reliance on structural biology as a common tool during the iterative process of lead optimization.

Conformational dynamics within the drug binding pockets of enzymes is a common feature, dictated by the chemistry of enzyme catalysis. Hence, binding pocket structures are not static; rather, they often change in response to encounters with inhibitory molecules. Thus, as described in this chapter, there can often be a temporal component to enzyme-inhibitor affinity. Advances in kinetic methodologies and instrumentation have made the determination of inhibition kinetics more facile, so that such measurements can be a routine part of the SAR of lead optimization. It is therefore no longer necessary to rely solely on equilibrium measures of inhibitor binding affinity, such as IC_{50} and K_i values, for lead optimization. Instead, routine measurements of enzyme-inhibitor association and dissociation rates are becoming practical, with throughput that makes these measurements germane to drug discovery. Hence, increased attention is being paid to the importance of understanding these kinetic components of drug–target interactions, and their potential impact on clinical efficacy. For example, the duration of drug efficacy in vivo has been suggested to depend in part on the duration of the drug–target complex; this is experimentally measured as the residence time which is the reciprocal of the dissociation rate constant for the drug–target complex. Drugs that demonstrate long residence times, especially when this exceeds the pharmacokinetic half-life of the drug, may significantly extend the pharmacodynamic efficacy of a drug in vivo, and may also ameliorate the potential for off target-mediated adverse events. Future drug discovery efforts may thus be focused not merely on optimization of inhibitor affinity, but also on the extension of residence time (see Further Reading).

FURTHER READING

Copeland, R.A. 2000. *Enzymes: A practical Introduction to Structure, Mechanism and Data Analysis*, 2nd ed. Hoboken, NJ: John Wiley & Sons.

Copeland, R.A. 2013. *Evaluation of Enzyme Inhibitors in Drug Discovery: A Guide for Medicinal Chemists and Pharmacologists*, 2nd ed. Hoboken, NJ: John Wiley & Sons.

Copeland, R.A., Pompliano, D.L., and Meek, T.D. 2006. Drug-target residence time and its implications for lead optimization. *Nat. Rev. Drug Discov.* 5:730–739.

Nagar, B., Bornmann, W.G., Pellicena, P. et al. 2002. Crystal structures of the kinase domain of c-Abl in complex with the small molecule inhibitors PD173955 and imatinib (STI-571). *Cancer Res.* 62:4236–4243.

Roberts, N.A., Martin, J.A., Kinchington, D. et al. 1990. Rational design of peptide-based HIV proteinase inhibitors. *Science* 248:358–361.

Robertson, J.G. 2005. Mechanistic basis of enzyme-targeted drugs. *Biochemistry* 44:5561–5571.

Schramm, V.L. 2005. Enzymatic transition states: Thermodynamics, dynamics and analogue design. *Arch. Biochem. Biophys.* 433:13–26.

Wasserman, S. DOE X-ray light sources and pharmaceutical discovery. *AAAS Annual Meeting*, 2014.

Wood, E.R., Truesdale, A.T., McDonald, O.B. et al. 2004. A unique structure for epidermal growth factor receptor bound to GW572016 (Lapatinib): Relationships among protein conformation, inhibitor off-rate, and receptor activity in tumor cells. *Cancer Res.* 64:6652–6659.

12 Receptors
Structure, Function, and Pharmacology

Hans Bräuner-Osborne

CONTENTS

12.1 INTRODUCTION

Communication between cells is mediated by compounds such as neurotransmitters and hormones which upon release will activate receptors in the target cells. This communication is of pivotal importance for many physiological functions, and dysfunction in cell communication pathways often has severe consequences. Many diseases are caused by dysfunction in the pathways, and in these cases, drugs designed to act at the receptors can have beneficial effects. Thus, receptors are very important drug targets.

The first receptors were cloned in the mid-1980s and since then hundreds of receptor genes have been identified. Based on the sequence of the human genome, it is currently estimated that more than one thousand human receptors exist. Almost all receptors are heterogeneous, meaning that several receptor subtypes are activated by the same signaling molecule. One such example is the excitatory neurotransmitter glutamate. As shown in Figure 12.1, the amino acid sequence of the glutamate receptors varies and the receptors form subgroups, which, as will be discussed in Chapter 15, share pharmacology.

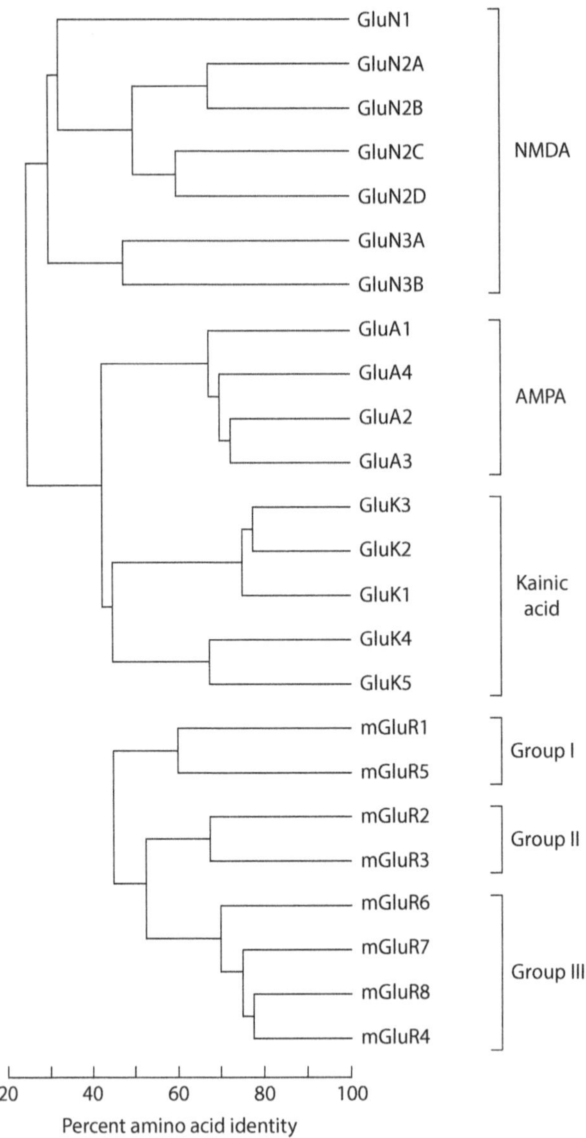

FIGURE 12.1 Phylogenetic tree showing the amino acid sequence identity between cloned mammalian glutamate receptors. The subgroups according to receptor pharmacology have been noted. The NMDA, AMPA, and kainic acid receptors belong to the superfamily of ligand-gated ion channels whereas the metabotropic glutamate receptors (mGluR1–8) belong to the superfamily of G protein-coupled receptors.

The same signaling molecule can act on both G protein-coupled receptors and ligand-gated ion channels (Figure 12.1). One of the reasons for the heterogeneity is that it allows cells to be regulated in subtle ways. For example, whereas the fast synaptic action potential is initiated by glutamate receptors of the ligand-gated ion channel family, these receptors are themselves regulated by slower and longer acting glutamate receptors from the G protein-coupled receptor family. The action on these two receptor families is shared by a number of other neurotransmitters such as GABA (Chapter 15), acetylcholine (Chapter 16), and serotonin (Chapter 18).

12.1.1 Synaptic Processes and Mechanisms

Receptors are located in a complex, integrated, and highly interactive environment which can be further illustrated by the processes and mechanisms of synapses (Figure 12.2). The synapses are key elements in the interneuronal communication in the peripheral and in the central nervous system (CNS). In the CNS, each neuron has been estimated to have synaptic contact with several thousand other neurons, making the structure and function of the CNS extremely complex.

The receptor is activated upon release of the signaling molecule and it is evidently equally important to be able to stop the signaling again. This is often achieved by transporters situated in the vicinity of the receptor, which move the signaling molecule from the extracellular to the intracellular space, where it is either stored or metabolized (see Chapters 14, 15, and 18). Blockade of a transporter or a metabolic enzyme will cause an elevation of the extracellular concentration of the signaling molecule and lead to increased receptor activation, and transporters and metabolic enzymes can thus be viewed as indirect receptor targets. Synaptic functions may also be facilitated by stimulation of the neurotransmitter biosynthesis, for example, by administration of a biochemical precursor. Transport mechanisms in synaptic storage vesicles (Figure 12.2) are also potential sites for pharmacological intervention. Autoreceptors normally play a key role as a negative feedback mechanism regulating the release of certain neurotransmitters, making this class of presynaptic receptors therapeutically interesting.

Pharmacological stimulation or inhibition of the above-mentioned synaptic mechanisms are, however, likely to affect the function of the entire neurotransmitter system. Activation of neurotransmitter receptors may, in principle, represent the most direct and selective approach to stimulate of a particular neurotransmitter system. Furthermore, activation of distinct subtypes of receptors operated

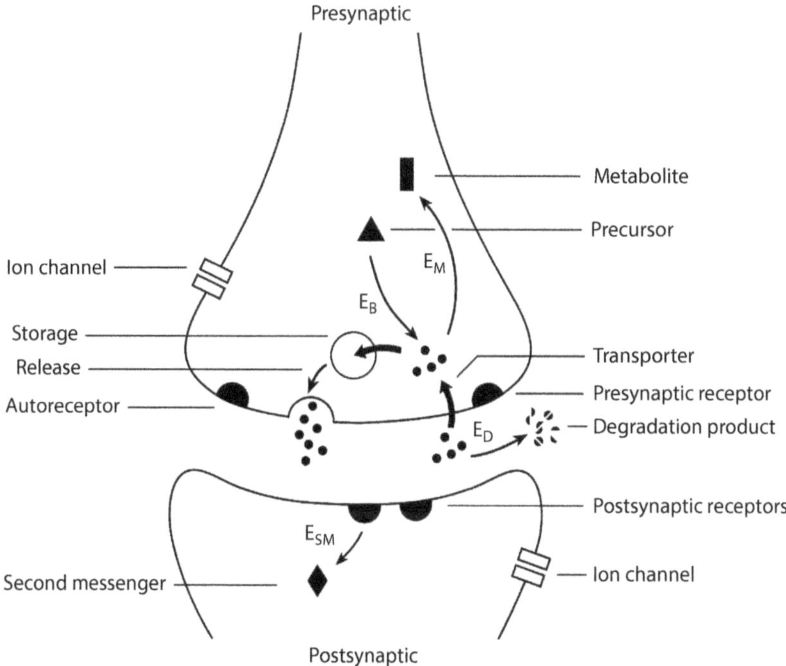

FIGURE 12.2 Generalized schematic illustration of processes and mechanisms associated with an axosomatic synapse in the CNS. E, enzymes, metabolic (E_M), biosynthetic (E_B), degradation (E_D), second messenger (E_{SM}); •, neurotransmitter.

by the neurotransmitter concerned may open up the prospect of highly selective pharmacological intervention. Nevertheless, indirect mechanisms of targeting receptors via regulation of the level of the endogenous agonist at the site-of-action remains an important pharmacological principle which has also been applied outside the synapse as exemplified by compounds increasing insulin release and preventing glucagon-like peptide-1 (GLP-1) breakdown.

Direct activation of receptors by full agonists may result in rapid receptor desensitization (insensitive to further activation). Partial agonists are less liable to induce receptor desensitization and may therefore be particularly interesting for neurotransmitter replacement therapies. Desensitization may be a more or less pronounced problem associated with therapeutic use of receptor agonists, whereas receptor antagonists which in other cases have proved useful therapeutic agents, may inherently cause receptor supersensitivity. The presence of allosteric binding sites at certain receptor complexes, which may function as physiological modulatory mechanisms, offer unique prospects of selective and flexible pharmacological manipulation of the receptor complex concerned. Whilst some receptors are associated with ion channels, others are coupled to second messenger systems. Key steps in such enzyme-regulated multistep intracellular systems (Figure 12.2), also including regulation of gene transcription by second messengers, represent novel targets for therapeutic interventions.

12.2 RECEPTOR STRUCTURE AND FUNCTION

Receptors have been divided into four major superfamilies: G protein-coupled receptors, ligand-gated ion channels, tyrosine kinase receptors, and nuclear receptors. The three first receptor superfamilies are located in the cell membrane and the latter family is located intracellularly.

Our understanding of ligand–receptor interactions and receptor structure has increased dramatically during the last decade, not least due to the rapidly growing number of 3D crystallographic structures that have been determined of either full receptors or isolated ligand binding domains. Thus, today, structures of partial or full receptors of all four receptor superfamilies have been determined. Clearly, the information obtained from 3D structures of ligand binding domains in the presence of ligands is very valuable for rational drug design (see Chapter 4). Likewise, knowledge about receptor mechanisms can be used to, e.g., design allosteric modulators interfering with receptor activation.

12.2.1 G Protein-Coupled Receptors

The G protein-coupled receptors (GPCRs) are the largest of the four superfamilies with some estimated 800 human receptor genes. Approximately, 50% of these are taste and odor-sensing receptors, which are not of immediate interest for the pharmaceutical industry, but are of interest for, e.g., tastant and fragrance manufactures. Nevertheless, it is estimated that 30% of all currently marketed drugs act on GPCRs and the superfamily thus remains a very important target for drug research. It is fascinating to note the very broad variety of signaling molecules or stimuli, which are able to act via this receptor superfamily, including tastes, odors, light (photons), ions, monoamines, nucleotides, lipids, amino acids, peptides, proteins, and pheromones.

The GPCRs are also referred to as seven-transmembrane (7TM) receptors due to the seven α-helical transmembrane segments found in all GPCRs (Figure 12.3) and the fact that the receptors can also signal via G protein independent pathways (see later). The GPCRs have been further subdivided into class A, B, and C based on their amino-acid sequence homology. Thus, receptors within class A are more closely related to each other than to receptors in class B and C, etc. This grouping also coincides with the way ligands bind to the receptors. Thus, as illustrated in Figure 12.3, the orthosteric binding sites (binding site of the endogenous agonist) are generally located in the transmembrane region of class A receptors (e.g., acetylcholine, histamine, dopamine, serotonin, opioid, and cannabinoid GPCRs, see Chapters 16 through 19), both in the extracellular loops and

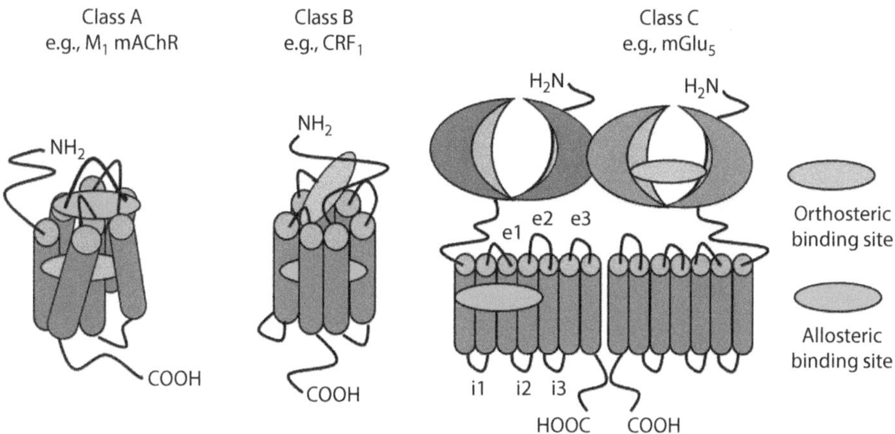

FIGURE 12.3 The superfamily of G protein-coupled receptors. All G protein-coupled receptors contain seven α-helical transmembrane segments and are thus also called seven-transmembrane (7TM) receptors. Cartoon of the three classes showing the typical orthosteric (endogenous agonist) and allosteric binding sites. (Reprinted with permission from Melancon, B.J., Hopkins, C.R., Wood, M.R. et al., *J. Med. Chem.*, 55, 1445–1464. Copyright 2012 American Chemical Society.)

amino-terminal domain of class B receptors (e.g., glucagon and GLP-1 GPCRs) and exclusively in the extracellular amino-terminal domain of class C receptors (e.g., glutamate and GABA GPCRs, see Chapter 15). Allosteric binding sites have also been located in all three receptor classes as outlined in Figure 12.3.

The intracellular loops of GPCRs interact with G proteins. As illustrated in Figure 12.4, the G protein is trimeric consisting of Gα, Gβ, and Gγ subunits. Receptor activation will cause an interaction of the receptor with the trimeric Gαβγ protein, catalyzing an exchange of GDP for GTP in the Gα subunit whereupon the G protein disassociates into activated Gα and Gβγ subunits. Both of these will then activate effector molecules such as adenylate cyclase or ion channels (Figure 12.4). 16 Gα, 5 Gβ, and 12 Gγ subunits have been identified in humans and like the receptors they form groups based on the amino acid homology and the effectors they interact with.

Most GPCRs desensitize quickly upon activation via phosphorylation of specific serine/threonine residues in the intracellular loops and/or C-terminal by kinases such as G protein-coupled receptor kinases (GRKs). Once phosphorylated, β-arrestin molecules will bind to the receptor and cause arrest of the G protein-mediated signaling and induce internalization. Recent evidence has shown that β-arrestins can activate the tyrosine kinase pathway directly leading to non-G protein-mediated cellular effects (Figure 12.5a). In some cases, it has even been possible to develop ligands that selectively activate or inhibit the β-arrestin pathway without activating the G proteins or vice versa. Such biased ligands will induce different cellular effects than ligands activating or inhibiting both signaling pathways with the potential to retain the desired clinical effect while diminishing unwanted side effects (Figure 12.5b and c). The molecular mechanisms of biased ligand action is a matter of debate but is likely either caused by ligand stabilization of different active or inactive receptor conformations, or by differences in ligand kinetics.

Recent evidence has shown that some if not all GPCRs exist as dimeric or even oligomeric complexes. As shown in Figure 12.3 class C receptors dimerize which leads to either homo- or heterodimers. The latter is, for example, the case for $GABA_B$ receptors, which are formed by heterodimerization of $GABA_{B1}$ and $GABA_{B2}$ receptor subunits whereas, e.g., metabotropic glutamate receptors (mGluRs) homodimerize. Whether class A and B receptors also homo- or heterodimerize have been heatedly debated in the literature and only a few examples have been convincingly shown

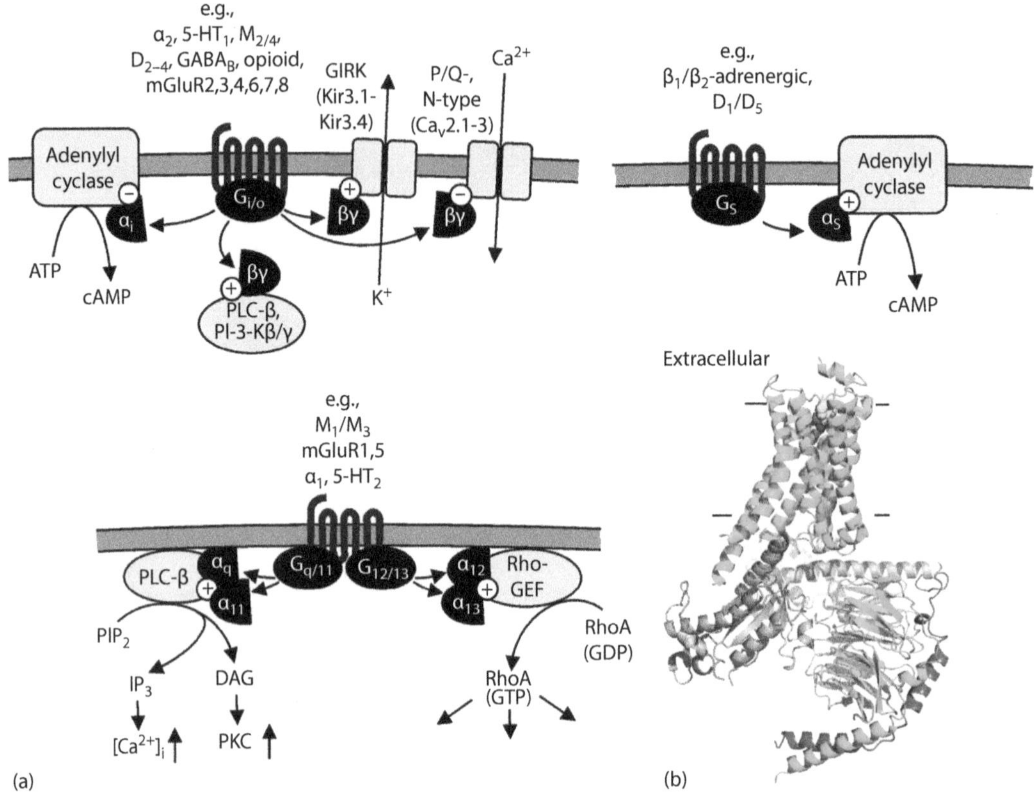

FIGURE 12.4 (a) Principal G protein coupling pathways for a range of G protein-coupled receptors discussed in further detail in Chapters 15 through 19. Receptor activation will catalyze an exchange of GDP for GTP in the α-subunit which leads to activation and separation of the α- and βγ-subunits. The G proteins are divided into four families based on the α-subunits, which upon activation lead to characteristic intracellular effects: Gi/o, inhibition of adenylyl cyclase and decreased cAMP level; Gs, activation of adenylyl cyclase and increased cAMP level; Gq/11, activation of PLC-β and increased IP_3 and Ca^{2+} levels; G12/13, activation of Rho-GEF and RhoA. It is important to note that the activated βγ-subunits also lead to cellular effects via, e.g., modulation of ion channels. (Adapted from Wettschreck, N. and Offermanns, S., *Physiol. Rev.*, 85, 1159, 2005. With permission.) (b) Structure of an active conformation of β2-adrenergic receptor (green) bound to the agonist BI-16707 (yellow sphere) and trimeric Gs protein (Gα, orange; Gβ, cyan; Gγ, purple). (Adapted by permission from Rasmussen S.G.F., DeVree B.T., Zou Y. et al., *Nature* 477, 549, Copyright 2011. Macmillan Publishers Ltd.). α_{1-2} and β_{1-2}, adrenergic receptor subtypes; D_{1-5}, dopamine receptor subtypes 1–5; GIRK, G protein-regulated inward rectifier potassium channel; $5\text{-HT}_{1,2}$, serotonin receptor subtypes 1 and 2; M_{1-5}, muscarinic acetylcholine receptor subtypes 1–5; mGluR1–8, metabotropic glutamate receptor subtypes 1–8; PLC-β, phospholipase C-β; PI-3-K, phosphoinositide-3-kinase; PIP2, phosphatidylinositol 4,5-bisphosphate; IP_3, inositol 1,4,5-trisphosphate; DAG, diacylglycerol; PKC, protein kinase C; Rho-GEF, Rho-guanine nucleotide exchange factor.

to be of physiological importance. One such case is in the field of opioid receptors (see Chapter 19), where it has been shown that the κ, μ, and δ subtypes can form pharmacologically distinct receptor subtypes by heterodimerization.

Collectively, the fact that one GPCR can activate several signaling pathways and heterodimerize to create additional subtypes has greatly complicated our view of receptor function. From a medicinal chemistry point of view, it is interesting to note that in some cases it has been possible to develop ligands selectively targeting a specific signaling pathway or heterodimer. This has opened up not only new possibilities but also new challenges in drug design.

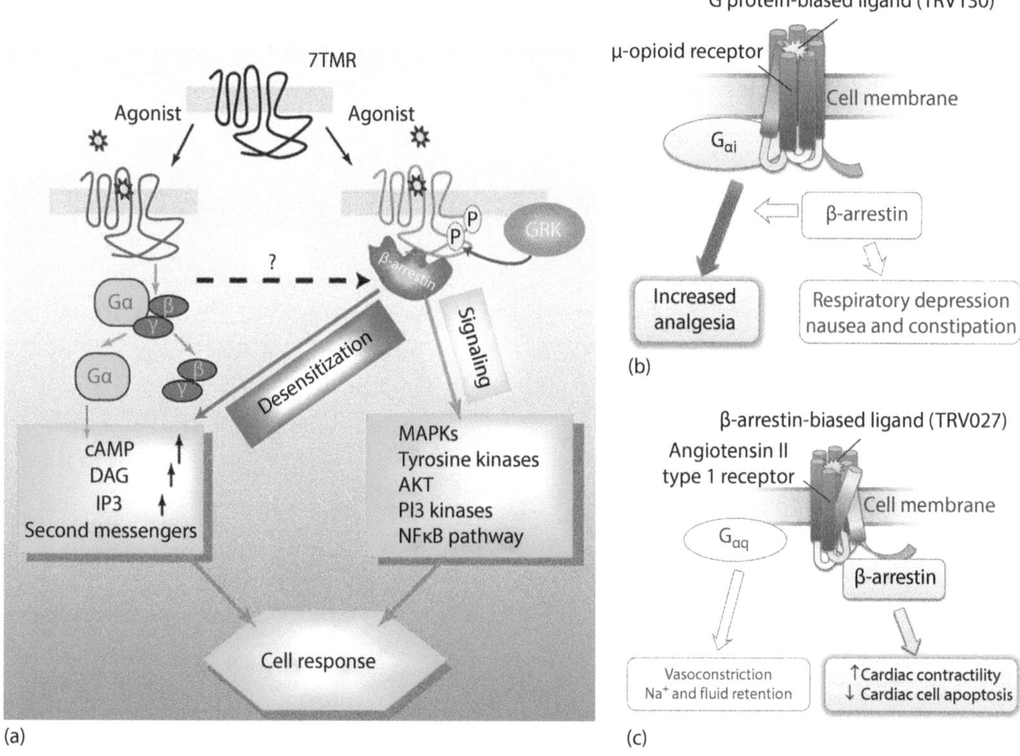

FIGURE 12.5 (a) The G protein and β-arrestin signaling pathway of 7TM receptors (7TMRs). Agonist activation of 7TMRs initiates the classical G protein cascade (see Figure 12.4 for further details) and rapid receptor phosphorylation by G protein-coupled receptor kinases (GRKs). The latter lead to recruitment of β-arrestins which cause desensitization and internalization of the receptor and activation of tyrosine kinase pathways. (b) Hypothesis that a G protein pathway biased agonist at the μ-opioid receptor would retain analgesic effect while lowering unwanted side effects mediated through the β-arrestin pathway. TRV130 ([(3-methoxythiophen-2-yl) methyl]({2-[(9R)-9-(pyridin-2-yl)-6-oxaspiro[4.5]decan-9-yl]ethyl})amine) is an example of such a G protein pathway biased agonist which has shown efficacy in pain in human trials. (c) Hypothesis that a β-arrestin pathway biased agonist at the angiotensin II type 1 receptor would retain desired cardiac effects while lowering unwanted side effects mediated through the G protein pathway. The octapeptide TRV027 (Sar-Arg-Val-Tyr-Ile-His-Pro-D-Ala-OH) has such a biased profile and have shown promising results in early human clinical trials of chronic heart failure. (a: From Lefkowitz, R.J. and Shenoy, S.K., *Science*, 308, 512, 2005. Reprinted with permission from AAAS; b and c: Reprinted from *Trends Pharmacol. Sci.*, 35, Violin, J.D., Crombie, A.L., Soergel, D.G. et al., Biased ligands at G protein-coupled receptors: Promise and progress, 308–316. Copyright 2014, with permission from Elsevier.)

12.2.2 Ligand-Gated Ion Channel Receptors

Ligand-gated ion channel receptors can be divided into three major groups termed the Cys-loop, ionotropic glutamate receptor, and purinergic P2X families, respectively. Whereas the two latter families are exclusively excitatory cation-permeable channels, the former are either excitatory (serotonin and nicotinic acetylcholine receptors) or inhibitory (glycine and GABA receptors) by influx of Na^+/Ca^{2+} or Cl^- ions which will hypo- or hyperpolarize the cell, respectively (see Chapter 13 for further details).

12.2.2.1 Cys-Loop Receptor Family

The nicotinic acetylcholine receptor, at the nerve-muscle synapse, is the best understood Cys-loop receptor which upon acetylcholine binding allows as many as 10,000 potassium and sodium ions per millisecond to pass through the channel. As shown in Figure 12.6, the receptor consists of two

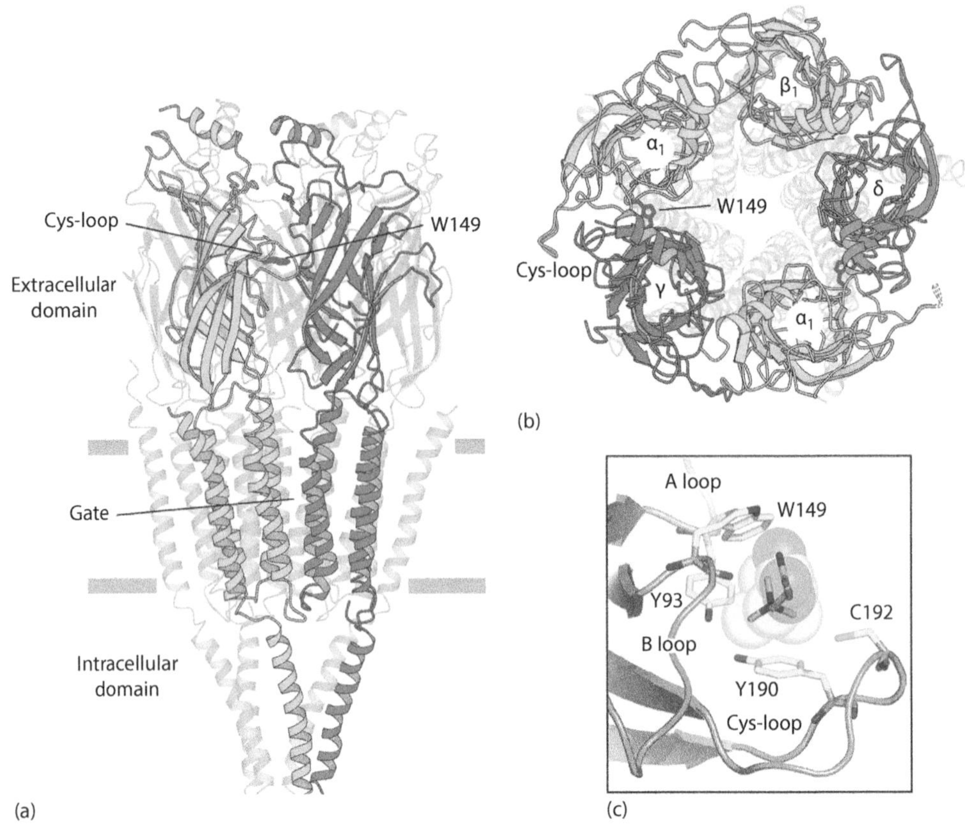

FIGURE 12.6 Structure of the family of Cys-loop ligand-gated ion channel receptors. (a) Side view and (b) top view of 3D structure of the neuromuscular nicotinic acetylcholine receptor which consist of five subunits (two α_1-, one β_1-, one γ-, and one δ-subunit) forming an ion channel in the center. The two agonist binding sites are located in the $\alpha_1\gamma$ and $\alpha_1\delta$ interfaces in the extracellular domain. Each subunit has four transmembrane α-helices of which the M2 helices of the five subunits line the pore. A gate, consisting of hydrophobic leucine and valine residues, is located in the most constricted part of the pore which tilts outward upon receptor activation leading to channel opening. (c) Close-up of the orthosteric binding site of the acetylcholine-binding protein (AChBP) with the agonist carbamylcholine bound highlighting the important cluster of aromatic residues and a cysteine forming key interactions with the agonist. (Adapted from Unwin, N., *Q. Rev. Biophys.*, 46, 283, 2013. With permission.)

α_1 subunits and three other subunits (β_1, γ and δ) which form a pentameric pore in the cell membrane. The agonist binding site is located in subunit interfaces in the extracellular domain. The pore itself is lined with five α-helices (termed M2), one from each of the five receptor subunits which contain hydrophobic leucine and valine residues at the most constricted part of the pore forming the gate for the hydrophilic ions. Agonist binding to the extracellular part of the α-subunits induces conformational changes, which are then relayed through the receptor subunits and ultimately leads to tilt of the pore-lining M2 helices, removal of the hydrophobic gate-lock, and channel opening.

Several high-resolution 3D structures of acetylcholine-binding protein (AChBP), a water-soluble homolog of the ligand-binding domain of nicotinic acetylcholine receptors from the snail *Lymnaea stagnalis*, have been solved in the presence of various ligands (Figure 12.6). These structures have shown that agonists bind in the interface between the subunits and provide detailed insight into the ligand–receptor interactions. For example, all endogenous agonists of the Cys-loop family contain an amine, which, according to the AChBP structures, is interacting with a cluster of aromatic residues via π-cation bonding.

Most Cys-loop receptors form heteropentamers (e.g., the neuromuscular nicotinic acetylcholine receptor described above), but some can form homopentamers (e.g., the nicotinic α_7 receptor). Numerous subunits for both nicotinic acetylcholine receptors and $GABA_A$ receptors have been cloned which can theoretically heteromize to a staggering high number of subunit combinations. However, in reality, only certain subunit combinations are formed in vivo and even fewer combinations have therapeutic interest. The glycine and serotonin Cys-loop receptors have fewer subunits, but heteromerization still exists and leads to distinct receptor subtypes. Interestingly, some subunits are unable to form their part of the agonist binding pocket in either one or both sides of the two interfaces they participate in. Depending on their subunit composition, Cys-loop receptors can bind from two to five agonist molecules. For example, the neuromuscular nicotinic acetylcholine receptor binds two agonist molecules whereas the nicotinic α_7 receptor and AChBP can bind five agonist molecules (Figure 12.6). Whether all agonist binding sites need to be occupied in order to achieve receptor activation has yet to be demonstrated.

12.2.2.2 Ionotropic Glutamate Receptor Family

The ionotropic glutamate receptor family comprises the 16 NMDA, AMPA, and kainic acid receptors listed in Figure 12.1 and two orphan receptors (termed δ1–2) with unknown function. The name of the receptor family is a bit misleading as GluN1 and GluN3A-B actually have glycine as ligand (see Chapter 15). Nevertheless, all 18 receptor subunits have the same overall structure: two large extracellular domains referred to as the amino-terminal domain (ATD) and ligand-binding domain (LBD), a transmembrane domain (TMD) consisting of three transmembrane segments and a re-entry loop and a C-terminal domain (CTD) (Figure 12.7). It is quite interesting to note

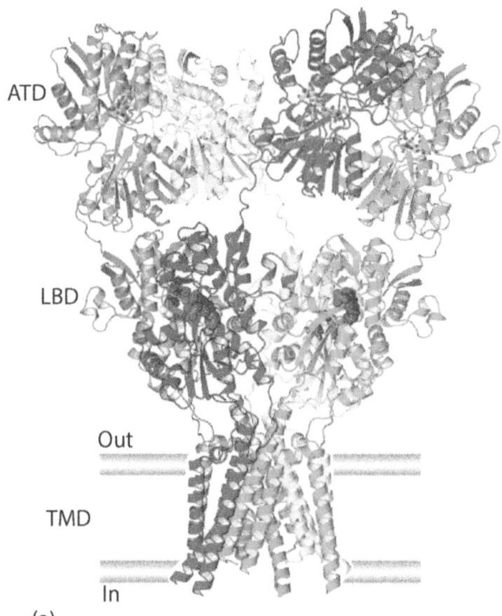

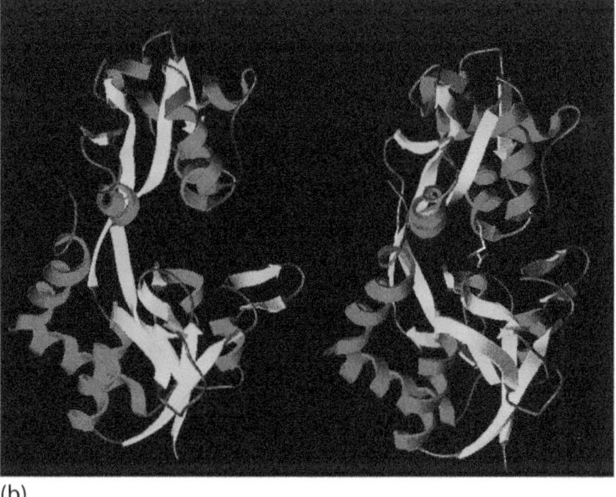

FIGURE 12.7 (a) 3D structure of the inactive conformation of full-length GluA2 ionotropic glutamate receptor with the antagonist ZK200775 (space filled) bound. Glycosylation moieties are shown as sticks. The location of the amino-terminal domain (ATD), ligand-binding domain (LBD), and transmembrane domain (TMD) is noted. The four subunits forming the receptor are shown in different colors. (Adapted with permission from Sobolevsky, A., Rosconi, M.P., and Gouaux, E., *Nature,* 462, 758, Copyright 2009, Macmillan Publishers Ltd.) (b) Structure of the LBD of GluA2 in the open inactive form (left) and the closed active form with glutamate bound in the cleft (right). The agonist-mediated closure of the LBD is thought to initiate the activation of the receptor via an outward pull of the M3 helices lining the ion channel. The structures were generated using the program "Swiss PDB viewer 3.5" with coordinates from Brookhaven Protein Data Base.

the resemblance of the structures of the ATD/LBD and TMD with the amino-terminal domain of mGluRs and potassium channels (see Chapter 13), respectively. Functional receptors are comprised of four subunits assembled around the ion channel lined by M3 helices from each subunit. All NMDA receptors are heteromeric assemblies as GluN1 together with either GluN2 and/or GluN3 subunits whereas AMPA and kainic acid receptors can either be homo- or heteromeric assemblies.

High-resolution 3D structures of the isolated LBD of the majority of ionotropic glutamate receptor subunits and of full-length GluA2 and GluN1/2B receptors have been determined in the absence of ligands and with full and partial agonists, antagonists, and/or allosteric modulators. Overall, these studies have shown that, like the mGluR receptors, activation is initiated by closure of the LBD around the ligand which leads to an outward pull of the M3 gating helices causing an opening of the channel pore (Figure 12.7). The plentitude of LBD structures has also provided a compelling insight into ligand–receptor interactions, molecular mechanisms of selectivity and efficacy. Such information is very valuable in the design of glutamate receptor subtype selective compounds as will be discussed in further detail in Chapter 15.

12.2.2.3 Purinergic P2X Receptor Family

ATP is primarily considered as an intracellular energy storage molecule, but also acts as an intercellular signaling molecule activating a group of P2Y GPCRs and seven $P2X_{1-7}$ ligand-gated ion-channels. Recent 3D structures of a full-length $P2X_4$ receptor has shown that the receptor consist of three homo- or heteromeric subunits which form a pore in the cell membrane (Figure 12.8). The orthosteric binding site is located in the three subunit interfaces in the extracellular domain. Comparison of the inactive apo (without ligand) and active ATP-bound structures have revealed that agonist binding leads to conformational changes in the extracellular domains which is relayed to the transmembrane domains causing opening of the nonselective cation channel (see Figure 12.8 for details).

12.2.3 Tyrosine Kinase Receptors

As illustrated in Figures 12.9 and 12.10, the tyrosine kinase receptors have a large extracellular agonist binding domain, one transmembrane segment, and an intracellular domain. The receptors can be divided into two groups: those that contain the tyrosine kinase as an integral part of the intracellular domain and those that are associated with a Janus kinase (JAK). Examples of the former group are the insulin receptor family and the epidermal growth factor (EGF) receptor family and examples of the latter are the cytokine receptor family such as the erythropoietin (EPO) receptor and the thrombopoietin (TPO) receptor. However, both groups share the same overall mechanism of activation: Upon agonist binding, two intracellular kinases are brought together which will initiate autophosphorylation of tyrosine residues of the intracellular tyrosine kinase domain (Figure 12.9). This will attract other proteins (e.g., Shc/Grb2/SOS and STAT for the two receptor groups, respectively), which are also phosphorylated and this will initiate protein cascades and ultimately lead to regulation of transcriptional factors (e.g., Elk-1, Figure 12.9) and thus regulation of genes involved in, e.g., cell proliferation and differentiation. As described for the GPCRs, all the proteins in the intracellular activation cascades are heterogeneous leading to individual responses (i.e., regulation of different subset of genes) in individual cell types.

Albeit the tyrosine kinase receptors share the overall activation mechanism, the family has turned out to be rather heterogeneous with respect to the structure and ligand–receptor interaction. Some of the receptors exist as monomers (e.g., the EGF receptor family) in the absence of agonist whereas others exist as covalently linked dimers (e.g., the insulin receptor family) or noncovalently linked dimers (e.g., the EPO receptor). In case of the monomers, agonist binding to either one or both subunits will bring the two receptor subunits together, and thereby initiate the autophosphorylation. In case of the preformed inactive dimers, agonist binding will cause a conformational change in the receptor which brings the two intracellular kinases together and thus initiate the

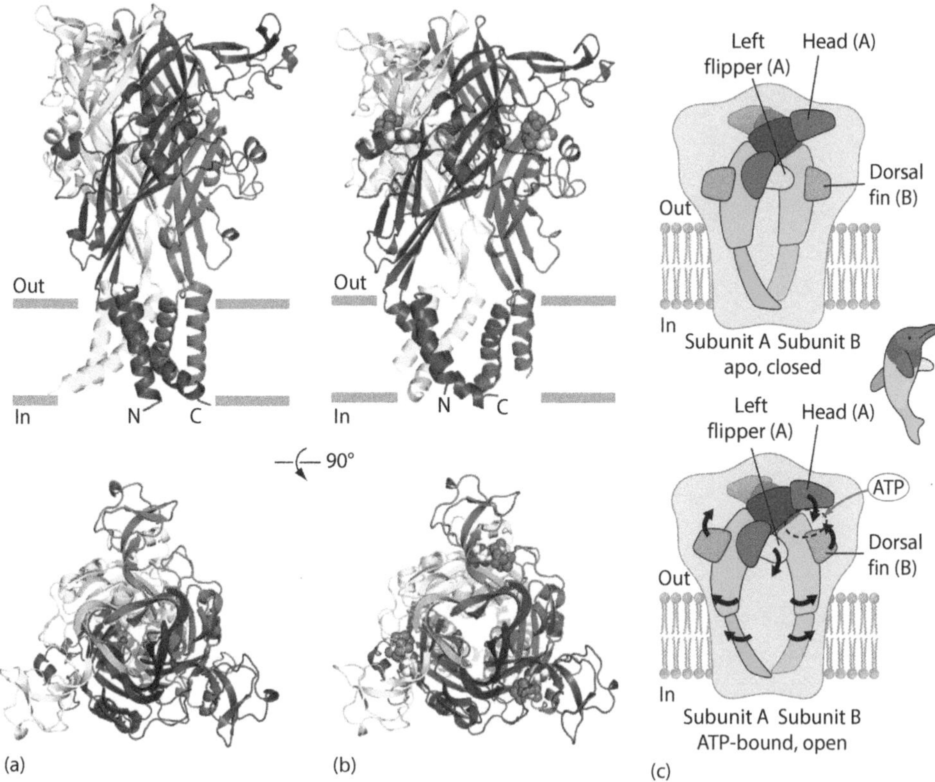

FIGURE 12.8 (a) 3D structure of a full-length purinergic $P2X_4$ receptor in its inactive apo (left) and (b) active ATP-bound (right) structure viewed from the side (top) and extracellular side (bottom). The receptor consists of three subunits shown in different colors with the orthosteric binding site located between the subunits. (c) Cartoon showing the movement when going from the inactive (top) to the active ATP bound (bottom) conformations (only two subunits shown for clarity). Each subunit adopts a dolphin-like structure. ATP binding in the inter-subunit cleft leads to downward movement of the "head" in one subunit and upward movement of the "dorsal fin" in the other subunit to encapsulate the agonist which simultaneously pushes the "left flipper" away. Through a lever motion, this leads to an outward movement of the transmembrane domains and opening of the nonselective cation channel. (Adapted with permission from Hattori, M. and Gouaux, E., *Nature*, 485, 207, Copyright 2012, Macmillan Publishers Ltd.)

autophosphorylation. One of the best understood examples in this regard is the EPO receptor of which the 3D structure of the extracellular agonist binding domain has been determined in the absence and presence of EPO (Figure 12.10). In the absence of EPO, the domain is a dimer in which the ends are too far apart for the JAKs to reach each other. EPO binds to the same amino acids on the receptor that forms the dimer interface and thereby tilts the two receptor subunits. This brings the JAKs close together and initiates the autophosphorylation (Figure 12.10).

12.2.4 Nuclear Receptors

Nuclear receptors are cellular proteins and are thus not embedded in the cell membrane like the previously described receptors. In contrast to the membrane bound receptors, they bind small lipophilic compounds and function as ligand-modulated transcription factors. The nuclear receptors have been classified into six subfamilies according to the type of hormone they bind and receptor sequence similarity. Ligands include steroid hormones (glucocorticoids, progestestins, mineralocorticoid androgens, and estrogens) and steroid derivatives (vitamin D_3 and bile acids), and nonsteroids (e.g., thyroid

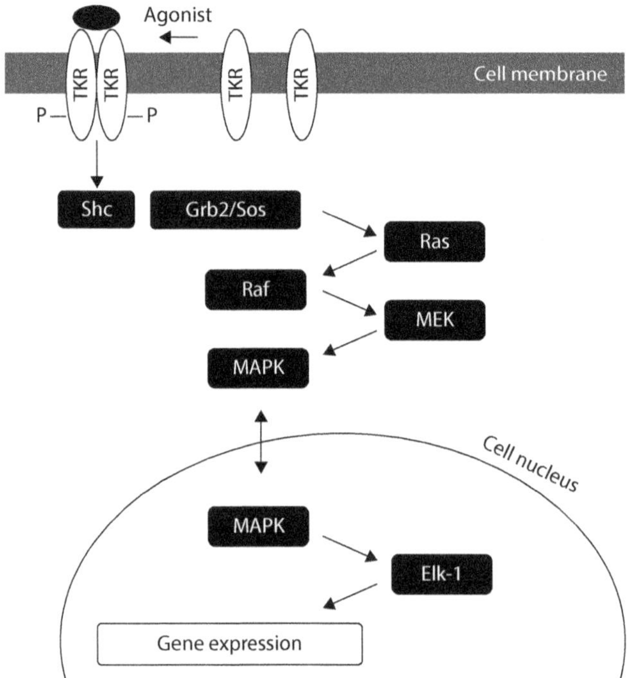

FIGURE 12.9 Cartoon of a protein cascade initiated by agonist binding to two tyrosine kinase receptors (TKR) causing autophosphorylation of the dimerized intracellular receptor domains. This causes activation of a cascade of intracellular proteins (abbreviated Shc, Grb2/SOS, Ras, Raf, MEK, and MAPK) which ultimately leads to activation of transcription factors (e.g., Elk-1) and thus regulation of gene expression.

hormone, retinoids, prostaglandins). The receptor family is relatively small (48 human subtypes), of which 50% still belongs to the group of orphan receptors with no known endogenous ligand.

The nuclear receptors consist of a ligand binding domain, a DNA binding domain, and a variety of other regulatory domains. Upon activation, two receptors dimerize, as homo- or heterodimers, and bind to specific recognition sites on the DNA. Co-activators will then associate with the dimeric receptor and initiate transcription of the target gene(s). Each receptor recognizes specific DNA sequences, also known as the hormone response elements which are located upstream of the genes that are regulated. Three-dimensional high-resolution structures of both ligand and DNA binding domains as well as full-length receptor bound to DNA have been determined. In drug research the main focus has been on the structures of the ligand binding domains which for several receptors have been determined in the absence and presence of ligands.

12.3 RECEPTOR PHARMACOLOGY

12.3.1 Recombinant versus In Situ Assays

The last decades have had a profound impact on how receptor pharmacology is performed. As mentioned in the introduction, receptor cloning was initiated in the mid-eighties and today the vast majority of receptors have been cloned. Thus, it is now possible to determine the effect of ligands on individual receptor subtypes expressed in recombinant systems rather than on a mixture of receptors in, e.g., an organ. This is very useful given that receptor selectivity is a major goal in terms of decreasing side effects of drugs and development of useful pharmacological tool compounds which can be used to elucidate the physiological function of individual receptor subtypes. Furthermore, recombinant assays allow one to assay cloned human receptors which would

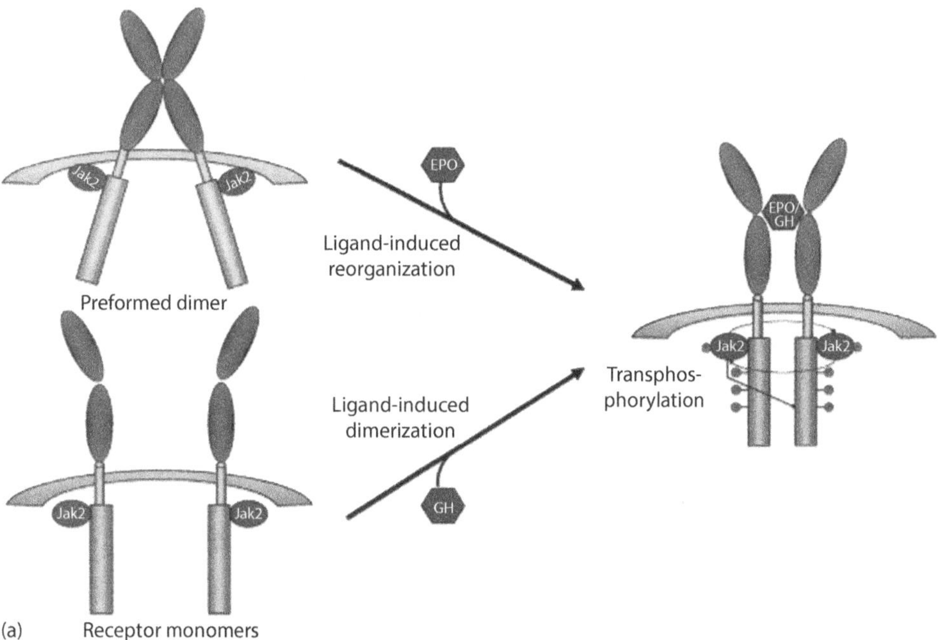

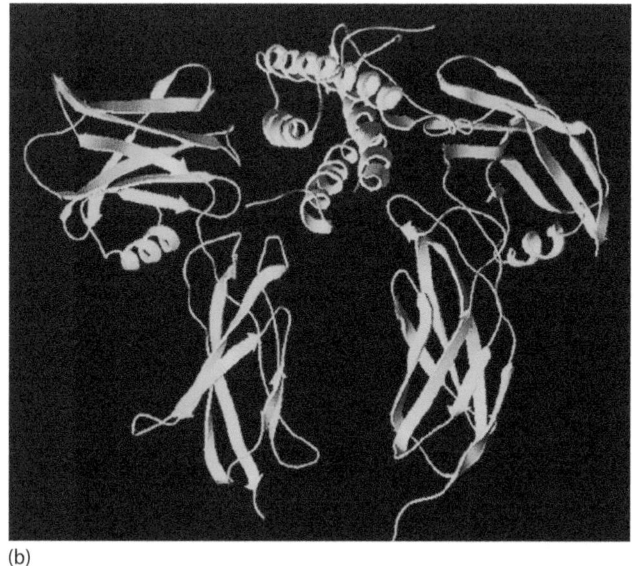

FIGURE 12.10 Cartoon of the activation mechanism of the erythropoietin (EPO) and growth hormone (GH) receptor which belong to the JAK/STAT receptor class of the superfamily of tyrosine kinase receptors. (a) The EPO receptor is dimerized in the inactive conformation by interaction of amino acids which are similar to those involved in binding of EPO and the intracellular JAK kinases are kept too far apart to initiate autophosphorylation. The GH receptor exists as monomers which are dimerized upon agonist binding. Upon agonist binding, intracellular JAK kinases of both receptors are brought in proximity and initiates transphosphorylation. (b) The actual structure of EPO (in cyan) bound to the extracellular receptor domains of the EPO receptor (in green). The structure was generated using the program "Swiss PDB viewer 3.5" with coordinates from Brookhaven Protein Data Base. (Reprinted [a] from *Handbook of Cell Signaling*, 2nd ed., Stauber, C.E., Busch, A., Naumann, A. et al., Academic Press, San Diego, CA, p. 245. Copyright 2010, with permission from Elsevier.)

otherwise not be possible. Most receptors have very similar human and rodent sequences, but due to the small differences in primary amino acid sequence there have been cases of drugs developed for rats rather than for humans, because the compounds were active on the rat receptor but not on the human receptor.

It should be noted that the use of organ and whole animal pharmacology is still required. As previously noted, the cellular effect of receptor activation depends on the intracellular contents of the proteins involved in, e.g., the signaling cascades. These effects can only be determined when the receptor is situated in its natural environment rather than in a recombinant system. In most situations, both recombinant and in situ assays are thus used to fully evaluate the pharmacological profile of new ligands. Furthermore, once a compound with the desired selectivity profile has been identified in the recombinant assays, it is important to confirm that this compound has the predicted physiological effects in, e.g., primary nonrecombinant cell lines, isolated organs, and/or whole animals.

12.3.2 Binding versus Functional Assays

Binding assays used to be the method of choice for primary pharmacological evaluation, mainly due to the ease of these assays compared to functional assays which generally required more steps than binding assays. However, several factors have changed this perception: (1) biotechnological functional assays have evolved profoundly and have decreased the number of assay steps and increased the throughput dramatically, (2) functional assay equipment has been automated, (3) ligand binding requires a high-affinity ligand, which for many targets identified in genome projects simply does not exist, (4) binding assays are generally unable to discriminate between agonists and antagonists, (5) binding assays will generally only identify compounds binding to the same site as the radioactively labeled tracer. One important aspect of binding assays is the ability to determine ligand–receptor kinetics (on-rate, off-rate, and ligand residence time) which are important pharmacological properties affecting drug efficacy in vivo.

The Fluorometric Imaging Plate Reader (FLIPR™) illustrates this development toward functional assays. Cells transfected with a receptor coupled to increase in intracellular calcium levels (e.g., a $G\alpha_q$-coupled GPCR or a Ca^{2+} permeable ligand-gated ion channel) are loaded with the dye Fluo-3 which in itself is not fluorescent. However, as shown in Figure 12.11, the dye becomes fluorescent when exposed to Ca^{2+} in the cell in a concentration-dependent manner. In this manner, ligand concentration–response curves can be generated on the FLIPR very fast as it automatically reads all wells of a 96-, 384-, or 1534-well tissue culture plate. Many other functional assays along these lines have been developed in recent years. Importantly, the majority of these assays can be applied on both recombinant and native receptor expressing cell lines.

12.3.3 Partial and Full Agonists

Agonists are characterized by two pharmacological parameters: potency and maximal response. The most common way of describing the potency is by measurement of the agonist concentration which elicit 50% of the compound's own maximal response (the EC_{50} value). The maximal response is commonly described as percent of the maximal response of the endogenous agonist. The maximal response is also often described as efficacy or intrinsic activity which were defined by Stephenson and Ariëns, respectively. Compounds, such as 2-Me-5-HT and MK-212 in Figure 12.11, show a lower maximal response than the endogenous agonist and are termed partial agonists. The parameters potency and maximal response are independent of each other and on the same receptor it is thus possible to have, e.g., a highly potent partial agonist and a low potent full agonist. Both parameters are important for drug research, and it is thus desirable to have a pharmacological assay system which is able to determine both the potency and the maximal response of the tested ligands.

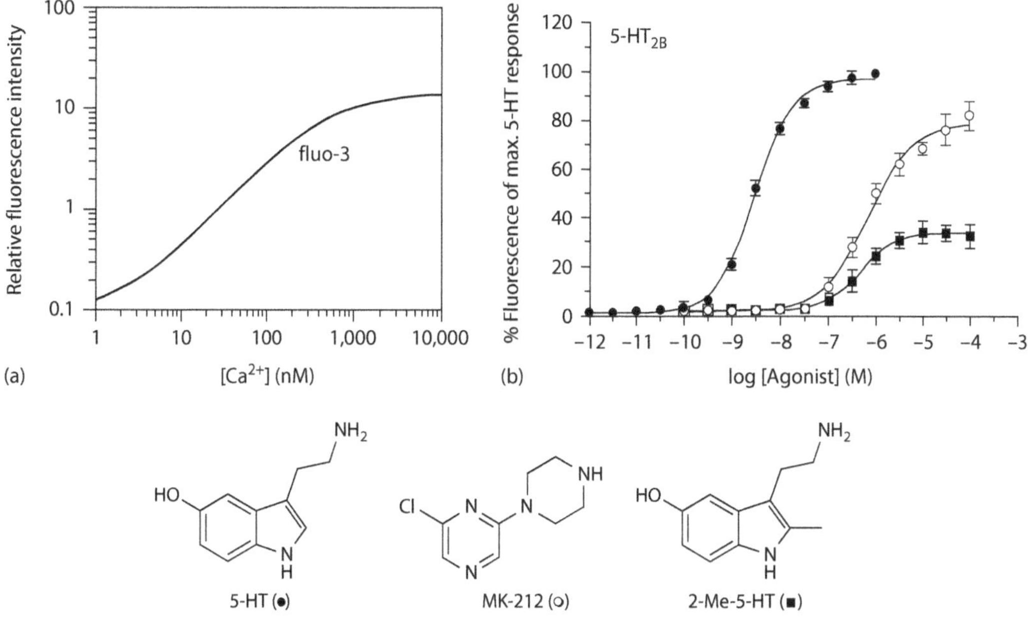

FIGURE 12.11 (a) Relation between Ca^{2+} concentration and relative fluorescence intensity of the fluorescent probe fluo-3. (b) The 5-HT_{2B} receptor subtype belongs to the superfamily of G protein-coupled receptors and is coupled to increase in inositol phosphates and intracellular Ca^{2+}. Cells expressing 5-HT_{2B} receptors were loaded with fluo-3 and the fluorescence was determined upon exposure to the endogenous agonist 5-HT (•) and the partial agonists MK-212 (○) and 2-Me-5-HT (■) on a FLIPR™. (Adapted from Jerman, J.C. et al., *Eur. J. Pharmacol.*, 414, 23, 2001. With permission.)

12.3.4 Antagonists

Antagonists do not activate the receptors but block the activity elicited by agonists and accordingly they are only characterized by the parameter affinity. The most common way of characterizing antagonists is by competition with an agonist (functional assay) or a radioactively labeled ligand (binding assay). In both cases, the antagonist concentration is increased and displaces the agonist or radioligand, which are held at a constant concentration. It is then possible to determine the concentration of antagonist, which inhibits the response/binding to 50% (the IC_{50} value). The IC_{50} value can then be transformed to affinity (K) by the Cheng–Prusoff equation:

Functional assay:

$$K = IC_{50}/(1 + [\text{Agonist}]/EC_{50}) \tag{12.1}$$

where

[Agonist] is the agonist concentration
EC_{50} is for the agonist in the particular assay

Binding assay:

$$K = IC_{50}/(1 + [\text{Radioligand}]/K_D) \tag{12.2}$$

where

[Radioligand] is the radioligand concentration
K_D is the affinity of the radioligand

It is important to observe that the Cheng–Prusoff equation is only valid for competitive antagonists.

The Schild analysis is often used to determine whether an antagonist is competitive or noncompetitive. In the Schild analysis, the antagonist concentration is kept constant while the agonist concentration is varied. For a competitive antagonist, this will cause a rightward parallel shift of the concentration–response curves without a reduction of the maximal response (Figure 12.12a). The degree of right-shifting is determined as the dose ratio (DR), which is the concentration of agonist giving a particular response in the presence of antagonist divided by the concentration of agonist that gives the same response in the absence of antagonist. Typically, one will chose the EC_{50} values to calculate the DR. In the Schild analysis, the log (DR-1) is depicted as a function of the antagonist concentration (Figure 12.12b). When the slope of the curve equals 1, it is a sign of competitive antagonism and the affinity can then be determined by the intercept of the abscissa. When the slope is significantly different from 1 or the curve is not linear, it is a sign of noncompetitive antagonism, which invalidates the Schild analysis.

As shown in the example in Figure 12.12, five concentration–response curves are generated to obtain one antagonist affinity determination, illustrating that the Schild analysis is rather work-intensive compared to, e.g., the transformation by the Cheng–Prusoff equation where one inhibition curve generates one antagonist affinity determination. However, the latter cannot be used to determine whether an antagonist is competitive or noncompetitive, which is the advantage of the Schild analysis. When testing a series of structurally related antagonists one would thus often determine the nature of antagonism with the Schild analysis for a couple of representative compounds. If these are competitive antagonists, it is reasonable to assume that all compounds in the series are competitive and thus determine the affinity of these by the use of the less work-intensive Cheng–Prusoff equation.

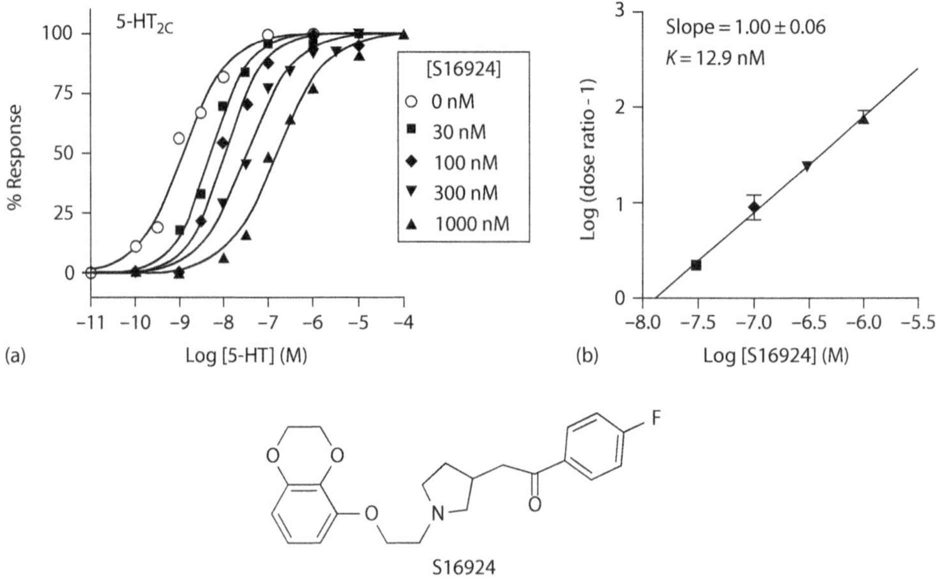

FIGURE 12.12 Schild analysis of the competitive antagonist S16924 on cells expressing the 5-HT_{2C} receptor. (a) Concentration–response curves of the agonist 5-HT were generated in the presence of varying concentrations of S16924. Note the parallel right shift of the curves and the same level of maximum response. (b) Dose ratios are calculated and plotted as a function of the constant antagonist concentration generating a straight line with a slope of 1.00 ± 0.012. These results and the observations from (a) are in agreement with a competitive interaction and the antagonist affinity can thus be determined by the intercept of the abscissa; $K = 12.9$ nM. (With kind permission from Springer Science+Business Media: *Naunyn Schmiedebergs Arch Pharmacol.*, Antagonist properties of the novel antipsychotic, S16924, at cloned, human serotonin 5-HT_{2C} receptors: A parallel phosphatidylinositol and calcium accumulation comparison with clozapine and haloperidol, 361, 2000, 549, Cussac, J.C., McCormick, D.J., Pang, Y.P. et al.)

12.3.5 Constitutively Active Receptors and Inverse Agonism

Most receptors display no or only minor basal activity but some receptors display increased basal activity in the absence of agonist, which has been referred to as constitutive activity. Interestingly, it has been shown that inverse agonists can inhibit this elevated basal activity, which contrast antagonists that inhibit agonist-induced responses but not the constitutive activity (Figure 12.13a).

Examples of important constitutively active receptors include the human ghrelin receptor and several viral receptors that display constitutive activity when expressed in the host cell. This latter group includes the ORF-74 7TM receptor from human herpesvirus 8 (HHV-8), which show a marked increased basal response when expressed in recombinant cells (Figure 12.13b). ORF-74 is homologous to chemokine receptors and does indeed bind chemokine ligands. As shown in Figure 12.13b, chemokines display a wide range of activities on the receptor from full agonism (e.g., GROα) to full inverse agonism (e.g., IP10), which correlates with the angiogenic/angiostatic effects of the chemokines.

Constitutive activity can also be caused by somatic mutations. Known examples include constitutively activating mutations in the thyrotropin receptor and the luteinizing hormone receptor which leads to adenomas, and the rhodopsin receptor which leads to night blindness.

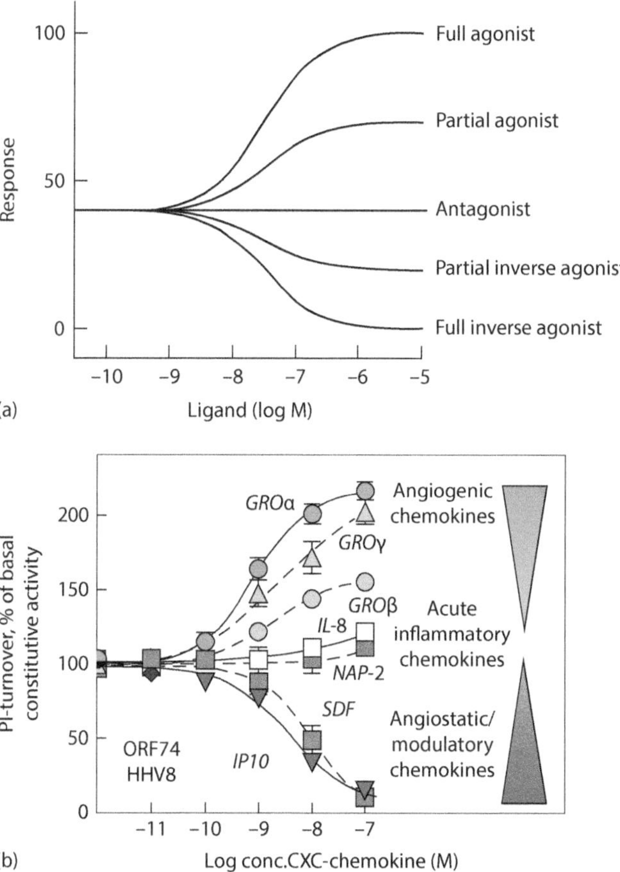

FIGURE 12.13 (a) The nomenclature of ligand efficacies and schematic illustration of their concentration-dependent effects on constitutive activity. (b) Ligand regulation of the constitutively active ORF-74 receptor from human herpesvirus 8 (HHV8). ORF-74 is a G protein-coupled receptor coupled to phosphatidylinositol (PI) turnover, which is regulated by a variety of human chemokines ranging from full agonism by GROα to full inverse agonism by IP10. (Reprinted from *Neuropharmacology*, 48, Rosenkilde, M.M., Virus-encoded chemokine receptors—Putative novel antiviral drug targets, 1–13. Copyright 2005, with permission from Elsevier.)

12.3.6 Allosteric Modulators

Allosteric modulators can both be stimulatory or inhibitory (noncompetitive antagonists) and typically these compounds bind outside the orthosteric binding site (Figure 12.3). Allosteric modulators have a number of potential therapeutic benefits compared to agonists and competitive antagonists which has led to significant increased pharmaceutical interest in recent years. This increased interest has also been fueled by the development of functional high-throughput screening assays which has made it possible to screen for allosteric modulators (see Section 12.3.2).

The allosteric modulators mentioned below act through allosteric mechanisms as evident from the fact that they do not displace radiolabeled orthosteric ligands. Furthermore, their activity is dependent on the presence of agonists as they do not activate the receptors by themselves. The fact that they bind outside of the orthosteric ligand binding pocket often leads to increased receptor subtype selectivity. Evolutionary pressure has led to conservation of the orthosteric binding site at different subtypes, as radical mutations would severely impact the binding properties. Thus, it is often seen that the orthosteric binding site is much more conserved than the remaining part of the receptor and accordingly, ligands binding to an allosteric site have a higher chance of being selective. Likewise, the allosteric ligands will have a different pharmacophore than the endogenous ligand which might improve, e.g., bioavailability. For example, ligands acting at the orthosteric site of the $GABA_A$ receptor need a negatively charged acid function and a positively charged basic function which greatly impairs the transport through biomembranes, whereas allosteric ligands such as the benzodiazepine diazepam (see Chapter 15) does not have any charged groups and show excellent bioavailability. It is well known that many agonists, particularly full agonists, lead to desensitization and internalization of receptors (Figure 12.5). Unlike agonists, the positive modulators should prevent the development of tolerance (as seen for, e.g., morphine), because they merely potentiate the endogenous temporal receptor activation pattern and avoid prolonged receptor activation leading to desensitization and internalization. The fact that the receptors are stimulated in a more natural way by positive modulators rather than the prolonged receptor activation caused by agonists may also lead to a difference in physiological effects which may or may not be an advantage.

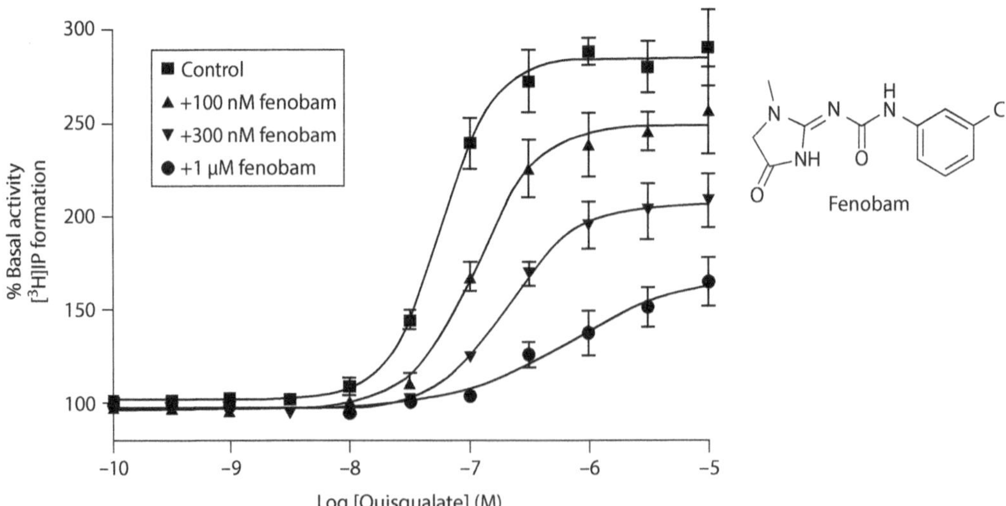

FIGURE 12.14 Schild analysis of the noncompetitive antagonist fenobam on cells expressing the metabotropic glutamate receptor subtype mGluR5. Concentration–response curves of the agonist quisqualate were generated in the presence of varying concentrations of fenobam. In contrast to the Schild analysis shown in Figure 12.12, a clear depression of the maximal response is seen with increasing antagonist concentrations. This shows that the antagonist is noncompetitive. The localization of the orthosteric and allosteric binding sites is depicted in Figure 12.3. (Adapted from Porter, R.H. et al., *J. Pharmacol. Exp. Ther.*, 315, 711, 2005.)

12.3.6.1 Negative Allosteric Modulators (Noncompetitive Antagonists)

As noted in the previous section, the Schild analysis is very useful to discriminate between competitive and noncompetitive antagonists, and an example of the latter is shown in Figure 12.14. Fenobam is a selective antagonist at the mGluR5 receptor, and the Schild analysis clearly demonstrates that the antagonism is noncompetitive due to the depression of the maximal response (compare Figures 12.12 and 12.14). As noted previously, glutamate binds to the large extracellular amino-terminal domain whereas fenobam has been shown to bind to the extracellular part of the 7TM domain. Fenobam does not hinder binding of glutamate to the extracellular domain, but hinder the conformational change leading to receptor activation.

12.3.6.2 Positive Allosteric Modulators

Positive allosteric modulation can be achieved through several mechanisms. For example, benzodiazepines positively modulate the $GABA_A$ receptor by increasing the frequency of channel opening (see Chapter 15). Positive modulation can also be obtained by blocking receptor desensitization as exemplified by cyclothiazide (see Chapter 15).

12.4 CONCLUDING REMARKS

The last decade of receptor research has provided many breakthroughs in our understanding of receptor structure, function, and pharmacology. The many new 3D structures of either full receptors or important domains have provided detailed knowledge about ligand–receptor interactions and receptor activation mechanisms. It has been shown that most receptors can activate several different signaling pathways which may also be selectively activated/inhibited by biased ligands. Finally, inverse agonism and allosteric modulation have pointed to novel ways that receptors can be regulated in vivo. Collectively, these new developments have created the foundation for structure-based drug design and new concepts of pharmacological intervention.

FURTHER READING

Alves, L., da Silva, J., Ferreira, D. et al. 2014. Structural and molecular modeling features of P2X receptors. *Int. J. Mol. Sci.* 15:4531–4549.

Burris, T.P., Solt, L.A., Wang, Y. et al. 2013. Nuclear receptors and their selective pharmacologic modulators. *Pharmacol. Rev.* 65:710–778.

Cusack, K.P., Wang, Y., Hoemann, M.Z., Marjanovic, J., Heym, R.G., and Vasudevan, A. 2015. Design strategies to address kinetics of drug binding and residence time. *Bioorg. Med. Chem. Lett.* 25:2019–2027.

Cussac, J.C., McCormick, D.J., Pang, Y.P. et al. 2000. Antagonist properties of the novel antipsychotic, S16924, at cloned, human serotonin 5-HT_{2C} receptors: A parallel phosphatidylinositol and calcium accumulation comparison with clozapine and haloperidol. *Naunyn Schmiedebergs Arch Pharmacol.* 361:549–554.

Hattori, M. and Gouaux, E. 2012. Molecular mechanism of ATP binding and ion channel activation in P2X receptors. *Nature* 485:207–212.

Jerman, J.C., Brough, S.J., Gager, T. et al. 2001. Pharmacological characterisation of human 5-HT2 receptor subtypes. *Eur. J. Pharmacol.* 414:23–30.

Kenakin, T. and Christopoulos, A. 2013. Signalling bias in new drug discovery: Detection, quantification and therapeutic impact. *Nat. Rev. Drug Discov.* 12:205–216.

Lefkowitz, R.J. and Shenoy, S.K. 2005. Transduction of receptor signals by β-arrestins. *Science* 308:512–517.

Melancon, B.J., Hopkins, C.R., Wood, M.R. et al. 2012. Allosteric modulation of seven transmembrane spanning receptors: Theory, practice, and opportunities for central nervous system drug discovery. *J. Med. Chem.* 55:1445–1464.

Porter, R.H., Jaeschke, G., Spooren, W. et al. 2005. Fenobam: A clinically validated nonbenzodiazepine anxiolytic is a potent, selective, and noncompetitive mGlu5 receptor antagonist with inverse agonist activity. *J. Pharmacol. Exp. Ther.* 315:711–721.

Rasmussen, S.G.F., DeVree, B.T., Zou, Y. et al. 2011. Crystal structure of the ß2 adrenergic receptor-Gs protein complex. *Nature* 477:549–555.

Rosenbaum, D.M., Rasmussen, S.G.F., and Kobilka, B.K. 2009. The structure and function of G-protein-coupled receptors. *Nature* 459:356–363.

Rosenkilde, M.M. 2005. Virus-encoded chemokine receptors—Putative novel antiviral drug targets. *Neuropharmacology* 48:1–13.

Sobolevsky, A., Rosconi, M.P., and Gouaux, E. 2009. X-ray structure, symmetry and mechanism of an AMPA-subtype glutamate receptor. *Nature* 462:745–756.

Stauber, C.E., Busch, A., Naumann, A. et al. 2010. *Handbook of Cell Signaling*, 2nd ed. San Diego, CA: Academic Press, p. 245–252.

Thompson, A.J., Lester, H.A., and Lummis, S.C.R. 2010. The structural basis of function in Cys-loop receptors. *Q. Rev. Biophys.* 43:449–499.

Traynelis, S.F., Wollmuth, L.P., McBain, C.J. et al. 2010. Glutamate receptor ion channels: Structure, regulation and function. *Pharmacol. Rev.* 62:405–496.

Unwin, N. 2013. Nicotinic acetylcholine receptor and the structural basis of neuromuscular transmission: Insights from Torpedo postsynaptic membranes. *Q. Rev. Biophys.* 46:283–322.

Violin, J.D., Crombie, A.L., Soergel, D.G. et al. 2014. Biased ligands at G-protein-coupled receptors: Promise and progress. *Trends Pharmacol. Sci.* 35:308–316.

Wettschreck, N. and Offermanns, S. 2005. Mammalian G proteins and their cell type specific functions. *Physiol. Rev.* 85:1159–1204.

13 Ion Channels
Structure and Function

Søren-Peter Olesen, Bo Hjorth Bentzen, and Daniel B. Timmermann

CONTENTS

13.1 INTRODUCTION

Ion channels form pores through cell membranes which are permeable to the small physiological ions Na^+, K^+, Ca^{2+}, and Cl^-. The channels can open and close and thereby turn the flux of the charged ions through the cell membrane on and off. By this mechanism, the ion channels govern the fast electrical activity of cells. Additionally, ion channels control Ca^{2+} influx and regulate responses as diverse as muscle contraction, neuronal signaling, hormone secretion, cell division, and gene expression. Moreover, ion channels are also found in organelles (endoplasmatic reticulum, nucleus, lysosomes, and mitochondria) where they control ion homeostasis and regulate organelle function. The opening of the channels is subject to regulation by physiological stimuli such as changes in membrane potential and ligand binding. Ion channels also lend themselves to pharmacological modulation and constitute important targets for drug treatment of diverse diseases including cardiac arrhythmia, arterial hypertension, diabetes, seizures, and anxiety.

13.1.1 Ion Channels Are Pores through the Cell Membrane

Ions polarize water molecules around them and carry a shell of hydration water rendering them insoluble in the hydrophobic phospholipid membrane. Therefore, ion transport in and out of cells and organelles has to occur through specialized molecules, allowing the cells to compose a specific intracellular ion milieu, which in many ways is different from the extracellular ion milieu, e.g., there is more than a 10-fold gradient in the Na^+ and K^+ concentrations and a 10.000-fold gradient for Ca^{2+} across the cell membrane (Table 13.1).

The transmembrane proteins establishing these gradients are transporters such as the Na^+–K^+ ATPase pumping three Na^+ out and two K^+ into the cell while consuming one ATP molecule. Other transporters are the Ca^{2+}-ATPases pumping Ca^{2+} out of the cell or into the endoplasmatic reticulum, and secondary active transporters such as the Na^+–Ca^{2+} exchanger not using energy themselves, but exploiting the ion gradients created by the ATPases. The transporters typically move 0.1–10 ions/ms each, they show saturation kinetics like enzymes, and they slowly build up the ion gradients.

The ion channels are different in many ways. They form water-filled pores through the cell membrane once they open, and permeation through the channels is only limited by diffusion. The transport is very fast—in the range of 10^4–10^5 ions/ms, and opening of ion channels may change the membrane potential by 100 mV within less than 1 ms. Ion channels are thus in an ideal position to govern the fast electrical activity of cells.

13.1.2 Ion Currents Change the Electrical Membrane Potential

In biological tissue, electrical currents are conducted by movement of ions. Bulk movements of ions in organs give rise to large currents resulting in voltage differences that can be measured on the body surface as was first done by Willem Einthovens in 1901 when he recorded human electrocardiograms. At the cellular level, the nature of excitable ion currents through the cell membrane was demonstrated by Hodgkin and Huxley in 1953 using a preparation of the squid axon. This giant nerve axon is about 1 mm in diameter, i.e., about 1000-fold thicker than human axons allowing electrodes to be positioned on either side of the membrane and the ion compositions on both sides to be controlled. Using this method, the authors showed that selective movement of Na^+ ions into the cell followed by an efflux of K^+ ions is the basis for the electrical activity in nerve cells. The ions move passively across the cell membrane when the permeability increases, and the direction of the movement is determined by the combined chemical and electrical forces acting on them. These two forces are generated by the concentration gradient and by the electrical field generated by the

TABLE 13.1
Typical Intracellular and Extracellular Ion Concentrations and Corresponding Equilibrium Potentials

Ion	Intracellular Concentration (mM)	Extracellular Concentration (mM)	Equilibrium Potential (mV)
Ca^{2+}	0.0001	1–2	+120
Na^+	10	145	+70
Cl^-	10	110	−62
K^+	140	4	−93

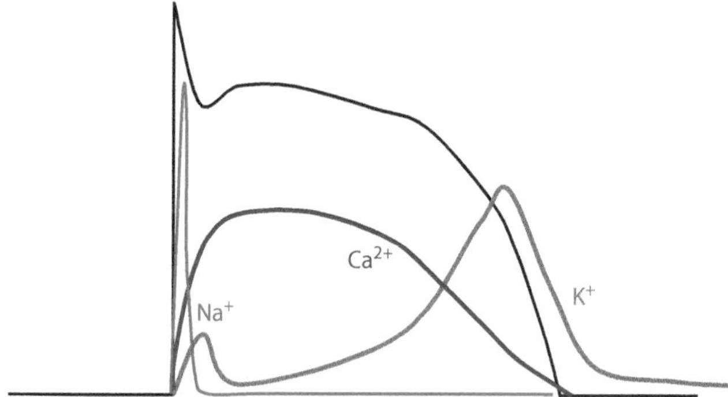

FIGURE 13.1 Cardiac action potential (curve in black) and time-course of selective Na^+, Ca^{2+}, and K^+ currents.

membrane potential, respectively. The net ion movement across the membrane is zero once the two forces equal each other which happens at the so-called equilibrium potential. This potential is determined by the ion distribution across the membrane, and for the typical intracellular and extracellular ion concentrations the equilibrium potentials are shown in Table 13.1.

The effects on the membrane potential of activation of selective ion channels are shown in Figure 13.1. The cardiac action potential (AP) is initialized by opening of voltage-gated Na^+ channels, and the influx of the positively charged Na^+ ions leads to a fast positive shift in the membrane potential (depolarization). Subsequently, voltage-gated Ca^{2+} channels are opened and the influx of Ca^{2+} ions keeps the membrane potential depolarized. Fast K^+ channels are activated early in the response and attenuate the depolarization, but the key role of the K^+ currents is to terminate the AP after about 350 ms when numerous K^+ channels open and the outflow of the positively charged K^+ ions mediate the repolarization.

The consequence of the sequential opening of Na^+–Ca^{2+} and K^+-selective channels is thus that the cell membrane potential will be pulled in the direction of the equilibrium potential for these ion species, i.e., about +70 mV for Na^+, +120 mV for Ca^{2+}, and −93 mV for K^+ (Table 13.1). Often the cell does not fully reach the equilibrium potential as shown for the cardiac AP, since several types of channels are usually open at the same time. Likewise the impact on the membrane potential of physiological or pharmacological ion channel block or activation depends on the presence of other simultaneous conductances and is not just linearly correlated to the number of ion channels being affected. Thus, it can be complicated to predict the functional effect of modulating ion channel function, and extensive target validation studies have to be conducted to establish the anticipated role of an ion channel subtype in an organ.

13.1.3 Gating of Ion Channels

Whereas it was clear to Hodgkin and Huxley that a sequential increase in Na^+ and K^+ membrane conductances underlies the neuronal AP, their method could not reveal the nature of the conductance pathway. This had to await another technological break through. In 1976, Neher and Sakmann reported the opening and closing of single acetylcholine-gated ion channels in striated muscle using a method by which they electrically isolated a patch of membrane in situ with a glass pipette. The method was called patch–clamp with reference to the patch of tissue and the clamp of the transmembrane voltage used to generate the electrical driving force. Since then the method has been extensively used to describe the characteristics and function of ion channels in all cells. Initially, endogenous currents in cells were measured, but following the cloning area the combination of this functional method and heterologous expression of cloned ion channels has been strong in the target-driven drug discovery process.

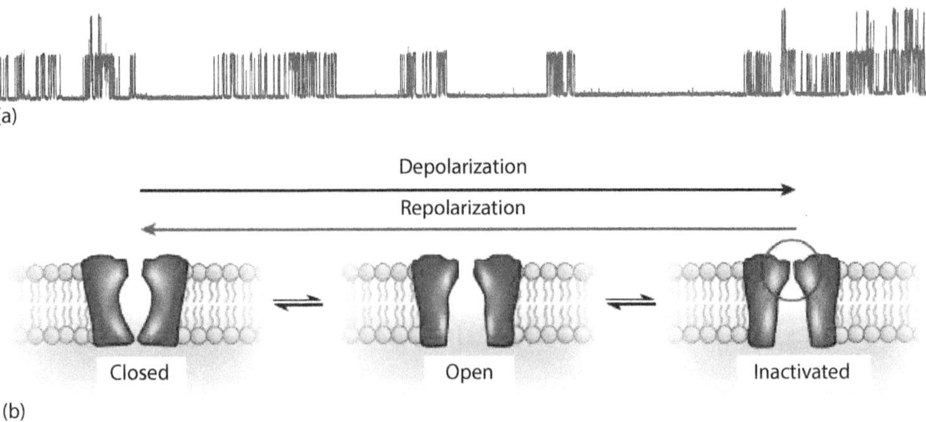

FIGURE 13.2 (a) Single channel recording of BK-type potassium channel. The baseline shows the closed state and the upward deflections are opening of single channels. The current through the channels is about 20 pA. (b) Opening, closing, and inactivation of ion channel. (From Sanguinetti, M.C. and Tristani-Firouzi, M., *Nature*, 440, 463, 2006.)

Patch–clamp studies of single ion channels have shown that the duration of channel opening ranges from a few μs to several hundred ms when exposed to a ligand or a voltage change (Figure 13.2). In the absence of a stimulus they either open less frequently or stay closed. The opening and closing of ion channels is called gating, and at the single channel level it is described by the distribution of open and closed times. Currents through all ion channels in the cell membrane can also be measured by ripping a hole in the cell which makes it possible to voltage-clamp the whole cell. This whole-cell current depicts the sum of hundreds of ion channels, and the kinetics of the current reflects the average open or closed times of the channels.

The gating is a dynamic process reflecting structural changes in the channel protein. The opening of a channel is preceded by conformational changes. Once the channel goes into the open state, the electrical current carried by ions through the channel can be recorded with a resolution of about 1 pA (10^{-12} A). The activation of single channels is a discrete event, and as seen from the recordings in Figure 13.2 the ion channels are either fully closed or fully open, although some channel types do show subconductance levels. Thus, it is possible to follow the movements between the two conformational states with an amazing time and current resolution.

The various ion channel types gate differently: some channels open only transiently whereas others stay open as long as the stimulus exists. Stimuli for ion channel activation are either (1) a change in the membrane potential, (2) a change in the concentration of extracellular ligands (neurotransmitters), (3) a change in the concentration of intracellular ligands (Ca^{2+}, H^+, cyclic nucleotides, or G-protein subunits), or (4) mechanical stimulation (e.g., stretch). Once the channels are exposed to the stimuli they open or activate, and when the stimulus is removed, the channels close in an opposite process called deactivation. A number of channel types do however also close in the presence of the stimulus. It is a general physiological phenomenon that continued stimulation of a signal process results in a decreasing output. This functional closure of ion channels in the presence of a stimulus is called inactivation and can occur either by parts of the channel protein plugging the open pore after a short delay, by collapse of the pore, or by decreased coupling between ligand binding and pore domains (Figure 13.2b).

13.1.4 Molecular Structures of Ion Channels

Ion channels are present in all cells, and they have been extensively characterized with respect to gating kinetics, voltage and ligand sensitivity, pharmacology, and other parameters. In addition, many ion channel types exhibit high affinity (pM or nM) to a number of toxins derived from scorpions, snakes, snails, or other animals, so toxins have been widely used to differentiate between the channel subtypes. The overall parameter used when describing an ion channel is its selectivity, i.e., whether it is selective to permeation of K^+, Na^+, Ca^{2+}, Cl^-, or divalent ions. Some channels are non-selective among cations. Since the selectivity is tightly coupled to the physiological function of the channels, this division is pragmatic and will be used in this chapter.

Following the sequencing of the human genome, 406 proteins with clear ion channel structure appeared. The characteristics of most of the cloned channels correspond well to the endogenous currents found in nerve, muscle, and other cells. The molecular constituents underlying other endogenous currents are however still debated, and these channels appear to be composed by several subunits from the same molecular family plus additional accessory proteins. For voltage-gated channels, the pore-forming subunits are denoted α-subunits, whereas the accessory subunits are called β, γ, or δ subunits. Moreover, it has become clear that ion channels are not solitary proteins, but are part of larger signaling networks. These ion channel protein complexes regulate ion channel activity, direct channels to their correct subcellular localization, and helps determine ion channel density. Pharmacological modulation of these interactions could provide an alternative strategy for regulating ion channel activity.

13.1.5 Ion Channels and Disease

The functional significance of specific ion channels in the body can be difficult to deduce from their molecular function, but it can be studied in organs or whole animals using pharmacological tools or selective toxins. Transgenic animals also provide valuable knowledge, but the most precise information about their role in humans has come from patients with diseases caused by dysfunctional ion channels. The diseases are typically caused by a point mutation in a single ion channel gene, and are jointly called channelopathies. By now more than 60 channelopathies are known. The most frequent and well-known disease is cystic fibrosis, arising from a point mutation in the Cl^- channel CFTR. In Northern Europe, 5% of the population is heterozygous for a mutation in the CFTR gene, and the prevalence of the disease is 0.5‰. Several types of cardiac arrhythmia (long and short QT syndromes, Brugada syndrome, Andersen syndrome, cardiomyopathy, nodal syndromes, and atrial fibrillation) are caused by mutations in cardiac K^+, Na^+, and Ca^{2+} channels. Mutations in neuronal and muscular ion channel subtypes cause epilepsy, ataxia, and myotonia. A number of endocrine, renal, as well as autoimmune diseases and syndromes are also caused by ion channel mutations. Luckily most channelopathies are rare, but their study has given invaluable information about the role of the ion channels in health and disease.

13.1.6 Physiological and Pharmacological Modulation of Ion Channels

In addition to the main mechanisms for ion channel activation (voltage, ligands, and mechanical stress), the channels may also in some cases be modulated by small organic molecules. The ligand-gated ion channels exhibit an endogenous ligand binding site, so compounds with similar functionalities can make potent drugs. The voltage-gated channels are not expected to naturally exhibit high-affinity binding sites, but may possess such as in the case of the dihydropyridine binding site on the Ca^{2+} channel. Most drugs act as positive or negative modulators of the channel gating, but some may also just plug the pore as the local anesthetics blocking the neuronal Na^+ channels or the neuromuscular blockers acting on the nicotinic channel in the neuromuscular junction.

Ion channels are the second largest drug target class, and drugs targeting ion channels are used as local anesthetics and for treating cardiovascular and neurological disorders. However, despite huge investments from the industry, relatively few drugs selectively targeting specific ion channels have entered the market in the last decade. Despite this, ion channel modulators are still eagerly pursued by the industry, perhaps stimulated by the increased biological and structural understanding of ion channels.

13.1.7 Drug Screening on Ion Channels

The center-stage role of ion channels in many physiological responses has been stressed by functional studies in cells, organs, and animals, by the emerging channelopathies as well as by the successful use of ion channel modulating drugs and naturally occurring toxins. Current drugs only target a dozen of the known channel members, while most of the other 400 types are currently all being investigated as potential drug targets by the pharmaceutical industry. Drug-discovery projects today depend strongly on large-scale blind-screening for finding new chemical lead molecules. High-throughput binding assays can be used for drug screening, but provides no functional information and is dependent on high-affinity ligands. Other options are ion flux assays which utilize specific ions such as rubidium, thallium, and lithium and measures their flux through K^+ and Na^+ channels, respectively. Fluorescent readouts from assays using voltage-sensitive dyes that indirectly measures ion channel activity by responding to changes in membrane potentials or by dyes responding to changes in ion concentrations (e.g., Ca^{2+}) can also be applied. Fluorescent, ion flux, and binding assays deliver high-throughput screening, but to some degree indirectly assay ion channel activity and do not allow direct control of voltage-dependent ion channel activity. The only "semi" high-throughput, high-quality technology to be used for screening on every ion channel type is the automated patch–clamp technique. With this method parallel recordings are performed by a robot on up to 384 arrays of ion channel expressing cells positioned on silicon chips, thereby enabling screening of large compound libraries (~30,000) in a matter of days.

13.1.8 Structure of Voltage-Gated Ion Channels

The superfamily of voltage-gated ion channels encompasses more than 140 members and is one of the largest families of signaling proteins, following the G-protein-coupled receptors and protein kinases. The pore-forming α-subunits of voltage-gated ion channels build upon common structural elements and come in four variations. The simplest version is composed of two transmembrane (TM) segments connected by a membrane-reentrant pore-loop with N- and C-termini on the inside (Figure 13.3). Four of such subunits form the channel. This architecture is typical for the so-called inward-rectifying K^+ channels (K_{ir}). It is found in a number of bacterial channels, suggesting it is the ancestor of the family. The second type, the two-pore potassium channel (K_{2P}) is made by a concatenation of two such subunits (4-TM), and the channel is formed by two double constructs. The third type is the 6-TM subunit, in which four extra membrane-spanning N-terminal domains including a voltage-sensor have been added to the basic 2-TM pore unit. Four of these 6-TM units form a channel. The group of 6-TM channels is rather large and includes the voltage-gated K^+ channels (K_V), the calcium-activated K^+ channels (K_{Ca}), the cyclic nucleotide-gated channels (CNG), the hyperpolarization-gated channels (HCN), the cation channel sperm-associated protein (CatSper), and the transient receptor potential channels (TRP). Finally, the fourth channel structure type is

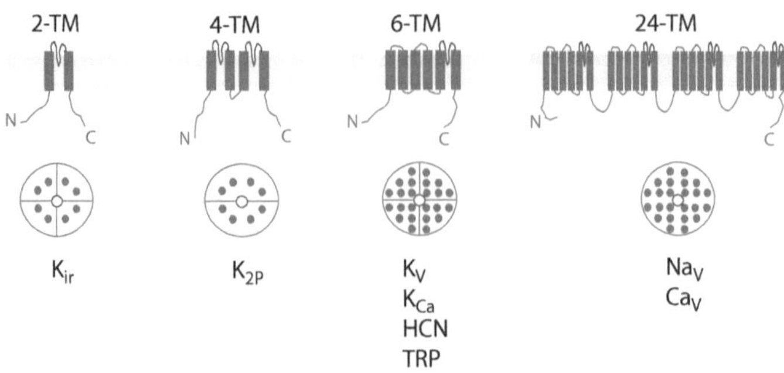

FIGURE 13.3 Topology of voltage-gated cation channels.

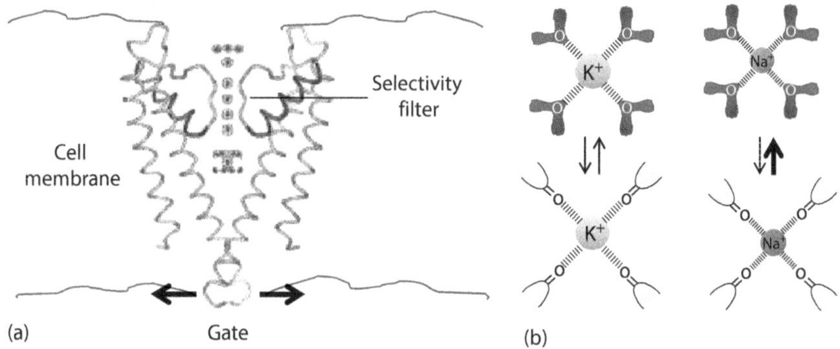

FIGURE 13.4 (a) Potassium channel structure with selectivity filter at the outer pore and gating mechanism at the inner pore. (b) Selectivity mechanism. The distance between the K^+ ions and the oxygen atoms is the same in water as in the selectivity filter enabling the K^+ ions to enter the pore at no energy cost. This is different for Na^+ ions, so they are excluded from the pore. (From MacKinnon, R., *FEBS Lett.*, 555, 62, 2003.)

made by concatenating four of the 6-TM subunits, making up a 24-TM subunit that forms the channel alone. This type is represented by the voltage-gated Na^+ and Ca^+ channels (Na_V and Ca_V). Within each of the four domains the six transmembrane segments are denoted S1–S6.

Three different parts of the channel are responsible for the functions: ion permeation, pore gating, and regulation. The narrow part of the pore is called the selectivity filter, and this has been studied by high-resolution X-ray in crystallized K^+ and Na^+ channels giving valuable insight into the selectivity mechanism (Figure 13.4). The residues in the pore loop line the K^+ channel selectivity filter and their peptide backbone carbonyl groups act as surrogate-water implying that the chemical energy of the dehydrated K^+ ions entering the pore is unchanged. By this means high selectivity and high permeability of the K^+ ions passing in single file are obtained. Although Na^+ ions are smaller than K^+ ions they will not enter the K^+ pore since it is energetically unfavorable. The Na^+ channel selectivity filter is larger and functions in a different way. At the extracellular end of the filter, the negatively charged side chains of four glutamate residues interact with Na^+ and partially remove its hydration. Following this two ion coordination sites formed by backbone carbonyls perfectly aligned to bind Na^+ with four planar waters of hydration are located. Thereby, Na^+ is conducted as a hydrated ion through the channel.

Crystal structure determination, molecular modeling, and crystal structures of voltage-gated Na^+ and K^+ channels explain how voltage sensing and gating occur. The membrane potential creates an electrical field across the membrane. Charged amino acids (usually arginine) are found in the fourth transmembrane segment (S4) of the channel, and move according to changes in the electrical field. Via the S4–S5 linker this movement bends the S6 segment (inner helix gate), and opens the pore.

13.2 PHYSIOLOGY AND PHARMACOLOGY OF VOLTAGE-GATED ION CHANNELS: POTASSIUM CHANNELS

The 2-TM K_{ir} channel family gives rise to six subtypes which play diverse roles in the body. Many K_{ir} channels are open at resting membrane potential and clamp the potential at −70 and −90 mV in nerve and heart cells, respectively (e.g., $K_{ir}4$ and $K_{ir}2$). The $K_{ir}3$ channels are gated by binding of the βγ subunit from the G_i protein. This mechanism is important in the atria of the heart, where stimulation of the para-sympathetic vagus nerve leads to release of acetylcholine, activating the G_i protein, and subsequently the $K_{ir}3$ channel to hyperpolarize the pacemaker cells. Crystal structures have revealed that K_{ir} channels have a longer ion permeation pathway as compared with other K^+ channels. It consists of both the transmembrane and a cytoplasmatic domain, that both contain gates. So, in addition to an inner helix gate, comparable to the one found in voltage-gated ion channels, $K_{ir}3$ has a cytosolic gate, the G loop gate that is important for regulation channel function.

The $K_{ir}6$ channels are expressed both in heart, vasculature, nerve, and in the pancreatic β-cells. This channel subtype can only be expressed in cells when it co-assembles with its accessory subunit, the so-called sulfonyl urea receptors (SUR) of the ABC transporter family. The β-cell subtype is composed of 4 $K_{ir}6.2$ + 4 SUR1. Like other K_{ir} channels it is activated by binding of phosphatidylinositol-4,5-bisphosphate (PIP_2). In contrast, the complex is blocked by ATP binding to the internal surface of $K_{ir}6.2$ and activated by MgADP binding to the nucleotide-binding domains of SUR1. The channel complex is also denoted the K_{ATP} channel and it is interesting for two reasons: it is a key regulatory protein in the β-cells coupling plasma glucose levels to insulin secretion, and the SUR has a well-exploited high-affinity drug binding site.

Briefly, insulin secretion is regulated by the following mechanism: an increase in plasma glucose leads, through an increased ATP level in the β-cells, to block of the K_{ATP} channel, depolarization, Ca^{2+}-influx, and insulin secretion (Figure 13.5). If this regulation is dysfunctional as in many type-2 diabetic patients, a similar functional effect can be obtained by directly blocking the K_{ATP} channel pharmacologically. The drug-binding site on SUR1 is on the inside of TM15 (plus partly on the inside of TM14), and the bulky substitution mutation S1237Y disrupts the site. Tolbutamide binds to this site only, whereas glibenclamide and metiglinide (Figure 13.6) bind to this as well as to a neighboring benzamido site. The latter low-affinity site is shared with the cardiac and vascular subunits SUR2A and SUR2B, respectively.

The cardiovascular side effects of the SUR-blockers are minimal whereas SUR-activators such as cromakalim and diazoxide which have been attempted primarily for the treatment of arterial hypertension, had to be abandoned since they cause orthostatic hypotension and reflex tachycardia.

The K_V channels fall into 12 subfamilies, which are all gated by changes in the membrane potential, but exhibit different kinetics. K_V channels can be composed of four different subunits from the same subfamily giving numerous possibilities for variation. Several K_V channel subfamilies are interesting drug targets. Retigabine is an activator of the $K_V7.2/3$ heteromultimeric channel and was in 2011 approved for the treatment of partial epilepsy, and XE-991 is a memory enhancing compound blocking the same channel (Figure 13.6) used in preclinical research.

Class III anti-arrhythmics block K_V channels in the heart (K_V1, K_V4, K_V11 subtypes) leading to a prolonged cardiac AP and termination of the so-called re-entry arrhythmia. Dofetilide, D-sotalol, and other anti-arrhythmics are selective for the K_V11 channels (hERG channels). These drugs show

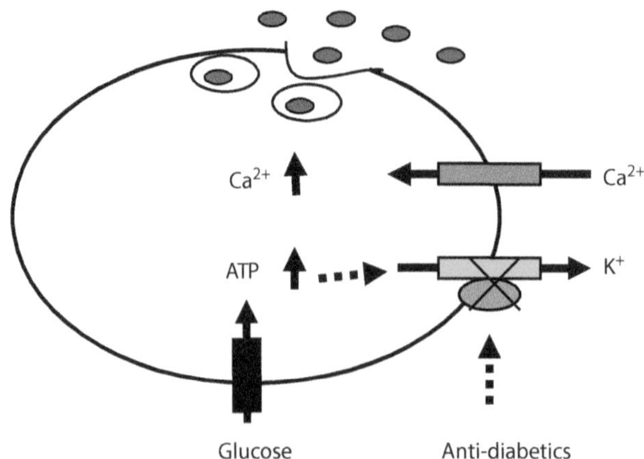

FIGURE 13.5 Ion channels in pancreatic β-cells and insulin secretion. The K channel subtype is called K_{ATP} and it is composed of the two molecular subunits $K_{ir}6.2$ and SUR1.

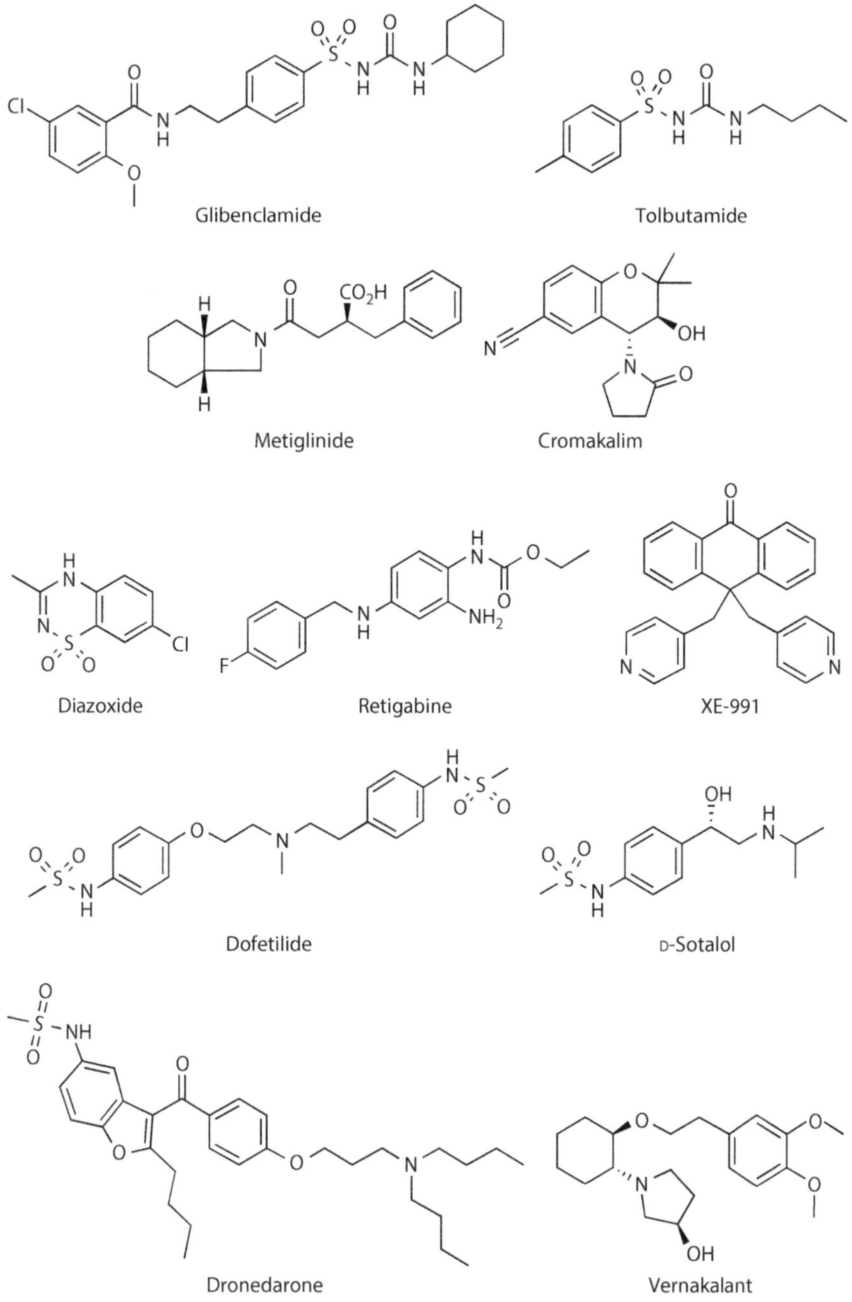

Glibenclamide

Tolbutamide

Metiglinide

Cromakalim

Diazoxide

Retigabine

XE-991

Dofetilide

D-Sotalol

Dronedarone

Vernakalant

FIGURE 13.6 Structures of the K_{ATP} channel blockers glibenclamide, tolbutamide, and metiglinide; the K_{ATP} channel openers cromakalim and diazoxide; the K_V7 channel opener retigabine; the K_V7 channel blocker XE-991; the K_V11 channel blockers dofetilide and D-sotalol; the multichannel blockers dronedarone and vernakalant.

anti-arrhythmic effects in some patients whereas they are pro-arrhythmic in others. The reason for the latter is that although the prolongation of the AP may terminate some arrhythmias, then blocking an important cardiac K^+ conductance being responsible for repolarizing the AP may destabilize the heart against triggered impulses (afterdepolarizations). In 2009 and 2012, two multi-channel blockers (dronedarone and vernakalant) were approved for the treatment of atrial fibrillation.

The K_V11 channel has a high-affinity binding site in the pore, which interacts with drugs of very different classes including antihistamines, antipsychotics, antidepressants, antibiotics, and many more. Pro-arrhythmia caused by drug binding to this site and channel block has been a major reason for withdrawal of drugs from the market and discontinued drug development projects, so the K_V11 channel has become a major cardiac safety pharmacology issue. The Ca^{2+}-activated K^+ channels, K_{Ca}, are divided into three families depending on their single-channel conductance. They are gated by Ca^{2+} binding either directly to the channel or indirectly to a constitutively bound calmodulin. The channels are generally involved in attenuating the activity of a given cell by hyperpolarizing this, when the internal Ca^{2+} concentration rises, thereby functioning as a molecular break.

Transient receptor potential (TRP) cation channels belong structurally to this group having six TM, and cation permeable pore loop between S5 and S6. Despite the structural similarity to K_V channels, the 28 different TRP subtypes can be either selective to Na^+/K^+, Mg^{2+} or Ca^{2+}, and functionally they may associate and be regulated by a number of G-protein-coupled receptors, kinases, and phospholipases. The mammalian TRP channels are divided into six subfamilies: TRPA for ankyrin (TRPA1), TRPC for canonical (TRPC1–7), TRPM for melastatin (TRPM1–8), TRPP for polycystin (TRPP2, TRPP3, TRPP5), TRPML for mucolipin (TRPML1–3), and TRPV for vanilloid (TRPV1–6) and TRPA (TRPA1). Because of the channels' many (patho)physiological roles, including roles in pain, cardiovascular, pulmonary and urinary system, cell proliferation, irritant and thermosensing, they have received a lot of attention from the pharmaceutical industry. Channel modulators could both target the activator ligand binding site, second messenger sites, the pore and possible allosteric sites. The TRP channel that has attracted the most attention as a potential drug target is the TRPV1 channel. This ion channel is activated not only by noxious heat but also by capsaicin, a constituent of chili pepper. TRPV1 has also been found to be up-regulated in various animal models of chronic pain and selective antagonists of TRPV1 reduce pain sensation in these models. Selective antagonists of TRPV1 are currently undergoing clinical trials in patients suffering from different types of chronic pain. Safety data however revealed that TRPV1 inhibition induced a mild increase in body temperature, wherefore the systemic use of TRPV1 inhibitors is currently debated. For now, the only TRP channel modulator on the market is a capsaicin patch (Qutenza) used for pain relief which likely works by inducing a desensitization of TRPV1-bearing nociceptive neurons.

13.3 VOLTAGE-GATED CALCIUM CHANNELS

13.3.1 Structure and Molecular Biology

The discovery of voltage-gated calcium channels (Ca_V) was originally made in the 1950s, through an investigation of crab leg muscle contraction. These experiments revealed that both membrane depolarization and muscle contraction depend on extracellular calcium ions, inferring that the muscle cells possess some membrane molecules enabling calcium to selectively permeate. By use of electrophysiological techniques, it was later found that a variety of functionally distinct Ca_Vs exist and that these ion channels are also expressed in nerve cells.

Functionally, Ca_Vs are closed at the resting membrane potential (i.e., −50 to −80 mV), but are activated by depolarization. Based on this and pharmacological properties, the 10 cloned α-subunits can be grouped in three families: $Ca_V1.x$ (L-type): high-voltage-activated dihydropyridine-sensitive calcium channels, requiring membrane potentials of ca. −20 to +10 mV to activate; $Ca_V2.x$: high-voltage activated dihydropyridine-insensitive channels; (T-type, $Ca_V3.x$): low-voltage activated currents which activate at much more negative membrane potentials, typically −50 to −40 mV. Following activation, Ca_Vs inactivate in the presence of sustained membrane depolarization, although the speed of inactivation can vary from ~50 ms to several seconds. Voltage-activated calcium currents, measured in native tissues, have traditionally been classified as L-, N-, P/Q- or R-type or T-type currents (see Table 13.2).

TABLE 13.2
Ca_V Channel Terminology and Properties

Channel Subtype	Ca_V1	Ca_V2	Ca_V3
Former names	L-type	$Ca_V2.1$ = P/Q type $Ca_V2.2$ = N type $Ca_V2.3$ = R type	T-type
Activation threshold	High voltage	High voltage	Low voltage
Blocker	Dihydropyridines Phenylalkylamines Benzothiazepines	$Ca_V2.1$ blockers: ω-conotoxin MVIIC, ω-agatoxin IVA $Ca_V2.2$: blockers: ω-conotoxin GVIA, ω-conotoxin MVIIA $Ca_V2.3$ blocker: SNX-482	Mibefradil (*R*)-Efonidipine Kurtoxin

The major component of the Ca_V is the large α_1 subunit, consisting of approximately 2000 amino acid residues. This subunit has 24 TM segments, arranged in four linked homologous domains (I–IV), each comprising six transmembrane α-helices (S1–S6), including the positively charged voltage-sensing S4 segments, and the S5–S6 pore loops, with the pore loops and S6 segments believed to line the channel lumen; the structure of the α_1- and other Ca_V subunits is schematically shown in Figure 13.7a.

Ca_Vs are several thousand-fold selective for Ca^{2+} ions over Na^+ and K^+ and this amazing selectivity is created by a ring of four negatively charged glutamic acid residues projecting into the ion channel pore, one such residue being contributed by each of the four pore loops.

When expressed alone, the α_1 subunit can form a functional ion channel. But native Ca_Vs are multi-subunit complexes in which the α_1 subunit interacts with a cytoplasmic β subunit, an extracellular membrane leaflet anchored α2δ subunit, and a 4-transmembrane spanning γ subunit (Figure 13.7b). The role of these subunits is to regulate surface expression, gating, and the pharmacological properties of Ca_Vs.

13.3.2 Physiological Roles of Voltage-Gated Calcium Channels

Ca^{2+} is an important second messenger molecule in eukaryotic cells where it initiates muscle contraction, neurotransmitter release, and activates many types of protein kinases. Many homeostatic mechanisms operate to keep intracellular $[Ca^{2+}] < 100$ nM under resting conditions. Outside the cell, $[Ca^{2+}]$ is 1–2 mM, creating a 10,000-fold concentration gradient. The Ca^{2+}-equilibrium potential is > +100 mV so Ca^{2+} always flows into a cell, when Ca_Vs are activated by depolarization. While the primary function of voltage-gated Na^+ and K^+ channels is to produce depolarization/re-polarization of the cell membrane, voltage-gated Ca^{2+} channels should be thought of as "gatekeepers" of calcium entry into excitable cells. Thereby, they constitute the principal entrance for calcium influx in nerves, endocrine, and muscle cells.

In muscle tissue, binding of Ca^{2+} to the protein troponin C allows myosin-mediated sliding of actin filaments, leading to shortening of muscle fibers. In skeletal muscle, the calcium necessary for this process actually comes from the sarcoplasmic reticulum and is released from this into the cytoplasm via ryanodine receptors. In this particular context, the Ca_V functions as a voltage sensor for the process—a direct interaction between the $Ca_V1.1$ α_1 subunit and the ryanodine receptors then activate the Ca^{2+} release.

Ca_Vs are also very important in cardiac and smooth muscle, where direct Ca^{2+}-influx through the Ca_V itself provides the Ca^{2+} necessary for muscular contraction. In cardiac muscle, $Ca_V1.2$ or $Ca_V1.3$ is responsible for the plateau phase of the cardiac AP which is important for cardiac muscle

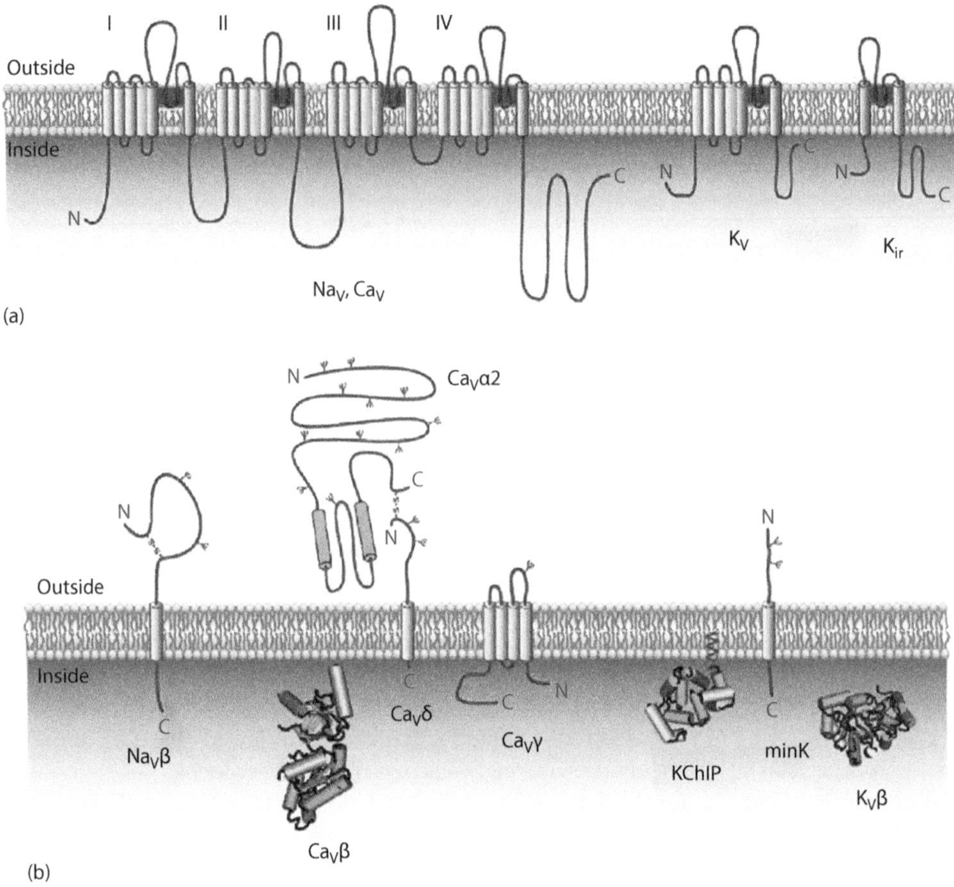

FIGURE 13.7 Overview of the membrane topology of voltage-gated ion channel α-subunits. (a) The voltage-sensing S4 transmembrane segments (*green*) contain several positively charged amino acid residues and the segments that constitute the ion channel pore (*shown in red*) are the S5, S6, and pore loop segments. (b) Membrane topology of auxiliary subunits of Na_V, Ca_V, and K_V ion channels. (From Catterall, W.A. et al., *Toxicon*, 49, 124, 2007.)

contraction and for regulation of the heart rate, so dihydropyridines are used for treatment of hypertension and cardiac arrhythmia. $Ca_V3.1$ and $Ca_V3.2$ subunits are found in the sino-atrial nodes where they play important roles for cardiac pacemaking.

Release of neurotransmitters from synaptic nerve terminals is triggered by influx of Ca^{2+} ions via $Ca_V2.1$ (P/Q-type) or $Ca_V2.2$ (N-type) subunits which are expressed in all nerve terminals. When neuronal APs travel down the axon and reach the nerve terminal, they provide the depolarization necessary for activation of Ca_Vs leading to Ca^{2+} influx. The $Ca_V2.1$ and $Ca_V2.2$ subunits bind directly to proteins of the protein machinery involved in membrane fusion of neurotransmitter-containing vesicles.

A similar role of Ca_Vs is found in various endocrine cells such as the pancreatic β-cells in which ATP-mediated closing of K_{ATP} channels leads to cellular depolarization, activation of $Ca_V1.3$ channels, and release of insulin-containing vesicles (Figure 13.5).

13.3.3 Pharmacology of Voltage-Gated Calcium Channels

There are two types of inhibition of Ca_V function, namely, blockade of the ion channel pore and allosteric modulation of ion channel function. An example of pore blockade is cadmium (Cd^{2+}) which produces nonselective inhibition of all type of Ca_Vs. The mechanism behind this effect is that

Cd^{2+} binds to the ring of four glutamates in the selectivity filter of the pore with much higher affinity than Ca^{2+} itself and thus blocks the pore. Most of the peptide toxins which block Ca_V subtypes with high specificity, also act by producing pore block. Allosteric modulation, on the other hand, is exemplified by the dihydropyridines which selectively affect members of the Ca_V1 family. The binding site for these compounds is located away from the pore and their mechanism of action relies on modification of the gating characteristics of the channel.

13.3.3.1 Ca_V1 Family (L-Type Currents)

The best characterized group of Ca_V modulators is the so-called "organic calcium blockers" or "calcium antagonists," comprising phenylalkylamines (e.g., verapamil), benzothiazepines (e.g., diltiazem), and the dihydropyridines (e.g., nifedipine; Figure 13.8), with distinct drug bindings sites for the two classes on the α_1 subunit. Several dihydropyridines are widely used clinically for the treatment of cardiovascular disorders such as hypertension, angina pectoris, and cardiac arrhythmia.

The organic calcium channel blockers bind with high affinity and selectivity to α_1 subunits of the Ca_V1 family, and act as allosteric modulators. This is highlighted by the fact that among the dihydropyridine-type compounds, positive modulators of Ca_V1 have also been identified, e.g., the compound (*S*)-Bay-K-8644 (Figure 13.8).

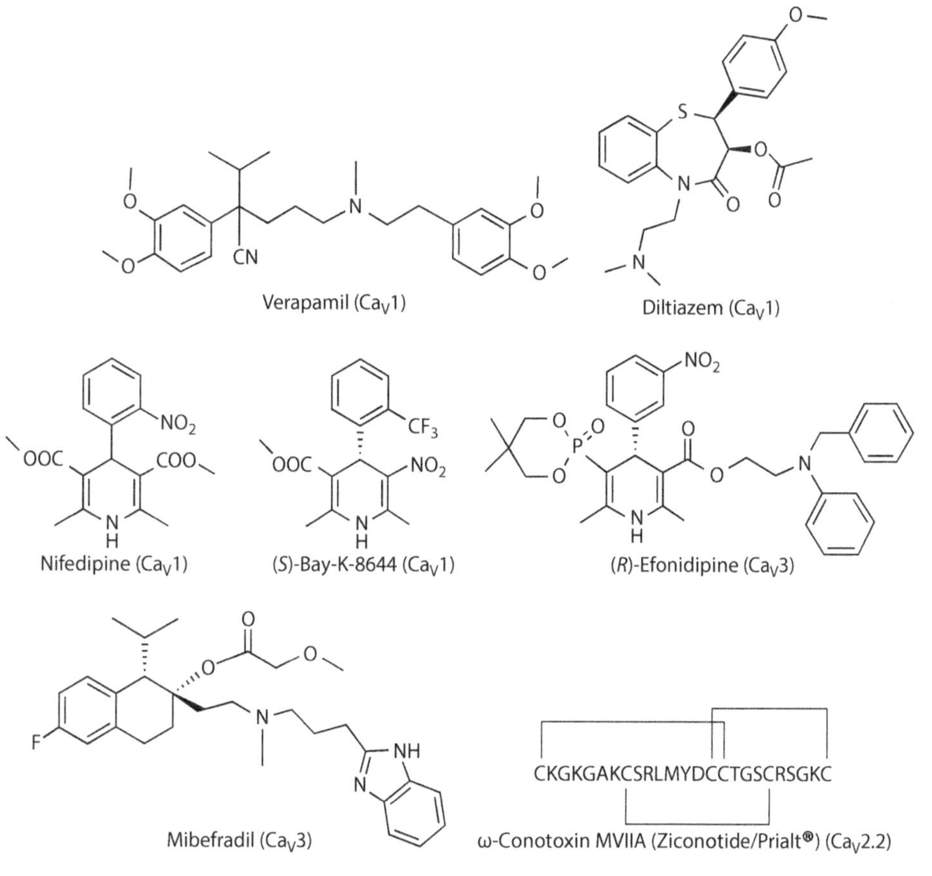

FIGURE 13.8 Chemical structure of drugs acting as blockers of Ca_V1 (L-type) and Ca_V3 (T-type) channels, the Ca_V1 channel activator (*S*)-Bay-K-8644, and the amino acid sequence of the highly specific peptide blocker of $Ca_V2.2$ (N-type) channels, ω-conotoxin MVIIA.

Amino acid residues important for binding of these compounds have been identified through mutagenesis studies and are located in the S5 and S6 segments of domains III and IV of the α_1 subunit (Figure 13.9).

Organic calcium blockers bind with much higher affinity to the inactivated conformations of the Ca_Vs, relative to the closed conformation, thereby trapping receptors in the inactivated state. Therefore, inhibition of Ca_Vs by these compounds has been termed "use-dependent": The rate and extent of Ca_V inhibition will increase with channel activation frequency. Use-dependence is generally considered to be an attractive quality of ion channel inhibitors, since only the highly active

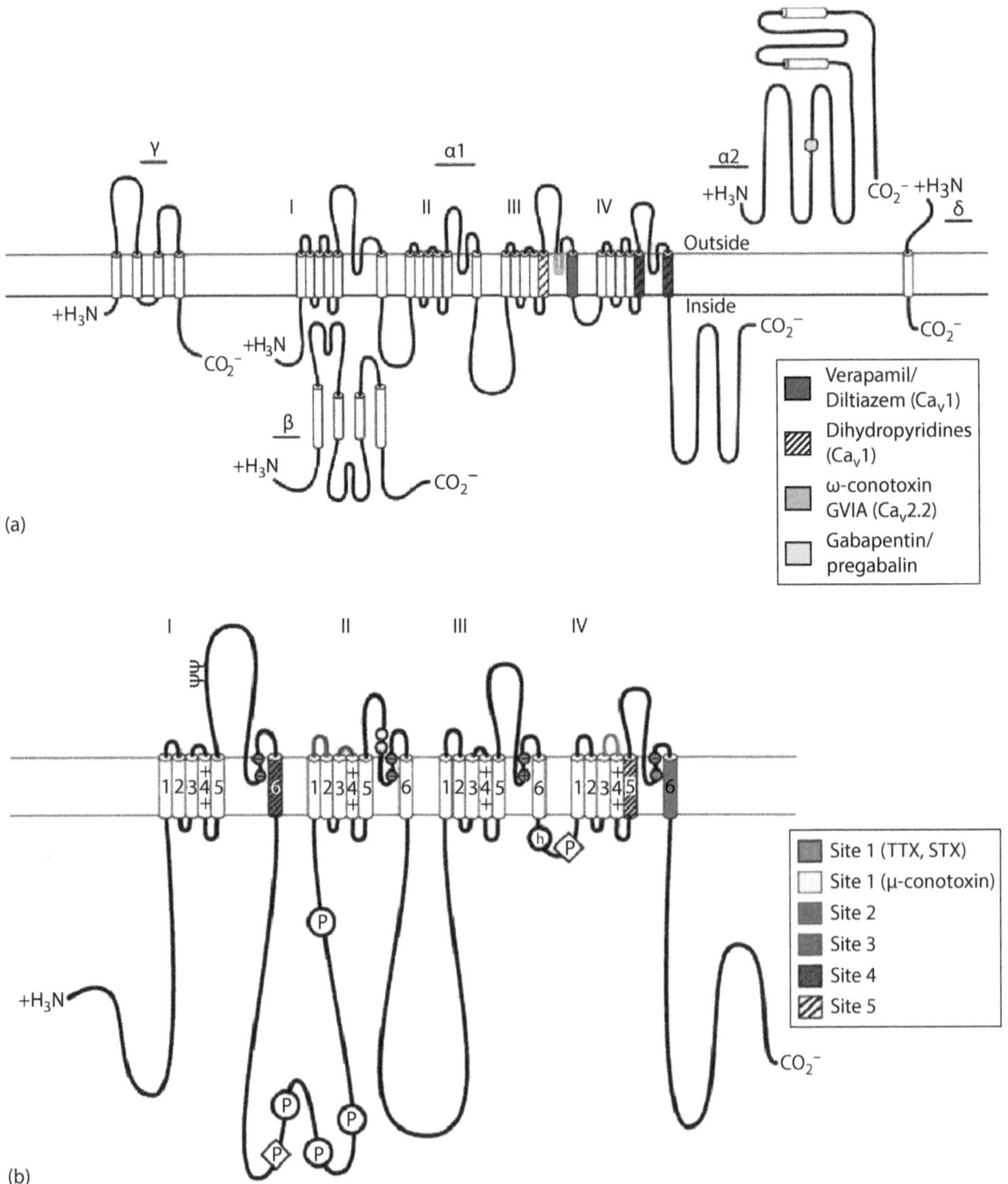

FIGURE 13.9 Overview of the binding sites of toxins and drugs acting at (a) Ca_V and (b) Na_V channels. (From Catterall, W.A. and Gutman, G., *Pharmacol. Rev.*, 57, 385, 2005; Catterall, W.A. et al., *Toxicon*, 49, 124, 2007.)

channels—presumably the ones responsible for a given disorder—will be inhibited, while less frequently activated channels are spared, thereby reducing the risk of adverse effects.

13.3.3.2 Ca_V2 Family (N-, P/Q-, and R-Type Current)

Within this family, the $Ca_V2.2$ subunit (N-type current) has attracted the most attention as potential drug target. The most efficient inhibitors of N-type currents are peptide toxins isolated from the venom of fish-eating marine snails that use these toxins to paralyze their prey. The category includes the 25–30 amino acid residue peptides ω-conotoxin GVIA and ω-conotoxin MVIIA (Figure 13.8) which bind to $Ca_V2.2$ with very high affinity and selectivity. Binding of ω-conotoxin GVIA mainly occurs to residues located in the pore loop region of domain III, suggesting that this toxin acts as a pore blocker of the $Ca_V2.2$ subunit.

The reason for the pharmacological interest in $Ca_V2.2$ is that these channels are responsible for neurotransmitter release in neural pathways relaying pain signals to the brain. Although ω-conotoxins are poorly suited for use as drugs because of their lack of biomembrane permeability, ω-conotoxin MVIIA (Ziconotide/Prialt®) was recently approved for use in humans. Since the drug has to be given through an intrathecal catheter to circumvent the blood–brain barrier, the clinical use of ω-conotoxin MVIIA is limited to severe chronic pain in particular patients such as those suffering from terminal cancer or AIDS. A search for selective, nonpeptide $Ca_V2.2$ blockers that can be administered orally is still ongoing.

$Ca_V2.1$ channels (P/Q-type current) are generally involved in neurotransmitter release in most synapses throughout the brain. $Ca_V2.1$ can be blocked by peptide toxins from either *Conus* snails (ω-conotoxin MVIIC) or from spider venom (ω-agatoxin IVA) (Table 13.2). From a drug discovery point of view their widespread role in neurotransmitter release represents a major safety liability.

The function(s) and pharmacology of $Ca_V2.3$ channels (R-type current) are not well understood. As is the case for $Ca_V2.1$ no selective inhibitor of $Ca_V2.3$ has been declared. A peptide toxin, SNX-482, isolated from tarantula venom was initially thought to be a selective blocker of $Ca_V2.3$ channels, but was later found to block other Ca_V, Na_V, and for K_V channels even at lower concentrations.

13.3.3.3 Ca_V3 Family (T-Type Current)

Certain small-molecule compounds appear to act as moderately selective blockers of Ca_V3. The vasodilating compound mibefradil (Figure 13.8) which has been used widely for treatment of hypertension and angina pectoris, inhibits $Ca_V3.1$–$Ca_V3.3$ channels in a use-dependent way with ~10-fold selectivity over $Ca_V1.2$ channels. Moreover, certain novel dihydropyridine compounds (e.g., (*R*)-efonidipine, Figure 13.8) inhibit Ca_V3 channels up to ~100-fold more potently compared to Ca_V1 channels. It is not yet known exactly how these compounds interact with Ca_V3, but this family of ion channels could have the potential as drug targets for treatment of cardiovascular disease. Certain classical antiepileptic compounds, such as ethosuximide, phenytoin, and zonisamide exert their antiepileptic action at least partly via inhibition of Ca_V3 channels. Substances such as nickel ions (Ni^{2+}), *n*-octanol, and the diuretic amiloride display moderate selectivity for Ca_V3 channels over the other Ca_V channel types. Kurtoxin is a scorpion venom toxin which produces potent blockade of Ca_Vs containing $Ca_V3.1$ and $Ca_V3.2$ but not $Ca_V3.3$ subunits.

13.3.3.4 Auxiliary Subunits

The drugs gabapentin and the more recently developed pregabalin are used clinically for the treatment of epilepsy and neuropathic pain. Their mechanism of action was not understood before the discovery that gabapentin binds with extremely high affinity to the $\alpha_2\delta$ subunit of Ca_Vs. This impairs the trafficking function of $\alpha_2\delta$ subunit which normally increases Ca_V channel cell surface expression. Thereby, gabapentin decreases the amplitude of calcium currents partially without producing the complete blockade seen with Ca_V inhibitors targeting the α_1 subunit. Both $Ca_V2.1$ and $Ca_V2.2$ are involved in mediating the effects of gabapentin/pregabalin. Both drugs are nontoxic which may be related to their partial blocking effect.

13.4 VOLTAGE-GATED SODIUM CHANNELS

13.4.1 Structure and Molecular Biology of Voltage-Gated Sodium Channels

Functionally, Na_Vs are closed at the resting membrane potential and open when the membrane becomes depolarized, activation requiring membrane potentials of −70 to −30 mV, with some variation between different Na_V types. Most Na_Vs inactivate within ~1–10 ms in the presence of sustained depolarization. In certain types of neurons, a more persistent Na_V current with slow inactivation has also been identified.

At the molecular level, Na_Vs are composed of a large (~2000 amino acid residues) α-subunit which is structurally similar to the α_1-subunit of Ca_Vs, and forms the ion conducting pore (Figure 13.7a). The rapid inactivation of most Na_Vs is explained by the cytoplasmic domain III–IV linker (Figure 13.9b h motiv) which functions as a "hinged lid," that simply swings in to occlude the intracellular mouth of the pore.

Nine different Na_V α-subunits (Na_V1.1–Na_V1.9) have been cloned and all these display >50% amino acid identity with each other, so they compose one subfamily. The Na_V1 family has most likely arisen from a single ancestral gene and that their present diversity reflects gene duplication events and chromosomal rearrangements occurring late in evolution.

By analogy to the Ca_Vs, functional Na_Vs can be formed from expression of α-subunits alone although native Na_Vs are protein complexes composed by α-subunits and auxiliary subunits. Only a single class of auxiliary Na_V subunits (β-subunits) has been identified. β-Subunits are composed of a large extracellular part, through which it interacts with the α-subunit (Figure 13.7b) and a small C-terminal portion consisting of a single transmembrane segment. The function of the β-subunits can be divided into (1) modulation of the functional properties of Na_Vs, (2) enhancement of membrane expression, and (3) mediating interactions between Na_Vs and extracellular matrix proteins as well as various signal transduction molecules. Moreover, the β-subunits are also suggested to serve a number of non-Na_V channel-modulating roles, including modulation of brain development.

13.4.2 Physiological Roles of Voltage-Gated Sodium Channels

The biological importance of Na_Vs relies on their ability to cause depolarization of cell membranes. Most of the Na_V α-subunits are capable of detecting even very small increases in membrane potential and this makes the Na_Vs activate, and subsequently inactivate, on a ms timescale. This combination of high sensitivity toward depolarization and very rapid gating kinetics make Na_Vs perfect for initiating and conducting APs.

Na_V1.1, Na_V1.2, Na_V1.3, or Na_V1.6 subunits are expressed in virtually all neurons within the CNS, in particular at the base and along the entire length of the axon. When an excitatory synaptic signal (e.g., glutamate, released by a neighboring neuron, acting on AMPA receptors, see Chapter 15) is received, this generates a small depolarization of the neuronal membrane in the dendrites and cell body. This rather modest depolarization is sufficient for activating Na_Vs at the initial segment of the axon, leading to the generation of an AP. Once the AP reaches the nerve terminal, this will activate Ca_Vs, leading to release of neurotransmitter. The importance of these Na_Vs for AP initiation and conduction is also highlighted by the fact that point mutations in the genes encoding Na_V1.1, Na_V1.2, and Na_V1.3 which alter their functional properties, have been linked to certain forms of epilepsy.

Dorsal root ganglion (DRG) neurons are important for transmitting sensory signals, including pain, from the periphery to the CNS. Sensory stimulation leads to generation and conduction of APs in DRG neurons and these APs are mediated by Na_Vs. Na_Vs of DRG neurons contain the Na_V1.7, Na_V1.8, and Na_V1.9 subunits which are almost exclusively expressed in these neurons. It has also been shown that expression of these α-subunits is altered in a complex fashion in animal models of inflammatory and neuropathic pain. Moreover, gain-of-function mutations of $Na_V1.7$ have been found in patients suffering from inherited pain disorders, whereas loss-of-function mutations result in congenital insensitivity to pain. This thereby tightly link $Na_V1.7$ function to capability of experiencing pain. Likewise, gain-of-function mutation in $Na_V1.8$ and $Na_V1.9$ lead to small-fiber neuropathy and congenital insensitivity to pain, respectively. From a therapeutic point of view, these α-subunits are therefore of particular interest, since compounds capable of selectively blocking $Na_V1.7$–$Na_V1.9$ channels could have great potential as analgesics.

13.4.3 Pharmacology of Voltage-Gated Sodium Channels

A large number of natural products (peptides and alkaloids) have been found to bind Na_Vs with high affinity. Radioligand binding, photoaffinity labeling, and mutagenesis techniques have been used to identify the regions of the α-subunit to which these substances bind. Six binding sites for these toxins are therefore used to provide the conceptual framework for understanding the pharmacology of Na_Vs (Table 13.3; Figure 13.9b). Given the high degree of homology between the Na_V1 subunits, very few examples of subunit-selective toxins are known. The substances mentioned in Table 13.3 thus bind to nearly all Na_V1 subunits. Most toxins act as gating modifiers, and only tetrodotoxin and saxitoxin binding to site 1 are pore blockers. The crystal structure of a homologous bacterial Na_V channel demonstrated how the binding sites create a drug site that when occupied blocks the pore. Access to this site by hydrophilic or larger molecules requires opening of the intracellular gate. This helps to explain how use-dependent block by anesthetics and other drugs arise, because they would bind more when the channel is opened frequently. The structure also revealed fenestrations in the sides of the pore penetrated by fatty acyl chains that extend into the central which could function as portals for the entry of small, hydrophobic resting state pore-blocking drugs.

In addition to toxins, a number of clinically used drug molecules are known to exert their pharmacological action through inhibition of Na_V function. Consistent with the physiological roles of Na_Vs, these drugs include antiepileptic compounds (carbamazepine, lamotrigine, phenytoin), local anesthetic and analgetic compounds (lidocaine), and drugs used to treat cardiac arrhythmia (class I anti-arrhythmics including quinidine, lidocaine, mexiletine, and flecainide). Recently, a lot of effort has been placed on finding subtype-selective Na_V blockers, especially targeting $Na_V1.7$ for treatment of pain, with several compounds demonstrating different levels of subtype selectivity and with several compounds having entered clinical trials. For now, the latest marketed drug is lacosamide which was approved in 2008 for the treatment of diabetic neuropathic pain and partial onset seizures.

TABLE 13.3
Toxin Binding Sites on Na_V Channels

Site No.	Site Location	Toxins Binding to Site	Mechanism of Action
1	Selectivity filter of pore	Tetrodotoxin, saxitoxin	Pore block
2	Interface between the S6 segments of domains I and IV	Plant alkaloid toxins: grayanotoxin, batrachotoxin, and veratridine	Inhibition of inactivation and channel opening at resting potential
3	Outer pore loop regions of domains I and IV	Sea anemone peptide toxins and α-scorpion toxins	Slow inactivation
4	Extracellular S3–S4 loop close to the voltage sensor	Large β-scorpion peptide	Enhance opening at negative membrane potential
5	Interface between the IS6 and IVS5 segments	Plant alkaloids ciguatoxins and brevetoxins	Enhance activation and inhibit inactivation
6	Unknown	δ-Conotoxins	Slow inactivation

13.5 CHLORIDE CHANNELS

For many years, the physiological role of Cl^- transport did not receive a lot of attention. However, this changed with the first cloning of chloride channels in the 1990s and with the discovery of more than a dozen inherited human disorders that was caused by Cl^- transport malfunction. Most cells have anion channels and the primary ion permeating these is Cl^-. The family is very diverse, and for now has not been found to share many properties other than that they are nonselective among anions. Excluding the ligand-gated Cl^- channels that will be covered in Chapter 12, the family so far consist of: the voltage-sensitive ClC subfamily, the cystic fibrosis transmembrane conductance regulator (CFTR), calcium-activated channels, high conductance channels, and volume-regulated channels. The nine ClC channel family members are involved in maintaining the resting membrane potential, transepithelial transport, acidification of intracellular compartments, and cell volume regulation. The ClC proteins are unique in the way that they are widely expressed in intracellular organelles. Surprisingly, although they share the same architecture, five of the nine ClCs are transporters, and mediate Cl^-/H^+ exchange rather than voltage-gated anion channel activity. Like it is the case for ClC, the CFTR also looks like an ion channel evolved from an ancestral transporter and has complicated gating mechanisms. The CFTR Cl^- channel is a 12-transmembrane cAMP-regulated ABC transport protein involved in transepithelial transport, and dysfunction of this Cl^- channel causes cystic fibrosis. The CFTR protein has been extensively used in attempts to establish a gene therapy for cystic fibrosis without success. Ca^{2+}-activated, high conductance channels, and volume-activated Cl^- currents have been characterized in native cells, but for most, molecular identities have not been completely resolved, although some new members have appeared such as the Anoctamin family (TMEM16) which includes 8-TM members of the Ca^{2+}-activated Cl^- channel family. The pharmacology of Cl^- channels is currently quite poor.

13.6 LIGAND-GATED ION CHANNELS

Na_V and K_V channels are essential for the generation and conduction of neuronal APs and Ca_V channels are essential for converting APs into neurotransmitter release—but it is the ligand-gated ion channels which receive the chemical signals of synaptic transmission and convert them into the electrical signals which initiate APs. See Chapter 12 for further details about ligand-gated ion channels.

FURTHER READING

Ashcroft, F. 2000. *Ion Channels and Disease*. London, U.K.: Academic Press.

Catterall, W.A. 2012. Voltage-gated sodium channels at 60: Structure, function and pathophysiology. *J. Physiol.* 590:2577–2589.

Catterall, W.A., Cestèle, S., Yarov-Yarovoy, V., Yu, F.H., Konoki, K., and Scheuer, T. 2007. Voltage-gated ion channels and gating modifier toxins. *Toxicon* 49:124–141.

Catterall, W.A. and Gutman, G. (eds.). 2005. International Union of Pharmacology. Compendium of voltage-gated ion channels. *Pharmacol. Rev.* 57:385–540.

Dalby-Brown, W., Hansen, H., Korsgaard, M.G., Mirza, N., and Olesen, S.-P. 2006. K_V7 channels: Function, pharmacology and channel modulators. *Curr. Top. Med. Chem.* 6:999–1023.

Doyle, D.A., Cabral, J.M., Pfuenzner, R.A. et al. 1998. The structure of the potassium channel: Molecular basis of K^+ conduction and selectivity. *Science* 280:69–77.

Duran, C. 2010. Chloride channels: Often enigmatic, rarely predictable. *Annu. Rev. Physiol.* 72:95–121.

Hille, B. 2001. *Ionic Channels of Excitable Membranes*, 3rd ed. Sunderland, MA: Sinauer Associates.

MacKinnon, R. 2003. Potassium channels. *FEBS Lett.* 555:62–65.

Nardi, A. and Olesen, S.-P. 2008. BK channel modulators: A comprehensive overview. *Curr. Med. Chem.* 15:1126–1146.

Nilius, B. and Szallasi, A. 2014. Transient receptor potential channels as drug targets: From the science of basic research to the art of medicine. *Pharmacol. Rev.* 66:676–814.

Sanguinetti, M.C. and Tristani-Firouzi, M. 2006. Herg potassium channels and cardiac arrhythmia. *Nature* 440:463–469.

Schmitt, N., Grunnet, M., and Olesen, S.-P. 2014. Cardiac potassium channel subtypes: New roles in repolarization and arrhythmia. *Physiol. Rev.* 94:609–653.

14 Neurotransmitter Transporters *Structure, Function, and Drug Binding*

Claus J. Løland and Ulrik Gether

CONTENTS

14.1 INTRODUCTION

Controlled transport of solutes across cell membranes is a prerequisite for maintaining viability and functionality of the cell. The process is controlled by proteins that are embedded to the cell membrane and capable of mediating fluxes of small molecules and/or ions either by generating small selective pores in the membrane (ion channels) or by carrying out a specific translocation processes (transporters). When a channel protein opens, thousands to millions of ions pass the membrane through a pore, down the electrochemical gradient of the individual ion. In contrast, transport proteins do not form pores. Rather they possess a binding site for the transported solute that is only accessible to one side of the membrane at any given time. For that reason, transport proteins—or carriers—facilitate the movement of only one or very few molecules per transport cycle. The transported substance can either be an ion or a small molecule and the movement can either occur with or against its electrochemical gradient. When the transport occurs against the gradient, the process is energetically coupled, either directly through the hydrolysis of ATP by the transport protein itself or indirectly by the use of transmembrane ion gradient. Not surprisingly, a vast amount of different transport proteins are found in both prokaryotic and eukaryotic organisms, transporting everything from nutrients, amino acids, and metabolites to ions, drugs, proteins, toxins, and transmitter molecules. It is generally assumed that about 10% of all human genes are transporter-related where the major classes in humans encompass ATP-driven ion pumps (e.g., the ubiquitously expressed Na-K ATPase), ATP binding cassette (ABC) transporters (e.g., the cystic fibrosis transmembrane conductance regulator and the multidrug resistance transporter p-glycoprotein), cytochrome B-like proteins, aquaporins (water transporters), and the solute carrier superfamily (SLC) (http://www.bioparadigms.org).

The immense functional heterogeneity among transporters is illustrated by the fact that the SLC gene family alone consists of nothing less than 52 different subfamilies (http://www.bioparadigms.org/slc/menu.asp). These include several types of plasma membranes transporters, for example, high-affinity glutamate transporters (SLC1), sodium-glucose co-transporters (SLC5), sodium-coupled neurotransmitter transporters (SLC6), amino acid and peptide transporters (SLC7, 15, 36, and 38), and a variety of ion exchangers (SLC8, 9, and 24). The SLC gene family also includes intracellular vesicular transporters, for example, the vesicular glutamate transporters (SLC17), the vesicular amine transporters (SLC18), and the vesicular inhibitory amino acid transporters (SLC32).

Notwithstanding the huge number of transport proteins present in the human body, relatively few of them are targets for the action of drugs. It might even be argued that transport proteins are relatively overseen as drug targets in spite of their critical physiological functions and some real "success stories," for example, inhibitors of the gastric ATP-driven proton pump, used against peptic ulcer, and inhibitors of monoamine transporters, used against depression/anxiety disorders (see Chapter 18). In this chapter, we will focus on the monoamine transporters and thus on neurotransmitter transporters belonging to the SLC6 family (also named neurotransmitter:sodium symporters or Na^+/Cl^--dependent transporters) (Table 14.1). Indeed, SLC6 transporters represent important targets for several drugs including not only medicines used against depression/anxiety but also against obesity, narcolepsy, and epilepsy, as well as drugs of abuse such as cocaine, amphetamine, "ecstasy," and the so-called bath salts. In addition, memory-enhancing drugs, or "study drugs," also target this class of transporters. Of interest, high-resolution X-ray crystal structures have become available of SLC6 transporter homologs opening up for entirely new possibilities for understanding of how these transporters operate at a molecular level and how their function can be altered by different types of drugs.

14.2 NEUROTRANSMITTER TRANSPORTERS BELONGING TO THE SLC6 FAMILY

The availability in the synaptic cleft of the neurotransmitters dopamine, serotonin, norepinephrine, glycine, and γ-amino butyric acid (GABA) is tightly regulated by specific transmembrane transport proteins belonging to the SLC6 family (Figure 14.1; Table 14.1). The transport proteins are either situated in the presynaptic membrane or on the surface of adjacent glial cells where they mediate rapid removal of the released neurotransmitters and thereby terminate their effect at the pre- and postsynaptic neurons. Inside the presynaptic nerve endings, specific vesicular transporters sequester the neurotransmitters into vesicles making them ready for subsequent release into the synaptic cleft upon arrival of the next stimulus. The plasma membrane neurotransmitter transporters serve three main purposes: First, the transport proteins increase the rate by which the released neurotransmitters are cleared from the synaptic cleft. This rapid removal of released neurotransmitters allows for hundreds of fold faster termination of neurotransmission than possible for simple diffusion. Second, reuptake may prevent diffusion of the neurotransmitter away from the synapse of their release, thereby minimizing chemical crosstalk between adjacent synapses. Third, transporters allow recycling by reuptake of transmitters into the nerve terminal with presumed savings in synthetic cost. The crucial physiological role of the neurotransmitter transporters have been cemented by gene knock-out experiments. In case of, e.g., the dopamine transporter (DAT), the disruption of the transporter gene in mice revealed the unequivocal importance of this carrier in control of locomotion, growth, lactation, and spatial cognitive function.

The SLC6 transporters do not only include transporters of neurotransmitters but also transporters of amino acids, metabolites (creatine), and osmolytes (betaine and taurine) (Table 14.1). Moreover, a large number of homologs have been identified in archae and bacteria. The function of the majority

TABLE 14.1
SLC6 Gene Family Neurotransmitter Transporters

	Endogenous Substrates	Synthetic Substrates	Potent Inhibitors	Therapeutic Use/ Potential
Dopamine transporter (DAT)	Dopamine Norepinephrine Epinephrine	Amphetamine MPP+	CFT, GBR12,909 benztropine, mazindol, RTI-55, Cocaine, Zn^{2+}	ADHD (amphetamines), Parkinsonism? (inhibitors)
Serotonin transporter (SERT)	5-HT	p-Cl-Amphetamine MDMA ("ecstasy")	Citalopram, escitalopram, fluoxetin, paroxetin, sertraline, imipramine, cocaine, RTI-55	Depression, anxiety, OCD (inhibitors)
Norepinephrine transporter (NET)	Dopamine Norepinephrine Epinephrine	Amphetamine MPP+	Nisoxetine, nortriptyline, desipramine, duloxetine, venlafaxine, mazindol	Depression (inhibitors)
Glycine transporter 1 (GlyT-1)	Glycine	—	(R)NFPS (ALX5407), NPTS, Org24598 Zn^{2+}	Schizophrenia? Psychosis? Dementia? (inhibitors)
Glycine transporter 2 (GlyT-2)	Glycine	—	ALX1393, ALX1405 Org25543	Anticonvulsant? Analgesic? (inhibitors)
GABA transporter 1 (GAT-1)	GABA	—	Tiagabine, SKF89976A, THPO *exo*-THPO	Epilepsy (tiagabine)
GABA transporter 2 (GAT-2 equivalent to mouse GAT-3)	GABA beta-Ala	—	—	?
GABA transporter 3 (GAT-3 equivalent to mouse GAT-4)	GABA beta-Ala	—	SNAP5114	?
GABA transporter 4 (GAT-4, equivalent to BGT-1 or mouse GAT-2)	GABA Betaine	—	EF1502, NNC052090, THPO Zn^{2+}	Epilepsy? (inhibitors) EF1502

Note: The known neurotransmitter transporters belonging to the SLC6 gene family with their respective substrates, inhibitors, and potential therapeutic use.

Abbreviations: CFT, (2β-carbomethoxy-3β-(4-fluorophenyl)tropane; MDMA, 3,4-methylenedioxymethamphetamine; MPP+, 1-methy l-4-phenylpyridinium; (*R*)NFPS, *N*-[3-(40-fluorophenyl)-3-(40-phenylphenoxy) propyl]sarcosine; NPTS, (*N*-[3-phenyl-3-(40-(4-toluoyl) phenoxy)propyl]sarcosine; THPO, (4,5,6,7-tetrahydroisoxazolo[4,5-c]pyridine-3-ol); ADHD, attention-deficit hyperactivity disorder; OCD, obsessive–compulsive disorder; *exo*-THPO, 4-amino-4,5,6,7-tetrahydrobenzo [d]isoxazol-3-ol.

of these transporters is still unknown; however, a few of them have been identified as amino acid transporters, for example, the leucine transporter (LeuT) from the *Aquifex aeolicus* bacterium and the multi-hydrophobic amino acid transporter (MhsT) from *Bacillus halodurans.*

At the molecular level, the SLC6 family transporters operate as Na^+-dependent co-transporters that utilize the transmembrane Na^+ gradient to couple "downhill" transport of Na^+ with "uphill" transport (against a concentration gradient) of their substrate from the extracellular to the intracellular environment. The transport process is so efficient that, for example, the serotonin transporter (SERT) can accumulate internal serotonin (5-HT) to concentrations hundreds of fold higher than the external medium when appropriate ion gradients are imposed. Most SLC6 transporters are also co-transporters of Cl^- and accordingly SLC6 transporters have been referred to as the family of Na^+/Cl^--dependent transporters.

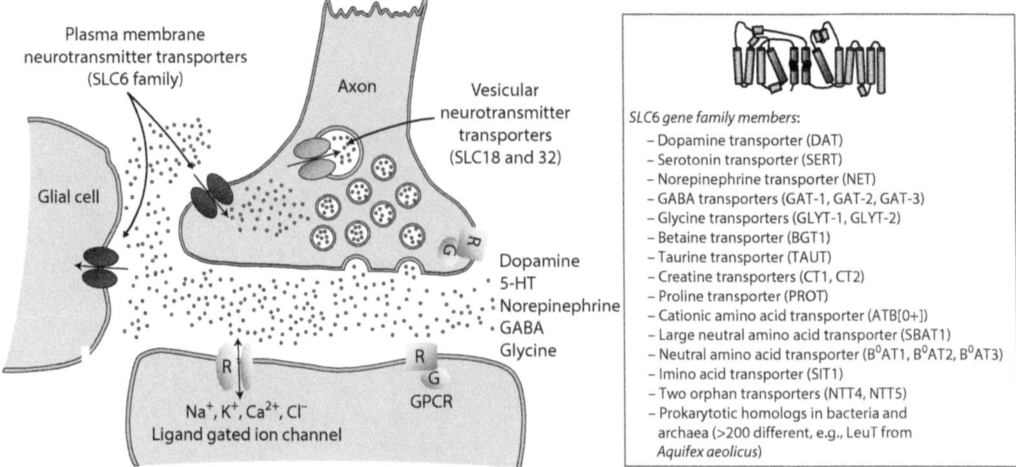

FIGURE 14.1 The role of neurotransmitter transporters in synaptic signaling. Neurotransmitters are sequestered into synaptic vesicles through the vesicular monoamine transporters (VMAT1-2) belonging to the SLC18 gene family or through the vesicular inhibitory amino acid transporter (VIAAT) belonging to the SLC32 gene family. Upon arrival of an axon potential, the synaptic vesicles release its content of neurotransmitter into the synaptic cleft by fusion of the vesicle with the plasma membrane. The neurotransmitter exerts its effects by activating ionotropic receptors (ligand-gated ion channels), such as $GABA_A$, glycine receptors and for $5\text{-}HT_3$ receptors or via G-protein-coupled receptors (GPCRs) such as dopamine, adrenergic, 5-HT, and for metabotropic $GABA_B$ receptors. The fast removal of neurotransmitter from the synaptic cleft is governed by neurotransmitter transporters belonging to the SLC6 family located on the presynaptic neuron (DAT, SERT, NET, GlyT2, GAT-1, and GAT-2) or on glia cells (GlyT-1, GAT-1, GAT-2, and GAT-3). Neurotransmitters taken up by the presynaptic neuron allow recycling with presumed savings in synthetic cost.

14.2.1 Structures and Mechanisms of SLC6 Transporters

It is generally believed that SLC6 transporters function according to an alternating access model first proposed by Peter Mitchell in 1957 and refined by Oleg Jardetzky in 1966. The model suggests a transport mechanism in which at any given time the substrate binding site only is accessible to either the intracellular or the extracellular side of the membrane. Thus, at all times, an impermeable barrier exists between the binding site and one side of the membrane, but the barrier can change from one side of the binding site to the other, giving the site alternate access to the two aqueous compartments that the membrane separates. A prerequisite for this model is the existence of both external and internal "gates," i.e., protein domains that are capable of completing the barrier by occluding access to the binding site of substrate from the external and internal domain, respectively (Figure 14.2). In order for a transport process to occur, the two gates must function in a coordinated manner so one is open when the other is closed. The coordination is likely triggered by the binding and unbinding of either substrate or ions or both in concert.

In the absence of any direct structural information on SLC6 transporters, the identification of substrate binding sites and gating domains is highly limited. Thus, structural information from homolog proteins is an important tool to obtain insight into the structure and function of this class of proteins. For the monoamine transporters, the structures of three homolog transporters have been solved: the previously mentioned bacterial NSS proteins, LeuT and MhsT as well as the dopamine transporter from *Drosophila melanogaster* (dDAT). LeuT which displays 20%–25% sequence identity to its mammalian counterparts was the first protein of this class to be successfully crystallized. Its structure was solved at high resolution (1.65 Å). Later the dDAT structure which bears more than 50% homology to its mammalian counterparts, was solved. It turned out to possess an overall structure very similar to LeuT (the LeuT-fold). The fact that the structural fold is conserved from

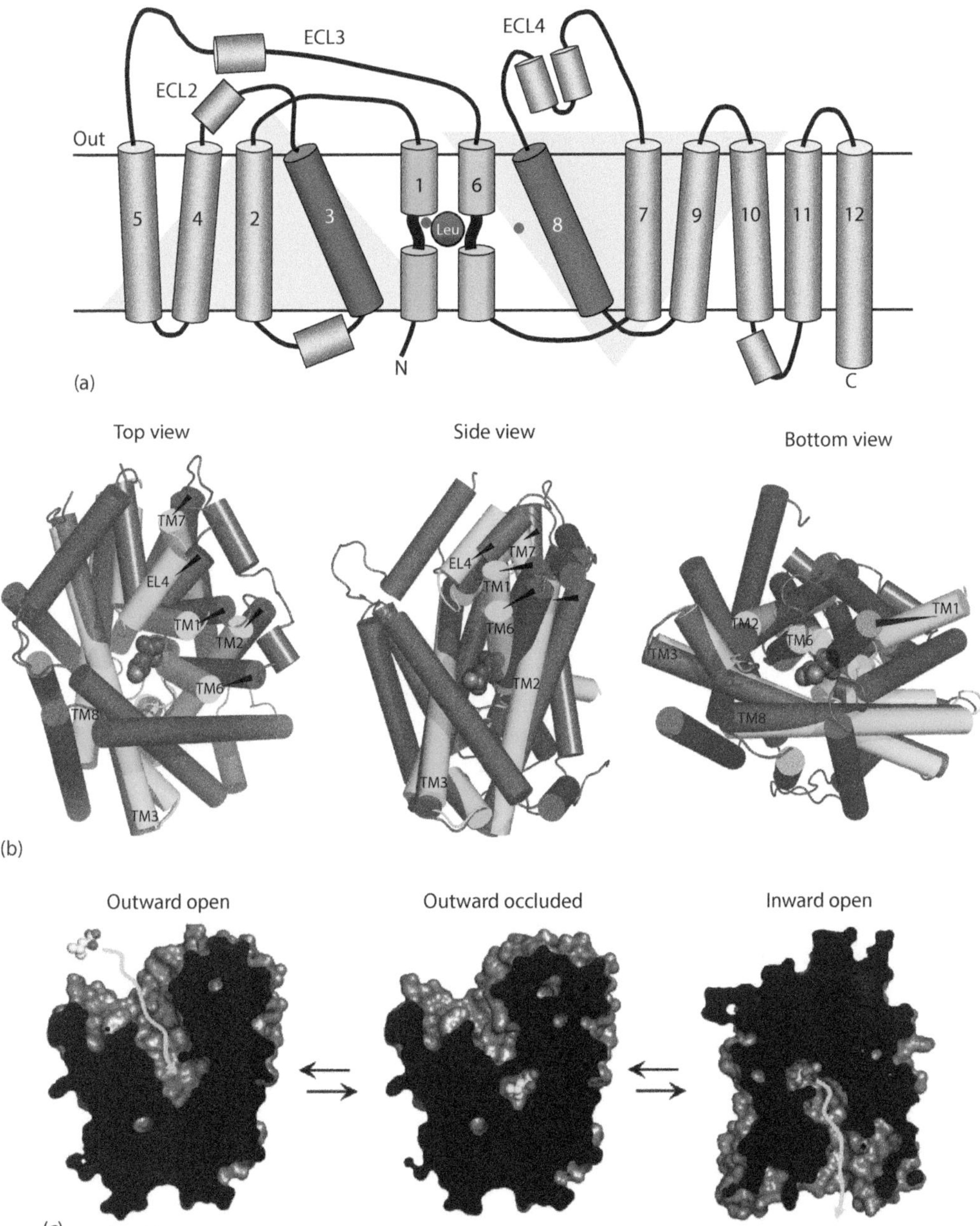

FIGURE 14.2 Structure and conformational changes in LeuT mediating the transitions from outward to inward facing structure. (a) Structural topology of LeuT. TM1, 3, 6, and 8 (colored) constitute the substrate binding site for leucine (purple). Na^+ is shown as blue circles. The yellow triangles depict the two inverted repeats consisting of TM1-5 and TM6-10. (b) Superimposing the outward facing (gray) and inward facing (cyan) structures of LeuT viewed from the extracellular side (left), in the plane of the membrane (middle) and from the intracellular side (right). Helices are shown as cylinders. The conformational changes are mainly governed by a bundle consisting of TM1, 2, 6, and 7 as well as ECL4. The TM3 and 8 which participate in leucine binding, are kept fixed relative to the bundle. For the inward facing conformation, only TM1, 2, 3, 6, 7, 8 and ECL4 are shown for clarity. (c) Cross-sectional space filling model of LeuT, shown from the plane of the membrane in the three crystallized conformations: outward open (left), outward occluded (middle), and inward open (right). The structures illustrate the alternating access model in all conformations by having formed a barrier to either side of the membrane.

Aquifex aeolicus, one of the oldest species of bacteria, to *Drosophila melanogaster*, argues strongly for a structural conservation all the way to mammals. This is further supported by biochemical and pharmacological observations made on the mammalian transporters. The transporter contains 12 transmembrane segments (TMs) organized in a shot glass-like shape with the binding site for substrate and Na^+ ions located in the very center of the protein (Figure 14.2). The first 10 TMs form a structural repeat that relates TM1–5 with TM6–10 around a pseudo-twofold axis of symmetry located in the plane of the membrane. The binding pockets for substrate and two bound Na^+ are formed by TM1, TM3, TM6, and TM8. TM3 and TM8 are long helices that are related by the twofold symmetry axis and are strongly tilted (~50°) (Figure 14.2). TM1 and TM6 are characterized by unwound breaks in the helical structure in the middle of the lipid bilayer. These breaks expose main chain carbonyl oxygen and nitrogen atoms for direct interaction with the substrate. Interestingly, the availability of an increasing number of high-resolution structures of membrane transporters have shown a preservation of this TM5+5 structural repeat, first observed in LeuT, within proteins having no apparent sequence homology to LeuT: The amino acid transporters ApcT and AdiC also possess 12 TMs with the first 10 TMs being part of a 5+5 repeat, but ApcT is proton coupled and AdiC is an exchanger. The repeat can also be located to different segments of the protein: the sodium-glucose transporter vSGLT has in total 14 TMs, one N-terminally and three C-terminally to the 5+5 repeat. The MhsT is thought to have only 11 TMs whereas others again have inserted 2 TMs before the repeat instead of after, as is the case for the betaine transporter BetP and carnitine transporter CaiT. The reuse of a common structural fold underlines the evolutionary advantage in functional simplicity, as well as it suggests that the overall transport mechanism could be preserved from archaebacterial amino acid transporters to mammalian neurotransmitter transporters.

Today, more than 50 X-ray crystal structures of LeuT have been posted in the NCBI database, but only in three distinct conformational states. These are, however, believed to be representing the most fundamental conformational transitions necessary for translocation to occur: the outward facing, the outward occluded, and the inward facing conformation. Thus, these three structures have provided the most compelling insight into the composition of gating domains and conformational transitions in NSS proteins.

In the outward facing conformation, the extracellular gate is open providing solvent accessibility to the substrate binding site. No substrate is bound here, but both Na^+ binding sites are occupied, suggesting that the binding of Na^+ ions opens the transporter placing it in a conformation ready to bind substrate. On the other hand, in the outward occluded conformation, substrate is bound along with Na^+ ions. When going from the outward open to outward occluded conformation, it becomes evident what parts of the protein constitute the extracellular gate. The transition is governed by the helical hinge-like movements of the extracellular halves of TM1, 2, and 6. The movements pivot at Val23 (TM1), Gly55 (TM2), and Leu257 (TM6) suggesting that substrate binding forges the interaction between the gating domains and poses constraints on TM1 and 6 (Figure 14.2). The TM11 and extracellular loop (ECL) 3 move toward the center of the LeuT as a consequence of TM1 and 6 closing in. The movement brings central gating residues in position: Phe253 comprises a 90° lateral rotation upon substrate binding functioning as a lid right on top of the binding site "locking" the substrate into the site. This movement is followed further "up" the protein by the formation of a salt bridge between Arg30 (TM1) and Asp404 (TM10). Finally, the large ECL4 closes the extracellular pathway by moving toward TM10 into the extracellular vestibule as cork in a bottle. The TM1–TM10 salt bridge and the movement of TM4 are in agreement with studies on the mammalian transporters, but here the ECL2 is much larger than in LeuT and mutations herein leave the transporter inactive, suggesting its implication in function. It contains several glycosylation sites and might only serve structural or regulatory purposes, but this is still to be elucidated.

The intracellular gate of LeuT in the outward occluded conformation is predicted to comprise ~20 Å of ordered protein structure, involving in particular the intracellular ends of TM1, TM6, and TM8. A key residue in the predicted gate is Tyr268 at the cytoplasmic end of TM6 which is conserved in all transporters of this class. The tyrosine is positioned below the substrate-binding site at the cytoplasmic surface of the protein and forms a cation–π interaction with an arginine in the N-terminus just below TM1 that forms a salt bridge with an aspartate at the cytoplasmic end of TM8 (Asp369). A likely possibility would be that opening of the gate to the inside will require disruption of this set of interactions. In agreement with this, recent experimental observations in the DAT in conjunction with computational simulations strongly support such a role of the interaction network and that the mechanism is highly conserved among all SLC6 transporters. From the crystal structure of the inward facing conformation of LeuT, it is evident that the transition away from the outward occluded form involves a tilting of the N-terminal part of TM1 by ~45° protruding into the predicted location of the membrane. The intracellular part of TM6 does also rotate away from the central binding site, albeit less drastic than TM1. In concordance with the intracellular part of TM1 and 6 opening, the extracellular part of the TMs closes in toward TM3 and 8 blocking access to the substrate binding site from the extracellular side. In contrast, on the intracellular side, the gate opens allowing access to the substrate binding side from the inside. TM2, 5, and 7 do also move, but they undergo a bending rather than a tilt as observed for TM1 and 6. The movement of TM7 causes ECL4 to plug into the extracellular vestibule creating an extracellular thick gate. To create the inward facing structure, it was necessary to mutate Tyr268 so its role in the transition is still not known. Also the Na^+ binding sites had to be mutated.

Taken together, it seems reasonable to conclude based on the LeuT structures and available functional data that SLC6 transporters follow an alternating access model. However, the mechanism of transport by LeuT is probably distinct from those suggested for other ion-coupled transporters, such as those involving movements of two symmetrical hairpins reaching from the extracellular and intracellular environment, respectively, that were offered for sodium-coupled glutamate transporters. Similarly, the suggested mechanism differs from the "rocker-switch" type mechanism proposed for Lac Permease and the glycerol-3-phosphate transporter, two other crystallized transport proteins of the major facilitator superfamily (MFS) that mediate proton-coupled secondary active transport.

14.2.2 Binding Sites for Na^+ and Cl^-: Importance in Substrate Binding and Translocation

The binding and co-transport of Na^+ is a feature which probably holds several purposes: First, Na^+ serves as a driving force for the translocation of substrate against its electrochemical gradient. Second, the ions coordinate the binding of substrate to the transporter; and third, the ions might function as conformational guides, ensuring that the transporter undergoes the proper conformational changes during the translocation cycle.

SLC6 transporters bind and translocate 1–3 Na^+ and one Cl^- during the translocation of one substrate molecule. SERT appears special by also mediating counter transport of one K^+ during one transport cycle. LeuT and dDAT crystal structures provided much insight into the localization of the ion binding sites. The two bound Na^+ ions have distinct roles: One Na^+ site (Na1) is connected to the bound substrate via two water molecules. In spite of this indirect coordination to substrate, it seems to be essential for substrate binding. The other Na^+, bound to the Na2 site, does not interact with the bound substrate; rather it could foster an interaction between conformational sensitive parts of the protein, thereby closing the cytoplasmic pathway by holding gating domains together. Its role is, however, still hypothetical. The binding site for Cl^- is located 5 Å away from the Na1 site through residues in TM2, 6, and 7. Substrate transport in LeuT does not appear to be Cl^--dependent and, hence, no Cl^- ion was found in that structure. Instead it has a negatively charged glutamate in the same position which seems to be protonated during the return of the empty transporter to the outward facing conformation, so, in that sense, the functionality of a negative charge is preserved. An elegant experiment on the GABA transporter (GAT-1) elucidated the functional role of chloride. They inserted the glutamate from LeuT in the binding site which resulted in a chloride independent GABA transporter. Instead, lowering pH on the intracellular side increased uptake suggesting that protonation is important for completing the return step. This result suggests that in the wild type transporter, the chloride ion is a substrate for GAT-1 and released to the cytosol in contrast to simply binding to the protein throughout the entire translocation cycle. The requirement of the negative charge during the translocation of GABA, but not during the return step, suggests that the role of chloride is mainly to compensate for the multiple positive charges that enable accumulation of substrate against huge concentration gradients.

14.2.3 Substrate Specificity and Binding Sites in SLC6 Neurotransmitter Transporters

The biogenic amine transporters include DAT, SERT, and the norepinephrine transporter (NET). Among these, DAT and NET display marked overlapping specificity for their substrates dopamine and norepinephrine; hence, DAT transports dopamine and norepinephrine with similar efficacy and the apparent affinity for dopamine is a few folds lower than that for norepinephrine (Figure 14.3). Moreover, NET transports dopamine with 50% of the efficacy seen for norepinephrine and the apparent affinity for norepinephrine is a few folds higher than that for dopamine. Accordingly, their classification as DAT and NET seems primarily determined by their localization to dopaminergic and noradrenergic neurons, respectively, rather than by their distinct substrate specificity. In contrast, SERT displays high specificity toward 5-HT although it has been shown that SERT can transport dopamine if present in very high concentrations. The GABA transporters (GAT-1 to 3) and the glycine transporters (GlyT-1 and -2) all display high specificity for their respective endogenous substrate; however, GAT-2 and GAT-3 can also transport beta-alanine (the only naturally occurring beta-amino acid) and GlyT-1 can transport the naturally occurring *N*-methyl-derivative of glycine, sarcosine.

The dDAT structure has been solved with dopamine bound in the binding site. This has provided compelling evidence for the binding mode and orientation of monoamines in the substrate binding site. The bound dopamine is surrounded by TM1, 3, 6, and 8. The amine group of dopamine interacts with the carboxylate of Asp46. This residue (Asp79 in hDAT) is conserved within monoamine transporters and has been proven crucial for both binding and transport mechanism. GATs, GlyTs, and LeuT contain glycine in this position which accommodates the carboxylate group in the substrates. The catechol group is harbored by TM3, 6, and 8. Most of the interactions between substrate and residues have been verified by molecular modeling and mutagenesis studies. The dDAT crystal structure and the experimental predictions differ at least at one central position: In the crystal structure, the binding pocket contains two water molecules that link the substrate to the Na1 site. In the models, the binding pocket is completely dehydrated. The dehydration proceeds further than the binding pocket, but results in an overall more closed-to-outside protein conformation. In the dDAT:dopamine complex, the conformation is rather open-to-out. However, a second X-ray structure in complex with the dopamine analog, 3,4-dichlorophenethylamine (DCP) showed a more occluded state of the transporter. DCP is less prone for oxidation than dopamine and thus more suitable for crystallographic studies.

Right above the dopamine molecule, a tyrosine in TM3 (Tyr124) forms via its hydroxyl group, a hydrogen bond with the main-chain carbonyl of Asp46 in TM1. This interaction could function as a latch to stabilize the irregular structure near the unwound region in TM1 and maybe even be the first determinant in the closure of the extracellular gate. This hypothesis is even more interesting in light of the fact that the tyrosine is strictly conserved among all NSS family members and has been implicated in substrate binding and transport in mammalian GAT-1, DAT, and SERT. Homology modeling studies of DAT and SERT have confirmed this interaction. In support of this hypothesis and thereby of role of the hydrogen bond in stabilizing the substrate binding site, mutation of the tyrosine in DAT (Tyr156) to phenylalanine decreases apparent dopamine affinity around 10-fold and decreased the maximum uptake capacity by approximately 50%. The interaction is present in both the dDAT:dopamine and dDAT:DCP structures, so if the hydrogen bond formation is dynamic, it must precede the dopamine bound structure.

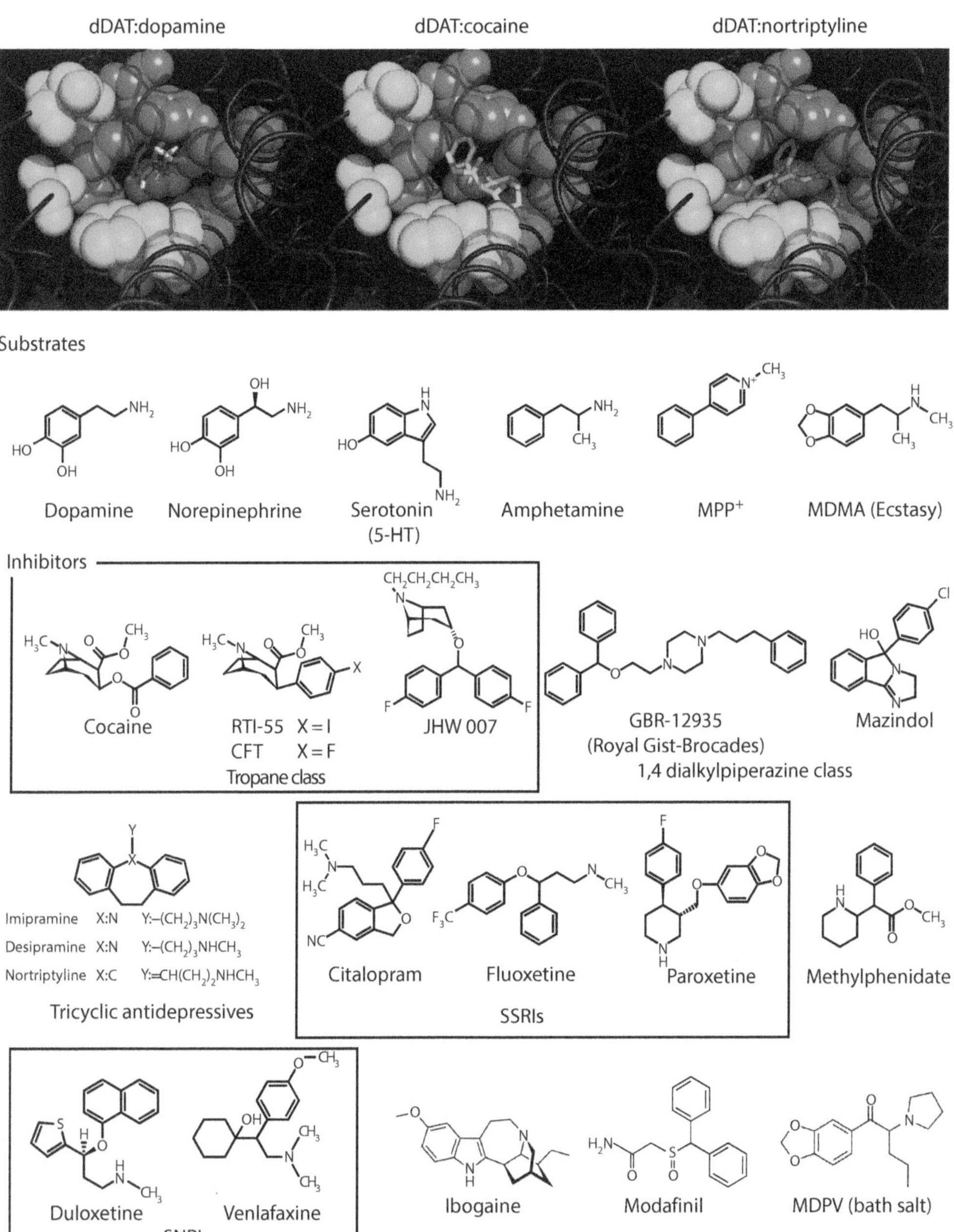

FIGURE 14.3 The binding site and structure of biogenic amine transporter substrates and inhibitors. Top panel: Crystal structure of dDAT solved in complex with dopamine, cocaine, and nortriptyline, respectively, with focus on the binding pocket hosting both substrates and inhibitors for the protein. The residues constituting the binding pocket are shown as spheres. They are colored according to their subsite position: Subsite A (red) consists of residues Phe43 and Asp46 from TM1 and Gly322 and Ser421 from TM6 and 8, respectively. Subsite B (blue) consists of Ala117, Val120, Asp121, and Tyr124 from TM3, Phe325 in TM6 and Gly425 from TM8. Subsite C (green) is Phe319 from TM6 and Asp475 and Ala479 from TM10. Middle and bottom panels: Chemical structure of most common substrates and inhibitors, respectively, for the biogenic amine transporters. SSRIs, selective serotonin reuptake inhibitors; SNRIs, serotonin norepinephrine reuptake inhibitors; MPP+, (1-methyl-4-phenylpyridinium); MDMA, 3,4-methylenedioxy-methamphetamine; MDPV, methylenedioxypyrovalerone.

The partial occlusion of the central site observed in the dDAT:DCP complex is mediated by an inward movement of TM1b, 2, and 6a. The movement is probably mediated by a movement of three conserved phenylalanine residues: Phe318, Phe319, and Phe325. Phe319 rotates to occlude the binding pocket, completely preventing solvent access and, thus, dehydrating the pocket. Also, Phe325 moves toward the catechol ring of DCP and stabilizes its binding. Phe318 rotates away from the binding site and participates in the inward movement of TM6b. To accommodate these side chain shifts, TM11 undergoes an outward movement and TM2 moves in. Interestingly, TM2, 7, and 11 form a binding site for cholesterol which has shown to be important for ligand binding.

14.3 DRUGS TARGETING BIOGENIC AMINE TRANSPORTERS: SPECIFICITY, USE, AND MOLECULAR MECHANISM OF ACTION

The biogenic amine transporters, DAT, NET, and SERT, are targets for a wide variety of drugs. Overall, these drugs can be classified as either pure inhibitors that block substrate binding and transport, or as substrates that in addition to compete with the endogenous substrate also are transported themselves.

The binding mode of inhibitors to the biogenic amine transporters has been derived from the crystal structures of dDAT and LeuT solved in complex with antidepressants and illicit drugs such as cocaine and amphetamine. Common for all the drugs is that they bind competitively with the substrate. The binding site can be divided into subsites A, B, and C. In the dDAT, subsite A consists of residues Phe43 and Asp46 from TM1 and Gly322 and Ser421 from TM6 and 8, respectively. Subsite B consists of Ala117, Val120, Asp121, and Tyr124 from TM3, Phe325 in TM6 and Gly425 from TM8. Subsite C consists of Phe319 from TM6 and Asp475 and Ala479 from TM10 (Figure 14.3). In general, subsite A accommodates the polar amine moiety of the inhibitors as it does for substrates. This is mainly mediated by Asp43, and also from the main chain carbonyl oxygen and nitrogen from the unwound regions of TM1 and 6 as potential hydrogen bonding partners. Ser421 does also contribute to the polarity in subsite A. Subsite B defines a nonpolar ridge (Phe325 and Val120) that accommodates the hydrophobic groups of the drugs. It also forms a groove (Ala117, Asp121, Ser421, and Gly425) that accommodates the polar groups in the drugs' rings, such as the chloro-, dichloro-, trifluoromethyl-, and benzodioxol groups. Subsite C is distal to the substrate binding site and located to the extracellular vestibule. This subsite interacts with bulky drugs that probably enhance affinity and specificity.

14.3.1 Cocaine, Benztropine, and Other Tropane Class Inhibitors

The most thoroughly studied class of inhibitors at the biogenic amine transporters is the tropane class with cocaine as the most well-known member (Figure 14.3). Cocaine is a moderately potent antagonist inhibiting the function of all three transporters nonselectively. However, correlative studies, as well as studies on genetically modified mice suggest the presynaptic DAT as the primary target for cocaine's stimulatory action. DAT knock-out mice are insensitive to the administration of cocaine and, moreover, knock-in mice expressing a functional DAT mutant incapable of binding cocaine shows insensitivity to cocaine administration. It is, therefore, the current view that the rapid increase in extracellular dopamine concentration elicited by cocaine inhibition of DAT produces the psychomotor stimulant and reinforcing effect that underlie cocaine abuse.

Some closely related cocaine analogs possess higher potency toward the biogenic amine transporters and, thus, have been more suitable than cocaine itself in experimental set-ups (e.g., radioligand binding assays) directed toward understanding the pharmacological properties of the transporters. Important examples include CFT (2β-carbomethoxy-3β-(4-fluorophenyl)tropane or WIN 35,428) and RTI-55 ((−)-2β-carbomethoxy-3β-(4-iodophenyl)tropane or β-CIT) (Figure 14.3). Both compounds display nanomolar affinity for the biogenic amine transporters; however, while CFT shows selectivity for DAT over NET and SERT, RTI-55 shows selectivity for SERT and DAT over NET.

The molecular mode of interaction of cocaine and analogs with DAT has long been the subject of speculation. In particular, it has been debated whether or not the cocaine-binding site in DAT overlaps with that of dopamine. If inhibition of dopamine uptake by cocaine is the result of an allosteric mechanism, it would be possible, at least in theory, to generate a cocaine antagonist for treatment of cocaine addiction that might block cocaine binding without affecting dopamine transport. Unfortunately, a crystal structure of dDAT in complex with cocaine, CFT or RTI-55 shows completely overlapping binding sites with dopamine. The binding is adjacent to both the Na^+ and the Cl^- ions. The tertiary amino group of cocaine forms a salt bridge with Asp48. Subsite B, which accommodates the aromatic moiety of the tropanes, has two nonconserved residues between hDAT and hSERT. This discrepancy could account for the differences in binding affinity of CFT and RTI-55. All tropane-based inhibitors induce an overall outward facing binding conformation, analogous to the dDAT:dopamine complex. This outward conformation as well as the majority of the interacting residues has been verified experimentally in both hDAT and hSERT and with molecular docking models. The major discrepancy between experiments and the crystal structure is that in the structure, the Asp46-Tyr124 hydrogen bond is still present. The experimental data and the docking models suggest that the 2β-methylester substituent of the tropane ring protrudes outward and disrupts this interaction.

For compounds of the tropane class, the tropane ring and the 2β-carbomethoxy group are crucial for their affinity. An exception for this rule is the benztropine class. This group of tropanes lacks the 2β-carbomethoxy group but still bind DAT with high affinity. Instead of the 2β-carbomethoxy group, the benztropines contain a diphenylmethoxy moiety. Recently, there has been increasing focus on benztropine analogs. Several of these compounds possess similar or even higher affinity and greater selectivity for DAT than cocaine. The compounds tested so far readily cross the blood–brain barrier and produce increases in extracellular levels of dopamine even for longer durations than cocaine. Nonetheless, several of these DAT inhibitors are less effective than cocaine as behavioral stimulants in rodent models. Furthermore, one BZT analog, JHW 007, was found to potently antagonize the behavioral effects of cocaine (Figure 14.3). Assuming a correlation between behavioral effects of cocaine in laboratory animals and abuse potential in humans, these findings suggest JHW 007 as a potential lead for development of cocaine abuse pharmacotherapeutics. The reason for this discrepancy in the stimulating effect has been suggested at least in part to be related to different pharmacodynamic properties of the compounds. Interestingly, experimental data suggest that in contrast to the cocaine-like compounds, the benztropine analogs bind and stabilize a more closed conformation of the transporter. It is conceivable that binding to the open and likely more prevalent outward facing conformation of DAT results in a faster on-rate which may facilitate faster inhibition of DAT function and thereby a more rapid rise in extracellular dopamine concentration. In contrast, binding to a more closed and predicted less prevalent conformation of the transporter may result in a slower on-rate of the compound and thereby a slower rise in dopamine levels and a less stimulatory effect. A structure of DAT:benztropine analog complex might clarify this issue.

14.3.2 Amphetamine and Other Nonendogenous Substrates

Several nonendogenous compounds are substrates of the biogenic amine transporters and are used either as medication, drugs of abuse, or biochemical tools. Amphetamine and derivatives thereof, for example, metamphetamine, *p*-chloroamphetamine, and 3,4-methylenedioxymetamphetamine (MDMA or ecstasy) are a class of psychostimulants that are transported by DAT, NET, and SERT (Figure 14.3). Methamphetamine preferentially acts on DAT and NET while *p*-chloroamphetamine and MDMA have higher specificity for SERT. This is supported by analyses of mice deficient in either DAT or SERT, i.e., DAT knock-out mice are hyperactive and do not respond to amphetamine, while SERT-deficient mice display locomotor insensitivity to MDMA. Interestingly, amphetamines do not only increase the synaptic concentration of dopamine by competing with dopamine for uptake via DAT but also by promoting reversal of transport resulting in efflux of dopamine via the transporter. This efflux dramatically increases levels of extracellular dopamine and is believed to be of major importance for the psychostimulatory properties of amphetamines. Increasing evidence supports that this efflux is not just the result of "facilitated exchange," but also might involve a channel mode of the transporter. Furthermore, studies suggest that the efflux is dependent on binding to the DAT C-terminus of Ca^{2+}/calmodulin-dependent protein kinase α (CaMKIIα) that in turn facilitates phosphorylation of one or more serines situated in the distal N-terminus of the transporter.

As for the inhibitors, dDAT has been crystallized in complex with D-amphetamine and (+)-methamphetamine. Relative to dopamine, the binding site for amphetamine is still overlapping, but moved more into subsite A by almost 3 Å and interacts directly with Asp46 and the main chain carbonyl of Phe319. In the crystal structure, the two amphetamines also appear to bind to an outward facing conformation. Experimental analysis and molecular docking models on hDAT suggested a more closed-to-outside conformation. Equal for both models is the intact hydrogen bond between Asp46 and Tyr124.

Of other types of stimulant drugs, we find the emerging group of bath salts. The name derives from instances in which the drugs have been sold disguised as true bath salts. The synthesis forms white granules or crystals which resemble the structure of true bath salts, but the content is entirely different. These drugs are designer drugs, derivatives of methamphetamine, and usually contain a benzoylethanamine also called a cathinone. Methylenedioxypyrovalerone (MDPV) is one of the more frequently used "bath salts." It is a potent inhibitor of both DAT and NET, but not a very potent inhibitor of SERT. The robust stimulation of dopamine transmission by MDPV predicts serious potential for abuse and may provide a mechanism to explain the adverse effects observed in humans taking high doses of "bath salts" preparations. However, the abuse potential of MDPV and similar fast evolving group of designer drugs is still not clarified.

14.3.3 Antidepressants

The biogenic amine transporters are also targets for medicines used against depression and anxiety, as also discussed in Chapter 18. The SSRIs (selective serotonin reuptake inhibitors), such as citalopram, fluoxetine, paroxetine, and sertraline are, as implicated by their name, potent and selective inhibitors of SERT (Figure 14.3). Another class of antidepressants includes the so-called SNRIs (serotonin norepinephrine reuptake inhibitors) or "dual action" antidepressants such as venlafaxine and duloxetine that are active at both SERT and NET (Figure 14.3). Finally, the classical and still often used tricyclic antidepressants (TCAs) are potent inhibitors of NET and/or SERT with imipramine and amitriptyline being approximately 10-folds more potent on SERT, while desipramine is a relative selective inhibitor of the NET. Interestingly, the anti-obesity drug sibutramine exerts its action via combined inhibition also of NET and SERT. However, a DAT component cannot be completely ruled out. Conceivably, this effect is achieved through a combination of an anorectic effect due to increased extracellular serotonin levels and increased thermogenesis due to increased norepinephrine levels. Sibutramine has now been withdrawn from the market in several countries including the EU and the United States due to cardiovascular side effects.

The binding sites for antidepressants at their main targets, NET and SERT, are poorly described. Also here, dDAT has been solved in complex with the TCA nortriptyline which might provide suggestions about the binding poses in its mammalian cousins (Figure 14.3). For the human transporters, nortriptyline seems only to bind to NET and SERT. However, dDAT seems to possess a binding pharmacology more similar to the human NET than human DAT. The structure is solved in the outward open conformation with the inhibitor bound to a site overlapping with the binding site for substrate and in close proximity to the Na^+ and Cl^-. The dibenzocycloheptene ring is binding in subsite B saddling around TM3. Parts of the tricyclic ring protrude into subsite C and interact with Ala479. The N-methylpropylamine group extends across the width of the drug binding site and forms a hydrogen bond with the main-chain carbonyl of Phe43 in subsite A. Mutagenesis studies performed in the human SERT supports a similar binding pose for the other TCAs clomipramine, imipramine, and amitriptyline. A molecular docking model of imipramine in SERT found a binding pose almost completely overlapping that of nortriptyline in dDAT.

The binding site for SSRIs is less conclusive. Here, dDAT crystal structures have so far not been of any assistance. Instead, LeuT has been engineered to harbor SERT-like pharmacology by mutating all key residues within 10 Å of the substrate binding pocket. Although the mutated LeuT was unable to transport, it was possible to produce diffracting crystal structures of this in complex with several inhibitors including the SSRIs sertraline, paroxetine fluoxetine, and fluvoxamine. In subsite A, Asp24, which corresponds to Asp46 in dDAT, forms a salt bridge with the amine group in the SSRIs. A similar interaction has been proposed with the aligned Asp98 in SERT. Subsite B has the most interactions which can be partitioned into two types of interactions: First, nonpolar residues interact with hydrophobic ring structures in the SSRIs. These are mediated by residues Val104 (dDAT: Val120) and Phe259 (dDAT Phe325). Second, a groove accommodates the polar groups in the rings of the drugs: the dichloro of sertraline, trifluoromethyl of fluoxetine and fluvoxamine, and the benzodioxol group of paroxetine. The groove is aligned by the residues Pro101, Ala105, Gly359, and Ser356 and capped by Tyr108 (Tyr124 in dDAT). Subsite C interacts with the bulky drugs paroxetine and fluoxetine but only partly with the smaller sertraline. Also the methoxy group of fluvoxamine protrudes into the groove. A similar pattern of interaction between SERT and the SSRIs has been suggested based on mutagenesis of selected residues within the three subsites. No structures have been solved in complex with citalopram. Mutagenesis studies suggests that it occupies all three subsites, but current docking models do not show a significant portion of S-citalopram protruding into subsite C. Rather it seems to have a pronounced subsite B component with its fluorophenyl protruding all the way in between TM3 and TM8. Also the cyano group appears to come in close proximity to the unwound part of TM6.

14.3.4 Other Atypical Biogenic Amine Transporter Inhibitors

Of other inhibitors that do not belong to the classes of psychostimulants or antidepressants, we find modafinil and ibogaine. Modafinil is a selective and low potency DAT inhibitor dominated by its diphenylmethyl moiety connected by a sulfinyl to an organic amide. Modafinil is used in the treatment of attention-deficit hyperactivity disorder (ADHD) and narcolepsy. It also has an off-label use for fatigue as well as a memory-enhancing drug (study drug) even though no clinical studies have reported positive effects of the latter. Experimental data suggest that modafinil have binding properties competitive to dopamine but its actual binding site has not been elucidated. However, it seems like the DAT binding conformation resembles that of the benztropines which might explain why it does not appear to possess stimulant properties. Ibogaine is a substance found in plants of the Apocynaceae family such as *Tabernanthe iboga*. It has a powerful psychedelic action and does also seem to interfere with the serotonergic system as reported for MDMA. Indeed, Ibogaine is an inhibitor of SERT, but in a noncompetitive manner. Compared to the other inhibitors of the biogenic amine transporters, the location of the binding site for ibogaine is probably the least characterized but could be completely unrelated to other known drugs. Apart from its psychedelic effects, ibogaine has been proposed to have anti-addictive properties and is used in medical subcultures as part of the treatment for cocaine addiction.

14.4 INHIBITORS OF GLYCINE AND GABA TRANSPORTERS: SPECIFICITY, USE, AND MOLECULAR MECHANISM OF ACTION

Other SLC6 family transporters than the biogenic amine transporters are targets for drugs or for drug discovery. GAT-1 is, for example, the target for the antiepileptic drug tiagabine; however, the molecular basis for its interaction with GAT-1 is not known. The *N*-dithienyl-butenyl derivative of *N*-methyl-*exo*-THPO (4-methylamino-4,5,6,7-tetrahydrobenzo[d]isoxazol-3-ol) (EF1502) has been shown to inhibit not only GAT-1 but also the betaine carrier (BGT-1) and to act as a very efficient anticonvulsant whose action is synergistic with that of tiagabine. Thus, BGT-1 is likely to be an important antiepileptic drug target. The explanation for the observations might be related to a differential distribution of BGT-1 and GAT-1. While GAT-1 is localized to synaptic sites, BGT-1 is localized to astrocytes and possibly extrasynaptic loci in the neurons; hence the efficacy of EF1502 owing to its interaction with BGT-1 could be explained by modulation of extracellular GABA concentrations at extrasynaptic sites.

The high-affinity glycine transporters (GlyT-1 and GlyT-2) might also represent interesting drug targets. Physiologically, GlyT-1 appears to play a role in astroglial control of glycine availability at NMDA receptors whereas GlyT-2 is likely to play a fundamental role in glycinergic inhibition as reflected in a lethal neuromotor deficiency in GlyT-2 knock-out mice. The putative role of GlyT-1 in regulating glycine availability at NMDA receptors has warranted attempts to develop high-affinity inhibitors of GlyT-1 as a novel class of antipsychotic drugs, i.e., blockade of the GlyT1 is envisioned to increase synaptic levels of glycine ensuring saturation of the glycine-B (GlyB) site at the NMDA receptor at which glycine acts as an obligatory co-agonist. Importantly, a derivative of sarcosine (3-(4-fluorophenyl)-3-(4′-phenylpheroxy))propylsarcosine (NFPS) has been shown to potentiate NMDA receptor-sensitive activity and produce an antipsychotic-like behavioral profile in rats. Several GlyT-1 and GlyT-2 inhibitors have now been described; however, details about their mode of interaction, e.g., from X-ray crystal structures, are still needed for these classes of transporters.

14.5 CONCLUDING REMARKS

The SLC6 neurotransmitter transporters represent a prototypical class of ion-coupled membrane transporters capable of utilizing the transmembrane Na^+ gradient to couple "downhill" transport of Na^+ with "uphill" transport (against a concentration gradient) of their substrate from the extracellular to the intracellular environment. The transporters play key roles in regulating synaptic transmission in the brain by rapidly sequestering transmitters such as dopamine, norepinephrine, serotonin, GABA, and glycine away from the extracellular space. Moreover, they are targets for a wide variety of drugs, including antidepressants, antiepileptics, and psychostimulants, as well as they are subject to current drug discovery efforts. High-resolution structural information has become available for this class of transporters through crystallization of the dopamine transporter from *Drosophila melanogaster* as well as from the bacterial homolog, LeuT. These structures serve as an important framework for future studies aimed at deciphering the precise molecular details and dynamics of the transport process for mammalian neurotransmitter transporters. The binding mode for a wide class of transporter inhibitors has been solved based on crystal structures of dDAT and LeuT in complex with substrates and inhibitors. The described structures serve as an important template for delineating the molecular determinants for drug binding to SLC6 neurotransmitter transporters.

FURTHER READING

Beuming, T., Kniazeff, J., Bergmann, M. et al. 2008. The binding sites for cocaine and dopamine in the dopamine transporter overlap. *Nat. Neurosci.* 11:780–789.

Kristensen, A.S., Andersen, J., Jørgensen, T.N. et al. 2011. SLC6 neurotransmitter transporters: Structure, function, and regulation. *Pharmacol. Rev.* 63:585–640.

Loland, C.J. 2015. The use of LeuT as a model in elucidating binding sites for substrates and inhibitors in neurotransmitter transporters. *Biochim. Biophys. Acta* 1850:500–510.

Shi, L., Quick, M., Zhao, Y., Weinstein, H., and Javitch, J.A. 2008. The mechanism of a neurotransmitter:sodium symporter—Inward release of Na^+ and substrate is triggered by substrate in a second binding site. *Mol. Cell* 30:667–677.

Wang, H., Goehring, A., Wang, K.H., Penmatsa, A., Ressler, R., and Gouaux, E. 2013. Structural basis for action by diverse antidepressants on biogenic amine transporters. *Nature* 503:141–145.

Wang, K.H., Penmatsa, A., and Gouaux, E. 2015. Neurotransmitter and psychostimulant recognition by the dopamine transporter. *Nature* 521:322–327.

Zhao, Y., Terry, D.S., Shi, L. et al. 2011. Substrate-modulated gating dynamics in a Na^+-coupled neurotransmitter transporter homologue. *Nature* 474:109–113.

Zomot, E., Bendahan, A., Quick, M., Zhao, Y., Javitch, J.A., and Kanner, B.I. 2007. Mechanism of chloride interaction with neurotransmitter:sodium symporters. *Nature* 449:726–730.

15 GABA and Glutamic Acid Receptor and Transporter Ligands

Bente Frølund and Lennart Bunch

CONTENTS

15.1 INTRODUCTION

γ-Aminobutyric acid (GABA [**15.1**]) is the major inhibitory neurotransmitter while (*S*)-glutamic acid (Glu [**15.2**]) is the major excitatory neurotransmitter in the central nervous system (CNS). Given the fact that the majority of central neurons are under inhibitory and excitatory control by GABA and Glu, the balance between the activities of the two is of utmost importance for CNS functions. Both the neurotransmitter systems are involved in regulation of a variety of physiological

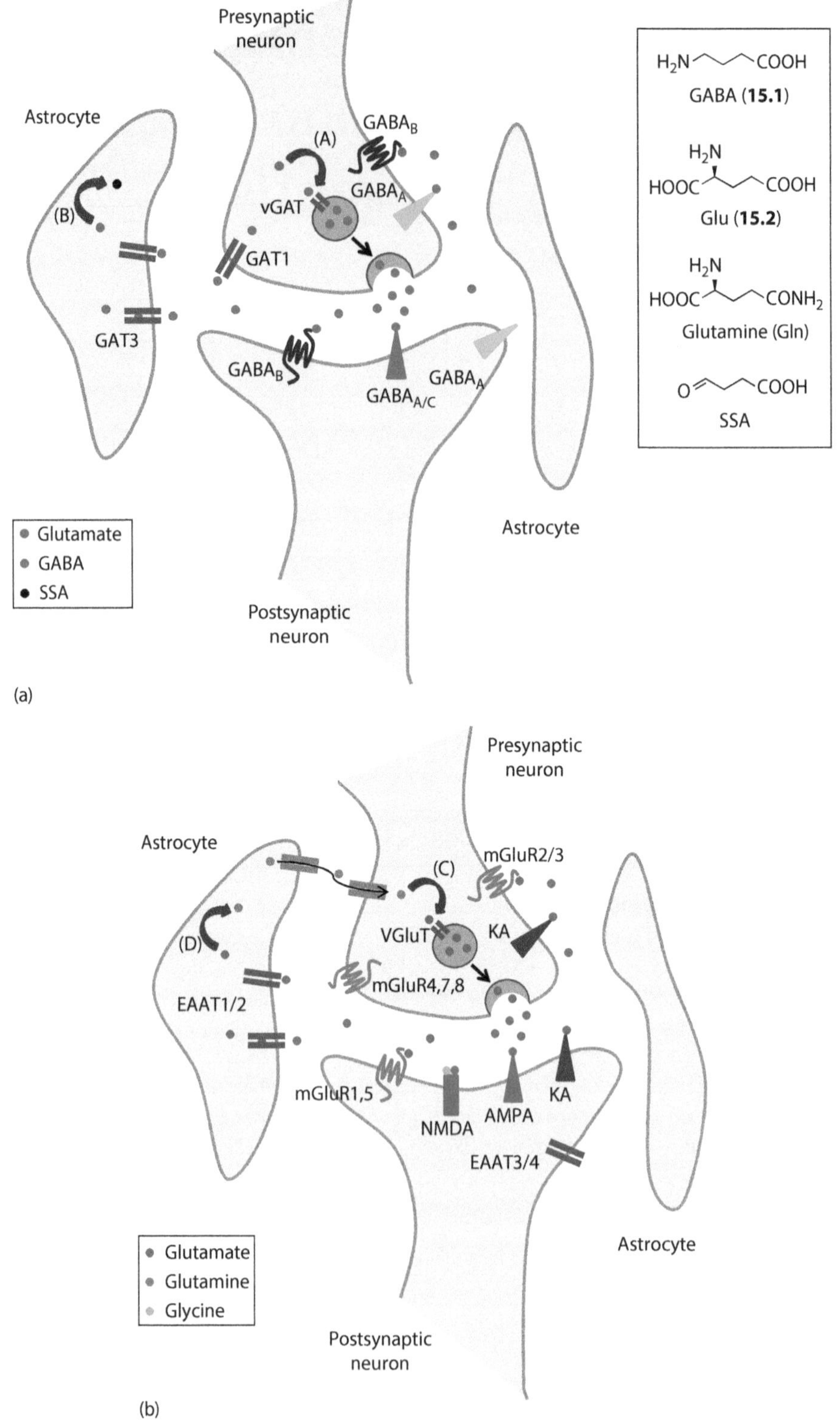

FIGURE 15.1 Schematic illustration of the biochemical pathways, transport mechanisms, and receptors at GABA (a) and Glu (b) operated neurons. Enzymes are indicated by the following: (A) L-glutamic acid decarboxylase (GAD); (B) GABA aminotransferase (GABA-AT); (C) glutaminase; and (D) glutamine synthase.

mechanisms, and dysfunctions of either of the two can be related to various neurological disorders in the CNS.

The transmission processes mediated by Glu and GABA are very complex and highly regulated. A general and simple model for the Glu and GABA neurotransmission is shown in Figure 15.1. Glu and GABA are formed in their respective presynaptic nerve terminals and upon depolarization released into the synaptic cleft in high concentration to activate postsynaptic ionotropic receptors that directly modify the membrane potential of the receptive neuron, generating an excitatory or inhibitory postsynaptic potential. This basic system is further modulated through G-protein-coupled receptors (GPCRs) for a variety of neuroactive substances, including Glu and GABA themselves. Subsequently, Glu and GABA are removed from the synaptic cleft into surrounding neurons and glia cells via specialized transporters to restore the neurotransmitter balance. The reuptaken Glu is enzymatically metabolized to form glutamine (Gln) or α-ketoglutarate, whereas GABA is converted to succinic acid semialdehyde.

15.2 THERAPEUTIC PROSPECTS FOR GABA AND GLUTAMIC ACID NEUROTRANSMITTER SYSTEMS

The therapeutic potential of manipulating the Glu/GABA neurotransmitter systems is large, and virtually all of the known components of the GABA and Glu neurotransmitter systems may be considered as *therapeutic targets*. Diseases include neurodegenerative disorders, e.g., Alzheimer's disease, Parkinson's disease, Huntington's chorea, epilepsy and stroke, and other psychiatric/neurological disorders, such as schizophrenia, depression, anxiety, and pain. Furthermore, cognitive enhancement, narcolepsy, spasticity, muscle relaxation, and insomnia are among the vast number of therapeutic possibilities.

GABA-based therapeutics have been in clinical use since the 1960s, where the most successful therapeutic application to date involves enhancement of the inhibitory GABA activity by modulation of the ionotropic $GABA_A$ receptor, notably by benzodiazepines (BZDs) and barbiturates. Vigabatrin, a suicide inhibitor for the enzyme GABA aminotransferase (GABA-AT) responsible for GABA degradation, is used clinically as an anticonvulsant. Elevation in extracellular GABA level by inhibition of the transport of GABA by the GABA transporter (GAT) is effected by tiagabine (**15.6**) (Figure 15.2) marketed for the treatment of epilepsy and in preclinical studies for the treatment of anxiety and insomnia. The *G*-protein-coupled $GABA_B$ receptor is the target for the antispastic drug baclofen (**15.24**) (Figure 15.8).

Drugs targeting the Glu neurotransmitter system have been slower to emerge. Memantine (see Section 15.6.3) is used with some success for the symptomatic treatment of Alzheimer's disease and a few compounds with mixed mechanisms of action, including reduction of Glu release (through blockade of Na channels), are used for the treatment of migraine, epilepsy, and amyotrophic lateral sclerosis.

Glu and GABA receptors and transporters are heterogeneous and may individually be involved in specific CNS disorders and disease conditions. These receptor/transporter subtypes are unevenly distributed in the CNS which opens up the prospect of developing ligands selective for receptor/transporter subtypes with predominant location in different brain regions of therapeutic relevance.

15.3 GABA TRANSPORT

The GATs belong to the family of Na^+/Cl^--dependent transporters (SLC-6 gene family) that also include transporters for the neurotransmitters dopamine, serotonin, norepinephrine, and glycine (see Chapter 14). Four subtypes of GATs have been cloned with inconsistent nomenclature among species. Thus, rat and human GAT-1, BGT-1, GAT-2, and GAT-3 correspond to mouse mGAT1, mGAT2, mGAT3, and mGAT4, respectively. The most abundantly expressed subtype in the mammalian brain is GAT-1 which accounts for 75%–85% of synaptosomal GABA uptake in mice.

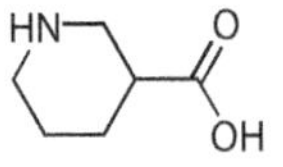

Nipecotic acid (**15.3**)

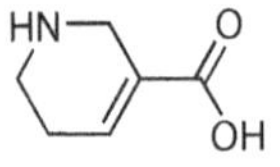

Guvacine (**15.4**)

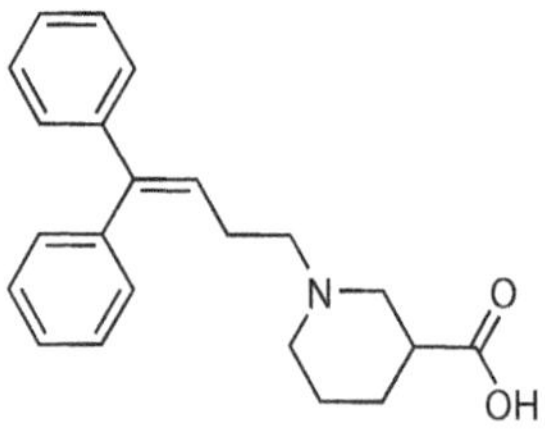

N-DPB-nipecotic acid (**15.5**)

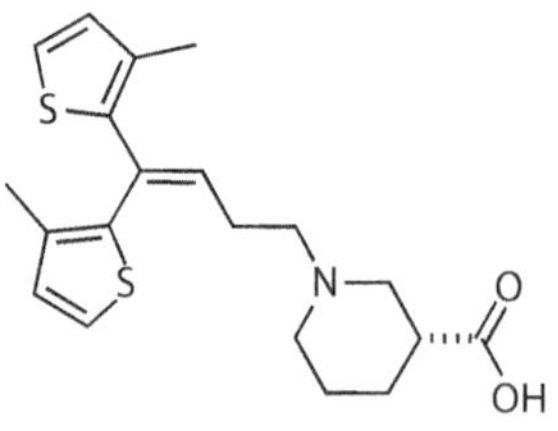

Tiagabine (**15.6**)

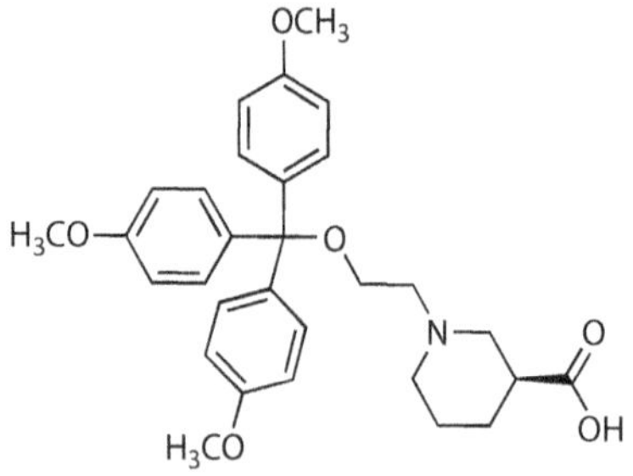

(S)-SNAP-5114 (**15.7**)

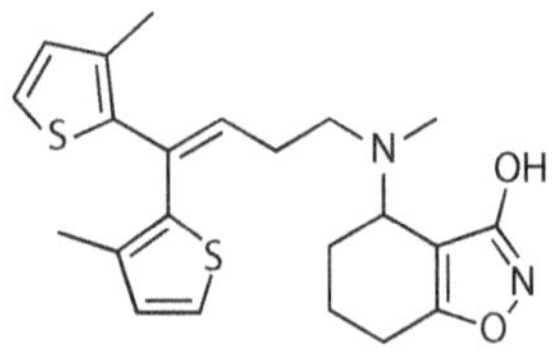

EF-1502 (**15.8**)

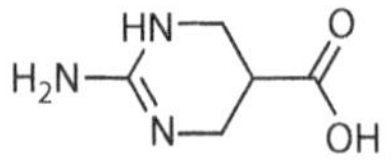

15.9

FIGURE 15.2 Structures of some GABA transport inhibitors.

15.3.1 Inhibitors of GABA Transport

Inhibition of GATs constitutes an attractive approach to increase overall GABA neurotransmission as it increases the synaptic GABA concentration. Due to its abundance, the focus of GAT research has been directed to GAT-1, leading to the development of the first antiepileptic drug, tiagabine (**15.6**), specifically targeting a GAT.

The cyclic GABA analogs nipecotic acid (**15.3**) and guvacine (**15.4**), competitive inhibitors and substrates for the GABA uptake, have been important lead structures for the development of a large number of lipophilic GABA uptake inhibitors. Introduction of a lipophilic moiety such as 4,4-diphenyl-3-butenyl (DPB) on the nitrogen atom led to *N*-DPB-nipecotic acid (**15.5**) and related analogs which are markedly more potent than the parent amino acids. These lipophilic compounds are able to cross the blood–brain barrier (BBB) and are potent anticonvulsants in animal models. Tiagabine (**15.6**), a structurally related compound, is now approved for the treatment of partial epileptic seizure. The clinical use of tiagabine is, however, limited owing to substantial side effects such as agitation, sedation, and psychotic-like episodes in patients predisposed to psychiatric illness. Interest in non-GAT-1 subtypes, especially GAT-3 and BGT-1, has therefore emerged.

Based on evolutionary relationship, BGT-1, GAT-2, and GAT-3 are mutually more closely related to each other than they are to GAT-1, challenging the development of non-GAT-1 subtype-selective inhibitors. Some inhibitors with partial and/or modest subtype selectivity have been developed: (*S*)-SNAP-5114 (**15.7**) shows selectivity toward GAT-3, although with some activity at BGT-1 and GAT-2, and the GAT-1/BGT-1-selective inhibitor EF-1502 (**15.8**), which has shown anticonvulsive effects in animal models of epilepsy, thereby indicating a possible role for BGT-1 as a therapeutic target. Bioisosteric replacement of the amino group in nipecotic acid by a guanidine group converted the GAT-1 selective nipecotic acid into a highly BGT-1 selective inhibitor (**15.9**), demonstrating that selectivity toward BGT-1 can be achieved.

15.4 GABA RECEPTORS AND THEIR LIGANDS

GABA exerts its effects on the CNS via activation of two different classes of GABA receptors: the ionotropic $GABA_A$ and $GABA_C$ receptors, mediating the fast synaptic transmission and the G-protein-coupled $GABA_B$ receptors, mediating a slower response via coupling to second messenger cascades.

15.4.1 Ionotropic GABA Receptors

The ionotropic GABA receptors belong to a superfamily of ligand-gated ion channels (Cys-loop receptors) that also includes the nicotinic acetylcholine, the glycine, and the serotonin (5-HT_3) receptors (see Chapter 12). The members of the Cys-loop superfamily share a common structural arrangement with five subunits encircling a central ion conducting pore. It contains an orthosteric binding site for GABA in each of the two α_1:β_2 interfaces and sites for allosteric modulators including the therapeutically important BZDs in the α_1:γ_2 interface. From a pool of 19 different human subunits (α_{1-6}, β_{1-3}, γ_{1-3}, δ, ε, π, θ and ρ_{1-3}), at least 26 native and mainly heteromeric $GABA_A$ receptor subtypes have been proposed. The expression of different subtypes varies greatly between brain regions, and the different functional properties of the individual receptor subtypes suggest involvement in a wide range of functions throughout the CNS. Also, the subunit compositions of the $GABA_A$ receptors appear to differ with different synaptic localizations and also seem to determine the type of inhibition mediated. Among synaptic $GABA_A$ receptors, which are predominantly composed of α_1, α_2, and/or α_3 in combination with β_2/β_3, and γ_2 subunits, the $\alpha_1\beta_2\gamma_2$ subtype is the predominant subtype. Synaptic $GABA_A$ receptors are suited to transduce the brief synaptic GABA transients into relatively short-lasting phasic GABA currents. Perisynaptic/extrasynaptic $GABA_A$ receptors are tailored to respond to the lower, slowly varying GABA concentrations surrounding them, leading to a persistent tonic current and are therefore more sensitive to GABA and less prone to desensitize than their synaptic counterparts. The bulk of tonic current is generated by $GABA_A$ receptors that contain either the δ subunit or the α5 subunit. $GABA_A$ receptors composed of ρ_{1-3} subunits assemble as homo- or pseudohomomers, also known as the $GABA_C$ receptors, and are expressed mainly in retina (Figure 15.3).

So far, the understanding of the molecular architecture of the orthosteric binding site at the ionotropic GABA receptors has to a large extent been based on crystal structures of acetylcholine binding

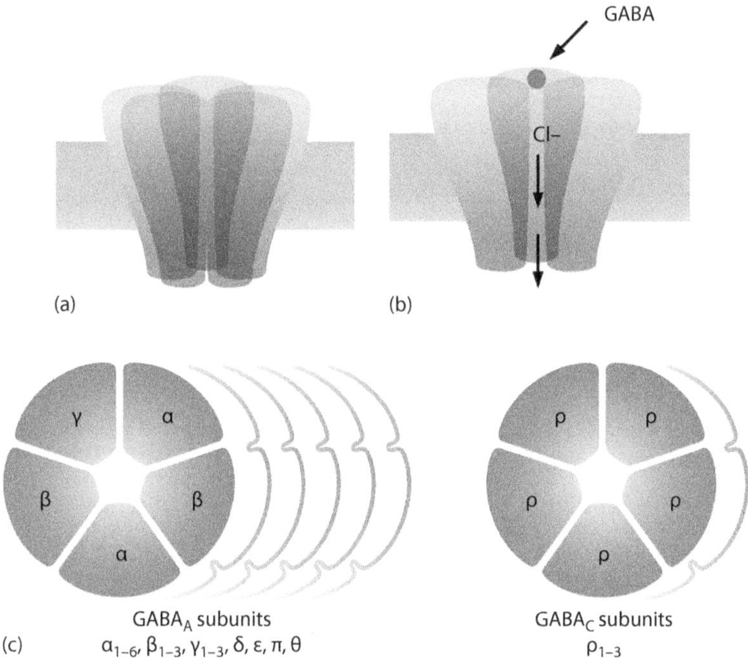

FIGURE 15.3 Schematic illustrations of (a) the pentameric structure of the ionotropic GABA receptors, (b) with indication of the GABA binding site and the chloride ion channel, (c) and the multiplicity of ionotropic $GABA_A$ and $GABA_C$ receptors.

proteins (AChBP) from snail and bacterial ion channels. Later, co-crystallization of glutamate with an eukaryotic glutamate-gated chloride channel (GluCl) was reported. The GluCl structure shares sequence identities with α_1, β_2, and γ_2 $GABA_A$ receptor subunits of 30%, 36%, and 31%, respectively. Locally, in the binding site region, the identity is as high as 48%. Recently, a high-resolution X-ray structure of a human β_3 homopentameric $GABA_A$ receptor was reported improving the structural basis for homology modeling.

Homology models have been updated continuously since the release of the first structures of the AChBP to recapitulate as much structural information from the available templates as possible. Such homology models offer an insight into the identities of the residues lining the binding pockets in the respective receptors. However, it is not straightforward to use these models for the prediction of ligand affinity.

15.4.2 Ionotropic GABA Receptor Ligands

The GABA binding site has very distinct and specific structural requirements for recognition and activation. Thus, only a few different structural classes of molecules have been reported. Within the series of compounds showing agonist activity at the $GABA_A$ receptor site are the selective agonists muscimol (**15.10**) and 4,5,6,7-tetrahydroisoxazolo[5,4-*c*]pyridine-3-ol (THIP) (**15.11**) which have been used for the pharmacological characterization of the $GABA_A$ receptors. Bicuculline methochloride (BMC) (**15.12**) and SR95531 (**15.13**) are the classical $GABA_A$ receptor antagonists where the structural diversity is very obvious (Figure 15.4).

So far, the structural insight into the agonist and antagonist binding at $GABA_A$ receptors has been guided by pharmacologically derived pharmacophore models and, more recently, these models have facilitated the development of receptor models compatible with structure-activity relationship (SAR) studies. On the basis of a hypothesis originating from the bioactive conformation of muscimol,

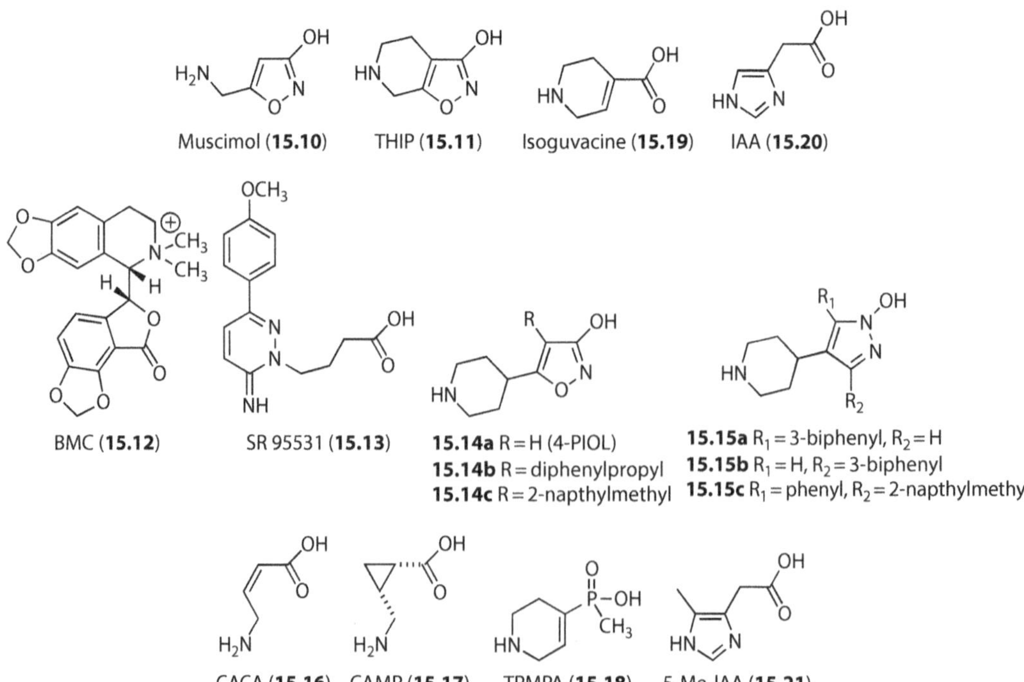

FIGURE 15.4 Structures of $GABA_A$ agonists (upper row), $GABA_A$ antagonists (middle row), and $GABA_C$ ligands (lower row).

the partial $GABA_A$ agonist 4-PIOL (**15.14a**), and on pharmacological data for an additional series of $GABA_A$ ligands, a simple 3D pharmacophore model for the orthosteric $GABA_A$ receptor ligands has been developed. The main features of this model are that the 3-hydroxyisoxazol rings of muscimol and 4-PIOL do not overlap in their proposed binding modes and that the two compounds interact with different conformations of an arginine residue (Arg66) located at the $GABA_A$ recognition site. The space surrounding the ligands has been defined and the existence of a cavity of considerable dimensions in the vicinity of the 4-position of the 3-hydroxyisoxazol moiety in the structure of 4-PIOL has been identified, whereas the corresponding position in muscimol is identified as sterically intolerable receptor region (Figure 15.5).

Based on this model, a series of selective and highly potent competitive antagonists have been developed including compounds **15.14b,c**. To optimize the antagonist model, 1-hydroxypyrazole analogs of 4-PIOL were used to guide placement in the binding pocket. In particular, 3-biphenyl analogs **15.15a** and **15.15b** were helpful as the placement of the large hydrophobic groups in the binding pocket essentially dictates orientation of the compounds. The structure of the 3-biphenyl compounds leads to a hypothesis suggesting hydrophobic cavities on either side of the core 1-hydroxypyrazole scaffold, which was supported by the pharmacological profile of a disubstituted analog **15.15c**, displaying retained activity compared to the monosubstituted analogs (Figure 15.6).

Since the β_3 and β_2 subunits are identical with respect to residues pointing toward the principal side of the binding pocket and as the complementary side of β_3 and α_1 only differ with respect to a few residues, a comparison of the binding sites of the optimized homology models and the human β_3 homopentameric $GABA_A$ receptor may provide a validation of the models. Despite the different ligands in models versus the X-ray structure, which makes direct comparison difficult given the known flexibility of the receptor, the models generally are in good agreement with the X-ray structure.

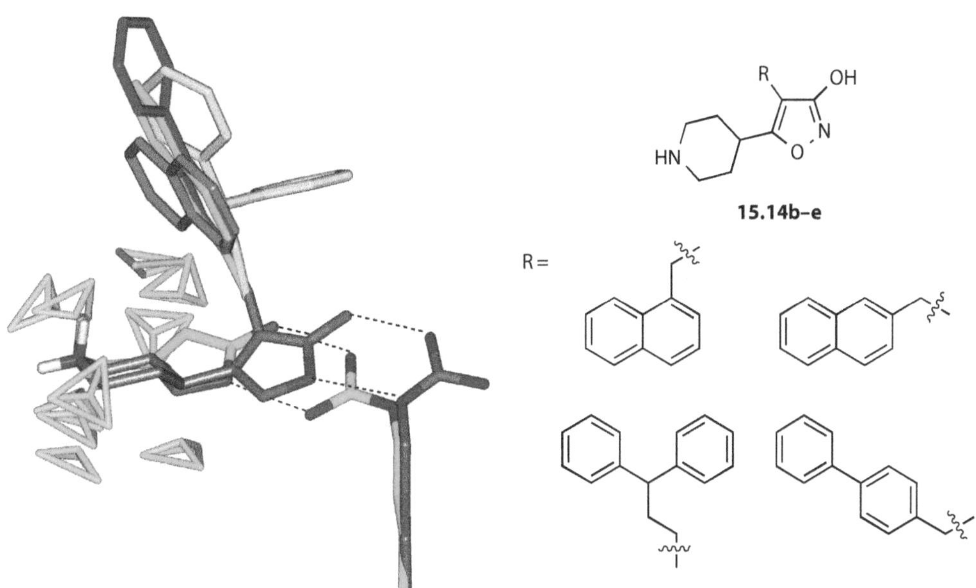

FIGURE 15.5 A superimposition of the proposed bioactive conformations of muscimol (**15.10**, green carbon atoms) and 4-PIOL (**15.14a**, gray carbon atoms) binding to two different conformations of an arginine residue at the orthosteric binding site. A series of 4-substituted 4-PIOL compounds (**15.14a–e**) are included illustrating the large space spanned by the 4-substituents. The tetrahedrons indicate receptor excluded volumes.

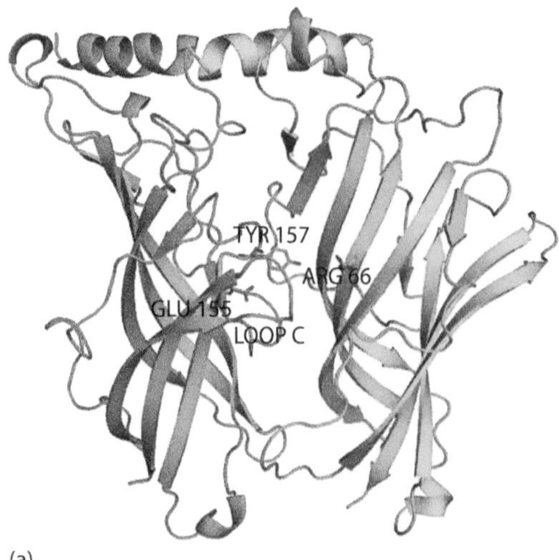

(a)

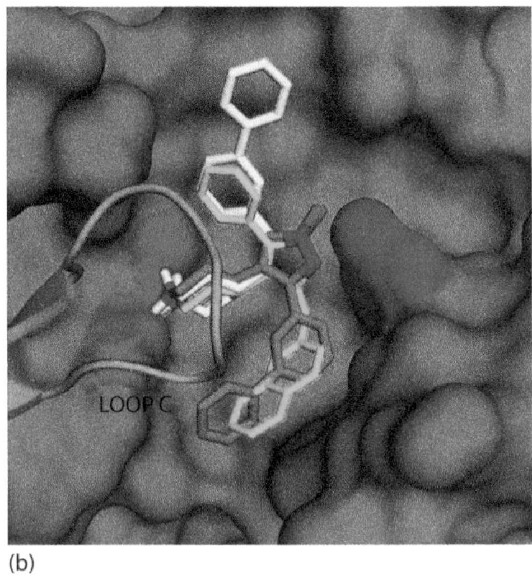

(b)

FIGURE 15.6 (a) Position of the orthosteric $GABA_AR$ binding site in the interface between the β (turquoise color) and α (olive green color) subunits with the important residues β_2-Glu155, β_2-Tyr157, α_1-Arg66, and loop C highlighted. (b) Overlaid binding poses for 1-hydroxypyrazole analogs of 4-PIOL: **15.15a** (pale yellow carbons), **15.15b** (dark salmon carbons), and **15.15c** (green carbons) shown within the surface of the orthosteric $GABA_AR$ binding site with negatively charged (hydrogen bond accepting) areas of the surface colored red and positively charged (hydrogen bond donating) areas of the surface colored blue.

In contrast, structure–activity studies of ligands targeting the $GABA_C$ receptors have been very limited. *cis*-4-Aminocrotonic acid (CACA [**15.16**]) (Figure 15.4) has been the key ligand for the identification of the $GABA_C$ receptors. The compound is a moderately potent partial $GABA_C$ agonist and inactive at $GABA_A$ receptors, but it has been shown to affect GABA transport as well. In the search for selective $GABA_C$ receptor ligands, the folded conformation of CACA has been used as a scaffold for new compounds such as *cis*-2-aminomethylcyclopropanecarboxylic acid (CAMP [**15.17**]). (+)-CAMP has been reported to be a selective $GABA_C$ receptor agonist with potency in the mid-micromolar range, displaying only weak activity at the $GABA_A$ receptors. Finally, the first antagonist capable of differentiating the $GABA_C$ receptors from both $GABA_A$ and $GABA_B$ receptors was (1,2,5,6-tetrahydropyridin-4-yl)methylphosphinic acid (TPMPA) (**15.18**).

The overall molecular architecture of the orthosteric sites at the $GABA_A$ and $GABA_C$ receptors appear to be quite similar as most $GABA_A$ agonists display some agonist/antagonist activities at $GABA_C$ receptors as well. THIP (**15.11**), the standard $GABA_A$ agonist, has been shown to be a partial agonist at $GABA_A$ receptors and a competitive antagonist at $GABA_C$ receptors. Likewise, the $GABA_A$ agonists muscimol (**15.10**), isoguvacine (**15.19**), and imidazol-4-acetic acid (IAA [**15.20**]) act as partial $GABA_C$ agonists. However, the fact that $GABA_A$ and $GABA_C$ receptors exhibit distinct antagonist selectivity clearly indicates that orthosteric sites of these receptors are not identical. This is further supported by the introduction of a methyl group in the 5-positon of IAA (5-Me-IAA [**15.21**]), which did separate the dual effect of IAA maintaining the agonist properties at ρ_1 $GABA_C$ receptors, whereas no activity at the $\alpha_1\beta_2\gamma_{2S}$ $GABA_A$ receptors was seen.

Enhancement of the inhibitory GABA activity would, in general, be beneficial in various conditions, including epilepsy, pain, anxiety, and insomnia. Direct activation of the ionotropic GABA receptors using $GABA_A$ agonists has for long not been anticipated as a useful therapeutic approach due to desensitization of the receptors. However, the partial $GABA_A$ agonist THIP has proven to be a potential drug in the treatment of insomnia, but a phase III clinical trial was discontinued due to unexpected side effects, including hallucinations and disorientation.

15.4.3 Modulatory Agents for the $GABA_A$ Receptor Complex

The $GABA_A$ receptor complex is the target for a large number of structurally diverse compounds, some of which are pharmacologically active and used clinically. These compounds include BZD, ethanol, general anesthetics, barbiturates, and neuroactive steroids, all of which act via a wide range of distinct allosteric binding sites within the pentameric receptor complex (Figure 15.7). The allosteric modulators exert their effects by binding to the $GABA_A$ receptor complex and affect GABA-gated chloride conductance. This modulation only takes place when GABA is present in the synaptic cleft which could be preferable rather than general receptor activation by exposure to a GABA agonist. Compounds within this group of modulators are marketed for the treatment of anxiety, epilepsy, insomnia, muscle relaxation, and anesthesia. Preclinical studies are going on with the focus on cognitive enhancement and schizophrenia as well.

Although multiple binding sites make $GABA_A$ receptor pharmacology complex, the complexity offers numerous possibilities of fine-tuned pharmacological interventions. The fact that receptor regions targeted by allosteric ligands typically are less conserved than the orthosteric sites in general opens up for development of subtype selective modulators with more specific therapeutic effects and reduced side effects.

Among the modulatory sites at the $GABA_A$ receptor complex, the BZD site is the best understood based upon not only the proven clinical efficacy of compounds acting at this site but also the availability of pharmacological tool compounds as well as genetically modified mice. The pharmacological profiles of ligands binding to the BZD site span the entire continuum from full and partial agonists, through antagonists, to partial and full inverse agonists (see Chapter 12). Antagonists do not influence the GABA-induced chloride flux but antagonizes the action of BZD site agonists as well as of inverse agonists.

Since the use of classical BZDs is associated with side effects and the risk of drug dependence, efforts have been put into development of more subtype-selective ligands for the modulatory sites with more specific pharmacological profiles compared to that of the classical BZDs. Knowledge obtained by correlating the pharmacological features of diazepam to the different $GABA_A$ receptor subtypes using genetically modified mice has formed the basis for the strategy taken in drug development within the BZD site area. At risk of oversimplification of the role of the individual receptor

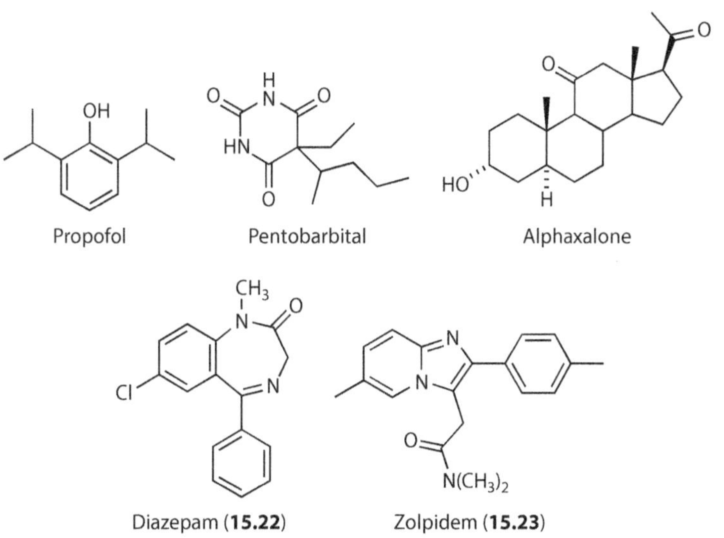

FIGURE 15.7 Structures of different classes of compounds known to bind to the various recognition sites associated with the $GABA_A$ receptor. Propofol (anesthetic), pentobarbital (barbiturate), alphaxalone (synthetic steroid), and the BZD site ligands diazepam and zolpidem.

subtypes, research based on this knowledge is focused on development of hypnotics (α_1 selective), nonsedating anxiolytics (α_2 and α_3 selective), antipsychotics (α_3 selective), and cognition-enhancement (α_5-selective inverse agonist) combined with the correlation of the α_1 subunit to sedation and addiction liability. In spite of intensive efforts in this area, where the critical issue seems to be the translation of the preclinical pharmacological profile into clinical utility, unselective BZD ligands like diazepam (**15.22**) and a few α_1-preferring ligands, including zolpidem (**15.23**) are still the most important BZD ligands on the market.

15.4.4 GABA$_B$ Receptor Ligands

The GABA$_B$ receptors belong to the subfamily C of GPCRs which also comprise the G-protein-coupled Glu receptors (see later sections and Chapter 12). The GABA$_B$ receptors exist as heterodimers consisting of two subunits, GABA$_{B1}$ and GABA$_{B2}$. The former contains the GABA binding domain, whereas GABA$_{B2}$ provides the G-protein-coupling mechanism. The diversity in this class of receptors arises from the two GABA$_{B1}$ splice variants, GABA$_{B1a}$ and GABA$_{B1b}$ which together with GABA$_{B2}$ form the two physiological receptors. Activation of the GPCR causes a decrease of calcium, an increase in potassium membrane conductance, and inhibition of cAMP formation. The resulting response is thus inhibitory and leads to hyperpolarization and decreased neurotransmitter release. However, activation of GABA$_B$ receptors also reduces neuronal excitability via activation of G-protein-regulated inwardly rectifying K^+ channels. Finally, GABA$_B$ receptors regulate intracellular signaling by inhibiting adenylyl cyclase activity. Predominant effects of GABA$_B$ agonists are muscle relaxation, but in addition various neurological and psychiatric disorders, including neuropathic pain, anxiety, depression, absence epilepsy, alcoholism, and drug abuse, are targets for GABA$_B$ agonist therapy.

Most of the existing structure–activity data for the GABA$_B$ receptor GABA binding site apply to understanding the role of substituents of the backbone of GABA as well as replacement of the carboxylic or the amine group. The GABA$_B$ receptors are selectively activated by baclofen (**15.24**), of which the (*R*)-form is the active enantiomer. Baclofen was developed as a lipophilic derivative of GABA, in an attempt to enhance the BBB penetrability of the endogenous ligand. Despite its approval for the treatment of muscle spasticity and a growing number of off-label and potential uses, baclofen is far from being an ideal drug molecule. Baclofen has a low penetration into the brain, a short duration of action, a narrow therapeutic window, and a rapid induction of drug tolerance.

In an attempt to improve the pharmacokinetic properties, bioisosteric replacement of the carboxylic acid in baclofen led to phaclofen (**15.25**) and saclofen (**15.26**), the phosphonic acid and sulfonic acid analogs of baclofen, respectively which were the first GABA$_B$ antagonists reported. In contrast, the phosphinic acid GABA bioisostere, CGP27492 (**15.27**), is a potent GABA$_B$ agonist, being approximately 10-fold more potent than GABA. Introduction of substituents in the 3-position of CGP27492 is well tolerated, where the 3-fluoro analog, lesogaberan (**15.28**), functions as a highly potent GABA$_B$ agonist 87-fold more active in human recombinant GABA$_B$ receptors compared to racemic baclofen. Interestingly, this is in contrast to its carboxylic acid equivalent which shows no GABA$_B$ receptor activity.

In an attempt to improve the pharmacology and pharmacokinetics of the phosphinic acid GABA$_B$ agonists mentioned earlier, a series of selective and highly potent GABA$_B$ antagonists was discovered. Among these, the compounds CGP35348 (**15.29**) and CGP55845 (**15.30**) were developed and were shown to be capable of penetrating the BBB after systemic administration.

In contrast to the GABA$_A$ receptor, allosteric modulators for the GABA$_B$ receptor are relatively less developed. The allosteric binding site(s) and structure–activity relationships are less well defined; however, multiple classes of molecules that act as positive allosteric modulators (PAMs) have been discovered. Three major structural classes of PAMs for the GABA$_B$ receptor have been described in the literature, illustrated by the pyrimidine GS39783 (**15.31**), the di-*tert*-butylbenzene

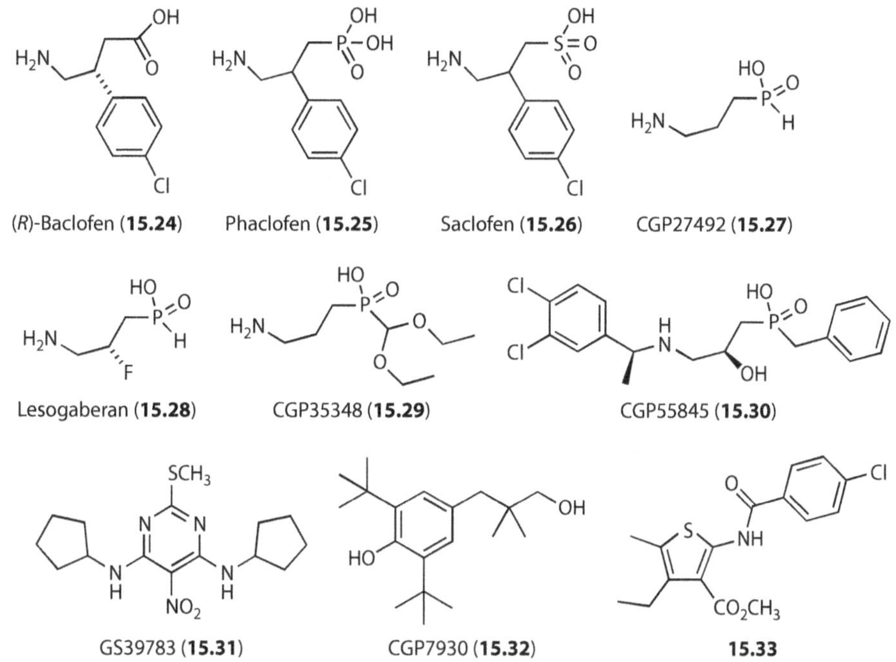

(*R*)-Baclofen (**15.24**) Phaclofen (**15.25**) Saclofen (**15.26**) CGP27492 (**15.27**)

Lesogaberan (**15.28**) CGP35348 (**15.29**) CGP55845 (**15.30**)

GS39783 (**15.31**) CGP7930 (**15.32**) **15.33**

FIGURE 15.8 Structures of $GABA_B$ receptor agonists (upper row), $GABA_B$ receptor antagonists (middle row), and allosteric modulators (lower row).

CGP7930 (**15.32**) and the thiophene **15.33**. The identification of new PAMs of the $GABA_B$ receptor continues to be an area of substantial interest due to the therapeutic potential and the openings for increased propensity for long-term use in relation to receptor desensitization and tolerance (Figure 15.8).

The release of the X-ray structure in 2013 of the $GABA_B$ receptor, along with co-crystallized agonists and antagonists, has established a platform for the use of state-of-the-art protein structure-based drug design tools to interpret the existing structure–activity data, to reveal the localization of allosteric binding site(s) and to discover new ligands.

15.5 GLUTAMATE: THE MAJOR EXCITATORY NEUROTRANSMITTER

Glu belongs to the group of 20 common α-amino acids and is thus a key constituent of peptides and proteins. In the CNS, it was recognized as the major excitatory neurotransmitter in the early 1980s. Since then, a plethora of Glu receptors and transporters have been identified and their expression in the CNS found to be high and ubiquitous. While glutamatergic neurotransmission is fundamental for all CNS activity, *excessive excitation* may lead to neuronal death. Given this incident, it is believed that either too low or too high glutamatergic activity may be implicated in a wide number of neurological as well as psychiatric diseases.

15.5.1 Receptor Classification

The Glu receptors (Figure 15.1) are divided into two main classes, the fast-acting ionotropic and the G-protein-coupled metabotropic Glu receptors (iGluRs and mGluRs, respectively). The iGluRs are further divided into three groups based on activation of distinct agonist *N*-methyl-D-aspartic acid (NMDA, **15.34**), 2-amino-3-(3-hydroxy-5-methyl-4-isoxazolyl)propionic acid (AMPA, **15.56**), and kainic acid (KA, **15.73**) receptors. Within each group, a number of subunits have been identified. For the NMDA receptors, these are abbreviated GluN1, GluN2A-D, and GluN3A,B; for the AMPA receptors, GluA1–4 and for the KA receptors, GluK1–5 (see phylogenetic tree in Figure 12.1). Functional iGluRs are formed by four of the aforementioned grouped subunits coming together to form an ion channel in the center. The iGluRs are referred to as being tetrameric in structure. Upon binding of Glu, opening of the ion channel leads to flux of Na^+, K^+, and Ca^{2+} ions and thus depolarization of the cell membrane and excitation of the neuron. The mGluRs modulates the activity of neurons and are GPCRs, also named 7TM receptors as described in Chapter 12. The mGluRs are divided into groups I, II, and III based on pharmacology, signal transduction mechanisms, and amino acid sequences. Group I consists of mGluR1,5 and stimulates phospholipase C, whereas groups II (mGluR2,3) and III (mGluR4,6–8) inhibit the formation of cAMP.

15.6 IONOTROPIC GLUTAMATE RECEPTOR LIGANDS

The development of selective ligands for the plethora of iGluRs is highly attractive as such compounds can be used to study the role and function of each iGlu receptor subtype. A key strategy is structure-based design due to the availability of a large number of X-ray structures of ligands crystallized with the iGlu ligand binding domain (LBD). X-ray structures deposited in the Protein Data Bank include not only the apo structure (without a ligand), Glu itself, but also a large number of agonists, antagonists, as well as partial agonists and modulatory ligands (see also Figures 2.7 and 12.7).

15.6.1 NMDA Receptor Ligands

NMDA receptors are unique in the sense that the concurrent binding of the co-agonist glycine (**15.49**) is required for receptor activation, and also membrane depolarization to release ion channel bound Mg^{2+}. In the brain, NMDA receptors are involved in synaptic plasticity and neuronal signaling processes, including mechanisms of learning and memory. Functional NMDA receptors are heteromeric, typically consisting of two GluN1 and two GluN2 subunits in a tetrameric structure. While the Glu binding site is located on GluN2A-D subunits, the glycine co-agonist binding site resides on the GluN1A,B subunits (Figure 15.9).

The synthetic agonist NMDA (**15.34**) is unique as it has the *R*-configuration at the α-carbon and also comprises an *N*-methyl group which plays a key role in its selectivity profile. Potent NMDA agonists have been developed, particularly by replacement of the distal acidic group and/or by conformational restriction of the three essential functional groups, namely, the α-amino group, the α-carboxyl group, and the ω-acidic moiety. Among the potent NMDA agonists with different distal acidic groups are tetrazolylglycine (**15.35**) and (*R*)-AMAA (**15.36**), exemplifying two widely used carboxyl group bioisosteric groups, the tetrazole and the 3-isoxazolol, respectively. (*R*)-AMAA (**15.36**) and other Glu ligands have been developed using the naturally occurring neurotoxin ibotenic acid (**15.37**) as a lead. Ibotenic acid (**15.37**) is, apart from being a potent NMDA agonist, a potent agonist of some mGluR subtypes and a somewhat weaker agonist at other Glu receptor types.

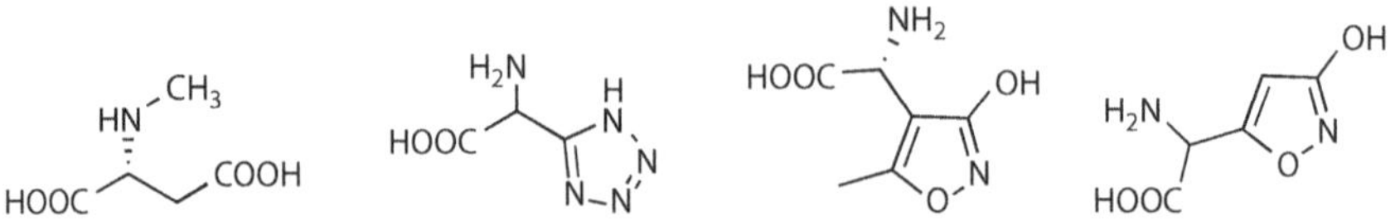

NMDA (**15.34**) Tetrazolylglycine (**15.35**) (*R*)-AMAA (**15.36**) Ibotenic acid (**15.37**)

FIGURE 15.9 Structures of Glu and some NMDA receptor agonists.

15.6.2 Competitive NMDA Receptor Antagonists

A large number of potent and selective competitive NMDA antagonists have been developed, and the availability of these compounds has greatly facilitated studies of the physiological and pathophysiological roles of NMDA receptors. The distance between the two acidic groups in NMDA antagonists is typically one or three C–C bonds longer than in Glu. Many potent ligands have successfully been developed using ω-phosphonic acid analogs such as (*R*)-APV (**15.38**) as lead structures. Combination of an ω-phosphonate group, a long carbon backbone, and conformational restriction has led to different series of potent antagonists. Conformational restriction has been achieved by use of double bonds (CGP 39653 [**15.39**]), ring systems (CGS19755 [**15.40**]), and bicyclic structures (LY235959 [**15.41**]). These antagonists have shown very effective neuroprotective properties in various in vitro models. However, many of these compounds suffer from poor BBB penetration. LY233053 (**15.42**) represents another class of antagonist with a tetrazole ring as the terminal acidic group. Substitution of the tetrazole for a phosphono group has limited effect on in vitro activity but results in improved bioavailability (Figure 15.10).

15.6.3 Uncompetitive and Noncompetitive NMDA Receptor Antagonists

The dissociative anesthetics phencyclidine (PCP) (**15.43**) and ketamine (**15.44**) block the NMDA receptor ion channel in a use-dependent manner. Thus, initial agonist activation of the channel is a prerequisite in order for such uncompetitive antagonists to gain access to the binding site which is found inside the ion channel. The antagonists are eventually trapped within the ion channel and may result in very slow kinetics. MK-801 (**15.45**) has been developed as a very effective uncompetitive NMDA antagonist and has been extensively investigated to probe the therapeutic utility of such compounds, notably for the treatment of ischemic insults such as stroke. MK-801 (**15.45**) and related high-affinity ligands have, however, shown severe side effects, including psychotomimetic effects, neuronal vacuolization, and impairment of learning and memory. Ligands with lower affinity, such as memantine (**15.46**), have shown improved therapeutic indexes. Memantine (**15.46**) is a marketed drug for the treatment of Alzheimer's disease and Parkinson's disease, and may also have potential in the treatment of AIDS dementia. Its fast kinetics and low affinity compared to MK-801 (**15.45**) may explain the absence of severe side effects (Figure 15.11).

Noncompetitive NMDA ligands tend not to bind to the same site as the uncompetitive ligands. Ifenprodil (**15.47**) and CP-101,606 (**15.48**) represent important series of noncompetitive NMDA receptor antagonists. These compounds have been shown positive effects in animal models of ischemia, and also as anticonvulsants and antinociceptive agents.

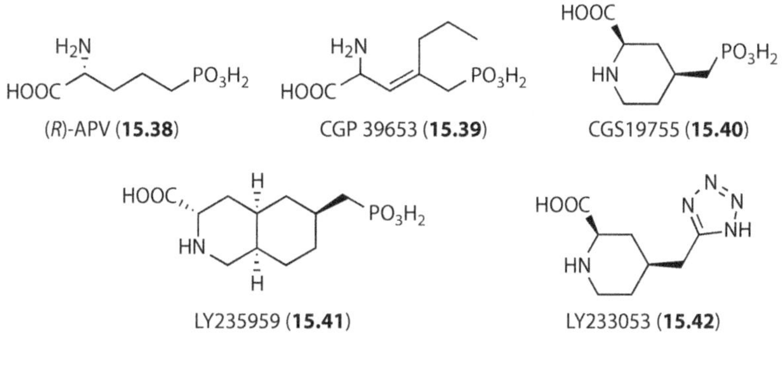

(*R*)-APV (**15.38**) CGP 39653 (**15.39**) CGS19755 (**15.40**)

LY235959 (**15.41**) LY233053 (**15.42**)

FIGURE 15.10 Structures of some competitive NMDA receptor antagonists.

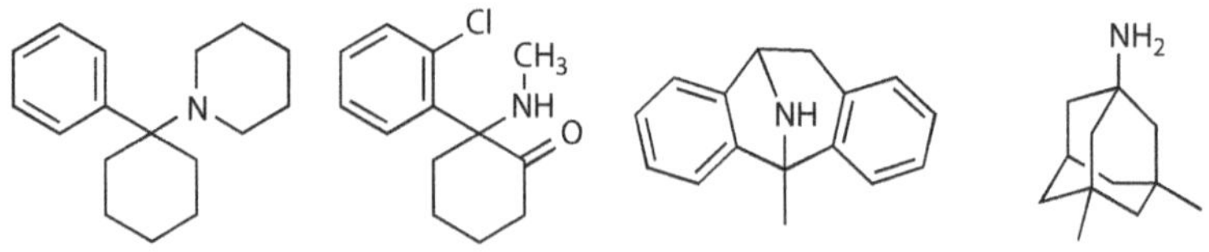

PCP (**15.43**) Ketamine (**15.44**) MK-801 (**15.45**) Memantine (**15.46**)

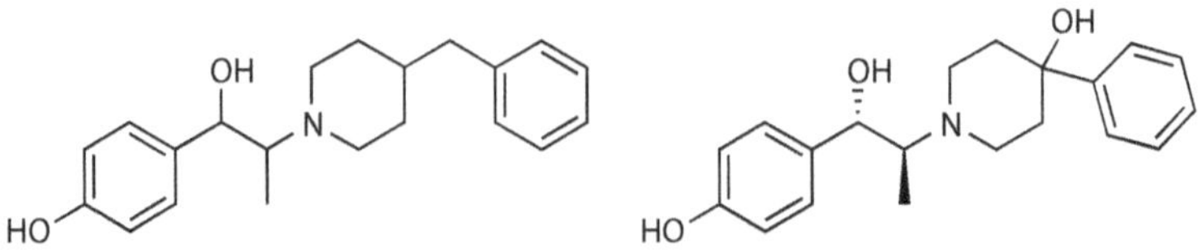

Ifenprodil (**15.47**) CP-101,606 (**15.48**)

FIGURE 15.11 Structures of some uncompetitive (**15.52–15.55**) and noncompetitive (**15.56** and **15.57**) NMDA receptor antagonists.

15.6.4 Glycine Co-Agonist Site

The excitatory co-agonist glycine (**15.49**) binds to the subunit GluN1 at the NMDA receptor. It is also named the glycine$_B$ receptor and is different from the inhibitory glycine receptors found primarily in the spinal cord of the mammalian CNS. The latter are selectively blocked by strychnine and named glycine$_A$ receptors. Glycine$_B$ receptors are believed to modulate the level of NMDA receptor activity. While glycine always is present in the synapse, it is the presynaptic release of Glu which leads to activation of the NMDA receptors. However, it is believed that the synaptic concentration of glycine can modulate NMDA receptor activity and possibly control receptor desensitization. Besides glycine, (*R*)-serine (D-serine) (**15.50**) is a potential endogenous agonist at glycine$_B$ receptors.

The limited success of competitive NMDA receptor antagonists as therapeutic agents has focused attention on the glycine$_B$ site. (*R*)-Cycloserine (**15.51**) and (*R*)-HA-966 (**15.52**) are both partial glycine agonists, capable of penetrating the BBB after systemic administration. (*R*)-Cycloserine (**15.51**) has shown promising effects in the treatment of schizophrenia and Alzheimer's disease, and partial agonists may have therapeutic advantages as compared to full antagonists in terms of reduced side effects. A number of glycine$_B$ antagonists have also been developed. L-689,560 (**15.54**) displays high potency and is derived from the endogenous compound kynurenic acid (**15.53**), the first glycine$_B$ antagonist reported (Figure 15.12).

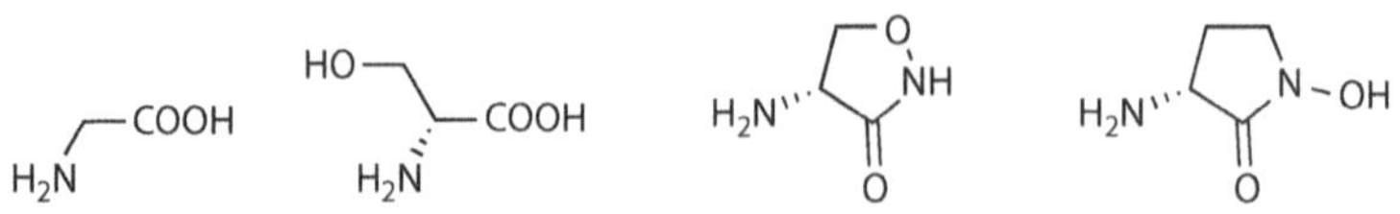

Glycine (**15.49**) (*R*)-Serine (**15.50**) (*R*)-Cycloserine (**15.51**) (*R*)-HA-966 (**15.52**)

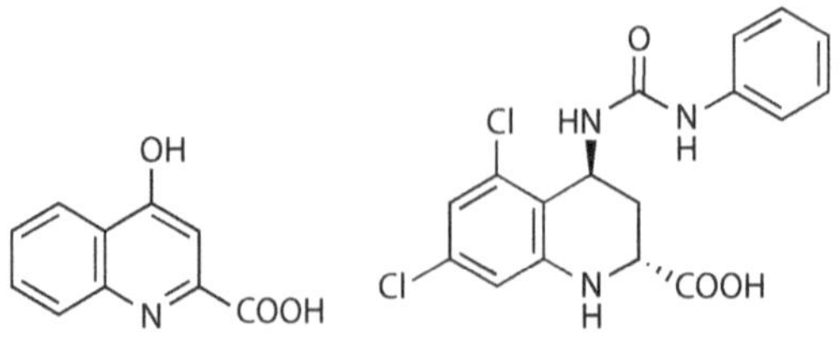

Kynurenic acid (**15.53**) L-689,560 (**15.54**)

FIGURE 15.12 Structures of glycine (**15.49**) and an agonist (**15.50**), two partial agonists (**15.51** and **15.52**) and two antagonists (**15.53** and **15.54**) at the glycine$_B$ receptor on the NMDA receptor complex.

15.6.5 AMPA Receptor Agonists

A large number of selective and potent AMPA receptor agonists have been developed by substituting a heterocyclic bioisosteric group for the distal carboxylate group of Glu. For example, the heterocycles 1,2,4-oxadiazole-3,5-dione, 3-isoxazolol, and uracil, as represented by quisqualic acid (**15.55**), (*S*)-AMPA (**15.56**), and (*S*)-willardiine (**15.57**), respectively, have been incorporated into numerous AMPA receptor agonists. The natural product quisqualic acid (**15.55**) was the first agonist in use for pharmacological characterization of AMPA receptors, but due to its nonselective action over the mGlu receptors, it was later replaced by (*S*)-AMPA (**15.56**) thereof giving the name to this class of receptors.

Conformational restriction of the skeleton of Glu has played an important role in the design of selective GluR ligands. However, only a few structurally rigid AMPA receptor-selective Glu analogs have been reported. One such example is the cyclized analog of AMPA, 5-HPCA (**15.58**). Interestingly, the pharmacological effects of (*R*)-5-HPCA (**15.58**) reside exclusively in the *R*-enantiomer which is in contrast to the *S*-form being the preferred enantiomer among AMPA receptor agonists (Figure 15.13).

15.6.6 Competitive and Noncompetitive AMPA Receptor Antagonists

Early pharmacological studies on AMPA and KA receptors were hampered by the lack of selective and potent antagonists. The discovery of the quinoxaline-2,3-diones such as CNQX (**15.59**) and DNQX (**15.60**) was a breakthrough, but it was not until the AMPA preferring and potent analog NBQX (**15.61**) was discovered that investigation of the pharmacological properties of the AMPA receptors accelerated. NBQX (**15.61**) showed good neuroprotective effect in models of cerebral ischemia; however, it failed in clinical trials because of nephrotoxicity due to a limited aqueous solubility. DNQX (**15.60**) has played a key role in elucidating the binding mode of competitive antagonists as it was the first antagonist to be co-crystallized with the GluR2 LBD. Attempts to improve the aqueous solubility of such antagonists without losing affinity for AMPA receptors by introducing appropriate polar substituents onto the quinoxaline-2,3-dione ring system have been highly successful and have resulted in very potent AMPA receptor antagonists as exemplified by ZK200775 (**15.62**) (Figure 15.14).

Another series of potent and selective competitive AMPA receptor antagonists based on the isantin oxime skeleton includes NS1209 (**15.63**) which shows long-lasting neuroprotection in animal models of ischemia and an increased aqueous solubility compared to NBQX (**15.61**). At least two classes of amino acid–containing compounds, based on decahydroisoquinoline-3-carboxylic acid

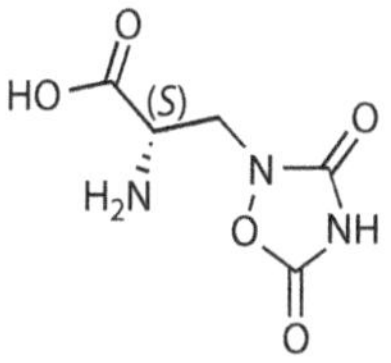

Quisqualic acid (**15.55**) 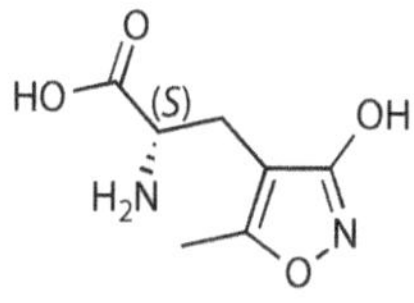 (*S*)-AMPA (**15.56**) 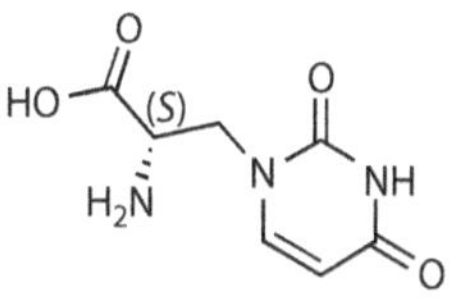 (*S*)-Willardiine (**15.57**)

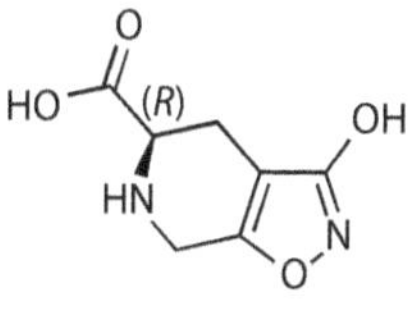

(*R*)-5-HPCA (**15.58**)

FIGURE 15.13 Structures of (*S*)-AMPA (**15.56**) and some AMPA receptor agonists.

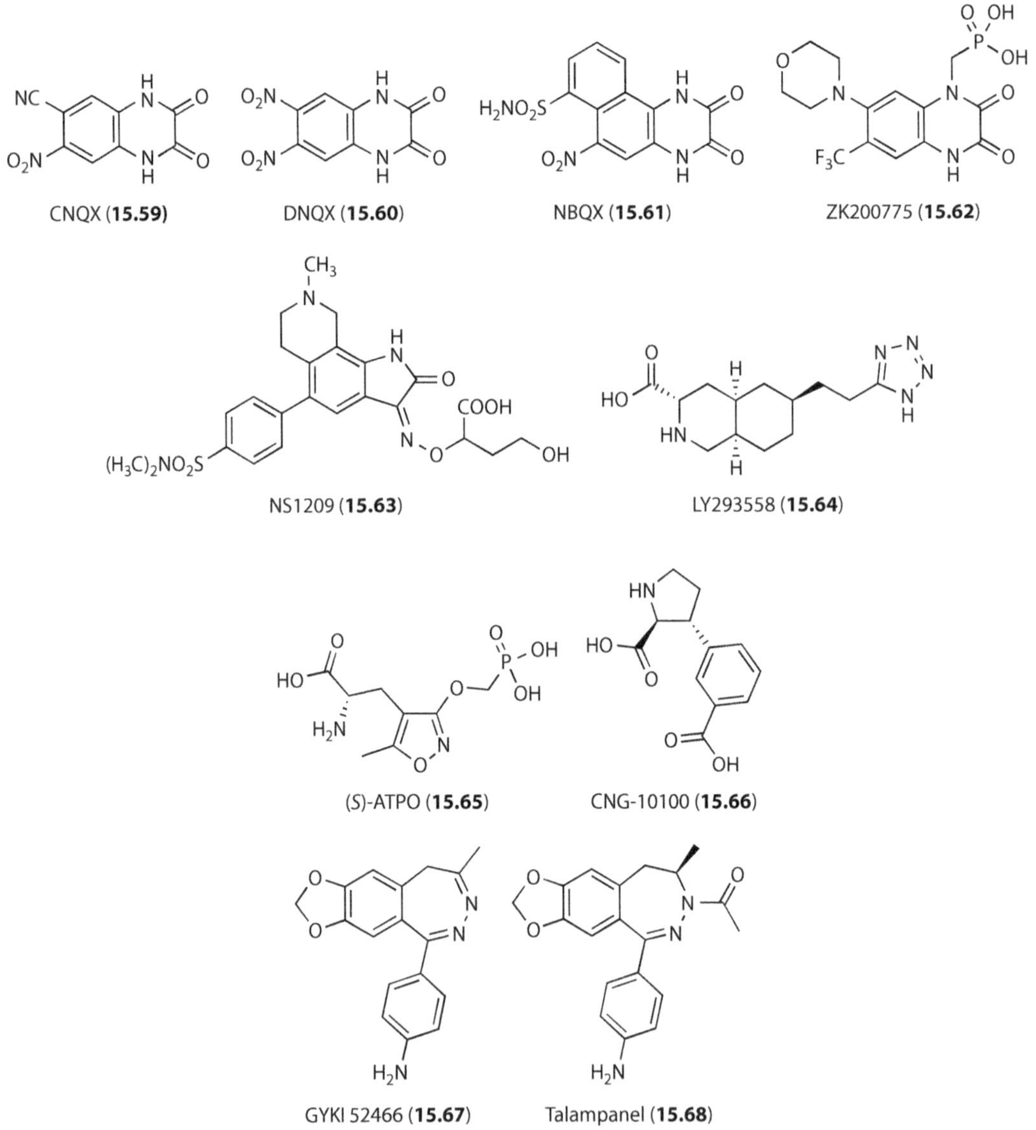

CNQX (**15.59**) DNQX (**15.60**) NBQX (**15.61**) ZK200775 (**15.62**)

NS1209 (**15.63**) LY293558 (**15.64**)

(*S*)-ATPO (**15.65**) CNG-10100 (**15.66**)

GYKI 52466 (**15.67**) Talampanel (**15.68**)

FIGURE 15.14 Structures of some competitive (**15.59–15.66**) and noncompetitive (**15.67** and **15.68**) AMPA receptor antagonists.

or AMPA, have been found to be competitive AMPA receptor antagonists. LY293558 (**15.64**), a member of the former class, is systemically active although it shows significant antagonist effects at KA receptors in addition to its potent AMPA receptor blocking effects. The AMPA receptor antagonist (*S*)-ATPO (**15.65**) which was designed using AMPA as a lead structure, has a carbon backbone longer than that which normally confers AMPA receptor agonism. The 3-phenyl-proline analog CNG-10100 (**15.66**) was designed as a broad-acting iGluR antagonist. The design was successful and CNG-10100 is now used as a lead structure for the discovery of subtype-selective iGluR antagonists.

The 2,3-benzodiazepines, such as GYKI 52466 (**15.67**) and talampanel (**15.68**), represent a class of noncompetitive AMPA receptor antagonists that have enabled effective pharmacological separation of AMPA and KA receptor-mediated neuroactivity. These compounds appear to bind to sites distinct from the agonist recognition site and are thus categorized as negative allosteric modulators.

Talampanel (**15.68**), currently under clinical development as a treatment for multiple sclerosis, epilepsy, and Parkinson's disease, may inhibit AMPA receptor function even in the presence of high levels of Glu.

15.6.7 Modulatory Agents at AMPA Receptors

The agonist-induced desensitization of AMPA receptors can be markedly inhibited by a number of structurally dissimilar AMPA receptor potentiators known as AMPA-kines, including aniracetam (**15.69**), cyclothiazide (CTZ) (**15.70**), and in particular, CX-516 (**15.71**) which have been shown to improve memory function in aged rats. These AMPA-kines positively modulate ion flux via stabilization of receptor subunit interface contacts and subsequent reduction in the degree of desensitization. A series of more potent arylpropylsulfonamide-based AMPA-kines has been identified, including LY395153 (**15.72**) (Figure 15.15).

15.6.8 KA Receptor Agonists and Antagonists

The pharmacology and pathophysiology of KA receptors are far less well understood than for AMPA receptors. However, identification of selective agonists and competitive antagonists has developed the field of KA receptor research during recent years and has provided insight into roles of these receptors in the CNS. For a number of years, KA (**15.73**) and domoic acid (**15.74**) have been used as standard KA receptor agonists despite their concurrent activities at AMPA receptors. (*S*)-ATPA (**15.75**) and (*S*)-5-I-willardiine (**15.76**) are more selective KA receptor agonists, and these compounds exhibit some selectivity for the low-affinity KA receptor subtype GluK1 compared to GluK2. (*S*)-ATPA (**15.75**) and (*S*)-5-I-willardiine (**15.76**) are structurally related to potent AMPA agonists discussed in earlier sections, illustrating that the structural characteristics required for activation of GluA1–4 and GluK1 receptors are quite similar. However, the presence of the relatively bulky and lipophilic *tert*-butyl- or iodo-substituents of these compounds is apparently the major determinant of the observed receptor selectivity (Figure 15.16).

Among the four possible stereoisomers of the 4-methyl substituted analog of Glu, only the 2*S*,4*R*-isomer (**15.77**) shows selectivity for KA receptors. Replacement of the 4-methyl group of (2*S*,4*R*)-Me-Glu (**15.77**) by a range of bulky, unsaturated substituents containing alkyl, aryl, or heteroaryl groups has yielded a number of interesting GluK1 receptor-selective compounds including LY339434 (**15.78**). LY339434 shows approximately 100-fold selectivity for GluK1 over GluK2 and no affinity for AMPA subtypes GluA1, 2, or 4 receptors.

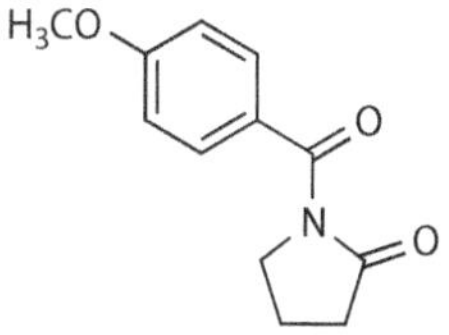

H_3CO Aniracetam (**15.69**)

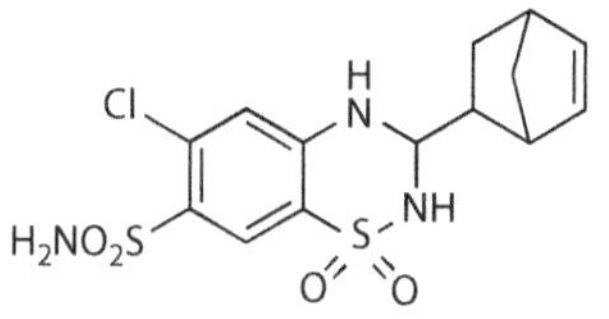

Cl, H_2NO_2S CTZ (**15.70**)

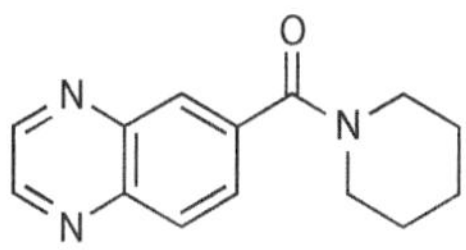

CX-516 (**15.71**)

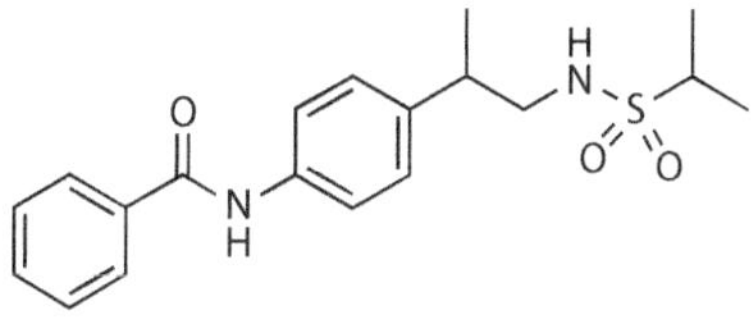

LY395153 (**15.72**)

FIGURE 15.15 Structures of some positive allosteric modulators of AMPA receptors.

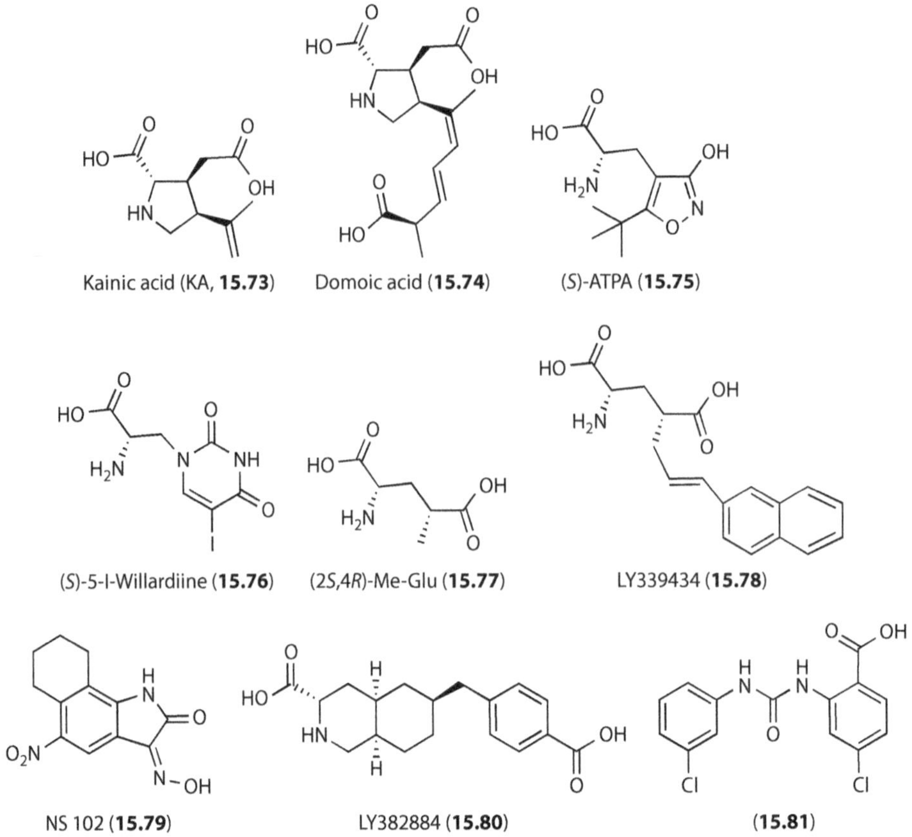

Kainic acid (KA, **15.73**) Domoic acid (**15.74**) (*S*)-ATPA (**15.75**)

(*S*)-5-I-Willardiine (**15.76**) (2*S*,4*R*)-Me-Glu (**15.77**) LY339434 (**15.78**)

NS 102 (**15.79**) LY382884 (**15.80**) (**15.81**)

FIGURE 15.16 Structures of KA (**15.73**) and some KA receptor agonists (**15.74–15.78**), two competitive (**15.79** and **15.80**) and one noncompetitive antagonist (**15.81**).

Whereas a large number of selective competitive AMPA receptor antagonists have been identified, only a few selective KA receptor antagonists have been reported. One of the first reported KA receptor-preferring antagonists was the isantin oxime, NS 102 (**15.79**) which shows some selectivity toward low-affinity [^{3}H]KA sites as well as antagonist effect at homomeric GluR6. However, low aqueous solubility has limited the use of NS 102 (**15.79**) as a pharmacological tool. A number of decahydroisoquinoline-based acidic amino acids, including LY382884 (**15.80**), have been characterized as competitive GluK1-selective antagonists that exhibit antinociceptive effects.

More recently, a series of arylureidobenzoic acids has been reported as the first compounds with noncompetitive antagonist activity at GluK1. The most potent ligands, exemplified by compound **15.81**, exhibit more than 50-fold selectivity for GluK1 over GluK2 or the AMPA receptor subtypes.

15.7 METABOTROPIC GLUTAMATE RECEPTOR LIGANDS

The cloning of the mGluRs and the evidence which has subsequently emerged on their potential utility as drug targets in a variety of neurological disorders, has encouraged medicinal chemists to design ligands targeted at the mGluRs. In analogy to the iGluRs, several X-ray structures of a mGluR ligand binding construct including different ligands have been obtained and afforded important structural knowledge of value, e.g., in the design of ligands.

15.7.1 Metabotropic Glutamate Receptor Agonists

The first agonist to show selectivity for mGlu receptors over to iGlu receptors was (1*S*,3*R*)-ACPD (**15.82**) which has been used extensively as a template for design of new mGlu receptor ligands. Introduction of a nitrogen atom in the C4 position of **15.82** gave (2*R*,4*R*)-APDC (**15.83**) which displays an increased potency for group II receptors compared to the parent compound while losing affinity for group I and III receptors (Figure 15.17).

LY354740 (**15.84**) displays low nanomolar agonist potency at mGluR2 and mGluR3, low micromolar agonist potency at mGluR6 and mGluR8, while showing no activity at the remaining mGluRs. ABHxD-I (**15.85**) displays potent agonist activity, comparable to Glu, at members of all three mGlu receptor groups. This observation has been of key importance in developing early models of the mGluR binding site. Compound **15.85** is quite a rigid molecule which adopts a conformation corresponding to an extended conformation of Glu. The observation that the compound is a potent agonist on all three mGluR groups led to the suggestion that Glu adopts the same extended conformation at all three receptor groups, and that group selectivity is thus not a consequence of different conformations but rather a consequence of other factors such as favored or disfavored interactions between the ligand and the host protein.

Apart from Glu itself (1*S*,3*R*)-ACPC (**15.82**), ibotenic acid (**15.37**) and quisqualic acid (**15.55**) were among the first potent metabotropic agonists, though fairly nonselective. Synthesis of homologs of these and other Glu analogs afforded compounds showing more selective activity at mGluRs. Thus, (*S*)-aminoadipic acid (**15.86**) was shown to be an mGluR2 and mGluR6 agonist, (1*S*,3*R*)-homo-ACPD (**15.87**) a group I agonist, whereas (*S*)-homo-AMPA (**15.88**) showed specific activity at mGluR6, and no activity at neither iGluRs nor at other mGluRs. A number of HIBO analogs including (*S*)-hexyl-HIBO (**15.89**) show group I antagonist activity and (*S*)-homo-Quis (**15.90**) is a mixed group I antagonist/group II agonist. The effect of backbone extension of different Glu analogs is often unpredictable, but chain length is nevertheless a factor of importance.

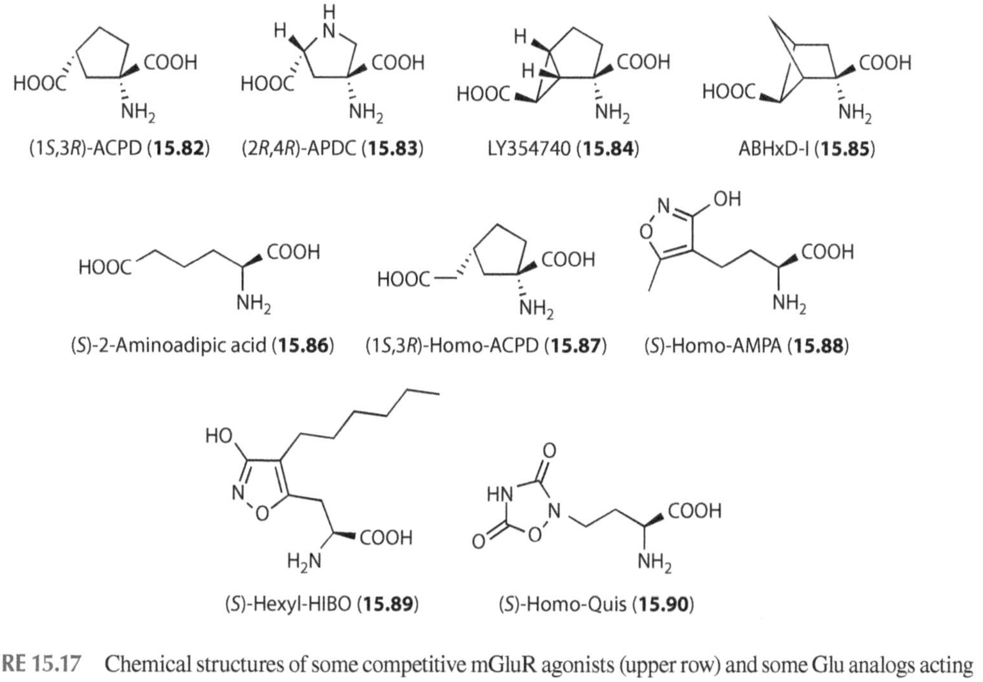

(1*S*,3*R*)-ACPD (**15.82**) (2*R*,4*R*)-APDC (**15.83**) LY354740 (**15.84**) ABHxD-I (**15.85**)

(*S*)-2-Aminoadipic acid (**15.86**) (1*S*,3*R*)-Homo-ACPD (**15.87**) (*S*)-Homo-AMPA (**15.88**)

(*S*)-Hexyl-HIBO (**15.89**) (*S*)-Homo-Quis (**15.90**)

FIGURE 15.17 Chemical structures of some competitive mGluR agonists (upper row) and some Glu analogs acting at iGluRs and/or mGluRs (middle row) and the corresponding homologs acting selectively at mGluRs (lower row).

15.7.2 Competitive Metabotropic Glutamate Receptor Antagonists

One of the first potent mGluR antagonists to be reported was (*S*)-4CPG (**15.91**), and it has been used extensively as a template for designing further potent and selective antagonists at mGluR1. The α-methylated analog, (*S*)-M4CPG (**15.92**), is an antagonist at both mGluR1 and mGluR2. It has been shown that the antagonist potency is increased by methylation at the 2-position of the phenyl ring. Thus, (+)-4C2MPG (**15.93**) is approximately fivefold more potent than the nonmethylated parent compounds. It is notable that most 4-carboxyphenylglycines show selectivity for the mGluR1 subtype with no or weak activities at the closely related mGluR5 subtype. One exception to this rule is (*S*)-hexyl-HIBO (**15.89**) which is equipotent as an antagonist at mGluR1 and mGluR5 (Figure 15.18).

α-Methylation has been widely used to derive mGluR antagonists from mGluR agonists. Maintaining the selectivity profiles as of their parent compounds, MAP4 (**15.94**) and MCCGI (**15.95**) antagonize mGluR2 and mGluR4, respectively, albeit with significantly reduced antagonist potency compared to the parent agonist.

Substituting agonists with bulky, lipophilic side chains has been a much more successful approach to the design of potent antagonists. Two of the early compounds in this class are 4-substituted analogs of Glu such as **15.97** and **15.98** which are potent and specific antagonists for mGluR2 and mGluR3. Interestingly, compounds with small substituents in the same position, such as (2*S*,4*S*)-Me-Glu (**15.96**), are more potent agonists at mGluR2 than Glu, with some activity at mGluR1 but without appreciable activity at mGluR4. Thus, by increasing the bulk and lipophilicity at the 4-position to give such "fly-swatter" substituents, the selectivity for group II is retained, and even increased, but the compounds are converted from agonists to antagonists. One of the most potent compounds of this type is LY341495 (**15.99**) with a xanthylmethyl substituent. However, LY341495 (**15.99**) also shows affinity for other subtypes, especially mGluR8.

It can be concluded that in their antagonized state, receptors from all three mGluR groups can accommodate quite large and lipophilic side chains in a variety of positions. Furthermore, compared with small α substituents such as methyl groups which most often confer antagonists with reduced potency, the large "fly-swatter" substituents in most cases confer antagonists with increased potency.

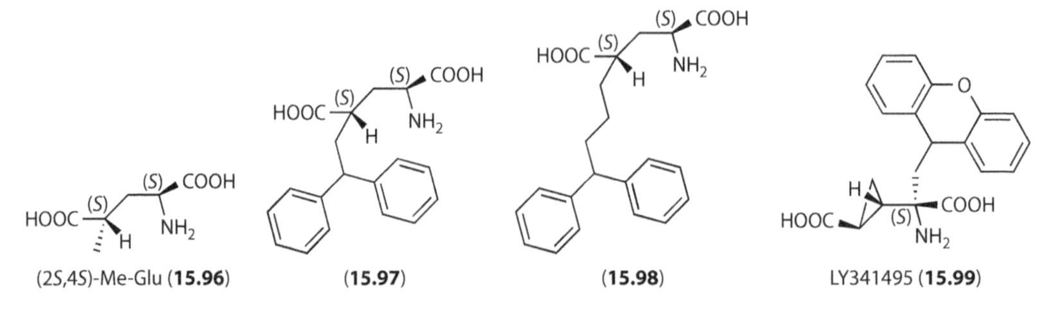

(*S*)-4CPG (**15.91**) (*S*)-M4CPG (**15.92**) (+)-4C2MPG (**15.93**) MAP4 (**15.94**) MCCGI (**15.95**)

(2*S*,4*S*)-Me-Glu (**15.96**) (**15.97**) (**15.98**) LY341495 (**15.99**)

FIGURE 15.18 Structures of some mGluR ligands.

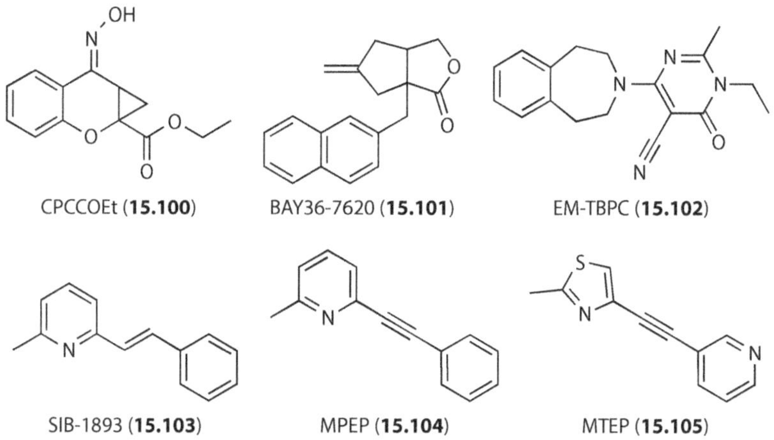

CPCCOEt (**15.100**) BAY36-7620 (**15.101**) EM-TBPC (**15.102**)

SIB-1893 (**15.103**) MPEP (**15.104**) MTEP (**15.105**)

FIGURE 15.19 Structures of some noncompetitive mGluR antagonists and positive allosteric modulators.

15.7.3 Allosteric Modulators of Metabotropic Glutamate Receptors

CPCCOEt (**15.100**) is a nonamino acid compound with no structural similarity with Glu which acts as a noncompetitive group I selective antagonist at the 7TM region rather than the agonist-binding site. A number of other nonamino acid mGluR antagonists have been discovered, e.g., BAY36-7620 (**15.101**) and EM-TBPC (**15.102**) which are potent mGluR1-specific antagonists acting at the 7TM domain (Figure 15.19).

The two compounds SIB-1893 (**15.103**) and MPEP (**15.104**) have been reported to be potent and selective, noncompetitive antagonists at mGluR5. Like CPCCOEt (**15.100**), MPEP (**15.104**) has been shown to act at the 7TM region rather than the agonist-binding site. MPEP (**15.104**) also antagonizes NMDA receptors with low micromolar potency which has led to the design of the analog MTEP (**15.105**) which is slightly more potent than **15.104** as an antagonist at mGluR5 and with no NMDA antagonist activity. SIB-1893 (**15.103**) and MPEP (**15.104**) also act as PAMs at mGluR4. The allosteric effect is dependent upon Glu activation, and the compounds are thus unable to activate the mGluR4 receptor directly. Instead, the compounds enhance the response mediated by Glu, causing a leftward shift of concentration–response curves and an increase in the maximum response.

15.8 EXCITATORY AMINO ACID TRANSPORTER LIGANDS

The excitatory amino acid transporters (EAATs) are responsible for maintaining a tolerably low concentration of Glu in the synapse. A total of five subtypes have been identified, of which the high-capacity transporters EAAT1,2 are expressed on glial cells and low-capacity transporters EAAT3–5 are expressed on neurons. Expression of the transporter subtypes in the CNS is diverse; EAAT1 is primarily expressed in the cerebellum, EAAT2 is the major transporter for Glu in the forebrain, and EAAT3 is the major neuronal transporter in the brain and spinal cord, EAAT4 in the cerebellum and EAAT5 is only expressed in the retina. Furthermore, three vesicular Glu transporter subtypes (VGLUT1–3) have been identified which serve to pack Glu into small vesicles for subsequent fusion with the presynaptic terminal membrane, thus releasing Glu into the synapse.

15.8.1 Inhibitors of the EAATs

The natural product KA (**15.73**) and its reduced analog dihydrokainic acid (**15.106**) both inhibit EAAT2 selectively in the mid-micromolar range. However, from a large series of asparagine analogs, WAY213613 (**15.107**) proved to be both potent (mid nanomolar) and selective for EAAT2. The nonselective inhibitor DL-TBOA (**15.108**) is an often used tool compound and its 3-benzyl

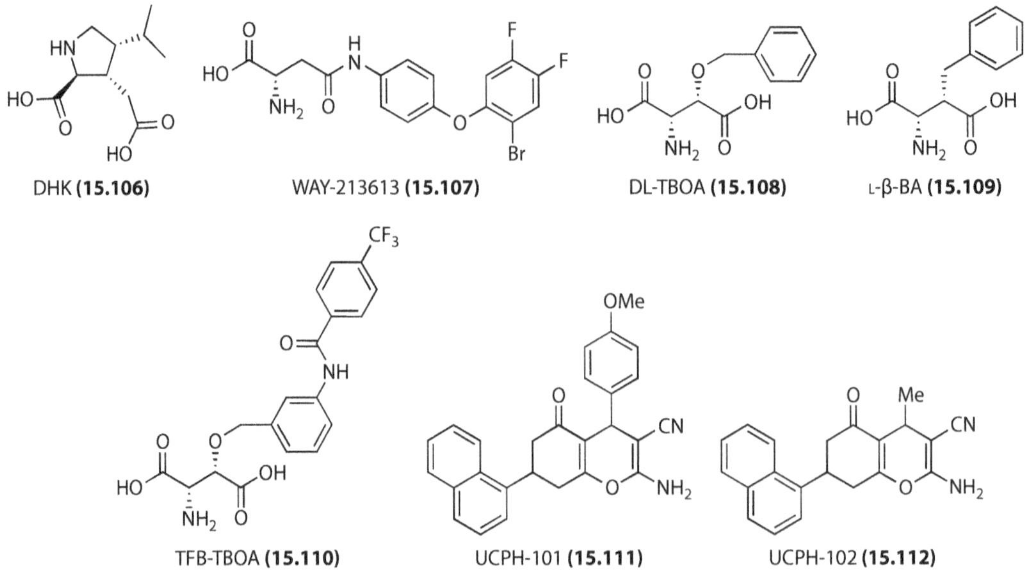

DHK (**15.106**) WAY-213613 (**15.107**) DL-TBOA (**15.108**) L-β-BA (**15.109**)

TFB-TBOA (**15.110**) UCPH-101 (**15.111**) UCPH-102 (**15.112**)

FIGURE 15.20 Chemical structures of reported EAAT inhibitors.

analog L-β-BA (**15.109**) displays a 10-fold preference for inhibition of EAAT3 over EAAT1,2. Interestingly, a large substituent in the 3-position resulted in a remarkable increase in potency (low nanomolar) to give the nonselective EAAT inhibitor TFB-TBOA (**15.110**). In summary, this suggests that the 3-position on asparagine is a "hot spot" for obtaining subtype selective and potent EAAT inhibitors.

The potent allosteric EAAT1 selective inhibitor UCPH-101 (**15.111**) was discovered after intensive SAR studies of a lead compound identified by a screening campaign. While UCPH-101 does not cross the BBB and is thus best suited as an in vitro pharmacological tool compound, oral administration of UCPH-102 (**15.112**) does indeed cross the BBB to give a favorable plasma:brain ratio (Figure 15.20).

15.9 DESIGN OF DIMERIC POSITIVE AMPA RECEPTOR MODULATORS

Many receptors, including the Glu receptors, exist as dimers or higher oligomers and this creates the possibility of designing ligand dimers which can bind to two binding sites simultaneously. Dimeric ligands have been developed in many different receptor areas and have led to compounds with, first of all, improved potency, but also improved selectivity, solubility, and pharmacokinetic properties can be observed. In spite of numerous examples of dimeric ligands with improved pharmacology compared to their monomeric analogs, no structural evidence has previously been presented for the simultaneous binding of such ligands to two identical binding sites. However, such evidence has been obtained for a dimeric PAM at AMPA receptors (Figure 15.21).

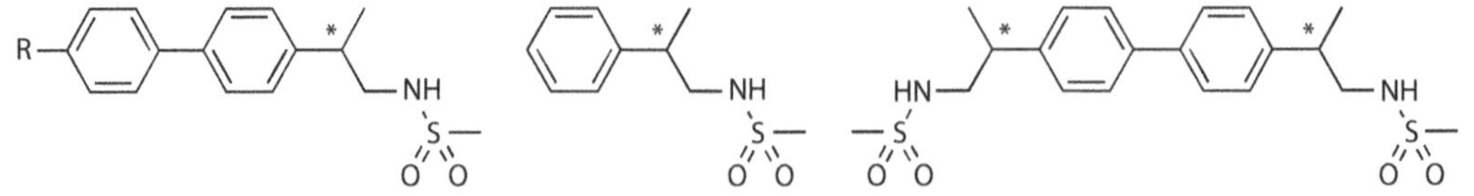

15.113 **15.114** **15.115**

FIGURE 15.21 Structure of CTZ (**15.70**) and other positive allosteric modulators at AMPA receptors including the symmetric dimeric analog **15.115**.

CTZ (**15.70**) (see Section 15.6.7) is a PAM at AMPA receptors and an X-ray structure of CTZ in complex with the GluR2 binding construct showed a symmetrical binding of two CTZ molecules in two identical binding sites close to each other. Another study showed a number of biarylpropylsulfonamide analogs (**15.113**) with good activity as positive modulators at the CTZ site, and these structures were used as templates for the design of a symmetrical dimeric ligand. By use of computer modeling, different symmetric dimeric ligands with two propylsulfonamide moieties, a biphenyl linker and different alkyl substituents were constructed and tested for binding by computer docking. This suggested dimer **15.115** as a potential ligand capable of binding in a symmetrical mode to two adjacent CTZ binding sites. Upon synthesis of the two enantiomers of monomer **15.114** and the three stereoisomers of the dimer **15.115** (*R,R*-, *S,S*- and mesoform), these were tested for activity at cloned AMPA receptors expressed in oocytes by electrophysiological experiments. (*R,R*)-**15.115** proved to be the most potent compound with $EC_{50} = 0.79$ μM at GluA2 compared to $EC_{50} = 1980$ μM for the monomer (*R*)-**15.114**. The maximal potentiation of Glu responses for the monomer as well as for the dimer was in the order of 800%–1000%, showing that they are both effective potentiators by blockade of AMPA receptor desensitization. Obviously, the dimeric compound was dramatically more potent than the monomer, more than three orders of magnitude, and a similar pattern was observed for the other less active enantiomer.

Eventually, (*R,R*)-**15.115** was co-crystallized with the GluA2 binding construct, and from the obtained X-ray structure is can be seen that (*R,R*)-**15.115** simultaneously binds to two identical modulatory binding sites at the AMPA receptors (Figure 15.22).

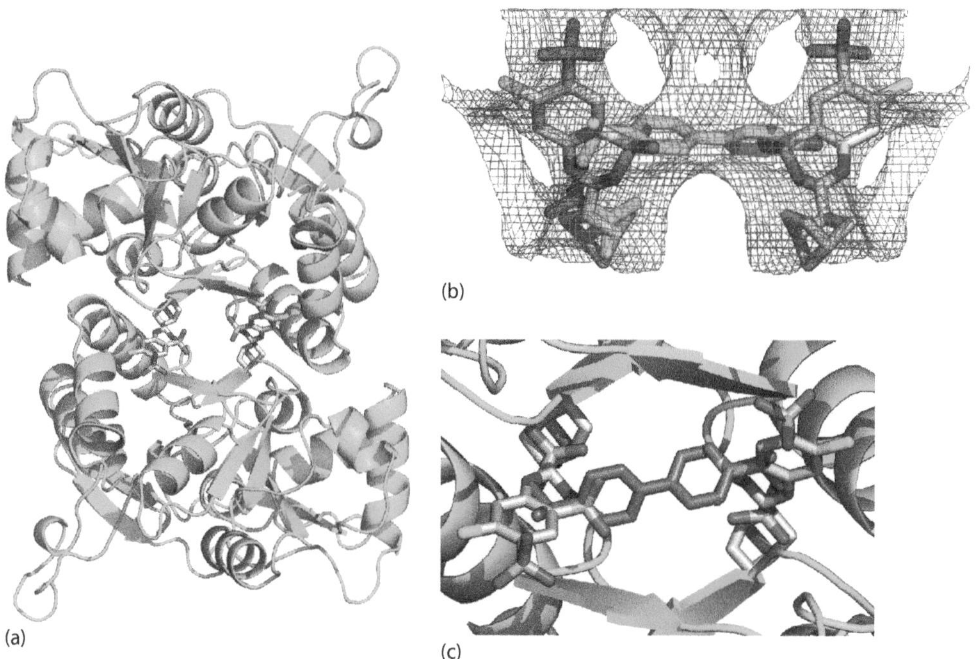

FIGURE 15.22 (a) X-ray structure of the upper part of the extracellular amino terminal domain of the GluR2 receptor in complex with CTZ (**15.70**) (white carbon atoms). (b) Illustration of the binding of dimer **15.115** (cyan carbon atoms) compared to the binding of CTZ (white carbon atoms) in the calculated cavity formed by the binding pocket surface (side view). (c) Close-up view of the X-ray structure of the GluR2 receptor in complex with dimer **15.115** (brown carbon atoms) compared to CTZ (white carbon atoms).

15.10 CONCLUDING REMARKS

The cloning of GABA and Glu receptor subtypes and their subsequent pharmacological characterization has been of great importance to the development of the field. Over the years, a large number of X-ray crystal structures have provided valuable information especially about the iGluRs and their mechanism of action. For the GABA receptors, structural models are now very recently supported by X-ray crystal structures forming a basis for structure-based studies. Selective ligands have only been discovered for a limited number of GABA and Glu receptor and transporter subtypes which impede a detailed understanding of the role and function of the subtypes in the healthy as well as diseased brain. Further insight into this would facilitate the development of new therapeutic agents within this important field.

FURTHER READING

GABA

Brown, K.M., Roy, K.K., Hockerman, G.H., Doerksen, R.J., and Colby, D.A. 2015. Activation of the γ-aminobutyric acid type B ($GABA_B$) receptor by agonists and positive allosteric modulators. *J. Med. Chem.* 58:6336–6347.

Krall, J., Balle, T., Krogsgaard-Larsen, N., Krogsgaard-Larsen, P., Kristiansen, U., and Frølund, B. 2015. $GABA_A$ receptor partial agonists and antagonists: Structure, binding mode and pharmacology. *Adv. Pharmacol.* 72:201–227.

Petersen, J.G., Bergmann, R., Krogsgaard-Larsen, P., Balle, T., and Frølund, B. 2014. Probing the orthosteric binding site of $GABA_A$ receptors with heterocyclic GABA carboxylic acid bioisosteres. *BioChem. Res.* 39:1005–1015.

Glutamate

Jensen, A.A., Fahlke, C., Bjørn-Yoshimoto, W.E., and Bunch, L. 2015. Excitatory amino acid transporters: Recent insights into molecular mechanisms, novel modes of modulation and new therapeutic possibilities. *Curr. Opin. Pharmacol.* 20:116–123.

Kaae, B.H., Harpsøe, K., Kastrup, J.S. et al. 2007. Structural proof of a dimeric positive modulator bridging two identical AMPA receptor-binding sites. *Chem. Biol.* 14:1294–1303.

Traynelis, S.F., Wollmuth, L.P., McBain, C.J. et al. 2010. Glutamate receptor ion channels: Structure, regulation, and function. *Pharmacol. Rev.* 62:405–496.

Vogensen, S.B., Greenwood, J.R., Bunch, L., and Clausen, R.P. 2011. Glutamate receptor agonists: Stereochemical aspects. *Curr. Top. Med. Chem.* 11:887–906.

16 Acetylcholine

Anders A. Jensen

CONTENTS

Cholinergic transmission mediated by the neurotransmitter acetylcholine (ACh) is found throughout the entire body. In the central nervous system (CNS), ACh plays key roles in processes underlying cognitive functions, arousal, reward, motor control, and analgesia, and in the periphery cholinergic transmission regulates a wide range of basic physiological functions, such as cardiac function, glandular secretion, gastrointestinal motility, and smooth muscle contraction. This chapter will focus on the therapeutic potential in central cholinergic neurotransmission, particularly when it comes to Alzheimer's disease (AD). Following a short introduction to the etiology of AD, the symptoms of the disease, and the different lines of research pursued in the field (Section 16.1), the complexity of cholinergic neurotransmission will be outlined (Section 16.2) and the three cholinergic protein classes predominantly pursued as drug targets for the treatment of AD and other CNS disorders will be presented (Sections 16.3 through 16.5).

16.1 ALZHEIMER'S DISEASE

AD is a progressive neurodegenerative disorder of the human CNS which in most cases manifests itself in mid to late adult life and usually leads to death within 7–10 years after the diagnosis. AD afflicts 2%–3% of the general population at age 65 with a doubling of incidence for every 5 years of age, and the disease accounts for ~60% of all cases of dementia in persons over 65 years of age. While age thus is the dominant risk factor in AD, genetic and epidemiological factors are also important determinants of the disorder.

Clinical diagnosis of AD is based on the progressive impairment of memory and at least one other cognitive dysfunction (such as difficulties with language, complex movements, or identifying objects) in the individual as well as on the exclusion of other diseases causing dementia. In mid or late stages of AD, the accuracy of the clinical diagnosis is very high (~90%). However, the diagnosis can only be made definitely based on direct examination of brain tissue derived from biopsy or autopsy. The AD brain is characterized by massive atrophy of cortical and hippocampal regions,

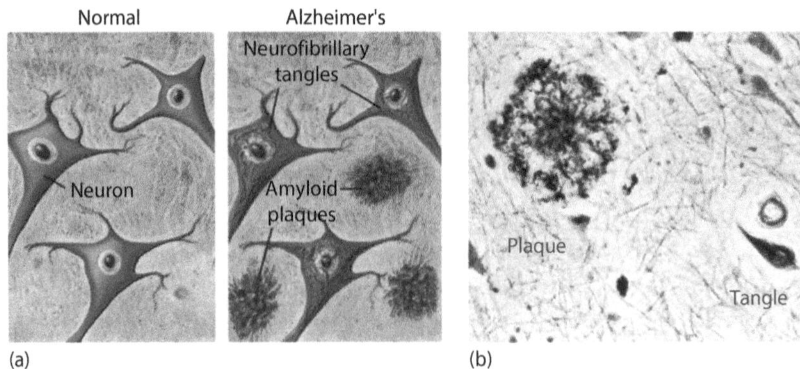

FIGURE 16.1 Amyloid plaques and neurofibrillary tangles. (a) Illustration of normal tissue and tissue affected by Alzheimer's disease. (b) Light micrograph of a section of hippocampal brain tissue affected by Alzheimer's disease. The presence of a neuritic amyloid plaque and neurons containing neurofibrillary tangles is indicated. ([a] Medical illustration courtesy of Alzheimer's Disease Research, a BrightFocus Foundation program. Copyright 2015. http://www.brightfocus.org/alzheimers; [b] courtesy of Scanpix. Photo by Dr. M. Goedert, Science Foto Library/ScanPix, Copenhagen, Denmark.)

and at the microscopic level, a widespread cellular degeneration and loss of neocortical neurons is observed together with the pathological hallmarks of the disease: the presence of *amyloid plaques* and *neurofibrillary tangles* (Figure 16.1).

The extracellular *amyloid plaques* are predominantly constituted by amyloid β ($A\beta_{40/42}$) peptides which are self-aggregating, 40/42 amino acids–long peptides formed from the proteolytic cleavage of the amyloid precursor protein (APP). In the nonamyloidogenic pathway, cleavage of APP by the protease α-secretase yields the N-terminal APPsα fragment and the transmembrane C-terminal fragment α-stub which is subsequently cleaved by γ-secretase yielding the nontoxic peptide p3 (Figure 16.2). In the amyloidogenic pathway prominent in the AD brain, however, the extracellular N-terminal domain of APP is cleaved at a different region by β-secretase to yield APPsβ and β-stub, and subsequent cleavage of β-stub by γ-secretase gives rise to the $A\beta_{40/42}$ peptides. In particular,

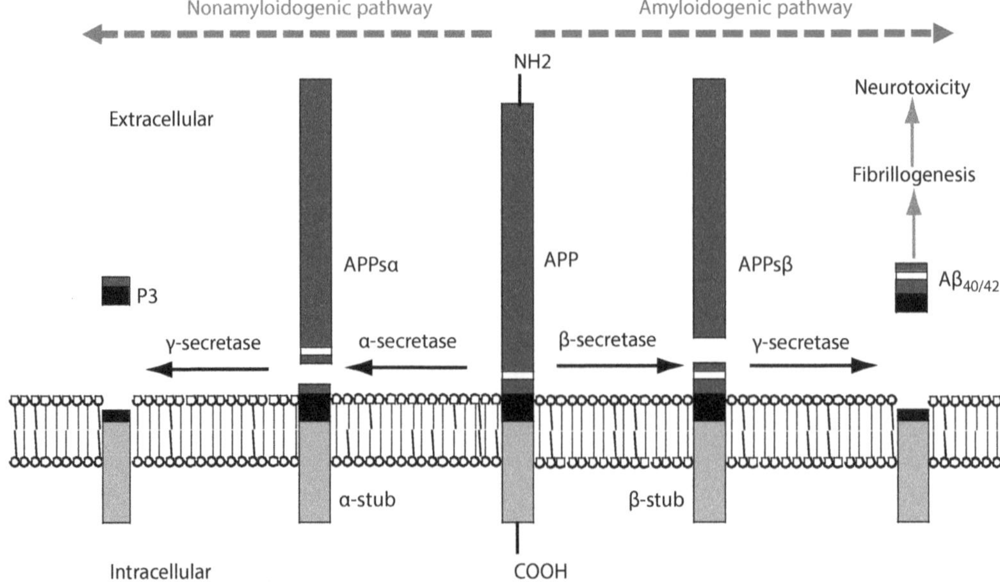

FIGURE 16.2 Proteolytic processing pathways of APP.

$A\beta_{42}$ undergoes oligomerization and deposition which leads to microglial and astrocytic activation, oxidative stress and progressive synaptic injury (Figure 16.2). The intracellular *neurofibrillary tangles* are bundles of paired helical filaments constituted mainly by hyperphosphorylated tau protein. Tau is abundantly expressed in neurons, where the protein under normal conditions maintains microtubule stability inside the cells. However, hyperphosphorylated tau aggregates into tangled clumps which leads to disintegration of the microtubules, cytoskeletal degeneration, and neuronal death.

The multifactorial pathogenesis of AD is not fully understood, and over the years several hypotheses about the origin of the disease have been proposed. The "cholinergic hypothesis" formulated in the 1970s–1980s was based partly on the emerging realization of the importance of ACh and cholinergic neurotransmission for cognitive functions and partly on the dramatic loss of cholinergic neurons in basal forebrain regions observed in the AD brain. However, the loss of cholinergic neurons is a relatively late event in AD development, and since subsequent studies have failed to demonstrate a direct causal relationship between cholinergic dysfunction and AD progression, the "cholinergic hypothesis" for the disease has largely been dismissed. The "amyloid cascade theory" stating that the aberrant production and deposition of plaques and tangles in AD may not only be disease markers but also play a causal role in the disease has become one of the dominant paradigms in the field, and it is supported by a substantial amount of histopathological, biochemical, genetic, and animal model data.

The complexity of the etiology of AD is reflected in the numerous strategies that have been pursued in the search for therapeutics. The "cholinergic hypothesis" for AD has fueled the development of drugs targeting this system, the rationale being to compensate for the progressive degeneration of cholinergic neurons by the administration of drugs augmenting cholinergic signaling in the remaining neurons. Glutamate is another highly important neurotransmitter for cognitive functions, and the therapeutic potential in glutamatergic mechanisms for AD is underlined by the uncompetitive NMDA receptor antagonist memantine being the only noncholinergic AD drug marketed (Namenda®) (Chapter 15). Finally, the key role of β-amyloid and tau aggregation in AD has inspired extensive research into these areas. The outcome of clinical trials with β-secretase and γ-secretase inhibitors, inhibitors of $A\beta_{40/42}$ aggregation/deposition and enhancers of amyloid clearance has so far been discouraging, whereas the potential in tau aggregation inhibitors still remains to be investigated.

16.2 THE CHOLINERGIC SYNAPSE

The cholinergic synapse and the events underlying cholinergic neurotransmission are outlined in Figure 16.3. Although cholinergic neurotransmission clearly can be affected by modulation of numerous different targets, the cholinesterases mediating the hydrolysis of ACh (**16.1**) into choline (**16.2**) upon its release from the presynaptic terminal into the synaptic cleft and the two receptor classes that mediate the signaling of the neurotransmitter, the muscarinic and nicotinic ACh receptors (mAChRs and nAChRs), have received most of the attention in cholinergic drug development. In the following sections, focus will be on these three protein classes.

16.3 CHOLINESTERASES

Most of the events underlying cholinergic neurotransmission essentially resemble those involved in other neurotransmitter systems, including the biosynthesis and storage of the neurotransmitter in presynaptic vesicles, the release of it into the synaptic cleft, and the activation of receptors located at presynaptic and postsynaptic terminals. The unique feature of cholinergic transmission compared to other neurotransmitter systems is the presence of two degradation enzymes that convert the neurotransmitter back into its precursor before this, not the neurotransmitter itself, is taken back up into the presynaptic terminal by a transporter (Figure 16.3). Acetylcholinesterase (AChE) and butyrylcholinesterase (BuChE) are present in cholinergic synapses in the CNS, in parasympathetic synapses in the periphery, and in the neuromuscular junction, with the AChE being responsible for ~80% of the total cholinesterase activity in the brain.

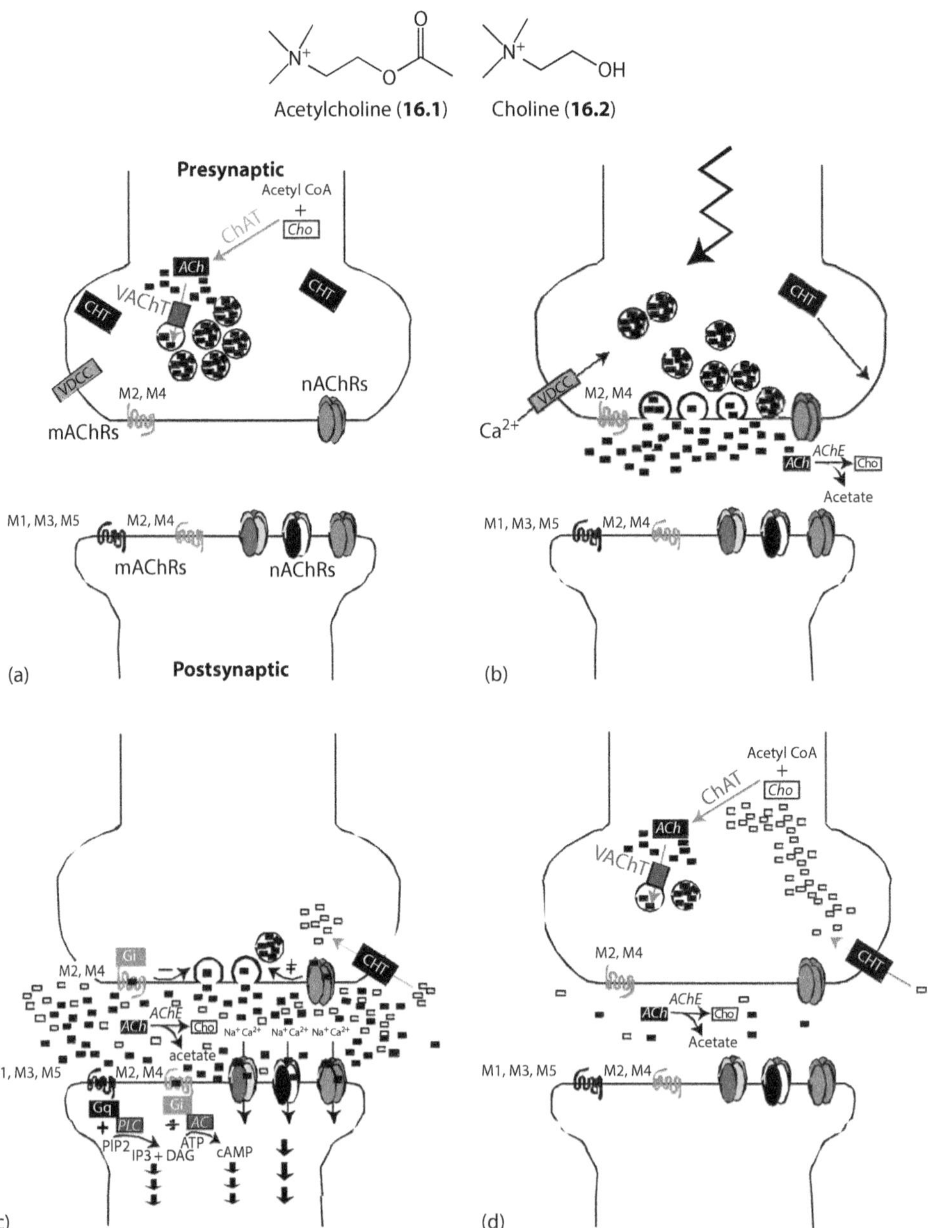

FIGURE 16.3 Cholinergic neurotransmission. (a) In the presynaptic terminal, ACh is synthesized by choline acetyltransferase (ChAT) through acetylation of its precursor Choline (Cho). The synthesized ACh is packaged into synaptic vesicles by uptake by the vesicular ACh transporter (VAChT). (b) Upon firing of the presynaptic neuron, the cytoplasmic Ca^{2+} concentration increases due to opening of voltage-dependent calcium channels (VDCCs), and this causes the synaptic vesicles to fuse with the plasma membrane and release ACh into the synaptic cleft. (c) ACh exerts its effects via activation of mAChRs and nAChRs. Activation of postsynaptic nAChRs and mAChRs elicits the fast and slow signaling of ACh, respectively, whereas activation of presynaptic receptors inhibits or augments synaptic release of ACh. The ACh-mediated signaling is terminated by acetylcholinesterase (AChE) which converts ACh back to Cho. The choline transporter (CHT) is transported to the plasma membrane of the presynaptic terminal. (d) CHT takes up Cho into the presynaptic terminal and with this synaptic transmission is terminated.

16.3.1 Cholinesterase Inhibitors

The physiological importance of cholinesterase activity is reflected by the fact that inhibitors of the enzymes include substances ranging from snake and insect toxins over pesticides and nerve gasses used in chemical warfare to clinical drugs against AD, glaucoma, and *Myasthenia Gravis* (a neuromuscular disease). This also accentuates that the mode and duration of AChE inhibition is of key importance for the induced effects. Inhibition of cholinesterase activity and the resulting amplification of the natural spatial and temporal tone of ACh-mediated signaling have proven efficacious in the treatment of AD symptoms, with four out of the five currently approved drugs against the disease being reversible AChE inhibitors (AChEIs). However, the clinical use of AChEIs is limited by their gastrointestinal side effects, and it is important to stress that the inhibitors predominantly are efficacious in the early stages of the disease and that their beneficial effects on the AD symptoms are modest at best. Despite these shortcomings, an important contribution of the AChEIs has been to demonstrate the potential in the cholinergic system as a target for cognitive enhancers which has served as an important proof of concept for many of the subsequent research activities in the mAChR and nAChR fields (Sections 16.4 and 16.5).

Natural product compounds have been a rich source of leads for drug discovery in the AChE field, and rational ligand design has been greatly facilitated by crystal structures of the enzyme in complex with inhibitors. Tacrine (**16.3**), the first AChEI to be marketed (Cognex®, 1993), is a reversible, nonselective AChE/BuChE inhibitor which also displays activities at monoamine oxidases, potassium channels, and ACh receptors, a "dirty" pharmacological profile that has been proposed to contribute to its therapeutic effects (Figure 16.4). However, tacrine is rarely used in the clinic today due to several reports of drug-induced liver damage. Donepezil (**16.4**), a rationally designed piperidine-based inhibitor (Aricept®, 1996), inhibits AChE in a noncompetitive manner and displays significant selectivity toward BuChE. Medicinal chemistry explorations of the carbamate-stigmine scaffold of physostigmine (**16.5**) from the Calabar bean (*Physostigma venenosum*) have yielded several "carbamoylating" AChEIs such as **16.6**, **16.7**, and the simplified analog rivastigmine (**16.8**, Exelon®, 2000). Galanthamine (**16.9**), an alkaloid originally isolated from *Galanthus nivalis*, was marketed for treatment of AD in 2001 (Razadyne®). The natural product compound huperzine A (**16.10**), isolated from the Chinese folk medicine *Huperzia serrata*, is a potent AChEI with no BuChE activity. Hybrid compounds combining structural components from two inhibitors, such as the tacrine/huperzine A hybrid huprine X (**16.11**) and the bivalent tacrine–indole ligand **16.12,** have displayed substantially higher inhibitory potencies, faster binding kinetics and/or higher AChE/BuChE selectivities than their parent compounds. In other hybrid compounds, the AChE activity has been supplemented with activities at other neurotransmitter systems, for example, in ladostigil (**16.13**), where the structure of rivastigmine (**16.8**) has been combined with the propylargyl group of the monoamine oxidase-B inhibitor rasagiline. Finally, several completely selective BuChE inhibitors have been developed, including cymserine analogs such as **16.14** (Figure 16.4).

16.3.2 Substrate Catalysis of and Ligand Binding to the AChE

The AChE is a 537 amino acids–long protein composed of a 12-stranded mixed β-sheet surrounded by 14 α-helices (Figure 16.5a). The hydrolysis of ACh and other substrates takes place in the catalytic site (or active site) located at the bottom of a long narrow gorge that penetrates half into the enzyme, ~20 Å from the enzyme surface. The ability of the substrate to efficiently reach this site is rooted in the existence of another ACh binding site, the "*peripheral anionic site*," located at the entrance to this gorge (Figure 16.5a). Binding of ACh to this site represents an important initial step in the ACh catalysis since this "capture" of the substrate from the surroundings and the subsequent translocation of it to the catalytic site greatly enhances the efficiency of the enzyme.

The catalytic site in AChE is comprised of two subsites, the "*catalytic anionic site*" and the "*esteratic subsite*." The catalytic anionic site is the key mediator of ACh binding to the catalytic site,

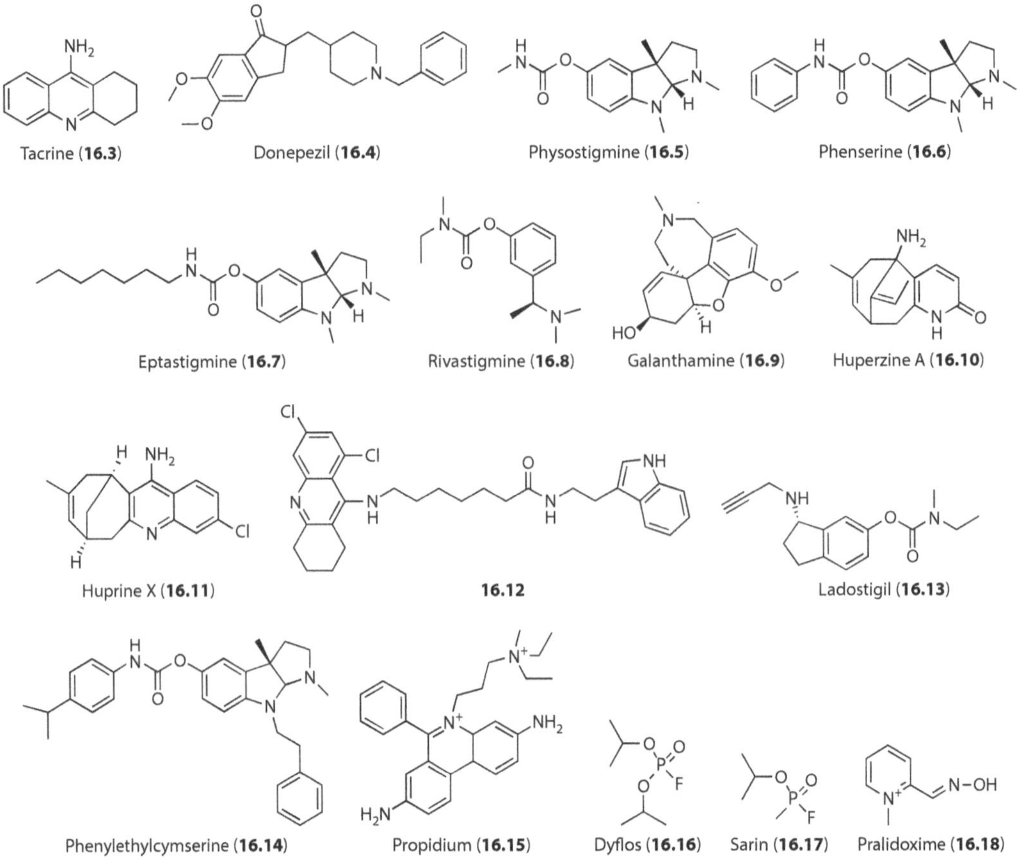

Tacrine (16.3)
Donepezil (16.4)
Physostigmine (16.5)
Phenserine (16.6)
Eptastigmine (16.7)
Rivastigmine (16.8)
Galanthamine (16.9)
Huperzine A (16.10)
Huprine X (16.11)
16.12
Ladostigil (16.13)
Phenylethylcymserine (16.14)
Propidium (16.15)
Dyflos (16.16)
Sarin (16.17)
Pralidoxime (16.18)

FIGURE 16.4 Chemical structures of cholinesterase inhibitors (**16.3–16.17**) and the AChE reactivator **16.18**.

Peripheral anionic site
Catalytic anionic site
Phe330
Trp84
Glu199
Esteratic subsite
Glu327
His440
Ser200
Rivastigmine
Galanthamine
Donepezil
Sarin
(a)
(b)

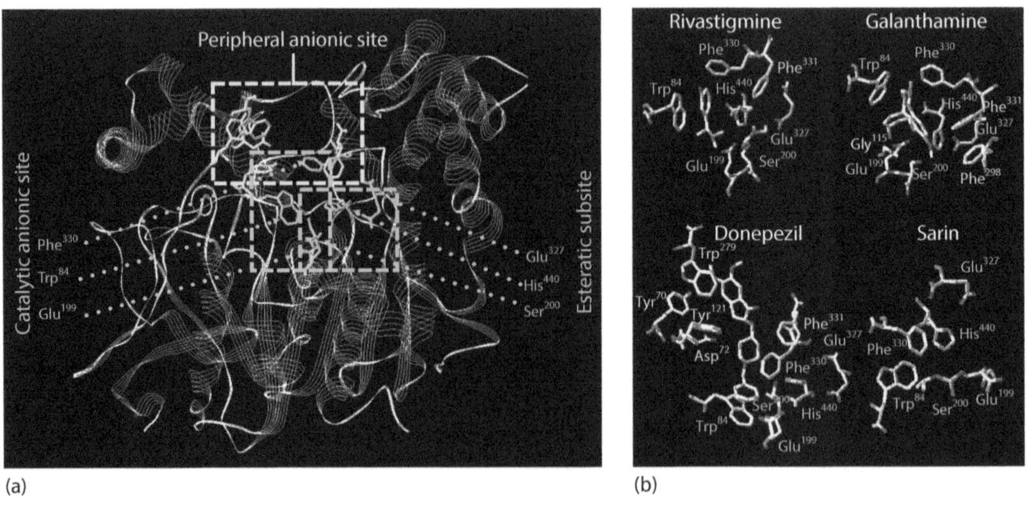

FIGURE 16.5 AChE structure and AChEI binding. (a) 3D structure of AChE. The localizations and residues constituting the peripheral anionic site, the catalytic anionic site, and the esteratic subsite are indicated in yellow, cyan, and green, respectively. (b) Crystal structures of the catalytic site of AChE in complexes with four AChEIs.

with a cation–π interaction being formed between the quaternized ammonium group of ACh and the aromatic ring of the Trp^{84} residue in this site (and additional contributions from the neighboring Glu^{199} and Phe^{330} residues). The anchorage of the choline moiety of ACh to this site presents the ester group of the substrate to the esteratic subsite which is made up by a serine-hydrolase catalytic triad consisting of the residues Ser^{200}, His^{440}, and Glu^{327} (Figure 16.5a). Although the mechanism for the ACh hydrolysis mediated by this catalytic triad may seem complex (Figure 16.6a), the catalytic turnover of the AChE is truly remarkable, with ~10,000 ACh molecules being hydrolyzed by each AChE molecule per second.

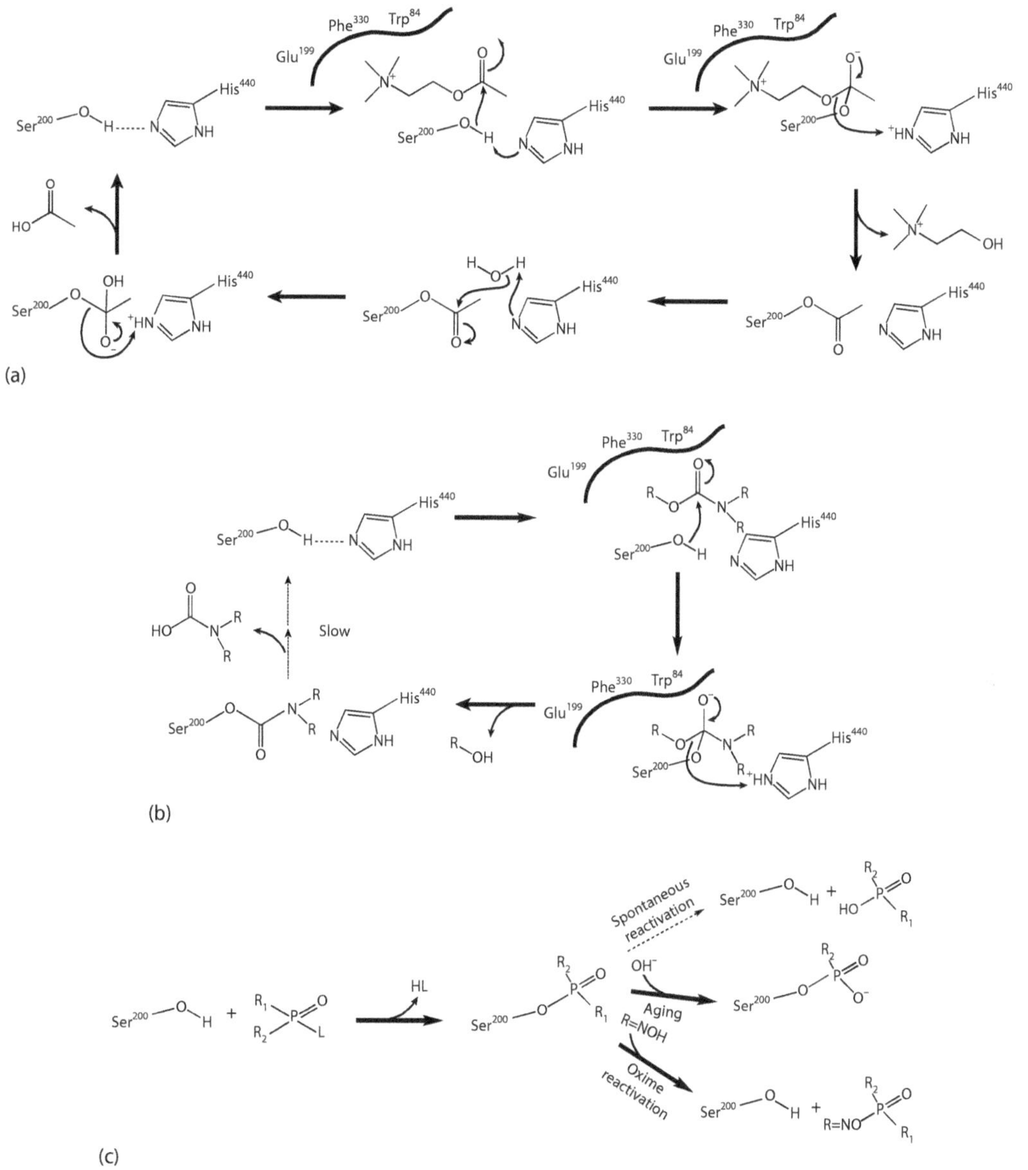

FIGURE 16.6 ACh hydrolysis in the AChE and the inhibition of it. (a) Hydrolysis of ACh in the AChE. (b) Inhibition of AChE by a carbamoylating AChEI. (c) Formation of the phosphorus AChE conjugate by organophosphorous AChEIs, and the following "aging" and oxime reactivation processes. R_1, O–alkyl or amid; R_2, alkyl, O–alkyl or amid; L, leaving group.

The complexity of AChE function is reflected in the diverse mechanisms by which different AChEIs exert their inhibition of the enzyme. The "carbamoylating" inhibitors (**16.5–16.8**) are actually substrates for AChE as they all contain carbamate groups that, analogously to the ester group in ACh, can be hydrolyzed by the enzyme, thus yielding the respective carbamate and stigmine (**16.5–16.7**) or dimethylamino-α-methylbenzyl (**16.8**) moieties of the compounds (Figure 16.6b). However, in contrast to the very fast (microseconds) cleavage of acetate from the Ser^{200} residue following the hydrolysis of ACh, the dissociation of the carbamate group from Ser^{200} is very slow (seconds–minutes). Thus, these AChE substrates exert *de facto* inhibition since the catalytic site is unable to bind and hydrolyze ACh as long as the carbamate group is covalently bound to Ser^{200} (Figure 16.6b). Thus, the kinetics of the inhibition mediated by these inhibitors is mainly determined by the stability of the formed Ser^{200}-carbamate conjugate which varies somewhat for the different carbamate groups.

In contrast to the carbamoylating inhibitors, the other reversible AChEIs (**16.3–16.4**, **16.9–16.15**) are "true" inhibitors in the sense that they are not hydrolyzed by AChE but inhibit its function through binding to various sites. Inhibitors like tacrine (**16.3**) and galanthamine (**16.9**) interact with residues both in the catalytic anionic site and the esteratic subsite and thus compete directly with ACh for binding to the catalytic site (Figure 16.5b). On the other hand, donepezil (**16.4**) targets the gorge connecting the catalytic site with surface of the enzyme, and the dimethoxy–indanone and benzyl piperidine moieties of the inhibitor interact with residues in peripheral anionic site and in the catalytic anionic site, respectively (Figure 16.5b). Propidium (**16.15**) and fasciculin-2, a peptide toxin found in the venoms of mamba species, inhibit AChE function noncompetitively by binding exclusively to the peripheral anionic site, thus prohibiting the substrate from gaining access to the catalytic site. Finally, the remarkably high inhibitory potencies displayed by bivalent ligands such as **16.12** have been attributed to their ability to form interactions with both the peripheral anionic site and the catalytic site in the AChE.

Irreversible organophosphorus AChEIs are esters or thiols derived from phosphoric, phosphonic, phosphinic, or phosphoramidic acids (e.g., **16.16** and **16.17**). Just as the carbamoylating inhibitors, these compounds are structural analogs to ACh and thus undergo similar initial interactions with the Ser^{200} residue in the esteratic subsite. This gives rise to the formation of an extremely stable covalent phosphorus-conjugate complex and a completely irreversible inhibition of enzyme function (Figures 16.5b and 16.6c). Thus, the toxicity of these inhibitors is rooted in the fact that normal AChE function in the body only can be restored by resynthesis of the enzyme. The formation of the phosphorus AChE conjugate can be reversed with oxime-based "reactivators," such as **16.18** (Figure 16.6c), and mAChR antagonists such as atropine have also been used to treat organophosphorate poisoning.

16.4 MUSCARINIC ACh RECEPTORS

The mAChRs belong to class A of the G-protein coupled receptor (GPCR) superfamily (Chapter 12). Thus, the mAChR component of cholinergic signaling is mediated by their coupling to G-proteins and other intracellular proteins (such as β-arrestins) and the resulting downstream effects arising from second messenger cascades, modulation of ion channels, and activation of various kinases. The five mAChR subtypes, termed M_1–M_5, are ubiquitously expressed throughout the CNS and in most peripheral tissues. The physiological functions mediated by the different subtypes have for decades been studied using knockout mice strains (where the expression of one or several mAChR subtypes has been eliminated), and in recent years these explorations have been further aided by the availability of truly subtype-selective pharmacological tools for the receptors (see later text). Whereas the M_1, M_4, and M_5 receptors predominantly are expressed in the CNS, M_2 and M_3 are found both centrally and in the periphery, including cardiac and smooth muscle tissues. An important realization to come from these studies is that many of the adverse effects produced by nonselective mAChR ligands or by AChEIs are attributable to the peripheral M_2 and M_3 receptors. These findings combined with the failure of many first-generation mAChR drugs in clinical development due to intolerable adverse effects have meant that focus in the field with respect to CNS indications have turned toward selective modulation of M_1, M_4, and M_5 receptors.

M_1 is the most predominant mAChR subtype in the CNS, where it primarily is located at postsynaptic densities. M_1 is expressed in several brain regions important for cognitive processes, and several lines of research have implicated augmentation of M_1 signaling as perhaps the most promising mAChR-based approach to treat the cognitive dysfunctions in AD and schizophrenia. M_4 has also emerged as an interesting target in connection with the psychosis associated with schizophrenia, a potential that has been ascribed to the modulation of dopaminergic signaling exerted by the receptor in striatum, hippocampus, and neocortex. Selective agonists or potentiators of M_4, M_1, or M_1/M_4 are currently being investigated for these indications. The high expression of M_5 receptors on dopaminergic neurons in the ventral tegmental area and regulation of dopamine release in nucleus accumbens exerted by these receptors has prompted an interest in selective M_5 antagonists for the treatment of drug addiction. Finally, despite the aforementioned adverse effects arising from peripheral M_2 receptors, the potential of selective antagonists at this receptor for cognitive enhancement is still being pursued. M_2 is the predominant presynaptic mAChR in the CNS, and antagonism of this autoreceptor attenuates the negative feedback on synaptic ACh release, thereby resulting in increased release of the neurotransmitter and augmented cholinergic signaling (Figure 16.3c).

The medicinal chemistry efforts invested for several decades to exploit some of the potential in mAChRs as drug targets for cognitive disorders have so far not paid off. While mAChR ligands have been and still are used for indications such as glaucoma, peptic ulcer, motion sickness, and asthma, the outcome of these efforts in terms of CNS drugs have been disappointing as compound after compound have been retracted from clinical development due to insignificant efficacy or unacceptable side effects. However, as discouraging as this has been, the recent development of truly subtype-selective allosteric and bitopic ligands for the receptors has sparked a renewed interest in mAChRs as CNS targets.

16.4.1 Orthosteric mAChR Ligands

None of the classical orthosteric mAChR agonists **16.19–16.23** in Figure 16.7 display significant selectivity for any of the five receptor subtypes. Muscarine (**16.19**), a constituent of *Amanita muscaria*, has given name to the mAChRs because of its selectivity for these receptors over nAChRs. Replacement of the ester moiety of ACh by a carbamate group yields carbachol (**16.20**) which is not only a nonselective agonist at the five mAChRs but also at the nAChRs. The heterocyclic agonist pilocarpine (**16.21**) from the leaves of South American *Pilocarpus* shrubs is widely used as topical miotic for the control of elevated intraocular pressure associated with glaucoma. Despite being a nonselective mAChR agonist, the low bioavailability of pilocarpine means that it does not induce systemic side effects when administered topically. The potent agonist oxotremorine (**16.22**) has been used extensively as a lead compound for structure–activity studies giving rise to mAChR ligands spanning the entire efficacy range from full agonists over partial agonists to competitive antagonists.

Development of subtype-selective orthosteric mAChR agonists has proven difficult. Arecoline (**16.24**), a constituent of areca nuts (the seeds of *Areca catechu*) and a cyclic "reverse ester" bioisostere of ACh, has constituted the lead for numerous analogs, including xanomeline (**16.25**) in which the metabolically labile ester moiety of **16.24** has been replaced by the more stable thiadiazole ring. Originally proposed to be M_1 selective, subsequent functional characterization of xanomeline has found it to be M_1/M_4-prefering (not selective). Just as several other agonists exhibiting varying degrees of M_1 preference (including **16.26** and **16.27**), xanomeline has been in clinical development for the treatment of AD, and it is the only one among these to have reached phase III trials. Although the clinical development eventually was discontinued, the pro-cognitive and antipsychotic efficacies displayed by xanomeline in these trials have been instrumental for the validation of M_1 and M_4 as putative AD and schizophrenia targets and for the current development of allosteric and bitopic ligands for these receptors (Section 16.4.2).

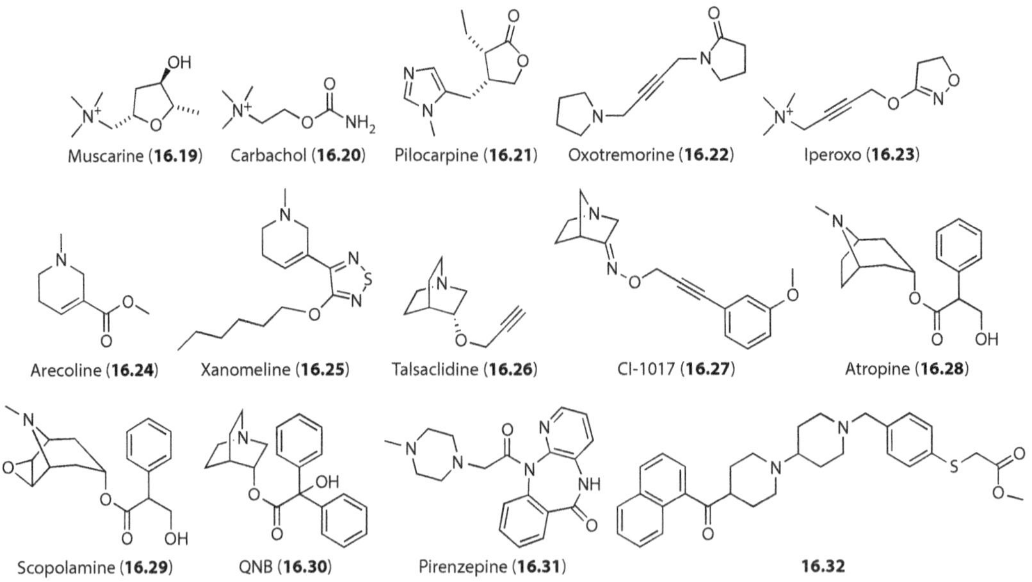

FIGURE 16.7 Chemical structures of orthosteric mAChR ligands.

Analogous to the lack of subtype-selective orthosteric agonists, it has also been difficult to develop truly subtype-selective competitive antagonists. The alkaloids atropine (**16.28**) and scopolamine (**16.29**) found in the berries of deadly nightshade (*Atropa belladonna*) and the close structural analog quinuclidinyl benzilate (QNB, **16.30**) are potent but nonselective mAChR antagonists that have been important pharmacological tools for the delineations of physiological functions of mAChRs over the years (Figure 16.7). Pirenzepine (**16.31**) is a M_1-preferring antagonist displaying significantly higher binding affinities for brain mAChRs (mainly M_1) over mAChRs in heart tissue (mainly M_2), although this M_1-over-M_2 selectivity has been less impressive when pirenzepine has been tested at recombinant mAChR subtypes. Conversely, truly selective M_2 receptor antagonists such as **16.32** have been developed and entered into clinical trials for cognitive enhancement but these have all been discontinued.

16.4.2 Allosteric and Bitopic Ligands of mAChRs

The renewed interest in mAChRs as CNS drug targets has been driven by the recently developed potent and subtype-selective allosteric ligands, but the concept of allosteric modulation (Chapter 12) is by no means new in the mAChR field. In fact, mAChRs have been the prototypic GPCRs in this respect since a wide range of diverse compounds capable of modulating their signaling have been known for decades. The first generation of these positive or negative allosteric modulators (PAMs and NAMs, respectively) had quite complex structures, as exemplified by **16.33–16.35**. Furthermore, the compounds were all characterized by having poor pharmacokinetic properties, low modulatory potencies at the mAChRs and activities at several other targets. Thus, it has been the emergence of high throughput (HTP) screening approaches enabling the search for novel structures from large compound libraries that has facilitated the identification of more potent and subtype-selective allosteric ligands suitable for in vivo studies.

The discovery of AC-42 (**16.36**) in a HTP screening represents the first example of a completely subtype-selective mAChR agonist (Figure 16.8). The compound is a fairly potent partial M_1 agonist exhibiting no activity at the other four mAChRs, and AC-42 analogs have been in clinical trials for the treatment of glaucoma. While AC-42 is not an orthosteric agonist, it does not have a completely allosteric mode of action either, and thus it is believed to be a bitopic ligand

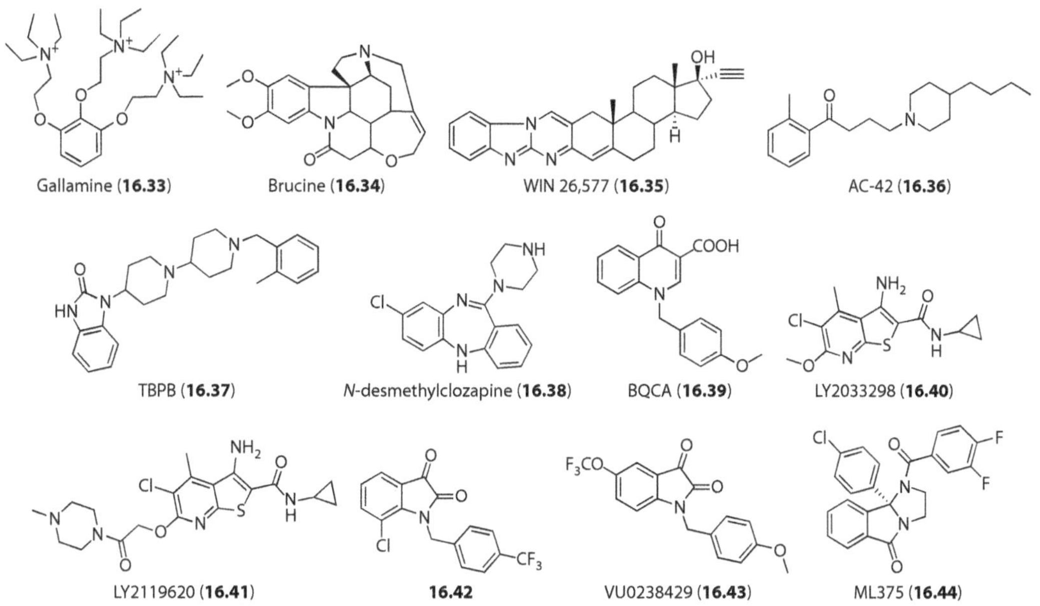

FIGURE 16.8 Chemical structures of allosteric and bitopic mAChRs ligands.

(see Section 16.4.3). Several equally selective M_1 agonists have followed in the trail of AC-42, including TBPB (**16.37**) and *N*-desmethylclozapine (**16.38**). Being one of two major in vivo metabolites of the prototypic atypical antipsychotic clozapine (Chapter 18), it has been speculated that the M_1 activity of *N*-desmethylclozapine may contribute to the clinical efficacy of clozapine in schizophrenic patients where other antipsychotics have limited effects.

In contrast to these M_1 agonists, BQCA (**16.39**) acts as a classical PAM exhibiting no intrinsic agonist activity at M_1 but potentiating ACh-mediated signaling through the receptor. BQCA is the most explored scaffold of the allosteric M_1 ligands to date, and interestingly BQCA-mediated potentiation of ACh-evoked signaling through M_1 has been shown to induce distinct intracellular effects compared to direct activation of the receptor by AC-42 or TBPB. This suggests that the cellular responses mediated by allosteric agonists and PAMs through M_1 could differ, and this could be important for their respective in vivo effects. A considerable number of PAMs exhibiting different degrees of selectivity for the M_4 receptor have also been published. Whereas LY2033298 (**16.40**) is a highly selective M_4 PAM that has been used as a pharmacological tool to explore the pro-cognitive and antipsychotic potential in the receptor, its analog LY2119620 (**16.41**) is an equipotent PAM at M_2 and M_4. Analogously, a series of *N*-methyl isatins exhibit very different selectivity profiles as mAChR PAMs, such as **16.42** that displays comparable EC_{50} values at M_1, M_2, M_3 and M_5 and the highly selective M_5 PAM VU0238429 (**16.43**). These examples illustrate how dramatic implications minor modifications on a modulator scaffold can have on its pharmacological properties. Finally, the selective M_5 receptor NAM ML375 (**16.44**), developed from a HTP screening hit, exhibits a high nanomolar IC_{50} value at this receptor and >1000-fold selectivity toward the four other receptors. ML375 is currently being applied in rodent studies probing the potential in selective M_5 antagonism for the treatment of drug addiction.

16.4.3 Ligand Binding to and Modulation of the mAChR

The binding modes of orthosteric and allosteric ligands to mAChRs have been studied extensively in mutagenesis and biochemical studies, and recent high-resolution crystal structures of the M_2 and M_3 receptors have provided detailed insight into these aspects (Figure 16.9). The orthosteric site in

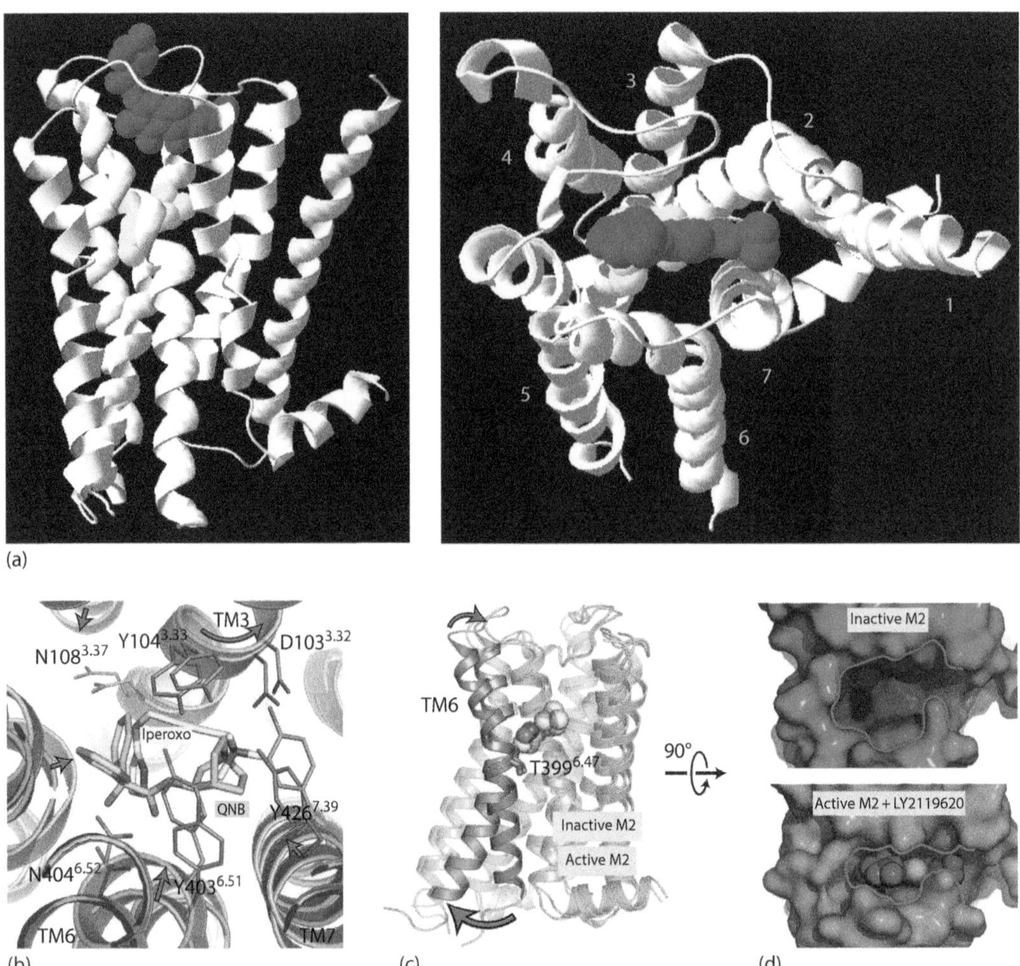

FIGURE 16.9 Orthosteric and allosteric ligand binding to the mAChR. (a) M_2 in complex with orthosteric agonist iperoxo (yellow, **16.23**) and the PAM LY2119620 (pink, **16.41**) (PDB: 4MQT). The receptor is viewed from the membrane (*left*) and from the extracellular side (*right*). (b) Overlay of the orthosteric binding sites in M_2 in the M_2/iperoxo (*orange*) and M_2/QNB (*blue*) crystal structures. Iperoxo and QNB (**16.30**) are given in yellow and green, respectively. (c) Overlay of the structures of M_2 in its inactive (*blue*) and active (*orange*) conformations. (d) Binding of the PAM LY2119620 to a narrow cavity above the orthosteric site facilitates the stabilization of the active conformation of M_2 (*orange*) mediated by iperoxo. Conversely, the PAM is unable to bind this cavity in the inactive receptor conformation (*blue*). (b, c, and d: Reprinted by permission from Macmillan Publishers Ltd. *Nature*, Kruse, A.C., Ring, A.M., Manglik, A., Hu, J., Hu, K., and Eitel, K., Activation and allosteric modulation of a muscarinic acetylcholine receptor, 504, 101–106. Copyright 2013.)

the mAChR (and in most other class A GPCRs) is located in a cavity formed by transmembrane helices (TMs) 3, 5, 6, and 7 (Figure 16.9a and b). All orthosteric mAChR ligands contain a quaternized ammonium group or an amino group (protonated at physiological pH), and the other parts of the molecules contain various heteroatoms capable of forming hydrogen bonds (Figure 16.7). The binding of the orthosteric ligand to the mAChR is centered in the ionic and cation–π interactions established by the positively charged group of the ligand with an aspartate residue in TM3 and aromatic residues in TM3, TM6, and TM7, respectively, and these interactions are supplemented by hydrogen bonds and van der Waals' interactions between the other part of the ligand and other residues in the binding site (Figure 16.9b). Agonist binding to the orthosteric site stabilizes an active receptor conformation which is characterized by inward movements of three of the four helixes forming the site compared to the inactive conformation. This structural difference explains why competitive antagonists of the mAChRs generally are larger molecules than orthosteric agonists (Figures 16.7 and 16.9b). The aforementioned difficulties obtaining subtype-selective orthosteric ligand are rooted in the complete conservation of the residues forming the orthosteric sites in the five subtypes. Conversely, the pronounced subtype-selectivity displayed by allosteric and bitopic ligands comes from these ligands targeting less conserved receptor regions.

The structural diversity of allosteric mAChR ligands (Figure 16.8) suggests that these ligands potentially could target several different sites in the receptor. However, extensive research into the modes of action of the first generation of allosteric modulators as well as AC-42 (**16.36**) and BQCA (**16.39**) indicates that most of these modulators target a common allosteric site located superficially to the orthosteric site in the mAChR. The existence of such a common allosteric site has been further supported by a recent crystal structure of M_2 in complex with iperoxo (**16.23**) and LY2119620 (**16.41**), in which the PAM is observed to bind to a site just above the orthosteric site and in this way facilitate the stabilization of the active mAChR conformation mediated by the orthosteric agonist (Figure 16.9a and d). Other PAMs acting through this common allosteric site most likely potentiate mAChR signaling through similar mechanisms, and conversely binding of a ligand to this site could be envisioned to favor the inactive rather than the active receptor conformation and in this way act as a NAM.

The mAChR PAMs (including **16.39–16.43**) obviously target site(s) in the receptors topographically distinct from the orthosteric site since they require concomitant binding of agonist to the receptor in order to mediate their effects. In contrast, several of the identified M_1-selective agonists (including **16.36–16.38**) have been proposed to be bitopic ligands, meaning that they target the orthosteric site as well as an allosteric site in close proximity to it. This hypothesis is supported by the structural resemblance that fragments of these agonists bear to orthosteric mAChR ligands, whereas other parts of the agonists are expected to protrude from the orthosteric site into neighboring receptor regions. In this scenario, these agonists obtain their pronounced subtype-selectivity from less conserved allosteric site while their interactions with the orthosteric site contribute to their binding affinity. The binding of some of the bitopic mAChR agonists have been speculated to bridge between the orthosteric site and the common allosteric site located just above it in the mAChR (Figure 16.9a).

16.5 NICOTINIC ACh RECEPTORS

The nAChRs are members of the Cys-loop receptor superfamily that also comprises ligand-gated ion channels for GABA, glycine, and serotonin (Chapter 12). The nAChR is a membrane-bound protein complex composed of five subunits forming an ion pore through which cations (Na^+, K^+, and Ca^{2+}) can enter the cell when the receptor is activated, resulting in depolarization of the neuron. The 16 human nAChR subunits assemble into "muscle-type" and "neuronal" receptor subtypes (Figure 16.10a and b). The muscle-type nAChR composed of α_1, β_1, δ, and γ/ε subunits is localized postsynaptically at the neuromuscular junction, where it mediates the electrical transmission across the anatomical gap between the motor nerve and the skeletal muscle, thus creating the skeletal muscle tone. The neuronal nAChRs are heteromeric or homomeric complexes assembled from α_2–α_7, α_9, α_{10}, and β_2–β_4 subunits (Figure 16.10b). The high heterogeneity of native neuronal nAChR populations thus arises from the differential expression of these subunits in the CNS combined with their ability to be incorporated into a substantial number of receptor combinations, each characterized by distinct functional and biophysical properties (Figure 16.10c). This molecular diversity extends down to the specific α/β combination, as exemplified by the two $\alpha_4\beta_2$ receptors stoichiometries $(\alpha_4)_2(\beta_2)_3$ and $(\alpha_4)_3(\beta_2)_2$, that exhibit very different agonist sensitivities and desensitization properties (Figure 16.10c). The neuronal nAChRs are located at presynaptic and postsynaptic densities in autonomic ganglia and in cholinergic neurons throughout the CNS, and equally important to their roles in cholinergic neurotransmission are their contributions as heteroreceptors regulating the synaptic release of several important neurotransmitters. In fact, much of the potential in neuronal nAChRs as drug targets arises from their regulation of dopaminergic, glutamatergic, or GABAergic neurotransmission (Figure 16.10d).

While it has been estimated that at least 20 different neuronal nAChR subtypes are expressed in native tissues, the $\alpha_4\beta_2$ and α_7 receptors are by far most abundant nAChR subtypes in vivo. The $\alpha_4\beta_2$ nAChRs constitute ~90% of the high-affinity binding sites for nicotine in the brain, and thus much of the therapeutic potential in nAChRs resides with this subtype (Figure 16.10d). The signaling characteristics of the homomeric α_7 nAChR differ significantly from those of heteromeric nAChRs as this receptor exhibits low ACh sensitivity, extremely fast desensitization kinetics, and a remarkably high Ca^{2+} permeability. The α_7 receptor has in recent years emerged as the most promising nAChR to target in disorders characterized by cognitive and/or sensory impairments, including AD and schizophrenia. Although these two major nAChRs naturally have attracted most attention in terms of drug development, several of the "minor" nAChR subtypes also constitute interesting therapeutic targets. For example, the α_6-containing nAChRs expressed almost exclusively in the midbrain have been shown to regulate synaptic dopamine release here which makes them interesting in relation to Parkinson's disease and nicotine addiction (Figure 16.10d).

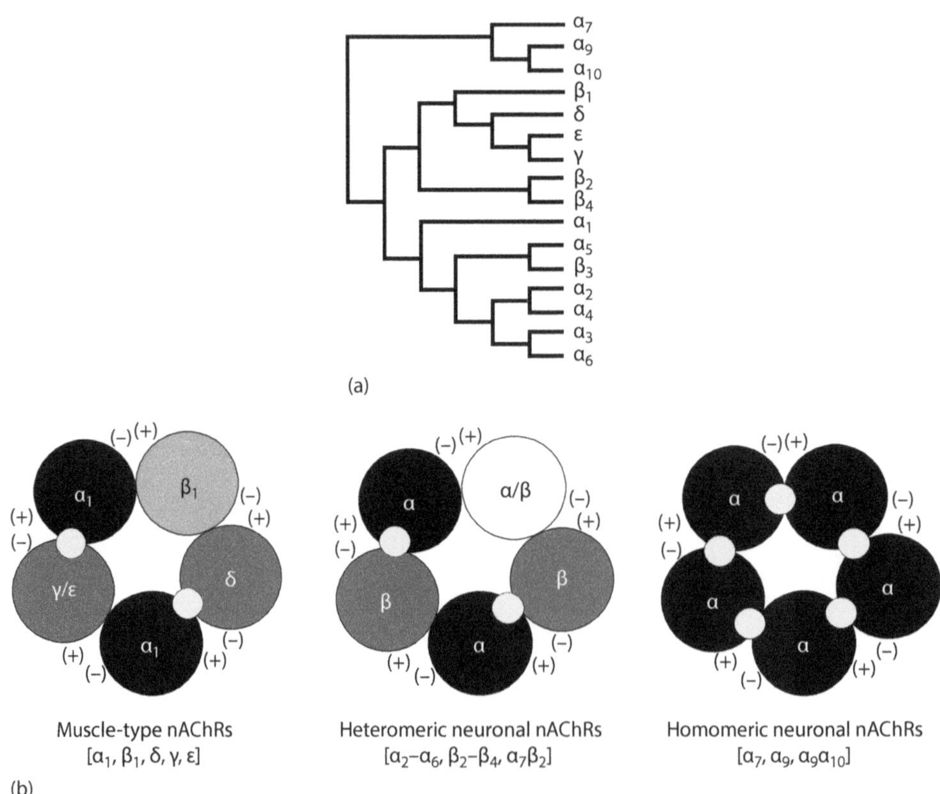

FIGURE 16.10 The nAChR family. (a) Phylogenetic tree of the 16 human nAChR subunits. The proximity of the outer branches between two subunits indicates how similar their amino acid sequences are (e.g., α_3 and α_6 are very similar). (b) The nAChR complexes formed by the 16 subunits. The localization of the orthosteric sites in the receptors is indicated with yellow circles. The (+)- and (−)-sides of the subunits in the pentameric complexes are indicated. *(Continued)*

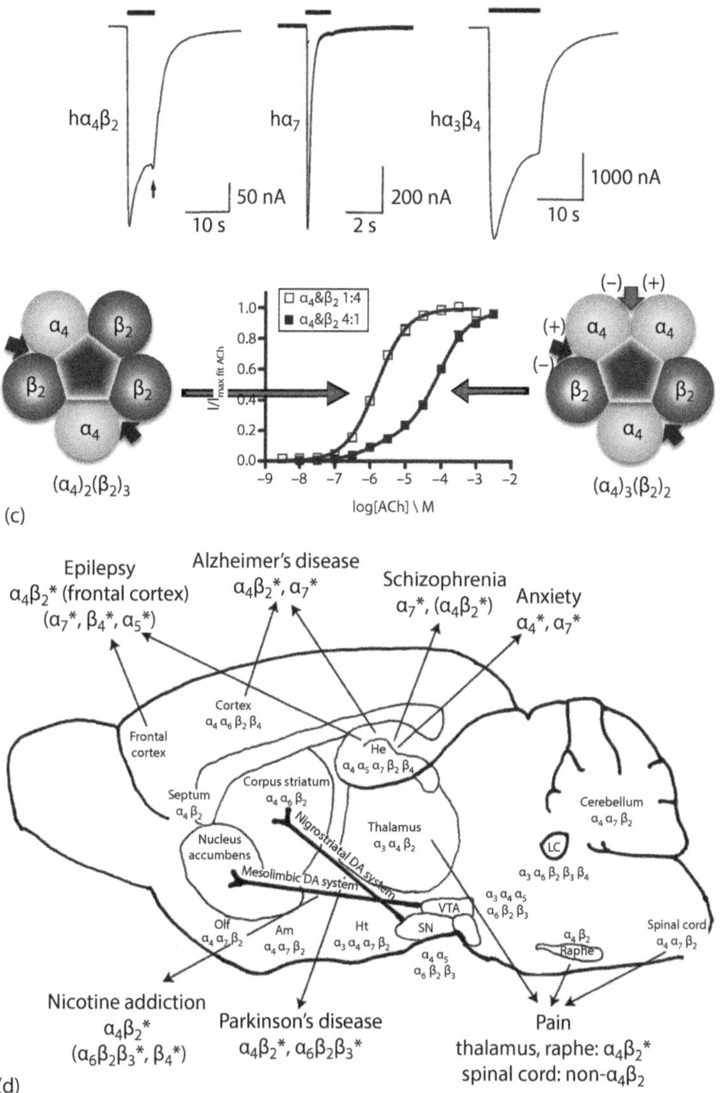

FIGURE 16.10 (*Continued*) The nAChR family. (c) Top: Traces from electrophysiological recordings of the responses evoked by maximally effective concentrations of ACh at human $\alpha_4\beta_2$, α_7, and $\alpha_3\beta_4$ nAChRs expressed in *Xenopus* oocytes. Black bars indicating the duration of ACh applications. Bottom: Concentration–response curves for ACh at $(\alpha_4)_2(\beta_2)_3$ and $(\alpha_4)_3(\beta_2)_2$ nAChRs expressed in *Xenopus* oocytes. (d) Distribution of nAChR subunits in the rodent brain. The predominantly expressed subunits in selected brain regions are shown, and the nAChRs proposed as potential therapeutic targets in various disorders are indicated. ([c, top] Reprinted from Chavez-Noriega, L.E. et al., *J. Pharmacol. Exp. Therapeut.*, 280, 46. Copyright 1997, with permission from ASPET; [c, bottom] reprinted with permission from Harpsøe, K. et al., *J. Neurosci.*, 31, 10759. Copyright 2011; [d] reprinted with permission from Jensen, A.A., Frølund, B., Liljefors, T., and Krogsgaard-Larsen, P., Neuronal nicotinic acetylcholine receptors: Structural revelations, target identifications, and therapeutic inspirations, *J. Med. Chem.*, 48, 4705–4745. Copyright 2005 American Chemical Society; Reprinted from *Pharmacol. Ther.*, 92, Picciotto, M.R., Caldarone, B.J., Brunzell, D.H., Zachariou, V., Stevens, T.R., and King, S.L., Neuronal nicotinic acetylcholine receptor subunit knockout mice: Physiological and behavioral phenotypes and possible clinical implications, 89–108. Copyright 2001, with permission from Elsevier.)

16.5.1 Orthosteric nAChR Ligands

As in the AChE and mAChR fields, medicinal chemistry development of nAChR ligands has been greatly assisted by natural sources. The classical nAChR agonists **16.45–16.50** are natural product compounds, which, albeit selective for nAChRs over mAChRs, are rather nonselective within the nAChR family (Figure 16.11). Because of their potent nAChR agonism and excellent pharmacokinetic properties, (*S*)-nicotine (**16.45**) and (±)-epibatidine (**16.46**) have been applied as lead compounds for the vast majority of nAChR agonists developed over the years. Several strategies have been applied to optimize the pharmacological properties of the two leads to different CNS indications. Modifications of the linker between the pyridine and pyrrolidine rings and/or introduction of new ring systems in (*S*)-nicotine (**16.45**) have resulted in several potent nAChR agonists, such as **16.51** and **16.52**, and opening of the pyrrolidine ring has given rise to ispronicline (**16.53**) and other potent analogs. The agonists display significant functional selectivity for $\alpha_4\beta_2$ over other nAChR subtypes, and several of them have been in clinical development as cognitive enhancers and analgesics. The 5-ethynyl nicotine analog altinicline (**16.54**) is a low-efficacious $\alpha_4\beta_2$ nAChR agonist exhibiting functional preference for β_4- over β_2-containing subtypes which has been in clinical trials for Parkinson's disease.

(±)-Epibatidine (**16.46**) was originally isolated from the skin of the Ecuadorian frog *Epipedobates tricolor*, and it is by far the most potent of the classical nAChR agonists. The therapeutic interest in epibatidine was sparked by studies showing that it is a highly efficacious and nonaddictive analgesic. However, because of the severe hypertension and neuromuscular paralysis also induced by the compound, several of its analogs have been developed in an attempt to optimize the analgesic effects compared to these side effects. (±)-UB-165 (**16.54**), a hybrid compound between epibatidine and anatoxin-a (**16.49**) from the algae *Anabaena flos-aquae*, is one of the few nAChR agonists that selectively activate "minor" heteromeric nAChRs without concomitant activation of the major $\alpha_4\beta_2$

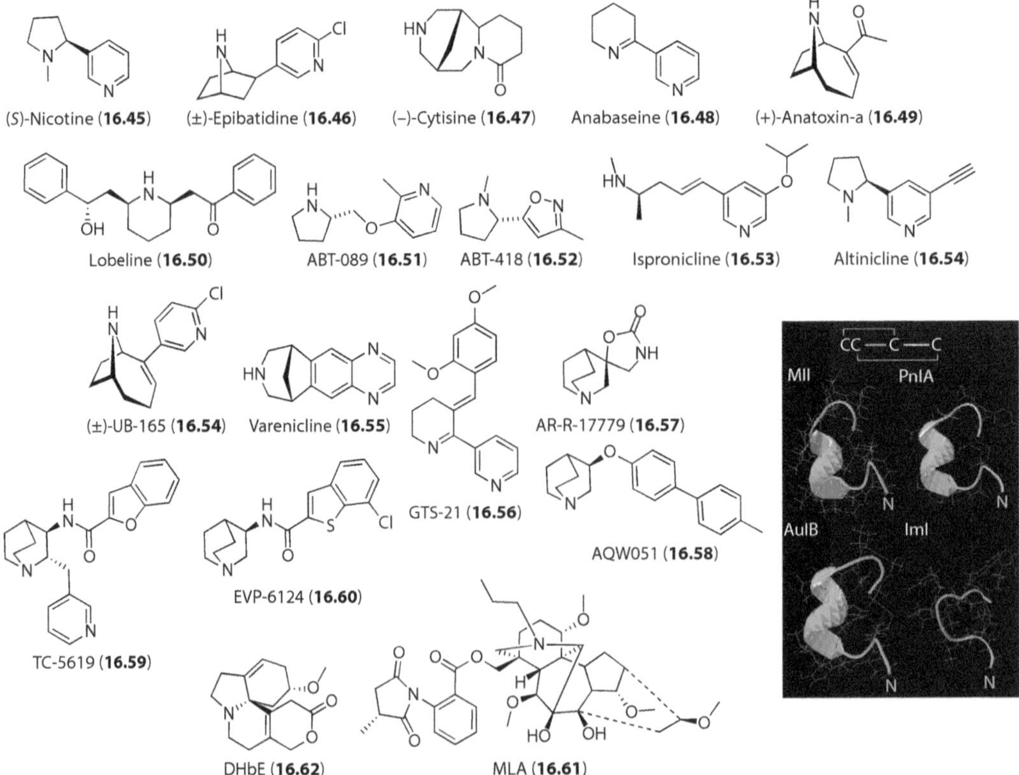

FIGURE 16.11 Chemical structures of orthosteric nAChR ligands and 3D structures of four α-conotoxins.

and α_7 subtypes. (–)-Cytisine (**16.47**) from *Laburnum anagyroides* is a potent partial agonist of the heteromeric nAChRs, and its analog varenicline (**16.55**) has recently been launched as a smoking cessation aid (Chantix®/Champix®). With both varenicline and ispronicline (**16.53**) presently in clinical development for mild to moderate AD, it remains to be seen whether $\alpha_4\beta_2$ nAChR agonists hold therapeutic potential for cognitive disorders.

Although the α_7 nAChR represents a low-affinity binding site for ACh and the natural source compounds **16.45–16.50**, potent and α_7-selective agonists have been developed from some of these leads (Figure 16.11). The toxin anabaseine (**16.48**) found in marine worms and ant species is a nonselective nAChR agonist displaying a somewhat higher efficacy at α_7 nAChR than at the heteromeric nAChRs. Introduction of conjugated aryl substituents in the 3-position of the pyridine ring of anabaseine has increased this functional selectivity substantially, for example, in the prototypic α_7 agonist GTS-21 (**16.56**). The quinuclidine (1-azabicyclo[2.2.2]octane) ring system has also been immensely important in this field as it can be found in several of the α_7-selective partial agonists currently undergoing clinical trials for cognitive disorders such as AD, schizophrenia and attention deficit hyper disorder (**16.57–16.60**). The frontrunner of these, EVP-6124 (**16.60**), is currently undergoing phase III trials in AD patients.

Very little medicinal chemistry development has been done on nAChR antagonists, even though recent studies suggest that these ligands possess antidepressant effects. Thus, the presently available nAChR antagonists predominantly originate from natural sources. The peptide toxin α-bungarotoxin from the Taiwan banded krait (*Bungarus multicinctus*) is a potent competitive antagonist of α_7 and muscle-type nAChRs, and methyllycaconitine (MLA, **16.61**), isolated from *Delphinum* and *Consolida* species, is a highly potent and selective α_7 antagonist. Conversely, the *Erythrina* alkaloid dihydro-β-erythroidine (DHβE, **16.62**) displays some preference for β_2- over β_4-containing heteromeric nAChRs (Figure 16.11). Finally, several highly subtype-specific competitive antagonists have been isolated from the venoms of a family of predatory cone snails, the so-called α-conotoxins. The 3D structures of these 12–20 amino acid peptides are established by the different arrangements of intramolecular disulfide bonds between four highly conserved cysteine residues, and their different subtype-selectivity profiles at the nAChRs arise from the differences in their nonconserved residues (Figure 16.11).

16.5.2 Allosteric Ligands of the nAChRs

Analogously to other ligand-gated ion channels such as $GABA_A$ and NMDA receptors (Chapter 15), nAChRs are highly susceptible to allosteric modulation. Several endogenous ligands such as steroids (**16.63**), 5-hydroxyindole (**16.64**), and Ca^{2+} and Zn^{2+} are known to modulate nAChR signaling through allosteric sites. Interestingly, the AChEIs physostigmine (**16.5**) and galanthamine (**16.9**) also act as PAMs at nAChRs, an activity component that has been proposed to contribute to the efficacy of the latter as an AD drug.

PAMs hold several advantages to regular agonists when it comes to augmentation of nAChR signaling. First, analogously to AChEIs and allosteric mAChR modulators they only exert their effect when ACh is present in the synapse, and thus enhancement of nAChR signaling will be mediated by the PAM in a physiological tone. Second, the fact that allosteric ligands typically target receptor regions less conserved than the orthosteric site means that they are more likely to be subtype-selective than orthosteric ligands. Finally, PAMs are generally less disposed to induce receptor desensitization than agonists. This latter point may be particularly important since the failure of previous clinical trials with nAChR agonists in some cases has been ascribed to the agonists acting as *de facto* antagonists due to their induction of receptor desensitization.

A selection of the nAChR PAMs published in recent years is presented in Figure 16.12. Whereas LY-2087101 (**16.65**) potentiates the signaling of several nAChRs, compounds **16.66–16.68** and **16.69–16.73** are selective $\alpha_4\beta_2$ and α_7 PAMs, respectively. NS 9283 (**16.67**) is an interesting $\alpha_4\beta_2$ PAM, since it acts specifically at the $(\alpha_4)_3(\beta_2)_2$ stoichiometry and thus is able to potentiate the signaling of low ACh-sensitivity $\alpha_4\beta_2$ nAChRs selectively (Figure 16.10c). α_7 nAChR PAMs are often divided

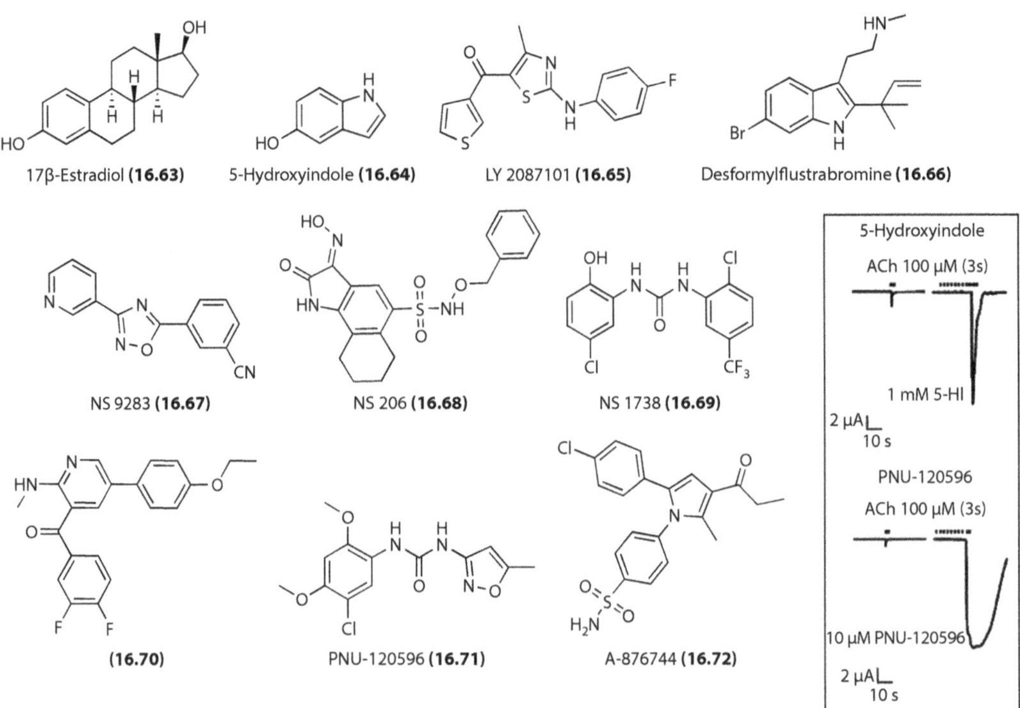

FIGURE 16.12 Chemical structures of nAChR PAMs and examples of the potentiation of ACh-evoked signaling through the α_7 nAChR mediated by a type I PAM (5-hydroxyindole) and a type II PAM (PNU-120596). (Reprinted from *Biochem. Pharmacol.*, 74, Bertrand, D. and Gopalakrishnan, M., Allosteric modulation of nicotinic acetylcholine receptors, 1155–1163. Copyright 2007, with permission from Elsevier.)

into two categories based on whether or not they modulate the extremely fast desensitization kinetics of this receptor as this has tremendous implications for the nature of their modulation. Type I PAMs like 5-hydroxyindole (**16.64**), NS 1738 (**16.69**), and **16.70** increase agonist sensitivity and enhance the amplitude of the receptor response, but having no or little effect on the desensitization they do not change the kinetics or the shape of the response (Figure 16.12). In contrast, type II PAMs like PNU-120596 (**16.71**) and A-876744 (**16.72**) slow down or block the desensitization of the receptor which produces dramatically different current response profiles (Figure 16.12). It remains to be seen whether the different degrees of α_7 modulation exerted by type I and II α_7 PAMs are important for their respective in vivo efficacies.

16.5.3 Ligand Binding to and Modulation of the nAChR

Numerous mutagenesis, biochemical and structural studies of nAChRs have provided a detailed insight into the molecular basis for orthosteric ligand binding. The orthosteric sites are located in the extracellular amino-terminal domain of the nAChR complex, more specifically at the two $\alpha^{(+)}/\beta^{(-)}$ interfaces in the heteromeric nAChRs and at the five $\alpha^{(+)}/\alpha^{(-)}$ interfaces in the homomeric receptors (Figures 16.10b and 16.13). Analogous to the orthosteric mAChR ligands, all orthosteric agonists and small-molecule competitive antagonists of nAChRs possess a quaternary amino group or a protonated amino group (Figure 16.11). This positively charged group docks into an "aromatic box" formed by five aromatic residues mainly from the $\alpha^{(+)}$ side of the orthosteric site, forming a strong cation–π interaction with one of these ring systems. This "primary binding component" is supplement by interactions between the other part of the orthosteric ligand and residues from the $\beta^{(-)}$ side and $\alpha^{(-)}$ side in the heteromeric and homomeric nAChR, respectively (Figure 16.13). Since the

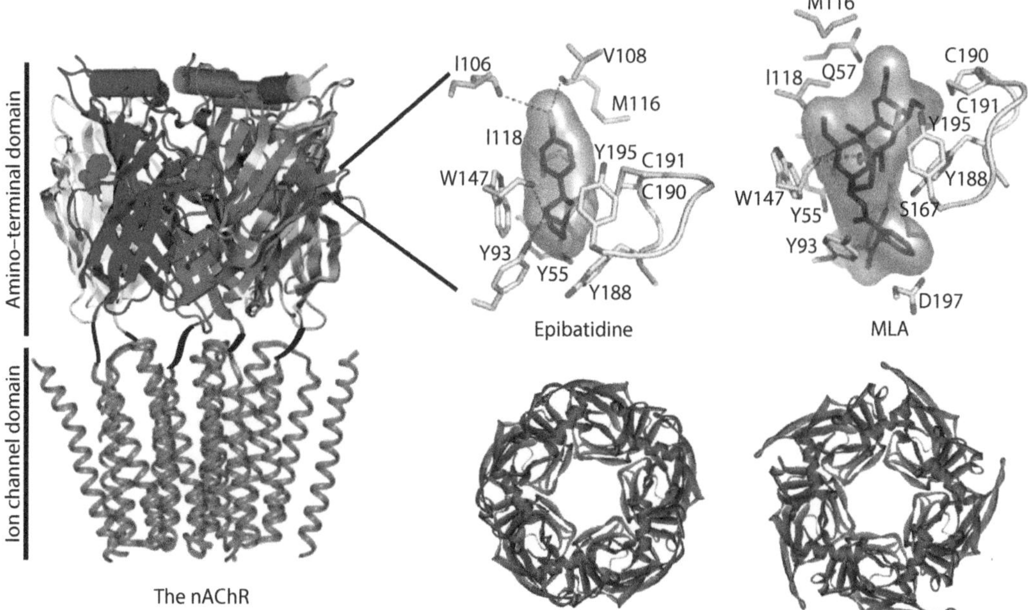

FIGURE 16.13 Orthosteric ligand binding to the nAChR. Left: the pentameric nAChR complex is given with the amino-terminal and ion channel domains indicated. Right: The binding modes of agonist (±)-epibatidine (**16.46**) and competitive antagonist MLA (**16.61**) to the orthosteric site in the receptor (top) and the differential degrees of closure of the binding pocket upon binding of the two ligands (bottom). ([middle and right] Reprinted from the Hansen, S.B., Sulzenbacher, G., Huxford, T., Marchot, P., Taylor, P., and Bourne, Y.: Structures of Aplysia AChBP complexes with nicotinic agonists and antagonists reveal distinctive binding interfaces and conformations. *EMBO J.* 2005. 24. 3635–3646. Copyright Wiley-VCH Verlag GmbH & Co. KGaA. Reproduced with permission; [left] reprinted with permission from Jensen, A.A., Frølund, B., Liljefors, T., and Krogsgaard-Larsen, P., Neuronal nicotinic acetylcholine receptors: Structural revelations, target identifications, and therapeutic inspirations, *J. Med. Chem.*, 48, 4705–4745. Copyright 2005 American Chemical Society.)

residues forming the aromatic box are highly conserved throughout nAChRs, subtype-selectivity in orthosteric ligands most often arise from this "complementary binding component." Agonists are typically smaller than competitive antagonists of nAChRs, and this size difference is the key determinant of the different intrinsic activities of the ligands (Figure 16.13). Whereas the orthosteric site is able to close up around the agonist when bound to the site, the presence of a larger ligand (the competitive antagonist) in the binding pocket does not allow for the same degree of closure and thus does not induce the conformational changes, leading to opening of the ion channel (Figure 16.13).

In contrast to the proposed common allosteric site in the mAChR, allosteric ligands of nAChRs have been demonstrated to act through several different binding sites (Figure 16.14a). Several PAMs of heteromeric nAChRs target allosteric sites located in those extracellular subunit interfaces that do not comprise the orthosteric sites. An example of such a PAM is NS 9283 (**16.67**) which binds to the extracellular $\alpha_4^{(+)}/\alpha_4^{(-)}$ subunit interface in the $(\alpha_4)_3(\beta_2)_2$ nAChR (Figures 16.10c and 16.14a). This explains the high specificity of this PAM for this $\alpha_4\beta_2$ stoichiometry since the $\alpha_4^{(+)}/\alpha_4^{(-)}$ interface is not present in the $(\alpha_4)_2(\beta_2)_3$ receptor or in other nAChRs. The nAChR signaling is also highly sensitive to ligands binding in the transmembrane receptor regions, where both intrasubunit sites (formed by the four TM helices in one subunit) and intersubunit sites (formed by TM helices from neighboring subunits) have been identified (Figure 16.14a). These sites are located directly behind the five TM2 helices forming the ion pore in the nAChR receptor, and it makes intuitively sense that ligand binding behind this ion pore can affect receptor gating, be it in an inhibitory (NAM) or stimulatory (PAM) manner. The degree and the nature of the modulation exerted through

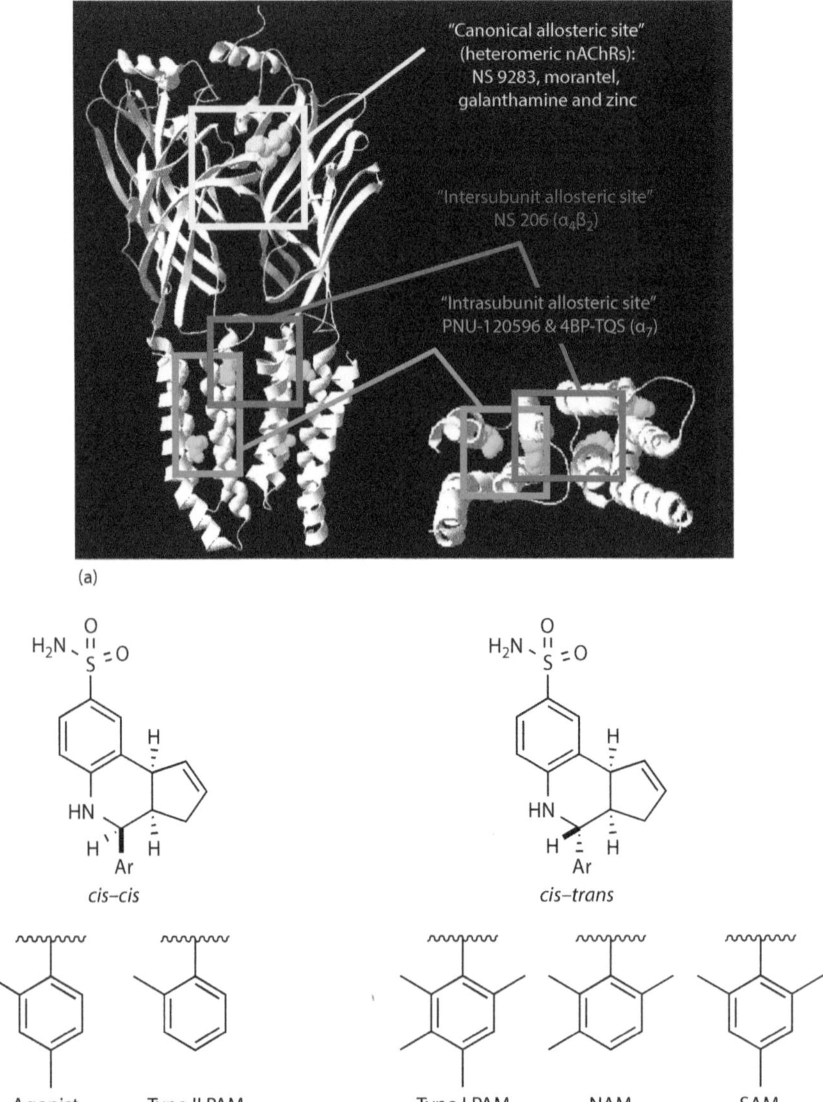

FIGURE 16.14 Allosteric ligand binding to the nAChR. (a) Selected allosteric sites in the nAChR complex and examples of modulators binding to them. For reasons of clarity, only two neighboring subunits of the pentameric receptor are shown. (b) The diverse effects on the α_7 nAChR signaling exerted by *cis–cis* and *cis–trans* isomers of 4-(naphtalen-1-yl)-3*a*,4,5,9*b*-tetrahydro-3*H*-cyclopenta[*c*]quinolone-8-sulfonamide with different methyl substitution patterns in the phenyl ring. SAM, silent allosteric modulator.

these sites can vary considerably between different nAChRs, and subtle structural modifications of the modulator can lead to dramatically different modulatory properties (Figure 16.14b). This can partly be ascribed to the different energy barriers between the resting, active, and desensitized conformations of different nAChR subtypes and partly to the mode of action of the specific modulator, that is which of the processes involved in nAChR signaling it affects and how (Figure 16.15a). An allosteric modulator can modulate receptor signaling through effects on the activation, deactivation, desensitization, and/or resensitization processes (Figure 16.15b). As exemplified for NS 9283 (**16.67**) in Figure 16.15c, several of these processes can be affected simultaneously by modulator binding, making the observed modulation of the sum of all these effects.

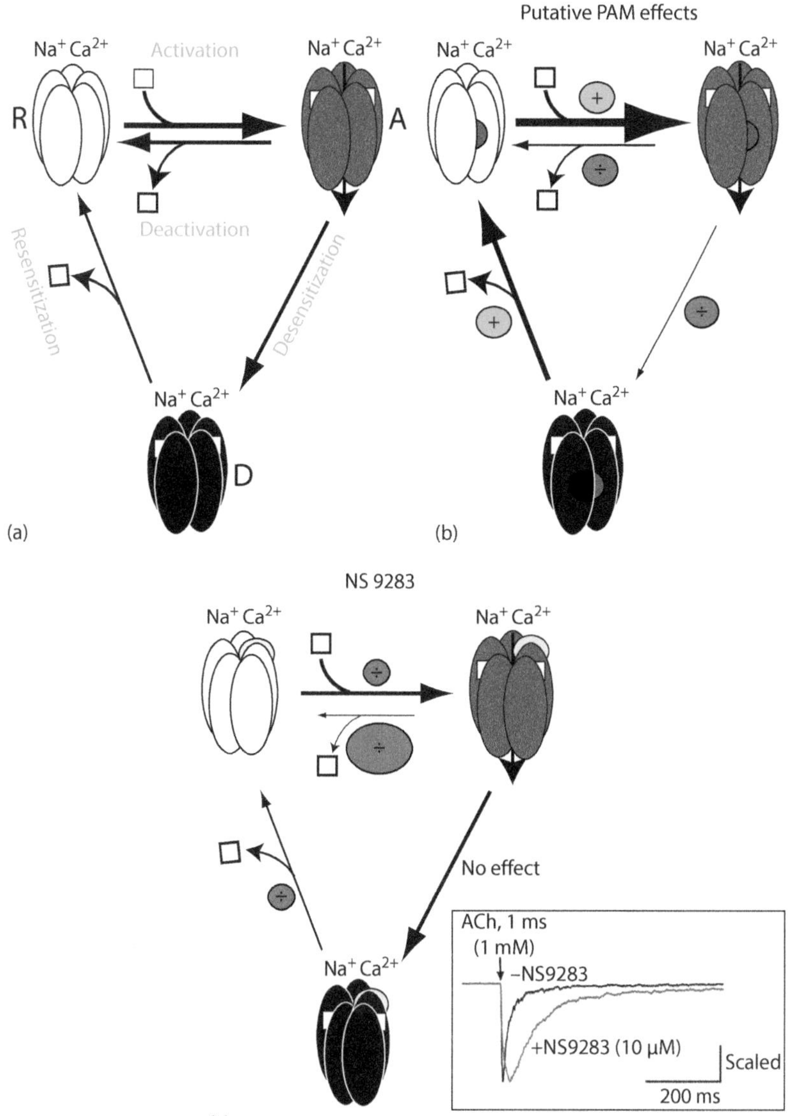

FIGURE 16.15 Allosteric modulation of the nAChR. (a) Simplified scheme outlining the equilibriums between resting [R, white], active [A, gray] and desensitized [D, black] nAChR conformations and the four major processes underlying receptor signaling. Binding of the agonist (white square) stabilizes the active receptor conformation and enables ion flux (*activation*). The agonist-bound, activated receptor then either releases the agonist and converts back to the unbound, resting conformation (*deactivation*) or converts to the agonist-bound, desensitized receptor (*desensitization*). The desensitized receptor eventually releases the agonist and returns to the resting, unbound conformation (*resensitization*). (b) The putative effects mediated by a PAM on agonist-evoked nAChR signaling. The PAM (purple circle) may potentiate receptor signaling by increasing *activation* or *resensitization* kinetics or by slowing down the *deactivation* or *desensitization* kinetics. The putative effects mediated by a NAM will be, as per definition, the opposite (not shown). (c) The potentiation of $(\alpha_4)_3(\beta_2)_2$ nAChR signaling by NS 9283 (**16.67**) is the sum of several effects mediated by the PAM. NS 9283 (yellow circle) does not change the desensitization kinetics of the receptor, and it exerts inhibitory effects at the activation and resensitization processes. Thus, its potentiation of the $(\alpha_4)_3(\beta_2)_2$ receptor arises from its major effect, a significant reduction of the deactivation kinetics (inserted figure). (Reprinted from Grupe, M., Jensen, A.A., Ahring, P.K., Christensen, J.K., and Grunnet, M.: Unraveling the mechanism of action of NS9283, a positive allosteric modulator of $(\alpha_4)_3(\beta_2)_2$ nicotinic ACh receptors. *Br. J. Pharmacol.* 2013. 168. 2000–2010. Copyright Wiley-VCH Verlag GmbH & Co. KGaA. Reproduced with permission.)

16.6 CONCLUDING REMARKS

The importance of cholinergic transmission for a plethora of processes throughout the body means that pharmacological modulation of cholinergic targets holds indisputable potential for the development of drugs for numerous indications. However, this abundance of the neurotransmitter in vivo also means that numerous unacceptable adverse effects can arise from modulation of cholinergic signaling, and historically insufficient selectivity in cholinergic drugs has turned out to be a major obstacle to the exploitation of this therapeutic potential. Despite their relative poor clinical efficacy as AD drugs, the validation of augmentation of cholinergic neurotransmission as treatment strategy for cognitive dysfunctions provided by the AChEIs seems to constitute a promising foundation for drug development in a field that is in dire need of better drugs. In light of this, the availability of the new generations of subtype-selective agonists and allosteric modulators of mAChRs and nAChRs developed in recent years, in particular those augmenting M_1, M_4, and α_7 receptor signaling, present an exciting possibility to probe the potential in the cholinergic system for the treatment of AD and other cognitive disorders in a more specific and comprehensive manner than done previously.

FURTHER READING

Anand, R., Gill, K.D., and Mahdi, A.A. 2014. Therapeutics of Alzheimer's disease: Past, present and future. *Neuropharmacology* 76:27–50.

Bertrand, D. and Gopalakrishnan, M. 2007. Allosteric modulation of nicotinic acetylcholine receptors. *Biochem. Pharmacol.* 74:1155–1163.

Colovic, M.B., Krstic, D.Z., Lazarevic-Pasti, T.D., Bondzic, A.M., and Vasic, V.M. 2013. Acetylcholinesterase inhibitors: Pharmacology and toxicology. *Curr. Neuropharmacol.* 11:315–335.

Foster, D.J., Choi, D.L., Conn, P.J., and Rook, J.M. 2014. Activation of M1 and M4 muscarinic receptors as potential treatments for Alzheimer's disease and schizophrenia. *Neuropsychiatr. Dis. Treat.* 10:183–191.

Jensen, A.A., Frølund, B., Liljefors, T., and Krogsgaard-Larsen, P. 2005. Neuronal nicotinic acetylcholine receptors: Structural revelations, target identifications, and therapeutic inspirations. *J. Med. Chem.* 48: 4705–4745.

Kruse, A.C., Kobilka, B.K., Gautam, D., Sexton, P.M., Christopoulos, A., and Wess, J. 2014. Muscarinic acetylcholine receptors: Novel opportunities for drug development. *Nat. Rev. Drug Discov.* 13:549–560.

Taly, A., Corringer, P.J., Guedin, D., Lestage, P., and Changeux, J.P. 2009. Nicotinic receptors: Allosteric transitions and therapeutic targets in the nervous system. *Nat. Rev. Drug Discov.* 8:733–750.

Wess, J., Eglen, R.M., and Gautam, D. 2007. Muscarinic acetylcholine receptors: Mutant mice provide new insights for drug development. *Nat. Rev. Drug Disc.* 6:721–733.

17 Histamine Receptors

Iwan de Esch, Henk Timmerman, and Rob Leurs

CONTENTS

17.1 INTRODUCTION

Histamine is an aminergic neurotransmitter and a local hormone and plays major roles in the regulation of several (patho)physiological processes. In biological systems, histamine is synthesized from L-histidine by histidine-decarboxylase (Scheme 17.1). In the brain (where histamine acts as a neurotransmitter), the synthesis takes place in restricted populations of neurons that are located in the tuberomammillary nucleus of the posterior hypothalamus. These neurons project to most cerebral areas and have been implicated in several brain functions (e.g., sleep/wakefulness, hormonal secretion, cardiovascular control, thermoregulation, food intake, and memory formation). In peripheral tissues (where histamine acts as a local hormone), the compound is stored in mast cells, eosinophils, basophils, enterochromaffin cells, and probably also some specific neurons. Once released, histamine is rapidly metabolized via N-methylation of the imidazole ring by the enzyme *histamine N-methyltransferase* and by oxidation of the amine function by *diamine oxidase* (Scheme 17.1).

Some of the symptoms of allergic conditions in the skin and the airway system are known to result from histamine release after mast cell degranulation. In 1937, Bovet and Staub discovered the first compounds that antagonize these effects of histamine. Ever since, there has been intense research activity devoted toward finding novel ligands with (anti)histaminergic activity.

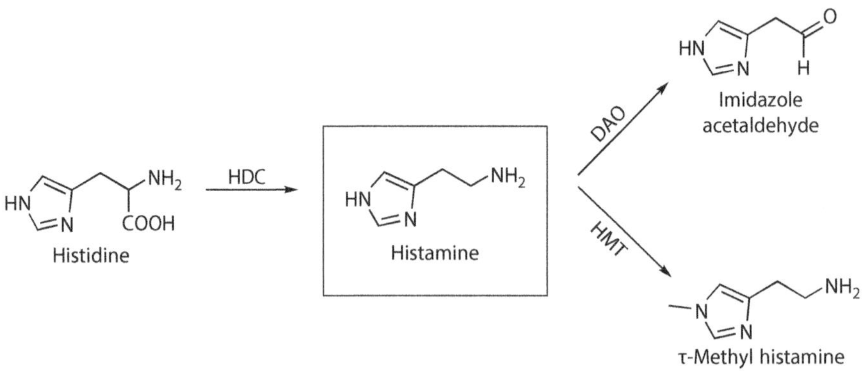

SCHEME 17.1 Synthesis and metabolism of histamine. HDC, histidine decarboxylase; DAO, diamine oxidase; HMT, histamine *N*-methyltransferase.

As more (patho)physiological processes that are mediated by histamine (e.g., stomach acid secretion and neurotransmitter release) were studied, it became apparent that the action of histamine is mediated by four distinct receptor subtypes. The first three histamine receptor subtypes were discovered by classical pharmacological means, i.e., using subtype selective ligands that were developed by medicinal chemists. The histamine H_4 receptor was discovered using the human genome, nicely illustrating the impact of molecular biology and genomics in drug discovery.

In this chapter, we will describe in detail the state-of-the-art knowledge on the molecular features of the histamine receptor proteins, the medicinal chemistry of the four histamine receptors, and the (potential) therapeutic applications of selective receptor ligands.

17.2 HISTAMINE H_1 RECEPTOR: MOLECULAR ASPECTS AND SELECTIVE LIGANDS

17.2.1 Molecular Aspects of the Histamine H_1 Receptor Protein

The H_1 receptor belongs to the large family of rhodopsin-like, G-protein-coupled receptors (GPCRs). In 1991, the cDNA encoding a bovine H_1 receptor protein was cloned after an expression cloning strategy in *Xenopus* oocytes by Fukui and coworkers. The human H_1 receptor gene resides on chromosome 3, and the deduced amino acid sequence revealed a 491 amino acid protein of 56 kDa. Using the cDNA sequence encoding the bovine H_1 receptor, the cDNA sequences and intronless genes encoding the rat, guinea pig, human and mouse H_1 receptor proteins were cloned soon thereafter. These receptor proteins are slightly different in length and highly homologous and do not show major pharmacological differences. In 2011, Iwata and colleagues solved the X-ray structure of a T4-lysozyme fusion construct of the human histamine H_1 receptor in complex with doxepin (see the section on H_1 antagonists).

The stimulation of the H_1 receptor leads to the phospholipase C-catalyzed formation of the second messengers inositol 1,4,5-triphosphate (IP_3) and 1,2-diacylglycerol (DAG) which in turn leads to the mobilization of intracellular calcium and the activation of protein kinase C, respectively.

17.2.2 H_1 Receptor Agonists

Modification of the imidazole moiety of histamine has been the most successful approach for obtaining selective H_1 agonists (Figure 17.1). The presence of the tautomeric N^{π}–N^{τ} system of the imidazole ring is not obligatory as reflected by the selective H_1 agonists 2-pyridylethylamine and 2-thiazolylethylamine. Substitution of the imidazole ring at the 2-position leads to relatively selective H_1 agonists. For example, 2-(3-bromophenyl)histamine is a relatively potent H_1 receptor agonist. Schunack and

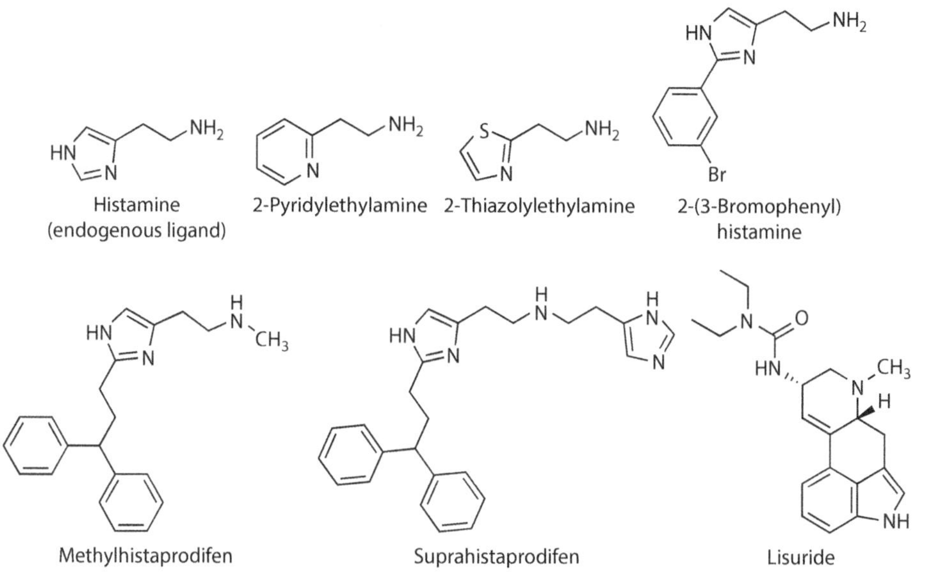

Histamine (endogenous ligand) 2-Pyridylethylamine 2-Thiazolylethylamine 2-(3-Bromophenyl) histamine

Methylhistaprodifen Suprahistaprodifen Lisuride

FIGURE 17.1 Histamine H_1 receptor agonists.

colleagues subsequently developed a series of the so-called histaprodifens on the basis of the hypothesis that the introduction of a diphenylalkyl substituent on the 2-position of the imidazole ring yields high-affinity agonists. This hypothesis was based on the realization that a diphenylmethyl group is a common feature of high-affinity H_1 antagonists (see Section 17.2.3). The introduction of the diphenylpropyl substituent at the 2-position of the imidazole ring and N-methylation of the ethylamine side chain results in the high potency agonist *N*-methyl-histaprodifen. Further modifications of the diphenylmethyl moiety were unsuccessful and indicated a considerable difference in SAR (and most likely receptor binding site) of the diphenyl moieties of the histaprodifens and the structurally related H_1 antagonists. A further increase in H_1 receptor agonist potency was obtained by a bivalent ligand approach. Suprahistaprodifen, a dimer of histaprodifen and histamine, is currently one of the most potent H_1 receptor agonists available. Surprisingly, recent high-throughput screening (HTS) of CNS-active drugs at the histamine H_1 receptor has identified the nonimidazole ergot derivative lisuride as another high-affinity H_1 receptor agonist.

17.2.3 H_1 Receptor Antagonists

The first antihistamines were identified and optimized by exclusively studying in vivo activities. This might be the explanation why several compounds originally reported as antihistamines were later on developed for other applications; e.g., the first so-called tricyclic antidepressants (e.g., doxepin, Figure 17.2) are often also very potent H_1 antagonists. More modern approaches, using genetically modified cells expressing the human H_1 receptor, currently provides more in-depth information on the molecular mechanism of actions. All therapeutically used H_1 antagonists in fact act as inverse agonists and favor an inactive conformation of the GPCR protein. In view of the detectable level of spontaneous activity of the H_1 receptor (i.e., receptor signaling without agonist, also known as constitutive GPCR activity), the H_1 antagonists tested so far all inhibit the constitutive activation of, e.g., nuclear factor-κB (NF-κB). Moreover, following the recent elucidation of the X-ray structure (resolution of 3.1 angstrom) of a T4-lysozyme stabilized human H_1 receptor, we now also have a clear idea on the binding interactions of the H_1 antagonists. In the X-ray structure, earlier predictions of the binding pocket of H_1 antagonists have proven reality (Figure 17.3); the tricyclic doxepin binds relatively deep in the binding pocket of the H_1 receptor via an ionic interaction of its

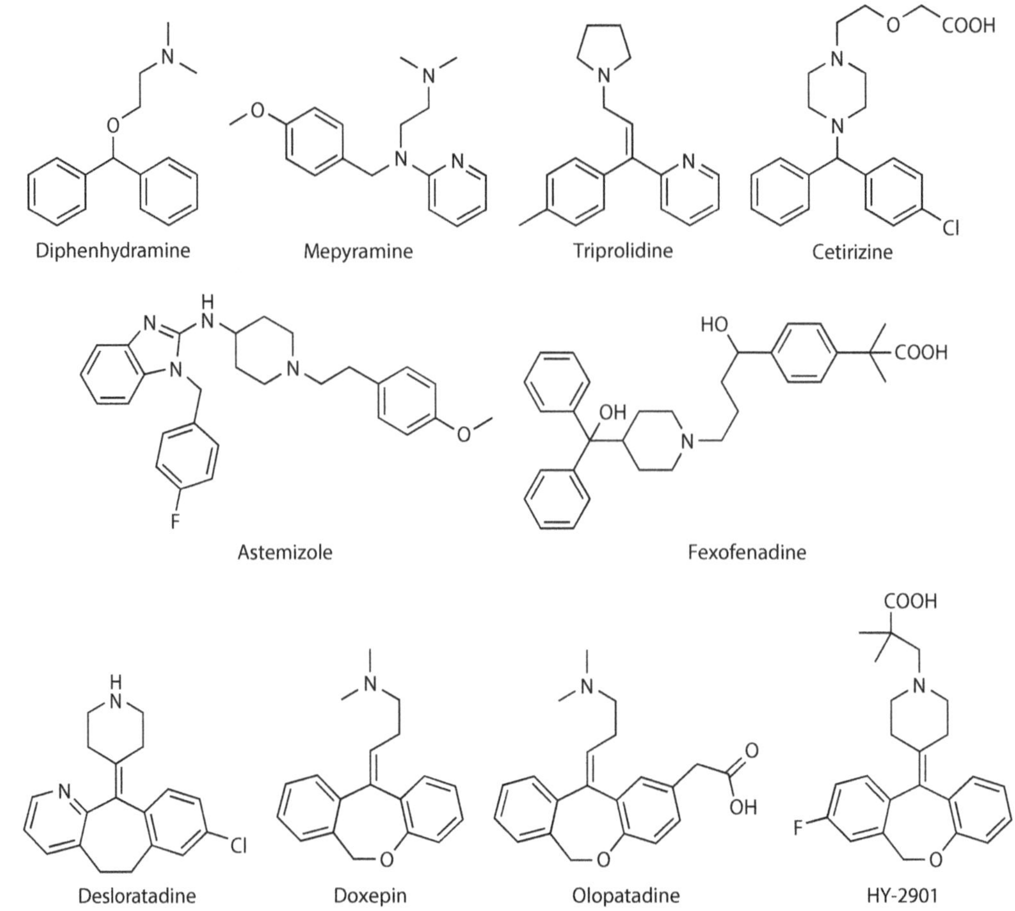

Diphenhydramine Mepyramine Triprolidine Cetirizine

Astemizole Fexofenadine

Desloratadine Doxepin Olopatadine HY-2901

FIGURE 17.2 Histamine H_1 receptor antagonists (inverse agonists).

protonated amine with the highly conserved negatively charged aspartate residue in transmembrane domain 3 (Asp170) and hydrophobic interactions of the tricyclic aromatic ring system with aromatic amino acids in transmembrane domains 5 and 6 (e.g., Phe435, Trp428, and Phe432).

Of the many first-generation histamine blockers, diphenhydramine (Figure 17.2) is considered as the archetype. The compound is known in medicine as Benadryl®, the first antihistamine successfully used in man. Other compounds of this class are, e.g., mepyramine and triprolidine. These compounds are highly potent H_1 antagonists and very useful both for pharmacological investigations and medicinal use. The so-called classical "antihistamines" easily penetrate the brain and are therefore also useful for in vivo CNS studies.

The early antihistamines had two major drawbacks: they were ligands for several targets; especially the antimuscarinic effects caused unpleasant side effects (e.g., dry mouth). By careful structural modifications, it is possible to obtain selective antihistamines. Another drawback of the first-generation H_1 antagonists was that the compounds show strong sedating effects, to such a level that some of them are still used as sleeping aids.

The notion that sedation is caused by a blockade of H_1 receptors in the brain sparked the search for nonbrain-penetrating compounds. Minor structural modifications resulted in a number of new, nonsedating H_1 antagonists (e.g., cetirizine, astemizole, fexofenadine, desloratadine, and olopatadine), also referred to as the second-generation H_1 blockers. Interestingly, the first of such compounds were more or less found by chance (e.g., fexofenadine being an active metabolite of terfenadine), and it took quite some time to understand why the compounds did not manifest CNS effects. It is now understood

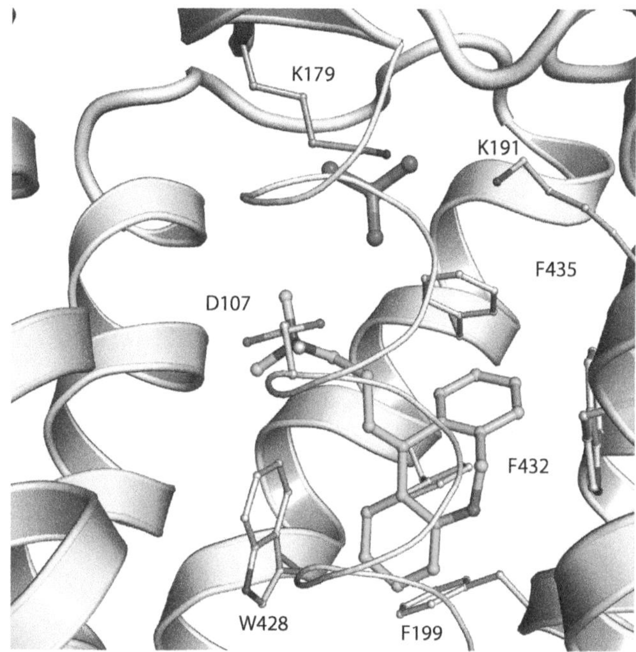

FIGURE 17.3 Doxepin binding site in the histamine H_1 receptor as determined by X-ray crystallography.

that besides the zwitterionic nature of many of these compounds, several of them also act as substrates of the PgP transport system in the blood–brain barrier. Consequently, these compounds are actively transported out of the brain and thereby are not able to occupy significant amounts of brain H_1 receptors. The new class of compounds, including terfenadine (later on due to HERG-blockade replaced by its metabolite fexofenadine), cetirizine (now replaced by the L-enantiomer), and loratadine (now replaced by its desoxy-active metabolite, desloratadine), reached as antiallergics the blockbuster status. The elucidation of the X-ray structure of the H_1 receptor in the complex also provided an interesting hypothesis for the interaction of zwitterionic antagonist with the H_1 receptor protein. In the receptor binding pocket, a phosphate ion can be found in the upper part of the binding pocket, interaction with, e.g., two lysine residues (Figure 17.3). Interestingly, one of these lysine residues has earlier been shown to affect the binding of zwitterions like cetirizine, suggesting that the "phosphate pocket" might be the binding site for the carboxylate moieties of the zwitterionic second-generation antagonists.

Although effective in treating allergic reactions, the second-generation H_1 antagonists do not display significant anti-inflammatory effects. Currently, research focuses on compounds also targeting inflammation; compounds having besides H_1 blocking properties antagonizing also, e.g., LTB_4 or blocking the synthesis of leukotrienes, have the interest of pharmaceutical companies. Recently, the combined blockade of H_1 and H_4 receptors (see later text) has also been indicated as an attractive new approach. Interestingly, since the turn of the century, the interest in the sleep promoting effects of histamine H_1 receptor antagonists has increased. Especially, the "old" derivative doxepin, a compound that blocks the H_1 receptor and also the H_2 receptor, is used as a sleep inducer. An analog of doxepin, HY-2901 (Figure 17.2), has been shown to have interesting properties for use as a sleep inducer and is currently under clinical evaluation as sleep aid.

17.2.4 Therapeutic Use of H_1 Receptor Ligands

The histamine H_1 receptor is a well-established drug target and has been thoroughly studied for decades. The first-generation but especially the second-generation antihistamines are clinically very successful and are widely available drugs. The main indications are hay fever, allergic rhinitis, and conjunctivitis, as well as comparable allergic affections; the application for asthmatic conditions does not seem to be of much use. The first-generation antihistamines are still used as in over-the-counter sleep aids or antiflu combination pills. As indicated before, currently, interest in sleep aids is increasing and new molecules are being developed (e.g., HY-2901).

17.3 HISTAMINE H_2 RECEPTOR: MOLECULAR ASPECTS AND SELECTIVE LIGANDS

17.3.1 Molecular Aspects of the Histamine H_2 Receptor Protein

The fact that the "antihistamines" did not antagonize histamine-induced effects at the stomach and the heart, led in 1966 to the proposal by Ash and Schild of two distinct histamine receptors: the H_1 and H_2 receptors. This hypothesis became generally accepted in 1972 when Black and his coworkers at Smith, Kline, and Beecham presented burimamide and related compounds. These ligands antagonize the effects of histamine on the stomach and the heart. Nowadays, the H_2 receptor is (as all histamine receptor subtypes) known to belong to the rhodopsin-like family of GPCRs. Using a polymerase chain reaction (PCR)–based method, based on the known sequence similarity of various GPCRs and gastric parietal mRNA, the H_2 receptor nucleotide sequence was elucidated. This DNA sequence encodes for a 359 amino acid GPCR receptor protein. Soon thereafter, the intronless genes encoding the rat, human, guinea pig, and mouse H_2 receptor were cloned by means of homology. The H_2 receptor proteins are slightly different in length and highly homologous and do not show major pharmacological differences. Interestingly, several polymorphisms have been found in the human H_2 receptor gene and one of the mutations might be linked to schizophrenia.

The histamine H_2 receptor is positively coupled to the adenylate cyclase system via G_s proteins in a variety of tissues (e.g., brain, stomach, heart, gastric mucosa, and lung). Moreover, cell lines recombinantly expressing the H_2 receptor show increases in cAMP following H_2 receptor activation.

17.3.2 H_2 Receptor Agonists

Simple modification of the histamine molecule has not been very successful to obtain selective and potent H_2 receptor agonists. A first step toward an H_2 receptor agonist was made with the discovery of dimaprit (Figure 17.4) which was found during a search for H_2 receptor antagonists in a series of isothiourea derivatives. Dimaprit is an H_2 receptor agonist that is almost as active as histamine at the H_2 receptor, but hardly displays any H_1 receptor agonism. Later, it was found that dimaprit is also a moderate H_3 receptor antagonist and a moderate H_4 receptor agonist. Using dimaprit as a template, amthamine (2-amino-5-(2-aminoethyl)-4-methylthiazole) was designed as a rigid dimaprit analog. Following the original suggestion of Green et al. that the sulfur atom of dimaprit might act as a

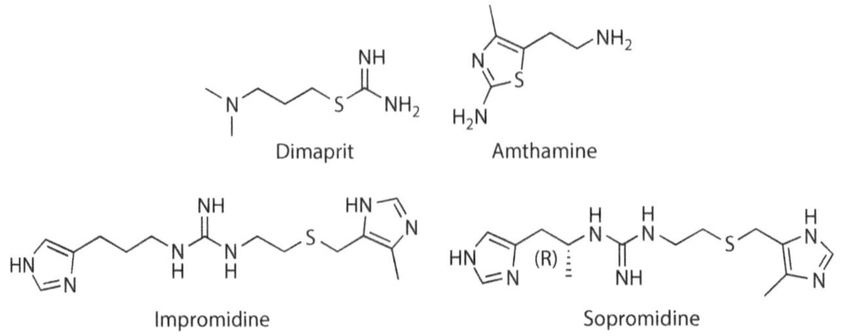

Dimaprit

Amthamine

Impromidine

Sopromidine

FIGURE 17.4 Reference histamine H_2 receptor agonists.

proton acceptor in a hydrogen bonding network with the H_2 receptor (in analogy of the idea of the interaction of the imidazole ring with the receptor protein), quantum chemical calculations by and synthesis of 2-aminothiazole analogs confirmed this idea. Amthamine combines a high H_2 receptor selectivity with a potency which is slightly higher compared to histamine, both in vitro and in vivo. An H_2 receptor agonist that is more potent than histamine is the guanidine derivative impromidine. This ligand actually combines a rather high H_2 receptor affinity with a reduced efficacy. Impromidine also shows moderate H_1- and potent H_3-receptor antagonistic and H_4-receptor agonistic activity. Interestingly, replacement of the propyl-imidazole moiety of impromidine with an α-methyl-ethylimidazole group results in the chiral analog (Figure 17.4). The *R*(−)-isomer, sopromodine, acts as a potent H_2 agonist, whereas the *S*(+)-isomer is a weak H_2 antagonist. Both compounds possess only weak H_3 and H_4 antagonistic activities, making *R*(−)-sopromodine one of the most potent and selective H_2 agonists to date.

17.3.3 H_2 Receptor Antagonists

The identification of N^{α}-guanylhistamine as a partial H_2 agonist in a gastric acid secretion model led to the development of the relatively weak H_2 antagonist burimamide (Scheme 17.2) following the replacement of the strong basic guanidine group by the noncharged, polar thiourea and side chain elongation. Years later, it was shown that burimamide is also an H_3 and H_4 receptor partial agonist. As H_2 receptor antagonist, burimamide lacked oral activity in man most likely due to its moderate potency. Nevertheless, burimamide was the lead for the development of selective and clinically useful H_2 receptor antagonists, such as cimetidine. Over time, many H_2 antagonists have been described; almost all of them possess 2 planar π-electron systems connected by a flexible chain. The 4-methylimidazole moiety of cimetidine can easily be replaced by other heterocyclic groups. Replacement by a substituted furan (e.g., ranitidine) or thiazole ring (e.g., famotidine) leads to compounds that are usually more potent at the H_2 receptor than cimetidine. Moreover, the replacement of

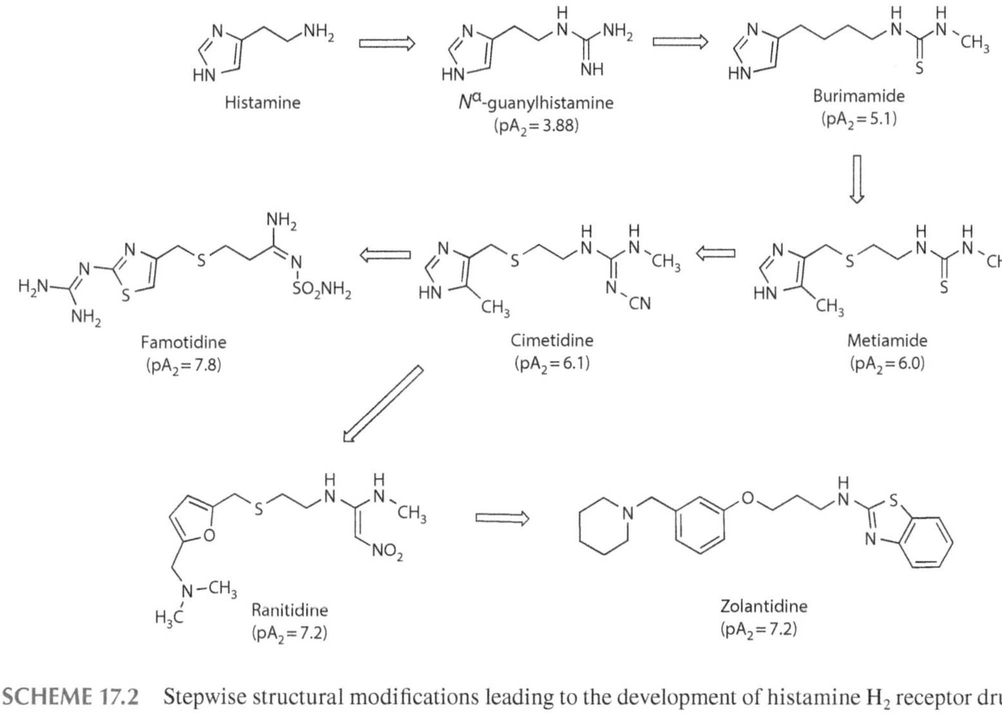

Histamine

N^{α}-guanylhistamine (pA_2 = 3.88)

Burimamide (pA_2 = 5.1)

Famotidine (pA_2 = 7.8)

Cimetidine (pA_2 = 6.1)

Metiamide (pA_2 = 6.0)

Ranitidine (pA_2 = 7.2)

Zolantidine (pA_2 = 7.2)

SCHEME 17.2 Stepwise structural modifications leading to the development of histamine H_2 receptor drugs to treat gastric ulcers.

the imidazole moiety also eliminates the undesired inhibition of cytochrome P-450. Most H_2 antagonists are rather polar compounds which do not readily cross the blood–brain barrier. The brain-penetrating H_2 antagonist zolantidine represents a rather nonclassical structure with the oxygen atom of the furan ring in ranitidine placed outside the aromatic ring and replacement of the polar group with a benzothiazole group (Scheme 17.2) and has become an important tool for in vivo CNS studies.

Like the H_1 receptor, the H_2 receptor was reported to be spontaneously active in transfected CHO cells. Based on this concept, many H_2 antagonists were reclassified: cimetidine, ranitidine, and famotidine are in fact inverse agonists, whereas burimamide acts in this model system as a neutral antagonist. This difference in pharmacological profile was also seen in a differential effect on H_2 receptor regulation, whereas long-term treatment with inverse agonists, like cimetidine, resulted in H_2 receptor upregulation, and exposure to the neutral antagonist burimamide did not affect the receptor expression. Such receptor upregulation was also observed in rabbit parietal cells, resulting in acid hypersecretion after H_2 antagonist withdrawal. These data present a mechanistic explanation for the known tolerance induction by H_2 antagonist that sometimes occurs in man.

17.3.4 Therapeutic Use of H_2 Receptor Ligands

At the moment, there is no clinical application of H_2 agonists, although sometimes histamine is used as diagnostic aid in patients with stomach problems. In contrast, H_2 antagonists have proven to be very effective drugs to alleviate the symptoms of duodenal ulcers, stomach ulcers, and reflux esophagitis. Nowadays, the blockbuster status of the H_2 antagonists has been strongly reduced with the introduction of the proton pump inhibitors, like omeprazole, to directly inhibit the gastric acid secretion and the eradication of *Helicobacter pylori* with antibiotics as actual cure, instead of symptomatic treatment.

17.4 Histamine H_3 Receptor: Molecular Aspects and Selective Ligands

17.4.1 Molecular Aspects of the Histamine H_3 Receptor Protein

The physiological role of histamine as a neurotransmitter became apparent in 1983, when Arrang and coworkers discovered the inhibitory effect of histamine on its own release and synthesis in the brain. This effect was not mediated by the known H_1 and H_2 receptor subtypes as no correlation with either the H_1 or the H_2 receptor activity of known histaminergic ligands was observed. Soon thereafter, the H_3 receptor agonist *R*-α-methylhistamine and the antagonist thioperamide (see Figures 17.5 and 17.6, respectively) were developed, thereby confirming that a new receptor subtype regulates the release and synthesis of histamine. In addition, the H_3 receptor regulates the release of other important neurotransmitters, such as acetylcholine, serotonin, noradrenaline, and dopamine. Next to its

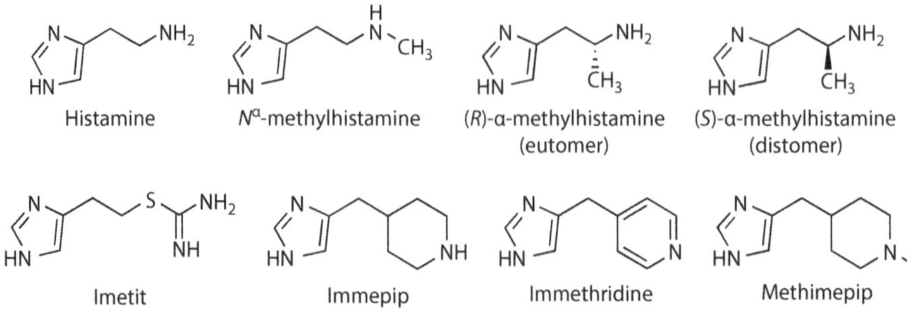

Histamine; N^{α}-methylhistamine; (*R*)-α-methylhistamine (eutomer); (*S*)-α-methylhistamine (distomer); Imetit; Immepip; Immethridine; Methimepip

FIGURE 17.5 Reference histamine H_3 receptor agonists.

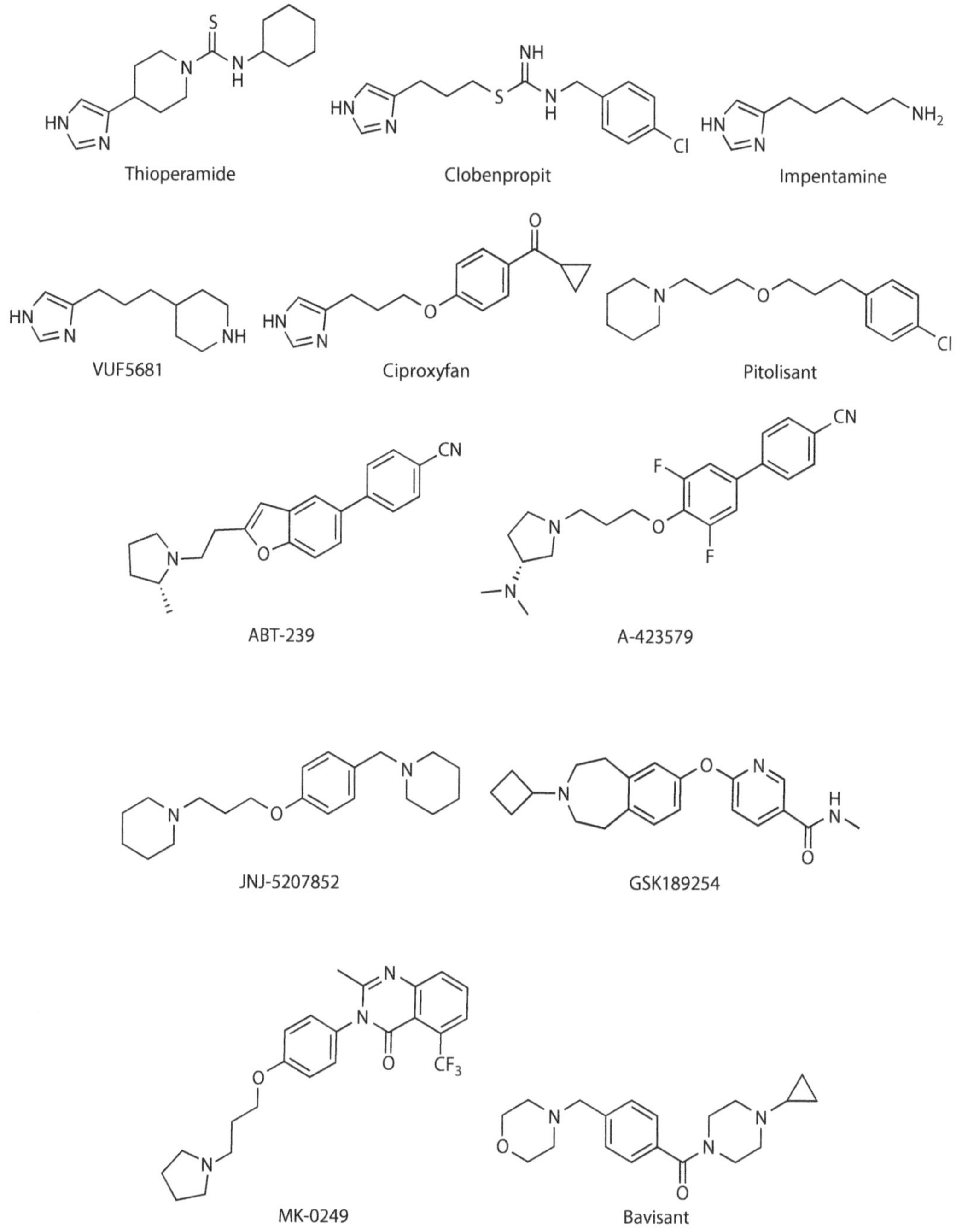

FIGURE 17.6 Selected histamine H_3 receptor antagonists and inverse agonists.

high expression in certain regions of the human brain (e.g., the basal ganglia, hippocampus, and cortical areas, i.e., the parts of the brain that are associated with cognition), the H_3 receptor is also present to some extent in the peripheral nervous system, e.g., in the gastrointestinal tract, the airways, and the cardiovascular system.

Initial efforts to identify the H_3 receptor gene using the anticipated homology with the previously identified H_1 and H_2 receptor genes all failed. Eventually, the human H_3 receptor cDNA was identified by Lovenberg and coworkers at Johnson & Johnson in 1999. In search for novel GPCR proteins using a homology search of commercial genome databases, a receptor with high similarity to the M_2 muscarinic acetylcholine receptor and high brain expression was identified. Expression of the gene and full pharmacological characterization established this protein as the histamine H_3 receptor. Cloning of the H_3 receptor genes of other species, including rat, guinea pig, and mouse, soon followed and important H_3 receptor species differences have been identified. The H_3 receptor mRNA undergoes extensive alternative splicing, resulting in many H_3 receptor isoforms that have different signaling properties and expression profiles. Moreover, the H_3 receptor displays particularly high constitutive activity which can also be observed in vivo, again leading to a reclassification of existing ligands into agonists, neutral antagonists, and inverse agonists.

The H_3 receptor signals via $G_{i/o}$ proteins as shown by the pertussis toxin-sensitive stimulation of [^{35}S]-GTPγS binding in rat cortical membranes. The inhibition of adenylyl cyclase after stimulation of the H_3 receptor results in lowering of cellular cAMP levels and modulation of cAMP response element-binding protein (CREB)-dependent gene transcription.

17.4.2 Histamine H_3 Receptor Agonists

At the H_3 receptor, histamine itself is a highly active agonist. Methylation of the terminal amino function results in N^{α}-methylhistamine (Figure 17.5), a compound that is H_3 selective and even more active than histamine. Methylation of the α-carbon atom of the ethylamine side chain also increases the potency at the H_3 receptor. This increased activity resides completely in the *R*-isomer; the corresponding *S*-isomer is approximately 100-fold less potent. Since the methylation leads to highly reduced activity at both the H_1- and H_2 receptor, but still substantial activity at the H_4 receptor, *R*-(α)-methylhistamine is a moderately selective agonist at the H_3 receptor. In combination with its less active *S*-isomer, this compound has proven to be highly useful for the pharmacological characterization of H_3 receptor-mediated effects. For potent H_3 agonism, the amine function of histamine can be replaced by an isothiourea group, as in imetit. This compound is also very active in vitro and in vivo, as is *R*-(α)-methylhistamine. The basic group in the imidazole side chain can also be incorporated in ring structures. For example, immepip is a potent H_3 agonist that is effective in vitro and in vivo. Although the described first-generation H_3 agonists were intensively used as reference ligands to study the H_3 receptor, all of them proved to have considerable activity for the recently discovered H_4 receptor. Therefore, a new generation of potent and selective H_3 agonists has been developed, most notably immethridine (pEC_{50} = 9.8; 300-fold selectivity over the H_4 receptor) and methimepip (pEC_{50} = 9.5; >10,000-fold selectivity over the H_4 receptor). These latter compounds are devoid of high H_4 receptor activity.

17.4.3 Histamine H_3 Receptor Antagonists

As with the first-generation H_3 agonists, the first-generation H_3 antagonists (all of them possessing an imidazole heterocycle) have considerable affinity for the more recently discovered histamine H_4 receptor. The first potent H_3 receptor antagonist (later reclassified as an inverse agonist) that was devoid of H_1 receptor and H_2 receptor activity was thioperamide (Figure 17.6). This compound has been used in many H_3 receptor studies as reference ligand and is active in vitro and in vivo (the compound is able to penetrate the CNS). However, thioperamide displays some 5-HT_3 receptor antagonism and also is an inverse agonist at the H_4 receptor. Moreover, a remarkable H_3 receptor species differences can be demonstrated with thioperamide, as the compound has a 10-fold higher affinity for the rat H_3 receptor than for the human H_3 receptor. Based on the H_3 receptor agonist imetit, the highly potent H_3 inverse agonist clobenpropit was developed (pA_2 = 9.9). This compound also has some 5-HT_3 receptor activity and displays partial agonist activity at H_4 receptors. Ciproxyfan is another potent and selective imidazole-containing antagonist with in vivo efficacy in a variety of preclinical animal models (e.g., models for attention). Impentamine is a potent histamine H_3 receptor

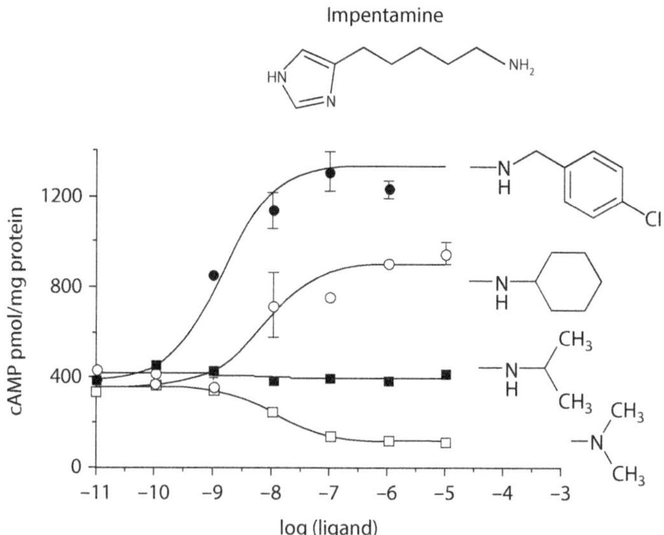

FIGURE 17.7 Alkylation of the primary amine function of impentamine leads to ligands that cover the complete spectrum of functional activity, i.e., agonism, neutral antagonism, and inverse agonism.

partial agonist in SK-N-MC cells expressing human H_3 receptors. It has also been shown that small structural modifications of impentamine, i.e., alkylation of the primary amine moiety of impentamine with, e.g., methyl-, isopropyl-, and *p*-chlorobenzyl groups, results in ligands that cover the complete spectrum of functional activity, i.e., agonism, neutral antagonism, and inverse agonism (Figure 17.7). The compound VUF5681 (Figure 17.6) was reported as a neutral H_3 antagonist, not affecting the basal signaling of the histamine H_3 receptor. As such, it has proven to be a useful molecular tool in H_3 receptor studies, for example, when studying H_3 constitutive activity in the rat brain.

The imidazole-containing compounds have been very important in characterizing the H_3 receptor and in setting the stage for the potential therapeutic use of H_3 antagonists. Furthermore, similarity studies resulted in pharmacophore models (Figure 17.8) that explain the SAR of the different classes of imidazole-containing ligands and indirectly describe the ligand binding site of the receptor. Yet, imidazole-containing ligands are associated with inhibition of cytochrome P-450 enzymes. Via this mechanism, the clearance of co-administrated drugs can be compromised, leading to severe drug–drug interactions and extrapyramidal symptoms. Classic medicinal chemistry work, elegantly conducted by the team of Ganellin (already involved in the development of the H_2 antagonist Cimetidine, *vide supra*) at the University College London, led to a first breakthrough in the search of nonimidazole H_3 antagonists, as illustrated in Scheme 17.3. The endogenous agonist histamine was once again taken as a lead structure. Attachment of a lipophilic group to the amine moiety led to N^{α}-(4-phenylbutyl)histamine, a compound with H_3 antagonist activity. Replacement of the imidazole heterocycle, initially deemed essential for H_3 affinity, led to *N*-ethyl-*N*-(4-phenylbutyl)amine with merely a twofold drop in affinity. Subsequent stepwise optimization of the structure for H_3 affinity by systemically modifying the basic group, the linker, and the aromatic moiety ultimately led to UCL 2190, a potent nonimidazole H_3 antagonist. Structural features of this compound, e.g., the amino-proxyphenyl substructure, reoccur in most H_3 medicinal chemistry programs that have been reported since.

Especially since the cloning of the H_3 receptor gene in 1999, the pharmaceutical industry has been actively exploring the potential of H_3 receptor ligands and many new antagonists/inverse agonists have been described and a variety has entered clinical trials. Typical examples are the UCL-derivative pitolisant, A-423579, the benzofuran ABT-239, GSK-189254, and JNJ-5207852, both of which contain two basic amines, and MK-0249 (Figure 17.6). Interestingly, JNJ-5207852 is

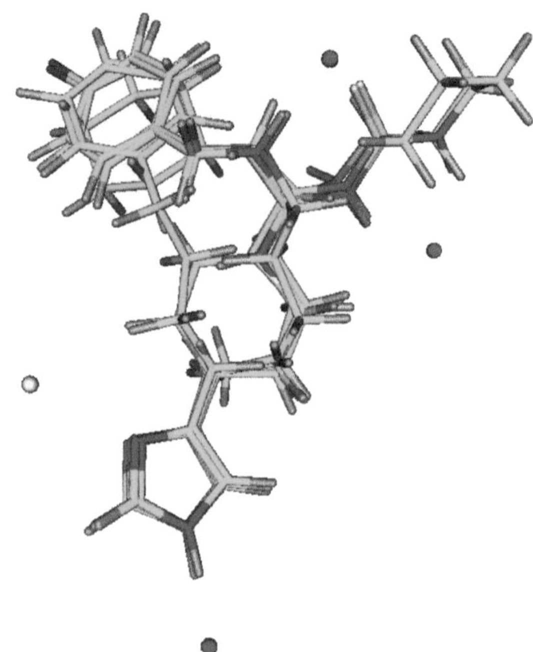

FIGURE 17.8 Pharmacophore model for imidazole-containing H_3 antagonists. Superposed are ten different compounds. Carbon atoms in green, nitrogen atoms in blue, sulfur atoms in yellow, and hydrogen atoms in white. All imidazole rings are perfectly superposed. The aromatic heterocycle and the basic groups in the imidazole side chain can interact with a total of four predicted H-bonding groups of the receptor site (yellow sphere indicating a H-bonding donor atom of the site and purple indicating H-bonding acceptor atoms of the site). The lipophilic groups at the terminus of the side chain of the ligands are located in two distinct positions, suggesting that the H_3 receptor has two lipophilic pockets for ligand binding. These findings were later validated by several groups using receptor homology modeling.

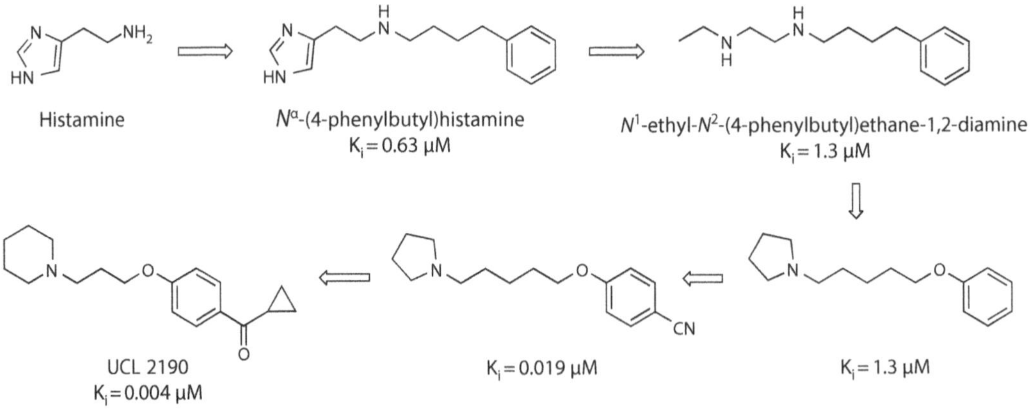

Histamine

N$^{\alpha}$-(4-phenylbutyl)histamine
$K_i = 0.63\ \mu M$

N^{1}-ethyl-*N*^{2}-(4-phenylbutyl)ethane-1,2-diamine
$K_i = 1.3\ \mu M$

UCL 2190
$K_i = 0.004\ \mu M$

$K_i = 0.019\ \mu M$

$K_i = 1.3\ \mu M$

SCHEME 17.3 Illustration of the stepwise development of UCL 2190 as one of the first potent nonimidazole H_3 receptor antagonists.

active in several models for cognition, but does not act as an appetite suppressant and has no effect on food intake. Other compounds, such as A-423579, have good efficacy in obesity models, but lack clear procognitive effects. At present, the differences in efficacy for distinct clinical applications of the different classes of H_3 ligands are not understood (e.g., involvement of different H_3 receptor isoforms) and subject of intense research.

17.4.4 Therapeutic Use of Histamine H_3 Receptor Ligands

Multiple lines of evidence indicate that the H_3 receptor is involved in numerous physiological processes and that this receptor bears potential as a promising drug target. A handful of applications for H_3 agonists have emerged from preclinical studies in the areas of migraine (modulating release of neurogenic peptides) and ischemic arrhythmias (modulating noradrenaline release). In migraine, the H_3 agonistic properties of N^{α}-methylhistamine have been reported to be beneficial in a phase II trial. H_3 antagonists and inverse agonists have been successfully used in animal models for narcolepsy, sleep/wakefulness, and cognitive disorders, neuropathic pain, and others. Full clinical validation of the benefit of H_3 receptor antagonists is still awaited, but a number of molecules (some with undisclosed structures) have in the meantime been evaluated in a number of clinical trials. Importantly, after obtaining an orphan drug status in the EU in 2010, pitolisant (Wakix from Bioprojet, France) has recently filed for EMA registration for the treatment of narcolepsy. Moreover, pitolisant is also tested for alleviation of sleeping disorders, related to, e.g., Parkinson's disease. Next to sleeping disorders, a lot of focus is also directed toward cognitive properties, but so far no spectacular results have been presented with, e.g., MK-0249 or bavisant (JNJ31001074). Outside the CNS, H_3 antagonists have shown positive clinical results in, e.g., nasal decongestion (JNJ3922075 and PF-03654746 in combination with the H_1 antagonist fexofenadine).

17.5 HISTAMINE H_4 RECEPTOR: MOLECULAR ASPECTS AND SELECTIVE LIGANDS

17.5.1 Molecular Aspects of the Histamine H_4 Receptor Protein

Immediately following the cloning of the H_3 receptor gene, several groups identified the homologous H_4 receptor sequence in the human genome databases. Indeed, the H_4 receptor has high sequence identity with the H_3 receptor (31% at the protein level, 54% in the transmembrane domains). The H_3 and H_4 receptors are also similar in gene structure. The human H_4 receptor gene is present on chromosome 18q11.2 and the gene contains three exons that are interrupted by two large introns (like the H_3 receptor gene). To date, two H_4 receptor isoforms have been identified, but no functional role has been reported so far. Cloning of the genes that encode the mouse, rat, guinea pig, and pig H_4 receptors reveal only limited sequence homology with the human H_4 receptor. The H_4 receptor is mainly expressed in bone marrow and peripheral leukocytes and mRNA of the human H_4 receptor is detected in, e.g., mast cells, dendritic cells, spleen, and eosinophils. The H_4 receptor has a pronounced effect on the chemotaxis of several cell types that are associated with immune and inflammatory responses. The H_4 receptor has also been reported to be expressed it the CNS, but final proof is awaited for a clear role of the H_4 receptor in brain function.

The H_4 receptor couples to $G_{i/o}$ proteins, thereby leading to a decrease in cAMP production and the regulation of CREB gene transcription. Furthermore, H_4 receptor stimulation affects the $G_{i/o}$ protein-mediated activation of mitogen-activated protein kinase. Studying the increased [^{35}S] GTPγS levels in H_4 transfected cells, it has been shown that the H_4 receptor is also constitutively active.

17.5.2 Histamine H_4 Receptor Agonists

Most of the first-generation imidazole-containing H_3 ligands have reasonable affinity for the H_4 receptor as well. The first imidazole-containing ligand that was reported to have some selectivity for the H_4 (40-fold) over the H_3 receptor is OUP-16 (Figure 17.9). This compound acts as a full H_4 agonist. More recently, the potent H_4 agonist 4-methylhistamine was discovered after the screening of a large number of histaminergic compounds. This compound was originally developed for an H_2 research program, but appears to be more than 100-fold more potent on the H_4 receptor than on any

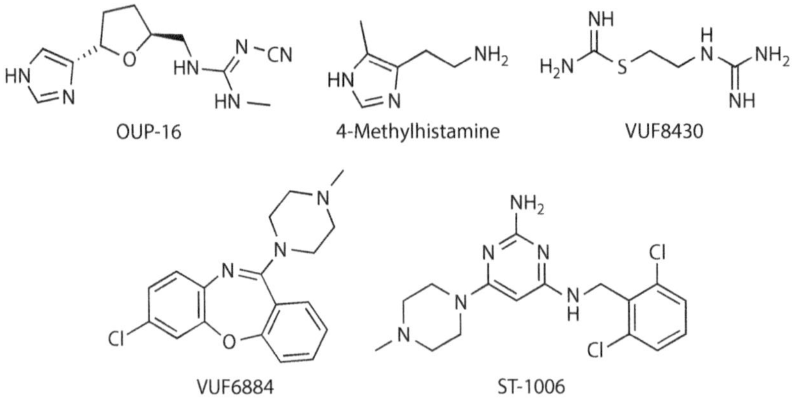

FIGURE 17.9 Reference histamine H_4 receptor agonists.

other histamine receptor subtype, including the H_2 receptor. VUF8430 was also reported as a potent H_4 agonist (pEC_{50} = 7.3) with a complementary selectivity profile, being 33-fold selective over the H_3 receptor. Again, VUF8430 was originally developed as a dimaprit analog in an H_2 research program. VUF6884 was developed as a clozapine analog with optimized histamine H_4 receptor affinity. This rigid compound is particularly useful for pharmacophore modeling studies (see later text). Interestingly, VUF6884 is a full agonist on histamine H_4 receptors and an even more potent histamine H_1 receptor inverse agonist. Clozapine derivatives are well known as promiscuous GPCR ligands. Finally, from a series of aminopyrimidines ST-1006 was discovered as a partial agonist in recombinant systems, but in human monocytes this molecule seems to behave as a full agonist.

17.5.3 Histamine H_4 Receptor Antagonists

Potent and selective H_4 receptor antagonists are also emerging. For this histamine receptor subtype, the first nonimidazole ligands were found by successful HTS campaigns. The first reported neutral antagonist was derived from an indole-containing hit structure that was efficiently converted into JNJ7777120 (Scheme 17.4). This indole carboxamide is currently considered as an H_4 receptor

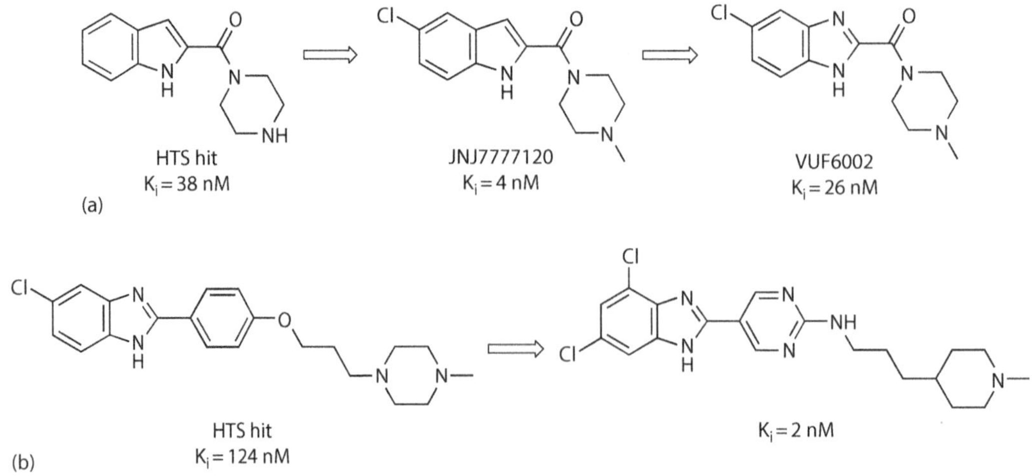

SCHEME 17.4 Illustration of two histamine H_4 receptor high-throughput screening hits (a and b) and subsequent hit optimization.

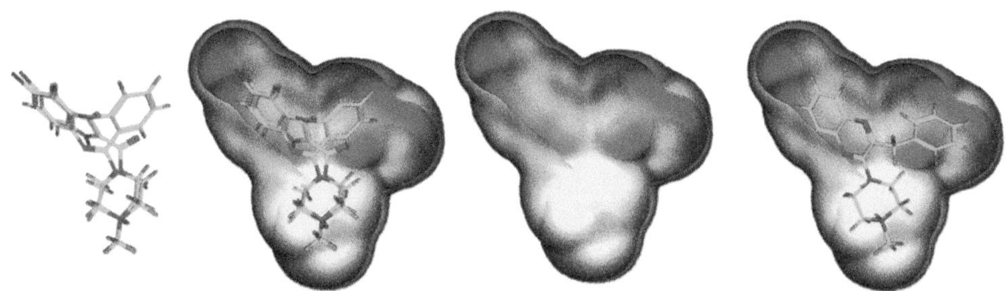

FIGURE 17.10 Pharmacophore modeling leading to the design of new ligands. Two reference histamine H_4 ligands (VUF6884 and JnJ7777120) were used to construct a pharmacophore model. This model indirectly describes the histamine H_4 receptor binding site. Based on this model, novel and potent ligands that fit the binding pocket could be designed, e.g., VUF10148). Carbon atoms in green, nitrogen atoms in blue, oxygen atoms in red, and hydrogen atoms in white. Color coding surface: hydrogen-bonding region in purple, hydrophobic regions in yellow, and mild polar regions in blue.

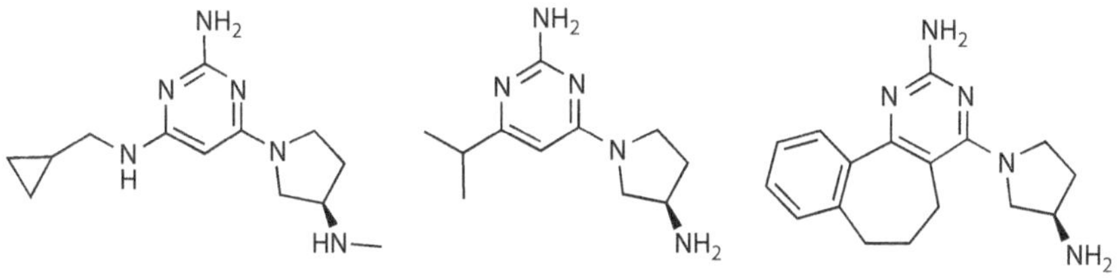

ZPL-38937887 JNJ-39758979 A-943931

FIGURE 17.11 Selected histamine H_4 receptor antagonists.

reference ligand with good selectivity over the other human histamine receptors. Unfortunately, the compound has a poor stability in human and rat liver microsomes and a half-life of only 2 hours in rats. The subsequently developed benzimidazole derivative JNJ10191584 is also a neutral H_4 antagonist. This compound is orally active in vivo and has improved liver microsomes stability but still a limited half-life. Also derived from a HTS hit, a series of 2-arylbenzimidazoles have been described as ligands with low nanomolar affinity for the H_4 receptor.

Following the discovery of the histamine H_4 receptor, a variety of academic and private laboratories have been developing therapeutically interesting H_4 antagonists. A significant number of H_4 receptor-related patent applications have recently been disclosed and some molecules even have reached the stage of clinical testing. Using a fragment screening, followed by approaches like pharmacophore modeling and subsequent ligand design (Figure 17.10), novel H_4 receptor ligands, like VUF 10148, have been developed. Other approaches have led to the discovery of aminopyrimidines as privileged structure for H_4 antagonists. Quite a number of companies and academic labs published on this scaffold with ZPL-38937887 (Ziarco, compound formerly known as PF-3893787), JNJ-39758979, and A-943931 as typical examples (Figure 17.11). Compounds are active in a variety of preclinical models of inflammation, asthma, dermatitis, and neuropathic pain. Moreover, both ZPL-38937887 and JNJ-39758979 have been tested in man (*vide infra*).

17.5.4 Therapeutic Use of Histamine H_4 Receptor Ligands

The presence of the H_4 receptor on immunocompetent cells and cells of hematopoietic lineage suggests that this new histamine receptor subtype plays an important role in the immune system. This hypothesis is supported by the fact that IL-10 and IL-13 modulate H_4 receptor expression and that binding sites for cytokine-regulated transcription factors, like interferon-stimulated response element, interferon regulatory factor-1, NF-κB, and nuclear factor-IL6, are present upstream of the H_4 gene. Considering the physiological role of the H_4 receptor, several applications, including allergy and asthma, chronic inflammations such as inflammatory bowel disease and rheumatoid arthritis have been investigated preclinically. The H_4 receptor is also being associated with pruritus (itch) and has been suggested to be involved in the progression of colon cancer.

The preclinical findings of the effects of H_4 antagonists like JNJ7777120 against itch have been validated in a clinical study with JNJ-39758979. Moreover, this antagonist is also being evaluated in clinical studies in asthma and allergic rhinitis, just like ZPL-3893787 and UR-63325 (structure still undisclosed).

17.6 CONCLUDING REMARKS

The medicinal chemistry of histamine receptors has so far been a very rewarding arena. Major blockbuster drugs have been developed on the basis of H_1 and H_2 receptor targeting and the first ligand for the H_3 receptor is hitting the market. Expectations for ligands targeting the two latest additions to the histamine receptor family are currently high.

Interestingly, for each of these receptor subtypes highly selective agonists and antagonists have been developed. The wide chemical diversity of the various selective receptor ligands reflects the relatively low homology between the various receptors (only the H_3 and H_4 receptors resemble each other to some extent). Moreover, it offers today's medicinal chemists an attractive arena for highly effective drug discovery efforts. This will be further aided by the recent elucidation of the X-ray structure of a variety of GPCRs, including the H_1 receptor, hopefully allowing even more effective future structure-based drug design.

FURTHER READING

Hancock, A.A. 2006. The challenge of drug discovery of a GPCR target: Analysis of preclinical pharmacology of histamine H_3 antagonists/inverse agonists. *Biochem. Pharmacol.* 71:1103–1113.

Leurs, R., Bakker, R.A., Timmerman, H., and de Esch, I.J. 2005. The histamine H_3 receptor: From gene cloning to H_3 receptor drugs. *Nat. Rev. Drug Discov.* 4:107–120.

Panula, P., Chazot, P.L., Cowart, M. et al. 2015. International union of pharmacology. XIII. Classification of histamine receptors. *Pharmacol. Rev.* 67:601–655.

Shimamura, T., Shiroishi, M., Weyand, S. et al. 2011. Structure of the human histamine H1 receptor complex with doxepin. *Nature* 475:65–70.

Thurmond, R.L., Gelfand, E.W., and Dunford, P.J. 2008. The role of histamine H_1 and H_4 receptors in allergic inflammation: The search for new antihistamines. *Nat. Rev. Drug Discov.* 7:41–53.

Zhang, M.Q., Leurs, R., and Timmerman, H. 1997. Histamine H_1-receptor antagonists. In *Burger's Medicinal Chemistry and Drug Discovery*, 5th ed., ed. M.E. Wolff, p. 495. John Wiley & Sons, Inc, New York.

18 Dopamine and Serotonin

Benny Bang-Andersen, Lena Tagmose, and Klaus P. Bøgesø

CONTENTS

18.1 INTRODUCTION

Dopamine (DA), serotonin (5-hydroxytryptamine [5-HT]), and norepinephrine (NE) are important neurotransmitters in the human brain. These neurotransmitters activate postsynaptic and presynaptic receptors, and their concentration is regulated by active reuptake into presynaptic terminals by transporters.

DA and 5-HT receptors are found in multiple subtypes that are divided into subclasses based on structural and pharmacological similarities. The DA and 5-HT receptors are all seven transmembrane (TM) G protein-coupled receptors (GPCRs) except for the 5-HT$_3$ receptor which is a ligand-gated ion channel, regulating the permeability of sodium and potassium ions. Five subtypes of DA receptors are known and grouped into the D$_1$-like receptors (D$_1$ and D$_5$) and the D$_2$-like receptors (D$_2$, D$_3$, and D$_4$), whereas 14 subtypes of 5-HT receptors are known and grouped into seven subclasses, namely, 5-HT$_1$ (5-HT$_{1A}$, 5-HT$_{1B}$, 5-HT$_{1D}$, 5-HT$_{1E}$, and 5-HT$_{1F}$), 5-HT$_2$ (5-HT$_{2A}$, 5-HT$_{2B}$, and 5-HT$_{2C}$), 5-HT$_3$, 5-HT$_4$, 5-HT$_5$ (5-HT$_{5A}$ and 5-HT$_{5B}$), 5-HT$_6$, and 5-HT$_7$. In addition, a variety of polymorphic and splice variants (functional and nonfunctional) have been described for subtypes of both DA and 5-HT receptors.

Transporters for DA (DAT), 5-HT (SERT), and NE (NET) belong to the same family, the so-called solute carrier 6 gene family of ion-coupled plasma membrane cotransporters. These transporters transport DA, 5-HT, and/or NE from the synapse and into the cell using the sodium gradient across the membrane. They are not specific for their substrates, and NET is, for example, important for the transport/clearance of DA in the cortex. This also fits with the fact that the highest homology among the cloned human transporters is found between DAT and NET. Recently, X-ray crystal structures of transporters have provided very important information about the molecular basis for antidepressant action and have expanded the understanding of the mechanism and regulation of neurotransmitter uptake at chemical synapses. Also, X-ray structures of D_3, 5-HT_{1B}, 5-HT_{2B}, and 5-HT_3 receptors are now available. Application of the new structures in drug design using computational methods will be discussed (Section 18.5). Several ligands have been described for many of these receptors and transporters. Antipsychotic drugs that are used in the treatment of schizophrenia will be discussed as examples of ligands binding to DA and 5-HT receptors (Section 18.2), whereas antidepressant drugs that are used for the treatment of depression and anxiety will be discussed as examples of ligands for transporters (Section 18.3 and Chapter 14). The multimodal antidepressant drugs will be discussed as examples of ligands that act via both transporters and receptors (Section 18.4).

18.2 RECEPTOR LIGANDS

18.2.1 ANTIPSYCHOTIC DRUGS

Antipsychotic drugs are primarily used to treat schizophrenia. Schizophrenia is distinguished from other psychotic disorders based on a characteristic cluster of symptoms, where the positive symptoms appear to reflect an excess or distortion of normal function (i.e., delusion, hallucinations, disorganized thinking, disorganized behavior, and catatonia), whereas the negative symptoms appear to reflect a diminution or loss of normal functions (i.e., affective flattening, poverty of speech, and an inability to initiate and persist in goal-directed activities). The cognitive symptoms (i.e., impairment of memory, executive function, and attention) have attracted more and more attention and much research is directed toward understanding the role of these symptoms in order to discover a treatment.

The antipsychotic drugs are typically divided into the classical and the atypical antipsychotic drugs. The classical antipsychotic drugs were discovered in the 1950s with chlorpromazine (**18.1**, Figure 18.1) as the first prominent example, whereas the atypical antipsychotic drugs were introduced into the treatment of schizophrenia during the 1990s. It is believed that the antipsychotic drugs exert their effect on positive symptoms by reducing DA hyperactivity in limbic areas of the brain.

The term classical antipsychotic drug is linked to compounds that show effect in the treatment of positive symptoms at similar doses that induce extrapyramidal symptoms (EPS, i.e., Parkinsonian symptoms, dystonia, akathisia, and tardive dyskinesia). It is believed that EPS is caused by the blockade of DA activity in striatal areas of the brain. The classical antipsychotic drugs are without effect on negative and cognitive symptoms, and these drugs may even worsen these symptoms. It has been argued that the deterioration of negative and cognitive symptoms by classical antipsychotic drugs may be a consequence of their EPS, and the separation of the antipsychotic effect and EPS is the foremost important property of the atypical antipsychotic drugs.

Thus, the term atypical antipsychotic drug is linked to a diverse group of drugs having antipsychotic effect at doses not giving EPS. However, all drugs from this group have their own compound-specific limitations, such as a strong tendency to increase weight for some of the compounds, whereas others have a tendency to prolong the QT interval (total duration of cardiac ventricular electrical activity) in the surface electrocardiogram. In the following, the classical as well as atypical antipsychotic drugs will be discussed with focus on their discovery, including structural considerations and pharmacological profiles of key compounds.

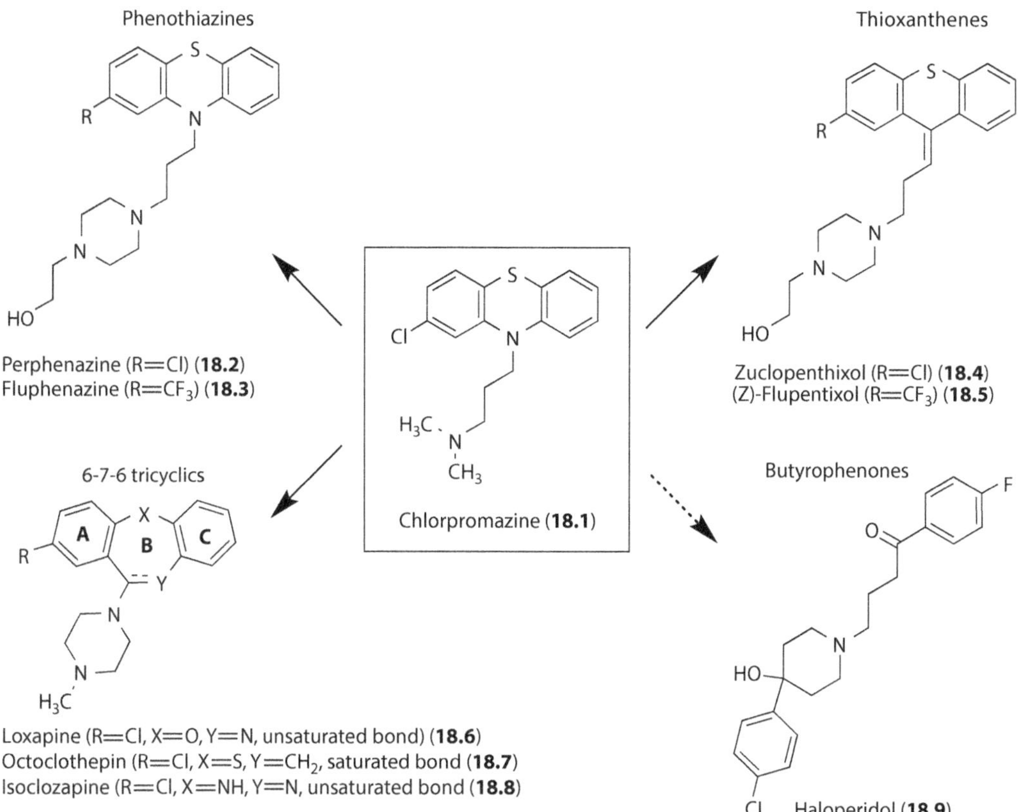

FIGURE 18.1 Classical antipsychotic drugs.

18.2.2 Classical Antipsychotic Drugs

Chlorpromazine was discovered in the beginning of the 1950s, and the structure of chlorpromazine with its phenothiazine backbone was an excellent lead for medicinal chemists. Thus, the modification of chlorpromazine without changing the phenothiazine backbone led to a number of drugs such as perphenazine (**18.2**) and fluphenazine (**18.3**) (Figure 18.1). Medicinal chemists also replaced the phenothiazine backbone with other tricyclic structures, and these modifications led to other classes of classical antipsychotic drugs such as the thioxanthenes and the 6–7–6 tricyclics.

Lundbeck in Denmark investigated in particular the thioxanthene backbone, and this work resulted in drugs such as zuclopenthixol (**18.4**) and (Z)-flupentixol (**18.5**) (Figure 18.1). The 6–7–6 tricyclic backbone has also led to a number of classical antipsychotic drugs such as loxapine (**18.6**), octoclothepin (**18.7**), and isoclozapine (**18.8**) (Figure 18.1). The R group found in all of these compounds is called the "neuroleptic substituent," and this substituent increases the D_2 receptor affinity/antagonism relative to unsubstituted molecules and is essential for potent neuroleptic effect.

In the late 1950s, researchers at Janssen discovered an entirely new class of classical antipsychotic drugs without a tricyclic structure, namely, the butyrophenones. Haloperidol (**18.9**, Figure 18.1) is the most prominent representative of this class of compounds, and today haloperidol is considered the archetypical classical antipsychotic drug.

The classical antipsychotic drugs were all discovered using in vivo animal models, as the current knowledge about receptor multiplicity and in vitro receptor-binding techniques were unknown at that time. However, many of the in vivo models which were used at that time as predictive for antipsychotic effect, are today considered more predictive of various side effects, e.g., EPS, and in hindsight it was difficult to find new antipsychotic drugs without the potential to induce EPS with the models available at that time.

An examination of the classical antipsychotic drugs by today's range of receptor-binding techniques and other more advanced biochemical methods has revealed that these drugs are predominantly postsynaptic D_2 receptor antagonists, and it is believed that this accounts for both their antipsychotic effect and their potential to induce EPS. However, these drugs also target several other receptors which may contribute to both their antipsychotic effect and their side effect profile.

18.2.3 Atypical Antipsychotic Drugs

Isoclozapine (**18.8**, Figure 18.1), which has the "neuroleptic chloro substituent" in benzene ring A, is a classical antipsychotic drug. On the contrary, clozapine (**18.10**, Figure 18.2), which has the chloro substituent in benzene ring C (Figure 18.1), has revolutionized the pharmacotherapy

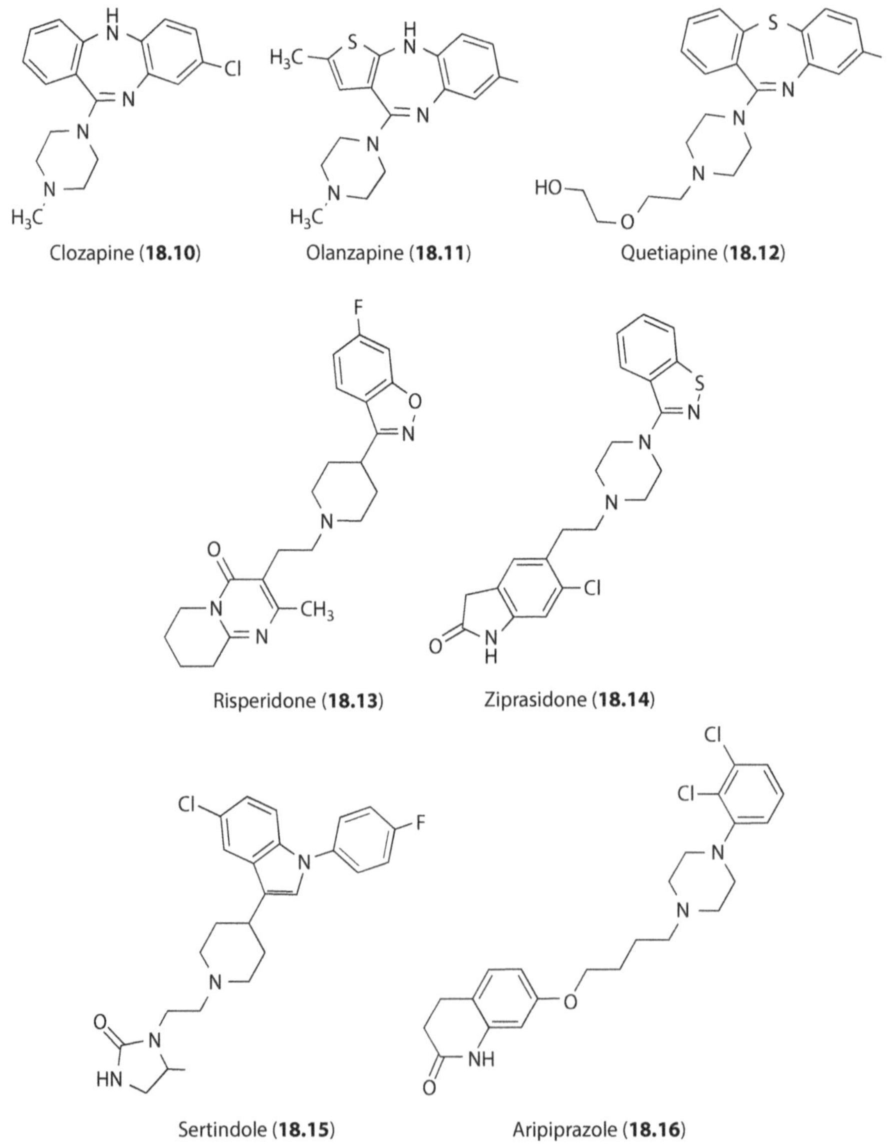

Clozapine (**18.10**) Olanzapine (**18.11**) Quetiapine (**18.12**)

Risperidone (**18.13**) Ziprasidone (**18.14**)

Sertindole (**18.15**) Aripiprazole (**18.16**)

FIGURE 18.2 Atypical antipsychotic drugs.

of schizophrenia. Thus, clozapine was the first antipsychotic drug that was effective in the treatment of positive symptoms of schizophrenia and free of EPS, but unfortunately clozapine can cause fatal agranulocytosis in a small percentage (1%–2%) of individuals, and much effort has been directed toward the identification of new antipsychotics with a clozapine-like clinical profile but without the risk of causing agranulocytosis.

This search has resulted in a number of atypical antipsychotics such as olanzapine (**18.11**), quetiapine (**18.12**), risperidone (**18.13**), ziprasidone (**18.14**), sertindole (**18.15**), and aripiprazole (**18.16**) (Figure 18.2). The structure of these compounds reveals that olanzapine and quetiapine were obtained by structural modification of clozapine, whereas risperidone and ziprasidone were obtained from the butyrophenones. Aripiprazole and sertindole are both quite different in chemical structure and not in a similar way related to the classical antipsychotics. Recently and after many years in preclinical and clinical development, iloperidone, lurasidone, and asenapine (structures not shown) were introduced in clinical practice, all presenting compound-specific advantages of limited impact as compared to the more established atypical antipsychotics.

In vitro binding data for selected DA, 5-HT, and NE receptors as well as data from the catalepsy model (in vivo rat model predictive of EPS in humans) are shown for haloperidol and key atypical antipsychotics (Table 18.1). All the compounds display mixed receptor profiles with affinity for even more receptors than included in the table (data not shown). The tendency is that classical antipsychotics display high affinity for D_2 receptors relative to 5-HT_{2A} receptors, whereas the atypical antipsychotics display an increased affinity for 5-HT_{2A} receptors relative to D_2. The relative ratio of D_2 versus 5-HT_{2A} receptor affinity has been suggested as a reason for atypicals not giving rise to EPS at therapeutic doses. A number of other factors may influence both the antipsychotic potential and the propensity to induce EPS, such as the affinity and efficacy at some of the other receptors, but also in vivo preference for limbic versus striatal regions of the brain might explain these differences.

Aripiprazole is a partial D_2 receptor agonist and therefore different from the other atypical antipsychotics. A partial D_2 receptor agonist is envisaged to stabilize a dysfunctioning DA system, inhibiting transmission in synapses with high tonus, and increasing function in those with low activity. This profile might explain why aripiprazole does not induce EPS. A number of pharmaceutical companies have had partial D_2 receptor agonist drug candidates in development, but a key

TABLE 18.1
Receptor Profile and Extrapyramidal Symptoms Potential of Antipsychotic Drugs

	Receptor Binding K_i (nM)							In Vivo ED_{50} (μmol/kg)
Compounds	D_1^a	D_2^a	D_3^b	D_4^b	5-HT_{2A}^a	5-HT_{2C}^a	α_1^a	Catalepsy max., sc[a]
Classical antipsychotic drug								
Haloperidol (**18.9**)	15	0.82	1.1	2.8	28	1,500	7.3	0.34
Atypical antipsychotic drugs								
Risperidone (**18.13**)	21	0.44	14	7.1	0.39	6.4	0.69	17
Olanzapine (**18.11**)	10	2.1	71	32	1.9	2.8	7.3	37
Quetiapine (**18.12**)	390	69	1,100	2,400	82	1,500	4.5	>80
Ziprasidone (**18.14**)	9.5	2.8	N/A	73	0.25	0.55	1.9	>48
Sertindole (**18.15**)	12	0.45	2.0	17	0.20	0.51	1.4	>91
Clozapine (**18.10**)	53	36	310	30	4.0	5.0	3.7	120

[a] Arnt, J. and Skarsfeldt, T., *Neuropsychopharmacology*, 18, 63, 1998.
[b] Lundbeck Screening Database, H. Lundbeck A/S, Valby, Denmark.
N/A, not available.

challenge has been to define and obtain the right level of functional activity at the D_2 receptor. Aripiprazole is currently the only marketed antipsychotic having this profile, and it has been very successful in treating schizophrenia and a number of other psychiatric diseases. It has recently been introduced as a depot formulation for long-term treatment. Brexpiprazole and cariprazine are in the preregistration phase and display partial D_2 receptor agonism as well as other DA and 5-HT receptor activities, illustrating that DA and 5-HT receptors are still very relevant targets for developing psychiatric drugs.

18.3 TRANSPORTER LIGANDS

18.3.1 Antidepressant Drugs

Antidepressant drugs represent ligands that target DAT, SERT, and NET to various degrees, and these include first-generation antidepressants (i.e., tricyclic antidepressants [TCAs]), selective serotonin reuptake inhibitors (SSRIs), combined serotonin and NE reuptake inhibitors (SNRIs), and the recently introduced multimodal antidepressants.

The SSRIs have been highly successful in the treatment of depression due to their high safety in use, and a number of additional indications (e.g., panic disorder, obsessive compulsive disorder, and social phobia) have been registered for many of these drugs in addition to major depressive disorder (MDD). However, there are still unmet medical needs in the treatment of depression, such as (1) efficacy in treatment-resistant MDD patients (33% of patients respond inadequately or not at all, even after consecutive treatment with up to four different antidepressants), (2) a rapid onset of antidepressant action (it is generally thought that 2–3 weeks of treatment is needed before a therapeutic response is detected), and (3) effective treatment of residual symptoms (e.g., cognitive symptoms including a diminished ability to think, concentrate, and plan). These unmet needs have driven drug discovery programs in pharmaceutical companies for years and some have been addressed, whereas other unmet needs are still very relevant. In the following, the discovery of the first-generation antidepressants and the SSRIs (exemplified by citalopram and escitalopram) will be discussed. The multimodal antidepressants vilazodone and vortioxetine will be discussed in Section 18.4.

18.3.2 First-Generation Drugs

The pharmacotherapy of depression started in the late 1950s with the introduction of the two drugs iproniazid (**18.17**) and imipramine (**18.19**) (Figure 18.3). Iproniazid was originally an antituberculosis drug, but it was noticed that the drug had an antidepressant effect. It was subsequently discovered that iproniazid was an unselective, irreversible inhibitor of the enzymes MAO-A and MAO-B that deaminate the monoamines NE, DA, and 5-HT. Structural modifications of the tricyclic antipsychotic drugs with chlorpromazine (**18.1**, Figure 18.1) as a prototype led to the 6–7–6 tricyclic compound imipramine (**18.19**) that was found to block the transporters for NE and 5-HT. Both these mechanisms led to an increase in the concentrations of NE and 5-HT in the synapse, which in turn led to the so-called monoamine hypothesis of depression, stating that there is a decreased availability of these neurotransmitters in depression.

Although the discovery of these two classes of drugs was of major therapeutic importance, it quickly turned out that both types had serious fatal side effects. Treatment with MAO inhibitors could induce a hypertensive crisis because of a fatal interaction with foodstuffs containing tyramine such as cheese. Dietary restrictions during treatment with MAO inhibitors were, therefore, required. Reversible MAO-A inhibitors (such as moclobemide [**18.18**]) have later been developed, but such drugs are still not completely devoid of the "cheese effect" because the tyramine potentiation is inherent to blockade of MAO-A in the periphery. MAO inhibitors are, therefore, only used to a lesser extent in antidepressant therapy.

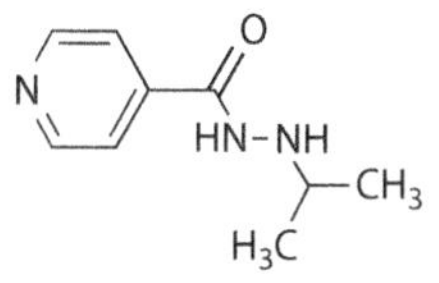

Iproniazid (**18.17**)

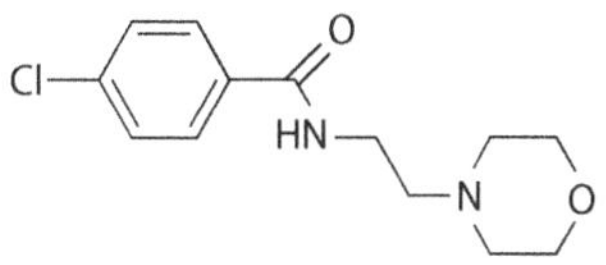

Moclobemide (**18.18**)

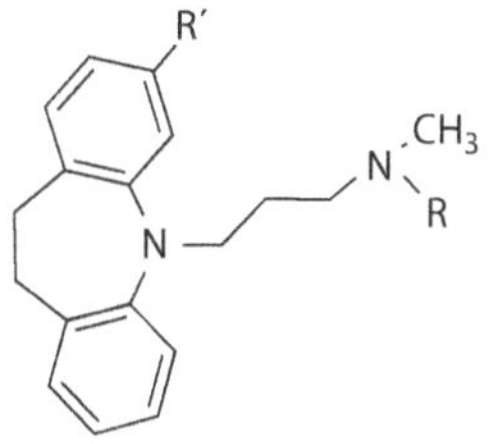

Imipramine (R′=H; R=CH_3) (**18.19**)
Desipramine (R′=H; R=H) (**18.20**)
Clomipramine (R′=Cl; R=CH_3) (**18.21**)

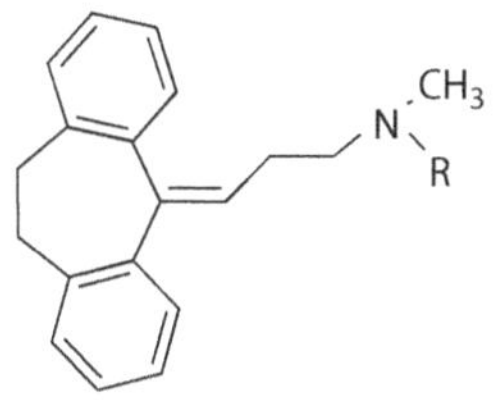

Amitriptyline (R=CH_3) (**18.22**)
Nortriptyline (R=H) (**18.23**)

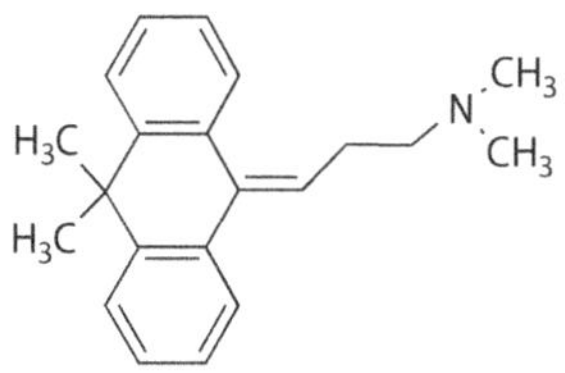

Melitracen (**18.24**)

FIGURE 18.3 Antidepressant drugs from MAO inhibitor and tricyclic classes.

18.3.3 Selective Serotonin Reuptake Inhibitors

Nortriptyline (**18.23**) is a relative selective NE reuptake inhibitor, while the corresponding dimethyl derivative, amitriptyline (**18.22**) (Figure 18.3), is a mixed 5-HT/NE reuptake inhibitor with concomitant high affinity for postsynaptic receptors as well. The same is true for the corresponding pair desipramine (**18.20**)/imipramine (**18.19**). Swiss psychiatrist Paul Kielholz coupled these observations to the clinical profiles of these drugs, and Swedish scientist Arvid Carlsson noticed that the tertiary amine drugs, which were mixed 5-HT and NE reuptake inhibitors, were "mood elevating," while the secondary amines, being primarily NE reuptake inhibitors, increased more "drive" in the depressed patients. As the foremost quality of an antidepressant drug should be mood elevation (elevation of drive before mood could induce a suicidal event), Carlsson advocated for the development of selective 5-HT reuptake inhibitors. Consequently, a number of pharmaceutical companies initiated drug discovery programs aiming at design of such drugs in the early 1970s.

18.3.4 Discovery of Citalopram

In the mid-1960s, chemists at Lundbeck were looking for more potent derivatives of the TCAs amitriptyline, nortriptyline, and melitracen (**18.24**) which the company had developed and marketed previously. The trifluoromethyl group had in other in-house projects proved to increase potency in thioxanthene derivatives with antipsychotic activity (see Figure 18.1), and it was therefore decided to attempt to synthesize the 2-CF_3 derivative of melitracen (**18.26**, Figure 18.4). The precursor molecule **18.25** was readily synthesized, but attempts to ring-close it in a manner corresponding to the existing melitracen method, using concentrated sulfuric acid, failed. However, another product was formed which through meticulous structural elucidation proved to be the bicyclic phthalane (or dihydroisobenzofuran) derivative **18.27**. Fortunately, this compound was examined in models for antidepressant activity and was very surprisingly found to be a selective NET inhibitor. Some analogs were synthesized, among them two compounds that later got the International Nonproprietary Names talopram (**18.28**) and talsupram (**18.29**). These compounds are still among the most selective NE reuptake inhibitors (SNIs) ever synthesized (Figure 18.4 and Table 18.2).

Both talopram and talsupram were investigated for antidepressant effect in clinical trials but were stopped in phase II for various reasons, among which was an activating profile in accordance with their potent NE reuptake inhibition. A project was, therefore, started in the beginning of 1971 with the aim of discovering an SSRI from the talopram structure.

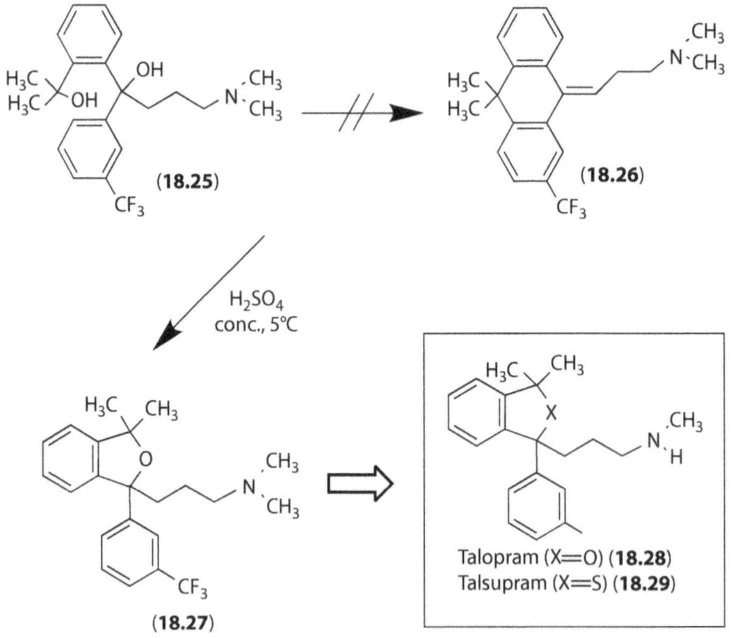

FIGURE 18.4 Discovery of phenylphthalane antidepressants.

TABLE 18.2
5-Hydroxytryptamine and Norepinephrine Reuptake Inhibition of Selected Talopram Derivatives

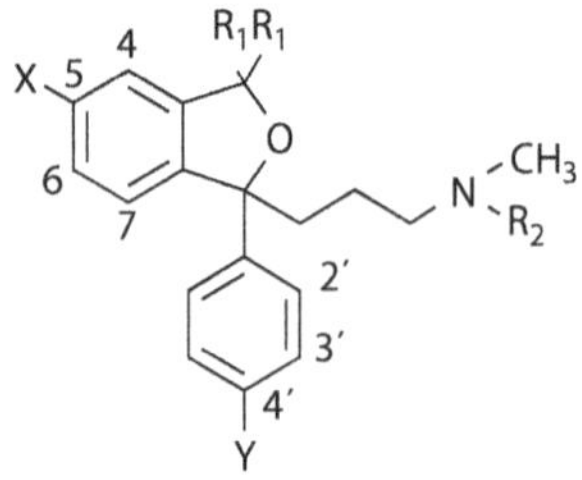

Compound	R_1	R_2	X	Y	5-HT Reuptake (In Vitro) Rabbit Blood pl. IC_{50} (nM)	NE Reuptake (In Vivo) Mouse Heart ED_{50} (μmol/kg)
Talopram (**18.28**)	CH_3	H	H	H	3,400	2.2
(**18.30**)	CH_3	CH_3	H	H	53,000	5
(**18.31**)	H	H	H	H	1,300	43
(**18.32**)	H	CH_3	H	H	600	66
(**18.33**)	H	CH_3	H	Cl	110	170
(**18.34**)	H	CH_3	Cl	H	220	>200
(**18.35**)	H	CH_3	Cl	Cl	24	>80
(**18.36**)	H	CH_3	H	Br	310	N/A
(**18.37**)	H	CH_3	H	CN	54	23
(**18.38**)	H	CH_3	CN	Cl	10	>80
Citalopram (**18.39**)	H	CH_3	CN	F	38	>40

Source: Data from Lundbeck Screening Database, H. Lundbeck A/S, Valby, Denmark.
N/A, not available.

It may not be obvious to use an SNI as template structure for an SSRI. However, in the first series synthesized, two compounds (**18.31** and **18.32**, Table 18.2) without the dimethylation of the phthalane ring showed a tendency for increased 5-HT reuptake, and in accordance with the structure–activity relationship (SAR) studies mentioned earlier for tricyclics, the *N,N*-dimethyl derivative **18.32** was the more potent compound. Therefore, compound **18.32** became a template structure for further structural investigation.

In this phase of the project, neuronal test models for measuring reuptake were not available, so 5-HT reuptake inhibition was measured as inhibition of tritiated 5-HT into rabbit blood platelets, while inhibition of NE reuptake was measured ex vivo as inhibition of tritiated NE into the heart of the mouse (Table 18.2). Although these models were not directly comparable, they were acceptable for the discovery of selective compounds.

The introduction of a chloro substituent into the template structure further increased 5-HT reuptake and decreased NE reuptake inhibition (**18.32** versus **18.34**), in accordance with observations by Carlsson that halogen substituents in both zimelidine (**18.43**, Figure 18.5) (see the following) and in imipramine derivatives (clomipramine, **18.21**, Figure 18.3) increased 5-HT reuptake. Indeed, the dichloro derivative **18.35** proved to be a selective 5-HT reuptake inhibitor. So the goal of obtaining an SSRI from an SNI was achieved very fast (in 1971), when less than 50 compounds had been synthesized.

The SAR were further explored, and it was established that high activity was generally found in 5,4′-disubstituted compounds where both substituents were halogen or other electron-withdrawing groups. Cyano-substituted compounds were obtained by the reaction of the bromo precursors

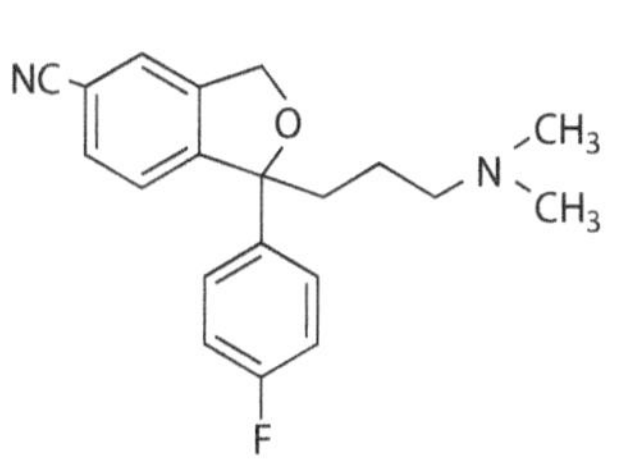

Citalopram (**18.39**)

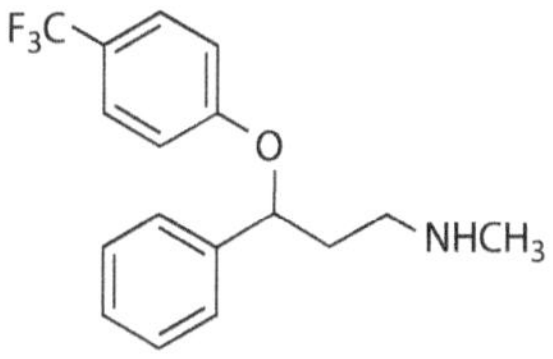

Fluoxetine (**18.40**)

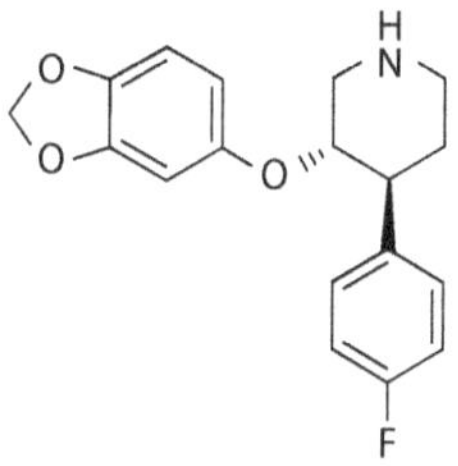

Paroxetine (**18.41**)

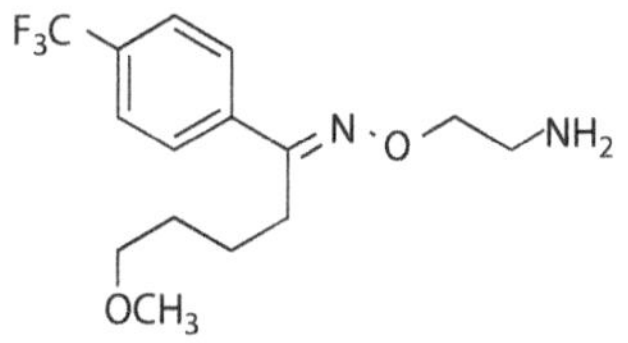

Fluvoxamine (**18.42**)

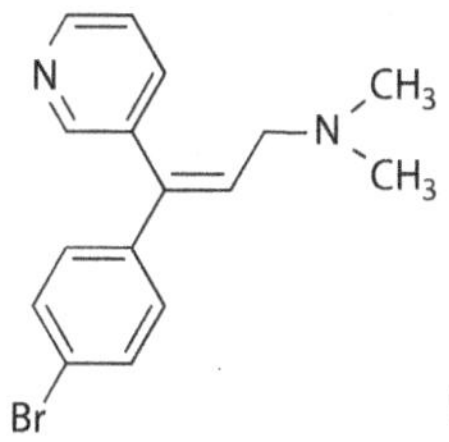

Zimelidine (**18.43**)

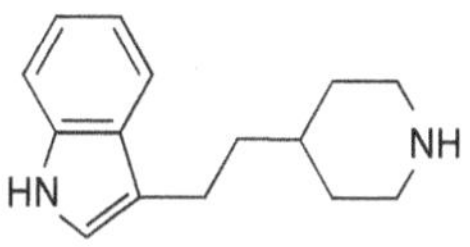

Indalpine (**18.44**)

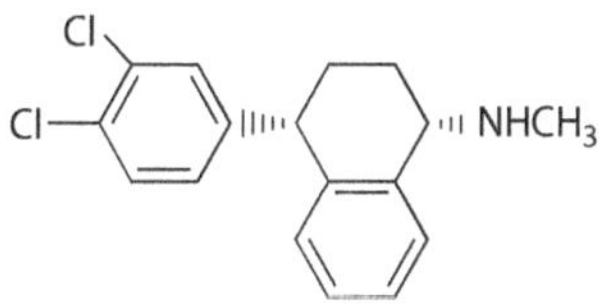

Sertraline (**18.45**)

FIGURE 18.5 Selective serotonin reuptake inhibitors.

(e.g., **18.36**) with CuCN. One of the cyano-substituted compounds was **18.39**, later known as citalopram. The compound was synthesized in August 1972. The cyano group could be metabolically labile, but it was subsequently shown not to be the case for citalopram neither in animals nor in humans. Citalopram displayed the best overall preclinical profile within this series and was consequently selected for development. The 5-cyano substituent in citalopram also proved to be chemically stable in a surprising manner; for example, it does not react with Grignard reagents which has led to a new and patentable process for its production.

Citalopram was launched for the treatment of MDD in 1989 in Denmark as Cipramil® and subsequently marketed worldwide, including the United States in 1998 under the trade name Celexa®. Overall, citalopram was on par with the other SSRIs with respect to efficacy, but had more favorable drug metabolism and pharmacokinetic (DMPK) properties (see Chapter 5), which are the likely reasons why citalopram became such a commercial success, attaining blockbuster status, even though it was the fifth SSRI introduced onto the US market.

18.3.5 Discovery of Escitalopram

Citalopram is a racemate, having an asymmetric carbon at the 1-position. When the first attempts to separate the enantiomers of citalopram were made in 1980, classical resolution via diastereomeric salt formation was the only option for separation of enantiomers. However, it is generally difficult to make salts of citalopram, and eventually direct resolution was given up. Chiral high-pressure liquid chromatography (HPLC) was in its infancy, and the available analytical columns were tried with negative results. Various attempts to asymmetric syntheses, avoiding acid ring closure of resolved intermediates due to the risk of racemization, failed.

Finally, attempts to resolve the so-called diol formed in the second last step was made. The diastereomeric esters with the enantiomers of α-methoxy-α-trifluoromethyl acetic acid (Mosher's acid) were prepared and separated on preparative (non-chiral) HPLC by repeated peak shaving. Small samples of the pure diastereomeric esters were obtained. Importantly, a seemingly spontaneous slow ring closure to citalopram of the mixture of diastereomeric esters (in the presence of triethylamine) was noticed. Encouraged by this observation, a stronger base was tried (potassium *tert*-butoxide) which surprisingly resulted in a stereoselective ring closure of the pure diastereomers to afford a small sample of the pure citalopram enantiomers. Later, it was realized that the diol could be resolved by diastereomeric salt formation with di-*p*-toluoyltartaric acid and, in this way, it became possible to produce larger quantities of the (*S*)- and (*R*)-enantiomers of citalopram.

The 5-HT reuptake inhibition of citalopram proved to reside in the (*S*)-enantiomer (escitalopram), whereas the (*R*)-enantiomer was about 100 times less potent. Escitalopram was launched as a single-enantiomer drug in 2002 and proved to be an effective antidepressant (trade names Cipralex® and Lexapro®) with several advantages as compared to citalopram and other SSRIs. In preclinical studies, escitalopram showed greater efficacy and faster onset of action than comparable doses of citalopram. This is attributed to the fact that the (*R*)-enantiomer of citalopram counteracts the activity of the (*S*)-enantiomer, possibly through the interaction at a second allosteric site on SERT (see further reading for a discussion of the allosteric site on SERT and its potential role in the effectiveness of escitalopram as an antidepressant drug). Further, in randomized, controlled clinical studies, escitalopram showed better efficacy than citalopram, with higher response and remission rates and faster onset of action. Indeed, a study from 2009 showed that escitalopram had the most favorable profile among all the antidepressants that were in clinical use at that time when taking efficacy and tolerability into consideration.

18.3.6 Other Selective Serotonin Reuptake Inhibitors

In Figure 18.5, the seven SSRIs that have reached the market, with the priority dates of the first patent application indicated, are shown. However, the two first compounds on the market were both withdrawn due to serious, although rare, side effects. Zimelidine (**18.43**) was found to

TABLE 18.3
Effect of Selective Serotonin Reuptake Inhibitors, Talopram, and Talsupram on the Inhibition of Reuptake of 5-Hydroxytryptamine, Norepinephrine, and Dopamine

	Uptake Inhibition IC_{50} (nM)			Ratio	
Compound	**5-HT**	**NE**	**DA**	**NE/5-HT**	**DA/5-HT**
Citalopram (**18.39**)	3.9	*6,100*	*40,000*	1,560	10,300
Escitalopram (*S*-**18.39**)	2.1	2,500	65,000	1,200	31,000
R-citalopram (*R*-**18.39**)	275	6,900	54,000	25	200
Indalpine (**18.44**)	2.1	2,100	1,200	1,000	570
Sertraline (**18.45**)	*0.19*	*160*	*48*	840	250
Paroxetine (**18.41**)	*0.29*	*81*	*5,100*	280	17,600
Fluvoxamine (**18.42**)	*3.8*	*620*	*42,000*	160	11,000
Zimelidine (**18.43**)	56	3,100	26,000	55	460
Fluoxetine (**18.40**)	*6.8*	*370*	*5,000*	54	740
Talopram (**18.28**)	1,400	2.5	44,000	0.0017	0.00006[a]
Talsupram (**18.29**)	770	0.79	9,300	0.0010	0.00008[a]

Source: Data in italics are from Hyttel, J., *Int. Clin. Psychopharmacol.*, 9(Suppl. 1), 19, 1994; Remaining data are from Lundbeck Screening Database, H. Lundbeck A/S, Valby.

[a] NE/DA.

induce an influenza-like symptom in 1%–2% of the patients which in rare cases (one of 10,000) resulted in the so-called Guillain–Barré syndrome. The drug was withdrawn in 1983 after 1½ years on the market. Indalpine (**18.44**) induced agranulocytosis in 1 of 20,000 patients and was withdrawn in 1984.

All the marketed SSRIs (except sertraline) were discovered in the first half of the 1970s (Figure 18.5), meaning that the companies lacked information regarding the structural classes their competitors were developing. Accordingly, this parallel development led to a rather diverse set of structural classes for the SSRIs. However, they were all selective 5-HT reuptake inhibitors (Table 18.3), although their selectivity ratios vary significantly, citalopram and escitalopram being the most SERT selective compounds. In general, the SSRIs have low affinity for DA, NE, and 5-HT receptors, although exceptions exist. With regard to interaction with cytochrome P450 enzymes, there are vital differences, e.g., paroxetine and fluoxetine have significant affinity for CYP2D6.

18.4 COMBINED RECEPTOR AND TRANSPORTER LIGANDS

18.4.1 Multimodal Antidepressant Drugs

The SSRI and SNRI antidepressant drugs made pharmacotherapy of depression safe and effective, but several medical needs remain to be addressed (see Section 18.3.1). Two recently approved multimodal antidepressants (vilazodone and vortioxetine) were designed to simultaneously modulate transporters and receptors (a multimodal drug is a compound interacting with ≥2 target classes, e.g., transporters and receptors) and thereby address some of those medical needs. Vortioxetine (Brintellix®) was approved for the treatment of MDD by the European Medicines Agency and the US Food and Drug Administration in 2013, whereas vilazodone (Viibryd®) is approved only in the United States. Vortioxetine was launched in the United States in January 2014 and subsequently in markets all over the world. The rationale behind the design of multimodal antidepressant drugs will be discussed, as will be the key steps in the development of vortioxetine.

18.4.2 Discovery of Vortioxetine

In the early 1990s, preclinical and clinical research indicated that the combination of a SERT inhibitor with an antagonist of somatodendritic 5-HT_{1A} autoreceptors resulted in a significantly larger increase in extracellular 5-HT levels in rat brain than seen for SSRIs alone. In addition, chronic dosing studies indicated that the maximum increase in 5-HT could be achieved faster for the drug combination than with the SSRI alone. Since the combination also led to a more rapid antidepressant effect and improved response rates in a clinical study, a search for single drugs modulating this combination of targets was initiated by many pharmaceutical companies. However, it turned out to be difficult to find such compounds, especially because it was not easy to define and obtain the right level of functional activity at the 5-HT_{1A} receptor. Vilazodone is the only approved antidepressant drug that so far has come out of this specific effort.

At Lundbeck, the effort toward a combined 5-HT_{1A} receptor antagonist and SERT inhibitor led to an increased understanding of which 5-HT receptors, apart from the 5-HT_{1A} receptor, might also augment the SSRI-induced increase in extracellular 5-HT. One such project was started in 2001 with the aim of finding a combined 5-HT_{2C} receptor antagonist and SERT inhibitor. A focused screen using the Lundbeck monoamine compound library led to the identification of the lead compound **18.46** (Table 18.4) which displayed the desired in vitro pharmacodynamic profile. Compound **18.46** was tested in a rat model that was established to determine the change in 5-HT levels in the brain after acute treatment (i.e., "acute microdialysis model"). Compound **18.46** significantly increased the extracellular 5-HT levels beyond that seen with an SSRI. However, compound **18.46** turned out to have an unsatisfactory in vitro DMPK profile due to its poor metabolic stability in human microsomes and potent inhibition of the cytochrome P450 isozyme CYP2D6.

Compound **18.47** (Table 18.4), later known as vortioxetine, was discovered during the lead optimization program and was initially characterized as having combined SERT inhibition and 5-HT_{2C} receptor activity (Table 18.4). As vortioxetine displayed a suitable in vitro DMPK profile, it was tested in the acute rat microdialysis model, where it increased the extracellular 5-HT level beyond

TABLE 18.4
From Lead Compound to Vortioxetine

H_3CO S NH N CH_3 ⇨ H_3C CH_3 S NH N

Assay	18.46	Vortioxetine (18.47)
SERT (IC_{50}, nM)	7.9	5.3/5.4[a]
5-HT_{2C} (K_i, nM)	13	180[b]
5-HT_{1A} (K_i, nM)	4000	39/15[a]
5-HT_{3A} (K_i, nM)	190	23/3.7[a]
Cl_{int} (L/min)[c]	2.8	0.5
CYP2D6 (IC_{50}, nM)	0.1	9.8

Source: Data are from Bang-Andersen, B. et al., *J. Med. Chem.*, 54, 3206, 2011.
[a] Data from different assays.
[b] Published value which is higher than that initially measured.
[c] Liver blood flow 1.4 L/min.

that seen for an SSRI. It later became clear that the 5-HT$_{2C}$ receptor activity was less pronounced than originally thought, but vortioxetine had pronounced 5-HT$_{1A}$ and 5-HT$_3$ receptor affinity. This was further investigated and a part of the lead optimization program was redirected toward compounds that combined SERT inhibition, 5-HT$_{1A}$ receptor agonism, and 5-HT$_{3A}$ receptor antagonism in a single molecule. Vortioxetine remained the overall best compound due to its superior combination of pharmacodynamic and DMPK properties Vortioxetine has later been shown to most likely mediate its pharmacological activities through SERT inhibition, 5-HT$_{1A}$ receptor agonism, and 5-HT$_3$ receptor antagonism, as well as antagonism at 5-HT$_7$ and 5-HT$_{1D}$ receptors and partial agonism at 5-HT$_{1B}$ receptors.

Vortioxetine has shown efficacy in the treatment of patients with MDD in a comprehensive clinical program. Furthermore, preclinical studies had indicated that vortioxetine could have beneficial effects on cognitive dysfunction in MDD, a residual symptom not well treated by the established antidepressants. This hypothesis has been confirmed in clinical studies which showed that vortioxetine can improve cognitive dysfunction in patients with MDD across a broad range of cognitive domains. Thus, vortioxetine's multimodal mechanism of action indicates a different antidepressant profile compared to other antidepressants in clinical use, and vortioxetine is the first antidepressant to include in its label an effect on certain aspects of cognitive dysfunction in patients with MDD.

18.5 MODELING TRANSPORTER AND RECEPTOR LIGAND BINDING

For a general introduction to structure-based drug design, the reader is referred to Chapter 4. In the following, a specific discussion of the modeling of key transporters and receptors relevant for antidepressant drug research is discussed.

In the previous decade, computational chemists had to rely solely on ligand-based design methods to guide medicinal chemistry in the design of novel ligands for GPCRs and biogenic amine transporters (BATs). These proteins are challenging targets for structure determination due to their instability outside of their natural membrane environment. However, over the previous decade the field has been revolutionized, especially for GPCRs, with more than 35 unique structures determined and some co-crystallized with over 10 different ligands. Structure-based drug design for GPCRs is now possible, and fragment screening is also a viable option. Application of these techniques has the potential to accelerate the development of novel therapeutic compounds, including for challenging or previously undruggable GPCRs, for the treatment of a wide range of disorders and diseases. Structure determination of neurotransmitter transporters including BATs has also become a reality with high-resolution X-ray structures of bacterial homologs of Na^+/Cl^--dependent neurotransmitter transporters such as the leucine transporter (LeuT) and, most recently, DAT. Structural information about the BATs and their interactions with antidepressant drugs is important for the understanding of their mechanism of action and for future drug development.

18.5.1 Modeling 5-HT Receptor-Ligand Binding

Multimodal drugs like vortioxetine were developed in the late 1990s to early 2000 when the only available GPCR X-ray structure was the rhodopsin receptor. This receptor has only little sequence identity to any of the 5-HT receptors that vortioxetine modulates and modeling efforts to guide chemistry were, therefore, limited to ligand-based design methods such as fingerprint searches, shape matching, and pharmacophore modeling using known SERT and 5-HT ligands. In 2013, the X-ray structures of the inactive states of the 5-HT$_{1B}$ and 5-HT$_{2B}$ receptors in complex with the antagonist ergotamine were published. The amino acid sequence identity between 5-HT$_{1A}$ and 5-HT$_{1B}$ is high (~40%), and there is now a good prognosis for building a fairly accurate homology model of the 5-HT$_{1A}$ receptor. It is, however, important that modeling efforts are supported by site-directed mutagenesis of residues suggested by the model to be situated in the binding site and that interact with the target ligands. Such studies will help validate the model and guide binding mode prediction for compounds like vortioxetine. The model can then be used for structure-based design purposes, including virtual screening to identify novel 5-HT_1 receptor ligands. With known GPCR protein structures, rational approaches become possible not only to identify and optimize the potency of new ligands, but to also control selectivity and kinetics (residence time). Drugability analyses of GPCR ligand binding sites have shown the importance of lipophilic hotspots in ligand binding, and knowing where these are located enables the design of ligands with drug-like physicochemical properties. Lipophilicity is a key property of most CNS drugs, and through structure-based design it can be used effectively for both potency and selectivity, avoiding less-productive presence that easily leads to drug candidates with too high a logP.

18.5.2 Modeling of Transporter-Ligand Binding at the Primary Site of BATs

Structure-based design approaches were not applicable when the first antidepressant drugs were developed because no protein structures sufficiently similar in sequence and function to human neurotransmitter transporters were available at the time. Thus, less powerful ligand-based methods like pharmacophore modeling were applied. In 2005, the crystal structure of a prokaryote (*Aquifex aeolicus*) LeuT with leucine bound within the protein core was published (see Chapter 14). The LeuT is a homolog of the BATs and belongs to the same transporter family. The overall sequence identity between LeuT and SERT is ~20% which is in the low end for "safe homology modeling," and, in particular, for ligand design. This structure was, nevertheless, a significant improvement over previous templates that do not belong to the same transporter family as the BATs. Publication of the structure initiated a cascade of homology modeling studies proposing 3D structural models for

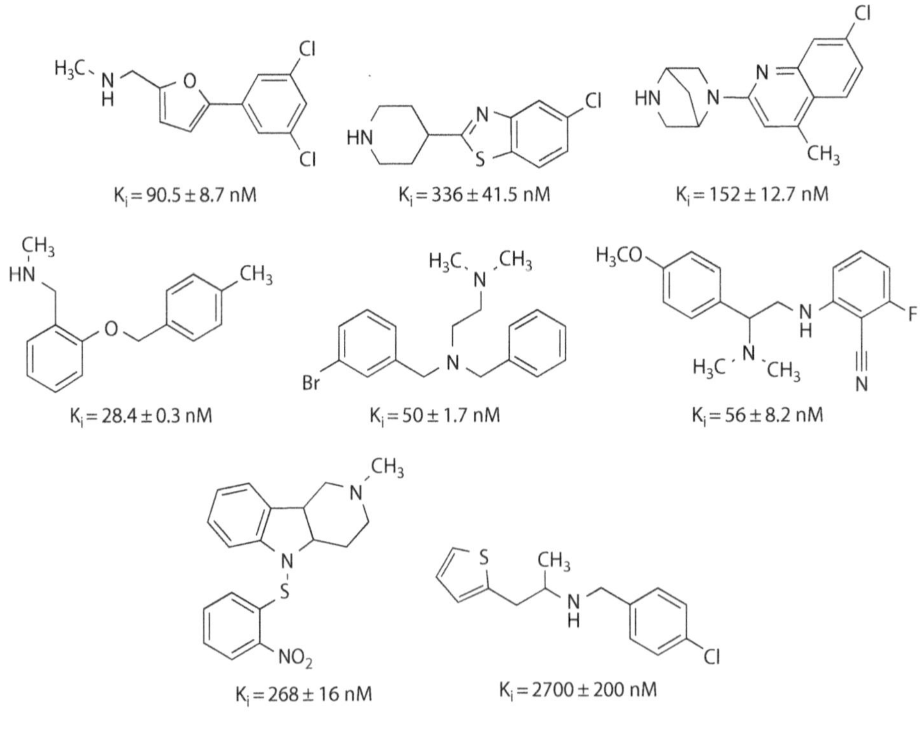

$K_i = 90.5 \pm 8.7$ nM

$K_i = 336 \pm 41.5$ nM

$K_i = 152 \pm 12.7$ nM

$K_i = 28.4 \pm 0.3$ nM

$K_i = 50 \pm 1.7$ nM

$K_i = 56 \pm 8.2$ nM

$K_i = 268 \pm 16$ nM

$K_i = 2700 \pm 200$ nM

FIGURE 18.6 Structures and binding affinities of novel SERT compounds found by virtual screening. The compounds have no or little structural resemblance to known SERT binders. LeuT-based SERT homology models were used in the virtual screening cascade, demonstrating the power of virtual screening as a method to identify new chemical matter. Binding affinities were measured by inhibition of specific [^{3}H]-citalopram binding to SERT. (Data and structures are from Gabrielsen, M. et al., *J. Chem. Inf. Model*, 54, 933, 2014.)

SERT as well as binding modes and bioactive conformations of known antidepressants when bound to SERT. Many of the homology modeling efforts were guided by site-directed mutagenesis studies to validate the model and support and guide the identified binding modes. Some groups used homology models to virtually screen vendor libraries of millions of compounds in order to identify SERT ligands with no or little structural resemblance to known SERT binders. A successful example of identification of such novel ligands can be found in Figure 18.6.

Recently, X-ray structures of a "surrogate" BAT in complex with four different classes of antidepressants (including the SSRIs sertraline, paroxetine, fluoxetine, and fluvoxamine) were published. The surrogate transporter was made by engineering LeuT to harbor human BAT pharmacology. Thirteen amino acid residues situated in the primary binding pocket of LeuT were mutated to the corresponding human (h) SERT residues. These LeuBAT complexes provide the first experimental structural information about SERT inhibitor binding to date (Figure 18.7a). At the same time, the crystal structure of *Drosophila melanogaster* DAT bound to the TCA nortriptyline was presented (Figure 18.7b). This construct should be an even better template for making a surrogate SERT structure since the sequence identity between *Drosophila* DAT and hSERT is much higher (~50%) than between *Aquifex* LeuT and hSERT (~20%).

For about three decades, it had been known that, in addition to the high-affinity primary binding site just described for SERT, the transporter also possesses a low-affinity allosteric binding site. It has been shown that binding of escitalopram, and to some extent paroxetine, at the low-affinity site can modulate the dissociation rate of other SERT ligands at the high-affinity site. The example of escitalopram is discussed in Section 18.3.1.4. Until recently, the location of the allosteric site in the transporter was unknown. The publication of X-ray structures of LeuT in complex with

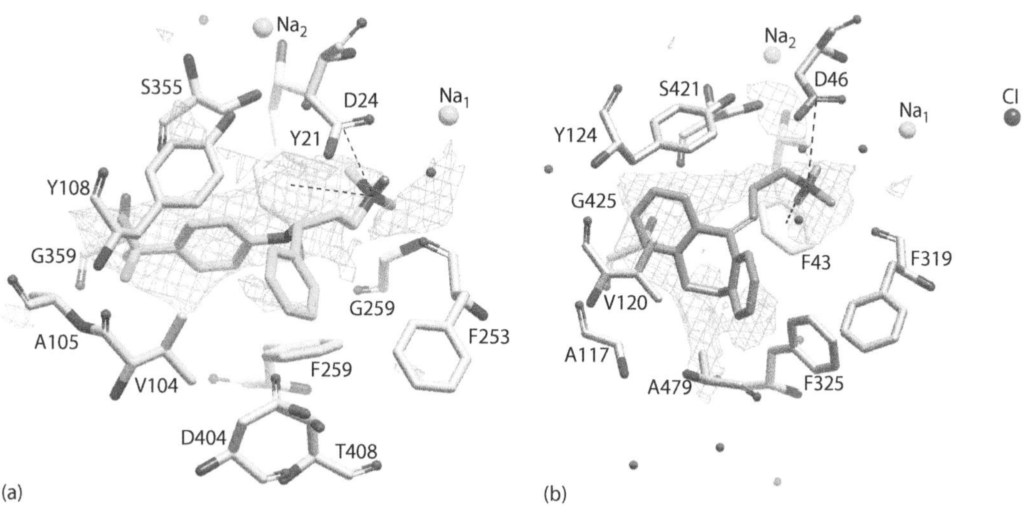

FIGURE 18.7 The topology and computational analysis of the primary binding sites of biogenic amine transporter (BATs) revealed by X-ray crystal co-complex structures. Shown are the structures of (a) a "surrogate" SERT in complex with (*R*)-fluoxetine (PDB:4MM8). (b) DAT in complex with nortriptyline (PDB:4M48). The yellow mesh depicts computed molecular interaction field from GRID calculations using the hydrophobic C1 = probe contoured at –2.5 kcal/mol. This yellow mesh represents areas of the binding site that are lipophilic and to which lipophilic groups of the ligands will bind well. Salt bridge and pi–cation interactions are shown with dotted lines. The driving forces for ligand binding are in both cases hydrophobic/π–π interactions between the ligand aromatic groups (and trifluoromethyl group of (*R*)-fluoxetine) and the hydrophobic amino acid side chains in the pocket (DAT/LeuT-BAT: F325/259, V120/104, Y124/108, F319/253) as well as formation of a salt bridge and a cation-π interaction between the positively charged amino group of the ligands and D46/24 and F43/Y21, respectively. (Courtesy of Ana Negri Martinez, H. Lundbeck A/S, Copenhagen-Valby, Denmark.)

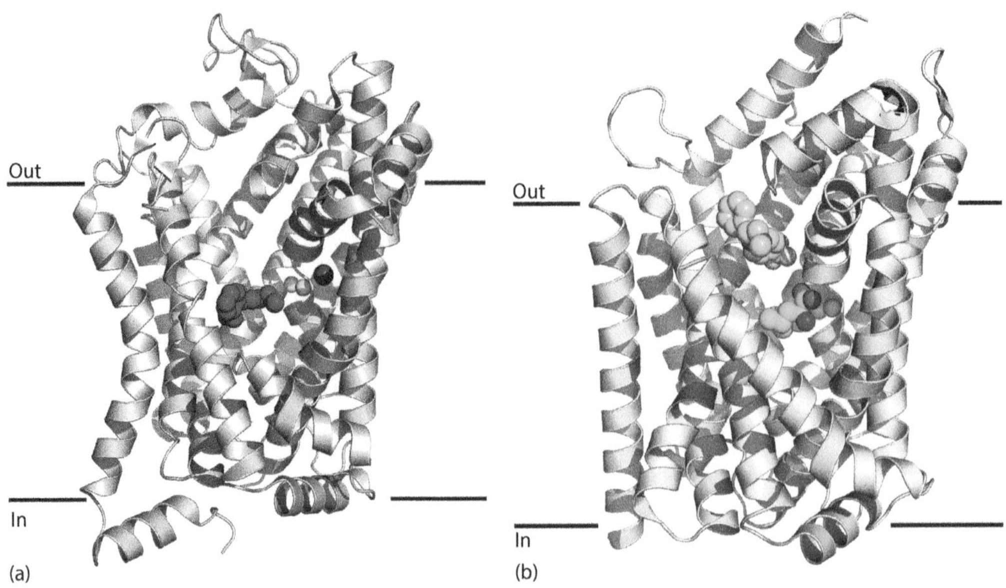

FIGURE 18.8 X-ray structures of (a) the DAT in complex with nortriptyline in the high-affinity primary binding site (PDB:4M48). The ligand, two Na^+ ions, and a Cl^- ion are shown as van der Waals spheres in magenta, yellow, and purple, respectively. The primary binding site is located at the core of the transporter which is locked in an outward open conformation with nortriptyline wedged between transmembrane helices 1, 3, 6, and 8, blocking the transporter from binding substrate and from isomerizing to an inward-facing conformation. (b) The LeuT in complex with substrate in the primary site and the selective serotonin reuptake inhibitor (*R*)-fluoxetine situated in an extracellular vestibule ~13 Å above the substrate binding pocket (PDB:3GWV). The substrate, the ligand, and two Na^+ ions are shown as van der Waals spheres in cyan, green, and yellow, respectively. The leucine transporter topology is shown in Chapter 14, Figure 2 (the DAT has the same topology). (Courtesy of Ana Negri Martinez, H. Lundbeck A/S, Copenhagen-Valby, Denmark.)

the substrate in the primary site and the SSRIs sertraline or fluoxetine situated in an extracellular vestibule ~13 Å above the substrate binding pocket has guided researchers to investigate whether the corresponding position in hSERT could be the allosteric site. Molecular modeling studies using a LeuT-based SERT homology model (induced fit docking to both the primary and allosteric sites in iteration with molecular dynamics simulations) were used to identify residues that could be involved in allosteric binding. The results of these studies have in turn guided steric hindrance mutagenesis studies of selected residues in hSERT that did indeed reduce allosteric binding of escitalopram and clomipramine. These studies support the hypothesis that the hSERT allosteric site is positioned in the extracellular vestibule and suggest that ligand binding to the allosteric site hinders both association and dissociation of antidepressants to and from the primary binding site (Figure 18.8).

18.6 CONCLUDING REMARKS

DA, 5-HT, and NE receptors and transporters have shown their relevance as drug targets over many years. The multimodal antidepressants vilazodone and vortioxetine further support the continued success of these targets, and cariprazine and brexpiprazole illustrate that this might continue. Recent years have led to an explosion in the understanding of the structure and function of these membrane-bound receptors and transporters which will enable a more rational design of new compounds. However, only the future can tell whether these new discoveries will result in novel and effective pharmacotherapies based on the DA and 5-HT systems.

FURTHER READING

Bang-Andersen, B., Ruhland, T., Jørgensen, M. et al. 2011. Discovery of 1-[2-(2,4-dimethylphenylsulfanyl) phenyl]piperazine (Lu AA21004): A novel multimodal compound for the treatment of major depressive disorder. *J. Med. Chem.* 54:3206–3221.

Bøgesø, K.P. and Sánchez, C. 2013. The discovery of citalopram and its refinement to escitalopram. In *Analogue-Based Drug Discovery III*, eds. J. Fischer, C.R. Ganellin, and D.P. Rotella, pp. 269–293. Wiley-VCH Verlag GmbH & Co. KGaA, Weinheim, Germany.

Bundgaard, C., Pehrson, A.L., Sánchez, C., and Bang-Andersen, B. 2015. Case Study 2: The discovery and development of the multimodal acting antidepressant vortioxetine. In *Blood-Brain Barrier in Drug Discovery: Optimizing Brain Exposure of CNS Drugs and Minimizing Brain Side Effects for Peripheral Drugs*, eds. L. Di and E.H. Kerns, pp. 505–520. John Wiley & Sons, Inc, Hoboken, NJ.

Penmatsa, A., Wang, K.H., and Gouaux, E. 2013. X-ray structure of dopamine transporter elucidates antidepressant mechanism. *Nature* 503:85–91.

Plenge, P., Shi, L., Beuming, T., Te, J., Newman, A.H., Weinstein, H., Gether, U., and Loland, C.J. 2012. Steric hindrance mutagenesis in the conserved extracellular vestibule impedes allosteric binding of antidepressants to the serotonin transporter. *J. Biol. Chem.* 287:39316–39326.

Wang, H., Goehring, A., Wang, K.H., Penmatsa, A., Ressler, R., and Gouaux, E. 2013. Structural basis for action by diverse antidepressants on biogenic amine transporters. *Nature* 503:141–146.

Wood, M. and Reavill, C. 2007. Aripiprazole acts as a selective dopamine D_2 receptor partial agonist. *Expert Opin. Invest. Drugs* 16:771–775.

Yamashita, A., Singh, S.K., Kawate, T., Jin, Y., and Gouaux, E. 2005. Crystal structure of a bacterial homologue of Na+/Cl–l–l--dependent neurotransmitter transport. *Nature* 437:215–223.

19 Opioid and Cannabinoid Receptors

Rasmus P. Clausen and Harald S. Hansen

CONTENTS

19.1 OPIOID RECEPTORS

> Presently she cast a drug into the wine of which they drank to lull all pain and anger and bring forgetfulness of every sorrow.
>
> *The Odyssey*, Homer (ninth century BC)

The history of opioids and its receptors span several millennia. The first evidence of uses of the seed pods of *Papaver somniferum* dates back to 4200 BC, and numerous findings and descriptions up through history witness the use of different parts of this plant for food, anesthesia, and ritual purposes. Opium (from *opos*, the Greek word for juice) refers to the liquid that appears on the unripe seed capsule when it is notched. This liquid contains as much as 16% of morphine, a compound that was isolated already in 1806 as the major active ingredient in opium. A few years later, codeine was also isolated. Morphine can now be produced and applied in its pure form for the treatment of pain and as an adjunct to general anesthetics, but it was quickly realized that morphine had the same potential of abuse as opium. In 1898, heroin was synthesized and claimed to be a safer, more efficacious, nonaddicting opiate as were several other analogs around that time; however, later they all proved to be unsafe. Heroin is an early example of a prodrug since the highly potent analgesic properties can be attributed to the rapid metabolism to 6-monoacetylmorphine and morphine, combined with higher blood–brain barrier penetration due to better lipid solubility compared to morphine (Figure 19.1).

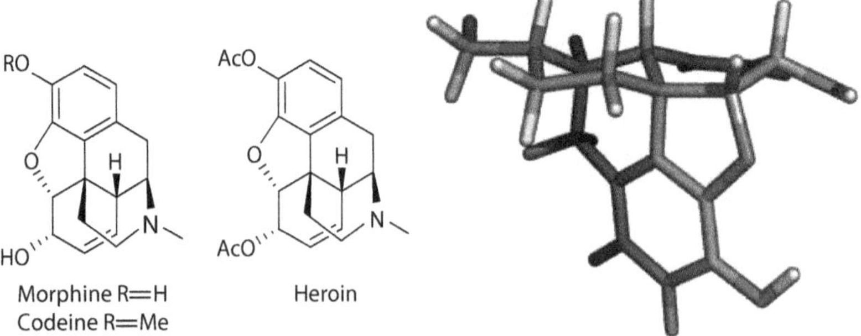

FIGURE 19.1 Chemical structures of morphine, codeine, and heroin. 3D structure of morphine.

19.1.1 Opioid Receptor Subtypes and Effector Mechanism

The idea that morphine and other opioids caused analgesia by interacting with a specific receptor arose around the 1950s. The observation 40 years earlier that the *N*-allyl analog of codeine antagonized the respiratory action of morphine was actually evidence of such a proposal. However, it was first fully realized when similar *N*-allyl analog of morphine (nalorphine) was shown to antagonize the analgesic effects of morphine.

Today, it is known that all of the opioid receptors are G-protein-coupled receptors (GPCRs) belonging to family A (Figure 19.2) that mediates its effects through G_i/G_o proteins (see Chapter 12). So far, four different opioid receptor subtypes have been cloned, sharing more than 60% sequence homology. These are termed μ, κ, and δ receptors (corresponding to MOR, KOR, and DOR, respectively) and an "orphan" receptor termed ORL_1 which was the first orphan GPCR to be cloned.

The different effects (Table 19.1) mediated by each receptor type (μ-euphoria versus κ-dysphoria; μ-supraspinal analgesia versus ORL_1-supraspinal antagonism of opioid analgesia) in the intact animal are the result of different anatomical localization and not due to different cellular responses. Each receptor type has been further subdivided into μ_1/μ_2, κ_1/κ_2, and δ_1/δ_2 receptors based on pharmacological and radioligand studies. However, the origin of this subdivision is not genetically based, and it is not known whether it arises from post-translational modification, cellular localization, or interactions with other proteins; however, it was recently shown that heterodimerization of the receptors could be important for some of these pharmacological differences.

Morphine has the ability to both excite and inhibit single neurons. Opioid inhibition of neuronal excitability occurs largely by the ability of opioid receptors to activate various potassium channels. Another well-established mechanism of action is the inhibition of neurotransmitter release. The observation in 1917 that morphine inhibited the peristaltic reflex in the guinea pig ileum (giving rise to constipation, one of the side effects of morphine) was 40 years later shown to result from inhibition of acetylcholine release. Also, glutamate, GABA, and glycine release throughout the central nervous system (CNS) can be inhibited by opioid receptor activation. In general, the CNS effects of opioids are inhibitory, but certain CNS effects (such as euphoria) result from excitatory effects.

19.1.2 Endogenous Opioid Receptor Ligands

It was proposed in the early 1970s that the physiological role of opioid receptors was not to be a target for opium alkaloids, but that endogenous agonists might exist as mediators of the opioid system. At that time, there were no hints of what kind of compounds to look for. After 2 years of collecting extracts from pig brain and applying them in a functional bioassay, Kosterlitz and coworkers in 1975 identified two closely related endogenous pentapeptide opioids (Table 19.2).

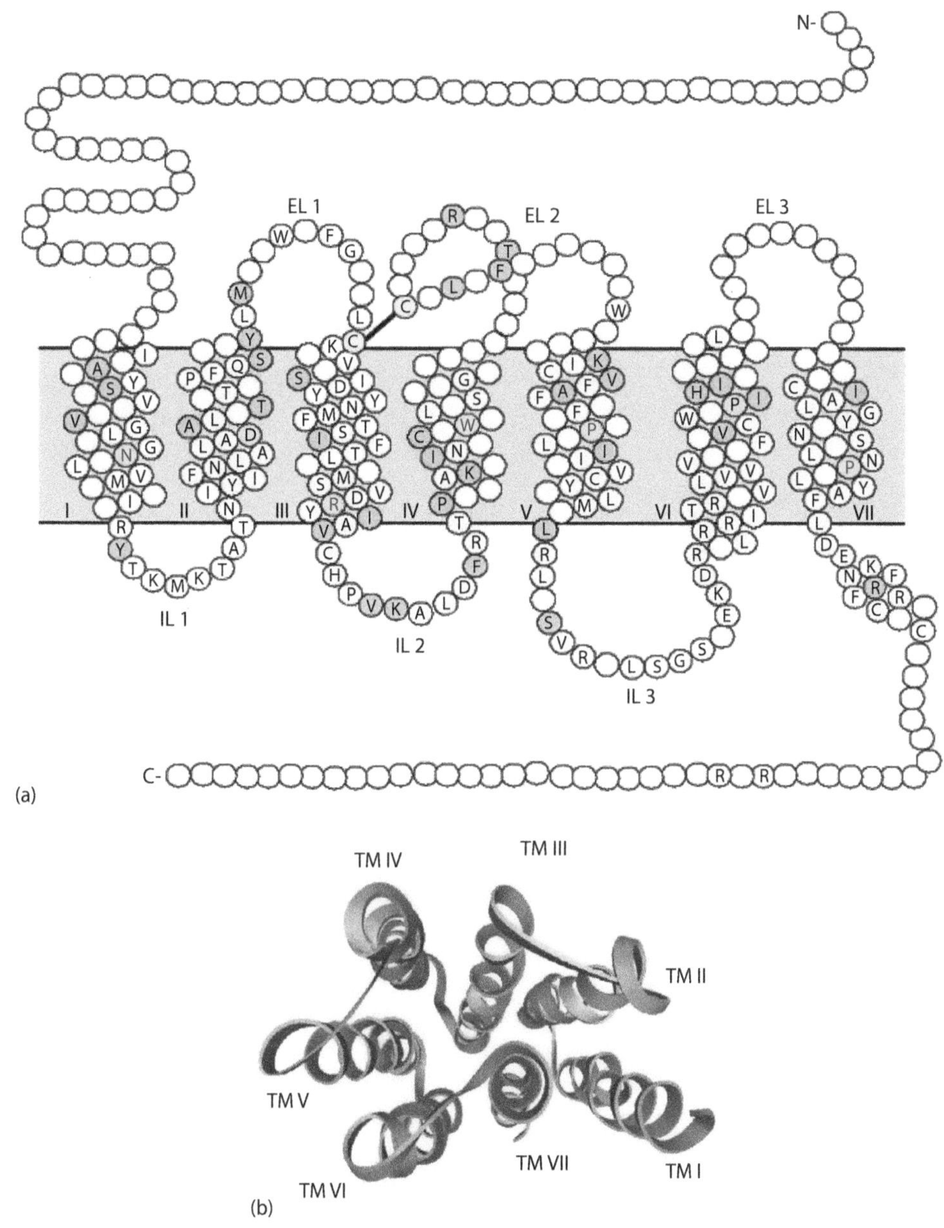

FIGURE 19.2 Structure of opioid receptors. (a) Serpentine model of the opioid receptor. Each transmembrane helix is labeled with a roman number. The white empty circles represent nonconserved amino acids and white circles with a letter represent identical amino acids among the four opioid receptors. Violet circles represent further identity between the MOR, DOR, and KOR. Green circles highlight the highly conserved fingerprint residues of family A receptors, Yellow circles depict the two conserved cysteines in EL loops 1 and 2, likely forming a disulfide-bridge. IL, intracellular loop; EL, extracellular loop. (b) Proposed arrangement of the seven transmembrane helices of opioid receptors as viewed from the top (extracellular side). (Reproduced with permission from Waldhoer, M. et al., *Ann. Rev. Biochem.*, 73, 953, 2004, Copyright Annual Reviews, http://ww.annualreviews.org.)

TABLE 19.1
Opioid Receptor Ligands

Receptor	Agonist	Antagonist	Agonist Effect(s)
μ	Morphiceptin	Naloxone	Analgesia
	DAGO		Respiratory depression
	Normorphine		Miosis
	Sufentanyl		Reduced gastrointestinal motility
			Nausea
			Vomiting
			Euphoria
δ	Deltorphin II	ICI 154,126	Supraspinal analgesia
	DPDPE	ICI 174,864	
	DADLE		
κ	U50,488	MR2266	Analgesia (spinal level)
			Trifluadom Miosis (weak)
			Respiratory depression (weak)
			Dysphoria

Note: DAGO, Tyr-D-Ala-Gly-MePhe-Gly-ol; DPDPE, [D-Pen^{2}, D-Pen^{5}]enkephalin; Pen, penicillamine; DADLE, [D-Ala^{2}, D-Leu^{5}]enkephalin; deltorphin II, Tyr-D-Ala-Phe-Glu-Val-Val-Gly-NH_2; morphiceptin, β-casomorphin-(1–4)-amide or Tyr-Pro-Phe-Pro-NH_2.

TABLE 19.2
Endogenous Opioid Peptides

Precursor	Opioid Peptide Product	Amino Acid Sequence
Pro-enkephalin	[Met]-enkephalin	**YGGFM**
	[Leu]-enkephalin	**YGGFL**
		YGGFMRF
		YGGFMRGL
	Peptide E	**YGGFM**RRVGRPEWWMDYQKR**YGGFM**
	BAM 22P	**YGGFM**RRVGRPEWWMDYQKRYG
	Metorphamide	**YGGFM**RRVNH_2
Pro-opiomelanocortin	β-Endorphin	**YGGFM**TSEKSQTPLVTLFKNAIIKNAYKKGE
Prodynorphin	Dynorphin A	**YGGFL**RRIRPKLKWDNQ
	Dynorphin A(1–8)	**YGGFL**RRI
	Dynorphin B	**YGGFL**RRQFKVVT
	α-Neoendorphin	**YGGFL**RKYPK
	β-Neoendorphin	**YGGFL**RKYP
Pronociceptin	Nociceptin	FGGFTGARKSARKLANQ
Orphanin-FQ	Orphanin-FQ	
	Endomorphin-1	YPWF-NH_2
	Endomorphin-2	YPFF-NH_2
Prodermorphin and prodeltorphin[a]	Dermorphin	Y(D)AFGYPS-NH_2
	Deltorphin	Y(D)MFHLMD-NH_2
	Deltorphin I	Y(D)AFDVVG-NH_2
	Deltorphin II	Y(D)AFEVVG-NH_2

Note: The pentapeptide sequences corresponding to [Met]-enkephalin and [Leu]-enkephalin contained in other opioid peptides are shown in bold. Note that β-endorphin and most of the opioid peptides derived from proenkephalin contain [Met]-enkephalin at their N-termini, whereas the sequence of [Leu]-enkephalin is present in those peptides derived from prodynorphin.

[a] Dermorphin and deltorphins are derived from multiple precursors and all have a naturally occurring D-amino acid in position 2.

The amino acid sequences are YGGFM and YGGFL and termed [Met]-and [Leu]-enkephalin, respectively. Since then, many other peptide opioids of varying lengths have been identified. They are all cleavage products of longer peptides and can be divided into four families based on their precursors. Three of these families all start with the [Met]- and [Leu]-enkephalin. The endogenous opioid peptides have varying affinities for the opioid receptor subtypes; however, none of them are specific for a single subtype, although the neuropeptide nociceptin is the endogenous ligand specific for ORL_1. The precursors are often made up of repeating copies of the opioid peptide products.

High-affinity opioid receptor peptides (dermorphins and deltorphins) have been isolated from frog skins and are quite unusual in having D-amino acids in the sequence. Also, milk-derived casomorphins, hemorphins from hemoglobin and cytochrophins (fragments of cytochrome B), have low affinity for the opioid receptors.

Besides the endogenous peptides, it has been shown that morphine is present in various tissues and body fluids and SH-SY5Y human neuroblastoma cells are capable of producing morphine. The biosynthetic route is similar to that found in *Papaver somniferum*.

19.1.3 Nonendogenous Opioid Receptor Ligands

The synthetic efforts in the opioid field over the last century have mainly been stimulated by the search for a safer alternative to morphine that maintained the analgesic effects but was devoid of respiratory depression and abuse potential. Different medicinal chemistry approaches have been followed in the development of opioid receptor ligands:

- Chemical modification of morphine and related structures (opiates)
- Simplification of the morphine structure
- Dimerization (bivalent ligands)
- Peptides and peptidomimetics

Early development was focused on the first two approaches. Examples of opiates that display similar affinity to all subtypes are shown in the upper part of Figure 19.3. Introduction of bulky substituents to the morphine structure generally yields antagonists, and naloxone and naltrexone are unselective antagonists. *N*-allyl analog nalorphine is an example of a mixed agonist–antagonist. It was originally characterized as an antagonist but later shown to have antagonist activity at MOR but agonist activity at KOR. Nalorphine was one of the first compounds to be extensively tested in the clinic in combination with morphine to find an ideal agonist–antagonist ratio for maximizing analgesics properties and minimizing adverse effects. Buprenorphine is a potent analgesic and partial agonist at the MOR and antagonist at DOR and KOR. Diprenorphine is reported as an unselective antagonist.

An increasing number of subtype-selective ligands has been reported, and a few examples are shown in the lower part of Figure 19.3. Compounds that are μ-selective include morphine that is an agonist and the irreversible antagonist β-FNA. SIOM and NTI are examples of δ-selective antagonists, and gNTI represents a κ-selective antagonist. Interestingly, gNTI was recently shown to have higher affinity toward HEK-293 cells expressing KOR together with DOR or MOR compared to cells expressing only KOR, and the agonist effect of gNTI at the heterodimers KOR/DOR and KOR/MOR could be blocked by antagonists selective against DOR and MOR, respectively. This underlines the importance of heterodimerization and the fact that gNTI is analgesic when injected into the spinal cord but not when injected into the brain could arise from different heteromeric populations.

The development of selective opiates has followed the "message-address" concept. This states that the amino and the aromatic group in the morphine determine the activity (the "message") of the opiates, whereas the lipophilic region around the allylic alcohol confers selectivity (the "address"). This is demonstrated by the conversion of NTI from a δ-selective antagonist into the κ-selective

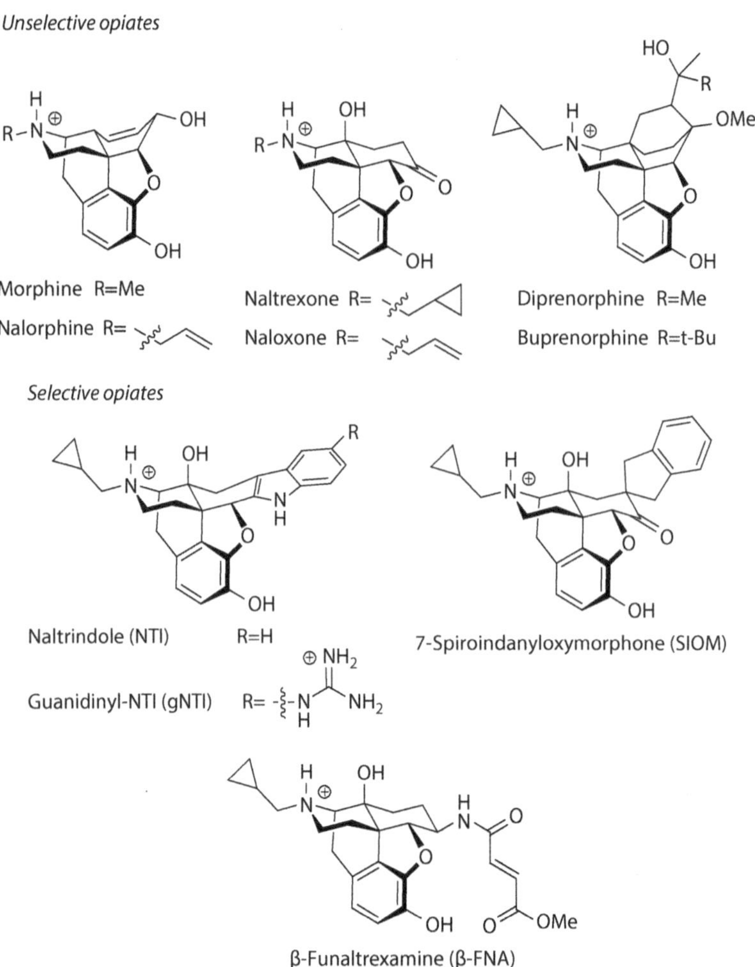

FIGURE 19.3 Chemical structures of classical unselective opiates (except morphine which is μ-selective) based on the morphine scaffold and chemical structures of selective opiates. Morphine and β-FNA are μ-selective agonist and irreversible antagonist, respectively. SIOM and NTI are examples of δ-selective antagonists, whereas the introduction of charged guanidinium group converts NTI into the κ-selective antagonist gNTI.

antagonist gNTI by the introduction of charged guanidinium group. Already in the 1960s, a 3D pharmacophore model was conceived stating the importance of the spatial placement of the amine, the aromatic group, and the lipophilic region for ligand affinity. The successive breakdown of the morphine structure has led to a number of simpler nonopiate structures obeying this early and simple 3D pharmacophore model. This breakdown is shown schematically in Figure 19.4 defining structural classes of opioid receptor ligands developed over the last century.

Examples of these classes are μ-selective agonist fentanyl (piperidine), ethyl-ketocyclazine (benzomorphane), methadone (phenylpropylamine), and meperidine (piperidine) (Figure 19.5). However, other structural classes have appeared, such as δ-selective agonist SNC-80 and κ-selective agonist U50,488, and, more recently, several new scaffolds come from screening compound libraries on the cloned opioid receptors including heteromeric combinations.

Dimerization of ligands is a popular strategy in medicinal chemistry to alter the pharmacological properties of a monomeric ligand. This strategy was advanced in the early 1980s by Portoghese and coworkers using opiates. Initially, the idea was to develop such bivalent ligands

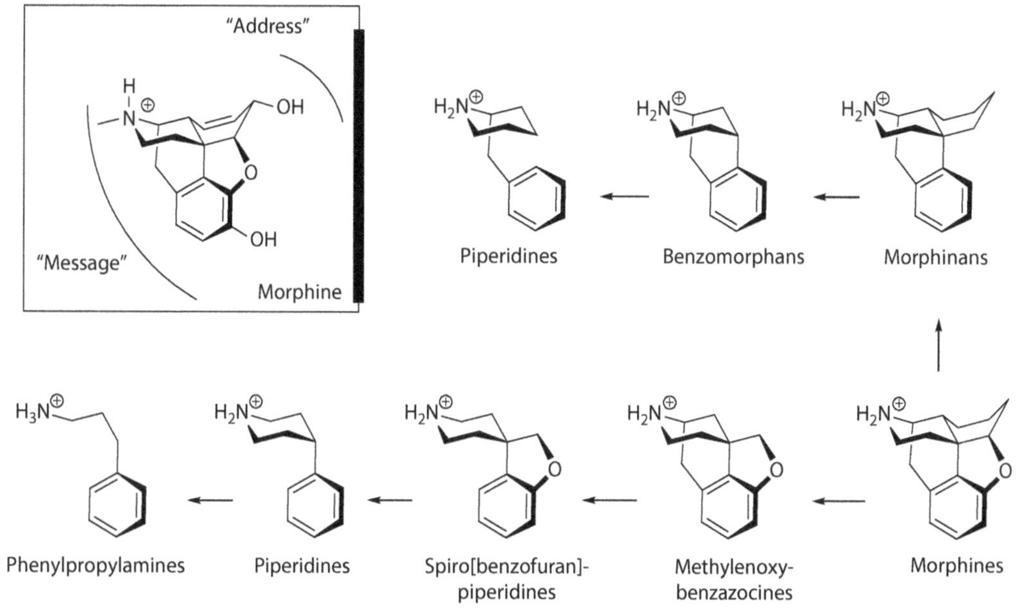

FIGURE 19.4 The message-address concept of the development of opiates is shown schematically in the box. The message region defines the activity of the compounds, whereas the address region defines the selectivity of the compounds. The structural development in the progressive simplification of the morphine scaffold via morphinans and benzomorphans to piperidines, and also via benzazocines, spiropiperidines to piperidines and phenylpropylamines.

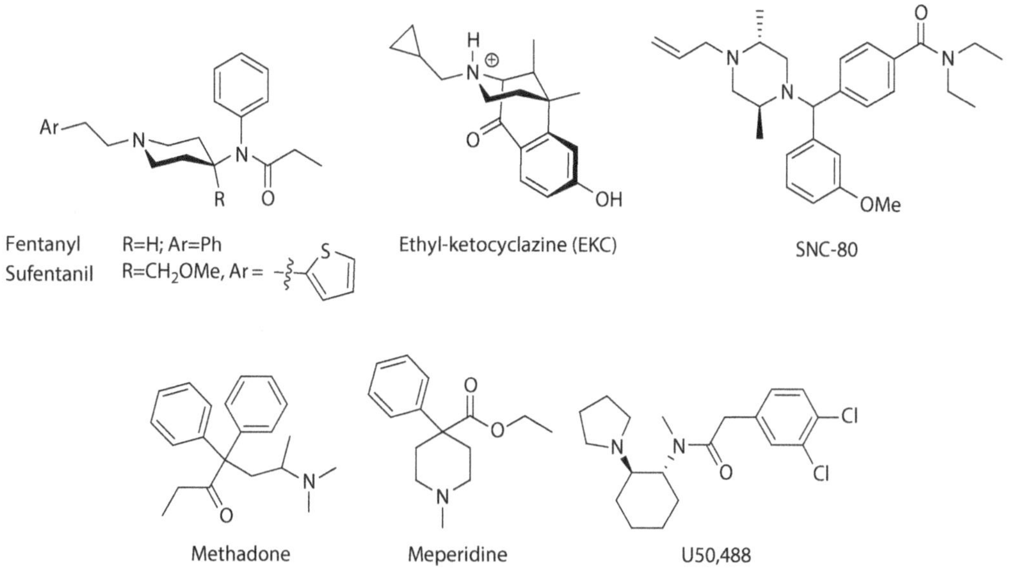

FIGURE 19.5 Examples of different structural classes of opioid receptor ligands.

with a spacer of optimal length that would exhibit a potency that is greater than that derived from the sum of its two monovalent pharmacophores. This would provide evidence that the receptors existed as dimers. One of the first series were compounds **19.1** (Figure 19.6, n = 0, 2, 4, 6, 8) dimerizing a naltrexone analog. The optimal spacer length was shown to be n = 4 giving the highest activity.

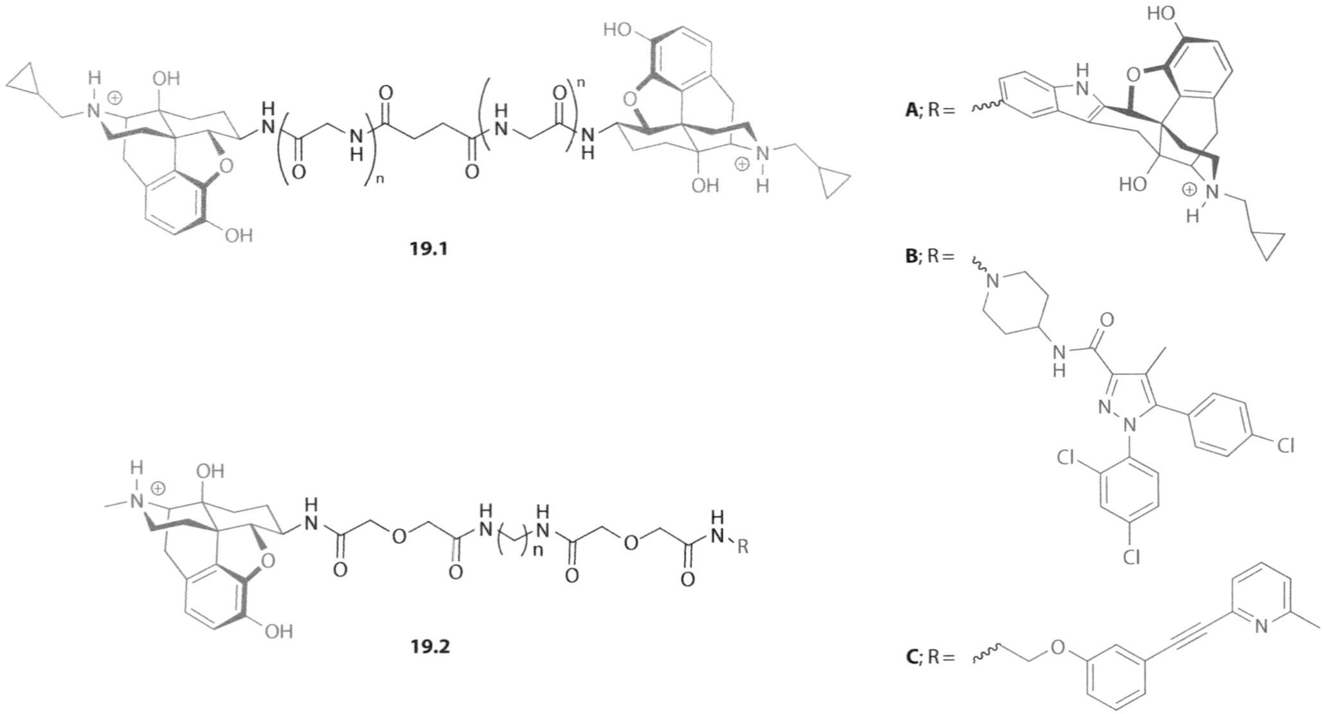

19.1

19.2

A; R=

B; R=

C; R=

FIGURE 19.6 Examples of dimeric or bivalent opioid receptor ligands. Structure **19.1** is a homodimeric ligand linking two identical analogs of naltrexone (green). Structure **19.2** represents heterodimeric ligands linking an analog of MOR agonist oxymorphone (red) with analogs (blue) of either DOR antagonist NTI (**A**), cannabinoid receptor CB_1 inverse agonist Rimonabant (**B**), or metabotropic glutamate receptor mGluR5 antagonist M-MPEP (**C**).

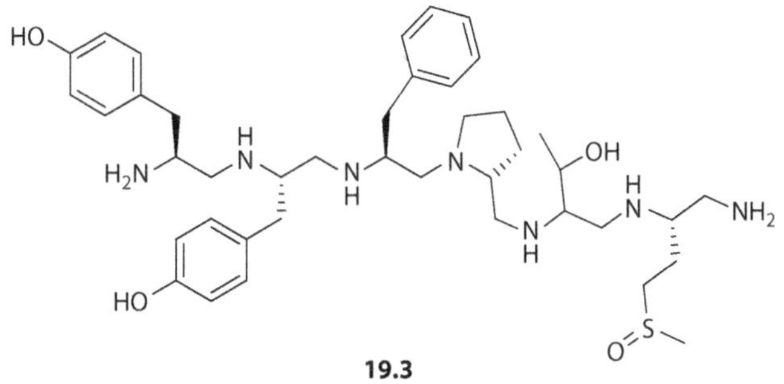

19.3

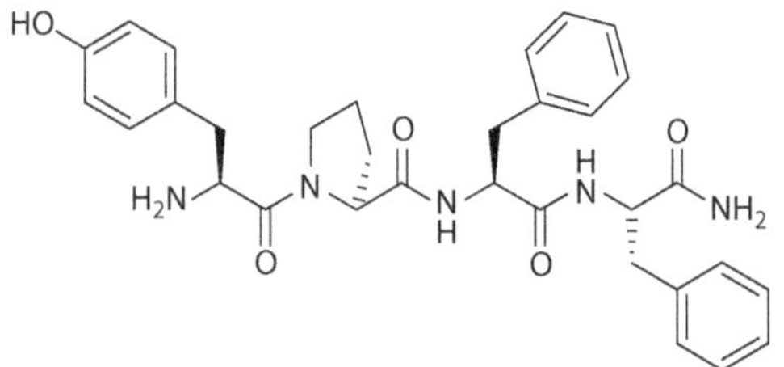

Endomorphin-2

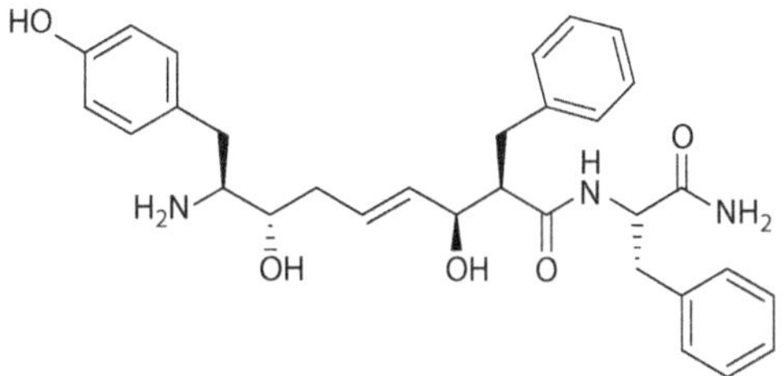

19.4

FIGURE 19.7 Examples of peptidomimetic opioid receptor ligands.

There is today substantial evidence that GPCRs exist as dimers. The concept of making bivalent ligands has been shown in many other areas to be able to modulate other pharmacological properties of a ligand such as degradation, uptake, etc. The concept has also been used to target heterodimeric receptor populations. For example, a series of heterobivalent ligands **19.2** (**A**, n = 2–7) were made by linking analogs of naltrexone and NTI, where tolerance and dependence was significantly reduced with increasing linker length, while agonist potency was increased. It is hypothesized that δ–κ heterodimers are targeted specifically with longer linkers. Also, ligands that selectively target δ–κ heterodimers that are localized in the spinal cord have been developed. More recently, heterobivalent ligands **19.2** have been developed linking an opioid ligand with ligands for other GPCRs such as the cannabinoid receptor or the metabotropic glutamate receptor (mGluR5). Thus, an analog of the opioid ligand naltrexone was covalently linked to a Rimonabant analog (**B**, cannabinoid receptor CB_1 inverse agonist) and M-MPEP (**C**, mGluR5 antagonist), respectively. These compounds also displayed reduced tolerance along with potent analgesic and antinociceptive properties. Furthermore, these compounds provide further evidence for the existence of such heteromeric receptor complexes.

The last approach that will be mentioned here is the use of peptides and peptidomimetics. New agonists and antagonists at opioid receptors have been obtained by making large combinatorial libraries of D- and L-amino acids and screening these compounds against MOR, KOR, and DOR. The sequences span from tetra- to deca-peptides. In this way, potent and selective peptides have been obtained that differ from the endogenous peptides. Furthermore, the modification of the peptide backbone has yielded potent peptidomimetics. The modifications include minor modifications such as backbone amide alkylation. But examples of more extensive modifications are the use of a polyamine backbone as in compound **19.3** or compound **19.4** which is a peptidomimetic analog of endomorphin-2, a potent agonist at MOR with high selectivity for MOR over DOR and KOR (Figure 19.7).

19.1.4 Therapeutic Applications and Prospects

Although development of opiates has been spurred primarily by the search for efficient analgesics with few side effects, other clinical applications of opioid receptor agonists and antagonists are known. Agonists are primarily applied as analgesic, anesthetic, antitussive, and in the treatment of diarrhea. Morphine and codeine are mostly used as analgesics. Fentanyl is a very potent analgesic used in anesthesia. Meperidine is used for acute pain. Methadone is applied to control withdrawal of heroin from addicts. Antagonists are used for reversal of some of the effects induced by agonists. Thus, naloxone has been used to reverse coma and respiratory depression of opioid overdose (methadone and heroin). It is also indicated as an adjunct agent to increase blood pressure under septic shock. Naltrexone has been approved as adjunctive therapy in the treatment of alcohol dependence and the treatment of narcotic addiction to opioids. However, there are also potential indications including obesity, obsessive compulsive disorder, and schizophrenia.

19.2 CANNABINOID RECEPTORS

The plant *Cannabis sativa* has for millennia been used for recreational and medicinal purposes, as it can be seen from old Chinese, Assyrian, and Roman literature. However, it was first in 1964 that the active principle causing the psychoactive effects was isolated and found to be Δ^9-tetrahydrocannabinol (THC) (Figure 19.8). Originally, it was thought that THC due to its lipophilicity somehow acted through fluidizing the cellular membranes, but in the early 1990s it was discovered that THC activates two receptors, cannabinoid receptor-1 (CB_1-receptor) and cannabinoid receptor-2 (CB_2-receptor). Cannabinoid effect in rodents is characterized by the so-called tetraed test. In this test, measurement of spontaneous activity, thermal pain sensation, catalepsy, and rectal temperature are made, and compounds with cannabinoid activity should produce hypomotility, analgesia, catalepsy, and hypothermia. Shortly after the discovery of the receptors, two endogenous compounds were identified that could activate these receptors, i.e., anandamide (arachidonoylethanolamide) and 2-arachidonoylglycerol (2-AG) (Figure 19.8), and they are called endocannabinoids. Both endocannabinoids are of lipid nature and thus not very water soluble. They associate with albumin in the extracellular space and endocannabinoids can function in an autocrine and paracrine fashion where they are formed "on demand" and then degraded, i.e., they are not stored in vesicles like neurotransmitters or peptide hormones. Tissue levels of anandamide and 2-AG are usually in the pmol/g and nmol/g tissue, respectively, but it is not clear whether these levels represent the ligand concentration available to the receptors. However, it is generally considered that

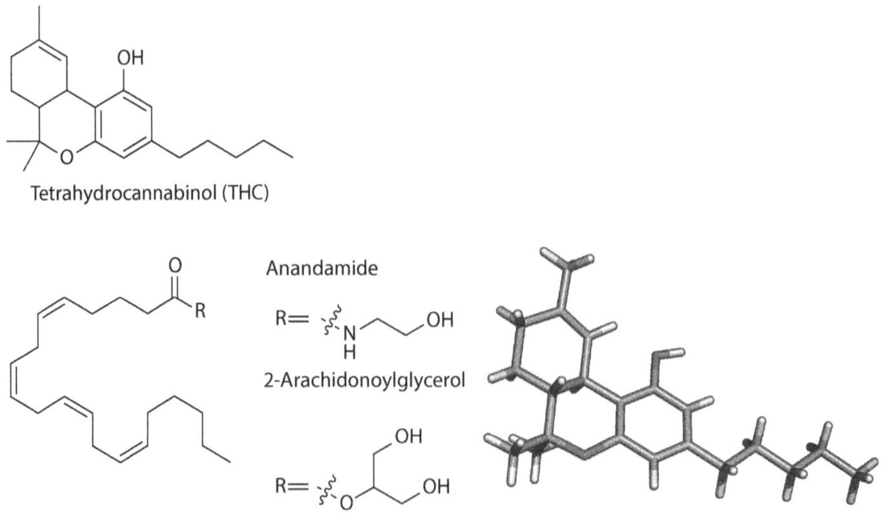

FIGURE 19.8 Plant cannabinoid (THC) and the two endocannabinoids, anandamide and 2-arachidonoyl glycerol. R = arachidonoyl.

there is an endogenous tone of endocannabinoid level in most tissues, especially of anandamide. Endocannabinoids are also found in very small concentrations in plasma, where they are thought to represent spillover from the tissues.

19.2.1 Endocannabinoid System

The endocannabinoid system comprises the two cannabinoid receptors, the two endocannabinoids and the enzymes that synthesize, and degrades the endocannabinoids. Biosynthesis of anandamide is complex, but it is formed from an unusual phospholipid having three fatty acids, *N*-arachidonoyl-phosphatidylethanolamine that again is formed from phosphatidylethanolamine by acylation of the amino group catalyzed by a poorly known calcium-stimulated *N*-acyltransferase. This enzyme uses acyl groups (e.g., arachidonoyl) from the *Sn*-1 position of phosphatidylcholine as substrate in the acylation process, and it will use whatever fatty acid is present. Thus, a number of *N*-acyl-phosphatidylethanolamines are apparently always formed having different fatty acids in the *N*-acyl position of which *N*-arachidonoyl is only a minor component and those with palmitic acid, stearic acid, or oleic acid are generally much more abundant. From this precursor phospholipid, *N*-acyl-phosphatidylethanolamine, a number of different enzyme-catalyzed pathways can result in the generation of acylethanolamides including anandamide that usually amounts to less than 5% of the acylethanolamides (Figure 19.9). Which pathway that is most relevant for a particular tissue or a particular physiological/pathophysiological setting is at present not known. The cellular localization

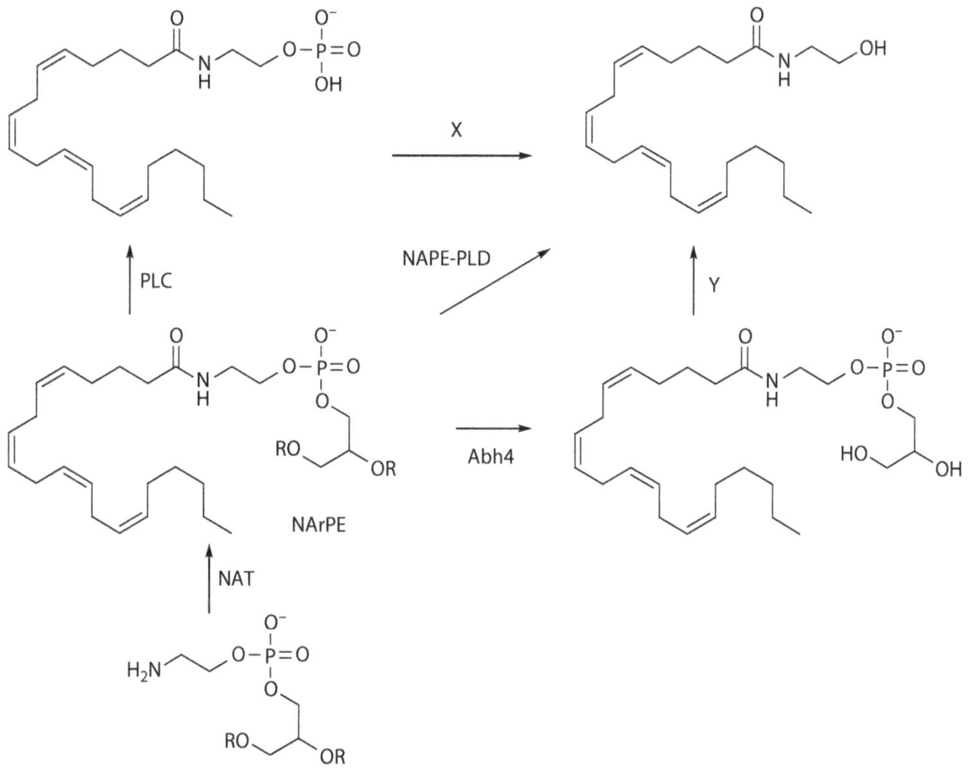

FIGURE 19.9 Biosynthesis of anandamide. The precursor phospholipids (NArPE) are generated from phosphatidylethanolamine by an *N*-acyltransferase (NAT). It can then be hydrolyzed by a phospholipase C (PLC), by *N*-acyl-phosphatidylethanolamine-hydrolyzing phospholipase D (NAPE-PLD), or by alpha-beta-hydrolase 4 (Abh4). Other acylethanolamides may be formed by the same enzymes. (R = fatty acids). The enzymes "X" and "Y" are not well characterized yet.

of anandamide formation is not known, and several of the involved enzymes have not been cloned yet. The different acylethanolamides have a number of more or less specific biological activities, e.g., palmitoylethanolamide is anti-inflammatory and oleoylethanolamide has anorexic and neuroprotective activity that may be mediated via activation of the transcription factor PPARα. These other acylethanolamides do not bind to the cannabinoid receptors. Anandamide is a partial agonist for the cannabinoid receptors, but it can also activate vanilloid receptor and several different ion channels, although it is uncertain to what degree it does this in vivo. All acylethanolamides are degraded by a fatty acyl ethanolamide hydrolase (FAAH), and FAAH-knockout mice has increased levels of anandamide and other acylethanolamides and an increased pain threshold. Acylethanolamides can also be degraded by some other hydrolases. Endogenous levels of anandamide and the other acylethanolamides can be increased several fold during tissue injury. It has been suggested that there exists an anandamide transporter responsible for the uptake of anandamide into cells before it is degraded by the FAAH enzyme which is located in the endoplasmic reticulum. However, a transporter protein has not been characterized and the concept of an uptake transporter for a lipophilic molecule is disputed.

2-AG is formed primarily from diacylglycerol, e.g., 1-stearoyl-2-arachidonoyl-glycerol, catalyzed by a *Sn*-1 specific diacylglycerol lipase and it is degraded by a monoacylglycerol lipase which is present in all tissues. The precursor, diacylglycerol, is known to be formed during receptor-stimulated turnover of inositol phospholipids where inositol-1,4,5-trisphosphate also is formed. Thus, this diacylglycerol formation is located to the cell membrane where the diacylglycerol lipase also is located. It is generally accepted that 2-AG is formed in postsynaptic neurons upon activation of neuronal inositol phospholipids turnover by phospholipase Cβ, whereupon it activates in a retrograde fashion the presynaptic CB_1-receptor that subsequently results in inhibition of neurotransmitter release (Figure 19.10). It is not exactly known how 2-AG travels through the aqueous fluid to reach the presynaptic neuron. This retrograde signaling can decrease the release of glutamate, GABA, acetylcholine, and other neurotransmitters. 2-AG is degraded by a monoacylglycerol lipase that is located

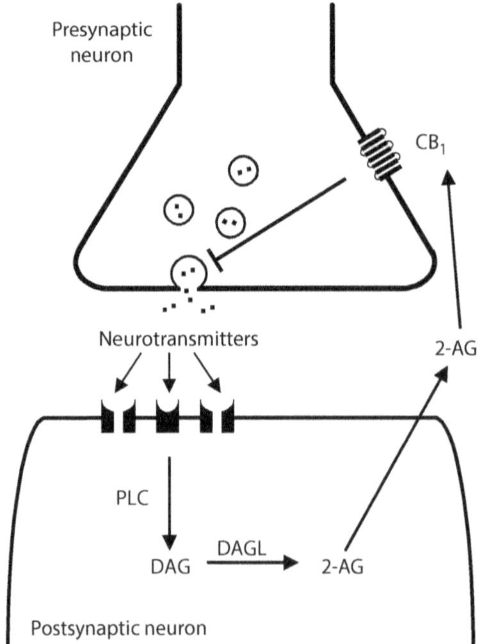

FIGURE 19.10 Synaptic endocannabinoid formation during neurotransmitter release. PLC, phospholipase C; DAG, diacylglycerol; DAGL, diacylglycerol lipase; 2-AG, 2-arachidonoyl glycerol; CB_1, cannabinoid receptor-1.

in the presynaptic neuron. In this way, 2-AG and the CB_1-receptors may contribute to homosynaptic plasticity of excitatory synapses and heterosynaptic plasticity between excitatory and inhibitory contacts that is part of the basic mechanism in learning and memory. This retrograde control is also called depolarization-induced suppression of inhibition and depolarization-induced suppression of excitation for GABAergic and glutamatergic synapses, respectively. Mice lacking monoacylglycerol lipase have an increased level of 2-AG in the brain causing tonic activation and consequently desensitization of the CB_1 receptor. These mice have enhanced learning performance.

19.2.2 Cannabinoid Receptor 1

CB_1-receptor belongs to the class A rhodopsin-like family of GPCRs, and it couples through $G_{i/o}$ proteins negatively to adenylate cyclase and positively to mitogen-activated protein kinase. In addition, CB_1-receptor can also couple to ion channels through these same G-proteins, positively to A-type and inwardly rectifying K^+-channels and negatively to N-type and P/Q type Ca^{2+}-channels. CB_1-receptor is primarily localized to the brain, where it is particularly abundant in cortex, hippocampus, amygdala, basal ganglia, cerebellum, and the emetic centers of the brain stem. CB_1-receptor can also be found in lower abundance in spleen, tonsils, white blood cells, gastrointestinal tissue, urinary bladder, adrenal gland, heart, lung, and reproductive organs. CB_1-receptor may form homodimeric complexes, and heterodimeric complexes with μ-opioid receptor and dopamine D2 receptor. Anandamide and 2-AG may reach the receptor from the lipid phase of the membrane and not from the aqueous site.

In vitro CB_1-receptor seems to have constitutive activity or it may be under endocannabinoid stimulatory tone. Several antagonists have also been shown to be inverse agonists, but it is unclear whether this has any in vivo significance.

Knockout mice that are lacking the receptor protein have been generated. They are generally healthy and fertile with no apparent gross anatomical defects. However, they do have a number of abnormalities, e.g., dysregulation of the hypothalamus–pituitary–adrenal axis, suggesting a role of endocannabinoids in modulating neuroendocrine functions. Furthermore, they have a lighter and leaner body phenotype and seem to have higher energy expenditure. In a number of experimental settings, these mice do also behave differently, e.g., in studies of alcohol dependence and of depression. CB_1-receptor knockout mice do not show hypothermia, hypoalgesia, and hypoactivity in response to THC. There is evidence of splice variation of CB_1-receptor of very low abundance, but their biological significance is unclear. Activation of GPR 55 (see Section 19.2.4) can be blocked by the CB_1 receptor antagonist Rimonabant (Figure 19.12), but not by the CB_2 receptor antagonist SR 144528 (Figure 19.13).

19.2.3 Cannabinoid Receptor 2

CB_2-receptor also belongs to the 7TM-receptors, and the human CB_2-receptor has 44% homology with human CB_1-receptor. It seems to couple to the same G-proteins and signaling pathways as does CB_1-receptor. However, CB_2-receptor is found primarily in the spleen, immune cells, tonsils, and brain microglial cells where its expression can be induced by transformation of microglia to macrophage-like cells. Furthermore, CB_2-receptor is found in osteoclasts, osteocytes and osteoblasts, where it plays a critical role in the maintenance of normal bone mass. CB_2 receptor can form heterodimeric complexes with GPR55 (see Section 19.2.4). CB_2-receptor knockout mice appear healthy and fertile, but they have low bone mass. In animal models of pain and inflammation, these CB_2-receptor knockout mice have indicated a clear role of this receptor in modulating acute pain, chronic inflammatory pain, postsurgical pain, cancer pain, and pain associated with nerve injury. Several CB_2 agonists for pain treatment have, however, failed in clinical trials. One reason could be that functionally selective agonists (e.g., biased agonists) may actually in vivo antagonize some signaling pathways stimulated by the endogenous unbiased agonist, e.g., 2-arachidonoyl glycerol.

19.2.4 Other Cannabinoid Receptors?

Since 2-AG and especially anandamide have a number of pharmacological effects that cannot be fully explained by activation of the known cannabinoid receptors, it has been suggested that there may exist more cannabinoid-like receptors. GPR55 has been suggested to be such a new cannabinoid receptor since it can be activated by both THC, 2-AG, anandamide, and the cannabinoid receptor agonist CP 55,940 (Figure 19.12). Another cannabinoid receptor agonist, WIN 55,212-2, can, however, not activate GPR55, and recently, the endogenous agonist for GPR 55 has been found to be lysophosphatidylinositol. A number of agonists (e.g., CP 55,940, WIN 55,212-2, and O-1812) and antagonists (e.g., Rimonabant, AM251, and LY 320135) (Figure 19.12) have been developed. Especially, anandamide can activate a number of other receptors, e.g., vanilloid receptor, and ion channels as mentioned earlier, and this may add to the confusion of the existence of additional cannabinoid receptors. 2-AG can at an apparently allosteric site activate GABA-A receptor.

19.2.5 Therapeutic Use and Potential

THC in capsules (Marinol/Dronabinol, Solway Pharmaceutical) is used for the treatment of nausea and vomiting that are common side effects of chemotherapy, and for the stimulation of appetite in AIDS patients. A synthetic THC-analog (Nabilone/Cesamet, Valeant Pharmaceuticals) (Figure 19.12) is also on the market for the same treatments. In Canada, THC in the form of an extract of cannabis sativa called Sativex (GW Pharmaceuticals) is provided as a mouthspray for multiple sclerosis patients, who can use it to alleviate neuropathic pain and spasticity. Sativex also contains other cannabinoids including cannabidiol that may add to its function. Medicinal cannabis, i.e., marihuana or hashish prescribed by a doctor for increased well-being and alleviation of pain, spasticity, or loss of appetite by patients having AIDS, cancer, and multiple sclerosis, has been approved in several countries, and there is a strong lobby for approval in certain U.S. states.

Sanofi-Aventis brought a CB_1-antagonist (SR141716A, Rimonabant, trade name Acomplia, Figure 19.12) on the European marked in 2006 for the treatment of obesity (body mass index above 30). Large clinical trials had shown that Rimonabant induced a weight loss of approximately 10% of initial body weight within 1 year. Discontinuation of Rimonabant treatment resulted in regain of lost weight. It was suspended by Sanofi in 2008 due to risk of depression and suicide. Many drug companies stopped their development of comparable CB_1 agonists for weight loss. A CB_1-receptor antagonist that does not cross the blood–brain barrier may possibly also have beneficial effects on energy metabolism.

Emerging evidence points to a possible participation of the endocannabinoid system in the regulation of the relapsing phenomenon of drug abuse in animal models. CB_1-receptor seems to be important in drug- as well as cue-induced reinstatement of drug-seeking behavior. Stimulation may elicit relapse not only to cannabinoid seeking but also to cocaine, heroin, alcohol, and methamphetamine, and this effect is significantly attenuated in animal experiments by pretreatment with CB_1-receptor antagonists. CB_2 agonists have shown positive results in preclinical studies on various forms of pain and atopic dermatitis.

Potential clinical application involves drugs that can increase or decrease endocannabinoid levels (i.e., FAAH-inhibitors and monoacylglycerol-lipase [MAGL] inhibitors or diacylglycerol-lipase inhibitors, respectively) or serve as agonists/antagonists for the two cannabinoid receptors (Table 19.3). Thus, the potential is large but so is the risk of side effects since the endocannabinoid system appears to be so ubiquitous.

19.2.6 FAAH Inhibitors and Anandamide Uptake Inhibitors

As discussed earlier, it is not clear whether an anandamide transporter exists and several compounds believed to be uptake inhibitors have turned out to be inhibitors of FAAH. As an example of an experimental FAAH-inhibitor, URB 597 is shown (Figure 19.11).

TABLE 19.3
Potential Clinical Applications of the Endocannabinoid System

	Therapeutic Target				
Clinical Condition	**CB_1 Agonists**	**CB_2 Agonists**	**CB_1 Antagonists**	**FAAH Inhibitors**	**MAGL Inhibitors**
AIDS and cancer Appetite stimulation and inhibition of nausea and vomiting	#			X	X
Inflammatory bowel diseases Stimulation of gastrointestinal mobility and reduction of inflammation	X	X			
Multiple sclerosis Inhibition of tremors and spasticity	X			X	X
Obesity Weight loss			X		
Osteoporosis Inhibition of bone loss		X			
Pain Chronic, inflammatory, and neuropathic	X	X		X	X
Depression					X

Note: X based on preclinical data from corresponding animal models; # in clinical use.

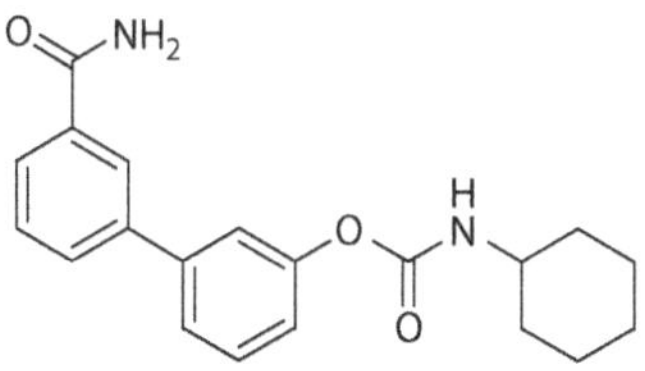

URB597

FIGURE 19.11 Inhibitor of FAAH, the enzyme degrading anandamide and other acylethanolamides.

19.2.7 CB_1-Receptor Agonist/Antagonist

Besides THC, anandamide and 2-AG, experimentally used synthetic CB_1-receptor agonists involve CP 55,940 and the aminoalkylindole WIN 55,212-2 that target both the cannabinoid receptors. More selective CB_1-receptor agonists are arachidonic acid derivatives like O-1812. Selective CB_1-receptor antagonist involves Rimonabant that for a short time was in clinical use and the experimental compounds AM251 and LY 320135 (Figure 19.12).

19.2.8 CB_2-Receptor Agonist/Antagonist

THC, 2-AG, and, to a lesser extent, anandamide activate the CB_2-receptor. Selective synthetic agonists include JHW 133 and GW 405833 (partial agonist) that are used experimentally (Figure 19.13). SR 144528 is a selective CB_2-receptor antagonist.

19.2.9 Diacylglycerol Lipase Inhibitors

There are no really specific inhibitors of diacylglycerol lipase, but RHC 80267, tetrahydrolipostatin (Orlistat), and O-3841 have been used in experimental settings (Figure 19.14).

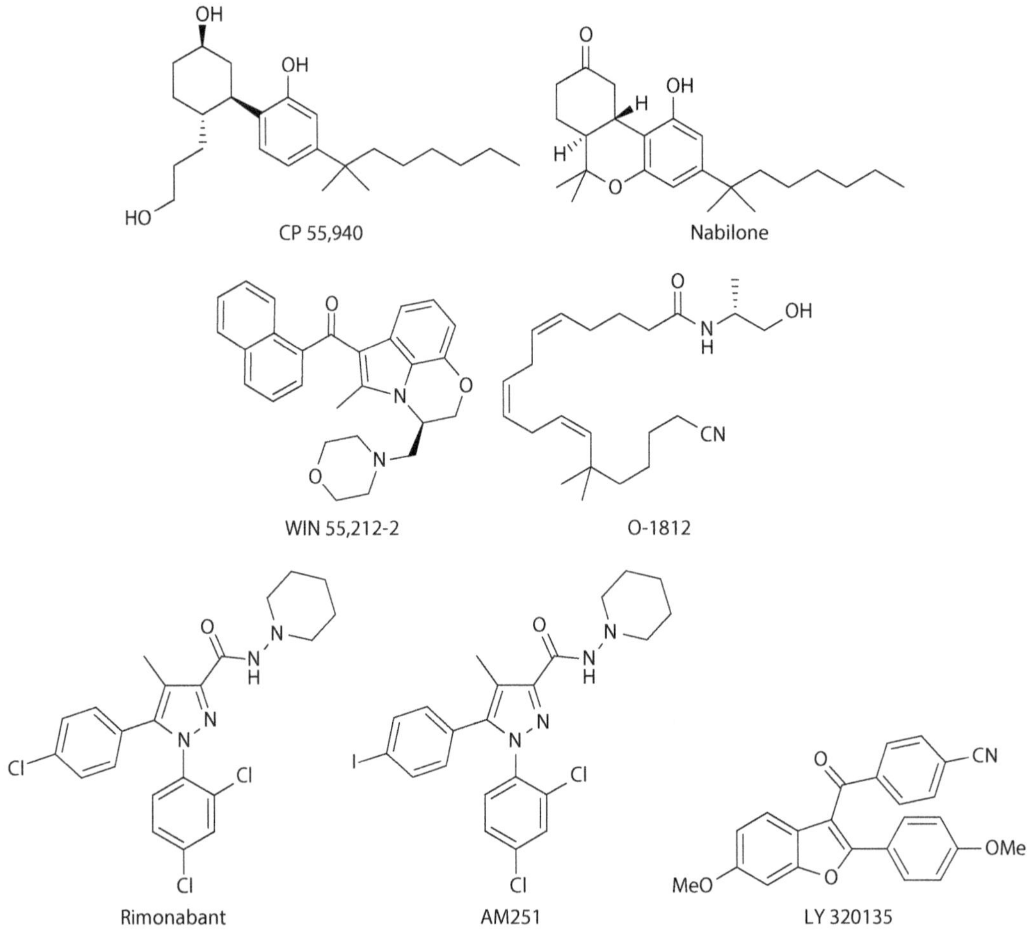

CP 55,940

Nabilone

WIN 55,212-2

O-1812

Rimonabant

AM251

LY 320135

FIGURE 19.12 Agonists and antagonists for CB_1-receptor.

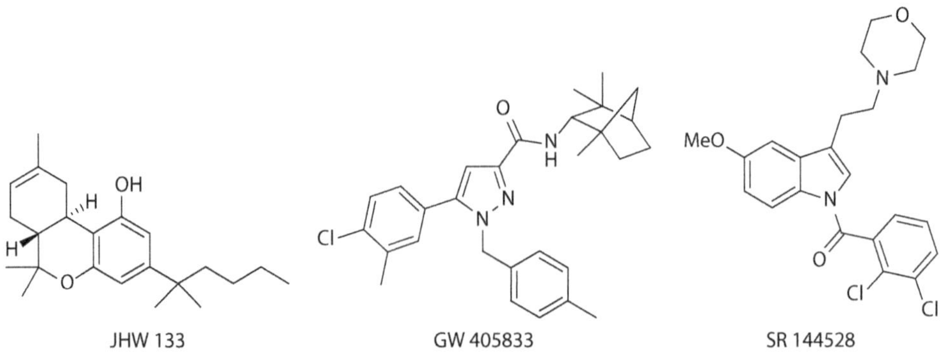

JHW 133

GW 405833

SR 144528

FIGURE 19.13 Agonists and antagonists for CB_2-receptor.

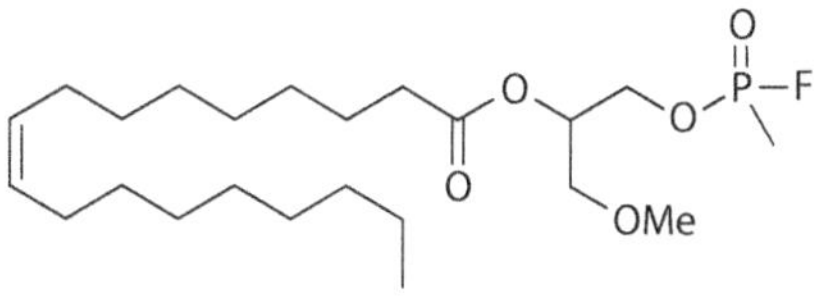

Tetrahydrolipostatin

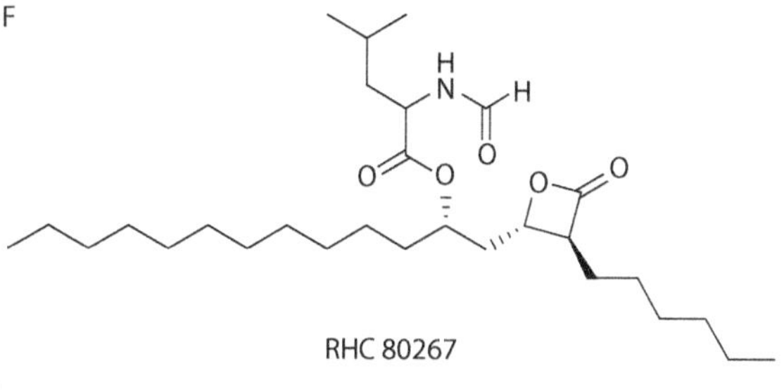

RHC 80267

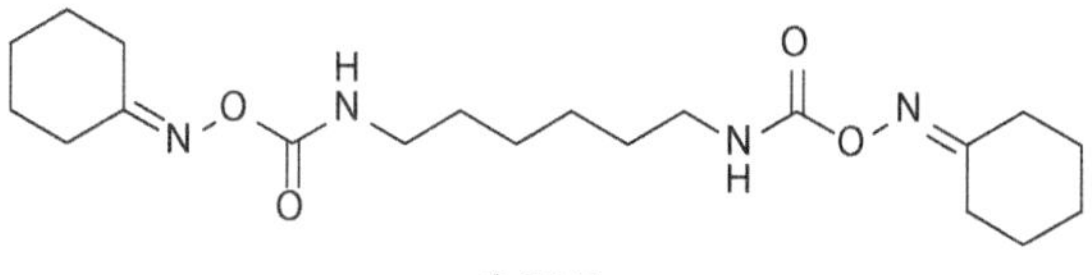

O-3841

FIGURE 19.14 Inhibitors of diacylglycerol lipase, the enzyme generating 2-arachidonoylglycerol.

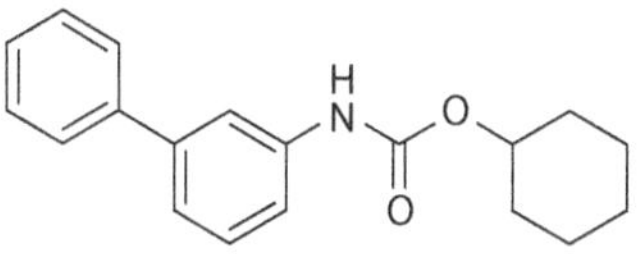

URB602

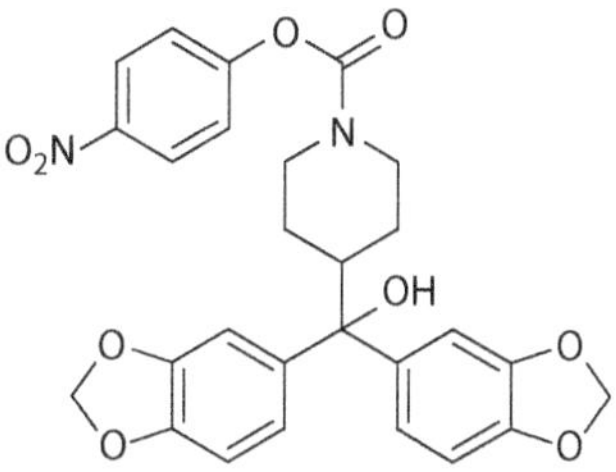

JZL184

FIGURE 19.15 Examples of inhibitors of monoacylglycerol lipase (MAGL), the enzyme degrading 2-arachidonoylglycerol.

19.2.10 Monoacylglycerol Lipase Inhibitors

URB602 and JZL 184 (Figure 19.15) are used as inhibitors of monoacylglycerol lipase in experimental settings, and especially JZL 184 is a selective inhibitor. There also exist some inhibitors which target both FAAH and monoacylglycerol lipase.

As can be seen from the description of the enzyme inhibitors earlier, this research area is at a very early stage of drug discovery, but if potent and specific inhibitors can be found, there are numerous clinical settings where such enzyme inhibitors may be alternatives to cannabinoid receptor agonists or antagonists by increasing or decreasing the formation of endocannabinoids.

FURTHER READING

Borgelt, L.M., Franson, K.L., Nussbaum, A.M., and Wang, G.S. 2013. The pharmacologic and clinical effects of medical cannabis. *Pharmacotherapy* 33:195–209.

Corbett, A.D., Henderson, G., McKnight, A.T., and Paterson, S.J. 2006. 75 years of opioid research: The exciting but vain quest for the Holy Grail. *Br. J. Pharmacol.* 147:S153–S162.

Dhopeshwarkar, A. and Mackie, K. 2014. CB2 cannabinoid receptors as a therapeutic target—What does the future hold? *Mol. Pharmacol.* 86:430–437.

Fujita, W., Gomes, I., and Devi, L.A. 2015. Revolution in GPCR signaling: Opioid receptor heteromers as novel therapeutic targets: IUPHAR review 10. *Br. J. Pharmacol.* 171:4155–4176.

Kaine, B.E., Svensson, B., and Ferguson, D.M. 2006. Molecular recognition of opioid receptor ligands. *AAPS J.* 8(1) Article 15:E126–E137.

Mechoulam, R. and Parker, L.A. 2013. The endocannabinoid system and the brain. *Ann. Rev. Psychol.* 64:21–47.

Pertwee, R.G. 2014. Elevating endocannabinoid levels: Pharmacological strategies and potential therapeutic applications. *Proc. Nutr. Soc.* 73:96–105.

Pertwee, R.G., Howlett, A.C., Abood, M.E. et al. 2010. International union of basic and clinical pharmacology. LXXIX. Cannabinoid receptors and their ligands: Beyond CB1 and CB2. *Pharmacol. Rev.* 62:588–631.

Portoghese, P.S. 2001. From models to molecules: Opioid receptor dimers, bivalent ligands and selective opioid receptor probes. *J. Med. Chem.* 44(14):2260–2269.

Vermuri, V.K. and Makriyannis, A. 2015. The medicinal chemistry of cannabinoids. *Clin. Pharmacol. Ther.* 97:553–558.

Waldhoer, M., Bartlett, S.E., and Whistler, J.L. 2004. Opioid receptors. *Ann. Rev. Biochem.* 73:953–990.

20 Neglected Diseases

Søren B. Christensen and Ib C. Bygbjerg

CONTENTS

20.1 INTRODUCTION

Infectious diseases including parasitic diseases account for approximately a third of the worldwide disease burden but only for 5% of the disease burden in high-income countries. Pharmaceutical companies assume that only a marginal profit can be made from drugs against diseases in low-income countries, and consequently, less than 2% of new chemical entities marketed in the last 30 years have been anti-infective drugs. The majority of the few marketed anti-infective agents were antiretroviral drugs, the development of which benefited from a serious political commitment from high-income countries for finding drugs against AIDS.

The limited interest for development of drugs against infectious diseases like malaria, African trypanosomiasis, Chagas disease, schistosomiasis, leishmaniasis, and tuberculosis has led to the use of the term "neglected diseases." Worldwide, these diseases are responsible for a heavier burden than cardiovascular or central nervous system (CNS) diseases. In the case of malaria, the number of mortalities was estimated to increase from 995,000 in all age groups in 1980 to 1,817,000 in 2004. A positive development, however, is that funding committed to malaria control has increased from $100 million (U.S.) in 2000 to 1.9 billion in 2012. As a consequence, the mortality has decreased to 1,238,000 in 2010 and to an even smaller number later on. This decrease is only partly a result of new drugs but mainly caused by intensified surveillance, treatment, and prophylaxis.

Medicine for Malaria Venture (MMV) is a nonprofit organization founded in 1999 which has set-up as a public private partnership in order to fund and manage the discovery, development, and delivery of new medicines for the treatment and prevention of malaria. Coartem Dispersible, a drug developed for the treatment of children, has emerged as a result of collaboration between MMV and Novartis. WHO has defined essential drugs as "those drugs that satisfy the health care needs of the majority of the population; they should therefore be available at all times in adequate amounts and in appropriate dosage forms, at a price the community can afford."

In the past, including the last decades, the major inspiration for the development of drugs against infectious diseases has come from natural products (see Chapter 7). Regarding antiparasitic drugs, some have primarily been developed for veterinary medicine, before subsequently being adopted by human medicine.

20.2 INFECTIONS CAUSED BY HELMINTHIC PARASITES

A parasite is an organism that lives in or on and takes its nourishment from another organism. A parasite cannot live independently. A helminth is a multicellular parasitic worm. In general, a helminth is visible to the naked eye in its adult stages. Parasites might have more hosts, e.g., intermediary and final hosts. The smaller host is generally called the vector. Many neglected diseases are caused by parasites. In contrast to bacteria which are prokaryotes, parasites are eukaryotes as are mammalians including humans.

20.2.1 Schistosomiasis

Schistosomiasis (bilharziasis) is caused by infection with flatworms belonging to the genus *Schistosoma*. *Schistosoma mansoni* is found in South America and Africa, *Schistosoma haematobium* is found throughout Africa, in particular in Egypt, and in parts of the Middle East, and *Schistosoma japonicum* is confined to the Far East. Approximately 200 million people in 74 developing countries suffer from the diseases, 20 million suffer severe consequences, such as colonic polyposis with bloody diarrhea (*S. mansoni*), splenomegaly and portal hypertension with hematemesis (vomiting of blood) (*S. japonicum* and *S. mansoni*), cystitis and ureteritis which may lead to bladder cancer (*S. haematobium*), and CNS lesions. The diseases are estimated to cause 280,000 deaths each year. Two safe, effective drugs, praziquantel (Figure 20.1, **20.1**) and oxamniquine (**20.2**), are now available for schistosomiasis and are included in the WHO Model List of Essential Drugs. Oxamniquine only affects infections with *S. mansoni*. Artemisinins (Figure 20.13), used for malaria, also have some effect on schistosomiasis. The mechanism of action is not understood for praziquantel and oxaminiquine.

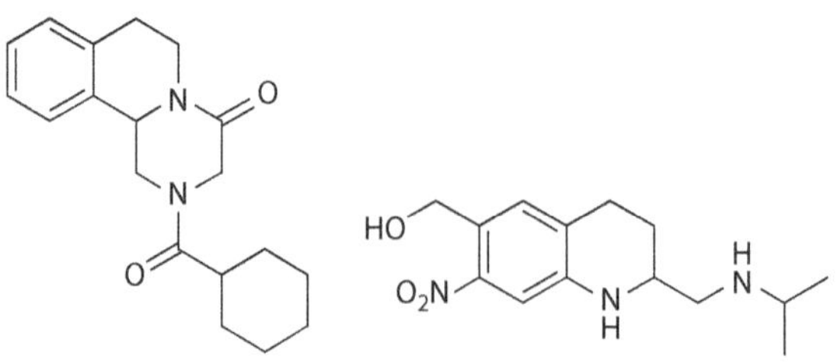

20.1 **20.2**

FIGURE 20.1 Structures of praziquantel (**20.1**) and oxamniquine (**20.2**).

20.2.2 Filariasis

Filariasis is caused by parasites belonging to the order Filaroidea. *Wuchereria bancrofti* and *Brugia malayi* both cause lymphatic filariasis (elephantiasis), *Loa loa* causes fugitive swelling in particular around the eyes, and *Onchocerca volvulus* causes onchocerciasis (river blindness). All the diseases are caused by helminthic worms transmitted through bites of insects belonging to the order Diptera. Some of the symptoms of the diseases are caused by the presence of parasites restricting the flow of lymph fluid, others by the tiny microfilaria invading the skin and eyes. Approximately 120 million people are infected with the parasites and 40 million severely disabled. Onchocerciasis is estimated to have infected 17.7 million peoples, of which 500,000 have visual impairment and 270,000 are blinded. The disease is limited to the vicinity of rivers where the vector, blackflies of the genus Simulium, is endemic. Unfortunately, the burden of the disease often forces the population to leave these areas uninhabited. The infection can be treated with ivermectin or with a combination of ivermectin with albendazole (Figures 20.2 through 20.4). Ivermectin, a dihydro derivative of avermectin B_{1a} and B_{1b} obtained from *Streptomyces avermitilis*, acts by opening invertebrate-specific glutamate-gated chloride ion channels in the nerve end and muscles of the parasites. This leads to death of microfilariae, the first larval stage. The drug does not cause immediate death of the adult parasite but reduces the worm's life-span. Human glutamate-gated chloride channels are not affected by ivermectin (**20.8**).

The observation that filarial nematodes live in symbiosis with *Endobacteria wolbachia* has initiated experimental treatment with antibiotics like tetracycline, rifampicin, and doxycycline.

20.3 INFECTIONS CAUSED BY PROTOZOAN PARASITES BELONGING TO OTHER GENERA THAN PLASMODIUM

Protozoan parasites are single-celled organisms that have an animal-like nutrition (they cannot perform photosynthesis). The life cycle of protozoan parasites involves two hosts; the smaller of which typically is named the vector. Important genera are *Plasmodium*, *Trypanosoma*, and *Leishmania*. Protozoans in the guts like *Entamoeba histolytica* will not be described in this chapter.

20.3.1 Trypanosomiasis

Two major tropical diseases, American trypanosomiasis (Chagas disease) and African trypanosomiasis (sleeping sickness), are caused by *Trypanosoma cruzi* and subspecies of *Trypanosoma brucei*, respectively. Two subspecies of *T. brucei* exist. *T. b. gambiense* is found in Central and West Africa, whereas *T. b. rhodiense* is present in East and South Africa. African trypanosomiasis is spread with tsetse flies (*Glossina* species). If untreated, the disease may be lethal: for *T. b. rhodiense* in the acute blood stage, for *T. b. gambiense* when the parasites enter the CNS causing coma (explaining the name sleeping disease) and death. In 1990, 300,000 new cases were estimated each year, whereas intensive measurements have led to a decrease to 30,000 new cases in 2009. Treatment of infections with *T. b. gambiense* is based on therapy using a combination of nifurtimox (Figure 20.2, **20.6**) and eflornithine (**20.7**), whereas the only drugs available for the treatment of the disease caused by *T. b. rhodiense* are suramin (**20.10**) in the acute stage and in the late stage the arsenical drug melarsoprol (**20.5**), both developed almost 100 years ago. Suramin was developed based on a misconception of Paul Ehrlich that the ability of trypan red to stain *Trypanosoma* parasites could be used to combat these organisms. The same misconception was used to develop chloroquine (Section 20.5.1).

The mechanism of action of nifurtimox (**20.6**) is not understood in detail, but it is assumed to increase the level of reactive oxygen species inside the parasite. Eflornithine (**20.7**), an analog of ornithine (**20.8**), is a suicide substrate for ornithine decarboxylase. Inhibition of this enzyme prevents the parasite from generating polyamines which are important for the survival. Scheme 20.1 illustrates a normal decarboxylation of an α-amino acid in an enzyme using pyridoxal phosphate as a coenzyme.

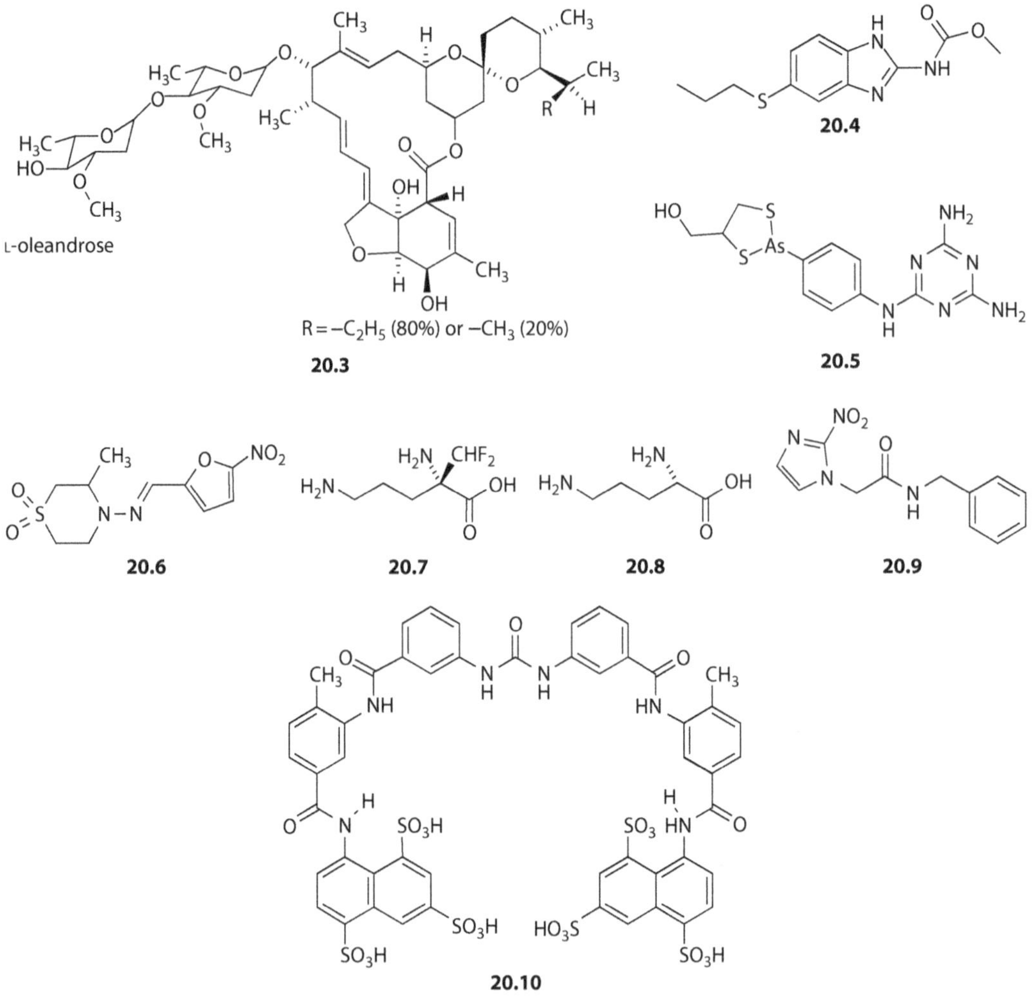

FIGURE 20.2 Structures of ivermectin (**20.3**), albendazole (**20.4**), melarsoprol (**20.5**), nifurtimox (**20.6**), eflornithine (**20.7**), ornithine (**20.8**), benznidazole (**20.9**), and suramin (**20.10**).

The presence of a difluoromethyl group in eflornithine enables the formation of a reactive intermediate which is attacked by the thiol group of cystein-360. The result is a covalent bond from ornithine decarboxylase to the substrate and thereby inactivation (Scheme 20.2).

Approximately 7.6 million persons are estimated to suffer from Chagas disease almost exclusively in Mexico, Central America, and South America. The preferred drugs are benznidazole (**20.9**) or nifurtimox (**20.6**), but there is still no effective cure. It is suggested that the two drugs act by provoking oxidative lesions of DNA in the nucleus and in the mitochondria, but other mechanisms of action are also suggested.

20.3.2 Leishmaniasis

Leishmaniasis is caused by parasites of the genus *Leishmania*. The diseases vary from simple self-healing skin ulcers (cutaneous leishmaniasis) found in South America, Africa, the Middle East, and Asia, severe disfiguring of nose, throat, and mouth cavities (mucocutaneous leishmaniasis)

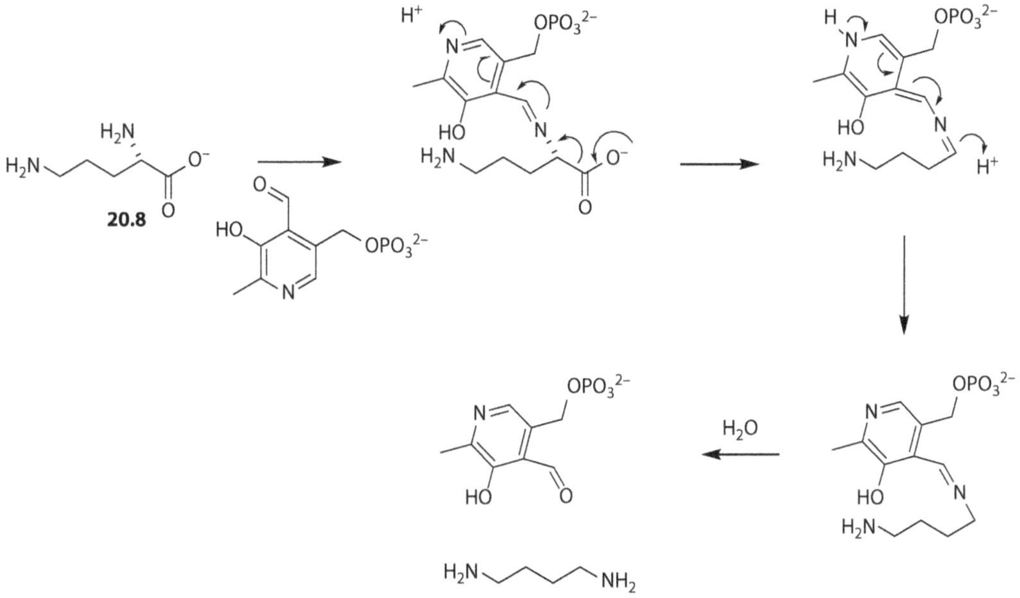

SCHEME 20.1 Decarboxylation of ornithine (**20.8**) by ornithine decarboxylase in which pyridoxal phosphate is a coenzyme.

found in South America, to life-threatening infections (visceral leishmaniasis). Visceral leishmaniasis can be fatal if untreated. Approximately 12 million humans are infected with leishmaniasis, and it is estimated that 59,000 die each year. The parasites nourish in the macrophages. The life cycle is illustrated in Figure 20.3.

Until an oral active drug became available, treatment of leishmaniasis was based on amphotericin B (Figure 20.4, **20.11**), pentamidine, or antimony-containing drugs like sodium stibogluconate and meglumine antimonate. Liposomal formulations of amphotericin B have increased the efficiency of the drug. Application of a drug in vesicles as liposomes will target the drug against cells performing phagocytosis like the macrophages. Since the macrophages host the parasites, some selectivity in activity is obtained. The main mechanism of action of amphotericin B is based on the amphiphilic nature of the molecule consisting of a lipophilic heptaene region and a hydrophilic polyol region. The polyene region complexes with steroids in the parasite membrane. The hydrophilic polyol region forms an ion channel permeable to small ions (Figure 20.4). Some selectivity is obtained because the drug has higher affinity for the double bonds of ergosterol dominating in the cell membrane of the parasites than for cholesterol in the membrane of mammalian cell.

Serendipitously, it was discovered that the cancer drug miltefosine (**20.12**) is an orally active drug against visceral leishmaniasis. The exact mechanism of action is still not understood, but it is assumed that accumulation in the cell membrane is important for the effect. The missing ester group in the molecule compared to phospholipids (e.g., see lysophosphatidylcholine **20.13**) causes resistance against metabolizing enzymes. In addition, miltefosine inhibits the formation of phosphatidylcholine which may predispose the cells for apoptosis (Figure 20.5).

20.4 MALARIA

Malaria is a leading cause of morbidity and mortality in the tropical world; some 300–500 million of the world population are infected with malaria parasites, presenting 120 million clinical cases each year. Among more than 100 species of *Plasmodium* parasites, only five can

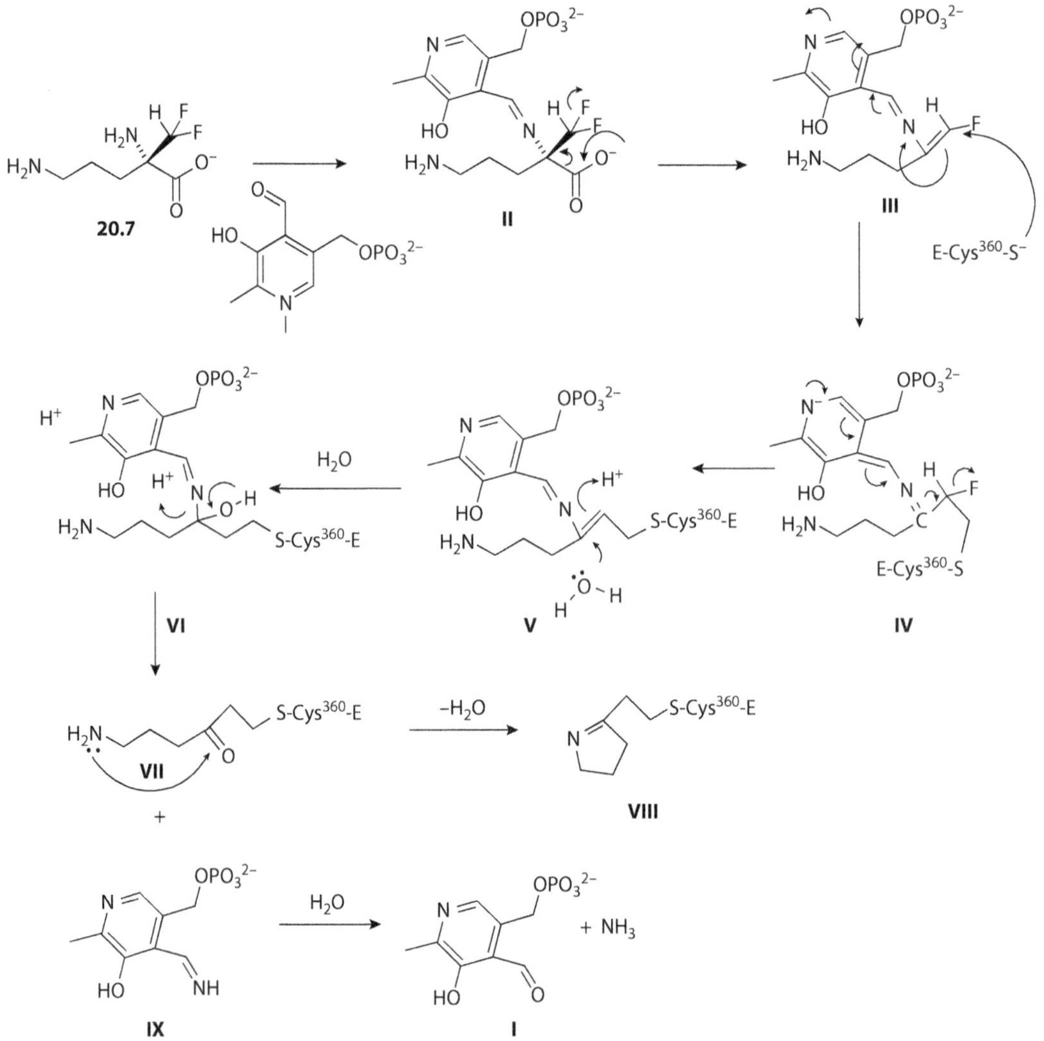

SCHEME 20.2 The suicide substrate eflornithine (**20.7**) is during the reaction with pyridoxyl phosphate converted into a reactive intermediate which reacts with the thiol group of cysteine-360 of ornithine decarboxylase. Pyridoxal phosphate reacts with **20.7** to give the imine **II** which is decarboxylated. However, the elimination of a fluoride gives the decarboxylation a new path affording a electrophile double bond which forms a covalent bond with a Cys360 (intermediate **V**) in the enzyme via **IV**. Notice the increased reactivity because of the resonance into the pyridine ring. A subsequent hydrolysis of the enamine affords the ketone **VII** which by reaction with the terminal amino group affords the pyrolidine **VIII**.

infect humans: *Plasmodium falciparum* (causing malignant tertian malaria), *Plasmodium malariae* (quartan malaria), *Plasmodium ovale* (ovale tertian malaria), and *Plasmodium vivax* (benign tertian malaria). *P. falciparum* is responsible for the major number of deaths. In addition to the four species which have been known to cause malaria for decades, *Plasmodium knowlesi* also infects humans.

The life cycle of the malaria parasite encompassing several stages is visualized in Figure 20.6. After entering a mosquito stomach with blood meal, the gametocytes will be activated to undergo another cycle and multiplication in the mosquitoes. Eventually, sporozoites will enter the salivary

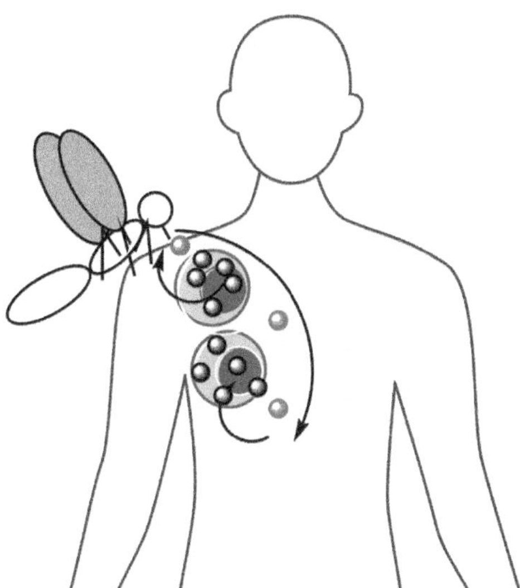

FIGURE 20.3 Life cycle of leishmania parasites. When the sand fly takes a blood meal, it regurgitates promastigotes (red spheres) into the skin. Macrophages take the promastigotes up by phagocytosis. In the macrophage, the promastigotes transform into amastigotes (black spheres) which proliferate. During another blood meal, the sand fly takes up the infected macrophages.

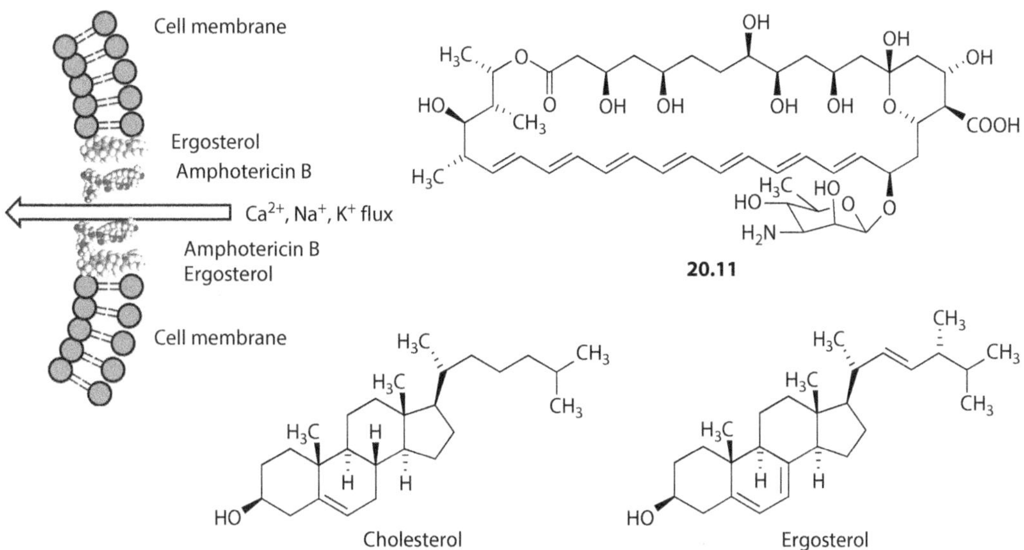

FIGURE 20.4 Mechanism of action of amphotericin B (**20.11**). The polyene region interacts with the double bonds of ergosterol which is found in the cell membrane of parasites. Mammalian cells contain cholesterol, in which the presence of only one double bond causes a weaker complex with amphotericin B. The orientation of the amphotericin B-ergosterol complexes creates an ion channel, through which an unregulated flux of small inorganic ion passes. Inability to control the concentration of inorganic ions eventually kills a cell.

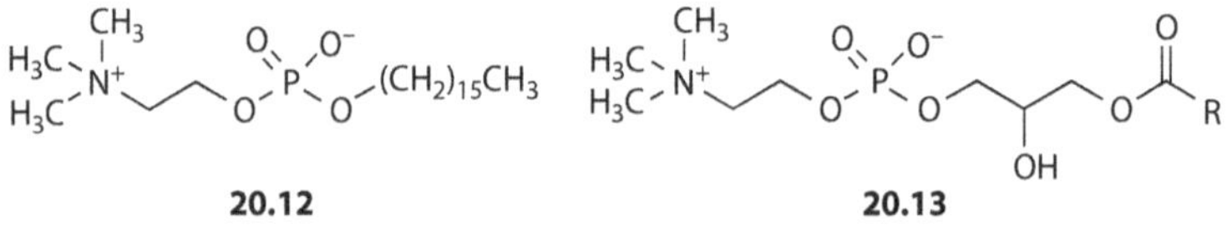

20.12 **20.13**

FIGURE 20.5 Miltefosine (**20.12**) and lysophosphatidylcholine (**20.13**).

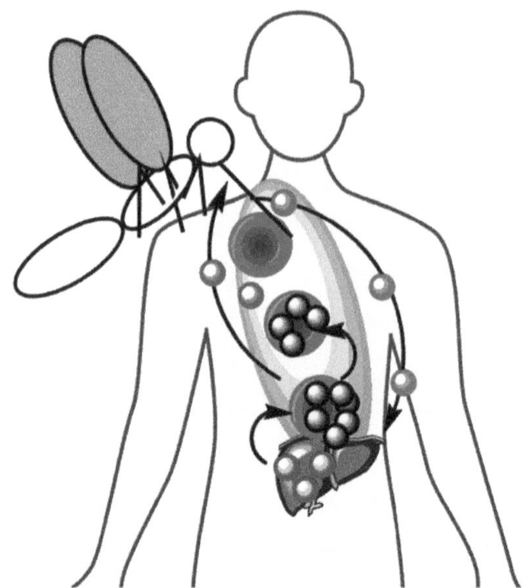

FIGURE 20.6 When an infected female Anopheles mosquito takes a blood meal, it regurgitates approximately 10 parasites as sporozoites (red spheres) into the blood. The sporozoites invade the liver where they proliferate asexually. After between 5 and 15 days, many thousand parasites return to the blood as merozoites which invade the erythrocytes. In the erythrocytes, the parasites (black spheres) go through the stages of rings, trophozoites, and schizonts. Finally, the parasites after 48 hours (*P. falciparum*, *P. ovale*, and *P. vivax*), tertian malaria, or 72 hours (*P. malaria*), quartan malaria, rupture the cell, and between 6 and 36 schizonts are liberated to the blood. These schizonts break down into merozoites which reinvade erythrocytes, and the life cycle inside the blood cells is repeated. The proliferation induces lysis of the erythrocytes and the clinical symptoms of malaria: chills with rising temperatures, followed by fever and intense sweating appear. In addition, there might be severe headache, fatigue, dizziness, nausea, lack of appetite, and vomiting. Since the trophozoites' major demand of amino acids is covered by catabolizing the hemoglobin, a heavy infection also will induce anemia. After some series of asexual proliferation, some parasites develop into female and male gametocytes (magenta spheres).

glands ready for being transferred into the human host. The falciparum parasites cause the erythrocytes to express a strain-specific protein which may adhere to the infected erythrocyte or to the walls of capillaries, resulting in reduced blood flow to vital organs. Reduced circulation to the brain contributes to the effects of cerebral malaria which can be fatal. Because of the symptoms of malaria, i.e., lack of blood (anemia) and fatigue, some persons chronically infected, such as many African school children, perform poorly. Studies suggest that national income in some African countries was suppressed as much as 18% because of malaria.

Today, the major burden of malaria is restricted to the tropical world: India, South East Asia, Central and South America, and, in particular, sub-Saharan Africa, but the endemic

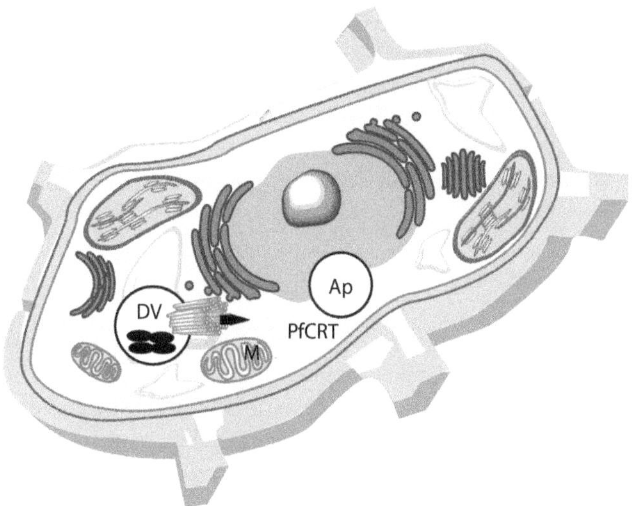

FIGURE 20.7 The malaria parasite with targets for drugs: drugs affecting the hemozoin formation (such as chloroquine and quinine) are active in the digestive vacuole (DV) and drugs affecting the electron flux influence the mitochondria (M). PfCRT: transporter taking drugs out of the digestive vacuole and thereby neutralizing the effect of the drug. Ap: apicoplast, in which metabolic pathways essential for the parasite occur.

area of malaria besides the tropics also encompasses the subtropics and some temperate zones. During the 1950s and 1960s, a combined use of dichlorodiphenyltrichloroethane (DDT), for the control of the vector mosquitoes, and chloroquine (Figure 20.9, **20.14**) for the control of the parasites almost eradicated malaria from the Indian subcontinent. Resistance of the parasites toward chloroquine and of the mosquitoes toward DDT and the environmental consequences of extended use of DDT led to discontinuation of the project. As a consequence, the malaria burden returned. It might be feared that the improved situation obtained though the last years by intense surveillance and treatment of malaria again can be lost if resistance against the preferred and affordable drugs develops like resistance against chloroquine developed during the 1960s and 1970s.

Malaria primarily is treated with chemotherapy targeting the organelles shown in Figure 20.7. In addition, reduction of human–vector contact contributes to reduction in infection rate. This may be obtained by reducing human–mosquito contact by sleeping under insecticide-treated mosquito nets (ITNs). Previously, ITNs had to be reimpregnated; nowadays, the insecticides are embedded in the nets so they last as long as the nets, LLITNs which has enhanced their effect considerably. Lowering the average lifetime, the local mosquito population may also be obtained by spraying the inner surface of dwellings with insecticides. This strategy will target mosquitoes which rest on walls after having taken a blood meal.

20.5 DRUGS AGAINST MALARIA

In the absence of vaccines, malaria therapy relies on small-molecule drugs. The more important drugs belong to three groups: (1) the quinolines (Section 20.5.1), (2) the antifols (Section 20.5.6), and (3) the artemisinins (Section 20.5.2). The quinolines are based on their structures divided into (1) the 4-aminoquinolines (Section 20.5.1.1), (2) the 4-quinolinemethanols (Section 20.5.1.2), and (3) 8-aminoquinolines (Section 20.5.5.1). In addition, a number of antibiotics are used successfully either individually or more commonly in combination with other drugs. Surprisingly, some antibiotics display considerable activity against the eukaryotic malaria parasite. This observation can be explained by similarities between the mitochondria of the parasite and prokaryotic mitochondria and the presence of the apicoplasts (Figure 20.7). There are several other drugs, including fosmidomycin, still being investigated as a potential drug.

The apicoplast probably is a remnant of endosymbiotic cyanobacteria which in plants have developed into the photosynthetic chloroplasts. Even though the apicoplasts do not perform photosynthesis, their metabolic pathways still are essential for the parasites. Both organelles have their own machinery for replication. Most antibiotics used in malaria therapy affect the apicoplasts.

20.5.1 Drugs Targeting Hemozoin Formation

The ultimate diagnosis of malaria is microscopic observation of parasites in the erythrocytes of a thick blood film. The presence of the malaria pigment, hemozoin, in the erythrocytes unequivocally reveals the presence of parasites. Hemozoin (**XIII**) is formed from the heme (ferroprotoporphyrin, **XI**) remaining after digestion of the peptide part of hemoglobin (Figure 20.8).

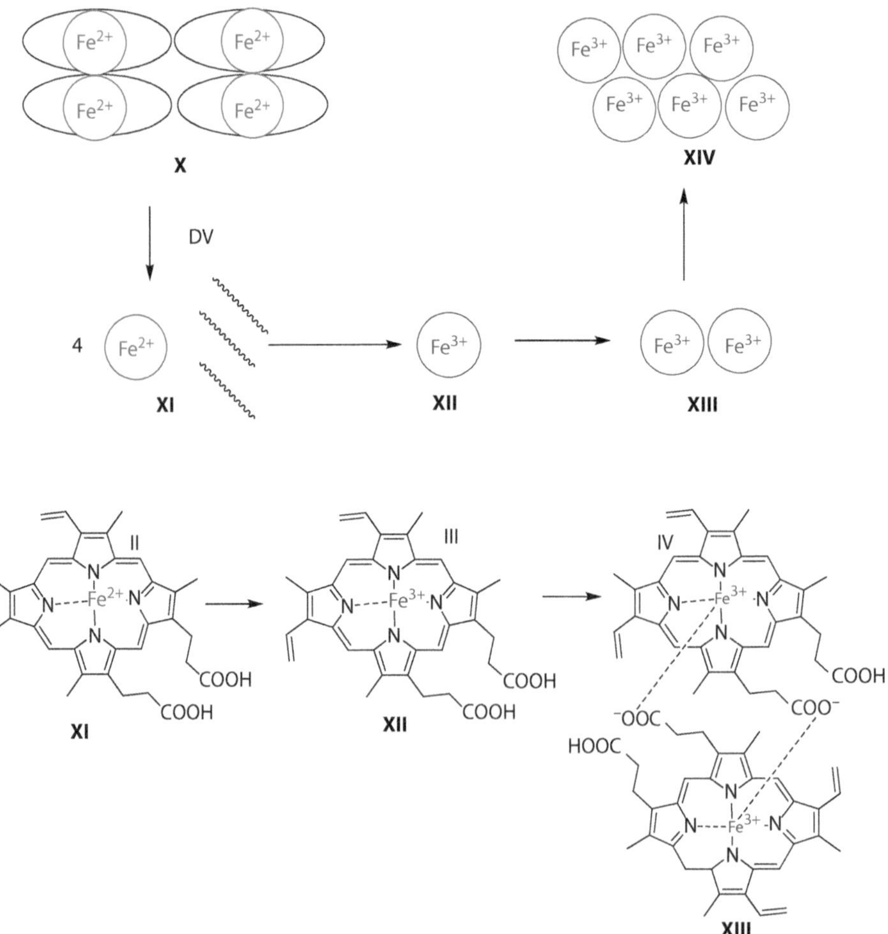

FIGURE 20.8 Hemoglobin (**X**) is a tetramer consisting of four protein chains (black ovals) and four iron(II) ions fixed in the hemoglobin by the protoporphyrin (red circles). After digestion of the protein, the unmasked heme (**XI**) is oxidized into hematin (**XII**). The hematin dimerizes and become a hemozoin (**XIII**) which falls out as a harmless precipitate (**XIV**).

The digestion proceeds in the digestive (food) vacuole of the parasite. The digestive vacuole is characterized by a pH between 5.0 and 5.5. The parasites use at least three types of proteases for the catabolism of hemoglobin. Since nude hematin (**XII**) is toxic for all kind of cells, the parasites have to detoxify it by converting it into an insoluble complex hemozoin (**XIV**). The mechanism of action behind a series of antimalarial drugs consists in blocking the formation of hemozoin by association with hematin. Two criteria have to be fulfilled for drugs acting by preventing detoxification of heme: (1) the drug must accumulate in the food vacuole of the parasite, and (2) the drug must bind to hematin. Thus, binding to hematin is a necessary, but not sufficient, requirement for an antimalarial drug targeting the hemazoin formation. Criterion 1 might be fulfilled by introduction of a basic aliphatic amine into the molecule, thus taking advantage of the low pH of the food vacuole. For a drug to be active in the human patient, several other factors including absorption, metabolism, and distribution need to be taken into account. It should be noted that the drugs associated with hematin work on the asexual erythrocytic stage of the parasite only, i.e., not on the liver stages or the gametocytes. Thus, a patient being treated for malaria may be cured but may remain infectious for up to 3 months due to surviving gametocytes.

20.5.1.1 4-Aminoquinolines

The 4-aminoquinolines are characterized by the presence of an amino group in the 4-position of the quinoline nucleus (Figure 20.9).

The 4-aminoquinolines possess an amino group in the 4-postion. As revealed in the resonance structures, this position of the amine group stabilizes the N-1 protonated ammonium ion (Figure 20.9).

In the late nineteenth century, Paul Ehrlich (1854–1915) noticed a selective uptake and staining of tissues with dyes such as methylene blue (**20.15**). Based on this, at that time pioneering, idea that this selective staining was caused by selective receptors for the dyes, he discovered that methylene blue had antimalarial activity, however not significant enough to become a drug. Elaboration on this idea led to the development of chloroquine (**20.16** R = H) and the 8-aminoquinolines, and also to the (unrelated) drug suramin (Figure 20.2, **20.10**) developed and still in use for African trypanosomiasis. It is claimed that chloroquine is the drug that, without comparison, has saved most lives. For more than 40 years, chloroquine was the first-line therapeutic and prophylactic agent for malaria.

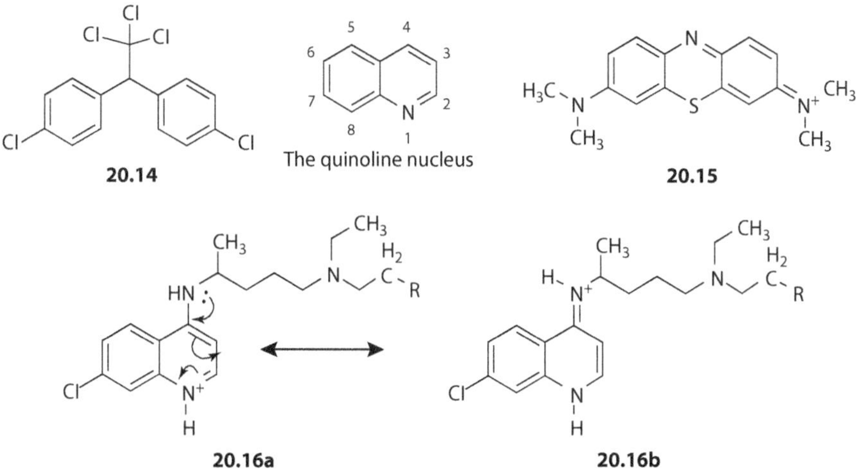

FIGURE 20.9 Dichlorodiphenyltrichloroethane (DDT, **20.14**), the quinoline nucleus, methylene blue (**20.15**), and protonated chloroquine (**20.16**, R = H). The resonance structures explain the high first pK_a (8.1) value of chloroquine. Hydroxychloroquine (**20.16**, R = OH) is still registered in some countries where chloroquine is no longer registered. The drug can be used for prophylaxis of malaria and long-term treatment or rheumatoid arthritis.

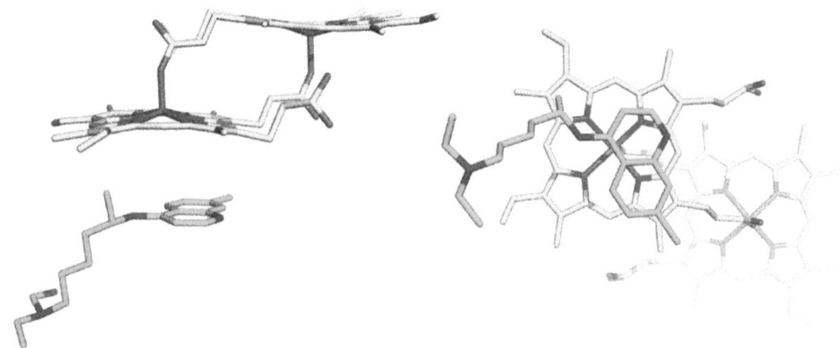

FIGURE 20.10 Two orthogonal views of a suggested binding mode of chloroquine (green) to hemozoin (yellow). Hetero-atoms are colored red, blue, and orange for oxygen, nitrogen, and iron, respectively. (Courtesy of Dr. F. S. Jørgensen.)

In the last three decades, however, chloroquine-resistant *P. falciparum* and *P. vivax* strains have developed, but the drug is still believed to be efficient toward infections with *P. ovale* and *P. malariae* and most strains of *P. vivax*. The resonance interaction between the electron pair of the exocyclic amino group and the quinoline nitrogen atom (Figure 20.9) will give the protonated 4- and 2-aminoquinolines higher pK_a values (8.1 and 10.2) than other aminoquinolines (for comparison, quinine pK_a of the corresponding acids are 5.1 and 9.7). The lipophilicity of the neutral molecule enables it to cross the membranes of the erythrocytes and the parasites. Having entered the food vacuole with a low pH, the two amino groups will become protonated, preventing chloroquine from leaving the vacuole by a passive penetration. Furthermore, the protonation of quinoline nitrogen enables a cation–π interaction between the quinoline and the hemozoin (Figure 20.10).

The complexation between chloroquine and hematin further contributes to the concentration of chloroquine in the food vacuole. Actually, the concentration of chloroquine is four orders of magnitudes higher inside the vacuole compared with that outside in sensitive parasites. The presented model also confirms the finding that variations in the side chain only to a minor extent influence the strength of the association. Notice the absence of stereocenters in the porphyrin nucleus. The achirality of this target explains that the two enantiomers of chloroquine have the same binding affinity. In agreement with the target being heme, chloroquine only affects the erythrocytic stages of the parasite, the only stages in which hemozoin is formed.

20.5.1.2 4-Quinolinemethanols

The quinolinemethanols may be considered as methanol substituted with 4-quinoline and an aliphatic amine (Figure 20.11).

The oldest representatives for the quinolinemethanols are the cinchona alkaloids (−)-quinine (**20.17**), (−)-cinchonidine (**20.18**), (+)-quinidine (**20.19**), and (+)-cinchonine (**20.20**). The aliphatic amine in these is a quinuclidine. All four alkaloids are isolated from bark of trees belonging to the genus *Cinchona* (Rubiaceae), often referred to as fever trees. Until isolation of quinine in large scale in 1820, the crude bark was the only efficient drug in Europe for the treatment of malaria and South America, from where it originated. The narrow therapeutic window of the cinchona alkaloids, in combination with different concentrations of the alkaloids in the barks depending on species and time of harvesting, made the access to the homogeneous compounds a major therapeutic improvement. Quinine in a decent pure state was isolated in 1820. In spite of the several severe side effects including cardiac arrhythmia, insulin release causing hypoglycemia, and peripheral vasodilatation, the cinchona alkaloids still are used for the treatment of multiresistant malaria, most frequently in combination with tetracyclines. The treatment, however, requires careful supervision.

The mechanism of action of the cinchona alkaloids is still debated. It has been argued that the lower pK_a value of the pyridine-nitrogen (pK_a 5.1) compared to that of the corresponding

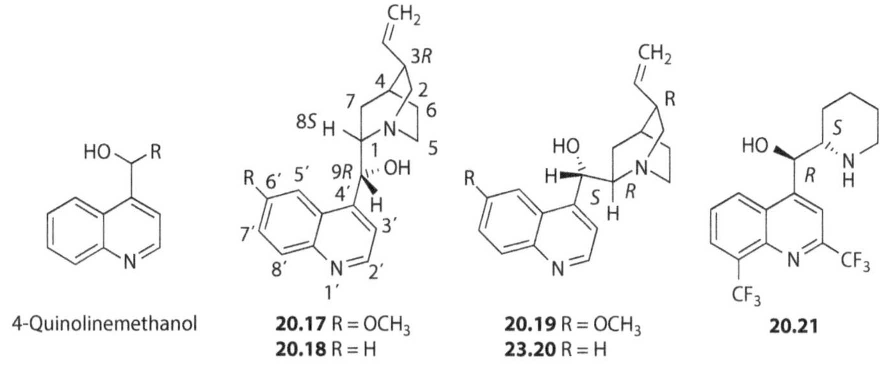

4-Quinolinemethanol

20.17 R = OCH_3
20.18 R = H

20.19 R = OCH_3
23.20 R = H

20.21

FIGURE 20.11 The 4-quinolinemethanol nucleus and (–)-quinine (**20.17**), (+)-cinchonidine (**20.18**), (–)-quinidine (**20.19**) and (+)-cinchonine (**20.20**); and the synthesized drug mefloquine (**20.21**)

nitrogen in the 4-aminoquinolines (pK_a 8.1) means that the compounds are less protonated at the pH of the digestive vacuole (pH 5.0–5.5). As a consequence, the complexation with hematin is less pronounced. It is established that quinine inhibits hemozoin formation like chloroquine. The mechanism, however, might be different, although both types of quinolines possess an aliphatic amino group ($pK_a \approx 10$) which will be protonated at pH below 6. Clinically, quinidine and cinchonine (both 8*R*,9*S*) are more active than quinine and cinchonidine (8*S*,9*R*). Quinine and quinidine have opposite configuration at both C8 and C9, and from a pharmacophore point of view, they may be considered to be enantiomeric even though they, from a chemical point of view, are diastereomeric since they have the same stereochemistry at C3. The same is true for cinchonine and cinchonidine. Quinidine and cinchonine, however, should only be used if there is shortage of quinine because of an even more narrow therapeutic window, particularly concerning cardiac toxicity. The importance of the relative stereochemistry at C8 and C9 is illustrated by the missing activity of 9-epiquinine, in which the relative configuration at C8 and C9 has been changed to 8*S*,9*S*.

Mefloquine (Figure 20.11, **20.21**) was selected among 300 quinolinemethanols prepared in order to develop agents effective against chloroquine-resistant malaria. In mefloquine, the aliphatic amine substituent on the methanol is a piperidine. In contrast to the naturally occurring alkaloids which only consist of one enantiomer, mefloquine is marketed as a racemic mixture. The pharmacokinetic of the two enantiomers, however, is different as illustrated by the finding that the half-life of the (–)-isomer is significantly longer than that of the (+)-isomer.

The 2,8-bis-trifluoromethyl arrangement proved to be the most active of the series. Compounds that possessed 2-aryl groups were found to have augmented antimalarial activity, but at the same time unacceptable phototoxic side effects. Mefloquine has been marketed as Lariam, a drug that has serious hallucinogenic side effects in some patients. Resistance against mefloquine has led to the use of combination therapy using mefloquine and arteminisinin derivatives.

20.5.1.3 Nonisoquinoline Drugs for Malaria: Phenanthrenemethanols

The dibutylaminopropyl groups of halofantrine (Figure 20.12, **20.22**) were found to give optimal antimalarial effect among the 9-phenanthrene analogs. Even though the evidence for the mechanism of action for this compound is less convincing than that for chloroquine, the findings that mefloquine only affects erythrocytic stages of the parasite and that some studies show association with hemazoin support the suggestion that halofantrine acts by preventing detoxification of hemazoin. Like quinine, halofantrine can induce cardiac arrhythmias. In addition, halofantrine can cause delusions.

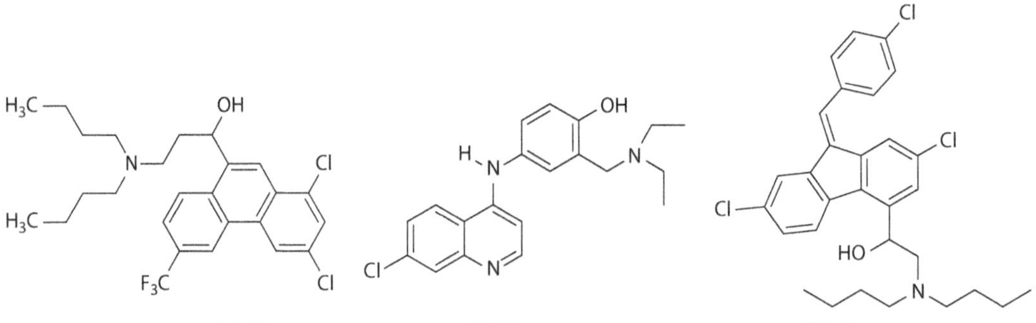

20.22 20.23 20.24

FIGURE 20.12 Halofantrine (**20.22**), amodiaquine (**20.23**), and lumefantrine (**20.24**).

A number of other drugs belonging to the quinoline group like amodiaquine (**20.23**) and lumefantrine (**20.24**) also are believed to prevent detoxification of heme. Amodiaquine and lumefantrine are among the most important drugs used in the so-called artemisinin combination therapy against falciparum malaria.

20.5.2 Artemisinins

Whereas European physicians did not get access to an effective antimalarial herbal drug until the sixteenth century, Chinese authors already 2000 years ago described the effect of *qing hao* and *cao hao* against intermittent fever. Since the Chinese way of describing symptoms of diseases is very different from the Western terminology, intermittent fever cannot in a simple way be translated into a well-known term. However, intermittent fever could include the fever caused by malaria. *Quing hao* has later been identified as *Artemisia apiacea* (Asteraceae) and *cao hao* as *Artemisia annua* (sweet wormwood). In the 1970s, Chinese scientists isolated artemisinin (Figure 20.13, **20.25**) from both of these species and showed that the compound was a potent antimalaria agent. Artemisinin is

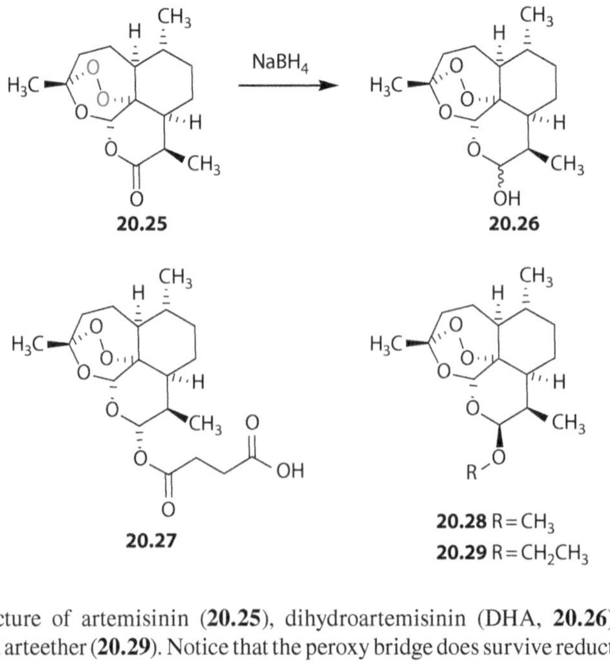

$NaBH_4$

20.25 20.26

20.27

20.28 R = CH_3
20.29 R = CH_2CH_3

FIGURE 20.13 Structure of artemisinin (**20.25**), dihydroartemisinin (DHA, **20.26**), artesunate (**20.27**), artemether (**20.28**), and arteether (**20.29**). Notice that the peroxy bridge does survive reduction with borohydride.

an irregular sesquiterpene lactone containing an endoperoxide bridge (marked with red). Synthetic analogs, in which the peroxide bridge has been removed or turned into an oxygen bridge, show no activity toward *Plasmodium* parasites. The amount of artemisinin that may be extracted from the wormwood varies between 0.01% and 0.8% of dry weight which is a serious limitation for the commercialization of the drug.

The mechanism of action, however, is still debated. Previously, it was believed that the lethal effect of the plasmodium parasite was caused by inhibition of the intracellular calcium pump (*Pf*ATP6) which is essential for the Ca^{2+} homeostasis in the parasite. After expression of this pump, it was realized that artemisinin does not affect this pump, not even in the presence of iron (III) ions.

One hypothesis is that artemisinin by reaction with the iron(II) ion of hem generates a C-radical (**20.30**, the Fenton reaction) (Scheme 20.3).

Mass spectrometric investigation of the product **20.31** has supported the theory. Although this idea still is favored in some laboratories, other studies indicate that the C-radical is not sufficiently reactive for this reaction to happen and give the antiplasmodial effect; instead, it may be caused by disturbance of the cellular redox homeostasis.

The poor solubility of artemisinin allows only oral or rectal administration of the drug. Since oral administration is contraindicated for severely ill patients, lipid-soluble derivatives (artemether, **20.28**, and arteether, **20.29**) and water-soluble derivatives (artesunate, **20.27**) have been developed by reacting dihydroartemisinin (**20.26**) with the corresponding alcohols or succinic anhydride (Figure 20.13). In the body artemether, arteether or artesunate are unmasked to give **20.26** which is the active drug. Fast glucuronidation causes a short half-life of **20.26**. The short half-time necessitates the use of combination therapy to avoid recrudescence (reappearance of a disease after it has been quiescent). Some examples of combination therapies are artesunate in combination with chlorproguanil and dapsone (Figure 20.20, **20.49**), artesunate in combination with amodiaquine

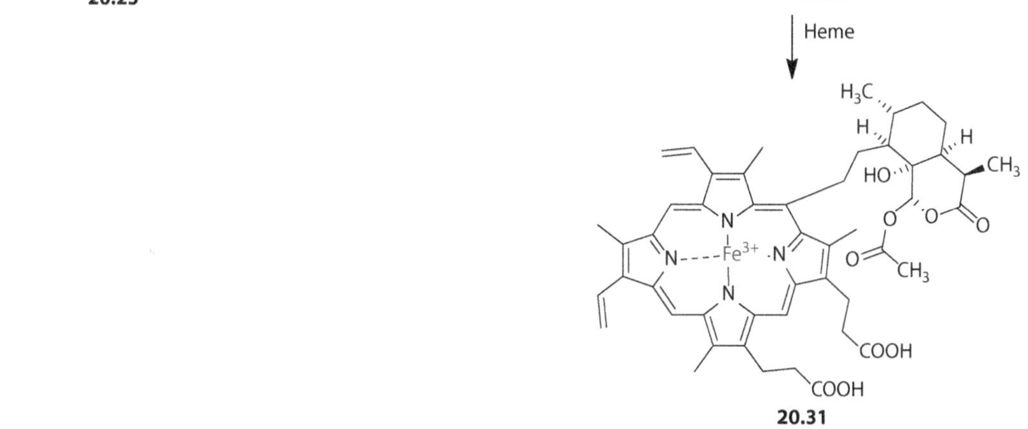

SCHEME 20.3 Ion(II) provoked radical formation of **20.25** and reaction with heme. The radical **20.30** is suggested to react with heme to afford alkylated **20.31**, thereby preventing precipitation.

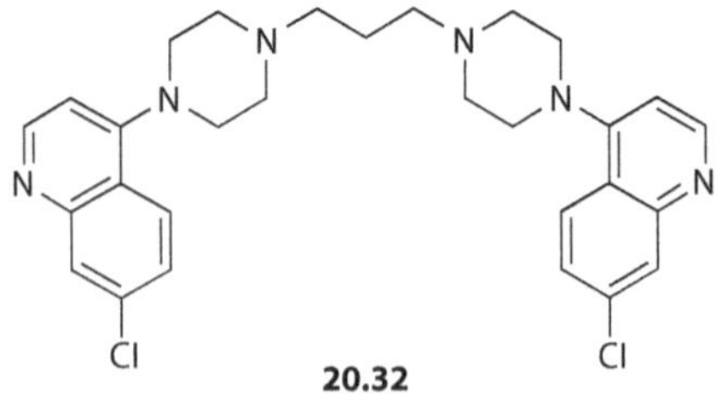

20.32

FIGURE 20.14 Structure of piperaquine (**20.32**).

(Figure 20.12, **20.23**) or mefloquine (Figure 20.11, **20.21**), and artesunate in combination with sulfadoxine (Figure 20.20, **20.48**) and pyrimethamine (Figure 20.18, **20.47**). Another artemisinin, artemether, is marketed in combination with lumefantrine, (Figure 20.12, **20.24**) and dihydroartemisinin is marketed in combination with piperaquine (Figure 20.14, **20.32**).

20.5.3 Drugs Targeting the Deoxyxylulose Pathway

In plants, the biosynthesis of isopentenyl diphosphate (Scheme 20.4, **20.36**) and dimethylallyl diphosphate (**20.37**), the precursor of all terpenoids, follows two independent pathways: (1) the mevalonate pathway and (2) the 1-deoxy-D-xylulose-5-phosphate pathway. In plants, the mevalonate pathway occurs in the cytosolic compartment, but the deoxyxylulose pathway takes place in the plastids which are analogous to apicoplasts. A search in the genome of *Plasmodium* parasites revealed the

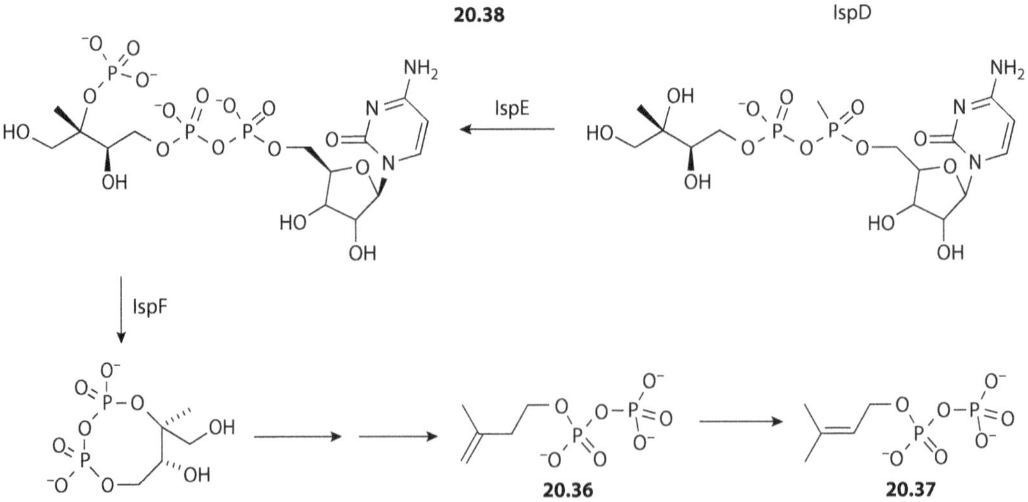

DXR DXR IspD IspE IspF

20.33 20.34 20.35 20.38 20.36 20.37

SCHEME 20.4 The deoxyxylulose pathway. DXR: 1-deoxy-xylulose 5-phophate reductoisomerase; IspD: 4-diphosphocytidyl-2-C-methyl-D-erythritol synthase; IspE: 4-diphosphocytidyl-2-C-ethyl-D-erythritol kinase; IspF: 2-C-methyl-D-erythritol-2,4-cyclodiphosphate synthase. Notice the structural similarities between the intermediate **20.34** and fosmidomycin (**20.38**).

presence of genes encoding the enzymes of the deoxyxylulose pathway including the genes encoding 1-deoxy-D-xylulose 5-phosphate reductoisomerase (DOXP-reductoisomerase). This enzyme catalyzes a crucial step in the deoxyxylulose pathway (Scheme 20.4): the reductive rearrangement of 1-deoxy-D-xylulose 5-phosphate (**20.33**) into 2-C-methyl-D-erythritol 4-phosphate (**20.35**). The pathway leads to the two building blocks isopentenyl diphosphate (**20.36**) and dimethylallyl diphosphate (**20.37**). Whereas plants can produce isopentenyl diphosphate and dimethylallyl diphosphate by both the mevalonate and the deoxyxylulose pathway, *Plasmodium* parasites apparently only can use the deoxyxylulose pathway and humans only the mevalonate pathway.

Fosmidomycin (Scheme 20.4, **20.38**), an antibiotic and herbicidal agent isolated from cultures of *Streptomyces lavendulae*, efficiently inhibits the plants' carotenoid and phytol synthesis. Fosmidomycin is a structural analog to 2-methylerythrose 4-phosphate (**20.34**). Fosmidomycin inhibits DXR and possibly other enzymes in the deoxyxylulose pathway. Since terpenoid formation in parasites is solely dependent on this pathway, inhibition is lethal for the parasites and some bacteria. In contrast, this pathway does not exist in humans, making us less sensitive to this drug. Combination therapy using fosmidomycin and clindamycin has revealed high antimalarial activity with only mild gastrointestinal side effects. The drug, however, is still in development since clinical trials have not given satisfactory cure rate.

20.5.4 Drugs Targeting Mitochondrial Functions

An important function of the mitochondria is to maintain an electron transport needed for nucleotide synthesis and for generating a pH gradient over the membrane. Parasites are dependent on de novo synthesis of the nucleotides. The mitochondrial cytochrome bc_1 complex is a part of the electron transport. The enzyme consists of a cytochrome and a Rieske protein bound to an iron sulfur subunit. In the ubiquinol binding pocket, two electrons are transferred from ubiquinol via the subunit to cytochrome c hem iron (Scheme 20.5).

20.5.4.1 Naphthoquinones

The antimalarial naphthoquinones are developed from naturally occurring naphthoquinones like lapachol (Figure 20.15, **20.39**). The problem of fast metabolism, however, prevented the clinical use. Among the several hundreds of naphthoquinones synthesized and tested, atovaquone (**20.40**)

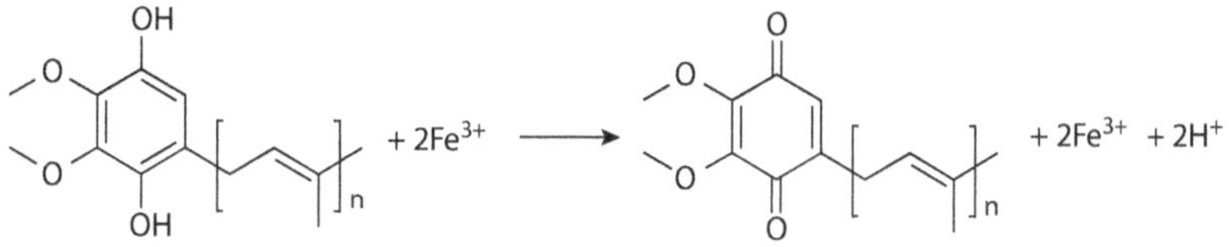

+ $2Fe^{3+}$ → + $2Fe^{3+}$ + $2H^+$

SCHEME 20.5 Ion(II) provoked oxidation of ubiquinol to give ubiquinone.

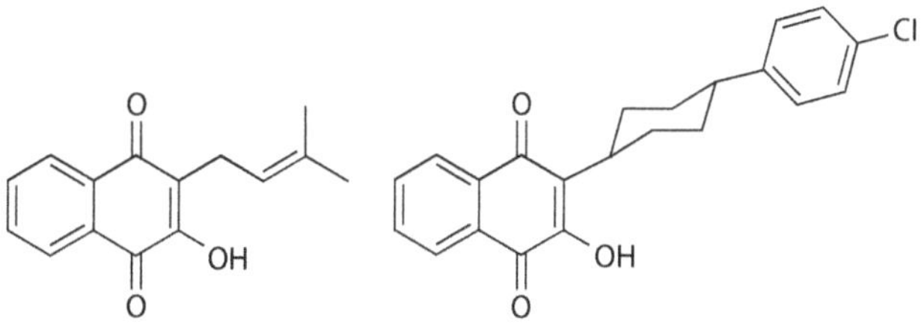

20.39 **20.40**

FIGURE 20.15 Structure of lapachol (**20.39**) and atovaquone (**20.40**).

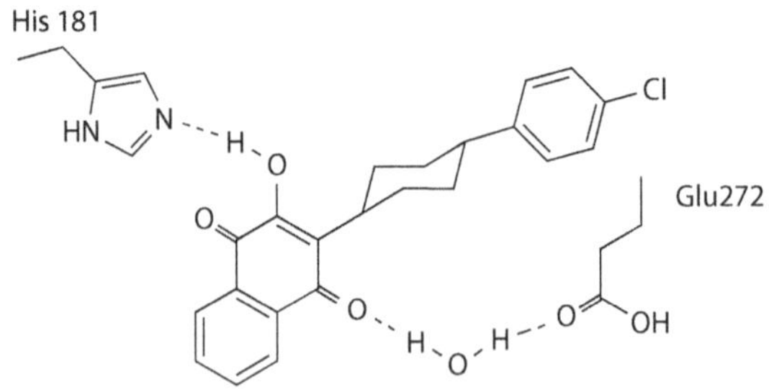

FIGURE 20.16 In the binding cavity, the complex between atovaquone and the protein is favored by a hydrogen bond from N-of histidine-181 and to glutamic acid-272.

was finally selected for use. Atovaquone is assumed to bind in the ubiquinol oxidation pocket of the parasite, thereby preventing electron transfer. Model studies performed on the yeast bc_1 complex suggest that a hydrogen bond between the hydroxy group of atovaquone and nitrogen of His181 of yeast Rieske-protein and a hydrogen bond from Glu272 of bc_1 complex via a water molecule to one of the carbonyls of atovaquone stabilize the complex and thereby prevent transfer of the electrons to the iron-sulfur complex (Figure 20.16). A problem using the yeast complex as a model appears from the observation that point mutation replacing leucine 275 with phenylalanine severely inhibits binding of atovaquone to the binding site. In the *Plasmodium* peptide, however, a phenylalanine is found in this position in the wild-type protein. Replacement in the yeast peptide of 11 amino acids (including L275F mutation) regained high-affinity binding of atovaquone. Point mutation to give a more bovine-like peptide made the peptide resistant to atovaquone. This resistance is similar to the resistance of the human cytochrome bc_1.

Rapid development of resistance and a high rate of recrudescence necessitated the use of combination therapy. Proguanil (Figure 20.18, **20.45**) combination with atovaquone (Malarone®) is at present an effective therapy for multidrug-resistant falciparum malaria. Until recently, however, the high costs of this treatment limited its use. This situation has changed after the expiration of the patent.

20.5.5 Drugs with Nonestablished Targets

20.5.5.1 8-Aminoquinolines

Primaquine (Figure 20.17, **20.41**) remains the only drug approved for cure of vivax malaria. The 8-aminoquinolines possess activity toward all stages of the parasite, including the hypnozoites in the liver and the gametocytes in the blood. Killing of hypnozoites prevents relapse which is caused

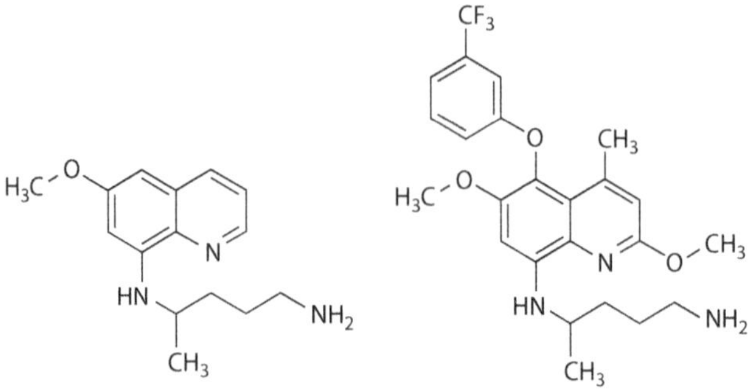

FIGURE 20.17 Primaquine (**20.41**) and tafenoquine (**20.42**).

by activation of hypnozoites resting in the liver. Relapse is pronounced for vivax malaria. Killing of gametocytes prevents transmission. The ability to affect all stages reveals that the mechanism of action of the 8-aminoquinolines must differ from that of the 4-aminoquinolines which only affect parasites digesting hemoglobin (Section 20.5.1.1). However, the mechanism of action of the 8-aminoquinolines has not yet been established. It is suggested that the compounds may affect the calcium homeostasis, may affect the mitochondria by oxidative stress, or disturb the distribution of phospholipids in the membranes.

Drawbacks of primaquine are a narrow therapeutic window, a short half-life (4–6 hours) which requires repeated administration for 14 days to achieve a cure, and hemolysis and methemoglobin formation. The latter side effect is particularly pronounced in patients with an inborn deficiency of glucose-6-phosphate dehydrogenase, a genetic abnormality especially common in areas where malaria is endemic. Structure–activity relationships have revealed that an appropriate substitution in the 2-position improved efficacy and decreased general systemic toxicity, a methyl group in the 4-position improved therapeutic activity but also toxicity, and that a phenoxy group in the 5-position decreased toxicity and maintained activity. The studies led to synthesis of tafenoquine (**20.42**) which has not yet been registered as a drug. The substituent in the 5-position provides tafenoquine with a half-life of 2–3 weeks. A long half-life is interesting for developing a single-dose oral cure for malaria. However, the long half-life could increase the side effects, in particular, for patients suffering from deficiency of glucose-6-phosphate dehydrogenase.

20.5.6 Drugs Targeting Folate Synthesis

Tetrahydrofolic acid is an important coenzyme in parasites as well as their hosts. The coenzyme is involved in the biosynthesis of thymine, purine nucleotide, and several amino acid syntheses. Malaria parasites are dependent on de novo folate synthesis (Scheme 20.6), whereas mammalian cells take up fully formed folic acid as vitamin B_9. Consequently, dihydropteroate synthase is totally absent in humans. In the mammalian as well as in parasitic cells, the precursors [folate or dihydrofolate (**20.43**), respectively] have to be reduced to the enzymatically active tetrahydrofolate, a reaction that is catalyzed by dihydrofolate reductase (DHFR). In mammalian, DHFR and thymidylate synthase are separate enzymes, whereas in protozoan parasites are covalently linked to one bifunctional enzyme (DHFR-TS). The binding site of dihydrofolate in DHFR-TS is sufficiently different from binding site in the human DHFR to allow selectivity. The binding site of DHFR-TS inhibitors like cycloguanil (Figure 20.18, **20.46**) and pyrimethamine (**20.47**) and the enzyme is illustrated in Figure 20.19 using pyrimethamine

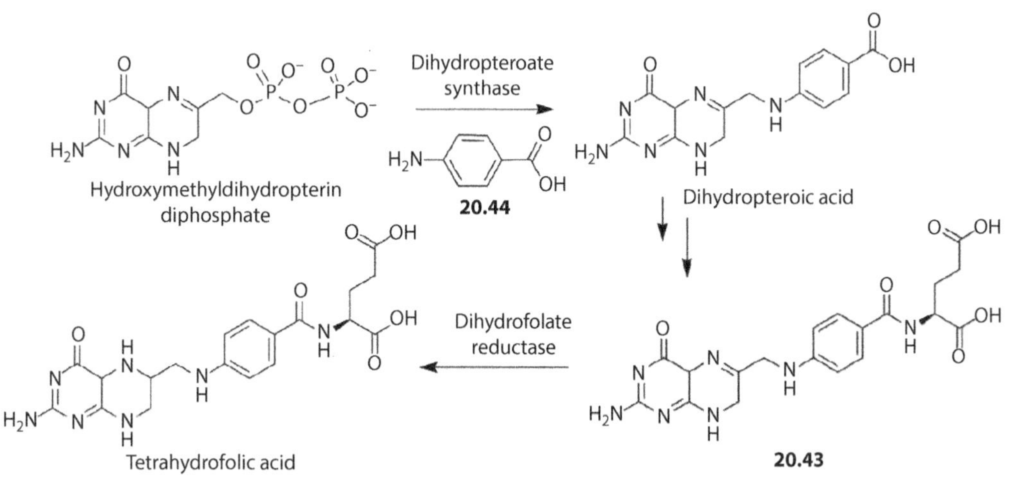

Hydroxymethyldihydropterin diphosphate

Dihydropteroate synthase

20.44

Dihydropteroic acid

Dihydrofolate reductase

Tetrahydrofolic acid

20.43

SCHEME 20.6 Simplified folate pathway. Dihydrofolic acid (**20.43**) and *p*-aminobenzoic acid (**20.44**).

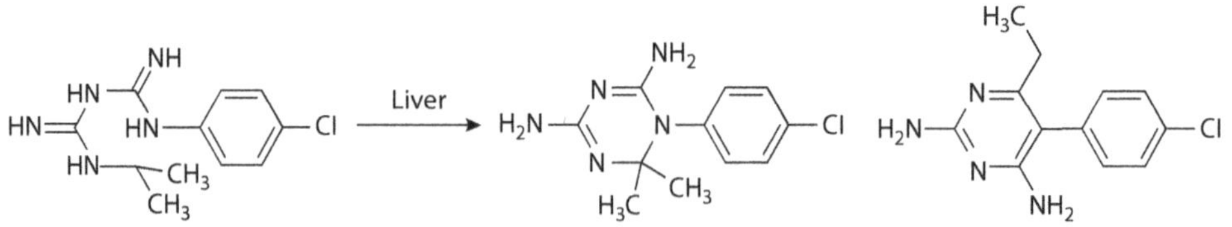

FIGURE 20.18 Metabolic conversion of proguanil (**20.45**) into cycloguanil (**20.46**). Structure of pyrimethamine (**20.47**).

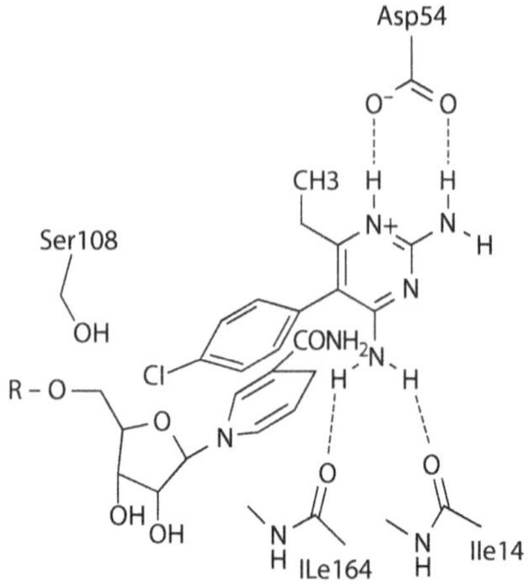

FIGURE 20.19 Binding of pyrimethamine (**20.47**) to the dihydrofolate binding site of dihydrofolate reductase (DHFR). The binding is stabilized through hydrogen bonds between the carboxylate of Asp54 and the positively charged NH group of the 2-amino group and the nitrogen of the pyrimidine ring, and of hydrogen bonds between 4-amino group and the backbone carbonyl groups of Ile14 and Ile164. In addition, a charge–transfer interaction between the chlorophenyl residue and the dihydropyridine ring of the NADPH and, finally, a hydrogen bond between the Ser108 and one of the hydrogen acceptors at the NADPH molecule stabilize the complex. (R = the remaining part of the NADPH molecule.)

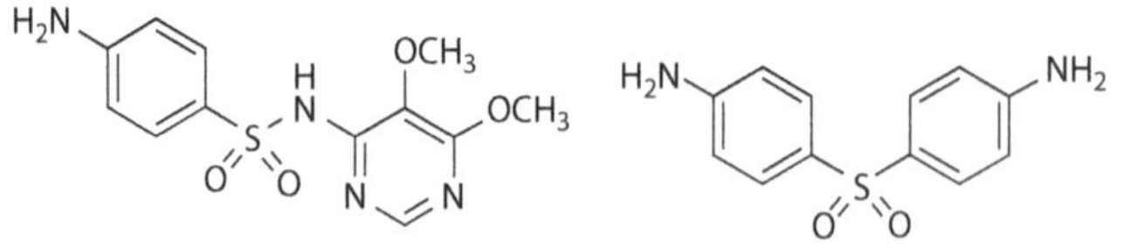

FIGURE 20.20 Structure of sulfadoxine (**20.48**) and dapsone (**20.49**). Compare these structures with para-aminobenzoic acid (Scheme 20.6, **20.44**).

as an example. Proguanil (**20.45**) will metabolically be converted into cycloguanil in the liver. The negatively charged carboxylate of Asp 54 of the enzyme binds to the positively charged amino group of pyrimethamine. The 4-amino group forms hydrogen bonds with the backbone carbonyl groups of Ile14 and Ile164. The coenzyme of DHFR, NADPH, is oriented through a hydrogen bond to Ser108.

Sulfadoxine (Figure 20.20, **20.48**) and dapsone (**20.49**) acts as antimetabolites of para-aminobenzoic acid (Scheme 20.6, **20.44**) which is a building block in the dihydrofolate synthesis. An antimetabolite is an agent which prevents incorporation a structural-related endogenic metabolite.

20.6 RESISTANCE

Treatment of malaria frequently fails because of the development of resistance. In fact, regular mutations of the parasites force continued development of new drugs. Resistance may develop either because of mutations in the target protein or because of introduction of a transport system which can decrease the concentration of the drug at the target.

20.6.1 Chloroquine Resistance

Chloroquine resistance has been correlated to a mutation in a wild-type digestive vacuolar membrane protein termed *Pf*CRT (*P*lasmodium *f*alciparum *c*hloroquine *r*esistance *t*ransporter). A mutation replacing Lys76 with Thr enables the protein in an energy-dependent manner to transport chloroquine out of the food vacuole decreasing the chloroquine concentration to below the pharmacological active concentration. Interestingly, some drugs like the antihistamine chlorpheniramine and the calcium channel antagonist verapamil may reverse this resistance probably by inhibiting the transporter. Clinically, the use of transporter inhibitors has only been used to a very limited extent. Other 4-aminoquinolines may also be substrates for the transporter explaining cross resistance. Chloroquine analogs with a changed side chain are developed in order to make analogs which are not substrates for the transporter.

A number of other hypotheses for resistance, including an increased value of the pH in the food vacuole or prevention of association of chloroquine and heme, cannot be fully excluded at present.

20.6.2 4-Quinolinemethanol Resistance

The membrane transport P-glycoprotein pump, *Pf*MDR1, which is an analog of the mammalian ABC multidrug-resistance transporter, plays a central role in resistance development of *P. falciparum* parasites. An increased number of *Pf*MDR1 transporters facilitate removal of the drug from the putative target in the cytosol.

20.6.3 Antifolate Resistance

Resistance against antifolate drugs is caused by mutations altering the active site, resulting in different binding affinities for different drugs. The resistance conferring mutations occur in a stepwise sequential fashion with a higher level of resistance occurring in the presence of multiple mutations. The decreased affinity for the drug is often followed by a decreased activity for the natural substrate, suggesting the parasites containing mutated forms of the enzyme may be selected against in the absence of a drug. Mutation of Ser108 into Asn causes steric interaction between the Asn side chain and the chlorophenyl group of pyrimethamine and thereby reduces the affinity of pyrimethamine (Figure 20.18, **20.47**) to *Pf*DHFR. This mutation, however, only causes a moderate loss of susceptibility to cycloguanil (**20.46**), in which the side chain is shorter. Additional replacement of Asn51 into Ile results in a higher pyrimethamine resistance but only a moderate effect of cycloguanil. On the contrary, replacement of Ser108 into Thr coupled with Ala16 into Val confers resistance to cycloguanil but only modest loss of susceptibility to pyrimethamine. However, the appearance of a quadruple mutant, Asn51 into Ile, Cys59 into Arg, Ser108 into Asn, and Ile164 into Leu has dramatically reduced the effect of dihydrofolate reductase inhibitors. Similarly, stepwise mutation in the DHPS enzyme leads to resistance toward sulfonamides which are often used in fixed combinations with antifolates, such as sulfadoxin-pyrimethamine.

20.6.4 Atovaquone Resistance

The most common mutation observed in atovaquone resistant parasites is a Tyr268 to Ser or Asn in the cytochrome bc_1-complex. This affords a 730-fold decrease of affinity for atovaquone. Similar mutation of Tyr279 to Ser affords an almost 300-fold reduction in affinity.

20.6.5 Artemisinin Resistance

Recent documentation (June 2015) from Southeast Asia reveals failure of artemisinin combination therapy (artemisinin–piperaquine) in some patients. The majority of the resistant parasites were found to have a triple mutation including a kelch13 Cys580Tyr substitution.

20.7 CONCLUDING REMARKS

A series of examples of drugs targeting biological systems only present in parasites has been given. In principle, by addressing targets not present in the host, but essential for the survival of the parasite should give a therapy without side effects. Unfortunately, very few drugs are truly selective. Quinine, as an example, has targets in the parasite cytosol, but does also induce a number of effects in the patient such as insulin release, inducing severe hypoglycemia, and cardiac arrhythmia. Another serious problem in the treatment of parasitic diseases is development of resistance which is addressed by giving a combination of drugs. Artemisinin and derivatives of artemisinin are given in combination with, e.g., lumefantrine. A further advantage of combination therapy is prevention of recrudescence, which, in particular, is a problem after treatment with drugs with a short half-life-like artemisinin.

In the last 10 years, considerable progress has been made in the battle against malaria. These achievements, however, are severely dependent on the use of artemisinin derivatives. The recent discovery of parasites resistant toward artemisinin–piperaquine combination therapy, therefore, indicates that a situation in which no effective and cheap drug against malaria is available may develop in the near future.

Some larger pharmaceutical companies have performed screening of their libraries for lead compounds and allowed access to the compounds for the development of new chemotherapeutics. Hopefully, these possibilities or other research results might give a new cheap drug against malaria which soon might be desperately needed.

It should be noticed that this chapter does not attempt to include all of the drugs or putative targets.

20.8 NOTE ADDED IN PROOF

The Nobel Prize in Physiology or Medicine 2015 was divided, one half jointly to William C. Campbell and Satoshi Ōmura "*for their discoveries concerning a novel therapy against infections caused by roundworm parasites*" and the other half to Youyou Tu "*for her discoveries concerning a novel therapy against Malaria*." The contribution of William C. Campbell and Satoshi Ōmura led to development of the drug ivermectin, the use of which has led to almost clearance of the diseases elephantiasis and river blindness (Section 20.2.2). Youyou Tu discovered artemisinin which has become an important tool in the fight against malaria (Section 20.5.2).

FURTHER READING

Azzouz, S., Maache, M., Garcia, R.G., and Osuma, A. 2005. Leishmanicidal activity of edelfosine, miltefosine and ilmofosine. *Basic Clin. Pharmacol. Toxicol.* 96:60–65.

Croft, S.L., Barrett, M.P., and Urbina, J.A. 2005. Chemotherapy of trypanosomiasis and leishmaniasis. *Trends Parasitol.* 21:508–512.

Egan, T.J. 2004. Haemozoin formation as a target for the rational design of new antimalarials. *Drug Des. Rev. Online* 1:93–110.

Farrar, J. Hotez, O.J., Junghanss, T., Kang, G., Lalloo, D., White, N.J., (eds.). 2014. *Manson's Tropical Diseases.* 23rd ed., Elsevier, Philadelphia, PA.

Gamo, F.-J., Sanz, L.M., Vidal, J. et al. 2010. Thousands of chemical starting points for antimalarial lead identification. *Nature* 465:305–310.

Rosenthal, P.J. (ed.). 2001. *Antimalarial Chemotherapy. Mechanism of Action, Resistance, and New Directions in Drug Discovery.* Totowa, NJ: Humana Press Inc.

Schlitzer, M. 2007. Malaria chemotherapeutics. *ChemMedChem* 2:944–986.

WHO Malaria Report: http://www.who.int/malaria/publications/world_malaria_report/en/.

21 Anticancer Agents

Fredrik Björkling, José Moreira, and Jan Stenvang

CONTENTS

21.1 THE DISEASE

Cancer constitutes an enormous burden on society, being the second most common cause of death, surpassed only by cardiovascular diseases. Based on GLOBOCAN estimates, in 2012 alone about 14.1 million new cancer cases and 8.2 million cancer-related deaths occurred worldwide, corresponding to 3 out of every 10 deaths. What is worse, due to the growth and aging of the population, as well as an increasing prevalence of lifestyle-related risk factors such as smoking, physical inactivity, and overweight, cancer incidence is ever increasing in a snowballing effect some refer to as "a rising cancer epidemic." On the upside, these dire statistics have partially been offset by the fact that more effective drugs and treatment regimens have been brought forth to clinical management of patients, and, as a result, cancer survival in most developed countries has effectively doubled in the last 40 years. The problem is that we need to continuously improve patient treatment and find more effective therapies to be able to cope with the increasing cancer incidence.

The Food and Drug Administration (FDA) has approved 58 new cancer drugs over the last 10 years compared to 29 in the previous 10 years. In 2015, the FDA approved a total of 45 novel drugs, of which 17 were approved for the treatment of cancer or cancer-related conditions. This number of new drugs (drugs that had not been approved previously for any indication) was the highest in nearly 20 years. An overview of some of the anticancer agents approved today indicating their primary target/mechanism of action and the indications they are used for is given in Table 21.1.

In this chapter, we will give an overview of the molecular and cellular alterations associated with the development of cancer and the challenges to be overcome for effective treatment of this disease. Finally, we will focus on the development and characterization of a small set of anticancer agents used in today's practice.

TABLE 21.1
Examples of Currently Used Anticancer Agents with Indication of Their Site of Interaction, Compound Classes, Their Name/Indications/Approval Year, and Their Specific Target/Mode of Interaction[a]

Site of Interaction	Compound Class	Examples of Launched and Experimental Drugs, Trade Name (Generic Drug Name, FDA Approval Year, Indication)	Specific Targets/Mode of Action
Nucleus	Nitrogen mustards	Cytoxan, Neosar (cyclophosphamide, 1959, broad; ovary, lung breast)	DNA cross-linking
	Nitrosoureas	Temodar (temozolomide, 1999, anaplastic astrocytoma)	DNA methylation
	Platinum compounds	Platinol (cisplatin, 1978, testicular)	DNA cross-linking
		Eloxatin (oxaliplatin, 2002, colorectal)	DNA cross-linking
	Epipodophyllotoxins		
	Topo I inhibitors	Cerubidine (daunorubicin, daunomycin, 1987, ovary)	Topo II-induced breaks
	Antimetabolites	Vepesid (etoposide, VP16, 1983, lung, testicular)	Topo II-induced breaks
	Antifolates	Camptosar (irinotecan, 1996, colon)	Topo I-induced breaks
	Anthracyclines	Hycamtin (topotecan, 1996, ovary)	Topo I-induced breaks
		Methotrexate (methotrexate, 1953, osteosarcoma)	DHFR
		Xeloda (capecitabine, 2001, breast, colon)	TS
		Gemzar (gemcitabine, 1996, pancreas, lung)	DNA polymerase
	Retinoids	Nolvadex (tamoxifen, 1977, breast)	Estrogen-receptor antagonist
	Antihormones	Lupron (leuprolide, 1985, prostate)	GnHR agonist reduce testosteron
	Steroids	Vesanoid (ATRA, 1995, APL)	Retinoic acid receptor
	Hydroxamic acids	Zolinza (vorinostat, 2006, CTCL)	Histone deacetylases
	Benzamide	Beleodaq (belinostat, 2014, PTCL)	Histone deacetylases
Microsomes	Small molecules	Zytiga (abiraterone acetate, 2011, prostate)	Inhibit testosteron production
		Aromasin (exemestane, 1999, breast)	Arometase/CYP-19 inhibitor
Cytoplasm	Small-molecule antagonists	Sutent (sunitinib maleate, 2006, gastrointestinal, renal)	Multiple tyrosine kinase inhibitor
		Afinitor (everolimus, 2011, pancreatic)	mTOR inhibitor
		Zydelig (idelalisib, 2014, CLL)	PI3Kδ inhibitor
	Cytidine analog	Decogen (decitabine, 2006, MDS)	DNA methyl transferases

(*Continued*)

TABLE 21.1 *(Continued)*

Examples of Currently Used Anticancer Agents with Indication of Their Site of Interaction, Compound Classes, Their Name/Indications/Approval Year, and Their Specific Target/Mode of Interaction[a]

Site of Interaction	Compound Class	Examples of Launched and Experimental Drugs, Trade Name (Generic Drug Name, FDA Approval Year, Indication)	Specific Targets/Mode of Action
Cell membrane	mAb	Rituxan (rituximab, 1997, non-Hodgkin's lymphoma)	CD20 antagonist
	Small molecule antagonists	Tarceva (erlotinib, 2004, lung, pancreas)	EGFR antagonist
		Herceptin (trastuzumab, 1998, breast)	Her-2 antagonist
		Opdivo (nivolumab, 2015, NSCLC)	Inhibit T-cell proliferation
Protein degradation	Proteasome inhibitor	Velcade (bortezomib, 2006, multiple myeloma)	Proteasome inhibitor
Tubulin modulator	Taxanes and vinca alkaloids	Oncovin (vincristine, 1963, broad leukemia, brain, breast)	
		Taxotere (docetaxel, 1996, breast, lung)	Hyperstabilization of microtubules
Endothelial cells	Small-molecule antagonists	Avastin (bevacizumab, 2006, lung, colon)	VEGFR antagonist
	mAb	Votrient (pazopanib, 2012, sarcoma)	VEGFR inhibitor
Lymphocytes, macrophages, and dendritic cells	Interferons	Roferon A (interferon alfa 2a, 1986, leukemias)	Stimulates the ability of immune cells to attack cancer cells
	Interleukin2	Cervarix (HVP vaccine, cervical cancer, 2009)	
	Cancer vaccines (DNA- or protein-based)	Gardasil (prophylactic HPV vaccine, 2006)	

[a] Compounds specifically discussed in this chapter are excluded from the table.

Abbreviations: APL, acute promyelocytic leukemia; ATRA, all-trans retinoic acid; CD20, nonglycosylated phosphoprotein expressed on the surface of all mature B-cells; CLL, cellular lymphatic leukemia; CTCL, cutaneous T-cell lymphoma; DHFR, dihydrofolic reductase; EGFR, epidermal growth factor receptor; GnHR, gonadotropin releasing hormone; Her-2, human epidermal growth factor receptor 2; HPV, human papilloma virus; MDS, myelodysplastic syndromes; mAb, monoclonal antibody; mTOR, mammalian target of rapamycin; NSCLC, nonsmall cellular lung cancer; PI3K, phosphoinositide 3 kinase; PTCL, peripheral T-cell lymphoma; Topo, topoisomerase; TS, thymidylate synthase; VEGFR, vascular endothelial growth factor receptor.

21.1.1 Hallmarks of Cancer

Following the completion of the Human Genome Project, numerous advances in DNA sequencing technology have enabled very large genome-wide studies of a number of cancers. These studies have resulted in groundbreaking discoveries, not least the identification of significantly mutated genes involved in a plethora of cellular processes in cancer. If one unified conclusion can be drawn from these studies, it is that cancer is a dauntingly complex disease, far more so than previously assumed. In a seminal article published in 2000, Weinberg and Hanahan proposed the concept of "the hallmarks of cancer," six biological capabilities underlying the multistep development of tumors. The hallmarks provide a conceptualizing rationalization of the complexity of cancer by reducing it to a limited number of underlying principles. Recently, building on research advances, the model was updated with two new hallmarks of cancer and two enabling characteristics (Table 21.2).

1. *Sustaining proliferative signaling*: The production of growth promoting signaling molecules, e.g., platelet-derived growth factor or transforming growth factor β/α by the cancer cell itself, is known as autocrine stimulation, thus creating a positive growth feedback loop. Another way of achieving self-sufficiency in growth signals is to constitutively activate downstream signal transduction pathways normally activated by pro-growth signals. The SOS-Ras-Raf-MAPK signal (Figure 21.1) transduction pathway is a good example, and the Ras proteins are indeed altered in 25% of all cancers. This pathway can be activated by overexpression of the cell surface receptor tyrosine kinase HER2/Neu overexpressed in breast cancers (Figure 21.1).
2. *Evading growth suppressors*: At the G1 cell cycle checkpoint, also termed the restriction point, normal cells detect the composition of growth/antigrowth signals including cytokines and nutrients in their environment, and then decide whether or not to enter another round of cell division. One key effector enzyme halting the cell cycle at G1 is the retinoblastoma protein (pRb). Hypophosphorylated pRb blocks cell proliferation by sequestering E2f family transcription factors, thereby inhibiting the binding of these to DNA so that the genes required for entry into the *S* phase of the cell cycle are not expressed. Deletion of pRb is frequent in cancers and thus lead to resistance to antigrowth signals.
3. *Enabling replicative immortality*: Normal cells have the capacity to divide only a finite number of times, typically 60–70 doublings before they die due to senescence, a cell fate different from apoptosis. This is because the telomeres, DNA sequences at the end of the chromosome arms, shorten a little at each mitotic cell division. Loss of telomeres results

TABLE 21.2
Hallmarks of Cancer and Associated Inhibitors

	Hallmarks of Cancer	Associated Inhibitors
1	Sustaining proliferative signaling	EGFR inhibitors
2	Evading growth suppressors	Cyclin-dependent kinase inhibitors
3	Enabling replicative immortality	Telomerase inhibitors
4	Activating invasion and metastasis	Inhibitors of HGF/c-Met
5	Inducing angiogenesis	VEGF signaling inhibitors
6	Resisting cell death	BH3 mimetics
7	Deregulating cellular energetics	Aerobic glycolysis inhibitors
8	Avoiding immune destruction	Immune activating anti-CTLA-4 mab
	Enabling characteristics	
A	Genome instability and mutation	PARP inhibitors
B	Tumor promoting inflammation	Selective anti-inflammatory drugs

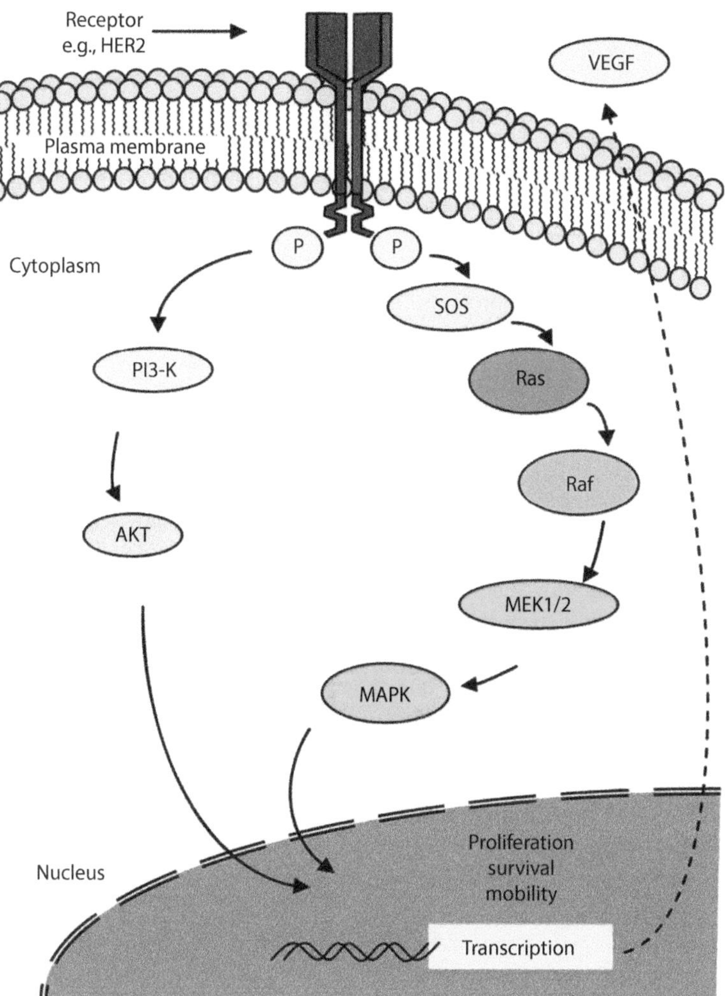

FIGURE 21.1 Cartoon depicting two signal transduction cascades often deregulated in cancer, namely, the PI3K-AKT and the SOS-Ras-Raf-Mek cascades. One way of activating these pathways is by signaling through the HER2 transmembrane tyrosine kinase receptor.

in DNA recombination and end to end fusions of chromosomes lead to cell senescence. More than 90% of human cancers avoid telomere shortening by expressing the catalytic subunit of telomerase hTERT, thereby avoiding death by senescence, with the remaining 10% of cancers using DNA recombination for this purpose.

4. *Activating invasion and metastasis*: Primary tumor growth is seldom the direct cause of death in cancer patients. Instead, the spreading of tumors known as metastasis is the most frequent cause of death (almost 90% of cases). Invasion is the penetration of tumor cells through basal membranes allowing a tumor to grow out from its original site into the surrounding tissue, whereas metastasis is the spreading of cancer cells to other sites in the body, and the subsequent establishment of secondary tumors in other tissues. Both processes involve changing the interaction of cancer cells with the extracellular environment/matrix.
5. *Inducing angiogenesis*: The initial growth phase of tumors is generally avascular as cells can be supplied by simple diffusion. However, in order to grow beyond at size of about 100 μm, tumors must develop neovascularization through the process of angiogenesis. To facilitate this, tumor cells produce the factors required for endothelial cells to perform

angiogenesis such as vascular endothelial growth factor (VEGF) or fibroblast growth factor (FGF1/2) which can be activated by different mechanisms as shown for VEGF in Figure 21.1. Angiogenesis is attractive in anticancer therapy because its inhibition is expected to have no implications for normal cells in the body regardless of their proliferation rate.

6. *Resisting cell death*: Resistance to apoptosis, or programmed cell death, is also a key acquired capacity of cancer cells. In normal cells, apoptosis targets cells for self-destruction when their DNA is damaged beyond repair or when the cells do not receive the proper pro-survival signals from its environment. A key family of apoptosis regulatory proteins is the Bcl-2 family containing pro-apoptotic (Bax, Bak, Bid, and Bim) and antiapoptotic (Bcl-2, Bcl-XL, and Bcl-W) proteins. The Bcl-2 protein itself is frequently upregulated in cancer cells, leading directly to resistance toward drug-induced apoptosis. Furthermore, deletion or mutational inactivation of the tumor suppressor gene *TP53* represents another way of evading apoptosis because p53 normally induces upregulation of pro-apoptotic Bax protein upon DNA damage. p53 is inactivated in 50% of all human tumors.
7. *Deregulating cellular energetics*: Cancer cells display an enhanced cellular energy metabolism through glycolysis, the so-called Warburg effect, a switch from normal respiration to anaerobic glycolysis. In order to support continuous cell growth and proliferation, cancer cells consume >20 times as much glucose and produce energy in the form of ATP almost 100 times faster as a normal cell. Hypoxia-inducible factor is an important factor that activates genes supporting anaerobic glycolysis when triggered by, e.g., DNA damage, protein signaling, and growth factors. This metabolic switch is controlled by phosphoinositide 3 kinase (PI3K) which in turn activates key regulatory proteins such as AKT and mTOR leading to the altered energy metabolism.
8. *Avoiding immune destruction*: An active immune system continuously recognizes and eliminates the vast majority of cancer cells before they establish themselves and form a tumor. The action of the immune system include three phases, elimination of cancer cells, immunological control of cancer cell growth, and the "escape" of tumor cell not affected by the immune system. A defect or suppressed immune system will lead to the "escape" of tumor cells that can continue to proliferate and cause tumors. In many cancers, malignant progression is accompanied by profound immune suppression; in addition, tumors have themselves the ability to promote the production of pro-inflammatory cytokines and other factors leading to suppression of immunity. Cancer immunotherapy, be it cellular-, antibody-, or cytokine-based, is at present one of the most promising new cancer treatment approaches. This is primarily due to the fact that whereas the benefits of molecular targeted therapies are each time negated by the rapid development of drug resistance, patients that respond to immunotherapies such as ipilimumab (Yervoy) or nivolumab (Opdivo) show durable long-term responses.

21.1.2 Enabling Characteristics of Cancer Cells

1. *Genome instability*: In addition to the hallmarks mentioned in the previous section, cancer cells display a number of additional alterations that can be construed as enabling cancer growth. One of the most important alterations is genome instability. The number of individual mutations required to induce all eight hallmarks of cancer would not normally accumulate in a single cell if not for genome instability. Genome instability is caused by mutations of DNA repair/checkpoint surveillance mechanisms which normally results in cell cycle arrest until the damaged DNA has been repaired, or in case DNA cannot be efficiently repaired results in elimination of the cell via apoptosis. The genome guarding protein p53 as well as proteins sensing DNA damage such as ATR, ATM, and their downstream signal kinases Chk1 and Chk2 play an important role in such checkpoints, and their loss often precedes genome instability.

One of the main obstacles in cancer therapy is tumor heterogeneity which is closely related to genome instability. Within one tumor, several different subpopulations of tumor stem cells often exist. Consequently, while first-line chemotherapy does often lead to good responses and tumor regression, tumor heterogeneity makes it very difficult to kill all tumor cells within a tumor. Thus, some cancer cells may be inherently resistant toward a given drug due to mutation of its primary target/receptor or due to dysregulation of the downstream signal transduction cascades. Such cells will often continue to divide and lead to a new tumor now resistant to the used drug. As a result, in modern anticancer therapy, combinations of two to three different drugs targeting different receptors/pathways are often used.

2. *Tumor promoting inflammation*: Recent evidence suggest that inflammation is critical in cancer development. Mechanistically, inflammation promotes tumor progression through cells of the immune system. The link between inflammation and cancer can be seen in, e.g., chronic infections, obesity, and smoking which are recognized as major risk factors in cancer development. Thus, the activation of the transcription factor nuclear factor κB promotes not only the generation of inflammatory mediators, e.g., interleukin-1 and tumor necrosis factor-α, but also, in immune cells, genes that promote cell growth, survival, migration, and angiogenesis. One of the most important immune cells associated with cancer are the tumor-associated macrophages which are found in most malignant tumors and correlate with poor prognosis.

21.2 ANTICANCER AGENTS

Despite intensive research aimed at understanding the molecular pathology of cancer, a great deal of anticancer agents currently in clinical use were discovered and even entered the clinic before their exact mechanism of action was determined. These drugs were often discovered in cellular screens of extracts from natural sources or in in vivo screens using a leukemic P388 mouse model. The drugs discovered typically inhibit DNA synthesis (antimetabolites), damage DNA (DNA alkylating agents and topoisomerase poisons), or inhibit the function of the mitotic microtubule-based spindle apparatus (taxanes) (Table 21.1). The reason for these agents still being in clinical use relates to the fact that they are often highly effective, although they have toxic properties toward normal fast proliferating cells as the intestinal- and gut lining, hair follicles, and the bone marrow cells, leading to the well-known effects of classical chemotherapy such as nausea, hair loss, and myelosuppression. The cytotoxics stands in contrast to the so-called targeted therapies that are developed in a totally different fashion by applying knowledge concerning the structure of a primary target with molecular *in silico* screening as well as high-throughput compound library screening or by designing protein-based medications which in the case of cancer treatment are often in the form of cell surface tyrosine kinase antagonistic antibodies. Examples of targeted therapeutic anticancer agents are kinase inhibitors imatinib (Gleevec) and gefitinib (Iressa), proteasome inhibitors bortezomib (Velcade), histone deacetylase (HDAC) inhibitors vorinostat (Zolinza) and belinostat (Beleodaq), and antibodies against cell surface receptors trastuzumab (Herceptin), several of which are extensively used today (Tables 21.1 and 21.3). It must be emphasized that the older cytotoxic compounds are in fact often highly specific and have well-defined targets in the cell, but this was not *a priori* knowledge, and only discovered once their mechanism of action had been worked out.

21.2.1 ANTICANCER AGENTS CURRENTLY USED

Reviewing the multitude of anticancer agents currently approved for clinical use is an overwhelming task and not within the scope of this chapter. Instead, we will describe the development and mechanism of action of two classic cytotoxics and four different targeted therapies. The former include

TABLE 21.3
Top Selling Anticancer Drugs in 2014

	Drug	Cancer Indication	Global Sales 2014 (billion $)	Company
1	Rituxan	Non-Hodgkin's lymphoma, CLL	7.6	Roche
2	Avastin	Colorectal, lung, ovarian, brain	7.0	Roche
3	Herceptin	Breast	6.9	Roche
4	Revlimid	Multiple myeloma	5.0	Celgene
5	Gleevec	Leukemia, GI	4.7	Novartis
6	Velcade	Multiple myeloma	3.1	Takeda/Johnson & Johnson
7	Alimta	Lung	2.8	Eli Lilly
8	Zytiga	Prostate	2.2	Johnson & Johnson
9	Erbitux	Colon, head and neck	1.9	Merck KGaA/BMS
10	Afinitor	Breast	1.6	Novartis

one antimetabolite, permetrexed (Alimta) which inhibit the production of precursors for DNA synthesis in the cell, and a taxane, paclitaxel (Taxol) which targets the structural protein β-tubulin involved in microtubule functioning. We will likewise review the development and mechanism of three new classes of anticancer therapeutics, the kinase inhibitors exemplified by the BCR-ALB tyrosine kinase inhibitor imatinib (Gleevec) and the ALK inhibitor crizotinib (Xalcori), the monoclonal antibody trastuzumab (Herceptin), and, finally, the HER2/Neu tyrosine kinase antagonist as a combined antibody-cytotoxic drug ado-trastuzumab emtansine (Kadcyla).

21.2.2 Alimta

The discovery of Alimta has its chemical origin in the early findings of the antimetabolites aminopteridines and thereafter methotrexate that both inhibit folate metabolism (Figure 21.2c). The impressive anticancer effects found for methotrexate validated folate antimetabolites in the earlier stages as antiproliferative agents. For decades, researchers have worked on the task of finding inhibitors of folate-dependent enzymes such as thymidylate synthase (TS), dihydrofolate reductase (DHFR), and glycinamide ribonucleotide formyltransferase (GARFT) which take part in the folic acid activation (Figure 21.3). The active form of folate is the reduced form tetrahydrofolate (Figure 21.2b) which plays an important role in the biochemical pathways to donate one carbon unit in the form of methyl, methylene, or formyl groups. These metabolic reactions are essential for the formation of DNA, RNA, ATP, and the catabolism of certain amino acids. Consequently, inhibiting this metabolic pathway abrogates cancer cell proliferation because cancer cells have high demands for ATP, and because they require high levels of nucleic acid precursors for DNA synthesis.

The pathway leading to the formation of tetrahydrofolate (THF) begins when folate (F) is reduced to dihydrofolate (DHF) which is then reduced to THF; DHFR catalyzes both steps (Figure 21.3). Methylene tetrahydrofolate (CH_2THF) is formed from tetrahydrofolate by the addition of methylene groups forming N^5, N^{10}-methylene tetrahydrofolate from one of the three carbon donors: formaldehyde, serine, or glycine (Figure 21.3).

The key reaction is the TS catalyzed methylation of deoxyuridine monophosphate to generate thymidylate which is needed for DNA synthesis. Methyl tetrahydrofolate (CH_3THF) is formed from methylene tetrahydrofolate by the reduction of the methylene group and formyl tetrahydrofolate (CHOTHF, folinic acid) results from the oxidation of the same precursor (Figure 21.2a).

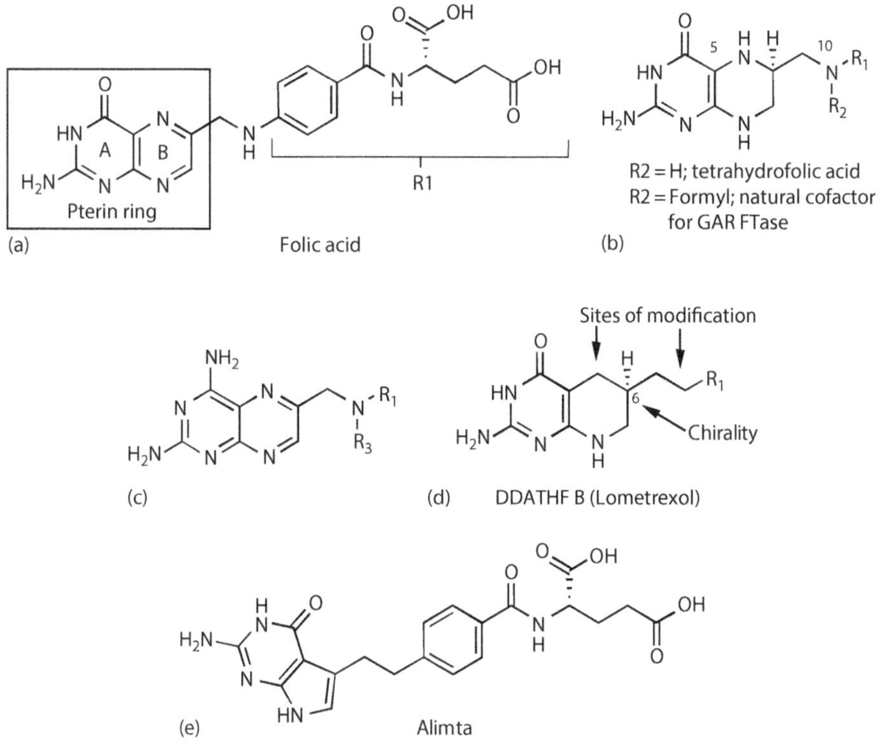

FIGURE 21.2 (a–e) Chemical structures of various antimetabolites preceding and founding the development of Alimta.

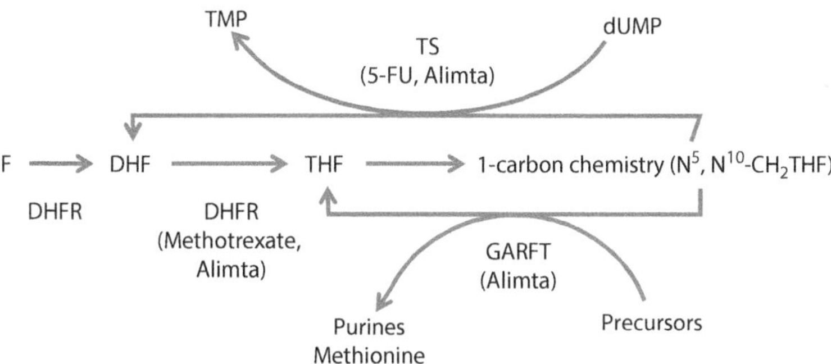

FIGURE 21.3 GARFT is one of the many enzymes involved in the biosynthesis of Purines.

Inspired by the active pterine structures (Figure 21.2a), many modifications were made in this ring system including modifications in ring A, such as substitution of NH_2 with methyl or hydrogen as well as exchange of the fused ring B for a fused phenyl ring. This resulted in compounds having high biological activity as TS inhibitors with concomitant antiproliferative activity. Some of these analogs were indeed taken into early clinical testing, but were stopped due to pharmacokinetic or toxicological problems.

An important new class of potent folate antimetabolites that are active as antitumor agents are represented by 5,10-dideaza-5,6,7,8-tetrahydrofolic acid (DDATHF, Lometrexol) in which the two nitrogens in positions 5 and 10 is exchanged for carbon and the B ring is reduced and as such mimic the structure of THF (Figure 21.2d). The target enzyme for DDATHF was shown to be GARFT

(Figure 21.3) catalyzing the first folate cofactor-dependent formyl transfer step in the de novo purine biosynthetic pathway instead of the DHFR enzyme which was the target for earlier folate inhibitors previously described.

The two diastereomeric forms of DDATHF were separated and their biological activity examined. Interestingly, they did not show any significant difference in activity and further work was, therefore, undertaken to remove this chiral center so as to obtain a stereochemically pure compound. After exploring many options such as opening of the ring B, the best candidate was found to be permetrexed (Alimta), developed and marketed by Eli Lilly (Figure 21.2e) which contains a fused pyrrole ring in place of ring B. While early studies with Alimta showed that its primary target was TS, more recent studies have demonstrated that following intracellular polyglutamation of Alimta the product affects folate metabolism dramatically by inhibiting several folate-dependent enzymes—TS, DHFR, GARFT (Figure 21.3)—in addition to 5-aminoimidazole-4-carboxamide ribonucleotide formyltransferase and C-1 tetrahydrofolate synthetase. This broad mechanism of action on folate activation may be responsible for the efficacy of Alimta. Today, Alimta is used in pleural mesothelioma and as a single agent for the treatment of patients with locally advanced or metastatic nonsmall cell lung cancer (NSCLC).

21.2.3 Taxol

Another example of a successful anticancer drug whose development goes far back in time is Taxol. Back around 1960, the National Cancer Institute launched a program of screening compounds from extracts of thousands of plants with the aim of identifying molecules of biomedical interest, hereunder compounds with anticancer activity. Paclitaxel (Taxol) was discovered in 1963. This complex polyoxygenated diterpenoid was originally isolated from the Pacific yew, *Taxus brevifolia*, and later found in several other species of *Taxus* including *Taxus wallichiana*, the Himalayan yew. By the early 1970s, the structure of Taxol was solved (Figure 21.4a). Yet, another decade had to pass before the molecular mechanism of the compound was elucidated which is stabilization of microtubules with concomitant cell cycle arrest at the G2M cell cycle phase. The binding of Taxol to polymerized β-tubulin is depicted in Figure 21.4b. Later on, availability problems associated with the limited supply of the Pacific yew were solved by applying a semisynthetic route starting with another natural product isolated from the English yew, *Taxus baccata* which is 10-deacetylbaccatin-III, avoiding the exhaustive complete synthesis of Taxol. This was achieved in the early 1980s and accomplished by acetylating 10-deacetylbaccatin-III and attaching a side chain to it. Taxol was finally approved by the FDA at the end of 1992.

Today, Taxol, developed and marketed by Bristol–Meyer Squibb, is an effective drug in the treatment of breast, ovarian, and lung cancer. In a relative short period of time, the taxane class of microtubule-targeting anticancer agents has become one of the most important chemotherapeutics available to treat cancer. Taxanes have a unique ability to be effective in the treatment of many different types of advanced cancers, such as carcinoma of the breast, lung, prostate, or bladder. Additionally, they have also demonstrated therapeutic benefit in the treatment of early-stage cancer. A total of more than 300 taxanes have been made and extensive structure–activity relationship (SAR) analyses have been carried out. The most important findings are summarized here (Figure 21.4a). In the lower part of the molecule, the activity is reduced by the removal of C1 hydroxy, C4 acetyl, 4,5,20-oxetane ring, and in the C2 benzoyl only limited substitutions are allowed. In the upper part of the molecule, derivatization of C7-hydroxyl or change in its stereochemistry has no significant effect on activity. Also, reduction of a C9-ketone slightly increases the activity, and both C10 hydroxyls and acetates retain activity. The C-13 side chain is essentially required for activity, even though some specific alteration is found to be useful for improving the parent compound. For example, can the C2′-hydroxyl group be used as an attachment point for a prodrug ester which results in a compound with in vivo activity, but not in vitro potency. Limited modifications of C2′ and C3′ are allowed, and the stereochemistry at these positions is important for

Modifications allowed

Essential groups

R1 = $-C_6H_5$, R2 = Ac; (Taxol)
R1 = O-tBu, R2 = H; (Taxotere)
(a)

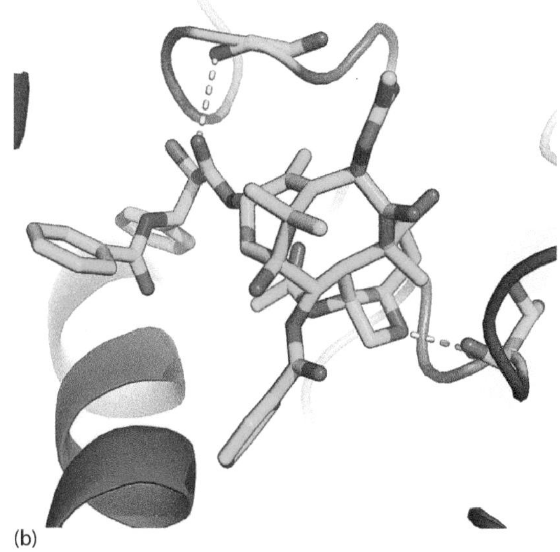

(b)

FIGURE 21.4 (a) Chemical structure of the taxanes with associated SAR information. (b) Crystal structure of Taxol bound to β-tubulin. PDB code: 1JFF. X-ray model prepared by Dr Lars Olsen.

high with preference for the natural 2′*R*,3′*S* isomer. Based on the SAR for Taxol, many derivatives have been prepared, in particular, with the aim of identifying new compounds with increased solubility since this is a major problem for Taxol having a solubility <0.3 μg/mL. Consequently, Taxol needs to be administered with a solubilizing carrier such as polyethoxylated castor oil. One example of an approved analog is docetaxel (Taxotere) (Figure 21.4a) which shows potent anticancer activity and improved solubility properties.

21.2.4 Gleevec

Imatinib or Gleevec is a tyrosine kinase inhibitor (see also Chapters 11 and 22) developed by Novartis which has greatly improved the treatment of chronic myeloid leukemia (CML). CML is a rather rare condition (prevalence is 1:100,000) which in about 90% of all cases is caused by a specific chromosomal translocation (t(29;22) (q34;q11)), also referred to as the Philadelphia chromosome. Through the creation of this new chromosomal breakpoint, this translocation creates a unique fusion protein BCR-ALB with oncogenic properties due to its constitutive tyrosine kinase activity which activates a number of cellular pathways including the JAK/STAT and the Ras-Raf-Mek-MAPK leading to the achievement of proliferation and antiapoptotic signaling as depicted in Figure 21.5.

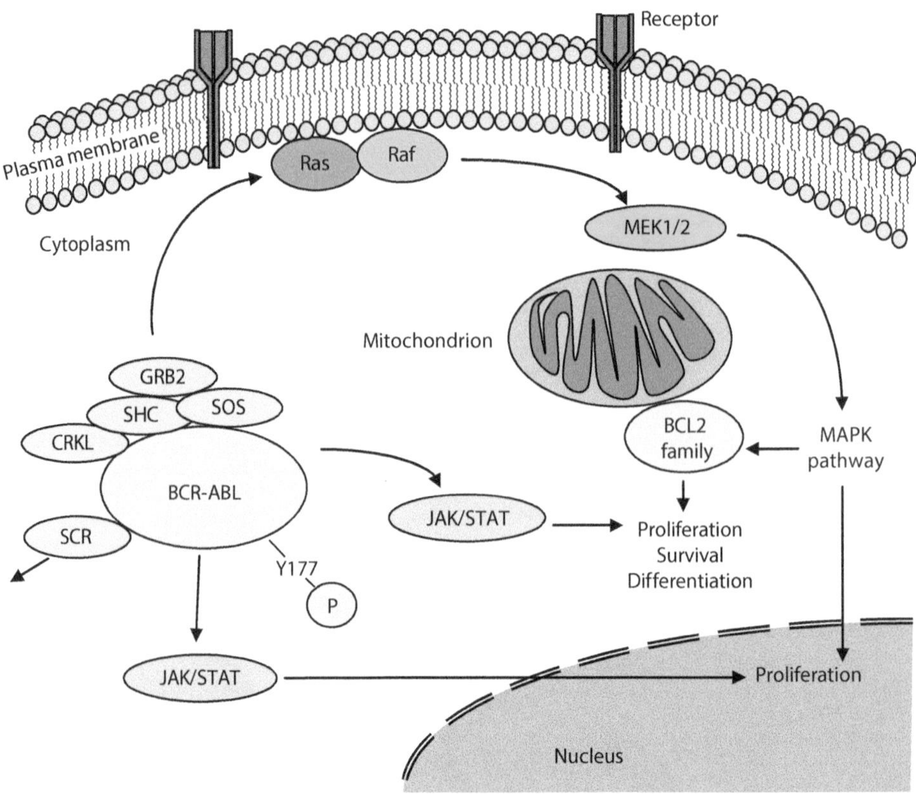

FIGURE 21.5 Cartoon highlighting pathways affected by the constitutive tyrosine kinase activity of the BCR–ALB oncoprotein and its cellular consequences.

The presence of the BCR-ALB protein exclusively in CML cells (cancer specificity) combined with the cellular dependence on this sole protein for malignant transformation and survival of transformed cells (also referred to as oncogene addiction) represents an unique situation where targeted cancer therapy is relatively easily achieved. Thus, the development of Gleevec is one of the best examples of a modern targeted anticancer therapeutics, and its design builds on a clear rationale for intervention (in this case, oncogene addiction), combined with detailed knowledge of the three-dimensional structure of the primary molecular target (the BCR-ALB oncoprotein). The successful launch of Gleevec has provided a great deal of inspiration and effort into the further development of small-molecule kinase inhibitors as anticancer drugs. The finding that Gleevec also inhibits c-KIT has promoted its recent use in the treatment of colorectal gastrointestina cancers.

The molecular starting point for the medicinal chemistry leading to the development of Gleevec was a phenylaminopyrimidine derivative (the blue core of compound A/B as contained in the box seen in Figure 21.6) found in a screen for protein kinase C (PKC) inhibitors. This compound showed good lead-like properties and could be further optimized with straightforward chemistry to improve activity, selectivity, as well as drug-like properties. Thus, the addition of a 3′-pyridyl group at the 3′-position of the pyrimidine gave compounds with superior activity as PKC inhibitors in cellular assays. It was further found that derivatives with an amide group in the phenyl ring provided inhibitory activity against tyrosine kinases, such as the BCR-ABL kinase (Figure 21.6, compound A). Subsequent SAR studies suggested that a substitution at position 6 of the diaminophenyl ring abolished PKC inhibitory activity while retaining activity for BRC-ALB kinase which was confirmed by the introduction of a methyl group in this position (Figure 21.6, compound B). Further modification was, however, needed due to the low solubility and bioavailability found

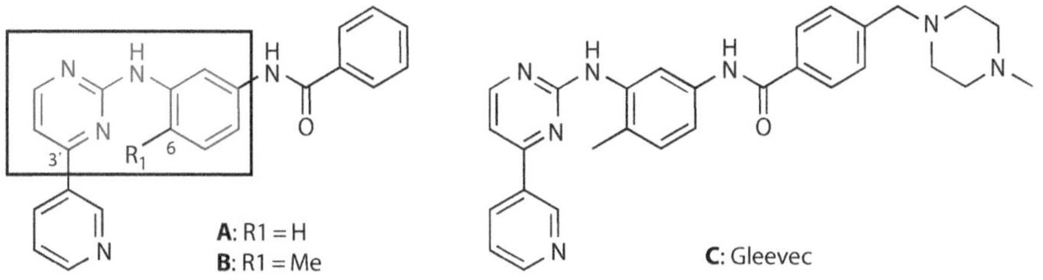

A: R1 = H
B: R1 = Me

C: Gleevec

Comp./Activity IC_{50} (μM)	v-Abl-K	PKCα	PKCβ
A	0.4	1.2	23
B	0.4	72	>500
C	0.038	>100	>100

FIGURE 21.6 Chemical structure of Gleevec and associated kinase inhibitors. Table shows the effect of SAR variations on the activity toward BCR–ALB and protein kinase C, α, and β isoforms.

for these compounds. Introduction of a highly polar side chain (an *N*-methylpiperazine) gave a marked improvement of both solubility and oral bioavailability. To avoid the mutagenic potential of aniline moieties, a spacer was introduced between the phenyl ring and the nitrogen atom which now afforded the best compound in the series, Gleevec (Figure 21.6, compound C). Computational chemistry, docking, and X-ray crystallography shows that Gleevec binds at the ATP-binding site of an inactive form of the kinase. The high specificity of Gleevec is explained by this unusual binding as well as some strong interactions of the *N*-methylpiperazine group with the ABL kinase backbone (Figure 21.7).

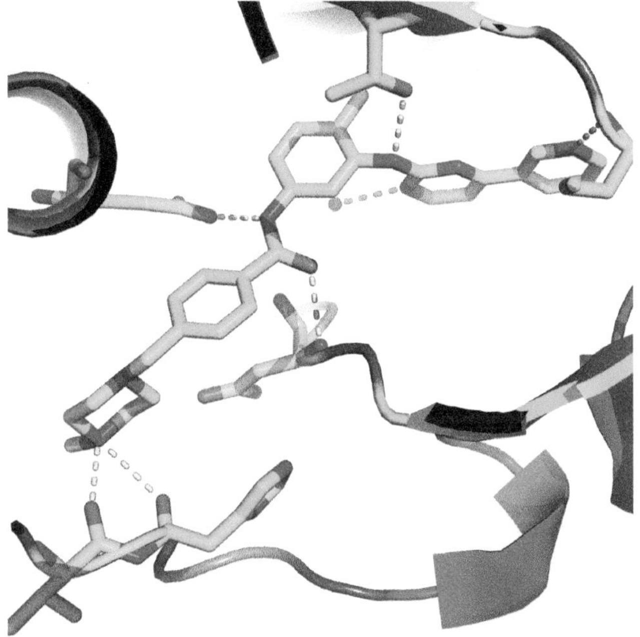

FIGURE 21.7 Crystal structure of the catalytic domain of ABL tyrosine kinase complexed with Gleevec. Hydrogen bonds with interacting amino acids are indicated. PDB code: 3GVU. X-ray model was prepared by Dr Lars Olsen.

21.2.5 Xalcori

Several protein kinases have emerged as attractive therapeutic targets for the treatment of cancer in part as a consequence of the launch of Gleevec. One of these, anaplastic lymphoma kinase (ALK), has oncogenic potential and plays an essential role in the pathogenesis of a wide variety of human cancers, such as NSCLC, breast, and colorectal cancer. NSCLC accounts for approximately 85% of lung cancers which is the leading cause of cancer death, with a 5-year survival in the range 1%–15%, mainly due to insensitivity to chemotherapy.

By large-scale high-throughput screening approaches, a number of genetic driver mutations have been identified in NSCLC, among them mutations in the kinase domain of the epidermal growth factor receptor (EGFR). This finding translated clinically to improved response rates in NSCLC patients treated with EGFR-tyrosine kinase inhibitors. In the early 1990s, the chromosomal translocation (t(2;5)(p23;q35)) causing the ALK fusion protein was identified as an oncogenic driver in anaplastic large cell lymphoma. In 2007, two research groups independently reported the discovery of ALK rearrangements in NSCLC. This was the first time ALK rearrangements were reported in a common solid tumor type and marks the beginning of a change in mindset in oncology drug development. Following the initial discovery of ALK rearrangements, Xalcori (crizotinib, Pfizer, Figure 21.8) went into Phase I clinical development in 2007. Xalcori is a multitarget tyrosine kinase inhibitor that blocks ALK, the c-Ros oncogene 1, the macrophage stimulating 1 receptor (c-MET-related tyrosine kinase), and the Met Proto-Oncogene (c-MET).

Xalcori was discovered starting from an initial screen of a potent class of kinase inhibitors, the 3-substituted indolin-2-ones. The target for inhibition was the c-MET kinase. The indolin-2-one core has previously been utilized for kinase inhibitors, and the anticancer compound sunitinib belong to this compound class (Figure 21.8). Kinase selectivity can be built in by substituents in indolin-2-one core, and initial optimization of this class led to compound A with low nM activity; however, the physicochemical properties were not favorable. This compound was co-crystallized with a kinase domain construct of c-MET and used to rationalize the further design of analogs with improved

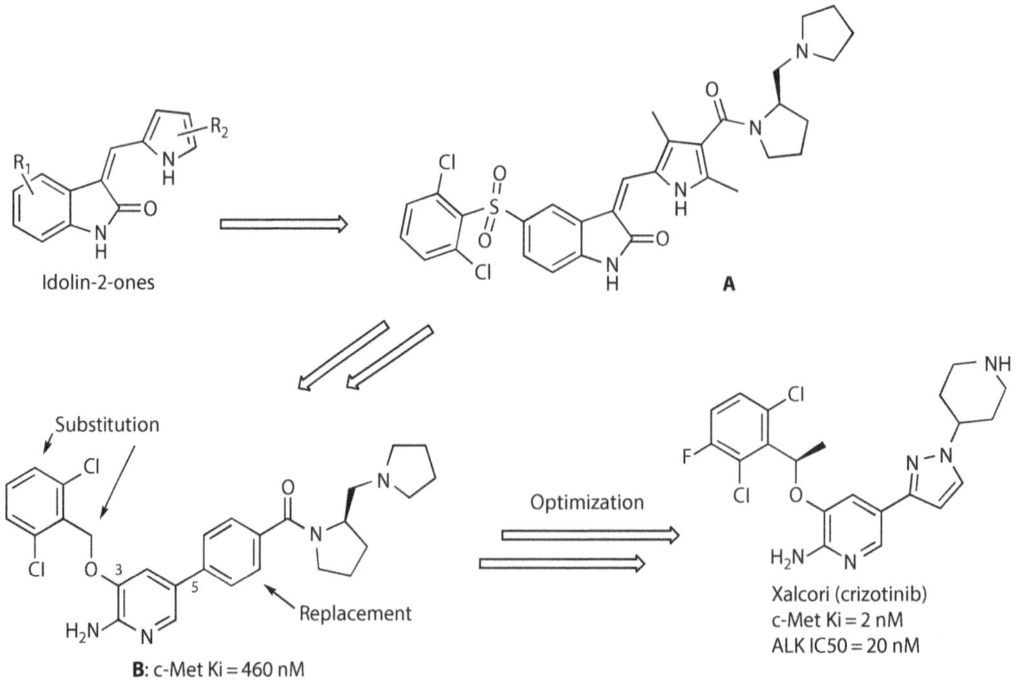

FIGURE 21.8 Discovery and chemical optimization toward Xalcori.

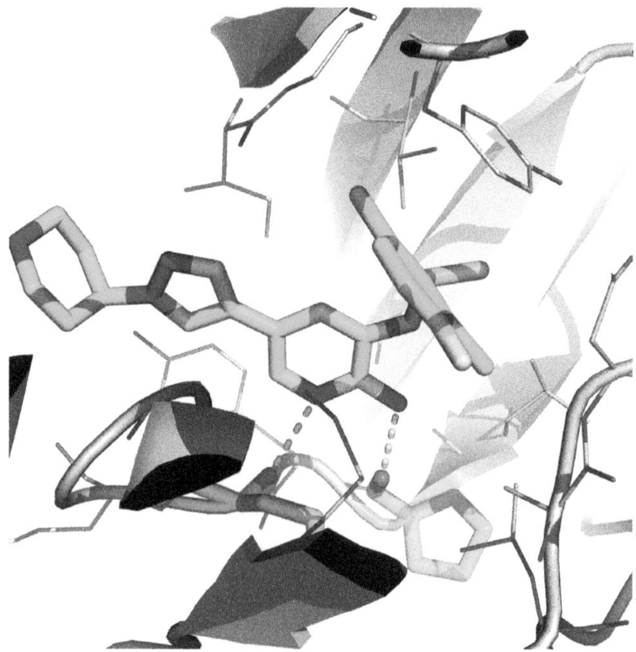

FIGURE 21.9 Xalcori co-crystallized with a kinase domain construct of c-MET. PDB code: 2WGJ. X-ray model was prepared by Dr Lars Olsen.

activity and physicochemical properties. A major modification was made by redesigning the central core rings to a novel 5-aryl-3-benzyloxy-2-aminopyridine scaffold (Figure 21.8, compound B). In this new series, the enzyme binding properties were optimized by substitution of the benzyloxy group at the 3-position of the pyridine ring that binds to the hydrophobic pocket. Also, guided by the structure a methyl group in the linker to fit a small hydrophobic cavity in the enzyme gave increased activities. Finally, the SAR could be optimized through the optimization of the substituents in the 5-position of the pyridine ring leading to the Xalcori structure (Figures 21.8 and 21.9).

In 2013, the FDA granted Xalcori approval for the treatment of patients with metastatic NSCLC whose tumors are ALK-positive. This rapid and successful development of Xalcori (crizotinib) as an inhibitor in ALK-rearranged NSCLC was enabled by the prior knowledge that NSCLC patients bearing tumors with activating EGFR mutations had benefit from EGFR tyrosine kinase inhibitors. Conditional approval of Xalcori for the treatment of ALK-rearranged NSCLC was granted based on objective response rates of 50%–60% and a median progression-free survival of 8.1 and 9.7 months, respectively, in two single-arm trials. To set this in context, standard of care single-agent chemotherapies in patients with NSCLC have response rates of less 10% and median progression-free survival of 3 months at most.

In addition to making the case for rapid and successful translation of preclinical molecular findings into the clinic, the development of Xalcori has also heralded another change in mindset. Until recently, clinical management of cancer was understood on a disease-specific manner. In other words, medical oncology has inherently been a disease-centric activity, where one determines whether a particular drug is effective in a specific disease setting. Xalcori specifically targets the ALK tyrosine kinase, thus cancers harboring the oncogenic EML4-ALK translocation will presumably be effectively inhibited by this drug, irrespective of the cancer type. Accordingly, Xalcori is also being tested in other ALK+ tumors such as ALCL and neuroblastoma, suggesting that eventually management of cancer may be driven more by which molecular alteration to target, rather than by which type of tissue the cancer originates from.

21.2.6 Herceptin

In the previous sections, we have looked at small-molecule inhibitors of various cancer pathways and targets. A totally different approach is to use natures' own building blocks and principles to target cancer calls selectively for destruction. One of the best examples is the humanized antibody trastuzumab or Herceptin targeting the HER2/Neu receptor that is often overexpressed in breast cancers. The development of antibody-based anticancer therapies goes back to the 1950s and the first experiments used polyclonal antisera, typically originating from immunization in rodents. The two main problems with these antibodies were lack of specificity and anaphylaxis due to immune response in humans. Today, these problems have been effectively solved by the development of monoclonal and the so-called humanized antibodies in which the invariable regions have been replaced with the homologous human sequences.

The HER2 receptor, also called the Neu or Erb2, is a member of the EGFR family of transmembrane tyrosine kinases. It is composed of an extracellular domain, a hydrophilic transmembrane domain, and a intracellular domain harboring its tyrosine kinase activity which when activated provides the upstream signal for activating cellular signaling pathways responsible for cell proliferation (the RAS-MAPK pathway) and pathways inhibiting cell death (the PI3K-AKT pathway) as depicted in Figure 21.1. In contrast to the other EGFR receptor members, HER2 has no identified binding ligand, but its tyrosine kinase activity can be activated upon dimerization of two HER2 receptors and also upon heterodimerization of HER2 with other EGFR family members. HER2 is often overexpressed in breast cancers which can be caused by gene duplication as well as transcriptional upregulation. This in turn leads to hyperactivation of its downstream pathways mentioned earlier, accelerating, and sustaining tumor growth (Figure 21.1). Therefore, there is a clear rationale for targeting the HER2/Neu receptor in breast cancers in cases where it is overexpressed (20%–30% of metastatic breast cancers). Large clinical trials were necessary to establish that the HER2/Neu receptor has to be overexpressed to high levels in order for Herceptin to provide a therapeutic benefit. This finding is now routinely used in diagnostics and in the classification of patients into groups amenable to Herceptin treatment which significantly benefits this well-defined subpopulation. Today, Herceptin is mostly used in combination regimens, together with chemotherapeutic drugs including topoisomerase II targeting anthracyclines or microtubule targeting paclitaxel in patient populations with highly elevated levels of HER2 expression. Although Herceptin showed some activity in preclinical models when administered as a monotherapy, this drug only yielded response rates of 12%–35% in a single-agent clinical setting. Conceivably because trastuzumabs' antitumor activity is dependent on essentially two modes of action: direct binding to HER2 blocks downstream signaling of several HER2-dependent cellular signaling pathways driving tumor growth. But the therapeutic activity of trastuzumab also critically depends on the involvement of the host's immune system. Animal studies demonstrated that like for most anticancer antibodies, antibody-dependent cell-mediated cytotoxicity (ADCC) is a major antitumor mechanism. Unfortunately, the factors that regulate the magnitude of ADCC are not well understood presently.

21.2.7 Kadcyla

Although unconjugated monoclonal antibodies have emerged as useful cancer therapeutics, recently, antibody-drug conjugate (ADC) drugs have attracted a lot of attention, and currently, more than 500 clinical trials have included drugs belonging to this class and more than 100 of these are within oncology. The ADCs belong to a drug class of therapeutics that exploits the powerful combination of the targeting specificity of monoclonal antibodies and cytotoxic action of other drugs. The antibodies can be "armed" with radioisotopes or cellular toxins, which are then effectively targeted to the surface of their target cells, or even activators of prodrugs, leading to high active concentrations of the active substances in targeted cells.

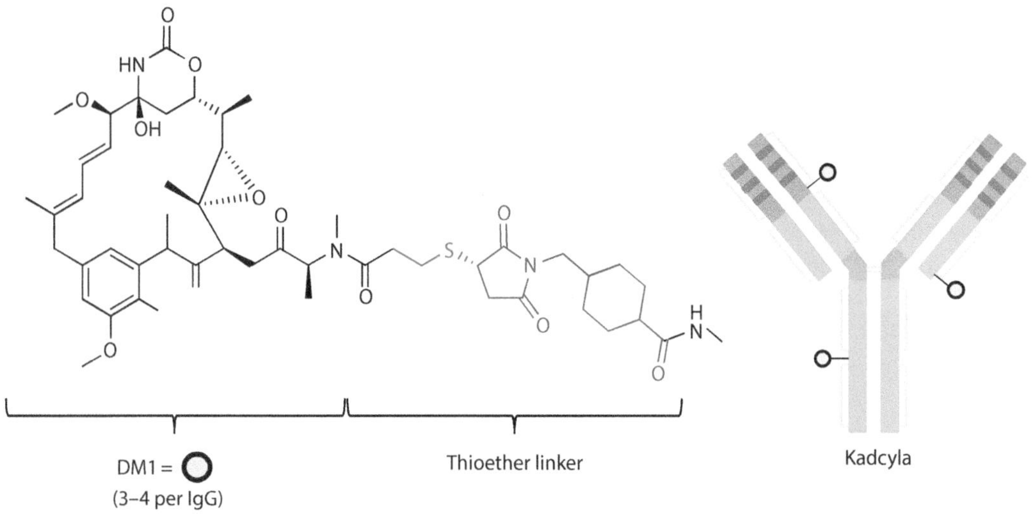

DM1 = O
(3–4 per IgG)
Thioether linker
Kadcyla

FIGURE 21.10 Chemical structure of cytotoxic DM1 and cartoon of the drug Kadcyla.

This dual mechanism of action is provided by direct linkage of a small-molecule cytotoxic agent to the antibody. This drug design has several advantages; the antibody exerts its normal therapeutic function by binding to targets on cell membranes, the cytotoxic agents is specifically delivered to the target cells and the target cells only will be exposed to the combined effects of the two drugs. One example of ADC is the T-DM1 (trastuzumab emtansine, ado-trastuzumab emtansine (in the United States), T-MCC-DM1, trade name Kadcycla, Figure 21.10) which is the first ADC to target the HER2 receptor. This ADC drug consists of a single HER2 targeting monoclonal antibody, trastuzumab, coupled with several molecules of the cytotoxic moiety DM1 (derivative of maytansine) which are linked by a nonreducible thioether linker. Maytansine is a cytotoxic agent that inhibits the assembly of microtubules by binding to tubulin. It is a natural product macrolide of the ansamycin type and can be isolated from plants of the genus Maytenus.

As mentioned earlier, the recombinant humanized monoclonal antibody trastuzumab (Herceptin) is an efficient drug in HER2-positive breast cancer patients and can, therefore, be exploited to deliver a cytotoxic cargo specifically to breast cancer cells. T-DM1 enters breast cancer cells by receptor-mediated endocytosis upon binding of T-DM1 to the extracellular subdomain IV of the HER2 receptor on the surface of the breast cancer cells. Release of the cytotoxic DM1 requires proteolytic degradation of the trastuzumab moiety which is executed in the lysosomes. Released DM1 from the lysosomes to the cytoplasma then inhibits microtubule assembly eventually causing cell death. Notably, coupling of DM1 to trastuzumab does not interfere with the binding affinity to HER2 or the anticancer effects. T-DM1 possesses strong antitumor activity in both preclinical experiments and in clinical trials. Furthermore, the activity of T-DM1 is superior to trastuzumab and T-DM1 remains active in trastuzumab-resistant breast cancer models. In summary, Kadcyla represent a new generation of anticancer compounds which combine specific targeting by a monoclonal antibody with the efficiency of a classical cytotoxic agent and mechanism.

21.3 CONCLUDING REMARKS

In this chapter, we have briefly discussed the molecular and cellular background underlying the development of cancer. Furthermore, we have focused on some anticancer agents in current clinical use. The anticancer agents reviewed represent a continuum in cancer drug development beginning with the antimetabolites used in cancer therapy for more than half a century (Alimta) over therapy targeting the structural function of the cancer cell (Taxol) to novel highly targeted small molecule (Gleevec and Xalcori), antibody based (Herceptin), and, finally, a combination of antibody and cytotoxic (Kadcyla) therapies.

This continuum not only represents a developmental timescale but also reflects today's clinical practice, attesting to the fact that in the field of cancer treatment old drugs continue to be used alongside much newer ones. This is because the older anticancer drugs, the cytotoxics, are generally highly effective although they are greatly hampered by their well-known side effects. On the other hand, while being more cancer specific, the new targeted therapies are often troubled by their rather restricted applications (with regard to indication) or lack of efficacy when administered as single agents. It is, therefore, often advantageous to combine the classic cytotoxics with new targeted molecules. Here, numerous preclinical and clinical research projects are beginning to indicate the directions toward effective combination regimens including both old and new drugs. We believe these approaches will be further developed in coming years and several new principles, e.g., in gene therapy, cancer immune therapy, and stem cell research, will find their way to the patient, possibly even as individually tailored personalized medications, but certainly also as traditional add-on to existing therapy.

FURTHER READING

Altmann, K.H. and Gertsch, J. 2007. Anticancer drugs from nature–natural products as a unique source of new microtubule-stabilizing agents. *Nat. Prod. Rep.* 24:327–357.

Costi, M.P., Ferrari, S., Venturelli, A., Calo, S., Tondi, D., and Barlocco, D. 2005. Thymidylate synthase structure, function and implication in drug discovery. *Curr. Med. Chem.* 12:2241–2258.

DeVita, V.T. Jr., Lawrence, T.S., and Rosenberg, S.A. 2014. Cancer: Principles and Practice of Oncology, 10th edn. Lipincott Williams & Wilkins; Philadelphia, PA. ISBN/ISSN: 9781451192940.

Hanahan, D. and Weinberg, R.A. 2011. Hallmarks of cancer: The next generation. *Cell* 144:646–674.

Hudis, C.A. 2007. Trastuzumab–mechanism of action and use in clinical practice. *New Engl. J. Med.* 357:39–51.

Kwak, E.L., Bang, Y.-J., Camidge, D.R. et al. 2010. Anaplastic lymphoma kinase inhibition in non-small-cell lung cancer. *New Engl. J. Med.* 363:1693–1703.

Weisberg, E., Manley, P.W., Cowan-Jacob, S.W., Hochhaus, A., and Griffin, J.D. 2007. Second generation inhibitors of bcr-abl for the treatment of imatinib-resistant chronic myeloid leukaemia. *Nat. Rev. Cancer* 7:345–356.

22 Targeting Receptor Tyrosine Kinases

Bo Falck Hansen and Steen Gammeltoft

CONTENTS

22.1 INTRODUCTION

Receptor tyrosine kinases (RTKs) are cell surface receptors that play an essential role in signal transduction from extracellular stimuli (see also Chapter 12). Tyrosine phosphorylation is a post-translational modification of proteins catalyzed by enzymes that transfer phosphoryl to tyrosine residues in protein substrates, using ATP as a phosphate donor. These enzymes are protein tyrosine kinases (TK), of which there are 58 RTKs and 32 nonreceptor types in the human genome. RTKs transduce signals of polypeptide and protein hormones, and growth factors which lead to regulation of critical cellular functions. Cytokine receptors are related to RTKs as their signaling is also propagated via tyrosine phosphorylation of intracellular proteins. Unlike RTKs, the cytokine receptors lack the intrinsic TK domain. Instead, the Janus kinases (JAK) which are noncovalently associated with the receptor, couple tyrosine phosphorylation with ligand binding to the receptor. The RTK and cytokine receptor families have a key role in numerous processes that affect cell proliferation and differentiation, cell migration, and cell cycle control, as well as stimulation of cell growth, modulation of cellular metabolism, and promotion of cell survival and apoptosis.

22.2 RTK FAMILY

The RTK family includes, among others, receptors for epidermal growth factor (EGF), platelet-derived growth factor, fibroblast growth factor, vascular endothelial growth factor, ephrin, insulin, and insulin-like growth factor 1 (IGF-1). Approximately 20 families of RTKs have been classified based on their primary and secondary structure. With most RTKs being composed of one subunit, some exists as multimeric complexes, for example, the insulin receptor (IR) that forms disulfide-linked dimers. All RTKs are single-pass membrane proteins which comprise an extracellular domain containing a ligand-binding site, a single hydrophobic transmembrane α helix and a cytosolic domain that includes a region with protein TK activity (Figure 22.1). Binding of a ligand leads to dimerization of most RTKs, leading to stimulation of the receptor's intrinsic TK activity. The TK of each receptor monomer then transphosphorylates a distinct set of tyrosine residues in the cytosolic domain of its dimer partner, a process termed autophosphorylation. The receptor kinase activity then phosphorylates other sites in the cytosolic domain and intracellular substrates; these phosphotyrosine residues serve as recruitment sites for a host of downstream signaling proteins including enzymes and adapter proteins, typically through Src homology-2 (SH2) or phosphotyrosine-binding domains which recognize phosphotyrosine residues in specific

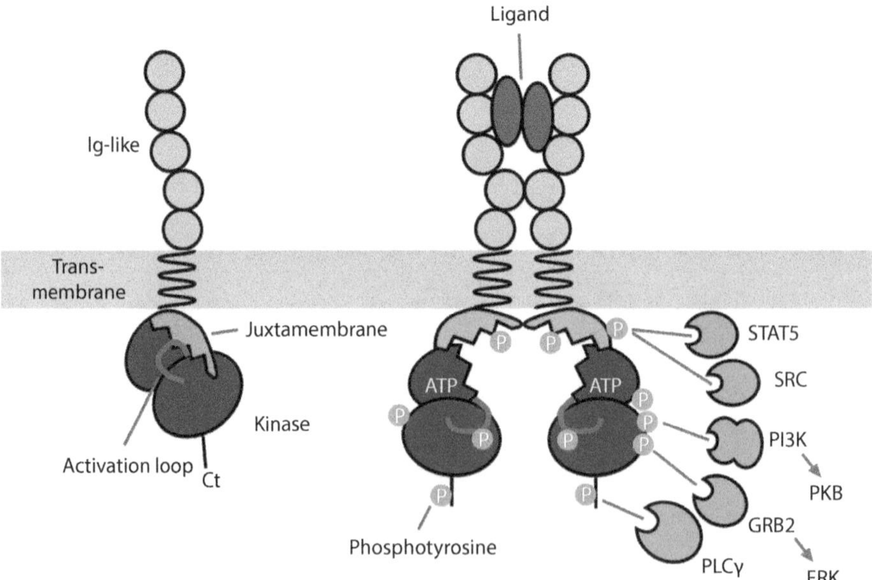

FIGURE 22.1 RTK structure and activation. The RTK prototype is a single-pass membrane protein with an extracellular Ig-like domain containing the ligand binding site, a single transmembrane domain, and a cytosolic domain with a juxtamembrane segment and a kinase containing the activation loop. Ligand binding leads to dimerization of RTK, stimulation of the receptor's intrinsic TK activity, and tyrosine phosphorylation of the activation domain. The TK of each receptor monomer then phosphorylates tyrosine residues in the cytosolic domain of its dimer partner. Phosphotyrosine residues are docking sites for several SH2 domain-containing proteins including STAT5, SRC kinase, PI3K, GRB2, and PLCγ. This leads to activation of a number of downstream signaling pathways including PKB and ERK. (Reproduced from Toffalini, F. and Demoulin, J.-B., *Blood*, 116, 2429, 2010. With permission from American Society of Hematology [HighWire Press].)

sequence contexts. RTK signaling pathways have a wide spectrum of functions that are elicited through regulation of gene transcription, protein synthesis, enzyme activity, and membrane carriers, leading to cell proliferation and differentiation, promotion of cell survival, and modulation of cellular metabolism.

22.2.1 Cytokine Receptors

The cytokine receptor family includes receptors for interleukins, interferon-γ, colony-stimulating factors (CSF), erythropoietin (EPO), thrombopoietin, growth hormone (GH), prolactin, and leptin. They are related to RTK as the mechanism of signal transduction involves TK activation and tyrosine phosphorylation of signaling proteins, but show differences in their molecular structure and cellular actions. Cytokine receptors and their ligands play a role in many physiological processes ranging from postnatal growth, reproduction, and lactation through the regulation of metabolism and body composition, bone formation, neural stem cell activation, to erythropoiesis, myelopoiesis, thrombopoiesis, and the inflammatory response. Cytokine receptors function as ligand-induced or -stabilized homodimers which interact with the JAK family of cytoplasmic TK to participate in signal transduction. EPO, GH, and other cytokine receptors are activated through hormone-induced receptor dimerization and autophosphorylation of JAK kinases that are associated with the cytoplasmic domain of the receptors. The receptor is then phosphorylated at specific tyrosine residues and serves as a docking site, via the SH2 domain, for transcription factors termed STATs (signal transducer and activators of transcription). STATs are then phosphorylated by the JAKs, dimerized, and translocated to the nucleus where they can carry out gene activation, leading to the cellular actions. These receptors primarily use JAK to activate the STAT regulators of gene transcription, but are also able to trigger the mitogen-activated protein kinase (MAPK) pathways, PI3K-AKT (also known as PKB), and other pathways. Cytokine receptors are involved in several pathological conditions including immunodeficiency, autoimmune disorders, myelopathies, anemia, thrombopenia, growth defects, reproductive and metabolic disturbances, as well as cancer.

22.2.2 RTKs as Drug Targets

RTKs as well as nonreceptor TK play a critical role in development and cell growth, and a wealth of information has accumulated implicating deregulated and dysfunctional RTKs in cancer and other diseases. Aberrant RTK activation in human cancers is mediated by four principal mechanisms: autocrine activation, chromosomal translocations, RTK overexpression, and gain-of-function mutations. Sequencing efforts in a wide variety of tumors have identified numerous mutations in RTKs. Consequently, the focus has been to develop RTK antagonists, rather than agonists which promote cell proliferation and survival. Nonetheless, the discovery and development of RTK antagonists have lagged behind the discovery and development of agents that target G-protein-coupled receptors. In part, this is because it has been difficult to discover analogs of naturally occurring RTK agonists that function as antagonists.

Several drugs have been developed for the treatment of cancers and other diseases caused by aberrant RTK activation. These drugs fall into two categories: small-molecule tyrosine kinase inhibitors (TKIs) that target the ATP-binding site of the intracellular TK domain, e.g., imatinib (Gleevec®, see also Chapters 11 and 21), and monoclonal antibodies that both interfere with RTK activation and target RTK-expressing cells for destruction by the immune system, e.g., bevacizumab (Avastin®, see also Chapters 7 and 21). Two generations of small-molecule TKIs targeting RTK as well as nonreceptor TK have been approved for the treatment of human malignancies since the breakthrough with imatinib in 2001. First-generation TKIs like imatinib bind to their target, the catalytic site in the TK domain, through classic competitive binding with ATP, and are fairly selective for their respective TK targets. Second generation of TKIs is now emerging and being introduced into clinical trials. The two most commonly employed strategies are introducing covalent (irreversible) binding of the drug to the RTK target, and kinase multitargeting by broadening the affected RTK targets of the drug within the cell. These novel TKIs appear promising as cancer treatments, and these drugs attempt to improve upon their first-generation predecessors.

Development of RTK agonists for pharmaceutical purposes has attracted sparse interest due to the growth promoting actions and role in cancer induction of many of the RTK family members. The IR is a special case among the RTK in the since it has much more pronounced effects toward metabolic responses, e.g., lowering of blood glucose. However, the discovery of analogs of endogenous RTK ligands that possess partial agonist activity heralds the possibility of clinically relevant ligand-based RTK antagonists. Endogenous ligands for the ErbB RTK family, including the EGF receptor, function as partial agonists which raised the possibility that they may act as antagonists with a therapeutic potential in cancer treatment. The full and partial agonists for a given ErbB receptor share a common binding site on that receptor. Thus, full and partial agonists are expected to compete with each other for receptor binding, and partial agonists are expected to function as competitive antagonists of agonist-induced ErbB receptor signaling. Preliminary data have shown that increasing concentrations of EGF partially antagonize stimulation of EGF receptor coupling to cell proliferation by a fixed concentration of the EGF receptor ligand amphiregulin. The partial agonist/antagonist paradigm may be generally applicable to all RTKs raising the potential of developing a new generation of RTK inhibitors based on their natural ligands.

22.2.3 Cytokine Receptors as Drug Targets

Due to their important role in health and disease, controlling cytokine action remains a major focus of drug discovery efforts. Many cytokines are themselves drugs, and interferons have been used to treat viral or bacterial infections and multiple sclerosis, as well as various forms of cancer. Cytokines activate inflammation and contribute to chronic inflammatory diseases such as rheumatoid arthritis, psoriasis, and Crohn's disease which has led to the development of cytokine antagonists including receptor-blocking antibodies. These are monoclonal antibodies which bind to interleukin receptors, and block their activation and function in several human inflammatory and autoimmune diseases like psoriasis, rheumatoid arthritis, and multiple sclerosis. Examples are daclizumab that binds the α-chain of interleukin-2 receptor, tocilizumab acting as an interleukin-6 receptor antagonist, and brodalumab, an anti-interleukin-17-receptor antibody. Finally, CSF, EPO, and thrombopoietin are used in the treatment of hematological disorders, and GH to treat short stature. Apart from bioactive mimicking peptides of EPO, GH, and interleukins, small-molecule drug design for cytokine receptors, at this time, remains elusive.

22.3 INSULIN RECEPTOR TYROSINE KINASE

Due to the limited pharmaceutical interest for RTK agonists, in this chapter we will use the IR as an example for ligand development for an RTK. Agonists for the IR have obviously been of much interest for the treatment of diabetes. The strategies applied for the IR can, in theory, be applied to all RTKs.

The cell surface receptor for insulin, the IR, was discovered 40 years ago, followed by its classification as an RTK, and the cloning and determination of its primary sequence, as well as of a related homologue, the IGF-1 receptor (IGF1R). These findings suggested that insulin shares its signaling pathways with IGF-1 and IGF-2 which exert their physiological effects via IGF1R, but can also cross-react with IR.

22.3.1 IR Gene

The IR is encoded by the single gene *INSR* located on chromosome 19. In mammals, the IR gene has acquired an additional exon 11, and alternate splicing during transcription results in either IR-A (exon 11 minus) or IR-B (exon 11 plus) isoforms. The most relevant functional difference between these two isoforms is the high affinity of IR-A for IGF-2. The IR isoforms form heterodimers, IR-A/IR-B, as well as hybrids with the IGF1R. The role of the hybrid receptors in physiology and diabetes is not fully understood.

22.3.2 IR Structure

The IR is a tetrameric protein composed of two alpha-subunits and two beta-subunits linked by disulfide bridges. The alpha-subunits are extracellular and contain the insulin-binding site, and the beta-subunits are transmembrane proteins with a cytosolic TK domain (Figure 22.2). The crystal structures of the human IR kinase domain in inactive and active forms have been determined. Similar inactive and active structures of the IGF1R kinase domain have since been determined.

The crystal structure of the extracellular domains of IR and IGF1R was solved over a period of 10 years through the pioneering work of Colin Ward and colleagues. The first breakthrough was the atomic crystal structure of the N-terminal three domains of the α-subunit IGF1R, followed by the X-ray crystal structures of IR ectodomain. Furthermore, a series of structures of insulin bound to various constructs of the IR ectodomain were reported. Finally, a solution structure of the transmembrane domain of human IR using NMR spectroscopy completed the picture. The structure of the IR ectodomain explains many features of insulin binding. Relative to the plasma membrane, the ectodomain has a folded-over conformation in which the two half-receptors lie antiparallel and surround an insulin-binding pocket. The binding of insulin occurs largely at two binding sites named sites 1 and 2. The most significant structural differences between IR and IGF1R are in the regions governing their ligand specificity.

22.3.3 IR Signaling and Actions

Upon binding of insulin to the disulfide-linked heterodimeric IR, the intracellular TK domain is activated. The exact mechanism by which the signal is transduced from ligand binding via the 23 amino acid transmembrane segment to activation of the intracellular kinase is not entirely clear. From X-ray studies, electron microscopic images and other structural information several different models for the activation mechanism have been suggested. The most plausible model is proposed by Dan Leahy and coworkers (Figure 22.3). They propose that in the nonligand bound state the two transmembrane regions of each receptor half are kept apart by more than 100 Å. Upon ligand binding on the outside of the cell at sites 1 and 2, a conformational change results in release of a steric constraint and leads to bending of the extracellular part so that the two transmembrane domains are brought close together. This in turn allows the kinase domains to trans-phosphorylate each other which ultimately allows opening of the autophosphorylation loop and access to the TK for IR substrates (IRS).

Activation of the IR TK mediates the pleiotropic actions of insulin. The two main signaling pathways by which IR and IGF1R regulate metabolism and gene expression, with central roles for the serine kinases AKT and ERK1/2 are well established: the PI3K-AKT pathway which is responsible for most of the metabolic actions of insulin, and the Ras-MAPK pathway which regulates gene expression and cooperates with the PI3K pathway to control cell growth and differentiation. Binding of insulin leads to tyrosine phosphorylation of several intracellular substrates including IRS (principally IRS-1 and IRS-2) and Shc. Each of these phosphorylated proteins serve as docking proteins for other signaling proteins that contain SH2 domains that specifically recognize distinct tyrosine-phosphorylated peptide motifs in proteins including the p85 regulatory subunit of PI3K, the adaptor Grb2, and the tyrosine phosphatase SHP2.

Activation of PI3K and the generation of phosphatidylinositol-(3,4,5)-triphosphate (PIP3), a lipid second messenger, stimulates the PIP3-dependent serine/threonine kinase, PDPK1 and subsequently AKT. The net effect of this pathway is to produce a translocation of the glucose transporter

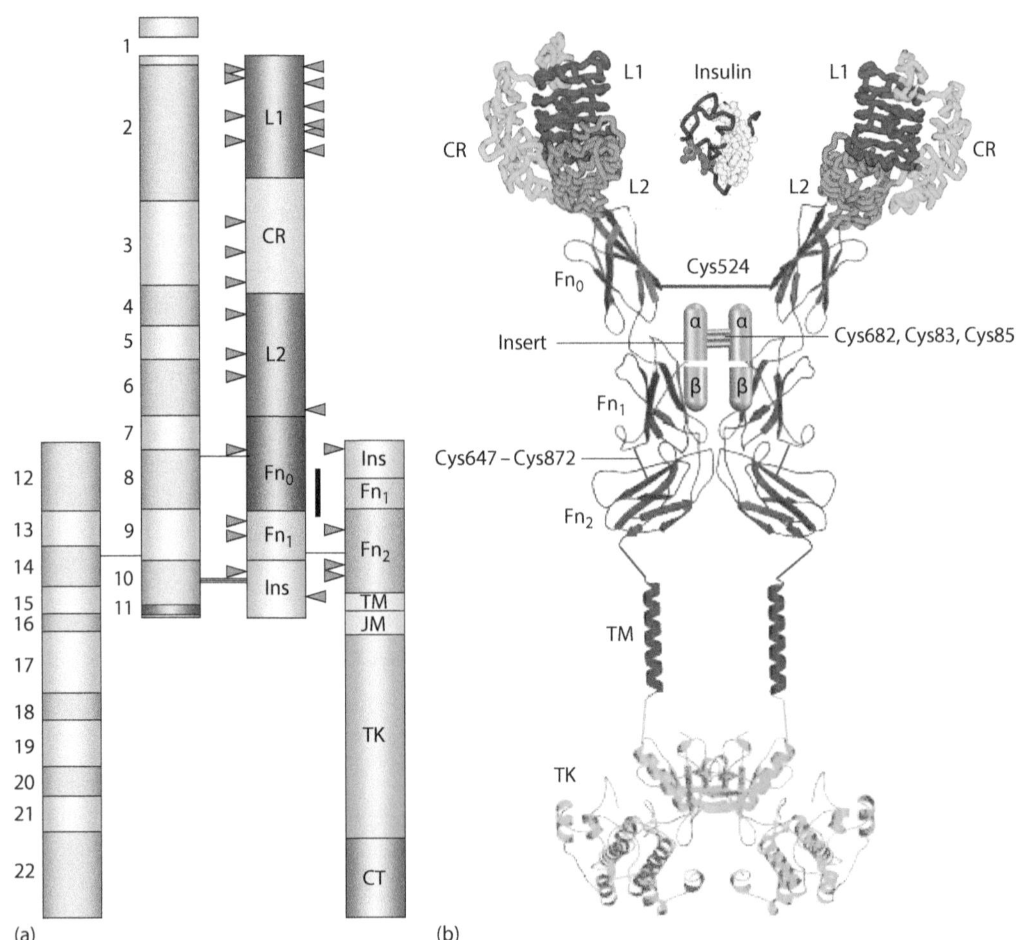

FIGURE 22.2 (a) A cartoon of the $\alpha_2\beta_2$ structure of the insulin receptor (drawn to scale). The left half of the diagram shows the boundaries of the 22 exons of the insulin-receptor gene. The right half of the diagram outlines the predicted boundaries of the protein modules. Module boundaries mainly correspond to exon boundaries. L1 and L2, leucine-rich domains 1 and 2; CR, cysteine-rich domain; Fn_0, Fn_1, Fn_2, fibronectin type III domains; Ins, insert in Fn_1; TM, transmembrane domain; JM, juxtamembrane domain; TK, tyrosine-kinase domain; CT, carboxy-terminal tail. The black bar along Fn_0 indicates the major immunogenic region. Orange arrowheads indicate N-glycosylation sites. Green arrowheads indicate ligand binding "hot spots" that have been identified by single-amino-acid site-directed mutagenesis. The two α-subunits are linked by a disulfide bond between the Cys524 residues in each Fn_0 domain. The three Cys residues at positions 682, 683, and 685 in the insert in the Fn_1 domain are also involved in α–α disulfide bridges. There is a single disulfide bridge between the α- and β-subunits which is formed between Cys647 in the Fn_1 domain and Cys872 in the Fn_2 domain. The numbering used is that of the B-isoform of the insulin receptor. Exon 11 is highlighted in orange. The physiological significance of the alternative splicing of exon 11 is still unclear; it is absent in the insulin-like growth factor 1 (IGF1) receptor which has no equivalent to exon 11. (b) The supra-domain organization of the insulin receptor. The diagram depicts a "stretched-out" model of predicted or actual modular structures (for the L1, CR, L2, and TK domains on the basis of X-ray analysis data). The three-dimensional structure of the insulin molecule is also shown. The classical binding surface is shown in yellow as van der Waals spheres, and Leu A13 and Leu B17 are shown in red. The insulin backbone is shown in blue. (Reprinted by permission from Macmillan Publishers Ltd. *Nat. Rev. Drug Discov.*, 1, 769–783. Copyright 2002.)

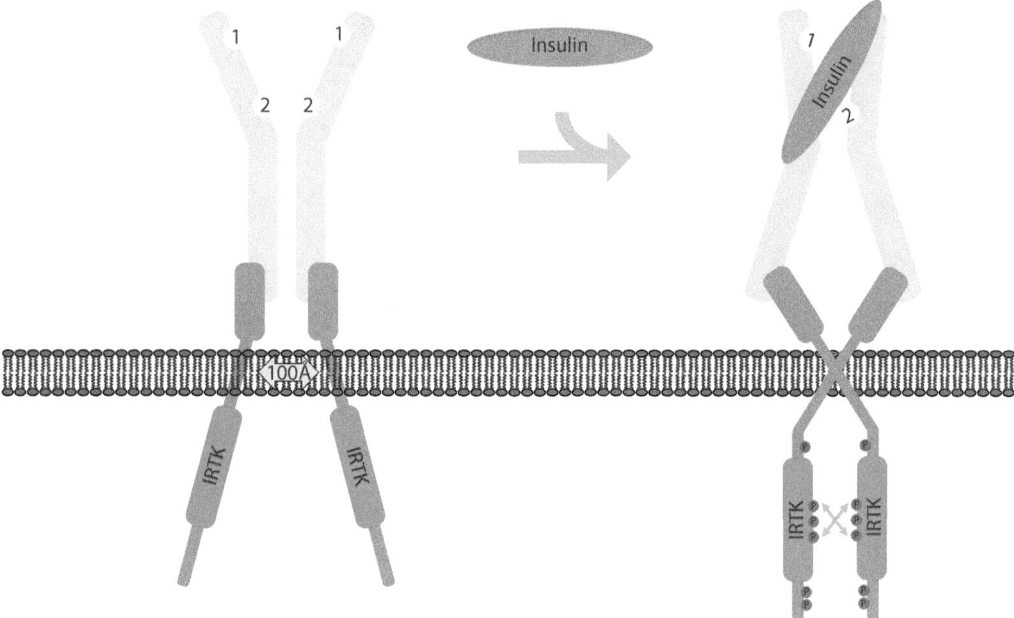

FIGURE 22.3 Upon binding of insulin to the IR, it is suggested that a conformational change takes places which brings the two transmembrane regions in close contact. This in turn brings the two kinase domains so close that the intramolecular trans-phosphorylation can take place.

GLUT4 from cytoplasmic vesicles to the cell membrane to facilitate glucose transport. Moreover, upon insulin stimulation, activated AKT is responsible for the antiapoptotic effect of insulin by inducing phosphorylation of BAD, and regulates the expression of gluconeogenic and lipogenic enzymes by controlling the activity of the winged helix or forkhead (FOX) class of transcription factors. Another pathway regulated by PI3K-AKT activation is mTORC1 which regulates cell growth and metabolism and integrates signals from insulin. AKT mediates insulin-stimulated protein synthesis by phosphorylating TSC2, thereby activating mTORC1 pathway. The Ras/RAF/MAP2K/MAPK pathway is mainly involved in mediating cell growth, survival, and cellular differentiation of insulin. Phosphorylated IRS1 recruits the GRB2/SOS complex which triggers the activation of the Ras/RAF/MAP2K/MAPK pathway.

In addition to binding insulin, the IR can bind IGF-1 and IGF-2. The short isoform IR-A has a higher affinity for IGF-2 binding, and IR-A present in a hybrid receptor with IGF1R binds IGF-1. Hybrid receptors composed of IGF1R and the long isoform IR-B are activated with a high affinity by IGF1, with low affinity by IGF-2, and not significantly activated by insulin, whereas hybrid receptors composed of IGF1R and IR-A are activated by IGF-1, IGF-2, and insulin. However, another study has reported that hybrid receptors composed of IGF1R and IR-A and hybrid receptors composed of IGF1R and IR-B have similar binding characteristics, both bind IGF-1 and have a low affinity for insulin.

22.4 PHARMACOKINETICS OF INSULIN ANALOGS

22.4.1 Why Make Insulin Analogs?

The driving force behind insulin analog development has been a desire to improve treatment of diabetes and increase the convenience for the patients. Even though insulin was a lifesaving drug when it was discovered in the early 1920s by Banting, Best, McCleod, and Collip, it subsequently became clear that insulin injected subcutaneously does not result in the same plasma profile as endogenously produced insulin (Figure 22.4). Endogenously produced insulin is secreted directly into the portal

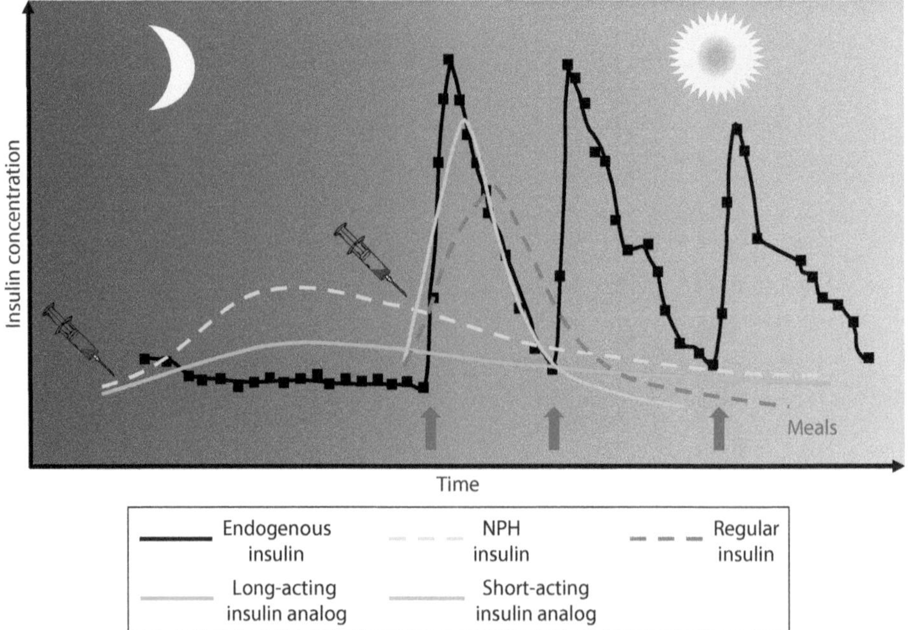

FIGURE 22.4 Endogenous insulin versus insulin injection. The development of insulin analogs (short- and long-acting) has allowed a more physiological profile of injected insulin compared to endogenously produced insulin.

vein and a substantial part of the produced insulin is removed in the liver at the first passage. The rise in plasma insulin in relation to a meal is quite sharp and also the decrease in plasma insulin occurs quite rapid. In contrast, when injected subcutaneously, native insulin results in a more sluggish rise and fall in the plasma concentration. In addition, during night, there is a tendency to have too much insulin resulting in an increased risk for hypoglycemia. In order to overcome these challenges related to treatment with native insulin, short-acting and long-acting insulin analogs have been developed to better mimic the profile of endogenously released insulin (see also Chapter 9).

22.4.2 Short-Acting Insulin Analogs

Native insulin is stable as hexamers in solution and upon injection into the subcutis it has to dissociate into dimers and monomers before entering the bloodstream (Figure 22.5). This process evidently takes some time, and the general principle behind fast action insulin analogs such as aspart, lyspro, and glulisine is to facilitate dimer and monomer formation after injection allowing for a more rapid absorption of monomers from the subcutis. This is done by substituting amino acids which are important for hexamer/dimer formation (Figure 22.6). All commercially available fast-acting insulin analogs are modified at the C-terminal of the B-chain. At the same time, it is important that the insulin analog otherwise retains the efficacy of native human insulin, i.e., ensuring that the safety parameters of the insulin analogs are unchanged. This is covered in more detail in Section 22.6.

22.4.3 Long-Acting Insulin Analogs

The first preparation with a longer duration of action was not an insulin analog, but native insulin formulated as suspension of crystalline zinc insulin combined with a positively charged polypeptide, protamine—NPH insulin. Upon injection, these crystalline preparations

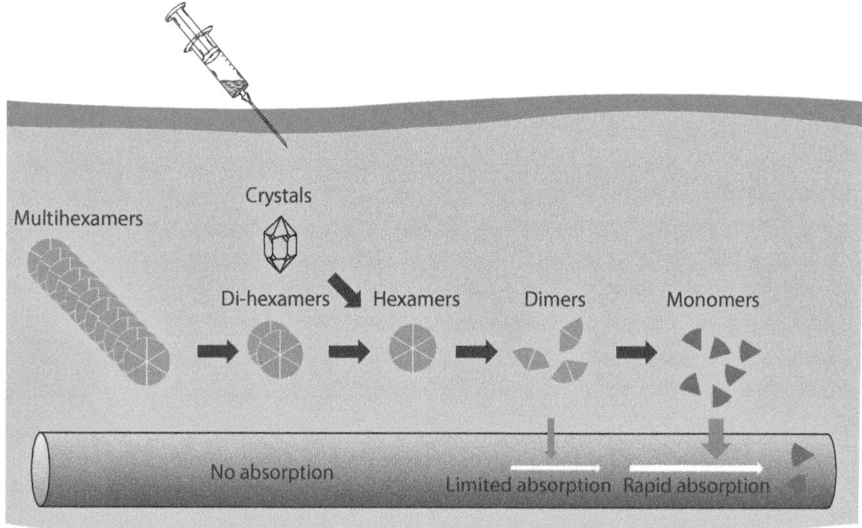

FIGURE 22.5 Absorption of insulin from subcutaneous tissue. Insulin is primarily absorbed in the monomeric form from subcutaneous tissue after injection. Fast-acting analogs are designed to be predominantly monomeric, whereas long-acting analogs make a depot after injecting, ensuring slower absorption.

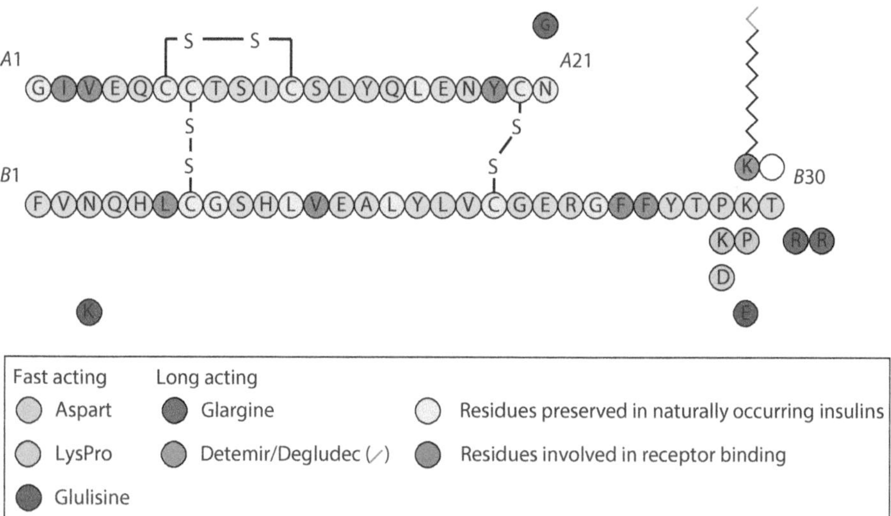

FIGURE 22.6 Insulin primary structure with changes in commercial insulin analogs. Primary structure of insulin and the substitutions in commercial available insulin analogs.

give rise to an intermediate-acting profile with a duration of action of approximately 10–16 hours (Figure 22.4). Although NPH insulin marked a clear improvement compared with native insulin with respect to duration of action, it is unable to provide 24 hours coverage throughout the day, and also there is a substantial risk of nighttime hypoglycemia due to the peaked insulin levels.

Subsequently, insulin analogs with a longer duration of action were developed. The main mechanism used to make long-acting insulin analogs is to protract the absorption from the injection site into the blood stream by increasing the size.

The first long-acting insulin developed took advantage of a relatively low isoelectric point (pH 5.4–6.7), making the molecule more soluble at an acidic pH and less soluble at physiological pH. In response to the increase in pH upon injection into the subcutis, insulin glargine forms crystals which results in a depot from which there is a slow release into the blood stream. The protraction mechanism for insulin detemir and insulin degludec is also depot formation at the injection site. In the case of these acylated insulin analogs, the depot is due to self-association and albumin binding at the injection site. Insulin detemir forms di-hexamers which also bind to albumin while insulin degludec's ultra-long duration of action is attributed primarily to its ability to form very long multihexamers from which monomers are continuously released into the blood circulation. In addition to this depot formation, the acylated insulin analogs also bind to serum albumin after release to the blood stream, making them even more long acting.

In summary, several studies with insulin analogs have shown that they improve not only the treatment outcome by giving a more physiological insulin profile and a more stable blood glucose levels, but also reducing the risk of hypoglycemic episodes. Finally, they have improved the convenience for the patients so they can control their blood glucose with less and more flexible injections.

22.4.4 Alternative Administration

Several trials exploring alternative administration have been conducted to increase the convenience for the patients in order to avoid subcutaneous injections. Thus, nasal, transdermal, buccal, and pulmonary administration have been tested, but have proven to have limited use.

Currently, various oral insulin preparations are undergoing clinical trials and ultimately, if successful, this would allow insulin treatment with a high degree of convenience for the patients. Oral administration of insulin is challenged by both the unfriendly milieu for peptides in the gastrointestinal (GI) tract (acid in the stomach, protein degrading enzymes) and by limited uptake of intact peptides from the GI lumen which makes the absorption of insulin extremely variable from time to time. Another major hurdle is interaction of oral insulin with ingestion of a meal. These hindrances have been tackled by the administration of insulin with absorption enhancers, inhibitors of peptidases, amino acid changes in the insulin molecule, to improve stability and solubility, the use of mucoadhesive polymeric systems (e.g., hydrogels), particulate carrier delivery systems (e.g., liposomes), or combinations of the these strategies.

With the use of these strategies, it is now possible to achieve effects of orally administered insulin in humans and it might be possible to develop insulin tablets in the future.

22.5 ACTIVATION OF THE IR

For almost 100 years, treatment of diabetes has been possible with insulin. The rather crude extracts from dog pancreas have now been substituted by highly purified preparations of genetically engineered human insulin analogs.

Despite these advances, the treatment is still not optimal. Pharmacokinetics can still be improved for type I diabetes patients to mimic endogenous insulin production and the risk of hypoglycemia is still a major problem for the patients. With respect to type II diabetes patients, the same challenges exists, but these patients display also insulin resistance and often require high insulin doses. This in turn increases the unwanted effects like weight gain and general anabolism. Thus, there is still considerable room for improvement in the treatment of both types of diabetes.

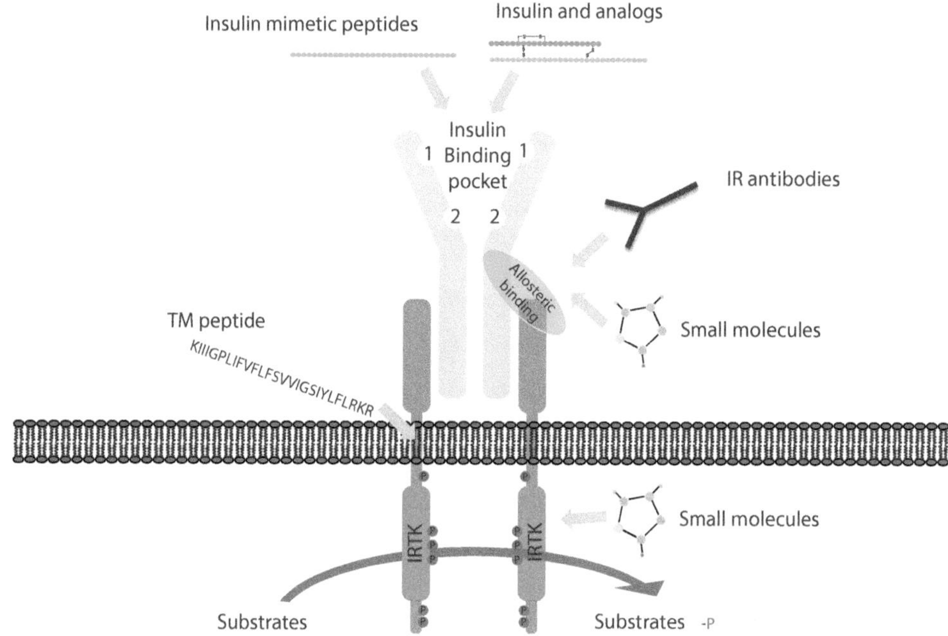

FIGURE 22.7 The possible strategies for activation of the insulin receptor.

Activation of the IR can be achieved by direct or allosteric modulation and/or activation. Since activation of the IR is attractive from a pharmaceutical point of view, much effort has been put into the development of potential drugs which can activate the receptor (Figure 22.7):

- Direct via the insulin binding domain
    - Insulin analogs
    - Small insulin mimetic peptides
- By allosteric modulation or activation
    - IR antibodies
    - Small molecules
    - Transmembrane peptides

22.5.1 Insulin Analogs

As described earlier, several insulin analogs have been produced to improve the pharmacokinetic properties of insulin, making them long- and short-acting, respectively, or to improve the ability to be absorbed through the epithelium. The common feature of these analogs is that, upon binding to the IR, they act as native insulin, giving rise to the same pattern of signals and biological effects.

However, during the last years, there has been an increased effort to design insulin analogs, which, in addition to pharmacokinetic changes, also display changed biological properties getting even closer to endogenous insulin distribution and physiology. Obviously, when the insulin molecule is modified, it is important to keep in mind that the IR and the related IGF1R are RTKs and as such capable of inducing mitogenic signals and promotion of growth. Thus, it is important to keep the safety aspect in mind when designing these new insulin analogs even though the treatment with insulin analogs can be regarded as hormone replacement (Figure 22.8). This issue will be covered in more detail in Section 22.6.

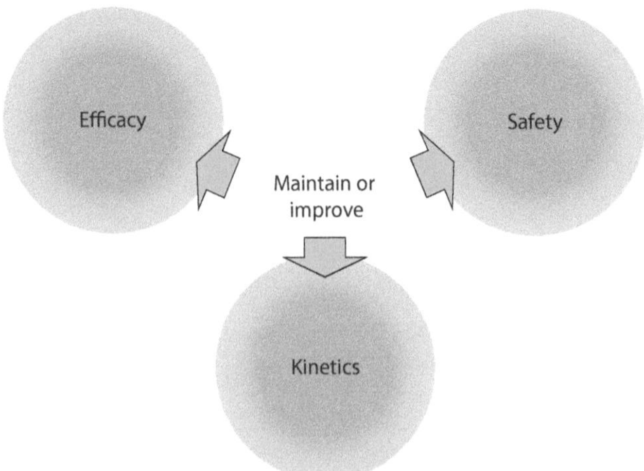

FIGURE 22.8 The relation between efficacy, kinetics, and safety when developing insulin analogs.

22.5.2 Insulin Mimetic Peptides

It was somewhat of a surprise when it turned out to be possible to identify single-chain peptides with no homology to insulin which could bind and activate the IR with high affinity. Initially, these synthetic peptides were identified by a semirational approach by investigation of phage display libraries (ranging in size from 20 to 40 amino acids, see Chapter 9). Peptides were screened for binding to three different hotspots: sites 1, 2, and 3 on the IR. IR sites 1 and 2 also bind insulin, whereas IR site 3 only binds some of the peptides. Subsequently, peptides binding to sites 1 and 2 were combined in various combinations as homo-dimers and hetero-dimers. When these peptides were analyzed further, it was possible to identify the peptide S597, a 27 amino acid single-chain peptide which binds to the human IR with an affinity in the same range as insulin, activates the IR, and lower blood glucose in rats with a potency similar to that of insulin.

Even though S597 binds to the same sites of IR as insulin (sites 1 and 2) and activates the TK, subsequent experiments showed that the downstream signaling is somewhat different. Thus, S597 is much better at inducing phosphorylation of AKT than phosphorylation of ERK compared to insulin. Furthermore, S597 was found to stimulate mitogenic effects quite poorly compared to metabolic effects, and, finally, significant differences between the gene expression patterns were seen between S597 and insulin. Thus, the peptide binds to the receptor via the insulin binding sites, but perhaps induce a somewhat altered conformational change upon binding which then results in a similar but not identical biological response. Potentially, such peptides could be useful in the treatment of diabetes in the future.

22.5.3 IR Antibodies

In the past, several antibodies directed against the IR have been described. Recently, XOMA Corporation has, by screening of a human phage display libraries for allosteric antibodies that activated the IR, identified 2 human monoclonal antibodies that bind allosterically to the IR. None of these antibodies bind to the traditional insulin binding site. The two antibodies differ in the way they activate the receptor. The first one, XMetA, is an allosteric activator and has effects on signaling from the IR by itself, while the second, XMetS, is an allosteric modulator which does not have any major effects in the absence of insulin, but increases insulin binding affinity mainly due to a decrease in the insulin dissociation rate. Interestingly, maximally effective concentrations of XMetA elicit phosphorylation patterns similar to 40–100 pM insulin which are sufficient for robust

Akt phosphorylation and full activation of glucose transport, but have little effect on ERK phosphorylation and mitogenic effects. This selectivity for the metabolic pathway of partial IR agonists like XMetA could turn out to be an advantageous feature of therapeutic agents designed to regulate blood glucose levels while minimizing undesirable outcomes of excessive insulin mitogenic stimulation. Recently, XOMA Corporation further showed that XMetA efficiently lowers blood glucose in animals and can improve glycemic control in mice with diet-induced obesity. In addition, they also showed that this antibody due to its size might be liver-preferential upon injection since the much larger size could lead to a more rapid access into liver, an insulin-sensitive tissue with well-fenestrated capillaries, when compared to other insulin-sensitive tissues with nonfenestrated capillaries, such as muscle and adipose. It is believed that insulin analogs with a preference for the liver could be advantageous due to a more physiological distribution of the insulin.

22.5.4 Small Molecules

Several small molecules with insulin mimetic properties have been identified. These include both compounds that bind to the extracellular part of the IR (e.g., a chaetochromin derivative 4548-G05) and compounds that by directly binding to the receptor kinase domain trigger its kinase activity (e.g., 5,8-diacetyloxy-2,3-dichloro-1,4-naphthoquinone). Several of these can, in addition to their insulin mimetic effects in vitro, also lower blood glucose in vivo by oral administration. In addition to compounds, which activates the IR by direct binding, it is theoretically possible to mimic the insulin response by inhibition of the protein tyrosine phosphatase (PTPase) which dephosphorylate the IR. Several researchers and companies have tried to develop antidiabetes drugs by identifying specific PTPase inhibitors. However, one major problem with this approach is that the specific PTPase responsible for the dephosphorylation of the IR is unknown.

These compounds have the possibility of being developed into orally available drugs for the treatment of diabetes with much increased convenience for the patients. However, the small-molecule approach for IR activation is hampered by the fact that tyrosine phosphorylation in general is related to growth and many cancer types are associated with increased tyrosine phosphorylation. Therefore, the requirement for specificity for this type of compounds is extremely high. Based on this observation, it would seem that compounds, which bind to the IR outside of the kinase domain, will have a better chance of being specific since the homology between the kinase domains of RTK as well as non-receptor TK is quite high.

22.5.5 Transmembrane Peptides

Finally, an alternative way of activating the IR has been suggested, in which peptides with a sequence identical to the transmembrane domain of the receptor can result in an insulin mimetic response. It has been proposed that the two transmembrane domains of the IR in the resting state are bound to each other and that upon ligand binding they are separated which then subsequently bring the two kinase domains into close proximity. The effect of the transmembrane (TM) peptides is proposed to be to induce a separation of the two transmembrane domains by binding to the domain itself. Since the transmembrane sequence differs substantially between the IR and other TK receptors, in theory it should be possible to obtain specificity with this approach. However, this model contradicts the well-documented model for IR activation suggested by Leahy and colleagues and it requires quite high concentrations of peptide before an effect is seen. However, it cannot be excluded that peptides that interact with the transmembrane domain could have a potential to activate the IR or other RTKs.

In summary, several approaches to activate the IR are feasible. The most promising from a pharmaceutical point of view are drugs that interact with the IR outside the TK domain in order to increase the specificity and avoid cross-activity with other RTKs. Currently, two possibilities exist: ligands which bind to the insulin-binding domain of the IR, or antibodies which interact with IR.

Whether any of these approaches will lead to new drugs to treat diabetes remains to be seen, but they certainly holds promise for developing new and improved treatments.

22.6 SAFETY

As discussed in Section 22.1, the IR belongs to the family of RTKs and as such has the potential to mediate growth signals. Certainly, cell biologists are aware of this, since many cell culture media are supplemented with insulin to aid cell growth.

22.6.1 Insulin X10 Lesson

The potential risk of promoting tumor growth with insulin treatment became clear when the first insulin analog insulin X10 was taken into the clinic almost 25 years ago. Upon subsequent toxicological testing in animals with very high doses, it turned out that this insulin analog resulted in a dose-dependent development of benign and malignant mammary tumors in female Sprague–Dawley rats. The finding was unexpected and the clinical program of insulin X10 was terminated accordingly.

The obvious question was how a single amino acid change (B10His -> B10Asp) could result in an insulin analog with such a dramatic change in the biological effects.

The purpose of the substitution in insulin X10 was to make the analog monomeric and therefore fast-acting (as described in Section 22.2), but it turned out that this substitution also changed the molecular characteristics of the insulin in several ways. First, it was shown that insulin X10 had an increased binding to the related IGF1R. In relative terms, it was approximately six times more potent than human insulin with respect to binding to the IGF1R. However, this was still much lower than binding of IGF-1 itself, and it was believed to be of minor importance. Another interesting feature of insulin X10 was the slower dissociation rate from the IR. This was accompanied by sustained signaling from the receptor. So, upon binding to the IR, insulin X10 sticks to the receptor for a much longer time and signals for a longer time. These two molecular mechanisms result in an insulin analog with increased mitogenic potency in both primary cell cultures and immortal cell cultures of mammary and other origins (Figure 22.9).

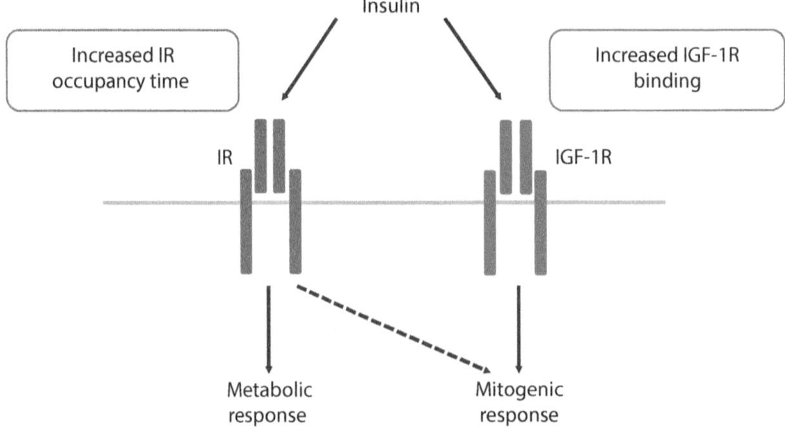

FIGURE 22.9 Increased mitogenic potency of insulin analogs (IR vs. IGF1R). Both increased binding to the IGF-1 receptor and sustained signaling from the insulin receptor can result in insulin analogs with increased mitogenic potency.

Interestingly, sustained signaling from the IGF1R is also seen with IGF-1 as a ligand. In other words, insulin X10 results in a signal from the IR which is similar to the signal from the IGF1R after stimulation with IGF-1. Since insulin X10 possesses both increased binding to the IGF1R and sustained signaling from the IR, it seems advisable to be cautious with respect to both characteristics when developing new insulin analogs since one or both mechanisms could be responsible (Figure 22.9). Insulin X10 is now recommended by the European Agency for the Evaluation of Medicinal Products (EMA) as the positive control in insulin analog mitogenicity studies.

22.6.2 Insulin Glargine Lesson

The issue of the potential risk of promoting tumor growth with insulin treatment remained primarily a scientific discussion for many years, but the interest was very much increased when four epidemiological studies published in June 2009 highlighted a potential link between insulin glargine (A21Gly, B31Arg, B32Arg) insulin and an increased incidence of cancer, stirring up an ongoing debate about the long-term clinical safety of insulin analogs. Insulin glargine has an increased binding to the IGF1R comparable to that of insulin X10 which then consequently fueled the discussion as to whether the increased IGF1R affinity could in fact play a role despite having a quite low affinity compared to the affinity of IGF-1 itself. On the other hand, insulin analogs do not bind to IGF-1 binding proteins. Consequently, the free and effective concentration of insulin analogs might be of relatively more importance than the free and effective concentration of IGF-1.

Fortunately, it turned out that insulin glargine upon injection rapidly is transformed in vivo into two metabolites: (A21Gly)insulin and (A21Gly, des-B30Thr)insulin with IGF1R binding comparable to human insulin and subsequent large-scale prospective studies have also revealed no increased risk of cancer with insulin glargine. Consequently, the original interpretations of the four epidemiological studies are now considered to be flawed and treatment with insulin glargine considered as safe. This, on the other hand, means that we still do not know whether an increase in IGF1R binding comparable to what is seen for insulin X10 holds a risk in the clinic since insulin glargine can be regarded as an insulin analog pro-drug which, upon in vivo transformation after injection merely is an insulin analog with IGF1R binding similar to that of human insulin.

22.6.3 Molecular Mechanisms behind Increased Mitogenicity

In order to be able to distinguish between the two possible mechanisms underlying the increased mitogenic potency of insulin analogs shown in Figure 22.7, several experimental insulin analogs were produced. Hence, insulin analogs with sustained signaling from the IR, but without the increased binding to the IGF1R, and insulin analogs with increased binding to the IGF1R with no sustained signaling from the IR, were made. With the use of these analogs, and different cell types with different IR and IGF1R expression levels, it was possible to show that each of these mechanisms independent of each other could lead to an insulin analog with increased in vitro mitogenic potency.

The relation between increased mitogenic potency in vitro and increased tumor growth in vivo is not clear. Theoretically, an insulin analog with an increased ability to stimulate cell division in vitro would potentially have a higher risk of stimulating tumor growth in vivo. However, the situation in vivo is obviously much more complex than in vitro with circulating growth factors and paracrine regulation. Also, when studying in vitro mitogenicity it is standard procedure to serum starve the cells for several hours/days and perform the measurement in the presence of a very low serum concentration to enhance the signal. Therefore, it is not entirely clear whether the mitogenic signals we can measure in vitro have any predictive value in vivo.

22.7 CONCLUDING REMARKS

Several attempts have been made to develop antagonists toward various RTKs to treat malignancies associated with mutations in RTKs or their signaling pathway. These endeavors have both been rewarding and disappointing. Several new drugs to treat various cancer forms with good therapeutic effect have been developed. In contrast, there have only been a relatively few successful attempts to make inhibitors for the IGF1R.

Agonists for RTKs have a limited therapeutic potential because of the inherent risk of initiating or promoting cancer. However, partial agonists might be a useful route since these could block the action of endogenous growth factors.

The IR is more of a special case when it comes to therapeutics since agonists of the IR will help lower blood glucose in diabetic patients. Thus, substantial efforts have been invested to develop IR activating ligands and several of these have the potential to be developed into new drugs in the future.

FURTHER READING

De Meyts, P. 2015. Insulin/receptor binding: The last piece of the puzzle? *Bioessays* 37:389–397.

Du, Y. and Wei, T. 2014. Inputs and outputs of insulin receptor. *Protein Cell* 5(3):203–213.

Evans, M., Schumm-Draeger, P.M., Vora, J., and King, A.B. 2011. A review of modern insulin analogue pharmacokinetic and pharmacodynamic profiles in type 2 diabetes: Improvements and limitations. *Diab. Obes. Metab.* 13:677–684.

Hubbard, S.R. and Miller, W.T. 2015. Closing in on a mechanism for activation. *eLife* 2014;3:e04909.

Issafras, H., Bedinger, D.H., Corbin, J.A. et al. 2014. Selective allosteric antibodies to the insulin receptor for the treatment of hyperglycemic and hypoglycemic disorders. *J. Diab. Sci. Technol.* 8(4):865–873.

Lemmon, M.A. and Schlessinger, J. 2010. Cell signaling by receptor tyrosine kinases. *Cell* 141:1117–1134.

Siddle, K. 2011. Signalling by insulin and IGF receptors: Supporting acts and new players. *J. Mol. Endocrinol.* 47:R1–R10.

Sousa, F., Castro, P., Fonte, P., and Sarmento, B. 2015. How to overcome the limitations of current insulin administration with new non-invasive delivery systems. *Ther. Deliv.* 6(1):83–94.

Waters, M.J. and Brooks, A.J. 2015. JAK2 activation by growth hormone and other cytokines. *Biochem. J.* 466:1–11.

23 Antibiotics

Smitha Rao C.V., Piet Herdewijn, and Anastassios Economou

CONTENTS

23.1 INTRODUCTION

23.1.1 History and Definition

Prior to Fleming's discovery of an antibacterial substance produced by the fungus *Penicillium rubens** in 1929, several authors discussed the production of antibacterial substances by microorganisms and their therapeutic potential. However, all these products were either too toxic for systemic application or low in activity (e.g., pyocyanase from *Pseudomonas aeruginosa* [1889]). H. Florey and coworkers isolated penicillin and demonstrated its activity and innocuity in animals and humans. This led S.A. Waksman to examine his *Streptomyces* collection, and in 1944, his group announced the discovery of streptomycin. Others quickly followed this "natural product approach" for finding antibiotics. A chemical approach had emerged in parallel, even earlier.

P. Ehrlich, while working in the late 1800s with dyes that selectively stained certain bacteria and not animal cells, conceived the idea of finding "a magic bullet"—a dye for selectively killing bacteria without harming the host. He discovered arsphenamine (Salvarsan) in 1910, the first chemical agent active against syphilis and trypanosomiasis, albeit with serious side effects. His work inspired G. Domagk in discovering the antibacterial property of a red dye, prontosil—the first chemotherapeutic agent for general infections (sulfa drug; 1936).

P. Vuillemin introduced the term "antibiotic" (from the Greek words *anti*, meaning "against," and *bios*, meaning "life") to describe antagonism between living organisms (1889). Later, Waksman defined antibiotic as a chemical substance produced by microorganisms that has the capacity to inhibit the growth and even to destroy bacteria and other microorganisms in dilute solutions (1942). Although Waksman's definition of antibiotics still remains largely accepted, it is also debated since "antibiotic" in contemporary use does not always adhere to the aforementioned criteria. Currently, many related definitions of antibiotics exist (Box 23.1).†

Here, we discuss both classical antibiotics and synthetic molecules introduced in human medicine as antibacterials. For simplicity, we use "antibiotics" broadly to describe natural/synthetic/semisynthetic antibacterial drugs.

23.1.2 Sources of Antibiotics

Microorganisms are a rich source of antibiotics with almost 29,000 classical antibiotics isolated from them. For clinical applications, microorganism-derived natural products or their synthetic mimics remain the primary source of antibiotics. Of all the FDA-approved antibacterials, 69% of the new molecular entities originate from natural products‡ with 97% isolated or derived from microbes. *Streptomyces*, *Penicillium*, and *Cephalosporium* are among the heavily represented genera.

Synthetic (man-made) antibiotics in therapeutic use—the sulfa drugs, quinolones, oxazolidinones, and diarylquinolines—were in fact discovered outside of antibacterial discovery programs (except diarylquinolines).

---

* *P. rubens* was previously misidentified as *P. chrysogenum*.

† Note: Antibiotics does not include antimicrobial compounds such as disinfectives (mostly used on inanimate surfaces), antiseptics (applied on the skin and mucous membranes), toxins/poisons (unsuitable for use). Antibiotics are generally systemic (i.e., have an effect on the whole body) and have specific molecular targets, whereas disinfectants are non-specific, act on multiple targets and are broadly toxic. However, "topical antibiotics" (applied on skin), fall in the borderline as they fulfil Waksman's definition but are toxic only when consumed internally.

‡ Natural product comprise unmodified natural materials or compounds, semisynthetic derivatives, or synthetic structures which were conceptually derived from a natural product.

BOX 23.1 DEFINITION OF TERMS

a. *Gram⁻/Gram⁺*: Gram staining, a staining procedure introduced by C. Gram in 1884, distinguishes bacteria into two types based on their cell wall structure. Bacteria that retain the dye crystal violet after solvent treatment are Gram-positive ($Gram^+$) while those that do not are Gram-negative ($Gram^-$). The ability of $Gram^+$ bacteria to retain the crystal violet dye after solvent treatment (in Gram staining) is due to their thick and multilayered peptidoglycan layer. A smaller group of $Gram^+$ bacteria, including *Mycobacterium tuberculosis*, retain the Gram stain even after acid washing and are called acid-fast bacteria.

b. *Antibiotics* (examples of the contemporary definitions; see text for Walksman and Ehrlich's definitions)
   1. The term antibiotic is used as a synonym for antibacterials used to treat bacterial infections in both people and animals (WHO, 2011).
   2. Antibiotics are drugs of natural or synthetic origin that have the capacity to kill or inhibit the growth of microorganisms (FAO, 2005).
   3. Antibiotics, also known as antimicrobial drugs, are drugs that fight infections caused by bacteria. They are not effective against viral infections like common cold, most sore throats, and the flu. (FDA, 2015).

c. *Bactericidal antibiotics* cause bacterial cell death.

d. *Bacteriostatic antibiotics* inhibit bacterial growth without any loss in viability.

e. *Broad-spectrum antibiotics* are active on a large number of bacterial species.

f. *Narrow-spectrum antibiotics* are active against a small number of bacterial species.

g. *Antibiotic resistance* is the acquired ability of bacteria to survive and multiply despite the presence of therapeutic levels of one or more antibiotics.

h. *Multidrug-resistant* (*MDR*) is defined as acquired nonsusceptibility to at least one agent in three or more antimicrobial categories.

i. *Extensively drug-resistant* (*XDR*) is defined as nonsusceptibility to at least one agent in all but two or fewer antimicrobial categories (i.e., bacterial isolates remain susceptible to only one or two categories).

j. *Pandrug-resistant* (*PDR*) is defined as nonsusceptibility to all agents in all antimicrobial categories.

k. *Antimicrobial resistance mechanisms*: Given the limited number of targets on which the known classes of antibiotics act, resistance mechanism are also well developed. The major types of clinically relevant resistance mechanisms are (1) modification of target structures/overproduction of target (2) enzymatic modification/inactivation of the antibiotic/elimination through the efflux pumps/decreased penetration of the antibiotic (3) bypassing a particular step in a pathway.

l. *Combination therapy* involves the use of two or more agents that (1) inhibit different targets in different pathways (e.g., the cocktail of antituberculosis drugs [see Table 23.3]), (2) inhibit different steps in the same pathway (e.g., sulfamethoxazole/trimethoprim), and (3) inhibit the same target in different ways (e.g., streptogramins).

23.1.3 Bacteria and the Need for New Drugs

Antibiotic resistance crept along ever since the wide and successful use of antibiotics. The WHO identified antibiotic resistance as one of the three greatest threats to human health.* Nosocomial (i.e., hospital-acquired) infections caused by multi- or pan-drug-resistant (Box 23.1) pathogens (e.g., *Staphylococcus aureus*, *Klebsiella pneumonia*, and *Pseudomonas aeruginosa*) are of serious concern; so is the emergence of extensively drug-resistant *Mycobacterium tuberculosis* strains.

23.2 ANTIBIOTICS IN CLINIC

From 1938 to 2013, the FDA had approved 155 antibacterial new molecular entities, but less than 62% of them are used today (Figure 23.1). Most antibiotics introduced in the past 30 years are semi-synthetic or synthetic natural product derivatives of previously approved antibiotics.

Currently, all the approved antibiotic classes target one of the five essential bacterial processes (Figure 23.1). As the time between the introduction of a new antibiotic into the clinic to the time when resistance is first observed is getting alarmingly short, there is a dire need for new antibiotics.

Usable antibiotics act on targets that are either absent or differ significantly from their eukaryotic counterparts. While the direct effect of the drug–target interaction is known, the associated cellular response mechanisms that contribute to the death of bacterial cells is a subject of ongoing research.

23.3 ANTIBIOTICS AFFECTING BACTERIAL CELL WALL FORMATION

Most prokaryotic cells are surrounded by a cell wall, responsible for their shape and their ability to survive in hypotonic environments. $Gram^+$ bacteria are surrounded by a plasma membrane and a thick cell wall consisting of peptidoglycan to which teichoic acids (polyol phosphate polymers) are linked. $Gram^-$ bacteria have a much thinner cell wall and an outer membrane comprising lipid, lipopolysaccharide, and protein (Figure 23.2). Mycobacterial cell wall is unique, more complex, and a formidable barrier to antibiotics.

The bacterial cell wall is a good target because it is an essential component, structurally and functionally conserved across bacterial species, absent from humans (no adverse effects), and is readily accessible to exogenously added compounds. Many of the antibiotics targeting the cell wall act by inhibiting enzymes or sequestering the substrates involved in peptidoglycan assembly and crosslinking.

Peptidoglycan forms an extensive polymer, also called "murein sacculus" (Latin murus: wall) which surrounds the entire cell. Peptidoglycan is a polymer of alternating *N*-acetylglucosamine (GlcNAc) and *N*-acetylmuramic acid (MurNAc), crosslinked by a short peptide bridge. MurNAc is the 3-*O*-D-lactylether of *N*-acetylglucosamine. GlcNAc and MurNAc are linked by β-(1,4) glycosidic bonds and form a linear structure. The carboxyl group of MurNAc is linked to a short peptide which forms a bridge with another GlcNAc-MurNAc strand (Figure 23.3).

23.3.1 β-Lactam Antibiotics

This class of antibiotics is the biggest (in terms of FDA-approved members as well as market sales), safe, and time-tested as reflected by widespread use for almost seven decades. Members of this class share the common β-lactam (four-membered cyclic amide) ring structure (Table 23.1) which is the active pharmacophore. Their mode of action is indicated in Table 23.2.

* http://www.who.int/medicines/areas/priority_medicines/BP6_1AMR.pdf.

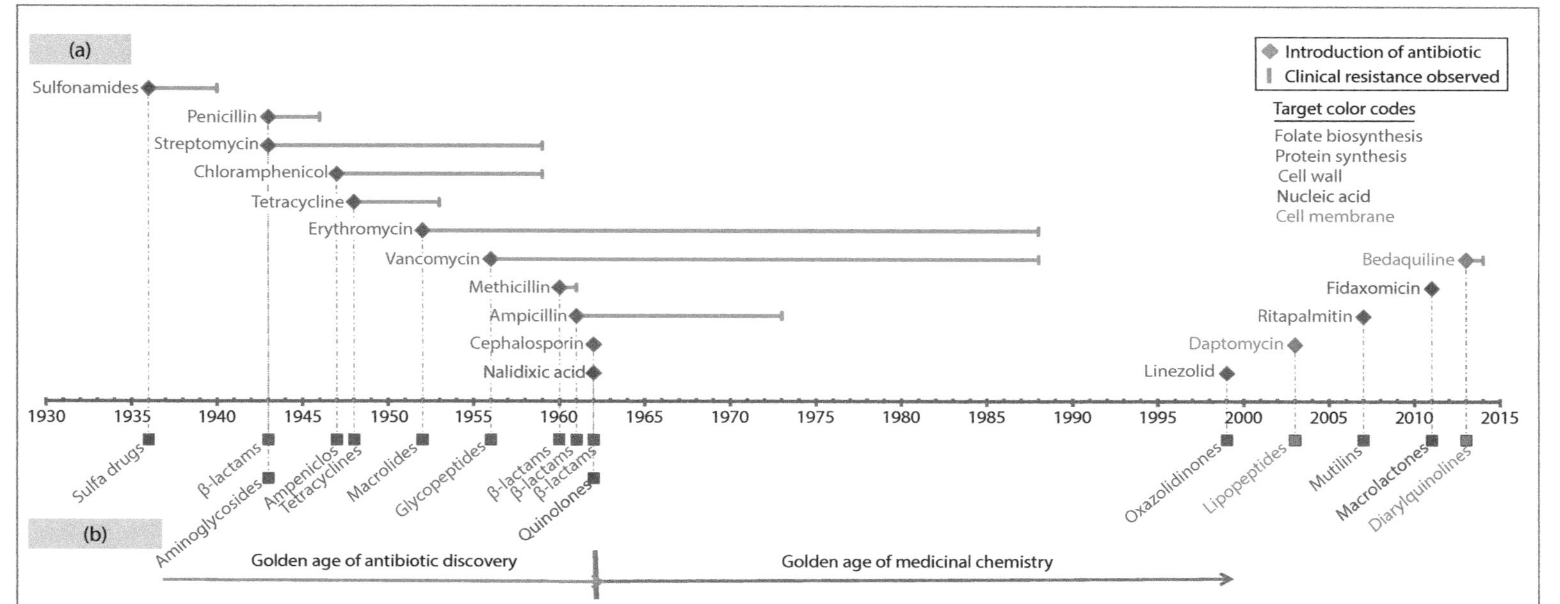

FIGURE 23.1 Antibiotic introduction and resistance timeline: The year of clinical introduction of key antibiotics is shown (a) along with their structural class (b). The horizontal bars indicate the duration between introduction and the first observation of clinical resistance. Absence of the horizontal bars indicates that resistance was observed in the year the antibiotic was introduced. Antibiotics are color-coded to indicate the essential bacterial process they target. (Data partly from Walsh, C.T. and Wencewicz, T.A., *J. Antibiot.*, 67, 7, 2014.)

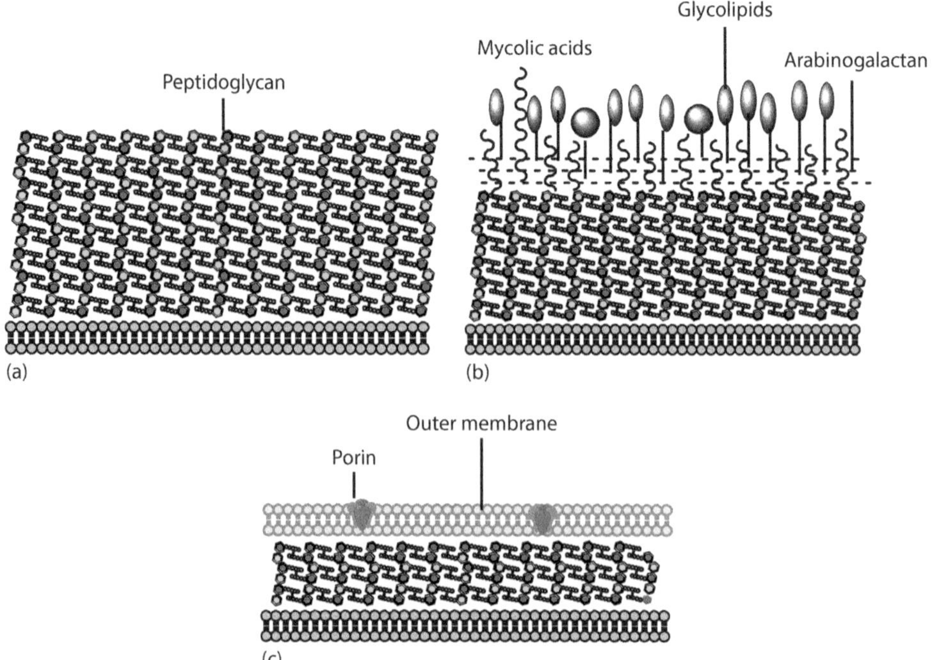

FIGURE 23.2 Schematic representation of the structure of the bacterial cell walls. (a) Gram$^+$ bacteria have a thick peptidoglycan layer surrounding the plasma membrane. (c) Gram$^-$ bacteria have a thin peptidoglycan layer that is surrounded by a lipid-rich bilayer outer membrane. Sandwiched between the plasma membrane and the outermembrane in Gram$^-$ bacteria is a concentrated gel-like matrix (periplasm) present in periplasmic space. (b) The mycobacterial cell wall is much thicker than that of other bacteria and is hydrophobic. It consists of peptidoglycan, arabinogalactan, and mycolic acids covalently linked together to form a complex that extends from the plasma membrane outward in layers, starting with the peptidoglycan and ending with mycolic acid.

23.3.1.1 Penicillins

England and the United States made significant efforts to produce sufficient quantities of penicillin and elucidate its structure (1942–1945). Addition of corn steep liquor increased penicillin production yielding penicillin G that has a benzyl side chain. When substituted with phenoxyethanol as precursor, phenoxymethyl penicillin (penicillin V) was obtained. Penicillin V was unexpectedly acid stable and was introduced as oral penicillin. Penicillin G and V are the first-generation penicillins of narrow spectrum (Gram$^+$ streptococci and staphylococci and Gram$^-$ gonococci but not against Gram$^-$ rods).

Preparation of other penicillins was vastly increased by the discovery of 6-aminopenicillanic acid (6-APA) which is the penicillin nucleus without any side chain attached. Batchelor et al. (1959) isolated 6-APA from fermentation media with no precursor added but the yield was very low. 6-APA could also be obtained by enzymatic removal of the side chain of penicillin by penicillin acylase (1960) or by chemical cleavage (1970).

Semisynthetic penicillins: Methicillin and oxacillin were stable to early penicillinase (an enzyme that opens the β-lactam ring and confers resistance to this class of drugs). Several derivatives followed. Amoxicillin (aminopenicillin; Table 23.1) and piperacillin (ureidopenicillin) are currently important members with improved antibacterial spectrum.

Although resistance to penicillins is widely prevalent, they continue to be the first-line antibiotics.

23.3.1.2 Cephalosporins

Brotzu discovered antibacterial activity of *Cephalosporium acremonium* (1948). Abraham et al. isolated cephalosporin P (1951), a steroid antibiotic related to fusidic acid and cephalosporin N (1954),

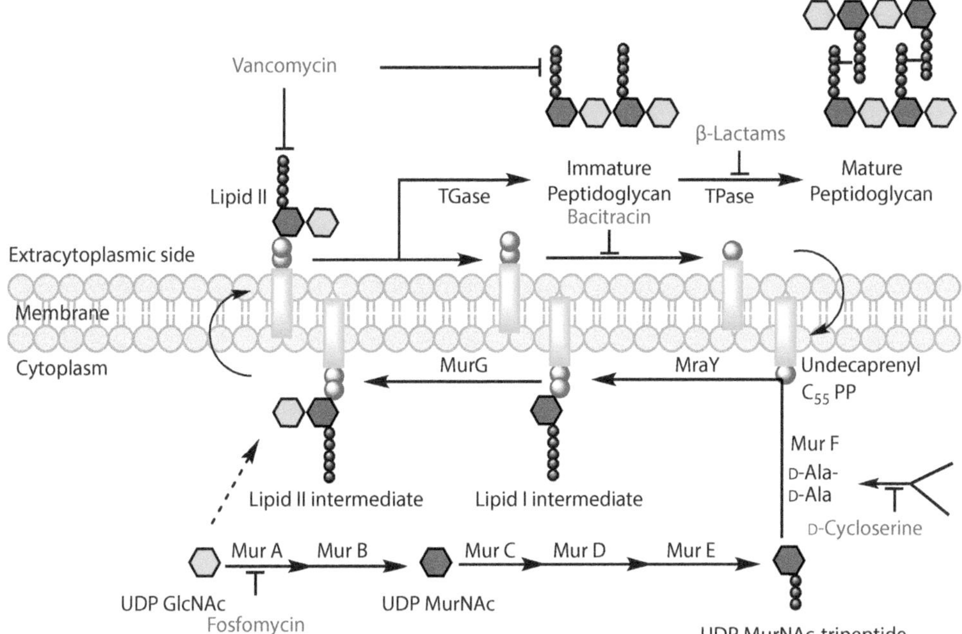

FIGURE 23.3 Stages in peptidoglycan biosynthesis inhibited by different antibiotics. The four stages of peptidoglycan biosynthesis are (1) synthesis of UDP-*N*-acetylmuramic acid (MurNAc) involving enzymes Mur A and Mur B and (2) building the pentapeptide side chain to the carboxyl group resulting in UDP-muramyl pentapeptide, accomplished by the enzymes Mur C-F, in the cytoplasm. (3) Membrane-bound reactions leading to a linear peptidoglycan polymer involve two lipid intermediates: lipid I, formed on the interior of the membrane, is converted to lipid II (by enzymes MraY and MurG, respectively) which is then flipped across the membrane and finally (4) peptide crosslinking of linear peptidoglycan by transpeptidases (TPase) occurs on the extracytoplasmic side. Antibiotics inhibiting the pathway are shown (blue text), their structures and mode of action are given in Tables 23.1 and 23.2, respectively. GlcNAc is *N*-acetylglucosamine; TGase is transglycosylase. Note that the figure shows the extracytoplasmic side in Gram$^+$ bacteria while this region is the periplasm in Gram$^-$ bacteria.

later called penicillin N which is a penicillin with D-α-aminoadipic acid as a side chain. Acid inactivation of this penicillin led to another antibiotic, cephalosporin C (1955). It has a D-α-aminoadipic acid side chain but a β-lactam-hydrothiazine ring system (instead of the thiazolidine cycle in penicillin). This product was acid stable and resistant to penicillinase. Chemical removal of the side chain from cephalosporin C resulted in 7-aminocephalosporinic acid (7 ACA), the cephalosporin nucleus.

Semisynthetic cephalosporins: Cephalosporins are often classified in generations. This classification is more or less related to the year of introduction and their properties. The initial focus of semisynthesis was to increase the pharmacokinetics and spectrum of activity of cephalosporins, but it later shifted to combating β-lactam resistance.

The first-generation cephalosporins (e.g., cefalothin, 1962) are narrow spectrum, used in Gram$^+$ infections. Second- and third-generation cephalosporins (e.g., cefuroxime and ceftazidime [Table 23.1], respectively) progressed toward broader spectrum. They are more effective against Gram$^-$ bacteria, but less against Gram$^+$ bacteria compared to their respective previous generations. Cephamycins, β-lactams with a methoxy group on C_7, were discovered in certain *Streptomyces* strains (1971). Fourth-generation cephalosporins (e.g., cefepime and cefpirome) possess potent broad-spectrum activity, though not against methicillin-resistant *Staphylococcus aureus* (MRSA). Due to their ability to penetrate into the cerebrospinal fluid, the fourth-generation drugs are effective for treating meningitis.

TABLE 23.1
Representative Structures of Antibiotics Affecting the Cell Wall

β-Lactams	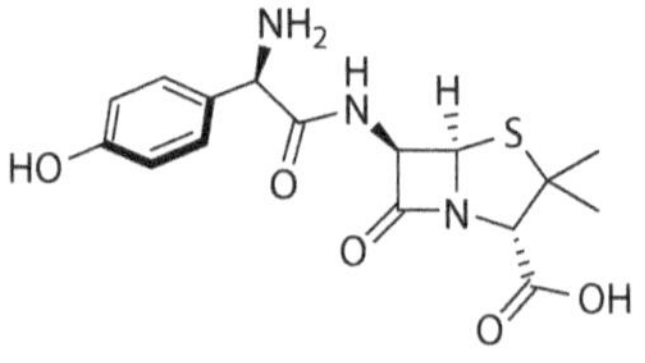 Amoxicillin (penicillin)	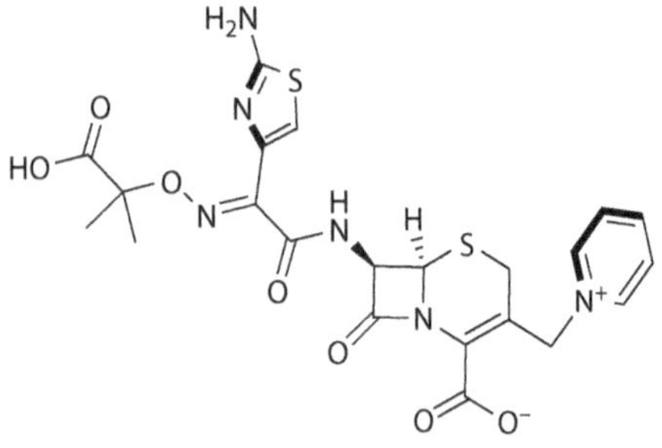 Ceftazidime (cephalosporin)
Nontraditional β-lactams	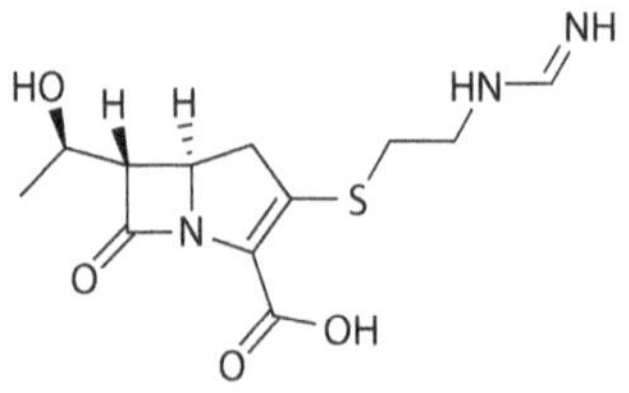 Imipenem (carbapenem)	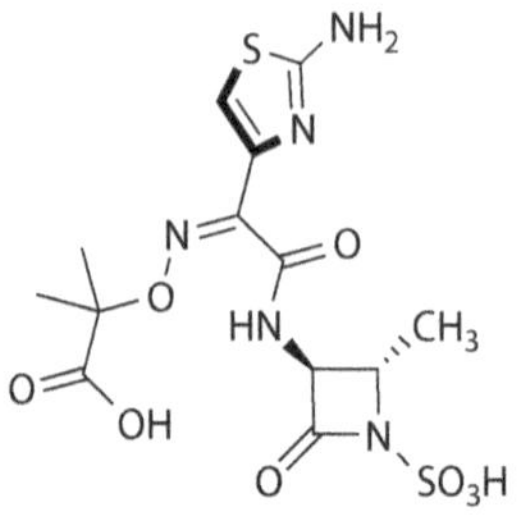 Aztreonam (monobactam)
β-Lactamase inhibitors	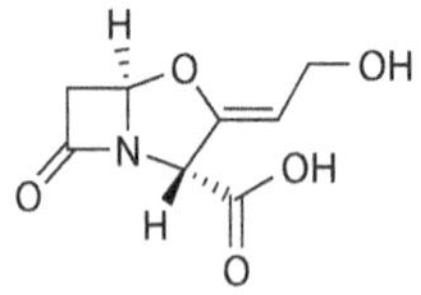 Clavulanic acid	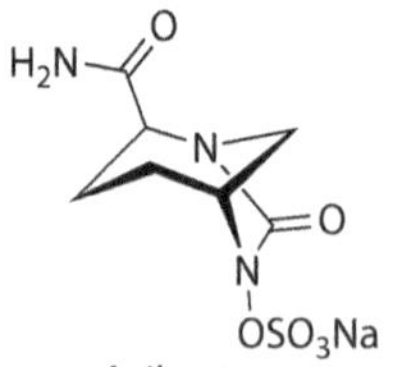 Avibactam
Polypeptide	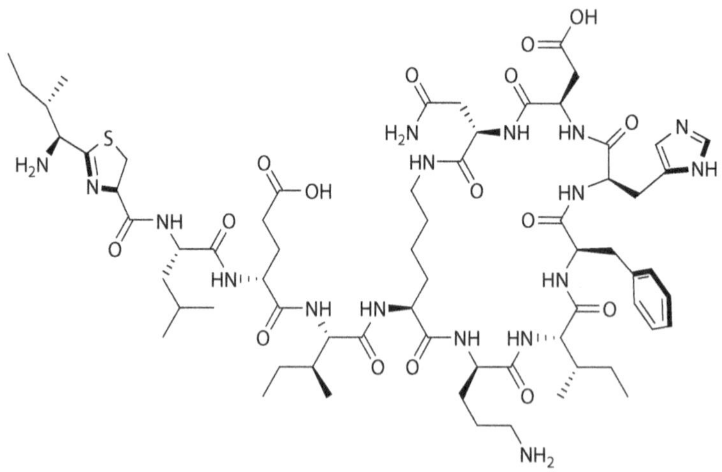 Bacitracin A	

(Continued)

TABLE 23.1 (*Continued*)
Representative Structures of Antibiotics Affecting the Cell Wall

Glycopeptide	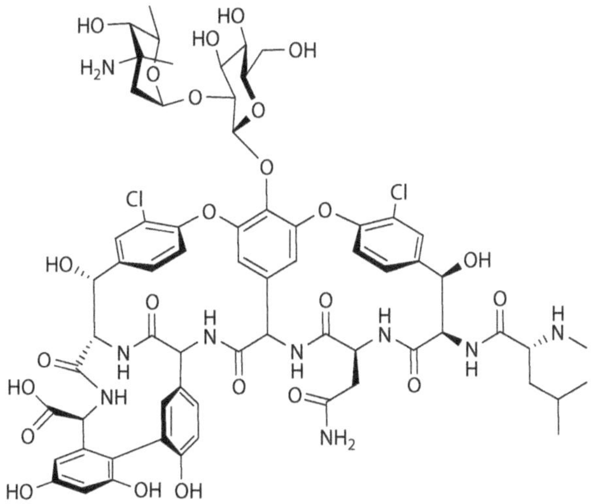 Vancomycin
Others	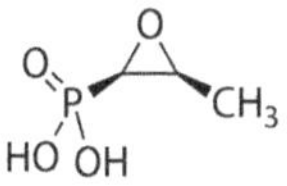 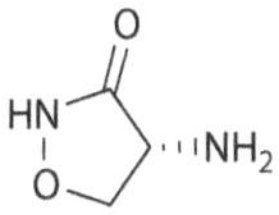 Fosfomycin (phosphonic acid derivative); D-Cycloserine (D-alanine analog)

Fifth-generation cephalosporin, ceftaroline (FDA approved, 2010), is highly effective against MRSA but not against a few multidrug-resistant (MDR) Gram⁻ bacteria. Nonetheless, its synergy with a β-lactamase inhibitor, tazobactam (see Section 23.3.1.3), restores its potency against many resistant strains.

One of the promising new derivatives is Ceftolozane (currently in late-stage clinical trials). Broad-spectrum activity and excellent safety profiles of cephalosporins make them one of the most widely used first-line drugs.

23.3.1.3 β-Lactamase Inhibitors

β-Lactamase inhibitors by themselves have weak antibacterial activity, but they enhance the action of conventional β-lactam antibiotics against β-lactamase-producing bacteria. Clavulanates and penicillin sulfones are irreversible "suicide substrates" that inactivate β-lactamases through secondary chemical reactions in their active site.

Clavulanates: Clavulanic acid (1976) (Table 23.1), the first β-lactamase inhibitor, is a β-lactam produced by the same actinomycete *Streptomyces clavuligerus* that produces cephamycin C. It is a potent inhibitor of many β-lactamases. The combination antibiotic co-amoxiclav (Table 23.3) is still one of the most successful antibiotics in the market. Clavulanic acid/ticarcillin combination is effective against aerobic Gram⁻ bacteria.

Penicillin sulfones: Penicillanic acid sulfone (sulbactam) and its prodrug pivsulbactam are combined with ampicillin, extending the antibacterial spectrum of the latter. Tazobactam possess broader spectrum of activity than clavulanic acid. A combination of tazobactam with piperacillin (e.g., Zosyn by Pfizer, U.S.) is particularly effective against some *P. aeruginosa* infections.

TABLE 23.2
Antibiotics: Mode of Action and Spectrum

Antibiotic Class	Member	Spectrum	Mode of Action
Antibiotics affecting cell wall synthesis			
β-Lactams	Amoxicillin	Broad spectrum	They block peptide crosslinking in the peptidoglycan polymer by inhibiting transpeptidases (penicillin binding proteins)
Polypeptides	Bacitracin	$Gram^+$	Intradicts the C_{55} lipid carrier cycle by interfering with dephosphorylation of the C_{55}-isoprenyl pyrophosphate
Phosphonic acid derivatives	Fosfomycin	Broad spectrum	Inactivates the enzyme MurA (enolpyruvyl transferase), blocking peptidoglycan precursor synthesis in the cytoplasm
Analogs of D-alanine	D-Cycloserine	Broad spectrum (second-line anti-TB drug)	Inhibits the enzymes D-Ala-D-Ala racemase (converts L-alanine to D-alanine) and alanine ligase (catalyzes ATP-dependent peptide bond formation between two molecules of D-alanine), involved in peptidoglycan precursor synthesis in the cytoplasm
Glycopeptides	Vancomycin	$Gram^+$	They block cell wall crosslinking by binding to the C-terminal D-Ala-D-Ala dipeptide of the nascent peptidoglycan
Antibiotics affecting cell membranes			
Lipopeptides (cationic)	Polymyxin B	$Gram^-$	Interacts with lipopolysaccharide on the outer membrane, permeabilizes the outer membrane and also damages the plasma membrane acting as a detergent
Lipopeptides (anionic)	Daptomycin	$Gram^+$	Detailed mode of action is still unclear Daptomycin possibly inserts into the lipid bilayer in a phosphatidylglycerol-dependent fashion, generating pores and leading to potassium efflux, membrane depolarization, and mislocalization of proteins
Diarylquinolines	Bedaquiline	Mycobacteria (*Mycobacterium tuberculosis*)	Decreases ATP levels by inhibiting mycobacterial ATP synthase enzyme on the membrane
Antibiotics affecting nucleic acid synthesis			
Rifamycins	Rifampicin	$Gram^+$	Inhibits bacterial transcription by binding to the bacterial RNA polymerase
Macrolactones	Fidaxomicin	$Gram^+$ (*Clostridium difficile* infections)	Inhibits the initiation of bacterial RNA synthesis by binding to RNA polymerase
Coumarins	Novobiocin	$Gram^+$, a few strains of $Gram^-$ bacteria	Inhibits DNA synthesis by inhibiting ATPase activity of DNA gyrase
Quinolones	Ciprofloxacin	Broad spectrum	Inhibits DNA synthesis by stabilizing the "cleavage complex" of DNA gyrase

(Continued)

TABLE 23.2 (*Continued*)
Antibiotics: Mode of Action and Spectrum

Antibiotic Class	Member	Spectrum	Mode of Action
Antibiotics affecting metabolic pathways			
Sulfa drugs	Sulfanilamide	$Gram^+$	Blocks nucleic acid precursor synthesis by inhibiting the enzyme dihydropteroate synthase in the folic acid pathway. Folic acid acts as a coenzyme in the synthesis of purines and pyrimidines
Antibiotics affecting protein biosynthesis[a]			
Aminoglycosides	Streptomycin	Broad spectrum	Binds to 30S ribosomal subunit; prevents formation of initiation complex, induces miscoding during protein synthesis
Tetracyclines/glycylglycines	Chlortetracycline/ tigecycline	Broad spectrum	Binds to 30S ribosomal subunit; inhibits aminoacyl-tRNA binding to the ribosome Tigecycline has a higher affinity for the ribosome; mode of action is same as tetracycline
Amphenicols	Chloramphenicol	Broad spectrum	Binds to 50S ribosomal subunit; targets the PTC[b] and inhibits peptide bond formation
Lincosamides	Clindamycin, lincomycin	$Gram^+$	(same as above; see Amphenicols)
Pleuromutilins	Retapamulin	$Gram^+$	(same as above; see Amphenicols)
Oxazolidinones	Linezolid	$Gram^+$	Binds to 50S ribosomal subunit; targets the PTC[b] and inhibits the first peptide bond formation
Macrolides	Erythromycin	Broad spectrum	Binds within the exit tunnel of 50S ribosomal subunit, thus blocking the exit of nascent polypeptides
Ketolides	Telithromycin	$Gram^+$ and some $Gram^-$ aerobes	(same as above; see Macrolides)
Streptogramins	Streptogramin B	$Gram^+$	(same as above; see Macrolides)
Fusidane	Fusidic acid	$Gram^+$ bacteria and $Gram^-$ cocci	Prevents dissociation of bacterial elongation factor G (EF-G:GDP) from the ribosome

Note: $Gram^+$, Gram-positive; $Gram^-$, Gram-negative bacteria.

[a] Mode of action of ribosome-targeting antibiotics.

[b] PTC denotes peptidyl-transferase center, the site of peptide-bond formation on the large subunit, comprising universally conserved residues of domain V of the 23S ribosomal RNA.

TABLE 23.3
Antibiotics in Combination Therapy

Antibiotic: Indication	Combination	Pathogen(s)
Co-amoxiclav: Lower respiratory tract infections, otitis media, skin and skin structure infections, urinary tract infections	Amoxicillin (β-lactam) + clavulanate potassium (β-lactamase inhibitor), restores bacterial susceptibility to the cell wall targeting β-lactam antibiotics	Gram$^+$ and Gram$^-$ aerobic bacteria and some anaerobes (e.g., *Escherichia coli*, *Enterococcus faecalis*, *Haemophilus influenza*, *S. aureus*, *Streptococcus peneumoniae*)
Co-trimoxazole: Pneumonia, bronchitis, urinary tract infections, acute otitis media, traveler's diarrhea	Trimethoprim + sulfamethoxazole (1:5 ratio), synergize in inhibiting tetrahydrofolate synthesis	Gram$^+$ and Gram$^-$ aerobic bacteria
Synercid (Pfizer): complicated skin and skin structure infections	Quinupristin (streptogramin B) + dalfopristin (streptogramin A), 30:70 w/w, synergizes in inhibiting bacterial protein synthesis	Methicillin-resistant *Staphylococcus aureus* (MRSA)
Antituberculosis (DOTS[a] regimen): Tuberculosis	Isoniazid (inhibits enoylreductase subunit of fatty acid synthase) + rifampicin (inhibits RNA polymerase) + ethambutol (inhibits arabinosyl transferases involved in cell wall biosynthesis) pyrazinamide (whose mechanism of action is not clear as yet); + streptomycin (protein synthesis inhibitor)	*Mycobacterium tuberculosis*

[a] DOTS, Directly observed treatment system.

Diazabicyclooctanes: The latest (third-generation) β-lactamase inhibitors are the non-β-lactam, diazabicyclooctanes (DBOs). An example is the synthetic molecule, Avibactam (Table 23.1) with γ-lactam core (a five-membered ring) bearing an O-sulfate substituent on the nitrogen. Avibactam/ceftazidime (third-generation cephalosporin) combination was approved (by the FDA in 2014) for the treatment of complicated urinary tract infections and intra-abdominal infections. Two other DBOs (MK-7655 and NXL105) are under clinical trials.

23.3.1.4 Carbapenems

2-Carbapenem is a bicyclic system, with a double bond between C_2 and C_3. Thienamycin was obtained from cultures of *Streptomyces cattleya* (in 1978). It is not stable and is used as the *N*-formimidoyl derivative (imipenem). Thienamycin is cleaved by dehydropeptidase present in the kidney. Hence, imipenem is coadministered with dehydropeptidase inhibitor, cilastatin. Imipenem (Table 23.1) is a broad-spectrum antibiotic, resistant to most β-lactamases. Meropenem, doripenem, and panipenem/betamipron are examples of this subclass. Although carbapenems remain an important subclass of antibiotics, no new member has progressed through the approval stage in the past decade.

23.3.1.5 Monobactams

Monobactams are monocyclic β-lactams. A product SQ26.180 was isolated from *Chromobacterium violaceum* and a related product, sulfazecin, was discovered in cultures of *Gluconobacter* sp. and *Pseudomonas* sp. (in 1981). Activity was improved by modification of the amide side chain. Introduction of the cefotaxime side chain gave an interesting new drug, Aztreonam (Table 23.1). It is the first and currently only FDA-approved monobactam. Aztreonam is specific to and highly active against Gram$^-$ bacteria, including *P. aeruginosa*. It is generally well tolerated and frequently used in patients unable to tolerate β-lactams. Baselia's triple combination (BAL30376) shows potent in vitro activity against Gram$^-$ bacteria including those resistant to aztreonam and related monobactams.

23.3.2 Bacitracin

Bacitracin was first isolated from a strain of *Bacillus subtilis* in 1945. The same product was discovered in the culture of *B. licheniformis* A5 in 1949 (but called Ayfivin). Bacitracin (or ayfivin) is a mixture in which the type A component predominates (Table 23.1). The mode of action of this nonribosomal decapeptide antibiotic is indicated in Table 23.2. Bacitracin is effective against Gram$^+$ pathogens. It is generally given topically and not orally due to high nephrotoxicity.

23.3.3 Fosfomycin

Fosfomycin, a phosphonic acid derivative isolated from *Streptomyces fradiae* and other *Streptomyces* species in 1969, is a three-carbon epoxy propyl phosphonate, with a low-molecular weight (138 g/mol) (Table 23.1). Fosfomycin is a broad-spectrum bactericidal, has low incidence of adverse reactions, and synergizes with β-lactams or aminoglycosides (see Table 23.2, mode of action).

23.3.4 Cycloserine

Of the two enantiomers of cycloserine, D-cycloserine (or seromycin), a natural product isolated from *Streptomyces garyphalus* and other *Streptomyces* species (in 1955), is used as an antibiotic. D-cycloserine ((*R*)-4-amino-1,2-oxazolidin-3-one) (Table 23.1) is a cyclic (or structural) analog of D-alanine. Although it inhibits two enzymes (see Table 23.2, mode of action), the activity of D-cycloserine is weak (bactericidal or bacteriostatic) and nonspecific. It is currently used as a second-line antibiotic for tuberculosis but has serious neurological side effects arising from its ability to penetrate the central nervous system.

23.3.5 Vancomycin

Glycopeptide antibiotics are glycosylated nonribosomal peptides produced by a diverse group of actinomycetes. They are also called "dalbaheptides" due to their mode of action (D-alanine-D-alanine binding) and their core structure (heptapeptides) (Tables 23.1 and 23.2).

Vancomycin and teicoplanin are the two first-generation (natural product) glycopeptides of which the former is in clinical use for over 50 years. They are used as drugs of last resort against life-threatening infections caused by Gram$^+$ pathogens. Teicoplanin has a greater potency against streptococci and enterococci but is not approved for use in the United States.

Vancomycin was isolated from *Streptomyces orientalis* in 1956. Although approved in 1958, its usage peaked during the 1980s (as MRSA emerged). Initially, vancomycin (known by the pseudonym "Mississippi mud") was associated with severe adverse effects—nephrotoxicity, ototoxicity, and hypersensitivity—attributed to impure preparations. These were alleviated as the purity of vancomycin increased (to ~95%).

Telavancin, a semisynthetic derivative of vancomycin has a lipophilic side chain (decylamino-ethyl), and a negatively charged phosphonomethyl aminomethyl group. Telavancin disrupts the plasma membrane by interacting with the lipid II precursor which explains its higher potency over vancomycin. Dalbavancin and oritavancin are the other second-generation products of vancomycin, in late-stage clinical trials.

Teicoplanins are lipoglycopeptides produced by *Actinoplanes teichomycelicus* (1978). Although similar to vancomycin in its activity, teicoplanin has increased potency against *Streptococcus* sp. and *Enterococcus* sp. It is effective against some vancomycin-resistant *S. aureus*, has better pharmacokinetic properties, and reduced occurrence of nephrotoxicity compared to vancomycin. The lipophilic nature of teicoplanin is the likely reason for its superior properties as also seen in the case of second-generation glycopeptides.

23.4 ANTIBIOTICS TARGETING THE PLASMA MEMBRANE

The plasma membrane of bacteria is a dynamic lipoprotein structure. It consists of major phospholipids phosphatidylethanolamine (PE), phosphatidylglycerol (PG), and their respective derivatives: mono-, bi-, or trimethylated forms of PE and a dimeric form of PG (diphosphatidylglycerol). Several proteins (also enzymes) are located in and around this membrane. While the plasma membrane is a good target that is easily accessible to exogenously added drugs (especially in Gram$^+$ bacteria), there are very few antibiotics that selectively target it without harming human membranes.

23.4.1 Polymyxins

Polymyxins are pentacationic cyclic lipodepsipeptides (Figure 23.4), discovered in 1947 from different strains of *Bacillus polymyxa.* Colistin, isolated in 1950 in Japan from *B. colistinus,* was identical with polymyxin E. Only polymyxin B and colistin (which appeared least toxic) have been used in medicine. Polymyxin B and E are mixtures containing several components differing in the structure of the fatty acid. The most important component is B_1 (or E_1), with some 15%–25% of B_2 (or E_2). Polymyxin and colistin are rapidly bactericidal on Gram$^-$ bacteria (see Table 23.2, mode of action). Polymyxin B and E, introduced for the treatment of infections caused by Gram$^-$ bacteria like *Pseudomonas*, were abandoned for systemic therapy 30 years ago due to neurotoxic and nephrotoxic side effects and were limited to topical use. Polymyxin was reinstated over a decade ago as the last-line intravenous therapy for infections caused by Gram$^-$ strains resistant to practically all other agents. The positive aspect is that, since polymyxin reintroduction, the observed nephrotoxicity is less severe in patients while neurotoxicity is rare and reversible when the medicine is discontinued. Improper dosage of polymyxin/colistin is one of the reasons cited for the observed discrepancy.

Novel polymyxin derivatives with three positive charges (NAB7061 and NAB741) instead of five are currently in phase I clinical trials. Although these are not very potent, they sensitize Gram$^-$ bacteria to other antibiotics by facilitating their entry inside the cell.

23.4.2 Daptomycin

Daptomycin (Figure 23.4), a fermentation product of *Streptomyces roseosporus* (discovered in 1985), is used for complicated skin and skin structure infections (FDA approved, 2003). Its precise mode of action is unclear (Table 23.2). Its ability to permeabilize membranes containing PG, abundant in bacterial plasma membrane and not in human cell membranes, is the most likely reason for its selective toxicity. Daptomycin is generally not active against Gram$^-$ bacteria due to its inability to penetrate the outer membrane effectively. Surotomycin (CB-183,315; phase III clinical trials), an analog of daptomycin, shows very good activity against anaerobic Gram$^+$ bacteria, including *Clostridium difficile.*

23.4.3 Diarylquinolines

Diarylquinolines are structurally and mechanistically different from both fluoroquinolones and other quinolines. A major structural difference is the specificity of the functionalized lateral (3′) chain borne by the diarylquinoline class (Figure 23.4). Bedaquiline (J & J and the TB Global Alliance) obtained fast-track approval from the FDA (2012) for combination therapy for the treatment of MDR TB in adults.

Bedaquiline was identified from a series of (synthetic) diarylquinolines in a whole-cell screen against *Mycobacterium smegmatis*. It is effective against both active as well as dormant Mycobacteria (see Table 23.2, mode of action). Balemans et al. (2012) have demonstrated that medicinal chemistry optimization can broaden the spectrum of diarylquinoline activity to include Gram$^+$ pathogens as well.

	R_1	R_2
Polymyxin B_1	$-C_2H_5$	$-C_6H_5$ (D-Phe)
Polymyxin B_2	$-CH_3$	$-C_6H_5$ (D-Phe)
Polymyxin E_1	$-C_2H_5$	$-CH(CH_3)_2$ (D-Leu)
Polymyxin E_2	$-CH_3$	$-CH(CH_3)_2$ (D-Leu)

Polymyxin (Lipopeptide)

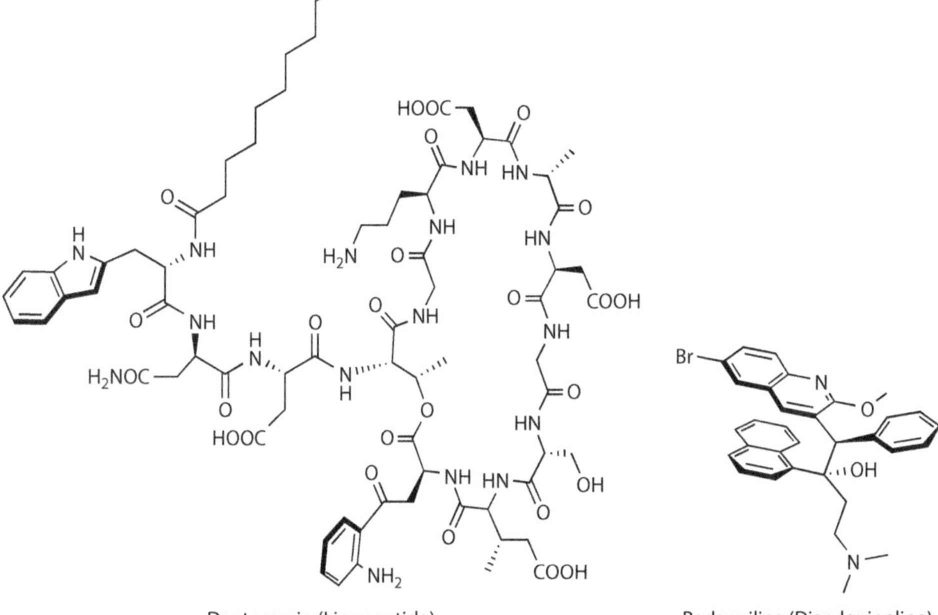

Daptomycin (Lipopeptide)

Bedaquiline (Diarylquinoline)

FIGURE 23.4 Representative structures of antibiotics targeting the plasma membrane.

23.5 ANTIBIOTICS AFFECTING NUCLEIC ACID SYNTHESIS

23.5.1 Antibiotics Inhibiting RNA Polymerase

RNA polymerase (RNAP) is a nucleotidyl transferase enzyme that catalyzes the formation of RNA from a DNA template. Bacterial RNAPs are highly conserved and differ from their eukaryotic/archaeal/viral counterparts, allowing selective inhibition. RNAP is essential for growth and survival of bacteria. Rifampicin and Fidaxomicin bind at different subsites on RNAP and act by different mechanisms (Figure 23.5; Table 23.2).

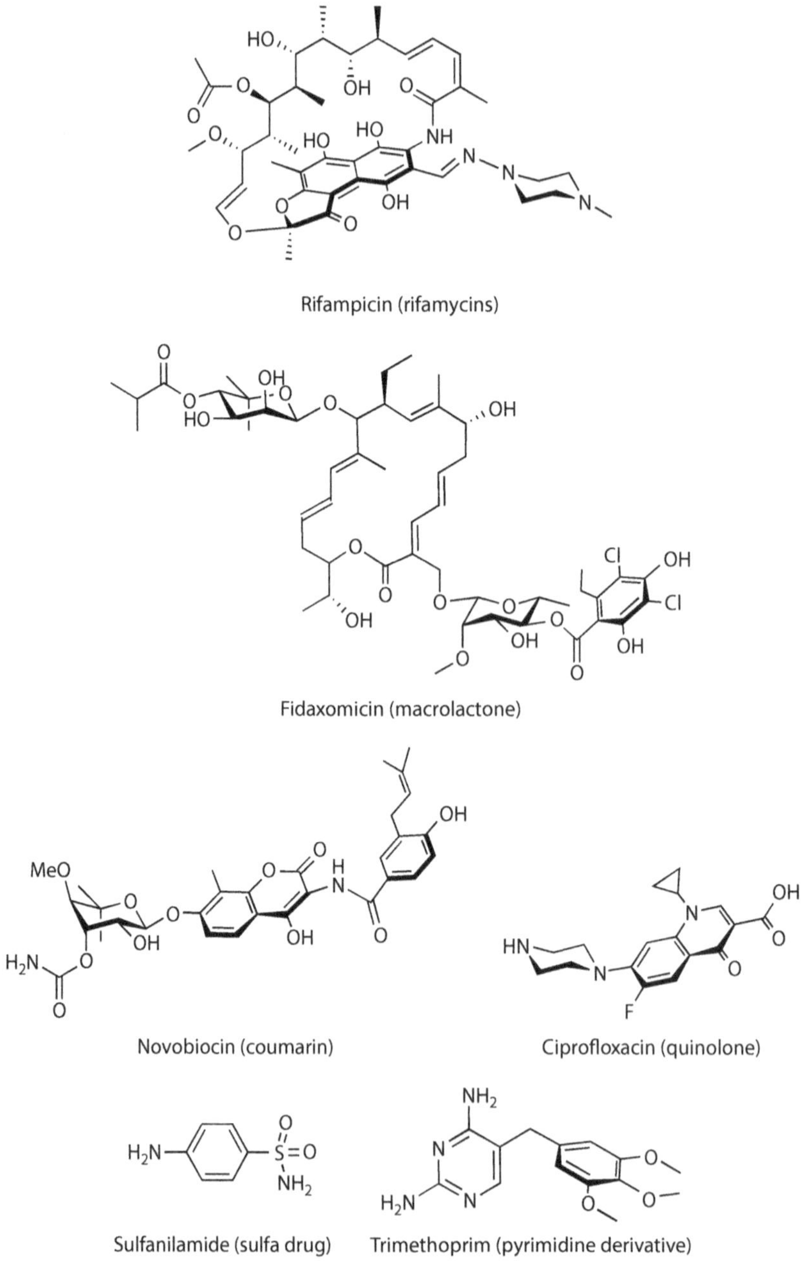

Rifampicin (rifamycins)

Fidaxomicin (macrolactone)

Novobiocin (coumarin)

Ciprofloxacin (quinolone)

Sulfanilamide (sulfa drug)

Trimethoprim (pyrimidine derivative)

FIGURE 23.5 Representative structures of antibiotics affecting nucleic acid and folate biosynthesis.

23.5.1.1 Rifamycins

Rifamycins consist of an aromatic naphthyl core and a polyketide-derived ansa-bridge. Rifamycin was isolated in a strain of *Streptomyces*, now reclassified as *Nocardia mediterranei* (1957). Rifampicin (Figure 23.5) was obtained by chemical modification of rifamycin B. Later, rifaximin was introduced (2004) whose spectrum is similar to that of rifampicin. They are very active against Gram$^+$ bacteria and some Gram$^-$ bacteria. Rifampicin gives a better, more regular absorption and is an excellent drug for tuberculosis and leprosy. Rifaximin is used for gastrointestinal infections.

23.5.1.2 Macrolactones

Macrolactone antibiotics have unsaturated lactone cores with deoxysugars and aromatic motifs. Fidaxomicin is an 18-membered polyketide macrolactone (Figure 23.5), discovered in 1975 from *Actinoplanes deccanensis*. It is bactericidal against Gram$^+$ (aerobic/anaerobic) bacteria but lacks activity against Gram$^-$ bacteria. Fidaxomicin was approved by the FDA (in 2011) for *C. difficile* infections. Narrow spectrum and poor absorption of fidaxomicin in the gastrointestinal tract has proved beneficial for this particular treatment.

23.5.2 Antibiotics Inhibiting DNA Gyrase

A molecule of DNA consists of two linear strands intertwined to form a double helix. In many bacteria, this double-stranded molecule closes into a covalent circle, while in others it is linear. When fully extended, a chromosome is 1000 times as long as the dimensions of the cell. Supercoiling is essential for packing the molecule inside the cell, while unwinding is necessary during replication and transcription.

Topoisomerases are enzymes that convert DNA from one topological form to another, that is, they promote or reverse supercoiling. Topoisomerase I cuts a single strand of the double helix, topoisomerase II cuts both simultaneously.

DNA gyrase is a topoisomerase II which is able to supercoil a relaxed DNA ring (reaction A) at the expense of ATP hydrolysis. While all topoisomerases can relax supercoiled DNA, negative supercoiling is unique to DNA gyrase. It is a validated target, essential in bacteria and absent in humans. DNA gyrase and its close relative topoisomerase IV (which plays an important role in partitioning DNA during cell division) have two subunits, A and B. In gyrase, the subunits are called GyrA and GyrB, respectively. The GyrA subunit is involved in interactions with DNA as it contains the active-site tyrosine responsible for DNA cleavage, while GyrB contains the ATPase active site. Two antibiotic classes, natural Coumarins and synthetic Quinolones, target DNA gyrase using different mechanisms (Table 23.2).

23.5.2.1 Coumarins

Novobiocin (1955) (Figure 23.5) is a coumarin derivative produced by *Streptomyces niveus* and *Streptomyces spheroides*. It is active against Gram$^+$ bacteria and *Haemophilus influenzae* and *Neisseria*. It was used in the treatment of infections caused by resistant staphylococci, but due to toxicity and solubility issues, it was replaced by other products.

23.5.2.2 Quinolones

The quinolone core typically has a N-linked cyclic moiety with various substituents at the C6 and/or C7 positions. The first member, Nalidixic acid (1962), is a by-product of chloroquine (an antimalarial drug) synthesis. Its original trade name Negram® indicates that it is active against Gram$^-$ bacteria. The first-generation quinolones, including oxolinic acid (1967), had weak activity. The synthesis of fluoroquinolones triggered multiple generations of modifications and optimizations for improving potency, spectrum of activity, and countering bacterial resistance. Examples include Ciprofloxacin (Figure 23.5), levofloxacin, and more recent, prulifloxacin.

Quinolones also target topoisomerase IV, especially in Gram$^+$ bacteria. Dual targeting has the advantage of low mutation frequencies for drug resistance. Their broad spectrum makes them a highly successful class of antibiotics. They are also used to treat intracellular pathogens due to their ability to enter the Gram$^-$ outer membrane via porins. Delafloxacin and finafloxacin are among the quinolones in late clinical trials, e.g., showing increased potency against Gram$^+$ bacteria including MRSA.

23.6 ANTIBIOTICS AFFECTING METABOLIC PATHWAYS

The folate biosynthesis pathway: Folic acid is the precursor for tetrahydrofolate, an important one-carbon unit donor in many biosynthetic pathways, including nucleic acid synthesis. Bacteria build up the folate skeleton de novo, unlike humans who assimilate folate from their diet. Sulfonamides (sulfa drugs) and Trimethoprim target the bacterial folate pathway at different stages (Figure 23.6). The effect of each individual antibiotic is bacteriostatic; the combination is synergistic and bactericidal.

23.6.1 Sulfa Drugs

The sulfa drugs have a common aryl sulfonamide moiety. The first prodrug, prontosil rubrum (1932), is metabolized in the liver to sulfanilamide. Sulfanilamide (1936) (Figure 23.5) proved effective in

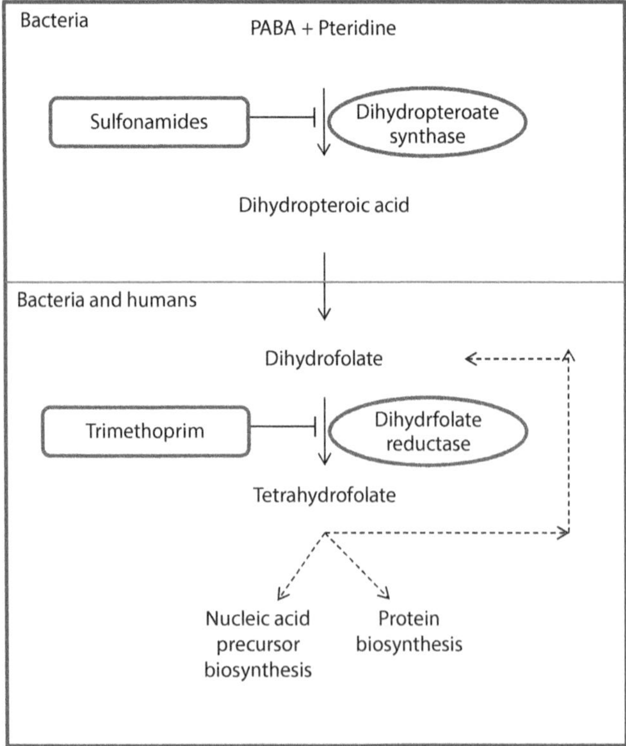

FIGURE 23.6 Sequential inhibition of folate biosynthesis by sulfonamide/trimethoprim. Bacteria build their folate skeleton from the start unlike humans. Sulfonamides competitively inhibit the enzyme dihydropteroate synthase (absent in humans) by mimicking *p*-aminobenzoic acid (PABA), one of the normal constituents of folic acid. Trimethoprim is a highly selective active site inhibitor of the bacterial dihydrofolate reductase enzyme (essential in humans but structurally different enough to allow selective inhibition). The overall effect is depletion of tetrahydrofolate, a carrier for one-carbon units, necessary for many biosynthetic pathways.

the treatment several infections. These broad-spectrum synthetic antimetabolites were eventually replaced by natural product antibiotics (e.g., penicillin) with greater potency. Renewed interest stems from their application in combination therapy with trimethoprim (see co-trimoxazole; Table 23.3).

23.6.2 Trimethoprim

Trimethoprim (Figure 23.5) is a pyrimidine derivative whose antibacterial activity was discovered in the early 1960s. In 1968, Bushby and Hitchings showed that trimethoprim is a sulfonamide potentiator. Although it can be used individually, it has up to 100-fold synergistic effect in combination with sulfonamide.

23.7 ANTIBIOTICS AFFECTING PROTEIN BIOSYNTHESIS

Ribosomes (cellular particles, 200–500 Å in diameter) are the protein-synthesizing factories of the cell. Up to 70,000 ribosomes are present in a fast-growing *E. coli* cell in rich growth media, and some of them associate with the plasma membrane on various proteinaceous receptors. All ribosomes comprise a small and a large subunit. In prokaryotes, the assembled ribosome (termed 70S; derived from the sedimentation coefficient) consists of 30 and 50S subunits. The 30S subunit contains a 16S RNA that traps messenger RNAs through their Shine–Dalgarno sequences. The 50S subunit contains a 23S and a 5S RNA molecule and is the site of peptide bond formation catalyzed by the 23S ribozyme. The ribosomes of eukaryotic cells are larger (80S) with 60 and 40S subunits. Several ribosomes can simultaneously translate one mRNA (polyribosome).

Many clinically useful antibiotics bind at or near the peptidyltransferase center (PTC) on the 50S subunit, where the peptide bond formation occurs. On the 30S subunit, the binding sites are clustered along the path of the mRNA and tRNAs (Table 23.2).

Puromycin, an antibiotic isolated from *Streptomyces alboniger* in 1953, although not in clinical use, has been very useful in the study of protein synthesis due to its resemblance to the aminoacyl-adenosine part of aminoacyl-tRNA.

23.7.1 Inhibitors of the 30S Ribosomal Subunit

23.7.1.1 Aminoglycosides

Aminoglycosides are amino sugars connected via glycosidic bonds typically to a 2-deoxystreptamine core. This is the only class of protein synthesis inhibitors that is broadly bactericidal. However, all aminoglycosides present a certain toxicity: vestibular disturbance, with problems of equilibrium, and cochleotoxicity which may result in partial and total loss of hearing. Hence, they are antibiotics of last resort. Many aminoglycosides are natural products, for example, streptomycin, gentamicin, and sisomicin.

Streptomycin (Figure 23.7), the first aminoglycoside discovered (1944; *Streptomyces griseus*), was active mainly against Gram$^-$ bacteria and served as a useful complement to early penicillins. Streptomycin and *p*-aminosalicylic acid were the first drugs used for tuberculosis.

Semisynthetic aminoglycosides were optimized for reduced toxicity (in particular, ototoxicity and nephrotoxicity). An example is amikacin, a semisynthetic derivative of kanamycin clinically introduced in 1976.

After a decline in interest due to toxicity and resistance, there is now a renewed interest in this group due to the need for antibiotics against multidrug and pandrug-resistant Gram$^-$ pathogens. Plazomicin, a synthetic derivative of sisomicin, is broad spectrum and lacks the toxic side effects as observed in clinical trials (phase II, 2012).

23.7.1.2 Tetracyclines

They derive their name from the tetracyclic ring system which is octahydronaphthacene. The first tetracyclines were chlortetracycline (1948; *Streptomyces aureofaciens*) and oxytetracycline (1950; *Streptomyces rimosus*). The structure was published in 1952 but complete stereochemistry was obtained later from X-ray diffraction analysis. These studies revealed that the removal of chlorine from chlortetracycline by hydrogenolysis led to an active product, tetracycline (Figure 23.7). Tetracycline was also obtained by fermentation of a medium poor in chloride using appropriate strains, e.g., *S. alboniger, S. viridifaciens*, etc.

Semisynthetic tetracyclines: Chemical manipulations of oxytetracycline led to the production of metacycline. Hydrogenation of metacycline under suitable conditions gave doxycycline (II generation). Doxycycline is very stable (no C6–OH group) and is lipophilic. It is more completely absorbed after oral administration. Minocycline (1972) is prepared by the chemical treatment of 6-deoxy-6-demethyltetracycline.

The tetracycline binding site in both prokaryotes and eukaryotes is structurally conserved. If not for the ability of bacteria to accumulate tetracyclines far more efficiently than mammalian cells, the drug does not differentiate between the eukaryotic and prokaryotic ribosomes. They are bacteriostatic, truly broad-spectrum antibiotics with low incidence of side effects which explains their position as first-line antibiotics.

Glycylcyclines have *N,N*-dimethylglycylamido moiety on the C9 position of ring D. Tigecycline (1998) (Figure 23.7), a glycyl derivative of minocycline, was approved 30 years after the last member from the tetracycline class. The modification resulted in broader spectrum of activity.

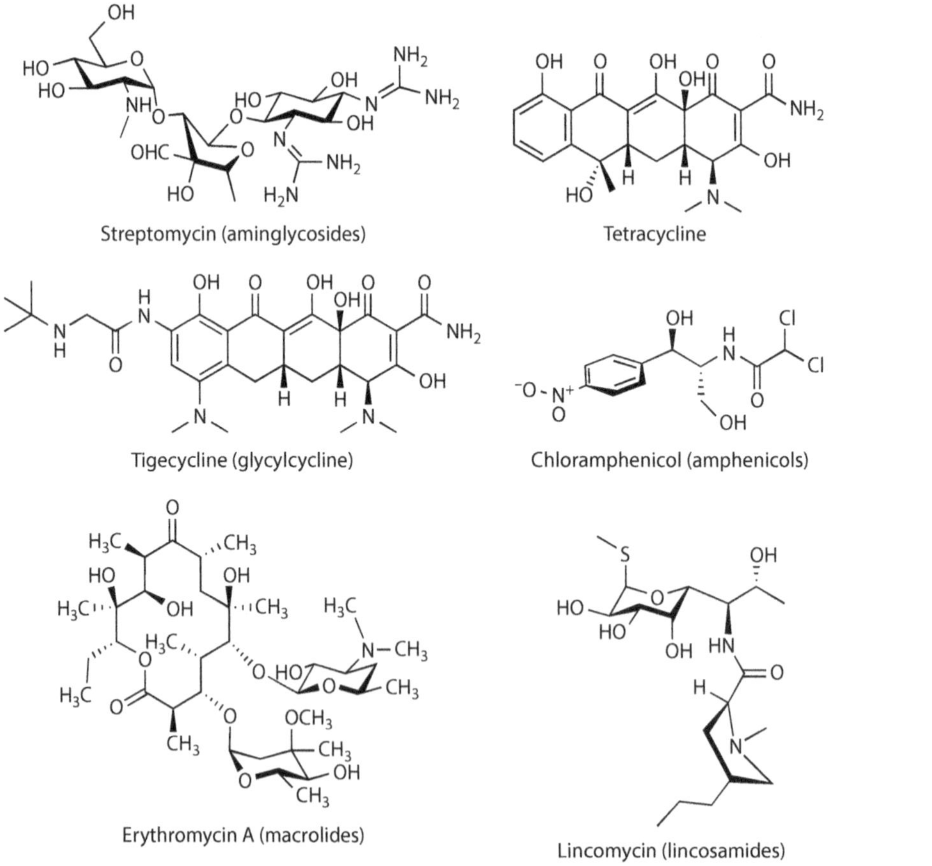

Streptomycin (aminglycosides)

Tetracycline

Tigecycline (glycylcycline)

Chloramphenicol (amphenicols)

Erythromycin A (macrolides)

Lincomycin (lincosamides)

FIGURE 23.7 Representative structures of antibiotics affecting protein biosynthesis. *(Continued)*

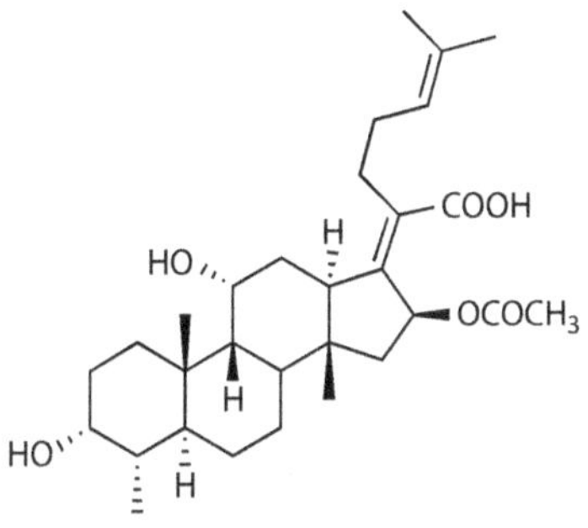

Fusidic acid (fusidane)

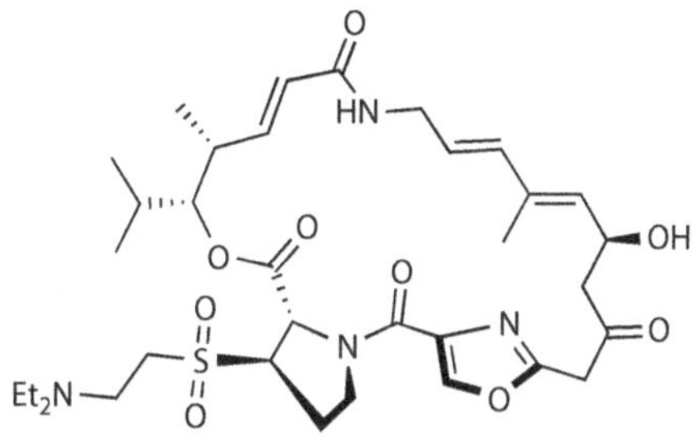

Dalfopristin (streptogramin A)

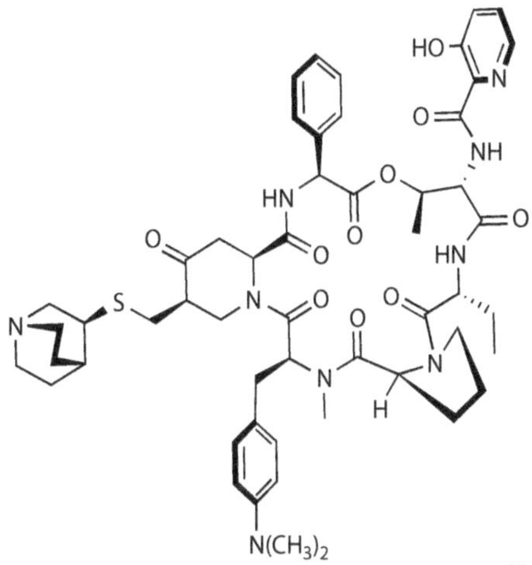

Quinupristin (streptogramin B)

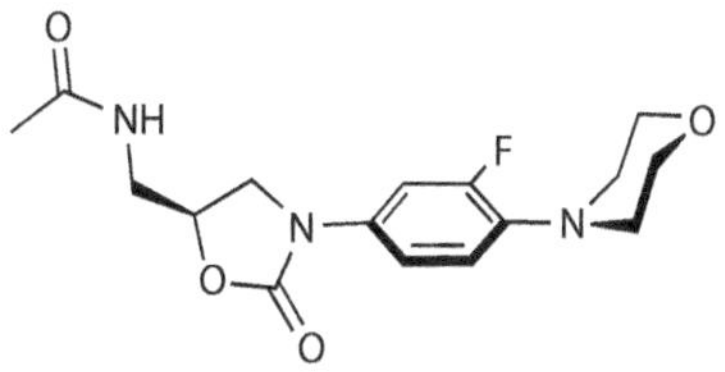

Linezolid (ozazolidinones)

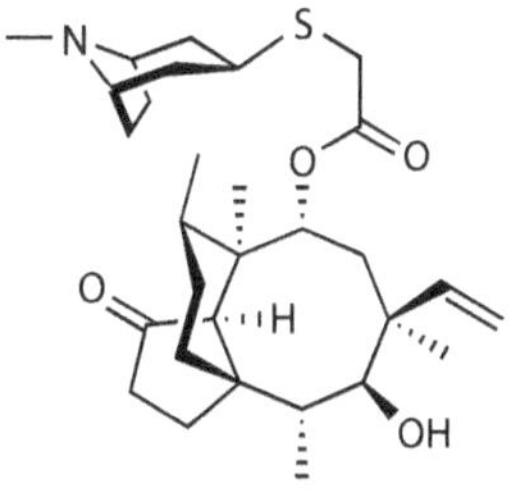

Retapamulin (pleuromutilins)

FIGURE 23.7 (*Continued*) Representative structures of antibiotics affecting protein biosynthesis.

Tigecycline is generally bacteriostatic (like tetracyclines) but is bactericidal against some pathogens. It is effective on strains resistant to the older tetracyclines and is active against MRSA, vancomycin-resistant enterococci, and many multidrug-resistant Gram$^-$ bacteria in addition to low activity against *P. aeruginosa*. Omadacycline, a minocycline derivative, and eravacycline, a fluorocycline, are in phase II and III clinical trials, respectively.

23.7.2 Inhibitors of the 50S Ribosomal Subunit

23.7.2.1 Amphenicols

They comprise phenylpropanoid antibiotics. The first broad-spectrum antibiotic was chloramphenicol (1947; *Streptomyces venezuelae*) (Figure 23.7). Its chemical structure was determined, and in 1949, a synthesis was described and used for commercial production. Of the four possible diastereoisomers, only the *R*,*R*-isomer is active and is separated during synthesis.

Many derivatives of chloramphenicol were prepared but only the sulfomethyl analog, thiamphenicol, is in clinical use. It is generally less active than chloramphenicol. The glycinate ester is used as a prodrug for injections. Chloramphenicol has moderate activity against Gram$^+$ and Gram$^-$ bacteria. It is not ideal for the treatment of these infections because of serious toxic reactions in the blood (aplastic anemia, thrombocytopenia). It is still used in the treatment of typhus and meningitis caused by *H. influenzae*.

23.7.2.2 Macrolides

The macrolide antibiotics have in common (1) a large lactone ring (hence the name macrolide), (2) a glycosidically linked aminosugar (sometimes two), and (3) usually a desoxysugar. The lactone ring may contain 12 (macrolides not used in medicine), 14 (erythromycin, oleandomycin), or 16 atoms (leucomycin, spiramycin, tylosin).

Erythromycin is the first clinically useful macrolide (1952; *Streptomyces erythreus*) (Figure 23.7). The structure was determined chemically (1954–1957); the stereochemistry and conformation by X-ray diffraction and NMR. Erythromycin is inactivated by acid. It is very active against Gram$^+$ bacteria.

Oleandomycin (1955; *Streptomyces antibioticus*) is usually administered as a triacetyl derivative which gives higher blood levels. Oleandomycin has a similar spectrum as erythromycin but the MIC is generally higher.

Semisynthetic macrolides include clarithromycin, roxithromycin, and azithromycin, whose spectra are comparable to those of erythromycin.

Telithromycin (1997), the first ketolide (semisynthetic derivative of erythromycin) approved by the FDA, had some rare but serious side effects and was partially withdrawn. Ketolides are effective against strains resistant to older macrolides. Ketolides currently under clinical trials are cethromycin and solithromycin, with better activity against some pathogens.

23.7.2.3 Lincosamides

Lincomycin (1962, *Streptomyces lincolnensis*) has a basic group in the proline part of the molecule and the sugar moiety contains a methylmercapto group (Figure 23.7). Replacement of a hydroxy group by chlorine, with inversion of configuration, resulted in Clindamycin (1967), with improved absorption and higher serum levels. Both are active against Gram$^+$ bacteria, with a spectrum similar to erythromycin. Clindamycin is used in the treatment of infections caused by anaerobic bacteria.

23.7.2.4 Fusidic Acid

Fusidic acid (1962, *Fusidium coccineum*) has a unique steroid-type (fusidane) structure (Figure 23.7) similar to cephalosporin P1. It is active against Gram$^+$ bacteria and Gram$^-$ cocci. Fusidic acid has been in clinical use outside the United States (since 1962) for skin infections. Safety concerns are related to gastrointestinal, allergic, hematological, and neurological adverse effects. Rapid emergence of resistant strains also limits its use. Currently, Cempra (Chapel Hill, U.S.) is evaluating a new dosage regimen in two phase II clinical trials.

23.7.2.5 Streptogramins

The streptogramin family (1950s; various *Streptomyces*) consists of two subgroups, type A and B, simultaneously produced in ~70:30 ratio. Group A streptogramins are cyclic polyunsaturated macrolactones that comprise a hybrid peptide/polyketide structure (e.g., pristinamycin II$_A$). Group B are cyclic hepta- or hexa-depsipeptides (e.g., quinupristin).

Streptogramins have been in use as feed additives in agriculture for decades. Driven by the need for new drugs to fight against resistant strains (e.g., vancomycin-resistant entercocci), Rhône-Poulenc Rorer improved the drug-like properties of pristinamycin, culminating in semisynthetic streptogramins—Dalfopristin (type A) and Quinupristin (type B) (Figure 23.7). The two streptogramins are bacteriostatic individually but bactericidal and synergistic when combined (see Synercid; Table 23.3). Efforts are in progress to evaluate new orally active streptogramins since synercid is not orally available and is administered intravenously.

23.7.2.6 Oxazolidinones

Members of this class (discovered in the mid-1980s) have a common oxazolidinone core with various N-linked aryl and heterocyclic rings and short C(5) side chains. Despite displaying bacteriostatic activity against Gram$^+$ pathogens, they were not pursued due to toxicity. Subsequent efforts by Pharmacia led to the antibiotic Linezolid (Figure 23.7) approved by the FDA (2000) for hard-to-treat Gram$^+$ bacterial infections (e.g., vancomycin-resistant enterococci).

Widespread resistance was not common to this class, probably due to the lack of such structures in nature. Prolonged therapy with linezolid has been linked to rare instances of lactic acidosis and liver injury. Continued efforts have focused on improving its spectrum, solubility, and pharmacological and toxicity profiles. Second-generation oxazolidinones are currently undergoing clinical trials.

23.7.2.7 Pleuromutilins

Pleuromutilins have a common fused cyclo-octane/pentanone with a bridged cyclohexane ring (Figure 23.7). Pleuromutilins, (1950s; *Pleurotus mutilis*), were used extensively in veterinary medicine. Ritapamulin (FDA approved, 2007), a semisynthetic pleuromutilin active against Gram$^+$ *S. aureus* and *S. pyrogens*, is used topically.

23.8 NEW DEVELOPMENTS IN ANTIBACTERIAL RESEARCH

Current antibiotic research is a massive line of work backed by advances in molecular structure determination, structure-based design, chemical synthesis, screening strategies, and genomic data, to mention a few. Given the breadth of the field, we confine our discussion to only a couple of interesting examples.

In clinical pipeline are new derivatives/molecules to address issues such as resistance, host toxicity, Gram-negative pathogens. Among the new molecules is POL7080 (Polyphlor Ltd., Basel, Switzerland; completed phase I clinical trials) which acts on an unconventional target LptD—an essential outermembrane protein, involved in exporting lipopolysaccharide molecules across the periplasm. POL7080 is a peptidomimetic molecule with potent antibacterial activity against *P. aeruginosa*. It has chemically evolved from a natural product, Protegrin I, an antimicrobial peptide from porcine leukocyte. Protegrin I disrupts the membrane via pore formation and has little clinical utility due to significant hemolytic activity.

One of the factors hampering natural product antibiotic discovery is that only a small minority of bacteria are cultivable in the lab. Uncultured bacteria accounts for approximately 99% of all species in external environments. Lewis K. and coworkers developed a novel method for growing uncultured bacteria which led to the discovery of the antibiotic Teixobactin (2015). It inhibits cell wall synthesis by binding to a highly conserved motif of lipid II (precursor of peptidoglycan) and lipid III (precursor of cell wall teichoic acid). Teixobactin (currently in preclinical stage) is potent against Gram$^+$ pathogens including *M. tuberculosis* with no detectable resistance.

23.9 CONCLUDING REMARKS

Development of antibiotics, one of the oldest fields in medicinal chemistry, continues to have a huge impact on human health, saving innumerable lives. This research domain was neglected during the last three decades, leaving a lean clinical pipeline and many difficult-to-treat bacterial infections. There is a pressing need for new antibiotics. The recent revival of interest in antibiotic research in some pharmaceutical companies and public–private partnerships will hopefully reverse the downward trend.

FURTHER READING

Brown, D.G., Troy, L., and Tricia, L.M. 2013. New natural products as new leads for antibacterial drug discovery. *Bioorgan. Med. Chem. Lett.* 24:413–418.

Fair, R.J. and Tor, Y. 2014. Antibiotics and bacterial resistance in the 21st century. *Perspect. Med. Chem.* 6:25–64.

Kohanski, M.A., Dwyer, D.J., and Collins, J.J. 2010. How antibiotics kill bacteria: From targets to networks. *Nat. Rev. Microbiol.* 8:423–435.

Marinelli, F. and Genilloud, O. (eds.). 2014. *Antimicrobials: New and Old Molecules in the Fight Against Multi-Resistant Bacteria.* Berlin, Germany: Springer-Verlag.

Pucci, M.J., Bush, K., and Page, M.G.P. 2014. Cautious optimism for the antibacterial pipeline. *Microbe* 9:147–152.

Sánchez, S. and Demain, A.L. (eds.). 2014. *Antibiotics: Current Innovations and Future Trends.* Norfolk, U.K.: Caister Academic Press.

Walsh, C.T. and Wencewicz, T.A. 2014. Prospects for new antibiotics: A molecule-centered perspective. *J. Antibiot.* 67:7–22.

Wright, P.M., Ian, B.S., and Andrew, G.M. 2014. The evolving role of chemical synthesis in antibacterial drug discovery. *Angew. Chem. Int. Ed.* 53(34):8840–8869.

Index

B

C

D

K

L

M

N

O

P